Properties of Absolute Value

$|a| \geq 0$

$|a| = |-a|$

$|a - b| = |b - a|$

$|a^2| = |a|^2 = a^2$

Multiplication Patterns

$(a + b)^2 = a^2 + 2ab + b^2$

$(a - b)^2 = a^2 - 2ab + b^2$

$(a + b)(a - b) = a^2 - b^2$

$(a + b)^3 = a^3 + 3a^2b + 3ab^2 + b^3$

$(a - b)^3 = a^3 - 3a^2b + 3ab^2 - b^3$

$(a + b)^n = \binom{n}{0}a^n + \binom{n}{1}a^{n-1}b + \binom{n}{2}a^{n-2}b^2 + \cdots + \binom{n}{n}b^n$

Properties of Exponents and Radicals

$b^n \cdot b^m = b^{n+m}$

$(b^n)^m = b^{mn}$

$(ab)^n = a^n b^n$

$\left(\dfrac{a}{b}\right)^n = \dfrac{a^n}{b^n}$

$\dfrac{b^n}{b^m} = b^{n-m}$

$\sqrt[n]{ab} = \sqrt[n]{a}\sqrt[n]{b}$

$\sqrt[n]{\dfrac{a}{b}} = \dfrac{\sqrt[n]{a}}{\sqrt[n]{b}}$

Interval Notation

(a, ∞)

$(-\infty, b)$

(a, b)

$[a, \infty)$

$(-\infty, b]$

$(a, b]$

$[a, b)$

$[a, b]$

Set Notation

$\{x \mid x > a\}$

$\{x \mid x < b\}$

$\{x \mid a < x < b\}$

$\{x \mid x \geq a\}$

$\{x \mid x \leq b\}$

$\{x \mid a < x \leq b\}$

$\{x \mid a \leq x < b\}$

$\{x \mid a \leq x \leq b\}$

Properties of Logarithms

$\log_b b = 1$

$\log_b 1 = 0$

$\log_b rs = \log_b r + \log_b s$

$\log_b\left(\dfrac{r}{s}\right) = \log_b r - \log_b s$

$\log_b r^p = p(\log_b r)$

Factoring Patterns

$a^2 - b^2 = (a + b)(a - b)$

$a^3 - b^3 = (a - b)(a^2 + ab + b^2)$

$a^3 + b^3 = (a + b)(a^2 - ab + b^2)$

Equations Determining Functions

Linear function: $\quad f(x) = ax + b$

Quadratic function: $\quad f(x) = ax^2 + bx + c$

Polynomial function: $\quad f(x) = a_n x^n + a_{n-1}x^{n-1} + \cdots + a_1 x + a_0$

Rational function: $\quad f(x) = \dfrac{g(x)}{h(x)}, \quad$ where g and h are polynomial functions

Exponential function: $\quad f(x) = b^x, \quad$ where $b > 0$ and $b \neq 1$

Logarithmic function: $\quad f(x) = \log_b x, \quad$ where $b > 0$ and $b \neq 1$

Turn the graphing calculator into a powerful tool for your success!

Explorations in College Algebra:
An Interactive Graphing Calculator Manual

by Deborah J. Cochener and Bonnie M. Hodge, both of Austin Peay State University

384 pages. Spiralbound. 8½" x 11". ISBN: 0-534-36423-3. ©1999. Published by Brooks/Cole.

You can quickly learn to use the graphing calculator to develop problem-solving and critical-thinking skills that will help improve your performance in your college algebra course!

Designed to help you succeed in your college algebra course, this unique and student-friendly workbook improves both your understanding *and* retention of college algebra concepts—using the graphing calculator. By integrating technology into mathematics, the authors help you develop problem-solving and critical-thinking skills.

To guide you in your explorations, you'll find:

- hands-on applications with solutions
- correlations charts that relate course topics to the workbook units
- key charts (specific to the TI-82, TI-83, TI-85, and TI-86) that show which units introduce keys on the calculator
- a "Troubleshooting Section" to help you avoid common errors

"This is the best graphing calculator lab manual supplement that I have seen. After reviewing it, I feel that it is time for a change . . . to a more subjective, comprehensive, exercise-packed lab manual . . . the lab manual being Explorations in College Algebra.*"*
> Chris Bendixen
> Sue Bennett College

"I did not find anything that I did not like about this manuscript. I thought it was well written and organized. The authors did a good job of describing the use of specific functions of the graphing calculators and following through with good examples that illustrated their use."
> David Ray
> University of Tennessee-Martin

Topics

This text contains 50 units divided into the following subsections:
> Basic Calculator Operations
> Graphically Solving Equations and Inequalities
> Graphing and Applications of Equations in Two Variables
> Conic Sections
> Miscellaneous (Matrices, Combinatorics and Probability, and Sequences and Series)
> Stat Plots
> Programming

Order your copy today!

To receive your copy of *Explorations in College Algebra: An Interactive Graphing Calculator Manual,* simply mail in the order form on the other side of this page, call (800) 354-9706, or visit us on the Internet: http://www.brookscole.com

ORDER FORM

Call our toll-free number (800) 354-9706 to purchase, or use the form below and mail or fax to (408) 375-6414.

No risk. All of our books and software are backed by our 30-day, money-back guarantee. Major credit cards accepted. We accept purchase orders from your company or institution. The cost of shipping will be added to your bill. To order simply fill out this coupon can return it to Brooks/Cole along with your check, money order, or credit card information.

_____ Yes! I would like to order **Explorations in College Algebra: An Interactive Graphing Calculator Manual,** by Deborah J. Cochener and Bonnie M. Hodge, ISBN: 0-534-36423-3 for $19.95.
<u>Note</u>: Pricing is subject to change. Please call (800) 354-9706 to receive current pricing information.

Residents of: AL, AZ, CA, CT, CO, FL, GA, IL, IN, KS, KY, LA, MA, MD, MI, MN, MO, NC, NJ, NY, OH, PA, RI, SC, TN, TX, UT, VA, WA, WI must add appropriate state sales tax.

Subtotal	_____
Tax	_____
Handling	$4.00
TOTAL	_____

Payment Options

____ Payment enclosed (check or money order)

____ Please charge the following credit card:

____ VISA ____ MasterCard ____ American Express

Card #: _____   Expiration Date: _____

Contact Name: _____   Phone: _____

Signature: _____

(Note: Credit card billing and shipping address must be the same.)

All orders under 250 pounds will be shipped via UPS, unless the customer references a specific carrier. Average shipping charges are estimated at 10% of the order. The cost of shipping will be added to your bill. Customer's requested shipper: _____

Prices are subject to change without notice. We will refund any pre-payments for unshipped, out-of-stock titles after 150 days and for not-yet-published titles after 180 days, unless an earlier date is requested in writing from you.

Please ship my order to: (please print)

Name _____

Street Address _____

City _____ State _____ Zip _____

Telephone (____) _____ e-mail: _____

Mail to:

Brooks/Cole Publishing Company
Source Code 9BCMA028
511 Forest Lodge Road
Pacific Grove, CA 93950-5098
Phone: (800) 354-9706 • Fax: (408) 375-6414

International Thomson Publishing
Education Group

COLLEGE ALGEBRA

FOURTH EDITION

Jerome E. Kaufmann

Brooks/Cole Publishing Company

I(T)P® *An International Thomson Publishing Company*

Pacific Grove • Albany • Belmont • Bonn • Boston • Cincinnati • Detroit • Johannesburg • London
Madrid • Melbourne • Mexico City • New York • Paris • Singapore • Tokyo • Toronto • Washington

Sponsoring Editor: *Margot Hanis*
Marketing Team: *Caroline Croley, Debra Johnston, and Christine Davis*
Editorial Assistant: *Jennifer Wilkinson*
Production Coordinator: *Laurel Jackson*
Production Service: *Susan Graham*
Manuscript Editor: *Connie Day*

Cover Design: *Lisa Henry*
Cover Photo: *Franklin, FPG, International*
Photo Editors: *Kathleen Olson and Terry Powell*
Typesetting: *American Composition & Graphics, Inc.*
Cover Printing: *Phoenix Color Corporation*
Printing and Binding: *World Color*

For more information, contact:

BROOKS/COLE PUBLISHING COMPANY
511 Forest Lodge Road
Pacific Grove, CA 93950
USA

International Thomson Editores
Seneca 53
Col. Polanco
11560 México, D. F., México

International Thomson Publishing Europe
Berkshire House 168-173
High Holborn
London WC1V 7AA
England

International Thomson Publishing GmbH
Königswinterer Strasse 418
53227 Bonn
Germany

Thomas Nelson Australia
102 Dodds Street
South Melbourne, 3205
Victoria, Australia

International Thomson Publishing Asia
60 Albert St.
#15-01 Albert Complex
Singapore 189969

Nelson Canada
1120 Birchmount Road
Scarborough, Ontario
Canada M1K 5G4

International Thomson Publishing Japan
Hirakawacho Kyowa Building, 3F
2-2-1 Hirakawacho
Chiyoda-ku, Tokyo 102
Japan

Printed in the United States of America

10   9   8   7   6   5   4   3

Library of Congress Cataloging-in-Publication Data

Kaufmann, Jerome E.
  College algebra / Jerome E. Kaufmann.— 4th ed.
    p.   cm.
  Includes index.
  ISBN 0-534-35751-2
  1. Algebra   I. Title.
QA154.2.K365   1998
512.9—dc21

98-24735
CIP

BRIEF CONTENTS

CONTENTS

5 POLYNOMIAL AND RATIONAL FUNCTIONS 365

6 SYSTEMS OF EQUATIONS 435

7 ALGEBRA OF MATRICES 491

PREFACE

College Algebra, Fourth Edition is written for students who need a college algebra course to serve as a prerequisite for the calculus sequence, for the finite math/calculus sequence, or to satisfy a liberal arts requirement. Sample outlines for these three types of courses are included at the end of this preface.

Four major ideas unify this text—namely, solving equations and inequalities, solving problems, developing graphing techniques, and developing and using the concept of a function.

College Algebra, Fourth Edition presents basic concepts of algebra in a simple, straightforward way. Examples motivate students and reinforce algebraic concepts by the application of these examples to real world situations that students can identify with. These examples also guide students to organize their work in a logical fashion and use meaningful shortcuts whenever appropriate.

In the preparation of this fourth edition, we made a special effort to incorporate improvements suggested by reviewers and users of the earlier editions without sacrificing the book's many successful features.

New in This Edition

- A **Chapter Test** has been added at the end of each chapter. Along with the Chapter Review Problem Sets, these practice tests should provide the students with ample opportunity to prepare for the "real" examinations. Cumulative Review Problem Sets appear at the end of Chapters 2 and 5. All answers for Chapter Review Problem Sets, Chapter Tests, and Cumulative Review Problem Sets appear at the back of the text.

- Problems called **Thoughts into Words** are now included in every problem set except the review exercises. These problems are designed to encourage students to express in written form their thoughts about various mathematical ideas. For examples, see Problem sets 0.5, 1.2, 1.3, and 5.6.

- Miscellaneous problem sections, now called **Further Investigations,** have been enhanced by adding more problems that lend themselves to small group work. These problems encompass a variety of ideas: some are proofs, some exhibit different approaches to topics covered in the text, some bring in supplementary topics and relationships, and some are more challenging problems. Note that, although these problems add variety and flexibility to the problem sets, they can be omitted entirely without disrupting the continuity of the text. For examples, see Problem Sets 1.1, 1.2, 2.1, and 2.3.

- **Applications** have been added to several problem sets, especially in Chapters 1, 4, and 6. Most of these new problems have replaced problems with number relationships but without any real world connections. Applications of radical equations have been added to Problem Set 1.5 and some applications of slope have been added to Section 2.3 and the accompanying problem set.

- **Alternate solutions** for some problem solving situations are presented to provide more flexibility for the development of problem solving techniques. For examples, see problems 3, 6, and 7 of Sections 1.4. Likewise, we sometimes show different approaches for skill development exercises. This provides more opportunity to accommodate different learning styles.

- Some **rewriting** and **reorganizing** has been done in this edition. Section 1.6 has been rewritten to explicitly include some work with conjunctions and disjunctions. Also, the conjunction-set intersection connection as well as the disjunction-set union connection are highlighted.

    In Section 3.3, zeros of a function are introduced. Then the connection between real number zeros of a function and the *x* intercepts of the graph of the function is emphasized.

    Old Section 3.6 (Inverse Functions) is now Section 4.3. This places the concept of inverse functions where it is needed to connect exponential and logarithmic functions. Furthermore, it should help eliminate the problem of confusing inverse variations and inverse functions.

    Reviewers suggested several places where a page, a paragraph, or a solution to a problem could be rewritten to further clarify the intended meaning. Sometimes we included a **Remark** to add a little flavor to the discussion.

Other Special Features

- **Chapter 0,** a review of intermediate algebra concepts, was written so that students can work through this material with a minimum of assistance from the instructor. The concepts are reviewed in enough detail to provide a good basis for understanding the numerous worked-out examples. The problem sets also provide ample practice material. The chapter test and/or the chapter review problem set could be used to determine problem areas.

- **Chapters 9 and 10,** if not covered as part of the regular course, could be used as enrichment material. The easy-to-read explanations along with carefully chosen examples make these chapters accessible to students with minimum help from the instructor.

- As recommended in the standards produced by NCTM and AMATYC, **problem solving** is an integral part of this text. With problem solving as its focus, Chapter 1 pulls together and expands on a variety of approaches to the process of solving equations and inequalities. Polya's four-phase plan is used as a basis for developing a variety of problem solving strategies. Functions are introduced in Chapter 3 and immediately used to solve problems. Exponential and logarithmic functions become problem solving tools in Chapter 4. Systems of equations provide more problem solving power in Chapter 6. Problem solving is the unifying theme of Chapters 9 and 10.

- **Specific graphing ideas** (intercepts, symmetry, restrictions, asymptotes, and transformations) are introduced and used throughout Chapters 2, 3, 4, 5, and 8. In Section 3.4, the extensive work with graphing parabolas from Sections 1.2 is used to motivate definitions for translations, reflections, stretchings, and shrinkings. These transformations are then applied to the graphs of $f(x) = x^3, f(x) = x^4, f(x) = \sqrt{x}$ and $f(x) = |x|$. Furthermore, in later chapters the transformations are applied to graphs of exponential, logarithmic, polynomial, and rational functions.

- The use of a **graphing utility** is introduced in Section 2.2. From then on, graphing calculator examples are incorporated, as appropriate, throughout the text. These examples usually reinforce ideas presented in the section and are written so that students without graphing calculators can read and benefit from them. The graphing ideas mentioned previously provide a sound basis for an efficient use of a graphing utility.

- Beginning with Problem Set 2.2, a group of problems called **Graphing Calculator Activities** is included in many of the problem sets. These activities, which are good for either individual or small group work, have been designed to reinforce concepts (see, for example, Problem Set 4.5) as well as lay groundwork for concepts about to be discussed (see, for example, Problem Set 2.2). Some of these activities ask students to predict shapes and locations of graphs based on previous graphing experiences and then to use a graphing utility to check their predictions (see, for example, Problem Set 3.4). The graphing calculator is also used as a problem solving tool (see, for example, Problem Set 5.5). When students do these activities, they should become familiar with the capabilities and limitations of a graphing utility.

- Please note the exceptionally pleasing design features of this text, including the **functional use of color.** The open format makes for a continuous and easy reading flow of material instead of working through a maze of flags, caution symbols, reminder symbols, etc.

The following are sample outlines for a college algebra course that (a) serves as a prerequisite for the standard calculus sequence, (b) serves as a prerequisite for a finite mathematics or business calculus course, or (c) satisfies a liberal arts requirement:

PRECALCULUS	BUSINESS	LIBERAL ARTS
0.1–0.8 Some Basic Concepts of Algebra	0.1–0.7 Some Basic Concepts of Algebra	0.1–0.6 Some Basic Concepts of Algebra
1.1–1.7 Equations, Inequalities, and Problem Solving	1.1–1.7 Equations, Inequalities, and Problem Solving	1.1–1.4 and 1.6 Equations, Inequalities, and Problem Solving
2.1–2.5 Coordinate Geometry and Graphing Techniques	2.1–2.4 Coordinate Geometry and Graphing Techniques	2.1–2.4 Coordinate Geometry and Graphing Techniques
3.1–3.6 Functions	3.1–3.6 Functions	3.1–3.6 Functions
4.1–4.6 Exponential and Logarithmic Functions	4.1–4.6 Exponential and Logarithmic Functions	4.1–4.6 Exponential and Logarithmic Functions
5.1–5.7 Polynomial and Rational Functions	5.1–5.5 Polynomial and Rational Functions	6.1–6.3 Systems of Equations
8.1–8.3 Conic Sections	6.1–6.5 Systems of Equations	7.1–7.2 Matrices
9.1–9.3 Sequences	7.4 Linear Programming	9.1–9.3 Sequences
10.6 The Binomial Theorem	9.1–9.3 Sequences	10.1–10.4 Counting Techniques and Probability
	10.1–10.5 Counting Techniques and Probability	

Instructors' Ancillaries

Instructor's Solutions Manual. Provides instructors with answers to all problems from the main text, plus selected solutions to the even-numbered problems.

Test Manual. Contains printed test forms, with answers, for the instructor.

Video Tutorial Series. These text-specific videos feature examples from every chapter and section of the text. All examples that are worked on the videos are cross-referenced with a video icon in the text.

Thomson World Class Learning ™ Testing Tools. This fully integrated suite of programs includes World Class Test 1.0, World Class Test On-Line 1.0, and World Class Manager 1.0. The program provides text-specific algorithmic testing options designed to offer instructors greater flexibility.

Students' Ancillaries

Student Solutions Manual (Author: Cheryl Cantwell). Provides stepped-out solutions to selected odd-numbered problems from the main text.

Explorations in College Algebra: An Interactive Graphing Calculator Manual (Authors: Deborah Jolly Cochener and Bonnie MacLean Hodge). A manual that teaches students the necessary skills to use the graphing calculator and provides supplementary exercises and instruction.

Tutorial CD. This CD contains a self-running tutorial with text-specific questions, which provides students with an opportunity for more practice.

Acknowledgments

I would like to thank the following reviewers for their valuable comments and suggestions:

Charles Demetriou
*State Technical Institute
of Memphis*

Sudhasree Gadam
University of North Texas

Linda Howard
*Elizabethtown Community
College*

Robert C. Limburg
*St. Louis Community College–
Florissant Valley*

Lynn Sutton
C.S. Mott Community College

I am very grateful to the staff of Brooks/Cole, especially Margot Hanis and Jennifer Wilkinson, for their continuous cooperation and assistance throughout this project. I would also like to express my sincere gratitude to Laurie Jackson and to Susan Graham. They continue to make my life as an author so much easier by carrying out the details of production in a dedicated and caring way.

Again, very special thanks are due to my wife, Arlene, who spends numerous hours typing and proofreading manuscripts.

Jerome E. Kaufmann

SOME BASIC CONCEPTS OF ALGEBRA: A REVIEW

The general nature of algebra makes it applicable to a large variety of occupations.

Algebra is often described as generalized arithmetic. That description may not tell the whole story, but it does indicate an important idea—namely, that a good understanding of arithmetic provides a sound basis for the study of algebra. Furthermore, a good understanding of some basic algebra concepts provides an even better basis for the study of more advanced algebraic ideas. Be sure that you can work effectively with the algebraic concepts we review in this first chapter.

0.1 SOME BASIC IDEAS

Let's begin by pulling together the basic tools we need for the study of algebra. In arithmetic, symbols such as 6, $\frac{2}{3}$, 0.27, and π are used to represent numbers. The operations of addition, subtraction, multiplication, and division are commonly indicated by the symbols $+$, $-$, $\times$, and $\div$, respectively. These symbols enable us to form specific **numerical expressions**. For example, the indicated sum of 6 and 8 can be written $6 + 8$.

In algebra, we use variables to generalize arithmetic ideas. For example, by using x and y to represent *any* two numbers, we can use the expression $x + y$ to represent the indicated sum of *any* two numbers. The x and y in such an expression are called **variables**, and the phrase $x + y$ is called an **algebraic expression**.

Many of the notational agreements we make in arithmetic can be extended to algebra, with a few modifications. The following chart summarizes those notational agreements regarding the four basic operations.

OPERATION	ARITHMETIC	ALGEBRA	VOCABULARY
Addition	$4 + 6$	$x + y$	The sum of x and y
Subtraction	$14 - 10$	$a - b$	The difference of a and b
Multiplication	7×5 or $7 \cdot 5$	$a \cdot b, a(b), (a)b,$ $(a)(b),$ or ab	The product of a and b
Division	$8 \div 4, \frac{8}{4}, 8/4,$ or $4\overline{)8}$	$x \div y, x/y, \frac{x}{y},$ or $y\overline{)x}$ $(y \neq 0)$	The quotient of x divided by y

Note the different ways of indicating a product, including the use of parentheses. The *ab* form is the simplest and probably the most widely used form. Expressions such as abc, $6xy$, and $14xyz$ all indicate multiplication. Notice the various forms used to indicate division. In algebra, the fraction forms $\frac{x}{y}$ and x/y are generally used, although the other forms do serve a purpose at times.

The Use of Sets

Some of the vocabulary and symbolism associated with the concept of sets can be effectively used in the study of algebra. A **set** is a collection of objects; the objects are called **elements** or **members of the set**. The use of capital letters to name sets and the use of set braces, { }, to enclose the elements or a description of the elements provide a convenient way to communicate about sets. For example, a set A that consists of the vowels of the alphabet can be represented as follows.

	$A = \{\text{vowels of the alphabet}\}$	Word description
or	$A = \{a, e, i, o, u\}$	List or roster description
or	$A = \{x \mid x \text{ is a vowel}\}$	Set-builder notation

A set consisting of no elements is called the **null set** or **empty set** and is written $\varnothing$.

    Set-builder notation combines the use of braces and the concept of a variable. For example, $\{x \mid x \text{ is a vowel}\}$ is read *The set of all x such that x is a vowel.* Note that the vertical line is read *such that.*

    Two sets are said to be **equal** if they contain exactly the same elements. For example, $\{1, 2, 3\} = \{2, 1, 3\}$ because both sets contain exactly the same elements; the order in which the elements are listed does not matter. A slash mark through an equality symbol denotes *not equal to*. Thus if $A = \{1, 2, 3\}$ and $B = \{3, 6\}$, we can write $A \neq B$, which is read *Set A is not equal to set B.*

Real Numbers

The following terminology is commonly used to classify different types of numbers.

$\{1, 2, 3\ 4, \ldots\}$	Natural numbers, counting numbers, positive integers
$\{0, 1, 2, 3, \ldots\}$	Whole numbers, nonnegative integers
$\{\ldots, -3, -2, -1\}$	Negative integers
$\{\ldots, -3, -2, -1, 0\}$	Nonpositive integers
$\{\ldots, -2, -1, 0, 1, 2, \ldots\}$	Integers

A **rational number** is defined as any number that can be expressed in the form a/b, where a and b are integers and b is not zero. The following are examples of rational numbers.

$$\frac{2}{3} \qquad -\frac{3}{4} \qquad\qquad\qquad 6\frac{1}{2} \quad \text{because } 6\frac{1}{2} = \frac{13}{2}$$

$$6 \quad \text{because } 6 = \frac{6}{1} \qquad\qquad\qquad -4 \quad \text{because } -4 = \frac{-4}{1} = \frac{4}{-1}$$

$$0 \quad \text{because } 0 = \frac{0}{1} = \frac{0}{2} = \frac{0}{3}, \text{etc.} \qquad 0.3 \quad \text{because } 0.3 = \frac{3}{10}$$

A rational number can also be defined in terms of a decimal representation. Before doing so, let's briefly review the different possibilities for decimal representations. Decimals can be classified as **terminating**, **repeating**, or **nonrepeating**. Here are some examples of each.

$$\begin{pmatrix} 0.3 \\ 0.46 \\ 0.789 \\ 0.2143 \end{pmatrix} \qquad \text{Terminating decimals}$$

$$\begin{pmatrix} 0.3333\ldots \\ 0.141414\ldots \\ 0.712712712\ldots \\ 0.24171717\ldots \\ 0.9675283283283\ldots \end{pmatrix} \qquad \text{Repeating decimals}$$

$$\begin{pmatrix} 0.472195631\ldots \\ 0.21411711191111\ldots \\ 0.752389433215333\ldots \end{pmatrix} \qquad \text{Nonrepeating decimals}$$

A **repeating decimal** has a block of digits that repeats indefinitely. This repeating block of digits may be of any size and may or may not begin immediately after the decimal point. A small horizontal bar is commonly used to indicate the repeating block. Thus $0.3333\ldots$ can be expressed as $0.\overline{3}$ and $0.24171717\ldots$ as $0.24\overline{17}$.

In terms of decimals, a rational number is defined as a number with either a terminating or a repeating decimal representation. The following examples illustrate some rational numbers written in $\frac{a}{b}$ form and in the equivalent decimal form.

$$\frac{3}{4} = 0.75 \qquad \frac{3}{11} = 0.\overline{27} \qquad \frac{1}{8} = 0.125 \qquad \frac{1}{7} = 0.\overline{142857} \qquad \frac{1}{3} = 0.\overline{3}$$

We define an **irrational number** as a number that cannot be expressed in $\frac{a}{b}$ form, where a and b are integers and b is not zero. Furthermore, an irrational number has a nonrepeating, nonterminating decimal representation. Following are some examples of irrational numbers. A partial decimal representation is given for each.

$$\sqrt{2} = 1.414213562373095\ldots$$
$$\sqrt{3} = 1.73205080756887\ldots$$
$$\pi = 3.14159265358979\ldots$$

The entire set of **real numbers** is composed of the rational numbers along with the irrationals. The following tree diagram can be used to summarize the various classifications of the real number system.

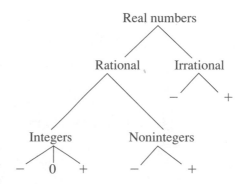

Any real number can be traced down through the tree. Here are some examples:

7 is real, rational, an integer, and positive.

$-\dfrac{2}{3}$ is real, rational, a noninteger, and negative.

$\sqrt{7}$ is real, irrational, and positive.

0.59 is real, rational, a noninteger, and positive.

The concept of a subset is convenient to use at this time. A set A is a **subset** of another set B if and only if every element of A is also an element of B. For example, if $A = \{1, 2\}$ and $B = \{1, 2, 3\}$, then A is a subset of B. This is written $A \subseteq B$ and is read A *is a subset of* B. The slash mark can also be used here to denote negation. If $A = \{1, 2, 4, 6\}$ and $B = \{2, 3, 7\}$, we can say A *is not a subset of* B by writing $A \not\subseteq B$. The following statements use the subset vocabulary and symbolism; they are represented in Figure 0.1.

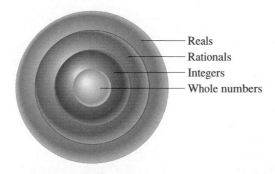

FIGURE 0.1

1. The set of whole numbers is a subset of the set of integers.

$$\{0, 1, 2, 3, \ldots\} \subseteq \{\ldots, -2, -1, 0, 1, 2, \ldots\}$$

2. The set of integers is a subset of the set of rational numbers.

$$\{\ldots, -2, -1, 0, 1, 2, \ldots\} \subseteq \{x \,|\, x \text{ is a rational number}\}$$

3. The set of rational numbers is a subset of the set of real numbers.

$$\{x \,|\, x \text{ is a rational number}\} \subseteq \{y \,|\, y \text{ is a real number}\}$$

Real Number Line and Absolute Value

It is often helpful to have a geometric representation of the set of real numbers in front of us, as indicated in Figure 0.2. Such a representation, called the **real number line**, indicates a one-to-one correspondence between the set of real numbers and the points on a line. In other words, to each real number there corresponds one and only one point on the line, and to each point on the line there corresponds one and only one real number. The number that corresponds to a particular point on the line is called the **coordinate** of that point.

FIGURE 0.2

Many operations, relations, properties, and concepts pertaining to real numbers can be given a geometric interpretation on the number line. For example, the addition problem $(-1) + (-2)$ can be interpreted on the number line as in Figure 0.3.

$(-1) + (-2) = -3$

FIGURE 0.3

FIGURE 0.4

The inequality relations also have a geometric interpretation. The statement $a > b$ (read *a is greater than b*) means that a is to the right of b, and the statement $c < d$ (read *c is less than d*) means that c is to the left of d (see Figure 0.4).

The property $-(-x) = x$ can be pictured on the number line in a sequence of steps. See Figure 0.5.

1. Choose a point having a coordinate of x.

2. Locate its opposite (written as $-x$) on the other side of zero.

3. Locate the opposite of $-x$ [written as $-(-x)$] on the other\ side of zero.

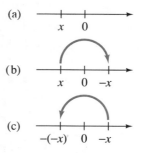

FIGURE 0.5

Therefore, we conclude that **the opposite of the opposite of any real number is the number itself,** and we symbolically express this by $-(-x) = x$.

REMARK   The symbol -1 can be read *negative one*, the *negative of one*, the *opposite of one*, or the *additive inverse of one*. The opposite-of and additive-inverse-of terminology is especially meaningful when working with variables. For example, the symbol $-x$, read *the opposite of x* or *the additive inverse of x*, emphasizes an important issue. Because x can be any real number, $-x$ (opposite of x) can be zero, positive, or negative. If x is positive, then $-x$ is negative. If x is negative, then $-x$ is positive. If x is zero, then $-x$ is zero.

The concept of absolute value can be interpreted on the number line. Geometrically, the **absolute value** of any real number is the distance between that number and zero on the number line. For example, the absolute value of 2 is 2, the absolute value of -3 is 3, and the absolute value of zero is zero (see Figure 0.6). Symbolically, absolute value is denoted with vertical bars. Thus we write $|2| = 2$, $|-3| = 3$, and $|0| = 0$. More formally, the concept of absolute value is defined as follows.

$$|-3| = 3 \qquad |2| = 2$$

$$\begin{array}{ccccccc} -3 & -2 & -1 & 0 & 1 & 2 & 3 \end{array}$$

$$|0| = 0$$

FIGURE 0.6

DEFINITION 0.1

For all real numbers a,

1. If $a \geq 0$, then $|a| = a$.

2. If $a < 0$, then $|a| = -a$.

According to Definition 0.1, we obtain

$|6| = 6$ by applying part 1

$|0| = 0$ by applying part 1

$|-7| = -(-7) = 7$ by applying part 2

Notice that the absolute value of a positive number is the number itself, but the absolute value of a negative number is its opposite. Thus the absolute value of any number except zero is positive, and the absolute value of zero is zero. Together, these facts indicate that the absolute value of any real number is equal to the absolute value of its opposite. All of these ideas are summarized in the following properties.

Properties of Absolute Value

The variables a and b represent any real number.

1. $|a| \geq 0$
2. $|a| = |-a|$
3. $|a - b| = |b - a|$ $a - b$ and $b - a$ are opposites of each other.

FIGURE 0.7

In Figure 0.7 we located points A and B at -2 and 4, respectively. The distance between A and B is 6 units and can be calculated by using either $|-2 - 4|$ or $|4 - (-2)|$. In general, if two points have coordinates x_1 and x_2, the distance between the two points is determined by using either $|x_2 - x_1|$ or $|x_1 - x_2|$, because by the third property, they are the same quantity.

Properties of Real Numbers

As you work with the set of real numbers, the basic operations, and the relations of equality and inequality, the following properties will guide your study. Be sure that you understand these properties, for they not only facilitate manipulations with real numbers but also serve as a basis for many algebraic computations. The variables a, b, and c represent real numbers.

Let's make a few comments about the properties of real numbers. The set of real numbers is said to be **closed** with respect to addition and multiplication. That is, the sum of two real numbers is a real number and the product of two real numbers is a real number. **Closure** plays an important role when we are proving additional properties that pertain to real numbers.

Addition and multiplication are said to be **commutative operations**. This means that the order in which you add or multiply two real numbers does not affect the result. For example, $6 + (-8) = -8 + 6$ and $(-4)(-3) = (-3)(-4)$. It is important to realize that subtraction and division are *not* commutative operations;

Properties of Real Numbers	
Closure properties	$a + b$ is a unique real number. ab is a unique real number.
Commutative properties	$a + b = b + a$ $ab = ba$
Associative properties	$(a + b) + c = a + (b + c)$ $(ab)c = a(bc)$
Identity properties	There exists a real number 0 such that $a + 0 = 0 + a = a.$ There exists a real number 1 such that $a(1) = 1(a) = a.$
Inverse properties	For every real number a, there exists a unique real number $-a$ such that $a + (-a) = (-a) + a = 0.$ For every nonzero real number a, there exists a unique real number $\dfrac{1}{a}$ such that $a\left(\dfrac{1}{a}\right) = \dfrac{1}{a}(a) = 1.$
Multiplication property of zero	$a(0) = (0)(a) = 0$
Multiplication property of negative one	$a(-1) = -1(a) = -a$
Distributive property	$a(b + c) = ab + ac$

order does make a difference. For example, $3 - 4 = -1$, but $4 - 3 = 1$. Likewise, $2 \div 1 = 2$, but $1 \div 2 = \dfrac{1}{2}$.

Addition and multiplication are associative operations. The associative properties are grouping properties. For example, $(-8 + 9) + 6 = -8 + (9 + 6)$; changing the grouping of the numbers does not affect the final sum. Likewise, for multiplication, $[(-4)(-3)](2) = (-4)[(-3)(2)]$. Subtraction and division are *not* associative operations. For example, $(8 - 6) - 10 = -8$, but $8 - (6 - 10) = 12$. An example showing that division is not associative is $(8 \div 4) \div 2 = 1$, but $8 \div (4 \div 2) = 4$.

Zero is the **identity element for addition**. This means that the sum of any real number and zero is identically the same real number. For example, $-87 + 0 = 0 + (-87) = -87$. One is the **identity element for multiplication**. The product of any real number and 1 is identically the same real number. For example, $(-119)(1) = (1)(-119) = -119$.

The real number $-a$ is called the **additive inverse of** a or the **opposite of** a. The sum of a number and its additive inverse is the identity element for addition. For example, 16 and -16 are additive inverses, and their sum is zero. The additive inverse of zero is zero.

The real number $1/a$ is called the **multiplicative inverse** or **reciprocal of** a. The product of a number and its multiplicative inverse is the identity element for multiplication. For example, the reciprocal of 2 is $\frac{1}{2}$, and $2\left(\frac{1}{2}\right) = \frac{1}{2}(2) = 1$.

The product of any real number and zero is zero. For example, $(-17)(0) = (0)(-17) = 0$. The product of any real number and -1 is the opposite of the real number. For example, $(-1)(52) = (52)(-1) = -52$.

The distributive property ties together the operations of addition and multiplication. We say that *multiplication distributes over addition*. For example, $7(3 + 8) = 7(3) + 7(8)$. Furthermore, because $b - c = b + (-c)$, it follows that *multiplication also distributes over subtraction*. This can be symbolically expressed as $a(b - c) = ab - ac$. For example, $6(8 - 10) = 6(8) - 6(10)$.

Algebraic Expressions

Algebraic expressions such as

$$2x \qquad 8xy \qquad -3xy \qquad -4abc \qquad z$$

are called **terms**. A term is an indicated product and may have any number of factors. The variables of a term are called **literal factors** and the numerical factor is called the **numerical coefficient**. Thus in $8xy$, the x and y are literal factors and 8 is the numerical coefficient. Because $1(z) = z$, the numerical coefficient of the term z is understood to be 1. Terms that have the same literal factors are called **similar terms** or **like terms**. The distributive property in the form $ba + ca = (b + c)a$ provides the basis for simplifying algebraic expressions by *combining similar terms*, as illustrated in the following examples.

$$
\begin{aligned}
3x + 5x &= (3 + 5)x & \quad -6xy + 4xy &= (-6 + 4)xy & \quad 4x - x &= 4x - 1x \\
&= 8x & &= -2xy & &= (4 - 1)x \\
& & & & &= 3x
\end{aligned}
$$

Sometimes an algebraic expression can be simplified by applying the distributive property to remove parentheses and combine similar terms, as the next examples illustrate.

$$4(x + 2) + 3(x + 6) = 4(x) + 4(2) + 3(x) + 3(6)$$
$$= 4x + 8 + 3x + 18$$
$$= 7x + 26$$

$$-5(y + 3) -2(y - 8) = -5(y) - 5(3) - 2(y) - 2(-8)$$
$$= -5y - 15 - 2y + 16$$
$$= -7y + 1$$

An algebraic expression takes on a numerical value whenever each variable in the expression is replaced by a real number. For example, when x is replaced by 5 and y by 9, the algebraic expression $x + y$ becomes the numerical expression $5 + 9$, which is equal to 14. We say that $x + y$ has a value of 14 when $x = 5$ and $y = 9$.

Consider the following examples, which illustrate the process of finding a value of an algebraic expression. The process is commonly referred to as **evaluating an algebraic expression.**

EXAMPLE 1

Find the value of $3xy - 4z$ when $x = 2$, $y = -4$, and $z = -5$.

Solution

$$3xy - 4z = 3(2)(-4) - 4(-5) \qquad \text{when } x = 2, y = -4, \text{ and } z = -5$$
$$= -24 + 20$$
$$= -4$$

EXAMPLE 2

Find the value of $a - [4b - (2c + 1)]$ when $a = -8$, $b = -7$, and $c = 14$.

Solution

$$a - [4b - (2c + 1)] = -8 - [4(-7) - (2(14) + 1)]$$
$$= -8 - [-28 - 29]$$
$$= -8 - [-57]$$
$$= 49$$

EXAMPLE 3

Evaluate $\dfrac{a - 2b}{3c + 5d}$ when $a = 14$, $b = -12$, $c = -3$, and $d = -2$.

Solution

$$\frac{a - 2b}{3c + 5d} = \frac{14 - 2(-12)}{3(-3) + 5(-2)}$$
$$= \frac{14 + 24}{-9 - 10}$$
$$= \frac{38}{-19} = -2$$

Look back at Examples 1–3 and note that we use the following **order of operations** when simplifying numerical expressions.

> **1.** Perform the operations inside the symbols of inclusion (parentheses, brackets, and braces) and above and below each fraction bar. Start with the innermost inclusion symbol.
>
> **2.** Perform all multiplications and divisions in the order in which they appear, from left to right.
>
> **3.** Perform all additions and subtractions in the order in which they appear, from left to right.

You should also realize that first simplifying by combining similar terms can sometimes aid in the process of evaluating algebraic expressions. The last example of this section illustrates this idea.

EXAMPLE 4

Evaluate $2(3x + 1) - 3(4x - 3)$ when $x = -5$.

Solution

$$
\begin{aligned}
2(3x + 1) - 3(4x - 3) &= 2(3x) + 2(1) - 3(4x) - 3(-3) \\
&= 6x + 2 - 12x + 9 \\
&= -6x + 11
\end{aligned}
$$

Now substituting -5 for x, we obtain

$$
\begin{aligned}
-6x + 11 &= -6(-5) + 11 \\
&= 30 + 11 \\
&= 41
\end{aligned}
$$

∎

PROBLEM SET 0.1

For Problems 1–10, identify each statement as *true* or *false*.

1. Every rational number is a real number.

2. Every irrational number is a real number.

3. Every real number is a rational number.

4. If a number is real, then it is irrational.

5. Some irrational numbers are also rational numbers.

6. All integers are rational numbers.

7. The number zero is a rational number.

8. Zero is a positive integer.

9. Zero is a negative number.

10. All whole numbers are integers.

For Problems 11–18, list those elements of the set of numbers $\left[0, \ \sqrt{5}, \ -\sqrt{2}, \ \dfrac{7}{8}, \ -\dfrac{10}{13}, \ 7\dfrac{1}{8}, \ 0.279, \ 0.4\overline{67}, \ -\pi, \ -14, \ 46, \ 6.75 \right]$ that belong to each of the following sets.

11. The natural numbers

12. The whole numbers

13. The integers

14. The rational numbers

15. The irrational numbers

16. The nonnegative integers

17. The nonpositive integers

18. The real numbers

For Problems 19–32, use the following set designations.

$N = \{x \mid x \text{ is a natural number}\}$

$W = \{x \mid x \text{ is a whole number}\}$

$I = \{x \mid x \text{ is an integer}\}$

$Q = \{x \mid x \text{ is a rational number}\}$

$H = \{x \mid x \text{ is an irrational number}\}$

$R = \{x \mid x \text{ is a real number}\}$

Place $\subseteq$ or $\not\subseteq$ in each blank to make a true statement.

19. N _____ R **20.** R _____ N

21. N _____ I **22.** I _____ Q

23. H _____ Q **24.** Q _____ H

25. W _____ I **26.** N _____ W

27. I _____ W **28.** I _____ N

29. $\{0, 2, 4, \ldots\}$ _____ W

30. $\{1, 3, 5, 7, \ldots\}$ _____ I

31. $\{-2, -1, 0, 1, 2\}$ _____ W

32. $\{0, 3, 6, 9, \ldots\}$ _____ N

For Problems 33–42, list the elements of each set. For example, the elements of $\{x \mid x \text{ is a natural number less than } 4\}$ can be listed $\{1, 2, 3\}$.

33. $\{x \mid x \text{ is a natural number less than } 2\}$

34. $\{x \mid x \text{ is a natural number greater than } 5\}$

35. $\{n \mid n \text{ is a whole number less than } 4\}$

36. $\{y \mid y \text{ is an integer greater than } -3\}$

37. $\{y \mid y \text{ is an integer less than } 2\}$

38. $\{n \mid n \text{ is a positive integer greater than } -4\}$

39. $\{x \mid x \text{ is a whole number less than } 0\}$

40. $\{x \mid x \text{ is a negative integer greater than } -5\}$

41. $\{n \mid n \text{ is a nonnegative integer less than } 3\}$

42. $\{n \mid n \text{ is a nonpositive integer greater than } 1\}$

43. Find the distance on the real number line between two points whose coordinates are as follows.

    a. 17 and 35   **b.** -14 and 12

    c. 18 and -21 **d.** -17 and -42

    e. -56 and -21 **f.** 0 and -37

44. Evaluate each of the following if x is a nonzero real number.

    a. $\dfrac{|x|}{x}$ **b.** $\dfrac{x}{|x|}$

    c. $\dfrac{|-x|}{-x}$ **d.** $|x| - |-x|$

In Problems 45–58, state the property that justifies each of the statements. For example, $3 + (-4) = (-4) + 3$ because of the commutative property of addition.

45. $x(2) = 2(x)$

46. $(7 + 4) + 6 = 7 + (4 + 6)$

47. $1(x) = x$

48. $43 + (-18) = (-18) + 43$

49. $(-1)(93) = -93$

50. $109 + (-109) = 0$

51. $5(4 + 7) = 5(4) + 5(7)$

52. $-1(x + y) = -(x + y)$

53. $7yx = 7xy$

54. $(x + 2) + (-2) = x + [2 + (-2)]$

55. $6(4) + 7(4) = (6 + 7)(4)$

56. $\left(\dfrac{2}{3}\right)\left(\dfrac{3}{2}\right) = 1$

57. $4(5x) = (4 \cdot 5)x$

58. $[(17)(8)](25) = (17)[(8)(25)]$

For Problems 59–79, evaluate each of the algebraic expressions for the given values of the variables.

59. $5x + 3y$; $x = -2$ and $y = -4$

60. $7x - 4y$; $x = -1$ and $y = 6$

61. $-3ab - 2c$; $a = -4, b = 7,$ and $c = -8$

62. $x - (2y + 3z)$; $x = -3, y = -4,$ and $z = 9$

63. $(a - 2b) + (3c - 4)$; $a = 6, b = -5,$ and $c = -11$

64. $3a - [2b - (4c + 1)]$; $a = 4, b = 6,$ and $c = -8$

65. $\dfrac{-2x + 7y}{x - y}$; $x = -3$ and $y = -2$

66. $\dfrac{x - 3y + 2z}{2x - y}$; $x = 4, y = 9, z = -12$

67. $(5x - 2y)(-3x + 4y)$; $x = -3$ and $y = -7$

68. $(2a - 7b)(4a + 3b)$; $a = 6$ and $b = -3$

69. $5x + 4y - 9y - 2y$; $x = 2$ and $y = -8$

70. $5a + 7b - 9a - 6b$; $a = -7$ and $b = 8$

71. $-5x + 8y + 7y + 8x$; $x = 5$ and $y = -6$

72. $|x - y| - |x + y|$; $x = -4$ and $y = -7$

73. $|3x + y| + |2x - 4y|$; $x = 5$ and $y = -3$

74. $\left|\dfrac{x - y}{y - x}\right|$; $x = -6$ and $y = 13$

75. $\left|\dfrac{2a - 3b}{3b - 2a}\right|$; $a = -4$ and $b = -8$

76. $5(x - 1) + 7(x + 4)$; $x = 3$

77. $2(3x + 4) - 3(2x - 1)$; $x = -2$

78. $-4(2x - 1) - 5(3x + 7)$; $x = -1$

79. $5(a - 3) - 4(2a + 1) - 2(a - 4)$; $a = -3$

80. You should be able to do calculations like those in Problems 59–79 with and without a calculator. Different types of calculators handle the priority-of-operations issue in different ways. Be sure you can do Problems 59–79 with your calculator.

THOUGHTS INTO WORDS

81. Do you think $3\sqrt{2}$ is a rational or an irrational number? Defend your answer.

82. Explain why $\dfrac{0}{8} = 0$ but $\dfrac{8}{0}$ is undefined.

83. The "solution" of the following simplification problem is incorrect. The answer should be -11. Find and correct the error.

$$8 \div (-4)(2) - 3(4) \div 2 + (-1) = (-2)(2) - 12 \div 1$$
$$= -4 - 12$$
$$= -16$$

84. Explain the difference between "simplifying a numerical expression" and "evaluating an algebraic expression."

0.2 EXPONENTS

Positive integers are used as **exponents** to indicate repeated multiplication. For example, $4 \cdot 4 \cdot 4$ can be written 4^3, where the raised 3 indicates that 4 is to be used as a factor three times. The following general definition is helpful.

DEFINITION 0.2

If n is a positive integer and b is any real number, then

$$b^n = \underbrace{bbb \cdots b}_{n \text{ factors of } b}$$

The number b is referred to as the **base** and n is called the **exponent**. The expression b^n can be read *b to the nth power*. The terms **squared** and **cubed** are commonly associated with exponents of 2 and 3, respectively. For example, b^2 is read *b squared* and b^3 as *b cubed*. An exponent of 1 is usually not written, so b^1 is simply written b. The following examples illustrate Definition 0.2.

$$2^3 = 2 \cdot 2 \cdot 2 = 8 \qquad\qquad \left(\frac{1}{2}\right)^5 = \frac{1}{2} \cdot \frac{1}{2} \cdot \frac{1}{2} \cdot \frac{1}{2} \cdot \frac{1}{2} = \frac{1}{32}$$

$$3^4 = 3 \cdot 3 \cdot 3 \cdot 3 = 81 \qquad\qquad (0.7)^2 = (0.7)(0.7) = 0.49$$

$$(-5)^2 = (-5)(-5) = 25 \qquad\qquad -5^2 = -(5 \cdot 5) = -25$$

We especially want to call your attention to the last two examples. Note that $(-5)^2$ means that -5 is the base used as a factor twice. However, -5^2 means that 5 is the base and that after it is squared, we take the opposite of the result.

Properties of Exponents

In a previous algebra course, you may have seen some properties pertaining to the use of positive integers as exponents. Those properties can be summarized as follows.

PROPERTY 0.1   *Properties of Exponents*

If a and b are real numbers and m and n are positive integers, then

1. $b^n \cdot b^m = b^{n+m}$
2. $(b^n)^m = b^{mn}$
3. $(ab)^n = a^n b^n$
4. $\left(\dfrac{a}{b}\right)^n = \dfrac{a^n}{b^n} \quad b \neq 0$
5. $\dfrac{b^n}{b^m} = b^{n-m} \qquad$ when $n > m, \quad b \neq 0$

   $\dfrac{b^n}{b^m} = 1 \qquad$ when $n = m, \quad b \neq 0$

   $\dfrac{b^n}{b^m} = \dfrac{1}{b^{m-n}} \qquad$ when $n < m, \quad b \neq 0$

Each part of Property 0.1 can be justified by using Definition 0.2. For example, to justify part 1, we can reason as follows.

$$b^n \cdot b^m = (\underbrace{bbb \cdots b}_{\substack{n \text{ factors} \\ \text{of } b}}) \cdot (\underbrace{bbb \cdots b}_{\substack{m \text{ factors} \\ \text{of } b}})$$

$$= \underbrace{bbb \ldots b}_{\substack{n + m \text{ factors} \\ \text{of } b}}$$

$$= b^{n+m}$$

Similar reasoning can be used to verify the other parts of Property 0.1. The following examples illustrate the use of Property 0.1 along with the commutative and associative properties of the real numbers. The steps enclosed in the dashed boxes can be performed mentally.

EXAMPLE 1

$$(3x^2y)(4x^3y^2) = \boxed{3 \cdot 4 \cdot x^2 \cdot x^3 \cdot y \cdot y^2}$$
$$= \boxed{12x^{2+3}y^{1+2}} \qquad b^n \cdot b^m = b^{n+m}$$
$$= 12x^5y^3 \qquad\qquad\qquad\qquad\qquad\qquad ■$$

EXAMPLE 2

$$\boxed{(-2y^3)^5 = (-2)^5(y^3)^5} \qquad (ab)^n = a^n b^n$$
$$= -32y^{15} \qquad\qquad (b^n)^m = b^{mn} \qquad\quad ■$$

EXAMPLE 3

$$\left(\frac{a^2}{b^4}\right)^7 = \boxed{\frac{(a^2)^7}{(b^4)^7}} \qquad \left(\frac{a}{b}\right)^n = \frac{a^n}{b^n}$$
$$= \frac{a^{14}}{b^{28}} \qquad (b^n)^m = b^{mn} \qquad\qquad ■$$

EXAMPLE 4

$$\frac{-56x^9}{7x^4} = \boxed{-8x^{9-4}} \qquad \frac{b^n}{b^m} = b^{n-m} \quad \text{when } n > m$$
$$= -8x^5 \qquad\qquad\qquad\qquad\qquad\qquad\qquad ■$$

Zero and Negative Integers as Exponents

Now we can extend the concept of an exponent to include the use of zero and negative integers. First let's consider the use of zero as an exponent. We want to use zero in a way that Property 0.1 will continue to hold. For example, if $b^n \cdot b^m = b^{n+m}$ is to hold, then $x^4 \cdot x^0$ should equal x^{4+0}, which equals x^4. In other words, x^0 *acts like* 1 because $x^4 \cdot x^0 = x^4$. Look at the following definition.

DEFINITION 0.3

If b is a nonzero real number, then

$$b^0 = 1$$

Therefore, according to Definition 0.3, the following statements are all true.

$$5^0 = 1 \qquad (-413)^0 = 1$$

$$\left(\frac{3}{11}\right)^0 = 1 \qquad (x^3 y^4)^0 = 1 \quad \text{if } x \neq 0 \text{ and } y \neq 0$$

A similar line of reasoning can be used to motivate a definition for the use of negative integers as exponents. Consider the example $x^4 \cdot x^{-4}$. If $b^n \cdot b^m = b^{n+m}$ is to hold, then $x^4 \cdot x^{-4}$ should equal $x^{4+(-4)}$, which equals $x^0 = 1$. Therefore, x^{-4} must be the reciprocal of x^4, because their product is 1. That is, $x^{-4} = 1/x^4$. This suggests the following definition.

DEFINITION 0.4

If n is a positive integer and b is a nonzero real number, then

$$b^{-n} = \frac{1}{b^n}$$

According to Definition 0.4, the following statements are true.

$$x^{-5} = \frac{1}{x^5} \qquad\qquad 2^{-4} = \frac{1}{2^4} = \frac{1}{16}$$

$$\left(\frac{3}{4}\right)^{-2} = \frac{1}{\left(\frac{3}{4}\right)^2} = \frac{1}{\frac{9}{16}} = \frac{16}{9} \qquad \frac{2}{x^{-3}} = \frac{2}{\frac{1}{x^3}} = 2x^3$$

The first four parts of Property 0.1 hold true *for all integers*. Furthermore, we do not need all three equations in part 5 of Property 0.1. The first equation,

$$\frac{b^n}{b^m} = b^{n-m}$$

can be used for *all integral exponents*. Let's restate Property 0.1 as it pertains to integers. We will include name tags for easy reference.

> ### PROPERTY 0.2
>
> If m and n are integers and a and b are real numbers, with $b \neq 0$ whenever it appears in a denominator, then
>
> **1.** $b^n \cdot b^m = b^{n+m}$ Product of two powers
>
> **2.** $(b^n)^m = b^{mn}$ Power of a power
>
> **3.** $(ab)^n = a^n b^n$ Power of a product
>
> **4.** $\left(\dfrac{a}{b}\right)^n = \dfrac{a^n}{b^n}$ Power of a quotient
>
> **5.** $\dfrac{b^n}{b^m} = b^{n-m}$ Quotient of two powers

Having the use of all integers as exponents allows us to work with a large variety of numerical and algebraic expressions. Let's consider some examples that illustrate the various parts of Property 0.2.

EXAMPLE 5

Evaluate each of the following numerical expressions.

a. $(2^{-1} \cdot 3^2)^{-1}$ **b.** $\left(\dfrac{2^{-3}}{3^{-2}}\right)^{-2}$

Solutions

a.
$$\begin{aligned}
(2^{-1} \cdot 3^2)^{-1} &= (2^{-1})^{-1}(3^2)^{-1} &&\text{Power of a product} \\
&= (2^1)(3^{-2}) &&\text{Power of a power} \\
&= (2)\left(\frac{1}{3^2}\right) \\
&= 2\left(\frac{1}{9}\right) = \frac{2}{9}
\end{aligned}$$

b.
$$\begin{aligned}
\left(\frac{2^{-3}}{3^{-2}}\right)^{-2} &= \frac{(2^{-3})^{-2}}{(3^{-2})^{-2}} &&\text{Power of a quotient} \\
&= \frac{2^6}{3^4} &&\text{Power of a power} \\
&= \frac{64}{81}
\end{aligned}$$

■

EXAMPLE 6

Find the indicated products and quotients and express the final results with positive integral exponents only.

a. $(3x^2y^{-4})(4x^{-3}y)$ **b.** $\dfrac{12a^3b^2}{-3a^{-1}b^5}$ **c.** $\left(\dfrac{15x^{-1}y^2}{5xy^{-4}}\right)^{-1}$

Solutions

a. $(3x^2y^{-4})(4x^{-3}y) = 12x^{2+(-3)}y^{-4+1}$ Product of powers

$\qquad\qquad\qquad\quad = 12x^{-1}y^{-3}$

$\qquad\qquad\qquad\quad = \dfrac{12}{xy^3}$

b. $\dfrac{12a^3b^2}{-3a^{-1}b^5} = -4a^{3-(-1)}b^{2-5}$ Quotient of powers

$\qquad\qquad\quad = -4a^4b^{-3}$

$\qquad\qquad\quad = -\dfrac{4a^4}{b^3}$

c. $\left(\dfrac{15x^{-1}y^2}{5xy^{-4}}\right)^{-1} = (3x^{-1-1}y^{2-(-4)})^{-1}$ First simplify inside parentheses

$\qquad\qquad\qquad\quad = (3x^{-2}y^6)^{-1}$

$\qquad\qquad\qquad\quad = 3^{-1}x^2y^{-6}$ Power of a product

$\qquad\qquad\qquad\quad = \dfrac{x^2}{3y^6}$ ■

The next three examples illustrate the simplification of numerical and algebraic expressions involving sums and differences. In such cases, Definition 0.4 can be used to change from negative to positive exponents so that we can proceed in the usual ways.

EXAMPLE 7

Simplify $2^{-3} + 3^{-1}$.

Solution

$$2^{-3} + 3^{-1} = \frac{1}{2^3} + \frac{1}{3^1}$$

$$= \frac{1}{8} + \frac{1}{3}$$

$$= \frac{3}{24} + \frac{8}{24}$$

$$= \frac{11}{24}$$ ■

EXAMPLE 8

Simplify $(4^{-1} - 3^{-2})^{-1}$.

Solution

$$
\begin{aligned}
(4^{-1} - 3^{-2})^{-1} &= \left(\frac{1}{4^1} - \frac{1}{3^2}\right)^{-1} \\
&= \left(\frac{1}{4} - \frac{1}{9}\right)^{-1} \\
&= \left(\frac{9}{36} - \frac{4}{36}\right)^{-1} \\
&= \left(\frac{5}{36}\right)^{-1} \\
&= \frac{1}{\left(\dfrac{5}{36}\right)^1} = \frac{36}{5}
\end{aligned}
$$

■

EXAMPLE 9

Express $a^{-1} + b^{-2}$ as a single fraction involving positive exponents only.

Solution

$$
\begin{aligned}
a^{-1} + b^{-2} &= \frac{1}{a^1} + \frac{1}{b^2} \\
&= \left(\frac{1}{a}\right)\left(\frac{b^2}{b^2}\right) + \left(\frac{1}{b^2}\right)\left(\frac{a}{a}\right) \\
&= \frac{b^2}{ab^2} + \frac{a}{ab^2} \\
&= \frac{b^2 + a}{ab^2}
\end{aligned}
$$

■

Scientific Notation

The expression $(n)(10)^k$ (where n is a number greater than or equal to 1 and less than 10, written in decimal form, and k is any integer) is commonly called **scientific notation** or the **scientific form** of a number. The following are examples of numbers expressed in scientific form.

$$(4.23)(10)^4 \qquad (8.176)(10)^{12} \qquad (5.02)(10)^{-3} \qquad (1)(10)^{-5}$$

Very large and very small numbers can be conveniently expressed in scientific notation. For example, a light year (the distance that a ray of light travels in one year) is approximately 5,900,000,000,000 miles, and this can be written as $(5.9)(10)^{12}$. The weight of an oxygen molecule is approximately 0.000000000000000000000053 of a gram, and this can be expressed as $(5.3)(10)^{-23}$.

To change from ordinary decimal notation to scientific notation, the following procedure can be used.

> Write the given number as the product of a number greater than or equal to 1 and less than 10, and a power of 10. The exponent of 10 is determined by counting the number of places that the decimal point was moved when going from the original number to the number greater than or equal to 1 and less than 10. This exponent is (a) negative if the original number is less than 1, (b) positive if the original number is greater than 10, and (c) 0 if the original number itself is between 1 and 10.

Thus we can write

$$0.00092 = (9.2)(10)^{-4}$$
$$872,000,000 = (8.72)(10)^8$$
$$5.1217 = (5.1217)(10)^0$$

To change from scientific notation to ordinary decimal notation, the following procedure can be used.

> Move the decimal point the number of places indicated by the exponent of 10. Move the decimal point to the right if the exponent is positive. Move it to the left if the exponent is negative.

Thus we can write

$$(3.14)(10)^7 = 31,400,000$$
$$(7.8)(10)^{-6} = 0.0000078$$

Scientific notation can be used to simplify numerical calculations. We merely change the numbers to scientific notation and use the appropriate properties of exponents. Consider the following examples.

Perform the indicated operations.

EXAMPLE 10

a. $\dfrac{(0.00063)(960,000)}{(3200)(0.0000021)}$ *b.* $\sqrt{90,000}$

Solution

a. $\dfrac{(0.00063)(960,000)}{(3200)(0.0000021)} = \dfrac{(6.3)(10)^{-4}(9.6)(10)^5}{(3.2)(10)^3(2.1)(10)^{-6}}$

$$= \dfrac{(6.3)(9.6)(10)^1}{(3.2)(2.1)(10)^{-3})}$$
$$= (9)(10)^4$$
$$= 90,000$$

b. $\sqrt{90{,}000} = \sqrt{(9)(10)^4}$
$= \sqrt{9}\sqrt{10^4}$
$= (3)(10)^2$
$= 3(100)$
$= 300$ ∎

Many calculators are equipped to display numbers in scientific notation. The display panel shows the number between 1 and 10 and the appropriate exponent of 10. For example, evaluating $(3{,}800{,}000)^2$ yields

$\boxed{\text{1.44E13}}$

Thus $(3{,}800{,}000)^2 = (1.444)(10)^{13} = 14{,}440{,}000{,}000{,}000$. Similarly, the answer for $(0.000168)^2$ is displayed as

$\boxed{\text{2.8224E}-8}$

Thus $(0.000168)^2 = (2.8224)(10)^{-8} = 0.000000028224$.

Calculators vary in the number of digits they display between 1 and 10 when they represent a number in scientific notation. For example, we used two different calculators to estimate $(6729)^6$ and obtained the following results.

$\boxed{\text{9.283316768E22}}$

$\boxed{\text{9.28331676776E22}}$

Obviously, you need to know the capabilities of your calculator when working with problems in scientific notation.

Many calculators also allow the entry of a number in scientific notation. Such calculators are equipped with an enter-the-exponent key often labeled $\boxed{\text{EE}}$. Thus a number such as $(3.14)(10)^8$ might be entered as follows.

ENTER	PRESS	DISPLAY
3.14	$\boxed{\text{EE}}$	3.14E
8		3.14E8

A $\boxed{\text{MODE}}$ key often appears on calculators; it enables you to choose the type of notation. Be sure you understand how to express numbers in scientific notation on your calculator.

PROBLEM SET 0.2

For Problems 1–42, evaluate each numerical expression.

1. 2^{-3}

2. 3^{-2}

3. -10^{-3}

4. 10^{-4}

5. $\dfrac{1}{3^{-3}}$

6. $\dfrac{1}{2^{-5}}$

7. $\left(\dfrac{1}{2}\right)^{-2}$

8. $-\left(\dfrac{1}{3}\right)^{-2}$

9. $\left(-\dfrac{2}{3}\right)^{-3}$

10. $\left(\dfrac{5}{6}\right)^{-2}$

11. $\left(-\dfrac{1}{5}\right)^{0}$

12. $\dfrac{1}{\left(\dfrac{3}{5}\right)^{-2}}$

13. $\dfrac{1}{\left(\dfrac{4}{5}\right)^{-2}}$

14. $\left(\dfrac{4}{5}\right)^{0}$

15. $2^5 \cdot 2^{-3}$

16. $3^{-2} \cdot 3^5$

17. $10^{-6} \cdot 10^4$

18. $10^6 \cdot 10^{-9}$

19. $10^{-2} \cdot 10^{-3}$

20. $10^{-1} \cdot 10^{-5}$

21. $(3^{-2})^{-2}$

22. $((-2)^{-1})^{-3}$

23. $(4^2)^{-1}$

24. $(3^{-1})^3$

25. $(3^{-1} \cdot 2^2)^{-1}$

26. $(2^3 \cdot 3^{-2})^{-2}$

27. $(4^2 \cdot 5^{-1})^2$

28. $(2^{-2} \cdot 4^{-1})^3$

29. $\left(\dfrac{2^{-2}}{5^{-1}}\right)^{-2}$

30. $\left(\dfrac{3^{-1}}{2^{-3}}\right)^{-2}$

31. $\left(\dfrac{3^{-2}}{8^{-1}}\right)^{2}$

32. $\left(\dfrac{4^2}{5^{-1}}\right)^{-1}$

33. $\dfrac{2^3}{2^{-3}}$

34. $\dfrac{2^{-3}}{2^3}$

35. $\dfrac{10^{-1}}{10^4}$

36. $\dfrac{10^{-3}}{10^{-7}}$

37. $3^{-2} + 2^{-3}$

38. $2^{-3} + 5^{-1}$

39. $\left(\dfrac{2}{3}\right)^{-1} - \left(\dfrac{3}{4}\right)^{-1}$

40. $3^{-2} - 2^3$

41. $(2^{-4} + 3^{-1})^{-1}$

42. $(3^{-2} - 5^{-1})^{-1}$

Simplify Problems 43–62; express final results without using zero or negative integers as exponents.

43. $x^3 \cdot x^{-7}$

44. $x^{-2} \cdot x^{-3}$

45. $a^2 \cdot a^{-3} \cdot a^{-1}$

46. $b^{-3} \cdot b^5 \cdot b^{-4}$

47. $(a^{-3})^2$

48. $(b^5)^{-2}$

49. $(x^3 y^{-4})^{-1}$

50. $(x^4 y^{-2})^{-2}$

51. $(ab^2 c^{-1})^{-3}$

52. $(a^2 b^{-1} c^{-2})^{-4}$

53. $(2x^2 y^{-1})^{-2}$

54. $(3x^4 y^{-2})^{-1}$

55. $\left(\dfrac{x^{-2}}{y^{-3}}\right)^{-2}$

56. $\left(\dfrac{y^4}{x^{-1}}\right)^{-3}$

57. $\left(\dfrac{2a^{-1}}{3b^{-2}}\right)^{-2}$

58. $\left(\dfrac{3x^2 y}{4a^{-1} b^{-3}}\right)^{-1}$

59. $\dfrac{x^{-5}}{x^{-2}}$

60. $\dfrac{a^{-3}}{a^5}$

61. $\dfrac{a^2 b^{-3}}{a^{-1} b^{-2}}$

62. $\dfrac{x^{-1} y^{-2}}{x^3 y^{-1}}$

For Problems 63–70, find the indicated products, quotients, and powers; express answers without using zero or negative integers as exponents.

63. $(4x^3 y^2)(-5xy^3)$

64. $(-6xy)(3x^2 y^4)$

65. $(-3xy^3)^3$

66. $(-2x^2 y^4)^4$

67. $\left(\dfrac{2x^2}{3y^3}\right)^{3}$

68. $\left(\dfrac{4x}{5y^2}\right)^{3}$

69. $\dfrac{72x^8}{-9x^2}$

70. $\dfrac{108x^6}{-12x^2}$

For Problems 71–80, find the indicated products and quotients; express results using positive integral exponents only.

71. $(2x^{-1} y^2)(3x^{-2} y^{-3})$

72. $(4x^{-2} y^3)(-5x^3 y^{-4})$

73. $(-6a^5 y^{-4})(-a^{-7} y)$

74. $(-8a^{-4} b^{-5})(-6a^{-1} b^8)$

75. $\dfrac{24x^{-1}y^{-2}}{6x^{-4}y^3}$

76. $\dfrac{56xy^{-3}}{8x^2y^2}$

77. $\dfrac{-35a^3b^{-2}}{7a^5b^{-1}}$

78. $\dfrac{27a^{-4}b^{-5}}{-3a^{-2}b^{-4}}$

79. $\left(\dfrac{14x^{-2}y^{-4}}{7x^{-3}y^{-6}}\right)^{-2}$

80. $\left(\dfrac{24x^5y^{-3}}{-8x^6y^{-1}}\right)^{-3}$

For Problems 81–88, express each as a single fraction involving positive exponents only.

81. $x^{-1} + x^{-2}$

82. $x^{-2} + x^{-4}$

83. $x^{-2} - y^{-1}$

84. $2x^{-1} - 3y^{-3}$

85. $3a^{-2} + 2b^{-3}$

86. $a^{-2} + a^{-1}b^{-2}$

87. $x^{-1}y - xy^{-1}$

88. $x^2y^{-1} - x^{-3}y^2$

For Problems 89–98, find the following products and quotients. Assume that all variables appearing as exponents represent integers. For example,

$$(x^{2b})(x^{-b+1}) = x^{2b+(-b+1)} = x^{b+1}$$

89. $(3x^a)(4x^{2a+1})$

90. $(5x^{-a})(-6x^{3a-1})$

91. $(x^a)(x^{-a})$

92. $(-2y^{3b})(-4y^{b+1})$

93. $\dfrac{x^{3a}}{x^a}$

94. $\dfrac{4x^{2a+1}}{2x^{a-2}}$

95. $\dfrac{-24y^{5b+1}}{6y^{-b-1}}$

96. $(x^a)^{2b}(x^b)^a$

97. $\dfrac{(xy)^b}{y^b}$

98. $\dfrac{(2x^{2b})(-4x^{b+1})}{8x^{-b+2}}$

For Problems 99–102, express each number in scientific notation.

99. 62,000,000

100. 17,000,000,000

101. 0.000412

102. 0.000000078

For Problems 103–106, change each number from scientific notation to ordinary decimal form.

103. $(1.8)(10)^5$

104. $(5.41)(10)^7$

105. $(2.3)(10)^{-6}$

106. $(4.13)(10)^{-9}$

For Problems 107–112, use scientific notation and the properties of exponents to help perform the indicated operations.

107. $\dfrac{0.00052}{0.013}$

108. $\dfrac{(0.000075)(4,800,000)}{(15,000)(0.0012)}$

109. $\sqrt{900,000,000}$

110. $\sqrt{0.000004}$

111. $\sqrt{0.0009}$

112. $\dfrac{(0.00069)(0.0034)}{(0.0000017)(0.023)}$

113. Explain how you would simplify $(3^{-1} \cdot 2^{-2})^{-1}$ and also how you would simplify $(3^{-1} + 2^{-2})^{-1}$.

114. How would you explain why the product of x^2 and x^4 is x^6 and not x^8?

0.3   POLYNOMIALS

Recall that algebraic expressions such as $5x$, $-6y^2$, $2x^{-1}y^{-2}$, $14a^2b$, $5x^{-4}$, and $-17ab^2c^3$ are called **terms**. Terms that contain variables with only nonnegative integers as exponents are called **monomials**. Of the previously listed terms, $5x$, $-6y^2$, $14a^2b$, and $-17ab^2c^3$ are monomials. The **degree** of a monomial is the sum of the exponents of the literal factors. For example, $7xy$ is of degree 2, whereas $14a^2b$ is of degree 3, and $-17ab^2c^3$ is of degree 6. If the monomial contains only one variable, then the exponent of that variable is the degree of the monomial. For example, $5x^3$ is of degree 3 and $-8y^4$ is of degree 4. Any nonzero constant term, such as 8, is of degree zero.

A **polynomial** is a monomial or a finite sum of monomials. Thus all of the following are examples of polynomials.

$$4x^2 \qquad 3x^2 - 2x - 4 \qquad 7x^4 - 6x^3 + 5x^2 - 2x - 1$$

$$3x^2y + 2y \qquad \frac{1}{5}a^2 - \frac{2}{3}b^2 \qquad 14$$

In addition to calling a polynomial with one term a monomial, we also classify polynomials with two terms as **binomials** and those with three terms as **trinomials**. The **degree of a polynomial** is the degree of the term with the highest degree in the polynomial. The following examples illustrate some of this terminology.

The polynomial $4x^3y^4$ is a monomial in two variables of degree 7.

The polynomial $4x^2y - 2xy$ is a binomial in two variables of degree 3.

The polynomial $9x^2 - 7x - 1$ is a trinomial in one variable of degree 2.

Addition and Subtraction of Polynomials

Both adding polynomials and subtracting them rely on basically the same ideas. The commutative, associative, and distributive properties provide the basis for rearranging, regrouping, and combining similar terms. Consider the following addition problems.

$$(4x^2 + 5x + 1) + (7x^2 - 9x + 4) = (4x^2 + 7x^2) + (5x - 9x) + (1 + 4)$$
$$= 11x^2 - 4x + 5$$

$$(5x - 3) + (3x + 2) + (8x + 6) = (5x + 3x + 8x) + (-3 + 2 + 6)$$
$$= 16x + 5$$

The definition of subtraction as *adding the opposite* $[a - b = a + (-b)]$ extends to polynomials in general. The opposite of a polynomial can be formed by taking the opposite of each term. For example, the opposite of $3x^2 - 7x + 1$ is $-3x^2 + 7x - 1$. Symbolically, this is expressed as

$$-(3x^2 - 7x + 1) = -3x^2 + 7x - 1$$

You can also think in terms of the property $-x = -1(x)$ and the distributive property. Therefore,

$$-(3x^2 - 7x + 1) = -1(3x^2 - 7x + 1) = -3x^2 + 7x - 1$$

Now consider the following subtraction problems.

$$(7x^2 - 2x - 4) - (3x^2 + 7x - 1) = (7x^2 - 2x - 4) + (-3x^2 - 7x + 1)$$
$$= (7x^2 - 3x^2) + (-2x - 7x) + (-4 + 1)$$
$$= 4x^2 - 9x - 3$$

$$(4y^2 + 7) - (-3y^2 + y - 2) = (4y^2 + 7) + (3y^2 - y + 2)$$
$$= (4y^2 + 3y^2) + (-y) + (7 + 2)$$
$$= 7y^2 - y + 9$$

Multiplying Polynomials

The distributive property is usually stated as $a(b + c) = ab + ac$, but it can be extended as follows.

$$a(b + c + d) = ab + ac + ad$$
$$a(b + c + d + e) = ab + ac + ad + ae \qquad \text{etc.}$$

The commutative and associative properties, the properties of exponents, and the distributive property work together to form a basis for finding the product of a monomial and a polynomial. The following example illustrates this idea.

$$3x^2(2x^2 + 5x + 3) = 3x^2(2x^2) + 3x^2(5x) + 3x^2(3)$$
$$= 6x^4 + 15x^3 + 9x^2$$

Extending the method of finding the product of a monomial and a polynomial to finding the product of two polynomials is again based on the distributive property.

$$(x + 2)(y + 5) = x(y + 5) + 2(y + 5)$$
$$= x(y) + x(5) + 2(y) + 2(5)$$
$$= xy + 5x + 2y + 10$$

Notice that each term of the first polynomial multiplies each term of the second polynomial.

$$(x - 3)(y + z + 3) = x(y + z + 3) - 3(y + z + 3)$$
$$= xy + xz + 3x - 3y - 3z - 9$$

Frequently, multiplying polynomials produces similar terms that can be combined, which simplifies the resulting polynomial.

$$(x + 5)(x + 7) = x(x+7) + 5(x + 7)$$
$$= x^2 + 7x + 5x + 35$$
$$= x^2 + 12x + 35$$

$$(x - 2)(x^2 - 3x + 4) = x(x^2 - 3x + 4) - 2(x^2 - 3x + 4)$$
$$= x^3 - 3x^2 + 4x - 2x^2 + 6x - 8$$
$$= x^3 - 5x^2 + 10x - 8$$

In a previous algebra course, you may have developed a shortcut for multiplying binomials, as illustrated by Figure 0.8.

STEP ①   Multiply $(2x)(3x)$.

STEP ②   Multiply $(5)(3x)$ and $(2x)(-2)$ and combine.

STEP ③   Multiply $(5)(-2)$.

REMARK   Shortcuts can be very helpful for certain manipulations in mathematics. But a word of caution: Do not lose the understanding of what you are doing. Make sure that you are able to do the manipulation without the shortcut.

$$(2x + 5)(3x - 2) = 6x^2 + 11x - 10$$

FIGURE 0.8

Exponents can also be used to indicate repeated multiplication of polynomials. For example, $(3x - 4y)^2$ means $(3x - 4y)(3x - 4y)$, and $(x + 4)^3$ means $(x + 4)(x + 4)(x + 4)$. Therefore, raising a polynomial to a power is merely another multiplication problem.

$$
\begin{aligned}
(3x - 4y)^2 &= (3x - 4y)(3x - 4y) \\
&= 9x^2 - 24xy + 16y^2
\end{aligned}
$$

[*Hint*: When squaring a binomial, be careful not to forget the middle term. That is, $(x + 5)^2 \neq x^2 + 25$; instead, $(x + 5)^2 = x^2 + 10x + 25$.]

$$
\begin{aligned}
(x + 4)^3 &= (x + 4)(x + 4)(x + 4) \\
&= (x + 4)(x^2 + 8x + 16) \\
&= x(x^2 + 8x + 16) + 4(x^2 + 8x + 16) \\
&= x^3 + 8x^2 + 16x + 4x^2 + 32x + 64 \\
&= x^3 + 12x^2 + 48x + 64
\end{aligned}
$$

Special Patterns

In multiplying binomials, some special patterns occur that you should learn to recognize. These patterns can be used to find products, and some of them will be helpful later when you are factoring polynomials.

$$
\begin{aligned}
(a + b)^2 &= a^2 + 2ab + b^2 \\
(a - b)^2 &= a^2 - 2ab + b^2 \\
(a + b)(a - b) &= a^2 - b^2 \\
(a + b)^3 &= a^3 + 3a^2b + 3ab^2 + b^3 \\
(a - b)^3 &= a^3 - 3a^2b + 3ab^2 - b^3
\end{aligned}
$$

The three following examples illustrate the first three patterns, respectively.

$$
\begin{aligned}
(2x + 3)^2 &= (2x)^2 + 2(2x)(3) + (3)^2 \\
&= 4x^2 + 12x + 9
\end{aligned}
$$

$$
\begin{aligned}
(5x - 2)^2 &= (5x)^2 - 2(5x)(2) + (2)^2 \\
&= 25x^2 - 20x + 4
\end{aligned}
$$

$$
(3x + 2y)(3x - 2y) = (3x)^2 - (2y)^2 = 9x^2 - 4y^2
$$

In the first two examples, the resulting trinomial is called a **perfect-square trinomial**; it is the result of squaring a binomial. In the third example, the resulting binomial is called the **difference of two squares**. Later, we will use both of these patterns extensively when factoring polynomials.

The cubing-of-a-binomial patterns are helpful primarily when you are multiplying. These patterns can shorten the work of cubing a binomial, as the next two examples illustrate.

$$(3x + 2)^3 = (3x)^3 + 3(3x)^2(2) + 3(3x)(2)^2 + (2)^3$$
$$= 27x^3 + 54x^2 + 36x + 8$$

$$(5x - 2y)^3 = (5x)^3 - 3(5x)^2(2y) + 3(5x)(2y)^2 - (2y)^3$$
$$= 125x^3 - 150x^2y + 60xy^2 - 8y^3$$

Keep in mind that these multiplying patterns are useful shortcuts, but if you forget them, simply revert to applying the distributive property.

Binomial Expansion Pattern

It is possible to write the expansion of $(a + b)^n$, where n is *any* positive integer, without showing all of the intermediate steps of multiplying and combining similar terms. To do this, let's observe some patterns in the following examples; each one can be verified by direct multiplication.

$$(a + b)^1 = a + b$$
$$(a + b)^2 = a^2 + 2ab + b^2$$
$$(a + b)^3 = a^3 + 3a^2b + 3ab^2 + b^3$$
$$(a + b)^4 = a^4 + 4a^3b + 6a^2b^2 + 4ab^3 + b^4$$
$$(a + b)^5 = a^5 + 5a^4b + 10a^3b^2 + 10a^2b^3 + 5ab^4 + b^5$$

First, note the patterns of the exponents for a and b on a term-by-term basis. The exponents of a begin with the exponent of the binomial and decrease by 1, term by term, until the last term, which has $a^0 = 1$. The exponents of b begin with zero ($b^0 = 1$) and increase by 1, term-by-term, until the last term, which contains b to the power of the original binomial. In other words, the variables in the expansion of $(a + b)^n$ have the pattern

$$a^n, \qquad a^{n-1}b, \qquad a^{n-2}b^2, \qquad \ldots, \qquad ab^{n-1}, \qquad b^n$$

where for each term, the *sum* of the exponents of a and b is n.

Next, let's arrange the *coefficients* in a triangular formation; this yields an easy-to-remember pattern.

$$
\begin{array}{ccccccccccc}
 & & & & 1 & & 1 & & & & \\
 & & & 1 & & 2 & & 1 & & & \\
 & & 1 & & 3 & & 3 & & 1 & & \\
 & 1 & & 4 & & 6 & & 4 & & 1 & \\
1 & & 5 & & 10 & & 10 & & 5 & & 1 \\
\end{array}
$$

Row number n in the formation contains the coefficients of the expansion of $(a + b)^n$. For example, the fifth row contains 1 5 10 10 5 1, and these numbers are the coefficients of the terms in the expansion of $(a + b)^5$. Furthermore, each can be formed from the previous row as follows.

1. Start and end each row with 1.

2. All other entries result from adding the two numbers in the row immediately above, one number to the left and one number to the right.

Thus from row 5, we can form row 6.

Now we can use these seven coefficients and our discussion about the exponents to write out the expansion for $(a + b)^6$.

$$(a + b)^6 = a^6 + 6a^5b + 15a^4b^2 + 20a^3b^3 + 15a^2b^4 + 6ab^5 + b^6$$

REMARK   The triangular formation of numbers that we have been discussing is often referred to as *Pascal's triangle*. This is in honor of Blaise Pascal, a seventeenth-century mathematician, to whom the discovery of this pattern is attributed.

Let's consider two more examples using Pascal's triangle and the exponent relationships.

EXAMPLE 1

Expand $(a - b)^4$.

Solution

We can treat $a - b$ as $a + (-b)$ and use the fourth row of Pascal's triangle to obtain the coefficients.

$$[a + (-b)]^4 = a^4 + 4a^3(-b) + 6a^2(-b)^2 + 4a(-b)^3 + (-b)^4$$
$$= a^4 - 4a^3b + 6a^2b^2 - 4ab^3 + b^4 \qquad ∎$$

EXAMPLE 2

Expand $(2x + 3y)^5$.

Solution

Let $2x = a$ and $3y = b$. The coefficients come from the fifth row of Pascal's triangle.

$$(2x + 3y)^5 = (2x)^5 + 5(2x)^4(3y) + 10(2x)^3(3y)^2 + 10(2x)^2(3y)^3 + 5(2x)(3y)^4 + (3y)^5$$
$$= 32x^5 + 240x^4y + 720x^3y^2 + 1080x^2y^3 + 810xy^4 + 243y^5 \qquad ∎$$

Dividing Polynomials by Monomials

In Section 0.5 we will review the addition and subtraction of rational expressions using the properties

$$\frac{a}{b} + \frac{c}{b} = \frac{a + c}{b} \qquad \text{and} \qquad \frac{a}{b} - \frac{c}{b} = \frac{a - c}{b}$$

These properties can also be viewed as

$$\frac{a + c}{b} = \frac{a}{b} + \frac{c}{b} \qquad \text{and} \qquad \frac{a - c}{b} = \frac{a}{b} - \frac{c}{b}$$

Together with our knowledge of dividing monomials, these properties provide the basis for dividing polynomials by monomials. Consider the following examples.

$$\frac{18x^3 + 24x^2}{6x} = \frac{18x^3}{6x} + \frac{24x^2}{6x} = 3x^2 + 4x$$

$$\frac{35x^2y^3 - 55x^3y^4}{5xy^2} = \frac{35x^2y^3}{5xy^2} - \frac{55x^3y^4}{5xy^2} = 7xy - 11x^2y^2$$

Therefore, to divide a polynomial by a monomial, we divide each term of the polynomial by the monomial. As with many skills, once you feel comfortable with the process, you may then choose to perform some of the steps mentally. Your work could take the following format.

$$\frac{40x^4y^5 + 72x^5y^7}{8x^2y} = 5x^2y^4 + 9x^3y^6$$

$$\frac{36a^3b^4 - 48a^3b^3 + 64a^2b^5}{-4a^2b^2} = -9ab^2 + 12ab - 16b^3$$

PROBLEM SET 0.3

In Problems 1–10, perform the indicated operations.

1. $(5x^2 - 7x - 2) + (9x^2 + 8x - 4)$

2. $(-9x^2 + 8x + 4) + (7x^2 - 5x - 3)$

3. $(14x^2 - x - 1) - (15x^2 + 3x + 8)$

4. $(-3x^2 + 2x + 4) - (4x^2 + 6x - 5)$

5. $(3x - 4) - (6x + 3) + (9x - 4)$

6. $(7a - 2) - (8a - 1) - (10a - 2)$

7. $(8x^2 - 6x - 2) + (x^2 - x - 1) - (3x^2 - 2x + 4)$

8. $(12x^2 + 7x - 2) - (3x^2 + 4x + 5) + (-4x^2 - 7x - 2)$

9. $5(x - 2) - 4(x + 3) - 2(x + 6)$

10. $3(2x - 1) - 2(3x + 4) - 4(5x - 1)$

In Problems 11–54, find the indicated products. Remember the special patterns that we discussed in this section.

11. $3xy(4x^2y + 5xy^2)$

12. $-2ab^2(3a^2b - 4ab^3)$

13. $6a^3b^2(5ab - 4a^2b + 3ab^2)$

14. $-xy^4(5x^2y - 4xy^2 + 3x^2y^2)$

15. $(x + 8)(x + 12)$

16. $(x - 9)(x + 6)$

17. $(n - 4)(n - 12)$

18. $(n + 6)(n - 10)$

19. $(s - t)(x + y)$

20. $(a + b)(c + d)$

21. $(3x - 1)(2x + 3)$

22. $(5x + 2)(3x + 4)$

23. $(4x - 3)(3x - 7)$

24. $(4n + 3)(6n - 1)$

25. $(x + 4)^2$

26. $(x - 6)^2$

27. $(2n + 3)^2$

28. $(3n - 5)^2$

29. $(x + 2)(x - 4)(x + 3)$

30. $(x - 1)(x + 6)(x - 5)$

31. $(x - 1)(2x + 3)(3x - 2)$

32. $(2x + 5)(x - 4)(3x + 1)$

33. $(x - 1)(x^2 + 3x - 4)$

34. $(t + 1)(t^2 - 2t - 4)$

35. $(t - 1)(t^2 + t + 1)$

36. $(2x - 1)(x^2 + 4x + 3)$

37. $(3x + 2)(2x^2 - x - 1)$

38. $(3x - 2)(2x^2 + 3x + 4)$

39. $(x^2 + 2x - 1)(x^2 + 6x + 4)$

40. $(x^2 - x + 4)(2x^2 - 3x - 1)$

41. $(5x - 2)(5x + 2)$

42. $(3x - 4)(3x + 4)$

43. $(x^2 - 5x - 2)^2$

44. $(-x^2 + x - 1)^2$

45. $(2x + 3y)(2x - 3y)$

46. $(9x + y)(9x - y)$

47. $(x + 5)^3$

48. $(x - 6)^3$

49. $(2x + 1)^3$

50. $(3x + 4)^3$

51. $(4x - 3)^3$

52. $(2x - 5)^3$

53. $(5x - 2y)^3$

54. $(x + 3y)^3$

For Problems 55–66, use Pascal's triangle to help expand each of the following.

55. $(a+b)^7$

56. $(a + b)^8$

57. $(x - y)^5$

58. $(x - y)^6$

59. $(x + 2y)^4$

60. $(2x + y)^5$

61. $(2a - b)^6$

62. $(3a - b)^4$

63. $(x^2 + y)^7$

64. $(x + 2y^2)^7$

65. $(2a - 3b)^5$

66. $(4a - 3b)^3$

In Problems 67–72, perform the indicated divisions.

67. $\dfrac{15x^4 - 25x^3}{5x^2}$

68. $\dfrac{-48x^8 - 72x^6}{-8x^4}$

69. $\dfrac{30a^5 - 24a^3 + 54a^2}{-6a}$

70. $\dfrac{18x^3y^2 + 27x^2y^3}{3xy}$

71. $\dfrac{-20a^3b^2 - 44a^4b^5}{-4a^2b}$

72. $\dfrac{21x^5y^6 + 28x^4y^3 - 35x^5y^4}{7x^2y^3}$

In Problems 73–82, find the indicated products. Assume all variables that appear as exponents represent integers.

73. $(x^a + y^b)(x^a - y^b)$

74. $(x^{2a} + 1)(x^{2a} - 3)$

75. $(x^b + 4)(x^b - 7)$

76. $(3x^a - 2)(x^a + 5)$

77. $(2x^b - 1)(3x^b + 2)$

78. $(2x^a - 3)(2x^a + 3)$

79. $(x^{2a} - 1)^2$

80. $(x^{3b} + 2)^2$

81. $(x^a - 2)^3$

82. $(x^b + 3)^3$

THOUGHTS INTO WORDS

83. Describe how to multiply two binomials.

84. Describe how to multiply a binomial and a trinomial.

85. Determine the number of terms in the product of $(x + y)$ and $(a + b + c + d)$ without doing the multiplication. Explain how you arrived at your answer.

0.4   FACTORING POLYNOMIALS

If a polynomial is equal to the product of other polynomials, then each polynomial in the product is called a factor of the original polynomial. For example, because

$x^2 - 4$ can be expressed as $(x + 2)(x - 2)$, we say that $x + 2$ and $x - 2$ are factors of $x^2 - 4$. The process of expressing a polynomial as a product of polynomials is called **factoring**. In this section we will consider methods of factoring polynomials with integer coefficients.

In general, factoring is the reverse of multiplication, so we can use our knowledge of multiplication to help develop factoring techniques. For example, we previously used the distributive property to find the product of a monomial and a polynomial, as the next examples illustrate.

$$3(x + 2) = 3(x) + 3(2) = 3x + 6$$
$$3x(x + 4) = 3x(x) + 3x(4) = 3x^2 + 12x$$

For factoring purposes, the distributive property (now in the form $ab + ac = a(b + c)$) can be used to reverse the process. (The steps indicated in the dashed boxes can be done mentally.)

$$3x + 6 = \boxed{3(x) + 3(2)} = 3(x + 2)$$
$$3x^2 + 12x = \boxed{3x(x) + 3x(4)} = 3x(x + 4)$$

Polynomials can be factored in a variety of ways. Consider some factorizations of $3x^2 + 12x$.

$$3x^2 + 12x = 3x(x + 4) \qquad \text{or} \qquad 3x^2 + 12x = 3(x^2 + 4x) \qquad \text{or}$$
$$3x^2 + 12x = x(3x + 12) \qquad \text{or} \qquad 3x^2 + 12x = \frac{1}{2}(6x^2 + 24x)$$

We are, however, primarily interested in the first of these factorization forms; we shall refer to it as the **completely factored form**. A polynomial with integral coefficients is in completely factored form if:

1. it is expressed as a product of polynomials with *integral coefficients*, and
2. no polynomial, other than a monomial, within the factored form can be further factored into polynomials with integral coefficients.

Do you see why only the first of the factored forms of $3x^2 + 12x$ is said to be in completely factored form? In each of the other three forms, the polynomial inside the parentheses can be factored further. Moreover, in the last form, $\frac{1}{2}(6x^2 + 24x)$, the condition of using only integers is violated.

This application of the distributive property is often referred to as **factoring out the highest common monomial factor**. The following examples further illustrate the process.

$$12x^3 + 16x^2 = 4x^2(3x + 4)$$
$$8ab - 18b = 2b(4a - 9)$$
$$6x^2y^3 + 27xy^4 = 3xy^3(2x + 9y)$$
$$30x^3 + 42x^4 - 24x^5 = 6x^3(5 + 7x - 4x^2)$$

Sometimes there may be a common *binomial* factor rather than a common monomial factor. For example, each of the two terms in the expression $x(y + 2) + z(y + 2)$ has a binomial factor of $y + 2$. Thus we can factor $y + 2$ from each term and obtain the following result.

$$x(y + 2) + z(y + 2) = (y + 2)(x + z)$$

Consider a few more examples involving a common binomial factor.

$$a^2(b + 1) + 2(b + 1) = (b + 1)(a^2 + 2)$$
$$x(2y - 1) - y(2y - 1) = (2y - 1)(x - y)$$
$$x(x + 2) + 3(x + 2) = (x + 2)(x + 3)$$

It may seem that a given polynomial exhibits no apparent common monomial or binomial factor. Such is the case with $ab + 3c + bc + 3a$. However, by using the commutative property to rearrange the terms, we can factor it as follows.

$$ab + 3c + bc + 3a = ab + 3a + bc + 3c$$

$\qquad\qquad\qquad = a(b + 3) + c(b + 3)$ **Factor *a* from the first two terms and *c* from the last two terms.**

$\qquad\qquad\qquad = (b + 3)(a + c)$ **Factor *b* + 3 from both terms.**

This factoring process is referred to as **factoring by grouping**. Let's consider another example of this type.

$ab^2 - 4b^2 + 3a - 12 = b^2(a - 4) + 3(a - 4)$ **Factor b^2 from the first two terms, 3 from the last two.**

$\qquad\qquad\qquad\quad = (a - 4)(b^2 + 3)$ **Factor the common binomial from both terms.**

Difference of Two Squares

In Section 0.3 we called your attention to some special multiplication patterns. One of these patterns was

$$(a + b)(a - b) = a^2 - b^2$$

This same pattern, viewed as a factoring pattern,

$$a^2 - b^2 = (a + b)(a - b)$$

is referred to as the **difference of two squares**. Applying the pattern is a fairly simple process, as these next examples illustrate. Again, the steps we have included in dashed boxes are usually performed mentally.

$$x^2 - 16 = \boxed{(x)^2 - (4)^2} = (x + 4)(x - 4)$$

$$4x^2 - 25 = \boxed{(2x^2) - (5)^2} = (2x + 5)(2x - 5)$$

Because multiplication is commutative, the order in which we write the factors is not important. For example, $(x + 4)(x - 4)$ can also be written $(x - 4)(x + 4)$.

You must be careful not to assume an analogous factoring pattern for the *sum* of two squares; *it does not exist*. For example, $x^2 + 4 \neq (x + 2)(x + 2)$ because $(x + 2)(x + 2) = x^2 + 4x + 4$. We say that a polynomial such as $x^2 + 4$ is not **factorable using integers**.

Sometimes the difference-of-two-squares pattern can be applied more than once, as the next example illustrates.

$$16x^4 - 81y^4 = (4x^2 + 9y^2)(4x^2 - 9y^2) = (4x^2 + 9y^2)(2x + 3y)(2x - 3y)$$

It may also happen that the squares are not just simple monomial squares. These next three examples illustrate such polynomials.

$$(x + 3)^2 - y^2 = [(x + 3) + y][(x + 3) - y] = (x + 3 + y)(x + 3 - y)$$

$$4x^2 - (2y + 1)^2 = [2x + (2y + 1)][2x - (2y + 1)]$$
$$= (2x + 2y + 1)(2x - 2y - 1)$$

$$(x - 1)^2 - (x + 4)^2 = [(x - 1) + (x + 4)][(x - 1) - (x + 4)]$$
$$= (x - 1 + x + 4)(x - 1 - x - 4)$$
$$= (2x + 3)(-5)$$

It is possible that both the technique of factoring out a common monomial factor and the pattern of the difference of two squares can be applied to the same problem. *In general, it is best to look first for a common monomial factor*. Consider the following examples.

$$2x^2 - 50 = 2(x^2 - 25)$$
$$= 2(x + 5)(x - 5)$$

$$48y^3 - 27y = 3y(16y^2 - 9)$$
$$= 3y(4y + 3)(4y - 3)$$

$$9x^2 - 36 = 9(x^2 - 4)$$
$$= 9(x + 2)(x - 2)$$

Factoring Trinomials

Expressing a trinomial as the product of two binomials is one of the most common factoring techniques used in algebra. As before, to develop a factoring technique we first look at some multiplication ideas. Let's consider the product $(x + a)(x + b)$, using the distributive property to show how each term of the resulting trinomial is formed.

$$(x + a)(x + b) = x(x + b) + a(x + b)$$
$$= x(x) + x(b) + a(x) + a(b)$$
$$= x^2 + (a + b)x + ab$$

Notice that the coefficient of the middle term is the *sum* of a and b and that the last term is the *product* of a and b. These two relationships can be used to factor trinomials. Let's consider some examples.

EXAMPLE 1

Factor $x^2 + 12x + 20$.

Solution

We need two integers whose sum is 12 and whose product is 20. The numbers are 2 and 10, and we can complete the factoring as follows.

$$x^2 + 12x + 20 = (x + 2)(x + 10)$$ ■

EXAMPLE 2

Factor $x^2 - 3x - 54$.

Solution

We need two integers whose sum is -3 and whose product is -54. The integers are -9 and 6, and we can factor as follows.

$$x^2 - 3x - 54 = (x - 9)(x + 6)$$ ■

EXAMPLE 3

Factor $x^2 + 7x = 16$.

Solution

We need two integers whose sum is 7 and whose product is 16. The only possible pairs of factors of 16 are $1 \cdot 16$, $2 \cdot 8$, and $4 \cdot 4$. A sum of 7 is not produced by any of these pairs, so the polynomial $x^2 + 7x + 16$ is *not factorable using integers*.

■

Trinomials of the Form $ax^2 + bx + c$

Now let's consider factoring trinomials where the coefficient of the squared term is not one. First, let's illustrate an informal trial-and-error technique that works quite well for certain types of trinomials. This technique is based on our knowledge of multiplication of binomials.

EXAMPLE 4

Factor $3x^2 + 5x + 2$.

Solution

By looking at the first term, $3x^2$, and the positive signs of the other two terms, we know that the binomials are of the form

$$(x + \_\_)(3x + \_\_)$$

Because the factors of the last term, 2, are 1 and 2, we have only the following two possibilities to try.

$$(x + 2)(3x + 1) \qquad \text{or} \qquad (x + 1)(3x + 2),$$

By checking the middle term formed in each of these products, we find that the second possibility yields the desired middle term of $5x$. Therefore,

$$3x^2 + 5x + 2 = (x + 1)(3x + 2) \qquad \blacksquare$$

EXAMPLE 5

Factor $8x^2 - 30x + 7$.

Solution

First, observe that the first term, $8x^2$, can be written as $2x \cdot 4x$ or $x \cdot 8x$. Second, because the middle term is negative and the last term is positive, we know that the binomials are of the form

$$(2x - \underline{})(4x - \underline{}) \qquad \text{or} \qquad (x - \underline{})(8x - \underline{})$$

Third, because the factors of the last term, 7, are 1 and 7, the following possibilities exist.

$$(2x - 1)(4x - 7) \qquad (2x - 7)(4x - 1)$$
$$(x - 1)(8x - 7) \qquad (x - 7)(8x - 1)$$

By checking the middle term formed in each of these products, we find that $(2x - 7)(4x - 1)$ produces the desired middle term of $-30x$. Therefore,

$$8x^2 - 30x + 7 = (2x - 7)(4x - 1) \qquad \blacksquare$$

EXAMPLE 6

Factor $5x^2 - 18x - 8$.

Solution

The first term, $5x^2$, can be written as $x \cdot 5x$. The last term, -8, can be written as $(-2)(4)$, $(2)(-4)$, $(-1)(8)$, or $(1)(-8)$. Therefore, we have the following possibilities to try.

$$(x - 2)(5x + 4) \qquad (x + 4)(5x - 2)$$
$$(x + 2)(5x - 4) \qquad (x - 4)(5x + 2)$$
$$(x - 1)(5x + 8) \qquad (x + 8)(5x - 1)$$
$$(x + 1)(5x - 8) \qquad (x - 8)(5x + 1)$$

By checking the middle terms, we find that $(x - 4)(5x + 2)$ yields the desired middle term of $-18x$. Thus

$$5x^2 - 18x - 8 = (x - 4)(5x + 2) \qquad \blacksquare$$

EXAMPLE 7

Factor $4x^2 + 6x + 9$.

Solution

The first term, $4x^2$, and the positive signs of the middle and last terms indicate that the binomials are of the form

$$(x + \underline{\quad})(4x + \underline{\quad}) \qquad \text{or} \qquad (2x + \underline{\quad})(2x + \underline{\quad})$$

Because the factors of the last term, 9, are 1 and 9 or 3 and 3, we have the following possibilities to try.

$$(x + 1)(4x + 9) \qquad (x + 9)(4x + 1)$$
$$(x + 3)(4x + 3) \qquad (2x + 1)(2x + 9)$$
$$(2x + 3)(2x + 3)$$

None of these possibilities yields a middle term of $6x$. Therefore, $4x^2 + 6x + 9$ is *not factorable using integers.* ∎

Certainly, as the number of possibilities increases, this trial-and-error technique for factoring becomes more tedious. The key idea is to organize your work so that all possibilities are considered. We have suggested one possible format in the previous examples. However, as you practice such problems, you may devise a format that works better for you. Whatever works best for you is the right approach.

There is another, more systematic technique that you may wish to use with some trinomials. It is an extension of the technique we used earlier with trinomials where the coefficient of the squared term was one. To see the basis of this technique, consider the following general product.

$$\begin{aligned}(px + r)(qx + s) &= px(qx) + px(s) + r(qx) + r(s) \\ &= (pq)x^2 + ps(x) + rq(x) + rs \\ &= (pq)x^2 + (ps + rq)x + rs\end{aligned}$$

Notice that the product of the coefficient of x^2 and the constant term is $pqrs$. Likewise, the product of the two coefficients of x (ps and rq) is also $pqrs$. Therefore, the coefficient of x must be a sum of the form $ps + rq$, such that the product of the coefficient of x^2 and the constant term is $pqrs$. Now let's see how this works in some specific examples.

EXAMPLE 8

Factor $6x^2 + 17x + 5$.

Solution

$$6x^2 + 17x + 5 \quad \text{Sum of 17}$$

Product of $6 \cdot 5 = 30$

We need two integers whose sum is 17 and whose product is 30. The integers 2 and 15 satisfy these conditions. Therefore, the middle term, $17x$, of the given trinomial can be expressed as $2x + 15x$, and we can proceed as follows.

$$6x^2 + 17x + 5 = 6x^2 + 2x + 15x + 5$$
$$= 2x(3x + 1) + 5(3x + 1)$$
$$= (3x + 1)(2x + 5) \blacksquare$$

EXAMPLE 9

Factor $5x^2 - 18x - 8$.

Solution

$$5x^2 - 18x - 8 \quad \text{Sum of } -18$$
$$\text{Product of } 5(-8) = -40$$

We need two integers whose sum is -18 and whose product is -40. The integers -20 and 2 satisfy these conditions. Therefore the middle term, $-18x$, of the trinomial can be written $-20x + 2x$, and we can factor as follows.

$$5x^2 - 18x - 8 = 5x^2 - 20x + 2x - 8$$
$$= 5x(x - 4) + 2(x - 4)$$
$$= (x - 4)(5x + 2) \blacksquare$$

EXAMPLE 10

Factor $24x^2 + 2x - 15$.

Solution

$$24x^2 + 2x - 15 \quad \text{Sum of 2}$$
$$\text{Product of } 24(-15) = -360$$

We need two integers whose sum is 2 and whose product is -360. To help find these integers, let's factor 360 into primes.

$$360 = 2 \cdot 2 \cdot 2 \cdot 3 \cdot 3 \cdot 5$$

Now by grouping these factors in various ways, we find that $2 \cdot 2 \cdot 5 = 20$ and $2 \cdot 3 \cdot 3 = 18$, so we can use the integers 20 and -18 to produce a sum of 2 and a product of -360. Therefore, the middle term, $2x$, of the trinomial can be expressed as $20x - 18x$, and we can proceed as follows.

$$24x^2 + 2x - 15 = 24x^2 + 20x - 18x - 15$$
$$= 4x(6x + 5) - 3(6x + 5)$$
$$= (6x + 5)(4x - 3) \blacksquare$$

Sum and Difference of Two Cubes

Earlier in this section we discussed the difference-of-squares factoring pattern. We pointed out that no analogous sum-of-squares pattern exists; that is, a polynomial such as $x^2 + 9$ is not factorable using integers. However, there do exist patterns for both the *sum* and the *difference of two cubes*. These patterns come from the following special products.

$$(x + y)(x^2 - xy + y^2) = x(x^2 - xy + y^2) + y(x^2 - xy + y^2)$$
$$= x^3 - x^2y + xy^2 + x^2y - xy^2 + y^3$$
$$= x^3 + y^3$$

$$(x - y)(x^2 + xy + y^2) = x(x^2 + xy + y^2) - y(x^2 + xy + y^2)$$
$$= x^3 + x^2y + xy^2 - x^2y - xy^2 - y^3$$
$$= x^3 - y^3$$

Thus we can state the following factoring patterns.

$$x^3 + y^3 = (x + y)(x^2 - xy + y^2)$$
$$x^3 - y^3 = (x - y)(x^2 + xy + y^2)$$

Note how these patterns are used in the next three examples.

$$x^3 + 8 = x^3 + 2^3 = (x + 2)(x^2 - 2x + 4)$$
$$8x^3 - 27y^3 = (2x)^3 - (3y)^3 = (2x - 3y)(4x^2 + 6xy + 9y^2)$$
$$8a^6 + 125b^3 = (2a^2)^3 + (5b)^3 = (2a^2 + 5b)(4a^4 - 10a^2b + 25b^2)$$

We do want to leave you with one final word of caution. **Be sure to factor completely.** Sometimes more than one technique needs to be applied, or perhaps the same technique can be applied more than once. Study the following examples very carefully.

$$2x^2 - 8 = 2(x^2 - 4) = 2(x + 2)(x - 2)$$
$$3x^2 + 18x + 24 = 3(x^2 + 6x + 8) = 3(x + 4)(x + 2)$$
$$3x^3 - 3y^3 = 3(x^3 - y^3) = 3(x - y)(x^2 + xy + y^2)$$
$$a^4 - b^4 = (a^2 + b^2)(a^2 - b^2) = (a^2 + b^2)(a + b)(a - b)$$
$$x^4 - 6x^2 - 27 = (x^2 - 9)(x^2 + 3) = (x + 3)(x - 3)(x^2 + 3)$$

$$3x^4y + 9x^2y - 84y = 3y(x^4 + 3x^2 - 28)$$
$$= 3y(x^2 + 7)(x^2 - 4)$$
$$= 3y(x^2 + 7)(x + 2)(x - 2)$$

$$x^2 - y^2 + 8y - 16 = x^2 - (y^2 - 8y + 16)$$
$$= x^2 - (y - 4)^2$$
$$= (x - (y - 4))(x + (y - 4))$$
$$= (x - y + 4)(x + y - 4)$$

PROBLEM SET 0.4

Factor each polynomial completely. Indicate any that are not factorable using integers.

1. $6xy - 8xy^2$

2. $4a^2b^2 + 12ab^3$

3. $x(z + 3) + y(z + 3)$

4. $5(x + y) + a(x + y)$

5. $3x + 3y + ax + ay$

6. $ac + bc + a + b$

7. $ax - ay - bx + by$

8. $2a^2 - 3bc - 2ab + 3ac$

9. $9x^2 - 25$

10. $4x^2 + 9$

11. $1 - 81n^2$

12. $9x^2y^2 - 64$

13. $(x + 4)^2 - y^2$

14. $x^2 - (y - 1)^2$

15. $9s^2 - (2t - 1)^2$

16. $4a^2 - (3b + 1)^2$

17. $x^2 - 5x - 14$

18. $a^2 + 5a - 24$

19. $15 - 2x - x^2$

20. $40 - 6x - x^2$

21. $x^2 + 7x - 36$

22. $x^2 - 4xy - 5y^2$

23. $3x^2 - 11x + 10$

24. $2x^2 - 7x - 30$

25. $10x^2 - 33x - 7$

26. $8y^2 + 22y - 21$

27. $x^3 - 8$

28. $x^3 + 64$

29. $64x^3 + 27y^3$

30. $27x^3 - 8y^3$

31. $4x^2 + 16$

32. $n^3 - 49n$

33. $x^3 - 9x$

34. $12n^2 + 59n + 72$

35. $9a^2 - 42a + 49$

36. $1 - 16x^4$

37. $2n^3 + 6n^2 + 10n$

38. $x^2 - (y - 7)^2$

39. $10x^2 + 39x - 27$

40. $3x^2 + x - 5$

41. $36a^2 - 12a + 1$

42. $18n^3 + 39n^2 - 15n$

43. $8x^2 + 2xy - y^2$

44. $12x^2 + 7xy - 10y^2$

45. $2n^2 - n - 5$

46. $25t^2 - 100$

47. $2n^3 + 14n^2 - 20n$

48. $25n^2 + 64$

49. $4x^3 + 32$

50. $2x^3 - 54$

51. $x^4 - 4x^2 - 45$

52. $x^4 - x^2 - 12$

53. $2x^4y - 26x^2y - 96y$

54. $3x^4y - 15x^2y - 108y$

55. $(a + b)^2 - (c + d)^2$

56. $(a - b)^2 - (c - d)^2$

57. $x^2 + 8x + 16 - y^2$

58. $4x^2 + 12x + 9 - y^2$

59. $x^2 - y^2 - 10y - 25$

60. $y^2 - x^2 + 16x - 64$

61. $60x^2 - 32x - 15$

62. $40x^2 + 37x - 63$

63. $84x^3 + 57x^2 - 60x$

64. $210x^3 - 102x^2 - 180x$

For Problems 65–74, factor each of the following, and assume that all variables appearing as exponents represent integers.

65. $x^{2a} - 16$

66. $x^{4n} - 9$

67. $x^{3n} - y^{3n}$

68. $x^{3a} + y^{6a}$

69. $x^{2a} - 3x^a - 28$

70. $x^{2a} + 10x^a + 21$

71. $2x^{2n} + 7x^n - 30$

72. $3x^{2n} - 16x^n - 12$

73. $x^{4n} - y^{4n}$

74. $16x^{2a} + 24x^a + 9$

75. Suppose that we want to factor $x^2 + 34x + 288$. We need to complete the following with two numbers whose sum is 34 and whose product is 288.

$$x^2 + 34x + 288 = (x + \underline{\quad})(x + \underline{\quad})$$

These numbers can be found as follows: Because we need a product of 288, let's consider the prime factorization of 288.

$$288 = 2^5 \cdot 3^2$$

Now we need to use five 2s and two 3s in the statement

$$(\quad) + (\quad) = 34$$

Because 34 is divisible by 2 but not by 4, four factors of 2 must be in one number and one factor of 2 in the other number. Also, because 34 is not divisible by 3, both factors of 3 must be in the same number. These facts aid us in determining that

$$(2 \cdot 2 \cdot 2 \cdot 2) + (2 \cdot 3 \cdot 3) = 34$$

or

$$16 \quad + \quad 18 \quad = 34$$

Thus we can complete the original factoring problem.

$$x^2 + 34x + 288 = (x + 16)(x + 18)$$

Use this approach to factor each of the following expressions.

a. $x^2 + 35x + 96$ **b.** $x^2 + 27x + 176$

c. $x^2 - 45x + 504$ **d.** $x^2 - 26x + 168$

e. $x^2 + 60x + 896$ **f.** $x^2 - 84x + 1728$

THOUGHTS INTO WORDS

76. Describe, in words, the pattern for factoring the sum of two cubes.

77. What does it mean to say that the polynomial $x^2 + 5x + 7$ is not factorable using integers?

78. What role does the distributive property play in the factoring of polynomials?

79. Explain your thought process when factoring $30x^2 + 13x - 56$.

80. Consider the following approach to factoring $12x^2 + 54x + 60$.

$$\begin{aligned} 12x^2 + 54x + 60 &= (3x + 6)(4x + 10) \\ &= 3(x + 2)(2)(2x + 5) \\ &= 6(x + 2)(2x + 5) \end{aligned}$$

Is this factoring process correct? What can you suggest to the person who used this approach?

0.5   RATIONAL EXPRESSIONS

Indicated quotients of algebraic expressions are called **algebraic fractions** or **fractional expressions**. The indicated quotient of two polynomials is called a **rational expression**. (This is analogous to defining a rational number as the indicated quotient of two integers.) The following are examples of rational expressions.

$$\frac{3x^2}{5} \qquad \frac{x - 2}{x + 3} \qquad \frac{x^2 + 5x - 1}{x^2 - 9} \qquad \frac{xy^2 + x^2 y}{xy} \qquad \frac{a^3 - 3a^2 - 5a - 1}{a^4 + a^3 + 6}$$

Because division by zero must be avoided, no values can be assigned to variables that will create a denominator of zero. Thus the rational expression $\frac{x - 2}{x + 3}$ is meaningful for all real number values of x except $x = -3$. Rather than making restrictions for each individual expression, we will merely assume that **all denominators represent nonzero real numbers**.

The basic properties of the real numbers can be used for working with rational expressions. For example, the property

$$\frac{a \cdot k}{b \cdot k} = \frac{a}{b}$$

which is used to reduce rational numbers, is also used to *simplify* rational expressions. Consider the following examples.

$$\frac{15xy}{25y} = \frac{3 \cdot \cancel{5} \cdot x \cdot \cancel{y}}{\cancel{5} \cdot 5 \cdot \cancel{y}} = \frac{3x}{5}$$

$$\frac{-9}{18x^2y} = -\frac{\overset{1}{\cancel{9}}}{\underset{2}{\cancel{18}}x^2y} = -\frac{1}{2x^2y}$$

Note that slightly different formats were used in these two examples. In the first one we factored the coefficients into primes and then proceeded to simplify; however, in the second problem we simply divided a common factor of 9 out of both the numerator and denominator. This is basically a format issue and depends upon your personal preference. Also notice that in the second example, we applied the property $\frac{-a}{b} = -\frac{a}{b}$. This is part of the general property that states

$$\frac{-a}{b} = \frac{a}{-b} = -\frac{a}{b}$$

The factoring techniques discussed in the previous section can be used to factor numerators and denominators so that the property $(a \cdot k)/(b \cdot k) = a/b$ can be applied. Consider the following examples.

$$\frac{x^2 + 4x}{x^2 - 16} = \frac{x\cancel{(x + 4)}}{(x - 4)\cancel{(x + 4)}} = \frac{x}{x - 4}$$

$$\frac{5n^2 + 6n - 8}{10n^2 - 3n - 4} = \frac{\cancel{(5n - 4)}(n + 2)}{\cancel{(5n - 4)}(2n + 1)} = \frac{n + 2}{2n + 1}$$

$$\frac{x^3 + y^3}{x^2 + xy + 2x + 2y} = \frac{(x + y)(x^2 - xy + y^2)}{x(x + y) + 2(x + y)}$$

$$= \frac{\cancel{(x + y)}(x^2 - xy + y^2)}{\cancel{(x + y)}(x + 2)} = \frac{x^2 - xy + y^2}{x + 2}$$

$$\frac{6x^3y - 6xy}{x^3 + 5x^2 + 4x} = \frac{6xy(x^2 - 1)}{x(x^2 + 5x + 4)} = \frac{6xy\cancel{(x + 1)}(x - 1)}{x\cancel{(x + 1)}(x + 4)} = \frac{6y(x - 1)}{x + 4}$$

Note that in the last example we left the numerator of the final fraction in factored form. This is often done if expressions other than monomials are involved. Either

$$\frac{6y(x - 1)}{x + 4} \qquad \text{or} \qquad \frac{6xy - 6y}{x + 4}$$

is an acceptable answer.

Remember that the quotient of any nonzero real number and its opposite is -1. For example, $6/-6 = -1$ and $-8/8 = -1$. Likewise, the indicated quotient of any polynomial and its opposite is equal to -1. For example,

$$\frac{a}{-a} = -1 \qquad \text{because } a \text{ and } -a \text{ are opposites}$$

$$\frac{a-b}{b-a} = -1 \qquad \text{because } a - b \text{ and } b - a \text{ are opposites}$$

$$\frac{x^2 - 4}{4 - x^2} = -1 \qquad \text{because } x^2 - 4 \text{ and } 4 - x^2 \text{ are opposites}$$

The next example illustrates how we use this idea when simplifying rational expressions.

$$\frac{4 - x^2}{x^2 + x - 6} = \frac{(2+x)}{(x+3)}\boxed{\frac{(2-x)}{(x-2)}}$$

$$= (-1)\left(\frac{x+2}{x+3}\right) \qquad \frac{2-x}{x-2} = -1$$

$$= -\frac{x+2}{x+3} \qquad \text{or} \qquad \frac{-x-2}{x+3}$$

Multiplying and Dividing Rational Expressions

Multiplication of rational expressions is based on the following property.

$$\frac{a}{b} \cdot \frac{c}{d} = \frac{ac}{bd}$$

In other words, we multiply numerators and we multiply denominators and express the final product in simplified form. Study the following examples carefully and pay special attention to the formats used to organize the computational work.

$$\frac{3x}{4y} \cdot \frac{8y^2}{9x} = \frac{\overset{2}{\cancel{3}} \cdot \cancel{8} \cdot \cancel{x} \cdot \overset{y}{\cancel{y^2}}}{\cancel{4} \cdot \underset{3}{\cancel{9}} \cdot \cancel{x} \cdot \cancel{y}} = \frac{2y}{3}$$

$$\frac{12x^2 y}{-18xy} \cdot \frac{-24xy^2}{56y^3} = \frac{\overset{2}{\cancel{12}} \cdot \overset{8}{\cancel{24}} \cdot \overset{x^2}{\cancel{x^3}} \cdot \cancel{y^3}}{\underset{3}{\cancel{18}} \cdot \underset{7}{\cancel{56}} \cdot \cancel{x} \cdot \underset{y}{\cancel{y^4}}} = \frac{2x^2}{7y} \qquad \frac{12x^2 y}{-18xy} = -\frac{12x^2 y}{18xy} \quad \text{and} \quad \frac{-24xy^2}{56y^3} = -\frac{24xy^2}{56y^3}$$

$$\text{so the product is positive.}$$

$$\frac{y}{x^2 - 4} \cdot \frac{x+2}{y^2} = \frac{\cancel{y}(\cancel{x+2})}{\underset{y}{\cancel{y^2}}(\cancel{x+2})(x-2)} = \frac{1}{y(x-2)}$$

$$\frac{x^2 - x}{x + 5} \cdot \frac{x^2 + 5x + 4}{x^4 - x^2} = \frac{\cancel{x}(\cancel{x-1})(\cancel{x+1})(x+4)}{(x+5)(\underset{x}{\cancel{x^2}})(\cancel{x+1})(\cancel{x-1})} = \frac{x+4}{x(x+5)}$$

To divide rational expressions, we merely apply the following property.

$$\frac{a}{b} \div \frac{c}{d} = \frac{a}{b} \cdot \frac{d}{c} = \frac{ad}{bc}$$

That is, the quotient of two rational expressions is the product of the first expression times the reciprocal of the second. Consider the following examples.

$$\frac{16x^2y}{24xy^3} \div \frac{9xy}{8x^2y^2} = \frac{16x^2y}{24xy^3} \cdot \frac{8x^2y^2}{9xy} = \frac{16 \cdot \overset{x^2}{\cancel{8}} \cdot \cancel{x^4} \cdot \cancel{y^3}}{\underset{3}{24} \cdot 9 \cdot \cancel{x^2} \cdot \underset{y}{\cancel{y^4}}} = \frac{16x^2}{27y}$$

$$\frac{3a^2 + 12}{3a^2 - 15a} \div \frac{a^4 - 16}{a^2 - 3a - 10} = \frac{3a^2 + 12}{3a^2 - 15a} \cdot \frac{a^2 - 3a - 10}{a^4 - 16}$$

$$= \frac{\cancel{3}\cancel{(a^2 + 4)}\cancel{(a - 5)}\cancel{(a + 2)}}{\cancel{3}a\cancel{(a - 5)}\cancel{(a^2 + 4)}\cancel{(a + 2)}(a - 2)}$$

$$= \frac{1}{a(a - 2)}$$

Adding and Subtracting Rational Expressions

The following two properties provide the basis for adding and subtracting rational expressions.

$$\frac{a}{b} + \frac{c}{b} = \frac{a + c}{b}$$

$$\frac{a}{b} - \frac{c}{b} = \frac{a - c}{b}$$

These properties state that rational expressions with a common denominator can be added (or subtracted) by adding (or subtracting) the numerators and placing the result over the common denominator. Let's illustrate this idea.

$$\frac{8}{x - 2} + \frac{3}{x - 2} = \frac{8 + 3}{x - 2} = \frac{11}{x - 2}$$

$$\frac{9}{4y} - \frac{7}{4y} = \frac{9 - 7}{4y} = \frac{2}{4y} = \frac{1}{2y}$$

Don't forget to simplify the final result.

$$\frac{n^2}{n - 1} - \frac{1}{n - 1} = \frac{n^2 - 1}{n - 1} = \frac{(n + 1)\cancel{(n - 1)}}{\cancel{n - 1}} = n + 1$$

If we need to add or subtract rational expressions that do not have a common denominator, then we apply the property $a/b = (a \cdot k)/(b \cdot k)$ to obtain equivalent fractions with a common denominator. Study the next examples and again pay special attention to the format we used to organize our work.

REMARK Remember that the **least common multiple** of a set of whole numbers is the smallest nonzero whole number divisible by each of the numbers in the set. When we add or subtract rational numbers, the least common multiple of the denominators of those numbers is the **least common denominator (LCD)**. This concept of a least common denominator can be extended to include polynomials.

EXAMPLE 1

Add $\dfrac{x+2}{4} + \dfrac{3x+1}{3}$.

Solution

By inspection we see that the LCD is 12.

$$\frac{x+2}{4} + \frac{3x+1}{3} = \left(\frac{x+2}{4}\right)\left(\frac{3}{3}\right) + \left(\frac{3x+1}{3}\right)\left(\frac{4}{4}\right)$$

$$= \frac{3(x+2)}{12} + \frac{4(3x+1)}{12}$$

$$= \frac{3x+6+12x+4}{12}$$

$$= \frac{15x+10}{12}$$ ∎

EXAMPLE 2

Perform the indicated operations.

$$\frac{x+3}{10} + \frac{2x+1}{15} - \frac{x-2}{18}$$

Solution

If you cannot determine the LCD by inspection, then use the prime-factored forms of the denominators.

$$10 = 2 \cdot 5 \qquad 15 = 3 \cdot 5 \qquad 18 = 2 \cdot 3 \cdot 3$$

The LCD must contain one factor of 2, two factors of 3, and one factor of 5. Thus the LCD is $2 \cdot 3 \cdot 3 \cdot 5 = 90$.

$$\frac{x+3}{10} + \frac{2x+1}{15} - \frac{x-2}{18} = \left(\frac{x+3}{10}\right)\left(\frac{9}{9}\right) + \left(\frac{2x+1}{15}\right)\left(\frac{6}{6}\right) - \left(\frac{x-2}{18}\right)\left(\frac{5}{5}\right)$$

$$= \frac{9(x+3)}{90} + \frac{6(2x+1)}{90} - \frac{5(x-2)}{90}$$

$$= \frac{9x+27+12x+6-5x+10}{90}$$

$$= \frac{16x+43}{90}$$ ∎

The presence of variables in the denominators does not create any serious difficulty; our approach remains the same. Study the following examples very carefully. For each problem we use the same basic procedure: (1) Find the LCD. (2) Change each fraction to an equivalent fraction having the LCD as its denominator. (3) Add or subtract numerators and place this result over the LCD. (4) Look for possibilities to simplify the resulting fraction.

EXAMPLE 3

Add $\dfrac{3}{2x} + \dfrac{5}{3y}$.

Solution

Using an LCD of $6xy$, we can proceed as follows.

$$\frac{3}{2x} + \frac{5}{3y} = \left(\frac{3}{2x}\right)\left(\frac{3y}{3y}\right) + \left(\frac{5}{3y}\right)\left(\frac{2x}{2x}\right)$$

$$= \frac{9y}{6xy} + \frac{10x}{6xy}$$

$$= \frac{9y + 10x}{6xy}$$

EXAMPLE 4

Subtract $\dfrac{7}{12ab} - \dfrac{11}{15a^2}$.

Solution

We can factor the numerical coefficients of the denominators into primes to help find the LCD.

$$\left.\begin{array}{l} 12ab = 2 \cdot 2 \cdot 3 \cdot a \cdot b \\ 15a^2 = 3 \cdot 5 \cdot a^2 \end{array}\right\} \qquad \text{LCD} = 2 \cdot 2 \cdot 3 \cdot 5 \cdot a^2 \cdot b = 60a^2b$$

$$\frac{7}{12ab} - \frac{11}{15a^2} = \left(\frac{7}{12ab}\right)\left(\frac{5a}{5a}\right) - \left(\frac{11}{15a^2}\right)\left(\frac{4b}{4b}\right)$$

$$= \frac{35a}{60a^2b} - \frac{44b}{60a^2b}$$

$$= \frac{35a - 44b}{60a^2b}$$

EXAMPLE 5

Add $\dfrac{8}{x^2 - 4x} + \dfrac{2}{x}$.

Solution

$$\left.\begin{array}{r} x^2 - 4x = x(x - 4) \\ x = x \end{array}\right\} \quad \text{LCD} = x(x - 4)$$

$$\frac{8}{x(x - 4)} + \frac{2}{x} = \frac{8}{x(x - 4)} + \left(\frac{2}{x}\right)\left(\frac{x - 4}{x - 4}\right)$$

$$= \frac{8}{x(x - 4)} + \frac{2(x - 4)}{x(x - 4)}$$

$$= \frac{8 + 2x - 8}{x(x - 4)}$$

$$= \frac{2\cancel{x}}{\cancel{x}(x - 4)}$$

$$= \frac{2}{x - 4} \qquad\blacksquare$$

EXAMPLE 6

Add $\dfrac{3n}{n^2 + 6n + 5} + \dfrac{4}{n^2 - 7n - 8}$.

Solution

$$\left.\begin{array}{r} n^2 + 6n + 5 = (n + 5)(n + 1) \\ n^2 - 7n - 8 = (n - 8)(n + 1) \end{array}\right\} \quad \text{LCD} = (n + 1)(n + 5)(n - 8)$$

$$\frac{3n}{n^2 + 6n + 5} + \frac{4}{n^2 - 7n - 8} = \left[\frac{3n}{(n + 5)(n + 1)}\right]\left(\frac{n - 8}{n - 8}\right) + \left[\frac{4}{(n - 8)(n + 1)}\right]\left(\frac{n + 5}{n + 5}\right)$$

$$= \frac{3n(n - 8)}{(n + 5)(n + 1)(n - 8)} + \frac{4(n + 5)}{(n + 5)(n + 1)(n - 8)}$$

$$= \frac{3n^2 - 24n + 4n + 20}{(n + 5)(n + 1)(n - 8)}$$

$$= \frac{3n^2 - 20n + 20}{(n + 5)(n + 1)(n - 8)} \qquad\blacksquare$$

Simplifying Complex Fractions

Fractional forms that contain rational expressions in the numerator and/or denominator are called **complex fractions**. The following examples illustrate some approaches to simplifying complex fractions.

<table>
<tr><td>

EXAMPLE 7

</td><td>

Simplify $\dfrac{\dfrac{3}{x} + \dfrac{2}{y}}{\dfrac{5}{x} - \dfrac{6}{y^2}}$.

</td></tr>
</table>

Solution A

Treating the numerator as the sum of two rational expressions and the denominator as the difference of two rational expressions, we can proceed as follows.

$$\frac{\dfrac{3}{x} + \dfrac{2}{y}}{\dfrac{5}{x} - \dfrac{6}{y^2}} = \frac{\left(\dfrac{3}{x}\right)\left(\dfrac{y}{y}\right) + \left(\dfrac{2}{y}\right)\left(\dfrac{x}{x}\right)}{\left(\dfrac{5}{x}\right)\left(\dfrac{y^2}{y^2}\right) - \left(\dfrac{6}{y^2}\right)\left(\dfrac{x}{x}\right)}$$

$$= \frac{\dfrac{3y}{xy} + \dfrac{2x}{xy}}{\dfrac{5y^2}{xy^2} - \dfrac{6x}{xy^2}} = \frac{\dfrac{3y + 2x}{xy}}{\dfrac{5y^2 - 6x}{xy^2}}$$

$$= \frac{3y + 2x}{\cancel{xy}} \cdot \frac{\overset{y}{\cancel{xy^2}}}{5y^2 - 6x}$$

$$= \frac{y(3y + 2x)}{5y^2 - 6x}$$

Solution B

The LCD of all four denominators $(x, y, x, \text{ and } y^2)$ is xy^2. Let's multiply the entire complex fraction by a form of 1—namely, $(xy^2)/(xy^2)$.

$$\frac{\dfrac{3}{x} + \dfrac{2}{y}}{\dfrac{5}{x} - \dfrac{6}{y^2}} = \left(\frac{\dfrac{3}{x} + \dfrac{2}{y}}{\dfrac{5}{x} - \dfrac{6}{y^2}}\right)\left(\frac{xy^2}{xy^2}\right)$$

$$= \frac{(xy^2)\left(\dfrac{3}{x}\right) + (xy^2)\left(\dfrac{2}{y}\right)}{(xy^2)\left(\dfrac{5}{x}\right) - (xy^2)\left(\dfrac{6}{y^2}\right)}$$

$$= \frac{3y^2 + 2xy}{5y^2 - 6x} \quad \text{or} \quad \frac{y(3y + 2x)}{5y^2 - 6x} \qquad \blacksquare$$

Certainly either approach (Solution A or Solution B) will work with a problem such as Example 7. We suggest that you study Solution B very carefully. This approach works effectively with complex fractions when the LCD of all the denominators is easy to find. Let's look at a type of complex fraction used in certain calculus problems.

EXAMPLE 8

Simplify $\dfrac{\dfrac{1}{x+h} - \dfrac{1}{x}}{h}$.

Solution

$$\frac{\dfrac{1}{x+h} - \dfrac{1}{x}}{\dfrac{h}{1}} = \left[\frac{x(x+h)}{x(x+h)}\right]\left[\frac{\dfrac{1}{x+h} - \dfrac{1}{x}}{\dfrac{h}{1}}\right]$$

$$= \frac{x(x+h)\left(\dfrac{1}{x+h}\right) - x(x+h)\left(\dfrac{1}{x}\right)}{x(x+h)(h)}$$

$$= \frac{x - (x+h)}{hx(x+h)} = \frac{x - x - h}{hx(x+h)}$$

$$= \frac{-h}{hx(x+h)} = -\frac{1}{x(x+h)} \qquad \blacksquare$$

Example 9 illustrates another way to simplify complex fractions.

EXAMPLE 9

Simplify $1 - \dfrac{n}{1 - \dfrac{1}{n}}$.

Solution

We first simplify the complex fraction by multiplying by n/n.

$$\left(\frac{n}{1 - \dfrac{1}{n}}\right)\left(\frac{n}{n}\right) = \frac{n^2}{n-1}$$

Now we can perform the subtraction.

$$1 - \frac{n^2}{n-1} = \left(\frac{n-1}{n-1}\right)\left(\frac{1}{1}\right) - \frac{n^2}{n-1} = \frac{n-1}{n-1} - \frac{n^2}{n-1}$$

$$= \frac{n-1-n^2}{n-1} \qquad \text{or} \qquad \frac{-n^2 + n - 1}{n-1} \qquad \blacksquare$$

Finally, we need to recognize that complex fractions are sometimes the result of applying the definition $b^{-n} = \dfrac{1}{b^n}$. Our final example illustrates this idea.

EXAMPLE 10

Simplify $\dfrac{2x^{-1} + y^{-1}}{x - 3y^{-2}}$.

Solution

First, let's apply $b^{-n} = \dfrac{1}{b^n}$.

$$\frac{2x^{-1} + y^{-1}}{x - 3y^{-2}} = \frac{\dfrac{2}{x} + \dfrac{1}{y}}{x - \dfrac{3}{y^2}}$$

Now we can proceed as in the previous examples.

$$\left(\frac{\dfrac{2}{x} + \dfrac{1}{y}}{x - \dfrac{3}{y^2}}\right)\left(\frac{xy^2}{xy^2}\right) = \frac{\dfrac{2}{x}(xy^2) + \dfrac{1}{y}(xy^2)}{x(xy^2) - \dfrac{3}{y^2}(xy^2)}$$

$$= \frac{2y^2 + xy}{x^2y^2 - 3x}$$

■

PROBLEM SET 0.5

For Problems 1–12, simplify each rational expression.

1. $\dfrac{14x^2y}{21xy}$

2. $\dfrac{-26xy^2}{65y}$

3. $\dfrac{-63xy^4}{-81x^2y}$

4. $\dfrac{x^2 - y^2}{x^2 + xy}$

5. $\dfrac{a^2 + 7a + 12}{a^2 - 6a - 27}$

6. $\dfrac{6x^2 + x - 15}{8x^2 - 10x - 3}$

7. $\dfrac{2x^3 + 3x^2 - 14x}{x^2y + 7xy - 18y}$

8. $\dfrac{3x - x^2}{x^2 - 9}$

9. $\dfrac{x^3 - y^3}{x^2 + xy - 2y^2}$

10. $\dfrac{ax - 3x + 2ay - 6y}{2ax - 6x + ay - 3y}$

11. $\dfrac{2y - 2xy}{x^2y - y}$

12. $\dfrac{16x^3y + 24x^2y^2 - 16xy^3}{24x^2y + 12xy^2 - 12y^3}$

For Problems 13–56, perform the indicated operations involving rational expressions. Express final answers in simplest form.

13. $\dfrac{4x^2}{5y^2} \cdot \dfrac{15xy}{24x^2y^2}$

14. $\dfrac{5xy}{8y^2} \cdot \dfrac{18x^2y}{15}$

15. $\dfrac{-14xy^4}{18y^2} \cdot \dfrac{24x^2y^3}{35y^2}$

16. $\dfrac{6xy}{9y^4} \cdot \dfrac{30x^3y}{-48x}$

17. $\dfrac{7a^2b}{9ab^3} \div \dfrac{3a^4}{2a^2b^2}$

18. $\dfrac{9a^2c}{12bc^2} \div \dfrac{21ab}{14c^3}$

19. $\dfrac{5xy}{x + 6} \cdot \dfrac{x^2 - 36}{x^2 - 6x}$

20. $\dfrac{2a^2 + 6}{a^2 - a} \cdot \dfrac{a^3 - a^2}{8a - 4}$

21. $\dfrac{5a^2 + 20a}{a^3 - 2a^2} \cdot \dfrac{a^2 - a - 12}{a^2 - 16}$

22. $\dfrac{t^4 - 81}{t^2 - 6t + 9} \cdot \dfrac{6t^2 - 11t - 21}{5t^2 + 8t - 21}$

23. $\dfrac{x^2 + 5xy - 6y^2}{xy^2 - y^3} \cdot \dfrac{2x^2 + 15xy + 18y^2}{xy + 4y^2}$

24. $\dfrac{10n^2 + 21n - 10}{5n^2 + 33n - 14} \cdot \dfrac{2n^2 + 6n - 56}{2n^2 - 3n - 20}$

25. $\dfrac{9y^2}{x^2 + 12x + 36} \div \dfrac{12y}{x^2 + 6x}$

26. $\dfrac{x^2 - 4xy + 4y^2}{7xy^2} \div \dfrac{4x^2 - 3xy - 10y^2}{20x^2y + 25xy^2}$

27. $\dfrac{2x^2 + 3x}{2x^3 - 10x^2} \cdot \dfrac{x^2 - 8x + 15}{3x^3 - 27x} \div \dfrac{14x + 21}{x^2 - 6x - 27}$

28. $\dfrac{a^2 - 4ab + 4b^2}{6a^2 - 4ab} \cdot \dfrac{3a^2 + 5ab - 2b^2}{6a^2 + ab - b^2} \div \dfrac{a^2 - 4b^2}{8a + 4b}$

29. $\dfrac{x + 4}{6} + \dfrac{2x - 1}{4}$

30. $\dfrac{3n - 1}{9} - \dfrac{n + 2}{12}$

31. $\dfrac{x + 1}{4} + \dfrac{x - 3}{6} - \dfrac{x - 2}{8}$

32. $\dfrac{x - 2}{5} - \dfrac{x + 3}{6} + \dfrac{x + 1}{15}$

33. $\dfrac{7}{16a^2b} + \dfrac{3a}{20b^2}$

34. $\dfrac{5b}{24a^2} - \dfrac{11a}{32b}$

35. $\dfrac{1}{n^2} + \dfrac{3}{4n} - \dfrac{5}{6}$

36. $\dfrac{3}{n^2} - \dfrac{2}{5n} + \dfrac{4}{3}$

37. $\dfrac{3}{4x} + \dfrac{2}{3y} - 1$

38. $\dfrac{5}{6x} - \dfrac{3}{4y} + 2$

39. $\dfrac{3}{2x + 1} + \dfrac{2}{3x + 4}$

40. $\dfrac{5}{x - 1} - \dfrac{3}{2x - 3}$

41. $\dfrac{4x}{x^2 + 7x} + \dfrac{3}{x}$

42. $\dfrac{6}{x^2 + 8x} - \dfrac{3}{x}$

43. $\dfrac{4a - 4}{a^2 - 4} - \dfrac{3}{a + 2}$

44. $\dfrac{6a + 4}{a^2 - 1} - \dfrac{5}{a - 1}$

45. $\dfrac{3}{x + 1} + \dfrac{x + 5}{x^2 - 1} - \dfrac{3}{x - 1}$

46. $\dfrac{5}{x} - \dfrac{5x - 30}{x^2 + 6x} + \dfrac{x}{x + 6}$

47. $\dfrac{5}{x^2 + 10x + 21} + \dfrac{4}{x^2 + 12x + 27}$

48. $\dfrac{8}{a^2 - 3a - 18} - \dfrac{10}{a^2 - 7a - 30}$

49. $\dfrac{5}{x^2 - 1} - \dfrac{2}{x^2 + 6x - 16}$

50. $\dfrac{4}{x^2 + 2} - \dfrac{7}{x^2 + x - 12}$

51. $x - \dfrac{x^2}{x - 1} + \dfrac{1}{x^2 - 1}$

52. $x - \dfrac{x^2}{x + 7} - \dfrac{x}{x^2 - 49}$

53. $\dfrac{2n^2}{n^4 - 16} - \dfrac{n}{n^2 - 4} + \dfrac{1}{n + 2}$

54. $\dfrac{n}{n^2 + 1} + \dfrac{n^2 + 3n}{n^4 - 1} - \dfrac{1}{n - 1}$

55. $\dfrac{2x + 1}{x^2 - 3x - 4} + \dfrac{3x - 2}{x^2 + 3x - 28}$

56. $\dfrac{3x - 4}{2x^2 - 9x - 5} - \dfrac{2x - 1}{3x^2 - 11x - 20}$

57. Consider the addition problem $\dfrac{8}{x - 2} + \dfrac{5}{2 - x}$. Note that the denominators are opposites of each other. If the property $\dfrac{a}{-b} = -\dfrac{a}{b}$ is applied to the second fraction, we obtain $\dfrac{5}{2 - x} = -\dfrac{5}{x - 2}$. Thus we can proceed as follows.

$$\dfrac{8}{x - 2} + \dfrac{5}{2 - x} = \dfrac{8}{x - 2} - \dfrac{5}{x - 2}$$

$$= \dfrac{8 - 5}{x - 2} = \dfrac{3}{x - 2}$$

Use this approach to do the following problems.

a. $\dfrac{7}{x - 1} + \dfrac{2}{1 - x}$

b. $\dfrac{5}{2x - 1} + \dfrac{8}{1 - 2x}$

c. $\dfrac{4}{a - 3} - \dfrac{1}{3 - a}$

d. $\dfrac{10}{a - 9} - \dfrac{5}{9 - a}$

e. $\dfrac{x^2}{x - 1} - \dfrac{2x - 3}{1 - x}$

f. $\dfrac{x^2}{x - 4} - \dfrac{3x - 28}{4 - x}$

For Problems 58–80; simplify each complex fraction.

58. $\dfrac{\dfrac{2}{x} + \dfrac{7}{y}}{\dfrac{3}{x} - \dfrac{10}{y}}$

59. $\dfrac{\dfrac{5}{x^2} - \dfrac{3}{x}}{\dfrac{1}{y} + \dfrac{2}{y^2}}$

60. $\dfrac{\dfrac{1}{x} + 3}{\dfrac{2}{y} + 4}$

61. $\dfrac{1 + \dfrac{1}{x}}{1 - \dfrac{1}{x}}$

62. $\dfrac{3 - \dfrac{2}{n-4}}{5 + \dfrac{4}{n-4}}$

63. $\dfrac{1 - \dfrac{1}{n+1}}{1 + \dfrac{1}{n-1}}$

72. $\dfrac{\dfrac{1}{(x+h)^3} - \dfrac{1}{x^3}}{h}$

73. $\dfrac{\dfrac{1}{x+h+1} - \dfrac{1}{x+1}}{h}$

64. $\dfrac{\dfrac{2}{x-3} - \dfrac{3}{x+3}}{\dfrac{5}{x^2-9} - \dfrac{2}{x-3}}$

65. $\dfrac{\dfrac{-2}{x} - \dfrac{4}{x+2}}{\dfrac{3}{x^2+2x} + \dfrac{3}{x}}$

74. $\dfrac{\dfrac{3}{x+h} - \dfrac{3}{x}}{h}$

75. $\dfrac{\dfrac{2}{2x+2h-1} - \dfrac{2}{2x-1}}{h}$

66. $\dfrac{\dfrac{-1}{y-2} + \dfrac{5}{x}}{\dfrac{3}{x} - \dfrac{4}{xy-2x}}$

67. $1 + \dfrac{x}{1 + \dfrac{1}{x}}$

76. $\dfrac{\dfrac{3}{4x+4h+5} - \dfrac{3}{4x+5}}{h}$

77. $\dfrac{x^{-1} + y^{-1}}{x - y}$

78. $\dfrac{x+y}{x^{-1} + y^{-1}}$

68. $2 - \dfrac{x}{3 - \dfrac{2}{x}}$

69. $\dfrac{a}{\dfrac{1}{a} + 4} + 1$

79. $\dfrac{x + x^{-1}y^{-2}}{x^{-1} - y^{-2}}$

80. $\dfrac{x^{-2} - 2y^{-1}}{3x^{-1} + y^{-2}}$

70. $\dfrac{\dfrac{3a}{2 - \dfrac{1}{a}} - 1}{}$

71. $\dfrac{\dfrac{1}{(x+h)^2} - \dfrac{1}{x^2}}{h}$

THOUGHTS INTO WORDS

81. What role does factoring play in the simplifying of rational expressions?

82. Explain in your own words how to multiply two rational expressions.

83. Give a step-by-step description of how to add $\dfrac{2x-1}{4} + \dfrac{3x+5}{14}$.

84. Look back at the two approaches shown in Example 7.

Which approach would you use to simplify $\dfrac{\dfrac{1}{4} + \dfrac{1}{6}}{\dfrac{1}{2} - \dfrac{3}{4}}$?

Which approach would you use to simplify $\dfrac{\dfrac{5}{8} + \dfrac{4}{9}}{\dfrac{5}{14} - \dfrac{2}{21}}$?

Explain the reason for your choice of approach for each problem.

0.6   RADICALS

Recall from our work with exponents that to **square a number** means to raise it to the second power—that is, to use the number as a factor twice. For example, $4^2 = 4 \cdot 4 = 16$ and $(-4)^2 = (-4)(-4) = 16$. A **square root of a number** is one of its two equal factors. Thus 4 and -4 are both square roots of 16. In general, a is a square root of b if $a^2 = b$. The following statements generalize these ideas.

1. Every positive real number has two square roots; one is positive and the other is negative. They are opposites of each other.

2. Negative real numbers have no real number square roots because the square of any nonzero real number is positive.

3. The square root of zero is zero.

The symbol $\sqrt{}$, called a **radical sign**, is used to designate the *nonnegative* square root, which is called the **principal square root**. The number under the radical sign is called the **radicand**, and the entire expression, such as $\sqrt{16}$, is referred to as a **radical**.

The following examples demonstrate the use of the square root notation.

$\sqrt{16} = 4$ $\quad$ $\sqrt{16}$ indicates the *nonnegative* or *principal square root* of 16.

$-\sqrt{16} = -4$ $\quad$ $-\sqrt{16}$ indicates the negative square root of 16.

$\sqrt{0} = 0$ $\quad$ Zero has only one square root. Technically, we could also write $-\sqrt{0} = -0 = 0$.

$\sqrt{-4}$ $\quad$ Not a real number

$-\sqrt{-4}$ $\quad$ Not a real number

To **cube a number** means to raise it to the third power—that is, to use the number as a factor three times. For example, $2^3 = 2 \cdot 2 \cdot 2 = 8$ and $(-2)^3 = (-2)(-2)(-2) = -8$. A **cube root of a number** is one of its three equal factors. Thus 2 is a cube root of 8, and as we will discuss later, it is the only real number that is a cube root of 8. Furthermore, -2 is the only real number that is a cube root of -8. In general, a is a cube root of b if $a^3 = b$. The following statements generalize these ideas.

1. Every positive real number has one positive real number cube root.

2. Every negative real number has one negative real number cube root.

3. The cube root of zero is zero.

REMARK Every nonzero real number has three cube roots, but only one of them is a real number. The other roots are complex numbers, which we will discuss in Section 0.8.

The symbol $\sqrt[3]{}$ is used to designate the cube root of a number. Thus we can write

$$\sqrt[3]{8} = 2 \qquad \sqrt[3]{-8} = -2 \qquad \sqrt[3]{\frac{1}{27}} = \frac{1}{3} \qquad \text{and} \qquad \sqrt[3]{-\frac{1}{27}} = -\frac{1}{3}$$

The concept of root can be extended to fourth roots, fifth roots, sixth roots, and in general, nth roots. If n is an *even positive integer*, then the following statements are true.

1. Every positive real number has exactly two real nth roots, one positive and one negative. For example, the real fourth roots of 16 are 2 and -2.

2. Negative real numbers do not have real nth roots. For example, there are no real fourth roots of -16.

If n is an *odd positive integer* greater than 1, then the following statements are true.

1. Every real number has exactly one real nth root.

2. The real nth root of a positive number is positive. For example, the fifth root of 32 is 2.

3. The real nth root of a negative number is negative. For example, the fifth root of -32 is -2.

In general, the following definition is useful.

DEFINITION 0.5

$\sqrt[n]{b} = a$ if and only if $a^n = b$

In Definition 0.5, if n is an even positive integer, then a and b are both nonnegative. If n is an odd positive integer greater than 1, then a and b are both nonnegative or both negative. The symbol $\sqrt[n]{}$ designates the principal root.

The following examples are applications of Definition 0.5.

$$\sqrt[4]{81} = 3 \qquad \text{because } 3^4 = 81$$
$$\sqrt[5]{32} = 2 \qquad \text{because } 2^5 = 32$$
$$\sqrt[5]{-32} = -2 \qquad \text{because } (-2)^5 = -32$$

To complete our terminology, the n in the radical $\sqrt[n]{b}$ is called the **index** of the radical. If $n = 2$, we commonly write $\sqrt{b}$ instead of $\sqrt[2]{b}$. In this text, when we use symbols such as $\sqrt[n]{b}$, $\sqrt[m]{y}$, and $\sqrt[r]{x}$, we will assume the previous agreements relative to the existence of real roots without listing the various restrictions, unless a special restriction is needed.

From Definition 0.5 we see that if n is any positive integer greater than 1 and $\sqrt[n]{b}$ exists, then

$$\left(\sqrt[n]{b}\right)^n = b$$

For example, $\left(\sqrt{4}\right)^2 = 4$, $\left(\sqrt[3]{-8}\right)^3 = -8$, and $\left(\sqrt[4]{81}\right)^4 = 81$. Furthermore, if $b \geq 0$ and n is any positive integer greater than 1 or if $b < 0$ and n is an odd positive integer greater than 1, then

$$\sqrt[n]{b^n} = b$$

For example, $\sqrt{4^2} = 4$, $\sqrt[3]{(-2)^3} = -2$, and $\sqrt[5]{6^5} = 6$. But we must be careful, because

$$\sqrt{(-2)^2} \neq -2 \qquad \text{and} \qquad \sqrt[4]{(-2)^4} \neq -2$$

Simplest Radical Form

Let's use some examples to motivate another useful property of radicals.

$$\sqrt{16 \cdot 25} = \sqrt{400} = 20 \qquad \text{and} \qquad \sqrt{16} \cdot \sqrt{25} = 4 \cdot 5 = 20$$
$$\sqrt[3]{8 \cdot 27} = \sqrt[3]{216} = 6 \qquad \text{and} \qquad \sqrt[3]{8} \cdot \sqrt[3]{27} = 2 \cdot 3 = 6$$
$$\sqrt[3]{-8 \cdot 64} = \sqrt[3]{-512} = -8 \qquad \text{and} \qquad \sqrt[3]{-8} \cdot \sqrt[3]{64} = -2 \cdot 4 = -8$$

In general, the following property can be stated.

PROPERTY 0.3

$$\sqrt[n]{bc} = \sqrt[n]{b}\,\sqrt[n]{c} \quad \text{if } \sqrt[n]{b} \text{ and } \sqrt[n]{c} \text{ are real numbers.}$$

Property 0.3 states that **the nth root of a product is equal to the product of the nth roots.**

The definition of nth root, along with Property 0.3, provides the basis for changing radicals to simplest radical form. The concept of **simplest radical form** takes on additional meaning as we encounter more complicated expressions, but for now it simply means that the radicand does not contain any perfect powers of the index. Consider the following examples of reductions to simplest radical form.

$$\sqrt{45} = \sqrt{9 \cdot 5} = \sqrt{9}\,\sqrt{5} = 3\sqrt{5}$$
$$\sqrt{52} = \sqrt{4 \cdot 13} = \sqrt{4}\sqrt{13} = 2\sqrt{13}$$
$$\sqrt[3]{24} = \sqrt[3]{8 \cdot 3} = \sqrt[3]{8}\sqrt[3]{3} = 2\sqrt[3]{3}$$

A variation of the technique for changing radicals with index n to simplest form is to factor the radicand into primes and then to look for the perfect nth powers in exponential form, as in the following examples.

$$\sqrt{80} = \sqrt{2^4 \cdot 5} = \sqrt{2^4}\sqrt{5} = 2^2\sqrt{5} = 4\sqrt{5}$$
$$\sqrt[3]{108} = \sqrt[3]{2^2 \cdot 3^3} = \sqrt[3]{3^3}\sqrt[3]{2^2} = 3\sqrt[3]{4}$$

The distributive property can be used to combine radicals that have the same index and the same radicand.

$$3\sqrt{2} + 5\sqrt{2} = (3 + 5)\sqrt{2} = 8\sqrt{2}$$
$$7\sqrt[3]{5} - 3\sqrt[3]{5} = (7 - 3)\sqrt[3]{5} = 4\sqrt[3]{5}$$

Sometimes it is necessary to simplify the radicals first and then to combine them by applying the distributive property.

$$3\sqrt{8} + 2\sqrt{18} - 4\sqrt{2} = 3\sqrt{4}\sqrt{2} + 2\sqrt{9}\sqrt{2} - 4\sqrt{2}$$
$$= 6\sqrt{2} + 6\sqrt{2} - 4\sqrt{2}$$
$$= (6 + 6 - 4)\sqrt{2}$$
$$= 8\sqrt{2}$$

Property 0.3 can also be viewed as $\sqrt[n]{b}\sqrt[n]{c} = \sqrt[n]{bc}$. Then, along with the commutative and associative properties of the real numbers, it provides the basis for multiplying radicals that have the same index. Consider the following two examples.

$$(7\sqrt{6})(3\sqrt{8}) = 7 \cdot 3 \cdot \sqrt{6} \cdot \sqrt{8}$$
$$= 21\sqrt{48}$$
$$= 21\sqrt{16}\sqrt{3}$$
$$= 21 \cdot 4 \cdot \sqrt{3}$$
$$= 84\sqrt{3}$$

$$(2\sqrt[3]{6})(5\sqrt[3]{4}) = 2 \cdot 5 \cdot \sqrt[3]{6} \cdot \sqrt[3]{4}$$
$$= 10\sqrt[3]{24}$$
$$= 10\sqrt[3]{8}\sqrt[3]{3}$$
$$= 10 \cdot 2 \cdot \sqrt[3]{3}$$
$$= 20\sqrt[3]{3}$$

The distributive property, along with Property 0.3, provides a way of handling special products involving radicals, as the next examples illustrate.

$$2\sqrt{2}(4\sqrt{3} - 5\sqrt{6}) = (2\sqrt{2})(4\sqrt{3}) - (2\sqrt{2})(5\sqrt{6})$$
$$= 8\sqrt{6} - 10\sqrt{12}$$
$$= 8\sqrt{6} - 10\sqrt{4}\sqrt{3}$$
$$= 8\sqrt{6} - 20\sqrt{3}$$

$$(2\sqrt{2} - \sqrt{7})(3\sqrt{2} + 5\sqrt{7}) = 2\sqrt{2}(3\sqrt{2} + 5\sqrt{7}) - \sqrt{7}(3\sqrt{2} + 5\sqrt{7})$$
$$= (2\sqrt{2})(3\sqrt{2}) + (2\sqrt{2})(5\sqrt{7}) - (\sqrt{7})(3\sqrt{2}) - (\sqrt{7})(5\sqrt{7})$$
$$= 6 \cdot 2 + 10\sqrt{14} - 3\sqrt{14} - 5 \cdot 7$$
$$= -23 + 7\sqrt{14}$$

$$(\sqrt{5} + \sqrt{2})(\sqrt{5} - \sqrt{2}) = \sqrt{5}(\sqrt{5} - \sqrt{2}) + \sqrt{2}(\sqrt{5} - \sqrt{2})$$
$$= (\sqrt{5})(\sqrt{5}) - (\sqrt{5})(\sqrt{2}) + (\sqrt{2})(\sqrt{5}) - (\sqrt{2})(\sqrt{2})$$
$$= 5 - \sqrt{10} + \sqrt{10} - 2$$
$$= 3$$

Pay special attention to the last example. It fits the special-product pattern $(a + b)(a - b) = a^2 - b^2$. We will use that idea in a moment.

More About Simplest Radical Form

Another property of *n*th roots is motivated by the following examples.

$$\sqrt{\frac{36}{9}} = \sqrt{4} = 2 \quad \text{and} \quad \frac{\sqrt{36}}{\sqrt{9}} = \frac{6}{3} = 2$$

$$\sqrt[3]{\frac{64}{8}} = \sqrt[3]{8} = 2 \quad \text{and} \quad \frac{\sqrt[3]{64}}{\sqrt[3]{8}} = \frac{4}{2} = 2$$

In general, the following property can be stated.

PROPERTY 0.4

$$\sqrt[n]{\frac{b}{c}} = \frac{\sqrt[n]{b}}{\sqrt[n]{c}} \quad \text{if } \sqrt[n]{b} \text{ and } \sqrt[n]{c} \text{ are real numbers and } c \neq 0.$$

Property 0.4 states that **the *n*th root of a quotient is equal to the quotient of the *n*th roots.**

To evaluate radicals such as $\sqrt{\dfrac{4}{25}}$ and $\sqrt[3]{\dfrac{27}{8}}$, where the numerator and the denominator of the fractional radicands are perfect *n*th powers, we can either use Property 0.4 or rely on the definition of *n*th root.

$$\sqrt{\frac{4}{25}} = \frac{\sqrt{4}}{\sqrt{25}} = \frac{2}{5} \quad \text{or} \quad \sqrt{\frac{4}{25}} = \frac{2}{5} \quad \text{because } \frac{2}{5} \cdot \frac{2}{5} = \frac{4}{25}$$

$$\sqrt[3]{\frac{27}{8}} = \frac{\sqrt[3]{27}}{\sqrt[3]{8}} = \frac{3}{2} \quad \text{or} \quad \sqrt[3]{\frac{27}{8}} = \frac{3}{2} \quad \text{because } \frac{3}{2} \cdot \frac{3}{2} \cdot \frac{3}{2} = \frac{27}{8}$$

Radicals such as $\sqrt{\dfrac{28}{9}}$ and $\sqrt[3]{\dfrac{24}{27}}$, where only the denominators of the radicand are perfect *n*th powers, can be simplified as follows.

$$\sqrt{\frac{28}{9}} = \frac{\sqrt{28}}{\sqrt{9}} = \frac{\sqrt{4}\sqrt{7}}{3} = \frac{2\sqrt{7}}{3}$$

$$\sqrt[3]{\frac{24}{27}} = \frac{\sqrt[3]{24}}{\sqrt[3]{27}} = \frac{\sqrt[3]{8}\sqrt[3]{3}}{3} = \frac{2\sqrt[3]{3}}{3}$$

Before we consider more examples, let's summarize some ideas about simplifying radicals. A radical is said to be in **simplest radical form** if the following conditions are satisfied.

1. No fraction appears within a radical sign.

Thus $\sqrt{\dfrac{3}{4}}$ violates this condition.

2. No radical appears in the denominator.

Thus $\dfrac{\sqrt{2}}{\sqrt{3}}$ violates this condition.

3. No radicand contains a perfect power of the index.

Thus $\sqrt{7^2 \cdot 5}$ violates this condition.

Now let's consider an example in which neither the numerator nor the denominator of the radicand is a perfect *n*th power.

$$\sqrt{\frac{2}{3}} = \frac{\sqrt{2}}{\sqrt{3}} = \frac{\sqrt{2}}{\sqrt{3}} \cdot \frac{\sqrt{3}}{\sqrt{3}} = \frac{\sqrt{6}}{3}$$

Form of 1

The process used to simplify the radical in the previous example is referred to as **rationalizing the denominator**. There is more than one way to rationalize the denominator, as illustrated by the next example.

EXAMPLE 1

Simplify $\dfrac{\sqrt{5}}{\sqrt{8}}$.

Solution A

$$\frac{\sqrt{5}}{\sqrt{8}} = \frac{\sqrt{5}}{\sqrt{8}} \cdot \frac{\sqrt{8}}{\sqrt{8}} = \frac{\sqrt{40}}{8} = \frac{\sqrt{4}\sqrt{10}}{8} = \frac{2\sqrt{10}}{8} = \frac{\sqrt{10}}{4}$$

Solution B

$$\frac{\sqrt{5}}{\sqrt{8}} = \frac{\sqrt{5}}{\sqrt{8}} \cdot \frac{\sqrt{2}}{\sqrt{2}} = \frac{\sqrt{10}}{\sqrt{16}} = \frac{\sqrt{10}}{4}$$

Solution C

$$\frac{\sqrt{5}}{\sqrt{8}} = \frac{\sqrt{5}}{\sqrt{4}\sqrt{2}} = \frac{\sqrt{5}}{2\sqrt{2}} = \frac{\sqrt{5}}{2\sqrt{2}} \cdot \frac{\sqrt{2}}{\sqrt{2}} = \frac{\sqrt{10}}{4}$$ ∎

The three approaches to Example 1 again illustrate the need to think first and then push the pencil. You may find one approach easier than another.

EXAMPLE 2

Simplify $\dfrac{\sqrt{6}}{\sqrt{8}}$.

Solution

$$\frac{\sqrt{6}}{\sqrt{8}} = \sqrt{\frac{6}{8}} \qquad \text{Remember that } \frac{\sqrt{a}}{\sqrt{b}} = \sqrt{\frac{a}{b}}.$$

$$= \sqrt{\frac{3}{4}} \qquad \text{Reduce the fraction.}$$

$$= \frac{\sqrt{3}}{\sqrt{4}}$$

$$= \frac{\sqrt{3}}{2} \qquad\qquad\qquad\qquad\qquad\qquad\qquad\qquad ■$$

EXAMPLE 3

Simplify $\dfrac{\sqrt[3]{5}}{\sqrt[3]{9}}$.

 Solution

$$\frac{\sqrt[3]{5}}{\sqrt[3]{9}} = \frac{\sqrt[3]{5}}{\sqrt[3]{9}} \cdot \frac{\sqrt[3]{3}}{\sqrt[3]{3}}$$

$$= \frac{\sqrt[3]{15}}{\sqrt[3]{27}}$$

$$= \frac{\sqrt[3]{15}}{3} \qquad\qquad\qquad\qquad\qquad\qquad\qquad\qquad ■$$

Now let's consider an example in which the denominator is of binomial form.

EXAMPLE 4

Simplify $\dfrac{4}{\sqrt{5} + \sqrt{2}}$ by rationalizing the denominator.

Solution

 Remember that a moment ago we found that $\left(\sqrt{5} + \sqrt{2}\right)\left(\sqrt{5} - \sqrt{2}\right) = 3$. Let's use that idea here.

$$\frac{4}{\sqrt{5} + \sqrt{2}} = \left(\frac{4}{\sqrt{5} + \sqrt{2}}\right)\left(\frac{\sqrt{5} - \sqrt{2}}{\sqrt{5} - \sqrt{2}}\right)$$

$$= \frac{4\left(\sqrt{5} - \sqrt{2}\right)}{\left(\sqrt{5} + \sqrt{2}\right)\left(\sqrt{5} - \sqrt{2}\right)} = \frac{4\left(\sqrt{5} - \sqrt{2}\right)}{3} \qquad ■$$

The process of rationalizing the denominator does agree with the previously listed conditions. However, for certain problems in calculus, it is necessary to **rationalize the numerator**. Again, the fact that $\left(\sqrt{a} + \sqrt{b}\right)\left(\sqrt{a} - \sqrt{b}\right) = a - b$ can be used.

EXAMPLE 5

Change the form of $\dfrac{\sqrt{x+h}-\sqrt{x}}{h}$ by rationalizing the *numerator*.

Solution

$$\frac{\sqrt{x+h}-\sqrt{x}}{h} = \left(\frac{\sqrt{x+h}-\sqrt{x}}{h}\right)\left(\frac{\sqrt{x+h}+\sqrt{x}}{\sqrt{x+h}+\sqrt{x}}\right)$$

$$= \frac{(x+h)-x}{h\left(\sqrt{x+h}+\sqrt{x}\right)}$$

$$= \frac{\cancel{h}}{\cancel{h}\left(\sqrt{x+h}+\sqrt{x}\right)}$$

$$= \frac{1}{\sqrt{x+h}+\sqrt{x}}$$

■

Radicals Containing Variables

Before we illustrate how to simplify radicals that contain variables, there is one important point we should call to your attention. Let's look at some examples to illustrate the idea.

Consider the radical $\sqrt{x^2}$ for different values of x.

Let $x = 3$; then $\sqrt{x^2} = \sqrt{3^2} = \sqrt{9} = 3$.

Let $x = -3$; then $\sqrt{x^2} = \sqrt{(-3)^2} = \sqrt{9} = 3$.

Thus if $x \geq 0$ then $\sqrt{x^2} = x$, but if $x < 0$ then $\sqrt{x^2} = -x$. Using the concept of absolute value, we can state that **for all real numbers, $\sqrt{x^2} = |x|$.**

Now consider the radical $\sqrt{x^3}$. Because x^3 is negative when x is negative, we need to restrict x to the nonnegative real numbers when working with $\sqrt{x^3}$. Thus we can write

if $x \geq 0$, then $\sqrt{x^3} = \sqrt{x^2}\sqrt{x} = x\sqrt{x}$

and no absolute value sign is needed.

Finally, let's consider the radical $\sqrt[3]{x^3}$.

Let $x = 2$; then $\sqrt[3]{x^3} = \sqrt[3]{2^3} = \sqrt[3]{8} = 2$.

Let $x = -2$; then $\sqrt[3]{x^3} = \sqrt[3]{(-2)^3} = \sqrt[3]{-8} = -2$.

Thus it is correct to write,

$\sqrt[3]{x^3} = x$ for all real numbers

and again, no absolute value sign is needed.

The previous discussion indicates that, technically, every radical expression with variables in the radicand needs to be analyzed individually to determine the necessary restrictions on the variables. However, to avoid having to do this on a problem-by-problem basis, we shall merely **assume that all variables represent positive real numbers.**

Let's conclude this section by simplifying some radical expressions that contain variables.

$$\sqrt{72x^3y^7} = \sqrt{36x^2y^6}\sqrt{2xy} = 6xy^3\sqrt{2xy}$$

$$\sqrt[3]{40x^4y^8} = \sqrt[3]{8x^3y^6}\sqrt[3]{5xy^2} = 2xy^2\sqrt[3]{5xy^2}$$

$$\frac{\sqrt{5}}{\sqrt{12a^3}} = \frac{\sqrt{5}}{\sqrt{12a^3}}\cdot\frac{\sqrt{3a}}{\sqrt{3a}} = \frac{\sqrt{15a}}{\sqrt{36a^4}} = \frac{\sqrt{15a}}{6a^2}$$

$$\frac{3}{\sqrt[3]{4x}} = \frac{3}{\sqrt[3]{4x}}\cdot\frac{\sqrt[3]{2x^2}}{\sqrt[3]{2x^2}} = \frac{3\sqrt[3]{2x^2}}{\sqrt[3]{8x^3}} = \frac{3\sqrt[3]{2x^2}}{2x}$$

PROBLEM SET 0.6

For Problems 1–8, evaluate.

1. $\sqrt{81}$ **2.** $-\sqrt{49}$ **3.** $\sqrt[3]{125}$

4. $\sqrt[4]{81}$ **5.** $\sqrt{\dfrac{36}{49}}$ **6.** $\sqrt{\dfrac{256}{64}}$

7. $\sqrt[3]{-\dfrac{27}{8}}$ **8.** $\sqrt[3]{\dfrac{64}{27}}$

For Problems 9–44, express each in simplest radical form. All variables represent positive real numbers.

9. $\sqrt{24}$ **10.** $\sqrt{54}$ **11.** $\sqrt{112}$

12. $6\sqrt{28}$ **13.** $-3\sqrt{44}$ **14.** $-5\sqrt{68}$

15. $\dfrac{3}{4}\sqrt{20}$ **16.** $\dfrac{3}{8}\sqrt{72}$ **17.** $\sqrt{12x^2}$

18. $\sqrt{45xy^2}$ **19.** $\sqrt{64x^4y^7}$ **20.** $3\sqrt{32a^3}$

21. $\dfrac{3}{7}\sqrt{45xy^6}$ **22.** $\sqrt[3]{32}$ **23.** $\sqrt[3]{128}$

24. $\sqrt[3]{54x^3}$ **25.** $\sqrt[3]{16x^4}$ **26.** $\sqrt[3]{81x^5y^6}$

27. $\sqrt[4]{48x^5}$ **28.** $\sqrt[4]{162x^6y^7}$ **29.** $\sqrt{\dfrac{12}{25}}$

30. $\sqrt{\dfrac{75}{81}}$ **31.** $\sqrt{\dfrac{7}{8}}$ **32.** $\dfrac{\sqrt{35}}{\sqrt{7}}$

33. $\dfrac{4\sqrt{3}}{\sqrt{5}}$ **34.** $\dfrac{\sqrt{27}}{\sqrt{18}}$ **35.** $\dfrac{6\sqrt{3}}{7\sqrt{6}}$

36. $\sqrt{\dfrac{3x}{2y}}$ **37.** $\dfrac{\sqrt{5}}{\sqrt{12x^4}}$ **38.** $\dfrac{\sqrt{5y}}{\sqrt{18x^3}}$

39. $\dfrac{\sqrt{12a^2b}}{\sqrt{5a^3b^3}}$ **40.** $\dfrac{5}{\sqrt[3]{3}}$ **41.** $\dfrac{\sqrt[3]{27}}{\sqrt[3]{4}}$

42. $\sqrt[3]{\dfrac{5}{2x}}$ **43.** $\dfrac{\sqrt[3]{2y}}{\sqrt[3]{3x}}$ **44.** $\dfrac{\sqrt[3]{12xy}}{\sqrt[3]{3x^2y^5}}$

For Problems 45–52, use the distributive property to help simplify each. For example,

$$3\sqrt{8} + 5\sqrt{2} = 3\sqrt{4}\sqrt{2} + 5\sqrt{2}$$
$$= 6\sqrt{2} + 5\sqrt{2}$$
$$= (6 + 5)\sqrt{2}$$
$$= 11\sqrt{2}$$

45. $5\sqrt{12} + 2\sqrt{3}$ **46.** $4\sqrt{50} - 9\sqrt{32}$

47. $2\sqrt{28} - 3\sqrt{63} + 8\sqrt{7}$ **48.** $4\sqrt[3]{2} + 2\sqrt[3]{16} - \sqrt[3]{54}$

49. $\dfrac{5}{6}\sqrt{48} - \dfrac{3}{4}\sqrt{12}$ **50.** $\dfrac{2}{5}\sqrt{40} + \dfrac{1}{6}\sqrt{90}$

51. $\dfrac{2\sqrt{8}}{3} - \dfrac{3\sqrt{18}}{5} - \dfrac{\sqrt{50}}{2}$ **52.** $\dfrac{3\sqrt[3]{54}}{2} + \dfrac{5\sqrt[3]{16}}{3}$

For Problems 53–68, multiply and express the results in simplest radical form. All variables represent nonnegative real numbers.

53. $(4\sqrt{3})(6\sqrt{8})$ **54.** $(5\sqrt{8})(3\sqrt{7})$

55. $2\sqrt{3}(5\sqrt{2} + 4\sqrt{10})$ **56.** $3\sqrt{6}(2\sqrt{8} - 3\sqrt{12})$

57. $3\sqrt{x}(\sqrt{6xy} - \sqrt{8y})$ **58.** $\sqrt{6y}(\sqrt{8x} + \sqrt{10y^2})$

59. $(\sqrt{3} + 2)(\sqrt{3} + 5)$ **60.** $(\sqrt{2} - 3)(\sqrt{2} + 4)$

61. $(4\sqrt{2} + \sqrt{3})(3\sqrt{2} + 2\sqrt{3})$

62. $(2\sqrt{6} + 3\sqrt{5})(3\sqrt{6} + 4\sqrt{5})$

63. $(6 + 2\sqrt{5})(6 - 2\sqrt{5})$ **64.** $(7 - 3\sqrt{2})(7 + 3\sqrt{2})$

65. $(\sqrt{x} + \sqrt{y})^2$ **66.** $(2\sqrt{x} - 3\sqrt{y})^2$

67. $(\sqrt{a} + \sqrt{b})(\sqrt{a} - \sqrt{b})$

68. $(3\sqrt{x} + 5\sqrt{y})(3\sqrt{x} - 5\sqrt{y})$

For Problems 69–80, rationalize the denominator and simplify. All variables represent positive real numbers.

69. $\dfrac{3}{\sqrt{5} + 2}$ **70.** $\dfrac{7}{\sqrt{10} - 3}$

71. $\dfrac{4}{\sqrt{7} - \sqrt{3}}$ **72.** $\dfrac{2}{\sqrt{5} + \sqrt{3}}$

73. $\dfrac{\sqrt{2}}{2\sqrt{5} + 3\sqrt{7}}$ **74.** $\dfrac{5}{5\sqrt{2} - 3\sqrt{5}}$

75. $\dfrac{\sqrt{x}}{\sqrt{x} - 1}$ **76.** $\dfrac{\sqrt{x}}{\sqrt{x} + 2}$

77. $\dfrac{\sqrt{x}}{\sqrt{x} + \sqrt{y}}$ **78.** $\dfrac{2\sqrt{x}}{\sqrt{x} - \sqrt{y}}$

79. $\dfrac{2\sqrt{x} + \sqrt{y}}{3\sqrt{x} - 2\sqrt{y}}$ **80.** $\dfrac{3\sqrt{x} - 2\sqrt{y}}{2\sqrt{x} + 5\sqrt{y}}$

For Problems 81–84, *rationalize the numerator.* All variables represent positive real numbers.

81. $\dfrac{\sqrt{2x + 2h} - \sqrt{2x}}{h}$

82. $\dfrac{\sqrt{x + h + 1} - \sqrt{x + 1}}{h}$

83. $\dfrac{\sqrt{x + h - 3} - \sqrt{x - 3}}{h}$

84. $\dfrac{2\sqrt{x+h} - 2\sqrt{x}}{h}$

THOUGHTS INTO WORDS

85. Is the equation $\sqrt{x^2 y} = x\sqrt{y}$ true for all real number values for x and y? Defend your answer.

86. Is the equation $\sqrt{x^2 y^2} = xy$ true for all real number values for x and y? Defend your answer.

87. Give a step-by-step description of how you would change $\sqrt{252}$ to simplest radical form.

88. Why is $\sqrt{-9}$ not a real number?

89. How could you find a whole number approximation for $\sqrt{2750}$ if you did not have a calculator or table available?

0.7 RELATIONSHIP BETWEEN EXPONENTS AND ROOTS

Recall that we used the basic properties of positive integral exponents to motivate a definition of negative integers as exponents. In this section, we shall use the properties of integral exponents to motivate definitions for rational numbers as exponents. These definitions will tie together the concepts of *exponent* and *root*. Let's consider the following comparisons.

From our study of radicals we know that

$$\left(\sqrt{5}\right)^2 = 5$$
$$\left(\sqrt[3]{8}\right)^3 = 8$$
$$\left(\sqrt[4]{21}\right)^4 = 21$$

If $(b^m)^n = b^{nm}$ is to hold when m is a rational number of the form $1/p$, where p is a positive integer greater than 1 and $n = p$, then

$$(5^{1/2})^2 = 5^{2(1/2)} = 5^1 = 5$$
$$(8^{1/3})^3 = 8^{3(1/3)} = 8^1 = 8$$
$$(21^{1/4})^4 = 21^{4(1/4)} = 21^1 = 21$$

Such examples motivate the following definition.

DEFINITION 0.6

If b is a real number, n is a positive integer greater than 1, and $\sqrt[n]{b}$ exists, then

$$b^{1/n} = \sqrt[n]{b}$$

Definition 0.6 states that $b^{1/n}$ means the nth root of b. We shall assume that b and n are chosen so that $\sqrt[n]{b}$ exists in the real number system. For example, $(-25)^{1/2}$ is not meaningful at this time because $\sqrt{-25}$ is not a real number. The following examples illustrate the use of Definition 0.6.

$$25^{1/2} = \sqrt{25} = 5 \quad 16^{1/4} = \sqrt[4]{16} = 2$$
$$8^{1/3} = \sqrt[3]{8} = 2 \quad (-27)^{1/3} = \sqrt[3]{-27} = -3$$

Now the following definition provides the basis for the use of *all* rational numbers as exponents.

DEFINITION 0.7

If m/n is a rational number expressed in lowest terms, where n is a positive integer greater than one, and m is any integer, and if b is a real number such that $\sqrt[n]{b}$ exists, then

$$b^{m/n} = \sqrt[n]{b^m} = \left(\sqrt[n]{b}\right)^m$$

In Definition 0.7, whether we use the form $\sqrt[n]{b^m}$ or $\left(\sqrt[n]{b}\right)^m$ for computational purposes depends somewhat on the magnitude of the problem. Let's use both forms on the following two problems.

$$8^{2/3} = \sqrt[3]{8^2} = \sqrt[3]{64} = 4 \quad \text{or} \quad 8^{2/3} = \left(\sqrt[3]{8}\right)^2 = (2)^2 = 4$$
$$27^{2/3} = \sqrt[3]{27^2} = \sqrt[3]{729} = 9 \quad \text{or} \quad 27^{2/3} = \left(\sqrt[3]{27}\right)^2 = (3)^2 = 9$$

To compute $8^{2/3}$, both forms work equally well. However, to compute $27^{2/3}$, the form $(\sqrt[3]{27})^2$ is much easier to handle. The following examples further illustrate Definition 0.7.

$$25^{3/2} = \left(\sqrt{25}\right)^3 = 5^3 = 125$$

$$(32)^{-2/5} = \frac{1}{(32)^{2/5}} = \frac{1}{\left(\sqrt[5]{32}\right)^2} = \frac{1}{2^2} = \frac{1}{4}$$

$$(-64)^{2/3} = \left(\sqrt[3]{-64}\right)^2 = (-4)^2 = 16$$

$$-8^{4/3} = \left(\sqrt[3]{8}\right)^4 = -(2)^4 = -16$$

It can be shown that all of the results pertaining to integral exponents listed in Property 0.2 (on page 18) also hold for all rational exponents. Let's consider some examples to illustrate each of those results.

$$\begin{aligned} x^{1/2} \cdot x^{2/3} &= x^{1/2+2/3} \\ &= x^{3/6+4/6} \\ &= x^{7/6} \end{aligned} \qquad\qquad b^n \cdot b^m = b^{n+m}$$

$$\begin{aligned} (a^{2/3})^{3/2} &= a^{(3/2)(2/3)} \\ &= a^1 = a \end{aligned} \qquad\qquad (b^n)^m = b^{nm}$$

$$\begin{aligned} (16y^{2/3})^{1/2} &= (16)^{1/2}(y^{2/3})^{1/2} \\ &= 4y^{1/3} \end{aligned} \qquad (ab)^n = a^n b^n$$

$$\begin{aligned} \frac{y^{3/4}}{y^{1/2}} &= y^{3/4-1/2} \\ &= y^{3/4-2/4} \\ &= y^{1/4} \end{aligned} \qquad\qquad \frac{b^n}{b^m} = b^{n-m}$$

$$\begin{aligned} \left(\frac{x^{1/2}}{y^{1/3}}\right)^6 &= \frac{(x^{1/2})^6}{(y^{1/3})^6} \\ &= \frac{x^3}{y^2} \end{aligned} \qquad\qquad \left(\frac{a}{b}\right)^n = \frac{a^n}{b^n}$$

The link between exponents and roots provides a basis for multiplying and dividing some radicals even if they have different indexes. The general procedure is one of changing from radical to exponential form, applying the properties of exponents, and then changing back to radical form. Let's apply these procedures in the next three examples.

$$\sqrt{2}\sqrt[3]{2} = 2^{1/2} \cdot 2^{1/3} = 2^{1/2+1/3} = 2^{5/6} = \sqrt[6]{2^5} = \sqrt[6]{32}$$

$$\begin{aligned} \sqrt{xy}\sqrt[5]{x^2y} &= (xy)^{1/2}(x^2y)^{1/5} \\ &= x^{1/2}y^{1/2}x^{2/5}y^{1/5} \\ &= x^{1/2+2/5}y^{1/2+1/5} \\ &= x^{9/10}y^{7/10} \\ &= (x^9y^7)^{1/10} = \sqrt[10]{x^9y^7} \end{aligned}$$

$$\frac{\sqrt{5}}{\sqrt[3]{5}} = \frac{5^{1/2}}{5^{1/3}} = 5^{1/2 - 1/3} = 5^{1/6} = \sqrt[6]{5}$$

Earlier we agreed that a radical such as $\sqrt[3]{x^4}$ is not in simplest form because the radicand contains a perfect power of the index. Thus we simplified $\sqrt[3]{x^4}$ by expressing it as $\sqrt[3]{x^3}\sqrt[3]{x}$, which in turn can be written $x\sqrt[3]{x}$. Such simplification can also be done in exponential form, as follows.

$$\sqrt[3]{x^4} = x^{4/3} = x^{3/3} \cdot x^{1/3} = x \cdot x^{1/3} = x\sqrt[3]{x}$$

Note the use of this type of simplification in the following examples.

EXAMPLE 1

Perform the indicated operations and express the answers in simplest radical form.

a. $\sqrt[3]{x^2}\sqrt[4]{x^3}$ **b.** $\sqrt{2}\sqrt[3]{4}$ **c.** $\dfrac{\sqrt{27}}{\sqrt[3]{3}}$

Solutions

a. $\sqrt[3]{x^2}\sqrt[4]{x^3} = x^{2/3} \cdot x^{3/4} = x^{2/3 + 3/4} = x^{17/12} = x^{12/12} \cdot x^{5/12} = x\sqrt[12]{x^5}$

b. $\sqrt{2}\sqrt[3]{4} = 2^{1/2} \cdot 4^{1/3} = 2^{1/2}(2^2)^{1/3} = 2^{1/2} \cdot 2^{2/3}$
$= 2^{1/2 + 2/3} = 2^{7/6} = 2^{6/6} \cdot 2^{1/6} = 2\sqrt[6]{2}$

c. $\dfrac{\sqrt{27}}{\sqrt[3]{3}} = \dfrac{27^{1/2}}{3^{1/3}} = \dfrac{(3^3)^{1/2}}{3^{1/3}} = \dfrac{3^{3/2}}{3^{1/3}} = 3^{3/2 \; 1/3} = 3^{7/6}$
$= 3^{6/6} \cdot 3^{1/6} = 3\sqrt[6]{3}$ ∎

The process of rationalizing the denominator can sometimes be handled more easily in exponential form. Consider the following examples, which illustrate this procedure.

EXAMPLE 2

Rationalize the denominator and express the answer in simplest radical form.

a. $\dfrac{2}{\sqrt[3]{x}}$ **b.** $\dfrac{\sqrt[3]{x}}{\sqrt{y}}$

Solutions

a. $\dfrac{2}{\sqrt[3]{x}} = \dfrac{2}{x^{1/3}} = \dfrac{2}{x^{1/3}} \cdot \dfrac{x^{2/3}}{x^{2/3}} = \dfrac{2x^{2/3}}{x} = \dfrac{2\sqrt[3]{x^2}}{x}$

b. $\dfrac{\sqrt[3]{x}}{\sqrt{y}} = \dfrac{x^{1/3}}{y^{1/2}} = \dfrac{x^{1/3}}{y^{1/2}} \cdot \dfrac{y^{1/2}}{y^{1/2}} = \dfrac{x^{1/3} \cdot y^{1/2}}{y} = \dfrac{x^{2/6} \cdot y^{3/6}}{y} = \dfrac{\sqrt[6]{x^2 y^3}}{y}$ ∎

Note in part (b) that if we had changed back to radical form at the step $\dfrac{x^{1/3}y^{1/2}}{y}$, we would have obtained the product of two radicals, $\sqrt[3]{x}\sqrt{y}$, in the numerator. Instead we used the exponential form to find this product and express the final result with a single radical in the numerator. Finally, let's consider an example involving *the root of a root*.

EXAMPLE 3

Simplify $\sqrt[3]{\sqrt{2}}$.

Solution

$$\sqrt[3]{\sqrt{2}} = (2^{1/2})^{1/3} = 2^{1/6} = \sqrt[6]{2}$$ ∎

PROBLEM SET 0.7

For Problems 1–16, evaluate.

1. $49^{1/2}$

2. $64^{1/3}$

3. $32^{3/5}$

4. $(-8)^{1/3}$

5. $-8^{2/3}$

6. $64^{-1/2}$

7. $\left(\dfrac{1}{4}\right)^{-1/2}$

8. $\left(-\dfrac{27}{8}\right)^{-1/3}$

9. $16^{3/2}$

10. $(0.008)^{1/3}$

11. $(0.01)^{3/2}$

12. $\left(\dfrac{1}{27}\right)^{-2/3}$

13. $64^{-5/6}$

14. $-16^{5/4}$

15. $\left(\dfrac{1}{8}\right)^{-1/3}$

16. $\left(-\dfrac{1}{8}\right)^{2/3}$

For Problems 17–32, perform the indicated operations and simplify. Express final answers using positive exponents only.

17. $(3x^{1/4})(5x^{1/3})$

18. $(2x^{2/5})(6x^{1/4})$

19. $(y^{2/3})(y^{-1/4})$

20. $(2x^{1/3})(x^{-1/2})$

21. $(4x^{1/4}y^{1/2})^3$

22. $(5x^{1/2}y)^2$

23. $\dfrac{24x^{3/5}}{6x^{1/3}}$

24. $\dfrac{18x^{1/2}}{9x^{1/3}}$

25. $\dfrac{56a^{1/6}}{8a^{1/4}}$

26. $\dfrac{48b^{1/3}}{12b^{3/4}}$

27. $\left(\dfrac{2x^{1/3}}{3y^{1/4}}\right)^4$

28. $\left(\dfrac{6x^{2/5}}{7y^{2/3}}\right)^2$

29. $\left(\dfrac{x^2}{y^3}\right)^{-1/2}$

30. $\left(\dfrac{a^3}{b^{-2}}\right)^{-1/3}$

31. $\left(\dfrac{4a^2x}{2a^{1/2}x^{1/3}}\right)^3$

32. $\left(\dfrac{3ax^{-1}}{a^{1/2}x^{-2}}\right)^2$

For Problems 33–48, perform the indicated operations and express the answer in simplest radical form.

33. $\sqrt{2}\sqrt[4]{2}$

34. $\sqrt[3]{3}\sqrt{3}$

35. $\sqrt[3]{x}\sqrt[4]{x}$

36. $\sqrt[3]{x^2}\sqrt[5]{x^3}$

37. $\sqrt{xy}\,\sqrt[4]{x^3y^5}$

38. $\sqrt[3]{x^2y^4}\sqrt[4]{x^3y}$

39. $\sqrt[3]{a^2b^2}\sqrt[4]{a^3b}$

40. $\sqrt{ab}\,\sqrt[3]{a^4b^5}$

41. $\dfrac{\sqrt[3]{4}}{\sqrt{8}}$

42. $\sqrt[3]{9}\sqrt{27}$

43. $\dfrac{\sqrt{2}}{\sqrt[3]{2}}$

44. $\dfrac{\sqrt{9}}{\sqrt[3]{3}}$

45. $\dfrac{\sqrt[3]{8}}{\sqrt[4]{4}}$

46. $\dfrac{\sqrt[3]{16}}{\sqrt[6]{4}}$

47. $\dfrac{\sqrt[4]{x^9}}{\sqrt[3]{x^2}}$

48. $\dfrac{\sqrt[5]{x^7}}{\sqrt[3]{x}}$

For Problems 49–57, rationalize the denominator and express the final answer in simplest radical form.

49. $\dfrac{5}{\sqrt[3]{x}}$

50. $\dfrac{3}{\sqrt[3]{x^2}}$

51. $\dfrac{\sqrt{x}}{\sqrt[3]{y}}$

52. $\dfrac{\sqrt[4]{x}}{\sqrt{y}}$ **53.** $\dfrac{\sqrt[4]{x^3}}{\sqrt[5]{y^3}}$ **54.** $\dfrac{2\sqrt{x}}{3\sqrt[3]{y}}$

55. $\dfrac{5\sqrt[3]{y^2}}{4\sqrt[4]{x}}$ **56.** $\dfrac{\sqrt{xy}}{\sqrt[3]{a^2b}}$

57. Simplify each of the following, expressing the final result as one radical. For example,

$$\sqrt{\sqrt{3}} = (3^{1/2})^{1/2} = 3^{1/4} = \sqrt[4]{3}$$

a. $\sqrt[3]{\sqrt{2}}$ **b.** $\sqrt[3]{\sqrt[4]{3}}$

c. $\sqrt[3]{\sqrt{x^3}}$ **d.** $\sqrt{\sqrt[3]{x^4}}$

THOUGHTS INTO WORDS

58. Your friend keeps getting an error message when evaluating $-4^{5/2}$ on his calculator. What error is he probably making?

59. Explain how you would evaluate $27^{2/3}$ without a calculator.

Further Investigations

60. Use your calculator to evaluate each of the following.

a. $\sqrt[3]{1728}$ **b.** $\sqrt[3]{5832}$

c. $\sqrt[4]{2401}$ **d.** $\sqrt[4]{65536}$

e. $\sqrt[5]{161051}$ **f.** $\sqrt[5]{6436343}$

61. In Definition 0.7 we stated that $b^{m/n} = \sqrt[n]{b^m} = \left(\sqrt[n]{b}\right)^m$. Use your calculator to verify each of the following.

a. $\sqrt[3]{27^2} = \left(\sqrt[3]{27}\right)^2$ **b.** $\sqrt[3]{8^5} = \left(\sqrt[3]{8}\right)^5$

c. $\sqrt[4]{16^3} = \left(\sqrt[4]{16}\right)^3$ **d.** $\sqrt[3]{16^2} = \left(\sqrt[3]{16}\right)^2$

e. $\sqrt[5]{9^4} = \left(\sqrt[5]{9}\right)^4$ **f.** $\sqrt[3]{12^4} = \left(\sqrt[3]{12}\right)^4$

62. Use your calculator to evaluate each of the following.

a. $16^{5/2}$ **b.** $25^{7/2}$ **c.** $16^{9/4}$

d. $27^{5/3}$ **e.** $343^{2/3}$ **f.** $512^{4/3}$

63. Use your calculator to estimate each of the following to the nearest thousandth.

a. $7^{4/3}$ **b.** $10^{4/5}$ **c.** $12^{2/5}$

d. $19^{2/5}$ **e.** $7^{3/4}$ **f.** $10^{5/4}$

Sometimes we meet the following type of simplification problem in calculus.

$$\frac{(x-1)^{1/2} - x(x-1)^{-(1/2)}}{[(x-1)^{1/2}]^2}$$

$$= \left(\frac{(x-1)^{1/2} - x(x-1)^{-(1/2)}}{(x-1)^{2/2}}\right) \cdot \left(\frac{(x-1)^{1/2}}{(x-1)^{1/2}}\right)$$

$$= \frac{x - 1 - x(x-1)^0}{(x-1)^{3/2}}$$

$$= \frac{x - 1 - x}{(x-1)^{3/2}} \qquad (x-1)^0 = 1$$

$$= \frac{-1}{(x-1)^{3/2}} \quad \text{or} \quad -\frac{1}{(x-1)^{3/2}}$$

For Problems 64–69, simplify each expression as we did in the previous example.

64. $\dfrac{2(x+1)^{1/2} - x(x+1)^{-(1/2)}}{[(x+1)^{1/2}]^2}$

65. $\dfrac{2(2x-1)^{1/2} - 2x(2x-1)^{-(1/2)}}{[(2x-1)^{1/2}]^2}$

66. $\dfrac{2x(4x+1)^{1/2} - 2x^2(4x+1)^{-(1/2)}}{[(4x+1)^{1/2}]^2}$

67. $\dfrac{(x^2+2x)^{1/2} - x(x+1)(x^2+2x)^{-(1/2)}}{[(x^2+2x)^{1/2}]^2}$

68. $\dfrac{(3x)^{1/3} - x(3x)^{-(2/3)}}{[(3x)^{1/3}]^2}$ **69.** $\dfrac{3(2x)^{1/3} - 2x(2x)^{-(2/3)}}{[(2x)^{1/3}]^2}$

0.8   COMPLEX NUMBERS

So far we have dealt only with real numbers. However, as we get ready to solve equations in the next chapter, there is a need for *more numbers*. There are some very simple equations that do not have solutions if we restrict ourselves to the set of real numbers. For example, the equation $x^2 + 1 = 0$ has no solutions among the real numbers. To solve such equations, we need to extend the real number system. In this section we will introduce a set of numbers that contains some numbers whose squares are negative real numbers. Then, in the next chapter and in Chapter 5, we will see that this set of numbers, called the **complex numbers**, provides solutions not only for equations such as $x^2 + 1 = 0$ but also for *any* polynomial equation in general.

Let's begin by defining a number i such that

$$i^2 = -1$$

The number i is not a real number and is often called the **imaginary unit**, but the number i^2 is the real number -1. The imaginary unit i is used to define a complex number as follows.

DEFINITION 0.8

A **complex number** is any number that can be expressed in the form

$a + bi$

where a and b are real numbers.

The form $a + bi$ is called the **standard form** of a complex number. The real number a is called the **real part** of the complex number, and b is called the **imaginary part**. (Note that b is a real number even though it is called the imaginary part.) Each of the following represents a complex number.

$6 + 2i$	is already expressed in the form $a + bi$. Traditionally, complex numbers for which $a \neq 0$ and $b \neq 0$ have been called imaginary numbers.
$5 - 3i$	can be written $5 + (-3i)$ even though the form $5 - 3i$ is often used.
$-8 + i\sqrt{2}$	can be written $-8 + \sqrt{2}i$. It is easy to mistake $\sqrt{2}i$ for $\sqrt{2i}$. Thus we commonly write $i\sqrt{2}$ instead of $\sqrt{2}i$ to avoid any difficulties with the radical sign.

−9i can be written $0 + (-9i)$. Complex numbers such as $-9i$, for which $a = 0$ and $b \neq 0$, traditionally have been called **pure imaginary numbers**.

5 can be written $5 + 0i$.

The set of real numbers is a subset of the set of complex numbers. The following diagram indicates the organizational format of the complex number system.

Complex numbers

$a + bi$, where a and b
are real numbers

Real numbers

$a + bi$, where $b = 0$

Imaginary numbers

$a + bi$, where $b \neq 0$

Pure imaginary numbers

$a + bi$, where $a = 0$ and $b \neq 0$

Two complex numbers $a + bi$ and $c + di$ are said to be *equal* if and only if $a = c$ and $b = d$. In other words, two complex numbers are equal if and only if their real parts are equal and their imaginary parts are equal.

Adding and Subtracting Complex Numbers

The following definition provides the basis for adding complex numbers.

$$(a + bi) + (c + di) = (a + c) + (b + d)i$$

We can use this definition to find the sum of two complex numbers.

$$(4 + 3i) + (5 + 9i) = (4 + 5) + (3 + 9)i = 9 + 12i$$

$$(-6 + 4i) + (8 - 7i) = (-6 + 8) + (4 - 7)i = 2 - 3i$$

$$\left(\frac{1}{2} + \frac{3}{4}i\right) + \left(\frac{2}{3} + \frac{1}{5}i\right) = \left(\frac{1}{2} + \frac{2}{3}\right) + \left(\frac{3}{4} + \frac{1}{5}\right)i$$

$$= \left(\frac{3}{6} + \frac{4}{6}\right) + \left(\frac{15}{20} + \frac{4}{20}\right)i = \frac{7}{6} + \frac{19}{20}i$$

$$\left(3 + i\sqrt{2}\right) + \left(-4 + i\sqrt{2}\right) = [3 + (-4)] + \left(\sqrt{2} + \sqrt{2}\right)i = -1 + 2i\sqrt{2}$$

Note the form for writing $2\sqrt{2}i$

The set of complex numbers is **closed with respect to addition**; that is, the sum of two complex numbers is a complex number. Furthermore, the commutative and associative properties of addition hold for all complex numbers. The additive identity element is $0 + 0i$, or simply the real number 0. The additive inverse of $a + bi$ is $-a - bi$ because

$$(a + bi) + (-a - bi) = [a + (-a)] + [b + (-b)]i = 0$$

Therefore, to *subtract* $c + di$ from $a + bi$, we add the additive inverse of $c + di$.

$$(a + bi) - (c + di) = (a + bi) + (-c - di)$$
$$= (a - c) + (b - d)i$$

The following examples illustrate the subtraction of complex numbers.

$$(9 + 8i) - (5 + 3i) = (9 - 5) + (8 - 3)i = 4 + 5i$$

$$(3 - 2i) - (4 - 10i) = (3 - 4) + [-2 - (-10)]i = -1 + 8i$$

$$\left(-\frac{1}{2} + \frac{1}{3}i\right) - \left(\frac{3}{4} + \frac{1}{2}i\right) = \left(-\frac{1}{2} - \frac{3}{4}\right) + \left(\frac{1}{3} - \frac{1}{2}\right)i = -\frac{5}{4} - \frac{1}{6}i$$

Multiplying and Dividing Complex Numbers

Because $i^2 = -1$, the number i is a square root of -1, so we write $i = \sqrt{-1}$. It should also be evident that $-i$ is a square root of -1 because

$$(-i)^2 = (-i)(-i) = i^2 = -1$$

Therefore, in the set of complex numbers, -1 has two square roots—namely, i and $-i$. This is symbolically expressed as

$$i = \sqrt{-1} \qquad \text{and} \qquad -i = -\sqrt{-1}$$

Let's extend the definition so that in the set of complex numbers, every negative real number has two square roots. For any positive real number b,

$$\left(i\sqrt{b}\right)^2 = i^2(b) = -1(b) = -b$$

Therefore, let's denote the **principal square root of** $-b$ by $\sqrt{-b}$ and define it to be

$$\sqrt{-b} = i\sqrt{b}$$

where b is any positive real number. In other words, the principal square root of any negative real number can be represented as the product of a real number and the imaginary unit i. Consider the following examples.

$$\sqrt{-4} = i\sqrt{4} = 2i$$
$$\sqrt{-17} = i\sqrt{17}$$
$$\sqrt{-24} = i\sqrt{24} = i\sqrt{4}\sqrt{6} = 2i\sqrt{6} \qquad \text{Note that we simplified the radical } \sqrt{24} \text{ to } 2\sqrt{6}.$$

We should also observe that $-\sqrt{-b}$, where $b > 0$, is a square root of $-b$ because

$$\left(-\sqrt{-b}\right)^2 = \left(-i\sqrt{b}\right)^2 = i^2(b) = (-1)b = -b$$

Thus, in the set of complex numbers, $-b$ (where $b > 0$) has two square roots: $i\sqrt{b}$ and $-i\sqrt{b}$. These are expressed as

$$\sqrt{-b} = i\sqrt{b} \qquad \text{and} \qquad -\sqrt{-b} = -i\sqrt{b}$$

We must be careful with the use of the symbol $\sqrt{-b}$, where $b > 0$. Some properties that are true in the set of real numbers involving the square root symbol do not hold if the square root symbol does not represent a real number. For example, $\sqrt{a}\sqrt{b} = \sqrt{ab}$ *does not hold* if a and b are both negative numbers.

Correct $\qquad \sqrt{-4}\sqrt{-9} = (2i)(3i) = 6i^2 = 6(-1) = -6$

Incorrect $\qquad \sqrt{-4}\sqrt{-9} = \sqrt{(-4)(-9)} = \sqrt{36} = 6$

To avoid difficulty with this idea, you should rewrite all expressions of the form $\sqrt{-b}$, where $b > 0$ in the form $i\sqrt{b}$ *before* doing any computations. The following examples further illustrate this point.

$$\sqrt{-5}\sqrt{-7} = \left(i\sqrt{5}\right)\left(i\sqrt{7}\right) = i^2\sqrt{35} = (-1)\sqrt{35} = -\sqrt{35}$$

$$\sqrt{-2}\sqrt{-8} = \left(i\sqrt{2}\right)\left(i\sqrt{8}\right) = i^2\sqrt{16} = (-1)(4) = -4$$

$$\sqrt{-2}\sqrt{8} = \left(i\sqrt{2}\right)\left(\sqrt{8}\right) = i\sqrt{16} = 4i$$

$$\sqrt{-6}\sqrt{-8} = \left(i\sqrt{6}\right)\left(i\sqrt{8}\right) = i^2\sqrt{48} = i^2\sqrt{16}\sqrt{3} = 4i^2\sqrt{3} = -4\sqrt{3}$$

$$\frac{\sqrt{-2}}{\sqrt{3}} = \frac{i\sqrt{2}}{\sqrt{3}} = \frac{i\sqrt{2}}{\sqrt{3}} \cdot \frac{\sqrt{3}}{\sqrt{3}} = \frac{i\sqrt{6}}{3}$$

$$\frac{\sqrt{-48}}{\sqrt{12}} = \frac{i\sqrt{48}}{\sqrt{12}} = i\sqrt{\frac{48}{12}} = i\sqrt{4} = 2i$$

Because complex numbers have a *binomial form*, we can find the product of two complex numbers in the same way that we find the product of two binomials. Then, by replacing i^2 with -1 we can simplify and express the final product in the standard form of a complex number. Consider the following examples.

$$
\begin{aligned}
(2 + 3i)(4 + 5i) &= 2(4 + 5i) + 3i(4 + 5i) \\
&= 8 + 10i + 12i + 15i^2 \\
&= 8 + 22i + 15(-1) \\
&= 8 + 22i - 15 \\
&= -7 + 22i
\end{aligned}
$$

$$
\begin{aligned}
(1 - 7i)^2 &= (1 - 7i)(1 - 7i) \\
&= 1(1 - 7i) - 7i(1 - 7i) \\
&= 1 - 7i - 7i + 49i^2 \\
&= 1 - 14i + 49(-1) \\
&= 1 - 14i - 49 \\
&= -48 - 14i
\end{aligned}
$$

$$(2 + 3i)(2 - 3i) = 2(2 - 3i) + 3i(2 - 3i)$$
$$= 4 - 6i + 6i - 9i^2$$
$$= 4 - 9(-1)$$
$$= 4 + 9$$
$$= 13$$

REMARK   Don't forget that when multiplying complex numbers, we can also use the multiplication patterns

$$(a + b)^2 = a^2 + 2ab + b^2$$
$$(a - b)^2 = a^2 - 2ab + b^2$$
$$(a + b)(a - b) = a^2 - b^2$$

The last example illustrates an important idea. The complex numbers $2 + 3i$ and $2 - 3i$ are called conjugates of each other. In general, the two complex numbers $a + bi$ and $a - bi$ are called **conjugates** of each other and **the product of a complex number and its conjugate is a real number**. This can be shown as follows.

$$(a + bi)(a - bi) = a(a - bi) + bi(a - bi)$$
$$= a^2 - abi + abi - b^2i^2$$
$$= a^2 - b^2(-1)$$
$$= a^2 + b^2$$

Conjugates are used to simplify an expression such as $3i/(5 + 2i)$, which *indicates the quotient of two complex numbers*. To eliminate i in the denominator and to change the indicated quotient to the standard form of a complex number, we can multiply both the numerator and denominator by the conjugate of the denominator.

$$\frac{3i}{5 + 2i} = \frac{3i}{5 + 2i} \cdot \frac{5 - 2i}{5 - 2i}$$
$$= \frac{3i(5 - 2i)}{(5 + 2i)(5 - 2i)}$$
$$= \frac{15i - 6i^2}{25 - 4i^2}$$
$$= \frac{15i - 6(-1)}{25 - 4(-1)}$$
$$= \frac{6 + 15i}{29}$$
$$= \frac{6}{29} + \frac{15}{29}i$$

The following examples further illustrate the process of dividing complex numbers.

$$\frac{2 - 3i}{4 - 7i} = \frac{2 - 3i}{4 - 7i} \cdot \frac{4 + 7i}{4 + 7i}$$

$$= \frac{(2 - 3i)(4 + 7i)}{(4 - 7i)(4 + 7i)}$$

$$= \frac{8 + 14i - 12i - 21i^2}{16 - 49i^2}$$

$$= \frac{8 + 2i - 21(-1)}{16 - 49(-1)}$$

$$= \frac{29 + 2i}{65} = \frac{29}{65} + \frac{2}{65}i$$

$$\frac{4 - 5i}{2i} = \frac{4 - 5i}{2i} \cdot \frac{-2i}{-2i}$$

$$= \frac{(4 - 5i)(-2i)}{(2i)(-2i)}$$

$$= \frac{-8i + 10i^2}{-4i^2}$$

$$= \frac{-8i + 10(-1)}{-4(-1)}$$

$$= \frac{-10 - 8i}{4} = -\frac{5}{2} - 2i$$

For a problem such as the last one, in which the denominator is a pure imaginary number, we can change to standard form by choosing a multiplier other than the conjugate of the denominator. Consider the following alternative approach.

$$\frac{4 - 5i}{2i} = \frac{4 - 5i}{2i} \cdot \frac{i}{i}$$

$$= \frac{(4 - 5i)(i)}{(2i)(i)}$$

$$= \frac{4i - 5i^2}{2i^2}$$

$$= \frac{4i - 5(-1)}{2(-1)}$$

$$= \frac{5 + 4i}{-2}$$

$$= -\frac{5}{2} - 2i$$

PROBLEM SET 0.8

For Problems 1–14, add or subtract as indicated.

1. $(5 + 2i) + (8 + 6i)$

2. $(-9 + 3i) + (4 + 5i)$

3. $(8 + 6i) - (5 + 2i)$

4. $(-6 + 4i) - (4 + 6i)$

5. $(-7 - 3i) + (-4 + 4i)$

6. $(6 - 7i) - (7 - 6i)$

7. $(-2 - 3i) - (-1 - i)$

8. $\left(\frac{1}{3} + \frac{2}{5}i\right) + \left(\frac{1}{2} + \frac{1}{4}i\right)$

9. $\left(-\frac{3}{4} - \frac{1}{4}i\right) + \left(\frac{3}{5} + \frac{2}{3}i\right)$

10. $\left(\frac{5}{8} + \frac{1}{2}i\right) - \left(\frac{7}{8} + \frac{1}{5}i\right)$

11. $\left(\frac{3}{10} - \frac{3}{4}i\right) - \left(-\frac{2}{5} + \frac{1}{6}i\right)$

12. $\left(4 + i\sqrt{3}\right) + \left(-6 - 2i\sqrt{3}\right)$

13. $(5 + 3i) + (7 - 2i) + (-8 - i)$

14. $(5 - 7i) - (6 - 2i) - (-1 - 2i)$

For Problems 15–30, write each in terms of i and simplify. For example,

$$\sqrt{-20} = i\sqrt{20} = i\sqrt{4}\sqrt{5} = 2i\sqrt{5}$$

15. $\sqrt{-9}$

16. $\sqrt{-49}$

17. $\sqrt{-19}$

18. $\sqrt{-31}$

19. $\sqrt{-\frac{4}{9}}$

20. $\sqrt{-\frac{25}{36}}$

21. $\sqrt{-8}$

22. $\sqrt{-18}$

23. $\sqrt{-27}$

24. $\sqrt{-32}$

25. $\sqrt{-54}$

26. $\sqrt{-40}$

27. $3\sqrt{-36}$

28. $5\sqrt{-64}$

29. $4\sqrt{-18}$

30. $6\sqrt{-8}$

For Problems 31–44, write each in terms of i, perform the indicated operations, and simplify. For example,

$$\sqrt{-9}\sqrt{-16} = \left(i\sqrt{9}\right)\left(i\sqrt{16}\right) = (3i)(4i)$$
$$= 12i^2 = 12(-1) = -12$$

31. $\sqrt{-4}\sqrt{-16}$

32. $\sqrt{-25}\sqrt{-9}$

33. $\sqrt{-2}\sqrt{-3}$

34. $\sqrt{-3}\sqrt{-7}$

35. $\sqrt{-5}\sqrt{-4}$

36. $\sqrt{-7}\sqrt{-9}$

37. $\sqrt{-6}\sqrt{-10}$

38. $\sqrt{-2}\sqrt{-12}$

39. $\sqrt{-8}\sqrt{-7}$

40. $\sqrt{-12}\sqrt{-5}$

41. $\frac{\sqrt{-36}}{\sqrt{-4}}$

42. $\frac{\sqrt{-64}}{\sqrt{-16}}$

43. $\frac{\sqrt{-54}}{\sqrt{-9}}$

44. $\frac{\sqrt{-18}}{\sqrt{-3}}$

For Problems 45–64, find each product, expressing the answers in standard form.

45. $(3i)(7i)$

46. $(-5i)(8i)$

47. $(4i)(3 - 2i)$

48. $(5i)(2 + 6i)$

49. $(3 + 2i)(4 + 6i)$

50. $(7 + 3i)(8 + 4i)$

51. $(4 + 5i)(2 - 9i)$

52. $(1 + i)(2 - i)$

53. $(-2 - 3i)(4 + 6i)$

54. $(-3 - 7i)(2 + 10i)$

55. $(6 - 4i)(-1 - 2i)$

56. $(7 - 3i)(-2 - 8i)$

57. $(3 + 4i)^2$

58. $(4 - 2i)^2$

59. $(-1 - 2i)^2$

60. $(-2 + 5i)^2$

61. $(8 - 7i)(8 + 7i)$

62. $(5 + 3i)(5 - 3i)$

63. $(-2 + 3i)(-2 - 3i)$

64. $(-6 - 7i)(-6 + 7i)$

For Problems 65–78, find each quotient, expressing the answers in standard form.

65. $\frac{4i}{3 - 2i}$

66. $\frac{3i}{6 + 2i}$

67. $\frac{2 + 3i}{3i}$

68. $\frac{3 - 5i}{4i}$

69. $\frac{3}{2i}$

70. $\frac{7}{4i}$

71. $\frac{3 + 2i}{4 + 5i}$

72. $\frac{2 + 5i}{3 + 7i}$

73. $\frac{4 + 7i}{2 - 3i}$

74. $\frac{3 + 9i}{4 - i}$

75. $\frac{3 - 7i}{-2 + 4i}$

76. $\frac{4 - 10i}{-3 + 7i}$

77. $\frac{-1 - i}{-2 - 3i}$

78. $\frac{-4 + 9i}{-3 - 6i}$

79. Using $a + bi$ and $c + di$ to represent two complex numbers, verify the following properties.

   a. The conjugate of the sum of two complex numbers is equal to the sum of the conjugates of the two numbers.

   b. The conjugate of the product of two complex numbers is equal to the product of the conjugates of the numbers.

THOUGHTS INTO WORDS

80. Is every real number also a complex number? Explain your answer.

81. Can the product of two nonreal complex numbers be a real number? Explain your answer.

Further Investigations

82. Observe the following powers of i.

$$i = \sqrt{-1}$$
$$i^2 = -1$$
$$i^3 = i^2 \cdot i = -1(i) = -i$$
$$i^4 = i^2 \cdot i^2 = (-1)(-1) = 1$$

Any power of i greater than 4 can be simplified to i, -1, $-i$, or 1 as follows.

$$i^9 = (i^4)^2(i) = (1)(i) = i$$
$$i^{14} = (i^4)^3(i^2) = (1)(-1) = -1$$
$$i^{19} = (i^4)^4(i^3) = (1)(-i) = -i$$
$$i^{28} = (i^4)^7 = (1)^7 = 1$$

Express each of the following as i, -1, $-i$, or 1.

   a. i^5 **b.** i^6 **c.** i^{11} **d.** i^{12}

   e. i^{16} **f.** i^{22} **g.** i^{33} **h.** i^{63}

83. We can use the information from Problem 82 and the binomial expansion patterns to find powers of complex numbers as follows.

$$(3 + 2i)^3 = (3)^3 + 3(3)^2(2i) + 3(3)(2i)^2 + (2i)^3$$
$$= 27 + 54i + 36i^2 + 8i^3$$
$$= 27 + 54i + 36(-1) + 8(-i) = -9 + 46i$$

Find the indicated power of each of the following.

   a. $(2 + i)^3$ **b.** $(1 - i)^3$ **c.** $(1 - 2i)^3$

   d. $(1 + i)^4$ **e.** $(2 - i)^4$ **f.** $(-1 + i)^5$

CHAPTER 0 SUMMARY

Be sure of the following key concepts from this chapter: set, null set, equal sets, subset, natural numbers, whole numbers, integers, rational numbers, irrational numbers, real numbers, complex numbers, absolute value, similar terms, exponent, monomial, binomial, polynomial, degree of a polynomial, perfect-square trinomial, factoring polynomials, rational expression, least common denominator, radical, simplest radical form, root, and conjugate of a complex number.

The following properties of the real numbers provide a basis for arithmetic and algebraic computation: closure for addition and multiplication, commutativity for addition and multiplication, associativity for addition and multiplication, identity properties for addition and multiplication, inverse properties for addition and multiplication, multiplication property of zero, multiplication property of -1, and distributive property.

The following properties of absolute value are useful.

1. $|a| \geq 0$
2. $|a| = |-a|$ a and b are real numbers
3. $|a - b| = |b - a|$

The following properties of exponents provide the basis for much of our computational work with polynomials.

1. $b^n \cdot b^m = b^{n+m}$
2. $(b^n)^m = b^{mn}$ m and n are rational numbers and a and b are
3. $(ab)^n = a^n b^n$ real numbers, except $b \neq 0$ whenever it
   appears in the denominator.
4. $\left(\dfrac{a}{b}\right)^n = \dfrac{a^n}{b^n}$
5. $\dfrac{b^n}{b^m} = b^{n-m}$

The following product patterns are helpful to recognize when multiplying polynomials.

1. $(a + b)^2 = a^2 + 2ab + b^2$
2. $(a - b)^2 = a^2 - 2ab + b^2$
3. $(a + b)(a - b) = a^2 - b^2$
4. $(a + b)^3 = a^3 + 3a^2b + 3ab^2 + b^3$
5. $(a - b)^3 = a^3 - 3a^2b + 3ab^2 - b^3$

Be sure you know how to do the following.

1. Factor out the highest common monomial factor.
2. Factor by grouping.
3. Factor a trinomial into the product of two binomials.
4. Recognize some basic factoring patterns:

$$a^2 + 2ab + b^2 = (a + b)^2;$$
$$a^2 - 2ab + b^2 = (a - b)^2;$$
$$a^2 - b^2 = (a + b)(a - b);$$
$$a^3 + b^3 = (a + b)(a^2 - ab + b^2);$$
$$a^3 - b^3 = (a - b)(a^2 + ab + b^2).$$

Be sure that you can simplify, add, subtract, multiply, and divide rational expressions using the following properties and definitions.

1. $\dfrac{a \cdot k}{b \cdot k} = \dfrac{a}{b}$

2. $\dfrac{-a}{b} = \dfrac{a}{-b} = -\dfrac{a}{b}$

3. $\dfrac{a}{b} \cdot \dfrac{c}{d} = \dfrac{ac}{bd}$

4. $\dfrac{a}{b} \div \dfrac{c}{d} = \dfrac{a}{b} \cdot \dfrac{d}{c} = \dfrac{ad}{bc}$

5. $\dfrac{a}{c} + \dfrac{b}{c} = \dfrac{a + b}{c}$

6. $\dfrac{a}{c} - \dfrac{b}{c} = \dfrac{a - b}{c}$

Be sure that you can simplify, add, subtract, multiply, and divide radicals using the following definitions and properties.

1. $\sqrt[n]{b} = a$ if and only if $a^n = b$
2. $\sqrt[n]{bc} = \sqrt[n]{b}\sqrt[n]{c}$
3. $\sqrt[n]{\dfrac{b}{c}} = \dfrac{\sqrt[n]{b}}{\sqrt[n]{c}}$

The following definition provides the link between exponents and roots.

$$b^{m/n} = \sqrt[n]{b^m} = \left(\sqrt[n]{b}\right)^m$$

This link, along with the properties of exponents, allows us (1) to multiply and divide some radicals with different indices, (2) to change to simplest radical form while in exponential form, and (3) to simplify expressions that are roots of roots.

A complex number is any number that can be expressed in the form $a + bi$, where a and b are real numbers and i is the imaginary unit such that $i^2 = -1$.

Addition and subtraction of complex numbers are defined as follows.

$$(a + bi) + (c + di) = (a + c) + (b + d)i$$
$$(a + bi) - (c + di) = (a - c) + (b - d)i$$

Because complex numbers have a binomial form, we can multiply two complex numbers in the same way that we multiply two binomials. Thus i^2 can be replaced with -1, and the final result can be expressed in the standard form of a complex number. For example,

$$(3 + 2i)(4 - 3i) = 12 - i - 6i^2$$
$$= 12 - i - 6(-1)$$
$$= 18 - i$$

The two complex numbers $a + bi$ and $a - bi$ are called conjugates of each other. The product $(a - bi)(a + bi)$ equals the real number $a^2 + b^2$, and this property is used to help with dividing complex numbers.

CHAPTER 0 REVIEW PROBLEM SET

For Problems 1–10, evaluate.

1. 5^{-3}

2. -3^{-4}

3. $\left(\dfrac{3}{4}\right)^{-2}$

4. $\dfrac{1}{\left(\dfrac{1}{3}\right)^{-2}}$

5. $-\sqrt{64}$

6. $\sqrt[3]{\dfrac{27}{8}}$

7. $\sqrt[5]{-\dfrac{1}{32}}$

8. $36^{-1/2}$

9. $\left(\dfrac{1}{8}\right)^{-2/3}$

10. $-32^{3/5}$

For Problems 11–18, perform the indicated operations and simplify. Express the final answers using positive exponents only.

11. $(3x^{-2}y^{-1})(4x^4y^2)$

12. $(5x^{2/3})(-6x^{1/2})$

13. $(-8a^{-1/2})(-6a^{1/3})$

14. $(3x^{-2/3}y^{1/5})^3$

15. $\dfrac{64x^{-2}y^3}{16x^3y^{-2}}$

16. $\dfrac{56x^{-1/3}y^{2/5}}{7x^{1/4}y^{-3/5}}$

17. $\left(\dfrac{-8x^2y^{-1}}{2x^{-1}y^2}\right)^2$

18. $\left(\dfrac{36a^{-1}b^4}{-12a^2b^5}\right)^{-1}$

For Problems 19–34, perform the indicated operations.

19. $(-7x - 3) + (5x - 2) + (6x + 4)$

20. $(12x + 5) - (7x - 4) - (8x + 1)$

21. $3(a - 2) - 2(3a + 5) + 3(5a - 1)$

22. $(4x - 7)(5x + 6)$

23. $(-3x + 2)(4x - 3)$

24. $(7x - 3)(-5x + 1)$

25. $(x + 4)(x^2 - 3x - 7)$

26. $(2x + 1)(3x^2 - 2x + 6)$

27. $(5x - 3)^2$

28. $(3x + 7)^2$

29. $(2x - 1)^3$

30. $(3x + 5)^3$

31. $(x^2 - 2x - 3)(x^2 + 4x + 5)$

32. $(2x^2 - x - 2)(x^2 + 6x - 4)$

33. $\dfrac{24x^3y^4 - 48x^2y^3}{-6xy}$

34. $\dfrac{-56x^2y + 72x^3y^2}{8x^2}$

For Problems 35–46, factor each polynomial completely. Indicate any that are not factorable using integers.

35. $9x^2 - 4y^2$

36. $3x^3 - 9x^2 - 120x$

37. $4x^2 + 20x + 25$

38. $(x - y)^2 - 9$

39. $x^2 - 2x - xy + 2y$

40. $64x^3 - 27y^3$

41. $15x^2 - 14x - 8$

42. $3x^3 + 36$

43. $2x^2 - x - 8$

44. $3x^3 + 24$

45. $x^4 - 13x^2 + 36$

46. $4x^2 - 4x + 1 - y^2$

For Problems 47–56, perform the indicated operations involving rational expressions. Express final answers in simplest form.

47. $\dfrac{8xy}{18x^2y} \cdot \dfrac{24xy^2}{16y^3}$

48. $\dfrac{-14a^2b^2}{6b^3} \div \dfrac{21a}{15ab}$

49. $\dfrac{x^2 + 3x - 4}{x^2 - 1} \cdot \dfrac{3x^2 + 8x + 5}{x^2 + 4x}$

50. $\dfrac{9x^2 - 6x + 1}{2x^2 + 8} \cdot \dfrac{8x + 20}{6x^2 + 13x - 5}$

51. $\dfrac{3x - 2}{4} + \dfrac{5x - 1}{3}$

52. $\dfrac{2x - 6}{5} - \dfrac{x + 4}{3}$

53. $\dfrac{3}{n^2} + \dfrac{4}{5n} - \dfrac{2}{n}$

54. $\dfrac{5}{x^2 + 7x} - \dfrac{3}{x}$

55. $\dfrac{3x}{x^2 - 6x - 40} + \dfrac{4}{x^2 - 16}$

56. $\dfrac{2}{x - 2} - \dfrac{2}{x + 2} - \dfrac{4}{x^3 - 4x}$

For Problems 57–59, simplify each complex fraction.

57. $\dfrac{\dfrac{3}{x} - \dfrac{2}{y}}{\dfrac{5}{x^2} + \dfrac{7}{y}}$

58. $\dfrac{3 - \dfrac{2}{x}}{4 + \dfrac{3}{x}}$

59. $\dfrac{\dfrac{3}{(x + h)^2} - \dfrac{3}{x^2}}{h}$

$6(x+2) - \dfrac{6x}{\sqrt{x+2}}$

$x+2$

60. Simplify the expression

$$\dfrac{6(x^2 + 2)^{1/2} - 6x^2(x^2 + 2)^{-1/2}}{[(x^2 + 2)^{1/2}]^2}$$

For Problems 61–68, express each in simplest radical form. All variables represent positive real numbers.

61. $5\sqrt{48}$

62. $3\sqrt{24x^3}$

63. $\sqrt[3]{32x^4y^5}$

64. $\dfrac{3\sqrt{8}}{2\sqrt{6}}$

65. $\sqrt{\dfrac{5x}{2y^2}}$

66. $\dfrac{3}{\sqrt{2} + 5}$

67. $\dfrac{4\sqrt{2}}{3\sqrt{2} + \sqrt{3}}$

68. $\dfrac{3\sqrt{x}}{\sqrt{x} - 2\sqrt{y}}$

For Problems 69–74, perform the indicated operations and express the answers in simplest radical form.

69. $\sqrt{5}\sqrt[3]{5}$

70. $\sqrt[3]{x^2}\sqrt[4]{x}$

71. $\sqrt{x^3}\sqrt[3]{x^4}$

72. $\sqrt{xy}\sqrt[5]{x^3y^2}$

73. $\dfrac{\sqrt{5}}{\sqrt[3]{5}}$

74. $\dfrac{\sqrt[3]{x^2}}{\sqrt[4]{x^3}}$

For Problems 75–86, perform the indicated operations and express the resulting complex number in standard form.

75. $(-7 + 3i) + (-4 - 9i)$

76. $(2 - 10i) - (3 - 8i)$

77. $(-1 + 4i) - (-2 + 6i)$

78. $(3i)(-7i)$

79. $(2 - 5i)(3 + 4i)$

80. $(-3 - i)(6 - 7i)$

81. $(4 + 2i)(-4 - i)$

82. $(5 - 2i)(5 + 2i)$

83. $\dfrac{5}{3i}$

84. $\dfrac{2 + 3i}{3 - 4i}$

85. $\dfrac{-1 - 2i}{-2 + i}$

86. $\dfrac{-6i}{5 + 2i}$

For Problems 87–92, write each in terms of i and simplify.

87. $\sqrt{-100}$

88. $\sqrt{-40}$

89. $4\sqrt{-80}$

90. $(\sqrt{-9})(\sqrt{-16})$

91. $(\sqrt{-6})(\sqrt{-8})$

92. $\dfrac{\sqrt{-24}}{\sqrt{-3}}$

For Problems 93 and 94, use scientific notation and the properties of exponents to help with the computations.

93. $\dfrac{(0.0064)(420000)}{(0.00014)(0.032)}$

94. $\dfrac{(8600)(0.0000064)}{(0.0016)(0.000043)}$

CHAPTER 0 TEST

1. Evaluate each of the following.

   a. -7^{-2} **b.** $\left(\dfrac{3}{2}\right)^{-3}$ **c.** $\left(\dfrac{4}{9}\right)^{\frac{3}{2}}$ **d.** $\sqrt[3]{\dfrac{27}{64}}$

2. Find the product $(-3x^{-1}y^{2})(5x^{-3}y^{-4})$ and express the result using positive exponents only.

For Problems 3–7, perform the indicated operations.

3. $(-3x - 4) - (7x - 5) + (-2x - 9)$ 4. $(5x - 2)(-6x + 4)$

5. $(x + 2)(3x^{2} - 2x - 7)$ 6. $(4x - 1)^{3}$

7. $\dfrac{-18x^{4}y^{3} - 24x^{5}y^{4}}{-2xy^{2}}$ $3x^3 - 2x^2 - 7x + 6x^2 - 4x - 14$

$3x^3 4x^2 - 11x - 14$

For Problems 8–11, factor each polynomial completely.

8. $18x^{3} - 15x^{2} - 12x$ 9. $30x^{2} - 13x - 10$

10. $8x^{3} + 64$ 11. $x^{2} + xy - 2y - 2x$

$18x^3 + 9x^2 - 24x^2 - 12x$

For Problems 12–16, perform the indicated operations involving rational expressions. Express final answers in simplest form.

12. $\dfrac{6x^{3}y^{2}}{5xy} \div \dfrac{8y}{7x^{3}}$

13. $\dfrac{x^{2} - 4}{2x^{2} + 5x + 2} \cdot \dfrac{2x^{2} + 7x + 3}{x^{3} - 8}$

14. $\dfrac{3n - 2}{4} - \dfrac{4n + 1}{6}$

15. $\dfrac{5}{2x^{2} - 6x} + \dfrac{4}{3x^{2} + 6x}$

16. $\dfrac{4}{n^{2}} - \dfrac{3}{2n} - \dfrac{5}{n}$

17. Simplify the complex fraction $\dfrac{\dfrac{2}{x} - \dfrac{5}{y}}{\dfrac{3}{x} + \dfrac{4}{y^{2}}}$.

For Problems 18–21, express each radical expression in simplest radical form. All variables represent positive real numbers.

18. $6\sqrt{28x^{5}}$ 19. $\dfrac{5\sqrt{6}}{3\sqrt{12}}$ 20. $\dfrac{\sqrt{6}}{2\sqrt{2} - \sqrt{3}}$ 21. $\sqrt[3]{48x^{4}y^{5}}$

For Problems 22–25, perform the indicated operations and express the resulting complex numbers in standard form.

22. $(-2 - 4i) - (-1 + 6i) + (-3 + 7i)$

23. $(5 - 7i)(4 + 2i)$

24. $(7 - 6i)(7 + 6i)$

25. $\dfrac{1 + 2i}{3 - i}$

$3 \times 5 =$

EQUATIONS, INEQUALITIES, AND PROBLEM SOLVING

We can use the equation $0.2(8) - 0.2x + x = 0.9(8)$ to determine the change of the mixture in a radiator from 20% antifreeze to 90% antifreeze. We would drain 7 liters of coolant from a radiator that holds 8 liters and replace them with pure antifreeze, which will change the protection against temperature from 12°F to −20°F.

A common thread throughout precalculus algebra courses is one of developing algebraic skills, then using the skills to solve equations and inequalities, and then using equations and inequalities to solve applied problems. In this chapter we shall review and extend a variety of concepts related to that thread.

1.1 LINEAR EQUATIONS AND PROBLEM SOLVING

An algebraic equation such as $5x + 2 = 12$ is neither true nor false as it stands; it is sometimes referred to as an **open sentence**. Each time that a number is substituted for x, the algebraic equation $5x + 2 = 12$ becomes a **numerical statement** that is either true or false. For example, if $x = 5$, then $5x + 2 = 12$ becomes $5(5) + 2 = 12$, which is a false statement. If $x = 2$, then $5x + 2 = 12$ becomes $5(2) + 2 = 12$, which is a true statement. **Solving an equation** refers to the process of finding the number (or numbers) that make(s) an algebraic equation a true numerical statement. Such numbers are called the **solutions** or **roots** of the equation and are said to **satisfy the equation**. The set of all solutions of an equation is called its **solution set**. Thus {2} is the solution set of $5x + 2 = 12$.

An equation that is satisfied by all numbers that can meaningfully replace the variable is called an **identity**. For example,

$$3(x + 2) = 3x + 6 \qquad x^2 - 4 = (x + 2)(x - 2) \qquad \text{and}$$

$$\frac{1}{x} + \frac{1}{2} = \frac{2 + x}{2x}$$

are all identities. In the last identity, x cannot equal zero; thus the statement

$$\frac{1}{x} + \frac{1}{2} = \frac{2 + x}{2x}$$

is true for all real numbers except zero. An equation that is true for some but not all permissible values of the variable is called a **conditional equation**. Thus the equation $5x + 2 = 12$ is a conditional equation.

Equivalent equations are equations that have the same solution set. For example,

$$7x - 1 = 20 \qquad 7x = 21 \qquad \text{and} \qquad x = 3$$

are all equivalent equations because {3} is the solution set of each. The general procedure for solving an equation is to continue replacing the given equation with equivalent but simpler equations until an equation of the form *variable = constant* or *constant = variable* is obtained. Thus, in the example above, $7x - 1 = 20$ was simplified to $7x = 21$, which was further simplified to $x = 3$, which gives us the solution set, {3}.

Techniques for solving equations revolve around properties of equality. The following list summarizes some basic properties of equality.

P R O P E R T Y 1 . 1 *Properties of Equality*

For all real numbers, a, b, and c,

1. $a = a$.   Reflexive property
2. If $a = b$, then $b = a$.   Symmetric property
3. If $a = b$ and $b = c$, then $a = c$.   Transitive property
4. If $a = b$, then a may be replaced by b, or b may be replaced by a, in any statement without changing the meaning of the statement.   Substitution property
5. $a = b$ if and only if $a + c = b + c$.   Addition property
6. $a = b$ if an only if $ac = bc$, where $c \neq 0$.   Multiplication property

---

The addition property of equality states that any number can be added to both sides of an equation to produce an equivalent equation. The multiplication property of equality states that an equivalent equation is produced whenever both sides of an equation are multiplied by the same nonzero real number.

Now let's consider how these properties of equality can be used to solve a variety of linear equations. A **linear equation** in the variable x is one that can be written in the form

$$ax + b = 0$$

where a and b are real numbers and $a \neq 0$.

EXAMPLE 1

Solve the equation $2x - 3 = 9$.

Solution

$$2x - 3 = 9$$
$$2x - 3 + 3 = 9 + 3 \qquad \text{Add 3 to both sides.}$$
$$2x = 12$$
$$\frac{1}{2}(2x) = \frac{1}{2}(12) \qquad \text{Multiply both sides by } \frac{1}{2}.$$
$$x = 6$$

Check   To check an apparent solution, we can substitute it into the original equation to see whether we obtain a true numerical statement.

$$2x - 3 = 9$$
$$2(6) - 3 \overset{?}{=} 9$$
$$12 - 3 \overset{?}{=} 9$$
$$9 = 9$$

Now we know that the solution set is $\{6\}$. ■

EXAMPLE 2

Solve the equation $-4x - 3 = 2x + 9$.

Solution

$$-4x - 3 = 2x + 9$$
$$-4x - 3 + (-2x) = 2x + 9 + (-2x) \qquad \text{Add } -2x \text{ to both sides.}$$
$$-6x - 3 = 9$$
$$-6x - 3 + 3 = 9 + 3 \qquad \text{Add 3 to both sides.}$$
$$-6x = 12$$
$$-\frac{1}{6}(-6x) = -\frac{1}{6}(12) \qquad \text{Multiply both sides by } -\frac{1}{6}.$$
$$x = -2.$$

 Check $-4x - 3 = 2x + 9$

$$-4(-2) - 3 \overset{?}{=} 2(-2) + 9$$
$$8 - 3 \overset{?}{=} -4 + 9$$
$$5 = 5.$$

Now we know that the solution set is $\{-2\}$. ■

EXAMPLE 3

Solve $4(n - 2) - 3(n - 1) = 2(n + 6)$.

Solution

First let's use the distributive property to remove parentheses and combine similar terms.

$$4(n - 2) - 3(n - 1) = 2(n + 6)$$
$$4n - 8 - 3n + 3 = 2n + 12$$
$$n - 5 = 2n + 12$$

Now we can apply the addition property of equality.

$$n - 5 + (-n) = 2n + 12 + (-n)$$
$$-5 = n + 12$$
$$-5 + (-12) = n + 12 + (-12)$$
$$-17 = n$$

Check $4(n - 2) - 3(n - 1) = 2(n + 6)$
$$4(-17 - 2) - 3(-17 - 1) \overset{?}{=} 2(-17 + 6)$$
$$4(-19) - 3(-18) \overset{?}{=} 2(-11)$$
$$-76 + 54 \overset{?}{=} -22$$
$$-22 = -22$$

The solution set is $\{-17\}$.   ■

As you study these examples, pay special attention to the steps shown in the solutions. Certainly, there are no rules about which steps should be performed mentally; this is an individual decision. We would suggest that you show enough steps so that the flow of the process is understood and so that the chances of making careless computational errors are minimized. We shall discontinue showing the check for each problem, but remember that checking an answer is the only way to be sure of your result.

EXAMPLE 4   Solve $\dfrac{1}{4}x - \dfrac{2}{3}x = \dfrac{5}{6}$.

Solution

$$\frac{1}{4}x - \frac{2}{3}x = \frac{5}{6}$$
$$12\left(\frac{1}{4}x - \frac{2}{3}x\right) = 12\left(\frac{5}{6}\right) \qquad \text{Multiply both sides by the LCD.}$$
$$12\left(\frac{1}{4}x\right) - 12\left(\frac{2}{3}x\right) = 12\left(\frac{5}{6}\right)$$
$$3x - 8x = 10$$
$$-5x = 10$$
$$x = -2$$

The solution set is $\{-2\}$.   ■

EXAMPLE 5   Solve $\dfrac{2y - 3}{3} + \dfrac{y + 1}{2} = 3$.

Solution

$$\frac{2y - 3}{3} + \frac{y + 1}{2} = 3$$

$$6\left(\frac{2y - 3}{3} + \frac{y + 1}{2}\right) = 6(3) \qquad \text{Multiply both sides by the LCD.}$$

$$6\left(\frac{2y - 3}{3}\right) + 6\left(\frac{y + 1}{2}\right) = 6(3) \qquad \text{Apply the distributive property on the left side.}$$

$$2(2y - 3) + 3(y + 1) = 18$$

$$4y - 6 + 3y + 3 = 18$$

$$7y - 3 = 18$$

$$7y = 21$$

$$y = 3$$

The solution set is $\{3\}$. (Check it!) ■

EXAMPLE 6

Solve $\dfrac{4x - 1}{10} - \dfrac{5x + 2}{4} = -3$.

Solution

$$\frac{4x - 1}{10} - \frac{5x + 2}{4} = -3$$

$$20\left(\frac{4x - 1}{10} - \frac{5x + 2}{4}\right) = 20(-3)$$

$$20\left(\frac{4x - 1}{10}\right) - 20\left(\frac{5x + 2}{4}\right) = -60$$

$$2(4x - 1) - 5(5x + 2) = -60$$

$$8x - 2 - 25x - 10 = -60$$

$$-17x - 12 = -60$$

$$-17x = -48$$

$$x = \frac{48}{17}$$

The solution set is $\left\{\dfrac{48}{17}\right\}$. ■

Problem Solving

The ability to use the tools of algebra to solve problems requires that we be able to translate the English language into the language of algebra. More specifically, at this time we need to translate *English sentences* into *algebraic equations* so that we can use our equation-solving skills. Let's work through an example and then comment on some of the problem-solving aspects of it.

PROBLEM 1

If 2 is subtracted from five times a certain number, the result is 28. Find the number.

Solution

Let n represent the number to be found. The sentence *If 2 is subtracted from five times a certain number, the result is 28* translates into the equation $5n - 2 = 28$. Solving this equation, we obtain

$$5n - 2 = 28$$
$$5n = 30$$
$$n = 6$$

The number to be found is 6.   ■

Now let's make a few comments about our approach to Problem 1. Making a statement such as *Let n represent the number to be found* is often referred to as **declaring the variable**. It amounts to choosing a letter to use as a variable and indicating what the variable represents for a specific problem. This may seem like an insignificant idea, but as the problems become more complex, the process of declaring the variable becomes more important. It is also a good idea to choose a *meaningful* variable. For example, if the problem involves finding the width of a rectangle, then a choice of w for the variable is reasonable. Furthermore, it is true that some people can solve a problem such as Problem 1 without setting up an algebraic equation. However, as problems increase in difficulty, the translation from English to algebra becomes a key issue. Therefore, even with these relatively easy problems, we suggest that you concentrate on the translation process.

To check our answer for Problem 1, we must determine whether it satisfies the conditions stated in the original problem. Because 2 subtracted from 5(6) equals 28, we know that our answer of 6 is correct. Remember, when you are checking a potential answer for a word problem, it is *not* sufficient to check the result in the equation used to solve the problem, because the equation itself may be in error.

Sometimes it is necessary not only to declare the variable but also to represent other unknown quantities in terms of that variable. Let's consider a problem that illustrates this idea.

PROBLEM 2

Find three consecutive integers whose sum is -45.

Solution

Let n represent the smallest integer; then $n + 1$ is the next integer and $n + 2$ is the largest of the three integers. Because the sum of the three consecutive integers is to be -45, we have the following equation.

$$n + (n + 1) + (n + 2) = -45$$
$$3n + 3 = -45$$
$$3n = -48$$
$$n = -16$$

If $n = -16$, then $n + 1$ is -15 and $n + 2$ is -14. Thus the three consecutive integers are -16, -15, and -14. ∎

Frequently, the translation from English to algebra can be made easier by recognizing a guideline that can be used to set up an appropriate equation. Pay special attention to the guidelines used in the solutions of the next two problems.

PROBLEM 3

Tina is paid time-and-a-half for each hour worked over 40 hours in a week. Last week she worked 45 hours and earned $380. What is her normal hourly rate?

Solution

Let r represent Tina's normal hourly rate. Then $\frac{3}{2}r$ represents $1\frac{1}{2}$ times her normal hourly rate (time-and-a-half). The following guideline can be used to help set up the equation.

$$
\begin{array}{ccccc}
\text{Regular wages} & & \text{Wages for 5} & & \\
\text{for first 40} & + & \text{hours of} & = & \text{Total wages} \\
\text{hours} & & \text{overtime} & & \\
\downarrow & & \downarrow & & \downarrow \\
40r & + & 5\left(\dfrac{3}{2}r\right) & = & \$380
\end{array}
$$

Solving this equation, we obtain

$$2\left[40r + 5\left(\frac{3}{2}r\right)\right] = 2(380)$$

$$2(40r) + 2\left[5\left(\frac{3}{2}r\right)\right] = 760$$

$$80r + 15r = 760$$

$$95r = 760$$

$$r = 8$$

Her normal hourly rate is thus $8 per hour. (Check the answer in the original statement of the problem!) ∎

PROBLEM 4

There are 51 students in a certain class. The number of females is 5 less than three times the number of males. Find the number of females and the number of males in the class.

Solution

Let m represent the number of males; then $3m - 5$ represents the number of females. The total number of students is 51, so the guideline is (number of males) plus (number of females) equals 51. Thus we can set up and solve the following equation.

$$m + (3m - 5) = 51$$
$$4m - 5 = 51$$
$$4m = 56$$
$$m = 14$$

Therefore, there are 14 males and $3(14) - 5 = 37$ females. ∎

PROBLEM SET 1.1

For Problems 1–42, solve each equation.

1. $9x - 3 = -21$

2. $-5x + 4 = -11$

3. $13 - 2x = 14$

4. $17 = 6a + 5$

5. $3n - 2 = 2n + 5$

6. $4n + 3 = 5n - 9$

7. $-5a + 3 = -3a + 6$

8. $4x - 3 + 2x = 8x - 3 - x$

9. $-3(x + 1) = 7$

10. $5(2x - 1) = 13$

11. $4(2x - 1) = 3(3x + 2)$

12. $5x - 4(x - 6) = -11$

13. $4(n - 2) - 3(n - 1) = 2(n + 6)$

14. $-3(2t - 5) = 2(4t + 7)$

15. $3(2t - 1) - 2(5t + 1) = 4(3t + 4)$

16. $-(3x - 1) + (2x + 3) = -4 + 3(x - 1)$

17. $-2(y - 4) - (3y - 1) = -2 + 5(y + 1)$

18. $\dfrac{-3x}{4} = \dfrac{9}{2}$

19. $-\dfrac{6x}{7} = 12$

20. $\dfrac{n}{2} - \dfrac{1}{3} = \dfrac{13}{6}$

21. $\dfrac{3}{4}n - \dfrac{1}{12}n = 6$

22. $\dfrac{2}{3}x - \dfrac{1}{5}x = 7$

23. $\dfrac{h}{2} + \dfrac{h}{5} = 1$

24. $\dfrac{4y}{5} - 7 = \dfrac{y}{10}$

25. $\dfrac{y}{5} - 2 = \dfrac{y}{2} + 1$

26. $\dfrac{x + 2}{3} + \dfrac{x - 1}{4} = \dfrac{9}{2}$

27. $\dfrac{c + 5}{7} + \dfrac{c - 3}{4} = \dfrac{5}{14}$

28. $\dfrac{2x - 5}{6} - \dfrac{3x - 4}{8} = 0$

29. $\dfrac{n - 3}{2} - \dfrac{4n - 1}{6} = \dfrac{2}{3}$

30. $\dfrac{3x - 1}{2} + \dfrac{x - 3}{4} = \dfrac{1}{2}$

31. $\dfrac{2t + 3}{6} - \dfrac{t - 9}{4} = 5$

32. $\dfrac{2x + 7}{9} - 4 = \dfrac{x - 7}{12}$

33. $\dfrac{3n - 1}{8} - 2 = \dfrac{2n + 5}{7}$

34. $\dfrac{x + 2}{3} + \dfrac{3x + 1}{4} + \dfrac{2x - 1}{6} = 2$

35. $\dfrac{2t - 3}{6} + \dfrac{3t - 2}{4} + \dfrac{5t + 6}{12} = 4$

36. $\dfrac{3y - 1}{8} + y - 2 = \dfrac{y + 4}{4}$

37. $\dfrac{2x + 1}{14} - \dfrac{3x + 4}{7} = \dfrac{x - 1}{2}$

38. $n + \dfrac{2n - 3}{9} - 2 = \dfrac{2n + 1}{3}$

39. $(x - 3)(x - 1) - x(x + 2) = 7$

40. $(3n + 4)(n - 2) - 3n(n + 3) = 3$

41. $(2y + 1)(3y - 2) - (6y - 1)(y + 4) = -20y$

42. $(4t - 3)(t + 2) - (2t + 3)^2 = -1$

Solve each of Problems 43–62 by setting up and solving an algebraic equation.

43. One number is 5 less than another number. Find the numbers if five times the smaller number is 11 less than four times the larger number.

44. The sum of three consecutive integers is 21 larger than twice the smallest integer. Find the integers.

45. Find three consecutive even integers such that if the largest integer is subtracted from four times the smallest, the result is 6 more than twice the middle integer.

46. Find three consecutive odd integers such that three times the largest is 23 less than twice the sum of the two smallest integers.

47. Find two consecutive integers such that the difference of their squares is 37.

48. Find three consecutive integers such that the product of the two largest is 20 more than the square of the smallest integer.

49. Find four consecutive integers such that the product of the two largest is 46 more than the product of the two smallest integers.

50. Over the weekend, Mario bicycled 69 miles. On Sunday he rode 9 miles more than $\frac{2}{3}$ of his distance on Saturday. Find the number of miles he rode each day.

51. For a given triangle, the measure of angle A is 10° less than three times the measure of angle B. The measure of angle C is one-fifth of the sum of the measures of angles A and B. Knowing that the sum of the measures of the angles of a triangle equals 180°, find the measure of each angle.

52. Jennifer went on a shopping spree, spending a total of $124 on a skirt, a sweater, and a pair of shoes. The cost of the sweater was $\frac{8}{7}$ of the cost of the skirt. The shoes cost $8 less than the skirt. Find the cost of each item.

53. The average of the salaries of Kelly, Renee, and Nina is $20,000 a year. If Kelly earns $4000 less than Renee, and Nina's salary is two-thirds of Renee's salary, find the salary of each person.

THOUGHTS INTO WORDS

63. Explain the difference between a numerical statement and an algebraic equation.

64. Are the equations $9 = 3x - 2$ and $3x - 2 = 9$ equivalent equations? Defend your answer.

54. Barry is paid double-time for each hour worked over 40 hours in a week. Last week he worked 47 hours and earned $378. What is his normal hourly rate?

55. Greg had 80 coins consisting of pennies, nickels, and dimes. The number of nickels was 5 more than one-third the number of pennies, and the number of dimes was 1 less than one-fourth of the number of pennies. How many coins of each kind did he have?

56. Rita has a collection of 105 coins consisting of nickels, dimes, and quarters. The number of dimes is 5 more than one-third of the number of nickels, and the number of quarters is twice the number of dimes. How many coins of each kind does she have?

57. In a class of 43 students, the number of males is 8 less than twice the number of females. How many females and how many males are there in the class?

58. A precinct reported that 316 people had voted in an election. The number of Republican voters was 6 more than two-thirds of the number of Democrats. How many Republicans and how many Democrats voted in that precinct?

59. Two years ago Janie was half as old as she will be 9 years from now. How old is she now?

60. The sum of the present ages of Eric and his father is 58 years. In 10 years, his father will be twice as old as Eric will be at that time. Find their present ages.

61. Brad is 6 years older than Pedro. Five years ago Pedro's age was three-fourths of Brad's age at that time. Find the present ages of Brad and Pedro.

62. Tina is 4 years older than Sherry. In 5 years the sum of their ages will be 48. Find their present ages.

65. How do you defend the statement that the equation $x + 3 = x + 2$ has no real number solutions?

66. How do you defend the statement that the solution set of the equation $3(x - 4) = 3x - 12$ is the entire set of real numbers?

Further Investigations

67. Verify that for any three consecutive integers, the sum of the smallest and the largest is equal to twice the middle integer.

68. Verify that no four consecutive integers can be found such that the product of the smallest and the largest is equal to the product of the other two integers.

69. Some algebraic identities provide a basis for shortcuts to do mental arithmetic. For example, the identity $(x + y)(x - y) = x^2 - y^2$ indicates that a multiplication problem such as $(31)(29)$ can be treated as $(30 + 1)(30 - 1) = 30^2 - 1^2 = 900 - 1 = 899$.

For each of the following, use the given identity to provide a way of mentally performing the indicated computations. Check your answers with a calculator.

a. $(x + y)(x - y) = x^2 - y^2$: $(21)(19)$; $(39)(41)$; $(22)(18)$; $(42)(38)$; $(47)(53)$

b. $(x + y)^2 = x^2 + 2xy + y^2$: $(21)^2$; $(32)^2$; $(51)^2$; $(62)^2$; $(43)^2$

c. $(x - y)^2 = x^2 - 2xy + y^2$: $(29)^2$; $(49)^2$; $(18)^2$; $(38)^2$; $(67)^2$

d. $(10t + 5)^2 = 100t^2 + 100t + 25 = 100t(t + 1) + 25$: $(15)^2$; $(35)^2$; $(45)^2$; $(65)^2$; $(85)^2$

1.2 MORE EQUATIONS AND APPLICATIONS

In the previous section we considered linear equations, such as

$$\frac{x - 1}{3} + \frac{x + 2}{4} = \frac{1}{6}$$

that have fractional coefficients with constants as denominators. Now let's consider equations that contain the variable in one or more of the denominators. Our approach to solving such equations remains essentially the same except **we must avoid any values of the variable that make a denominator zero.** Consider the following examples.

EXAMPLE 1 Solve $\dfrac{5}{3x} - \dfrac{1}{9} = \dfrac{1}{x}$.

Solution

First we need to realize that *x cannot equal zero*. Let's indicate this restriction so that it is not forgotten; then we can proceed as follows.

$$\frac{5}{3x} - \frac{1}{9} = \frac{1}{x}, \qquad x \neq 0$$

$$9x\left(\frac{5}{3x} - \frac{1}{9}\right) = 9x\left(\frac{1}{x}\right) \qquad \text{Multiply both sides by the LCD.}$$

$$9x\left(\frac{5}{3x}\right) - 9x\left(\frac{1}{9}\right) = 9x\left(\frac{1}{x}\right)$$

$$15 - x = 9$$

$$-x = -6$$

$$x = 6$$

The solution set is $\{6\}$. (Check it!)

EXAMPLE 2   Solve $\dfrac{65 - n}{n} = 4 + \dfrac{5}{n}$.

Solution

$$\frac{65 - n}{n} = 4 + \frac{5}{n}, \qquad n \neq 0$$

$$n\left(\frac{65 - n}{n}\right) = n\left(4 + \frac{5}{n}\right)$$

$$65 - n = 4n + 5$$

$$60 = 5n$$

$$12 = n$$

The solution set is $\{12\}$.   ∎

EXAMPLE 3   Solve $\dfrac{a}{a - 2} + \dfrac{2}{3} = \dfrac{2}{a - 2}$.

Solution

$$\frac{a}{a - 2} + \frac{2}{3} = \frac{2}{a - 2}, \qquad a \neq 2$$

$$3(a - 2)\left(\frac{a}{a - 2} + \frac{2}{3}\right) = 3(a - 2)\left(\frac{2}{a - 2}\right)$$

$$3a + 2(a - 2) = 6$$

$$3a + 2a - 4 = 6$$

$$5a = 10$$

$$a = 2$$

Because our initial restriction was $a \neq 2$, we conclude that this equation *has no solution*. The solution set is $\varnothing$.   ∎

Example 3 illustrates the importance of recognizing the restrictions that must be made to exclude division by zero.

Ratio and Proportion

A **ratio** is the comparison of two numbers by division. The fractional form is frequently used to express ratios. For example, the ratio of a to b can be written a/b. A statement of equality between two ratios is called a **proportion**. Thus, if a/b and c/d are equal ratios, the proportion $a/b = c/d$ $(b \neq 0$ and $d \neq 0)$ can be formed. There is a useful property of proportions.

If $\dfrac{a}{b} = \dfrac{c}{d}$ then $ad = bc$.

This property can be deduced as follows.

$$\frac{a}{b} = \frac{c}{d}, \qquad b \neq 0 \text{ and } d \neq 0$$

$$bd\left(\frac{a}{b}\right) = bd\left(\frac{c}{d}\right) \qquad \text{Multiply both sides by } bd.$$

$$ad = bc$$

This is sometimes referred to as the **cross-multiplication property of proportions.**

Some equations can be treated as proportions and solved by using the cross-multiplication idea, as the next example illustrates.

EXAMPLE 4   Solve $\dfrac{3}{3x - 2} = \dfrac{4}{2x + 1}$.

Solution

$$\frac{3}{3x - 2} = \frac{4}{2x + 1}, \qquad x \neq \frac{2}{3}, x \neq -\frac{1}{2}$$

$$3(2x + 1) = 4(3x - 2) \qquad \text{Apply the cross-multiplication property.}$$

$$6x + 3 = 12x - 8$$

$$11 = 6x$$

$$\frac{11}{6} = x$$

The solution set is $\left\{\dfrac{11}{6}\right\}$.   ■

Linear Equations Involving Decimals

To solve an equation such as $x + 2.4 = 0.36$, we can add -2.4 to both sides. However, as equations containing decimals become more complex, it is often easier to begin by *clearing the equation of all decimals*, which we accomplish by multiplying both sides by an appropriate power of 10. Let's consider two examples.

EXAMPLE 5   Solve $0.12t - 2.1 = 0.07t - 0.2$.

Solution

$$0.12t - 2.1 = 0.07t - 0.2$$
$$100(0.12t - 2.1) = 100(0.07t - 0.2) \qquad \text{Multiply both sides by 100.}$$
$$12t - 210 = 7t - 20$$
$$5t = 190$$
$$t = 38$$

The solution set is $\{38\}$.   ■

EXAMPLE 6

Solve $0.8x + 0.9(850 - x) = 715$.

Solution

$$0.8x + 0.9(850 - x) = 715$$
$$10[0.8x + 0.9(850 - x)] = 10(715) \qquad \text{Multiply both sides by 10.}$$
$$10(0.8x) + 10[0.9(850 - x)] = 10(715)$$
$$8x + 9(850 - x) = 7150$$
$$8x + 7650 - 9x = 7150$$
$$-x = -500$$
$$x = 500$$

The solution set is $\{500\}$.   ■

Changing Forms of Formulas

Many practical applications of mathematics involve the use of formulas. For example, to find the distance traveled in 4 hours at a rate of 55 miles per hour, we multiply the rate times the time; thus the distance is $55(4) = 220$ miles. The rule *distance equals rate times time* is commonly stated as a formula: $d = rt$. When using a formula, it is sometimes convenient first to change its form. For example, multiplying both sides of $d = rt$ by $1/t$ produces the equivalent form $r = d/t$. Multiplying both sides of $d = rt$ by $1/r$ produces another equivalent form, $t = d/r$. The following two examples further illustrate the process of obtaining equivalent forms of certain formulas.

EXAMPLE 7

If P dollars are invested at a simple rate of r percent, then the amount, A, accumulated after t years is given by the formula $A = P + Prt$. Solve this formula for P.

Solution

$$A = P + Prt$$
$$A = P(1 + rt) \qquad \text{Apply the distributive property to the right side.}$$
$$\frac{A}{1 + rt} = P \qquad \text{Multiply both sides by } \frac{1}{1 + rt}.$$
$$P = \frac{A}{1 + rt} \qquad \text{Apply the symmetric property of equality.} \qquad ■$$

| **EXAMPLE 8** | The area (A) of a trapezoid (see Figure 1.1) is given by the formula $A = \frac{1}{2}h(b_1 + b_2)$. Solve this equation for b_1. |

FIGURE 1.1

Solution

$$A = \frac{1}{2}h(b_1 + b_2)$$

$2A = h(b_1 + b_2)$ Multiply both sides by 2.

$2A = hb_1 + hb_2$ Apply the distributive property to the right side.

$2A - hb_2 = hb_1$ Add $-hb_2$ to both sides.

$\dfrac{2A - hb_2}{h} = b_1$ Multiply both sides by $\dfrac{1}{h}$.   ■

Notice that in Example 7, the distributive property was used to change from the form $P + Prt$ to $P(1 + rt)$. However, in Example 8 the distributive property was used to change $h(b_1 + b_2)$ to $hb_1 + hb_2$. In both examples the goal is to *isolate the term* containing the variable being solved for so that an appropriate application of the multiplication property will produce the desired result. Also note the use of *subscripts* to identify the two bases of the trapezoid. Subscripts allow us to use the same letter b to identify the bases, but b_1 represents one base and b_2 the other.

More on Problem Solving

Volumes have been written on the topic of problem solving, but certainly one of the best-known sources is George Polya's book *How to Solve It*.* In this book, Polya suggests the following four-phase plan for solving problems.

    1. *Understand the problem.*

    2. *Devise a plan* to solve the problem.

    3. *Carry out the plan* to solve the problem.

    4. *Look back* at the completed solution to review and discuss it.

We will comment briefly on each of the phases and offer some suggestions for using an algebraic approach to solve problems.

Understand the Problem   Read the problem carefully, making certain that you understand the meanings of all the words. Be especially alert for any technical terms used in the statement of the problem. Often it is helpful to sketch a figure, diagram, or chart to visualize and organize the conditions of the problem. Determine the known and unknown facts, and if one of the previously mentioned pictorial devices is used, record these facts in the appropriate places of the diagram or chart.

---

*George Polya, *How to Solve It* (Princeton: Princeton University Press), 1945.

Devise a Plan   This is the key part of the four-phase plan. It is sometimes referred to as the *analysis* of the problem. There are numerous strategies and techniques used to solve problems. We shall discuss some of these strategies at various places throughout this text; however, at this time we offer the following general suggestions.

1. Choose a meaningful variable to represent an unknown quantity in the problem (perhaps *t* if time is an unknown quantity) and represent any other unknowns in terms of that variable.

2. Look for a *guideline* that can be used to set up an equation. A guideline might be a formula, such as $A = P + Prt$ from Example 7, or a statement of a relationship, such as *the sum of the two numbers is 28*. Sometimes a relationship suggested by a pictorial device can be used as a guideline for setting up the equation. Also, be alert to the possibility that this *new* problem might really be an *old* problem in a new setting, perhaps even stated in different vocabulary.

3. Form an equation containing the variable so that the conditions of the guideline are translated from English into algebra.

Carry out the Plan   This phase is sometimes referred to as the *synthesis* of the plan. If phase two has been successfully completed, then carrying out the plan may simply be a matter of solving the equation and doing any further computations to answer all of the questions in the problem. Confidence in your plan creates a better working atmosphere for carrying it out. It is also in this phase that the calculator may become a valuable tool. The type of data and the amount of complexity involved in the computations are two factors that can influence your decision whether to use one.

Look Back   This is an important but often overlooked part of problem solving. The following list of questions suggests some things for you to consider in this phase.

1. Is your answer to the problem a *reasonable* answer?

2. Have you *checked* your answer by substituting it back into the conditions stated in the problem?

3. Looking back over your solution, do you now see another plan that could be used to solve the problem?

4. Do you see a way of generalizing your procedure for this problem that could be used to solve other problems of this type?

5. Do you now see that this problem is closely related to another problem that you have previously solved?

6. Have you tucked away for future reference the technique used to solve this problem?

Looking back over the solution of a newly solved problem can lay important groundwork for solving problems in the future.

Keep the previous suggestions in mind as we tackle some more word problems. Perhaps it would also be helpful for you to attempt to solve these problems on your own before looking at our approach.

PROBLEM 1

One number is 65 larger than another number. If the larger number is divided by the smaller, the quotient is 6 and the remainder is 5. Find the numbers.

Solution

Let n represent the smaller number. Then $n + 65$ represents the larger number. We can use the following relationship as a *guideline*.

$$\frac{\text{Dividend}}{\text{Divisor}} = \text{Quotient} + \frac{\text{Remainder}}{\text{Divisor}}$$

$$\frac{n + 65}{n} = 6 + \frac{5}{n}$$

We solve this equation by multiplying both sides by n.

$$n\left(\frac{n + 65}{n}\right) = n\left(6 + \frac{5}{n}\right), \qquad n \neq 0$$

$$n + 65 = 6n + 5$$

$$60 = 5n$$

$$12 = n$$

If $n = 12$, then $n + 65$ equals 77. The two numbers are 12 and 77.   ■

Sometimes we can use the concepts of ratio and proportion to set up an equation and solve a problem, as the next example illustrates.

PROBLEM 2

The ratio of male students to female students at a certain university is 5 to 7. If there is a total of 16,200 students, find the number of male and the number of female students.

Solution

Let m represent the number of male students; then $16200 - m$ represents the number of female students. The following proportion can be set up and solved.

$$\frac{m}{16200 - m} = \frac{5}{7}$$

$$7m = 5(16200 - m)$$

$$7m = 81000 - 5m$$

$$12m = 81000$$

$$m = 6750$$

Therefore, there are 6750 male students and $16200 - 6750 = 9450$ female students.   ■

The next problem has a geometric setting. In such cases, the use of figures is very helpful.

PROBLEM 3

If two opposite sides of a square are each increased by 3 centimeters and the other two sides are each decreased by 2 centimeters, the area is increased by 8 square centimeters. Find the length of the side of the square.

Solution

Let s represent the side of the square. Then Figures 1.2(a) and 1.2(b) represent the square and the rectangle formed by increasing two opposite sides of the square by 3 centimeters and decreasing the other two sides by 2 centimeters. Because the area of the rectangle is 8 square centimeters more than the area of the square, the following equation can be set up and solved.

FIGURE 1.2

$$(s + 3)(s - 2) = s^2 + 8$$
$$s^2 + s - 6 = s^2 + 8$$
$$s = 14$$

Thus the length of a side of the original square is 14 centimeters. ∎

Many consumer problems can be solved by using an algebraic approach. For example, let's consider a discount sale problem involving the relationship *original selling price minus discount equals discount sale price*.

PROBLEM 4

Jim bought a pair of slacks at a 30% discount sale for $28. What was the original price of the slacks?

Solution

Let p represent the original price of the slacks.

Original price − Discount = Discount sale price

$$(100\%)(p) \quad - (30\%)(p) = \qquad \$28$$

We switch this equation to decimal form to solve it.

$$p - 0.3p = 28$$
$$0.7p = 28$$
$$p = 40$$

The original price of the slacks was $40. ∎

Another basic relationship pertaining to consumer problems is *selling price equals cost plus profit*. Profit (also called markup, markon, and margin of profit), may be stated in different ways. It can be expressed as a percent of the cost, as a percent of the selling price, or simply in terms of dollars and cents. Let's consider a problem where the profit is stated as a percent of the selling price.

PROBLEM 5

A retailer of sporting goods bought a putter for $25. He wants to price the putter to make a profit of 20% of the selling price. What price should he mark on the putter?

Solution

Let s represent the selling price.

$$\text{Selling price} = \text{Cost} + \text{Profit}$$

$$s \qquad = \$25 \; + (20\%)(s)$$

Solving this equation involves using the methods we developed earlier for working with decimals.

$$s = 25 + (20\%)(s)$$
$$s = 25 + 0.2s$$
$$10s = 250 + 2s$$
$$8s = 250$$
$$s = 31.25$$

The selling price should be $31.25.   ∎

Certain types of investment problems can be solved by using an algebraic approach. As our final example of this section, let's consider one such problem.

PROBLEM 6

Cindy invested a certain amount of money at 10% interest and $1500 more than that amount at 11%. Her total yearly interest was $795. How much did she invest at each rate?

Solution

Let d represent the amount invested at 10%; then $d + 1500$ represents the amount invested at 11%. The following guideline can be used to set up an equation.

$$\text{Interest earned at 10\%} + \text{Interest earned at 11\%} = \text{Total interest}$$

$$(10\%)(d) \qquad + \quad (11\%)(d + 1500) \quad = \qquad \$795$$

We can solve this equation by multiplying both sides by 100.

$$0.1d + 0.11(d + 1500) = 795$$
$$10d + 11(d + 1500) = 79500$$
$$10d + 11d + 16500 = 79500$$
$$21d = 63000$$
$$d = 3000$$

Cindy invested $3000 at 10% and $3000 + $1500 = $4500 at 11%.   ∎

Don't forget phase four of Polya's problem-solving plan. We have not taken the space to look back over and discuss each of our examples. However, it would be beneficial for you to do so, keeping in mind the questions posed earlier regarding this phase.

PROBLEM SET 1.2

For Problems 1–32, solve each equation.

1. $\dfrac{x-2}{3} + \dfrac{x+1}{4} = \dfrac{1}{6}$

2. $\dfrac{5n-1}{4} - \dfrac{2n-3}{10} = \dfrac{3}{5}$

3. $\dfrac{5}{x} + \dfrac{1}{3} = \dfrac{8}{x}$

4. $\dfrac{5}{3n} - \dfrac{1}{9} = \dfrac{1}{n}$

5. $\dfrac{1}{3n} + \dfrac{1}{2n} = \dfrac{1}{4}$

6. $\dfrac{1}{x} - \dfrac{3}{2x} = \dfrac{1}{5}$

7. $\dfrac{35-x}{x} = 7 + \dfrac{3}{x}$

8. $\dfrac{n}{46-n} = 5 + \dfrac{4}{46-n}$

9. $\dfrac{n+67}{n} = 5 + \dfrac{11}{n}$

10. $\dfrac{n+52}{n} = 4 + \dfrac{1}{n}$

11. $\dfrac{5}{3x-2} = \dfrac{1}{x-4}$

12. $\dfrac{-2}{5x-3} = \dfrac{4}{4x-1}$

13. $\dfrac{4}{2y-3} - \dfrac{7}{3y-5} = 0$

14. $\dfrac{3}{2n+1} + \dfrac{5}{3n-4} = 0$

15. $\dfrac{n}{n+1} + 3 = \dfrac{4}{n+1}$

16. $\dfrac{a}{a+5} - 2 = \dfrac{3a}{a+5}$

17. $\dfrac{3x}{2x-1} - 4 = \dfrac{x}{2x-1}$

18. $\dfrac{x}{x-8} - 4 = \dfrac{8}{x-8}$

19. $\dfrac{3}{x+3} - \dfrac{1}{x-2} = \dfrac{5}{2x+6}$

20. $\dfrac{6}{x+3} + \dfrac{20}{x^2+x-6} = \dfrac{5}{x-2}$

21. $\dfrac{n}{n-3} - \dfrac{3}{2} = \dfrac{3}{n-3}$

22. $\dfrac{4}{x-2} + \dfrac{x}{x+1} = \dfrac{x^2-2}{x^2-x-2}$

23. $s = 9 + 0.25s$ **24.** $s = 1.95 + 0.35s$

25. $0.09x + 0.1(700 - x) = 67$

26. $0.08x + 0.09(950 - x) = 81$

27. $0.09x + 0.11(x + 125) = 68.75$

28. $0.08(x + 200) = 0.07x + 20$

29. $0.8(t - 2) = 0.5(9t + 10)$

30. $0.3(2n - 5) = 11 - 0.65n$

31. $0.92 + 0.9(x - 0.3) = 2x - 5.95$

32. $0.5(3x + 0.7) = 20.6$

For Problems 33–44, solve each formula for the indicated variable.

33. $P = 2l + 2w$ for w (Perimeter of a rectangle)

34. $V = \dfrac{1}{3}Bh$ for B (Volume of a pyramid)

35. $A = 2\pi r^2 + 2\pi rh$ for h (Surface area of a right circular cylinder)

36. $A = \dfrac{1}{2}h(b_1 + b_2)$ for h (Area of a trapezoid)

37. $C = \dfrac{5}{9}(F - 32)$ for F (Fahrenheit to Celsius)

38. $F = \dfrac{9}{5}C + 32$ for C (Celsius to Fahrenheit)

39. $V = C\left(1 - \dfrac{T}{N}\right)$ for T (Linear depreciation)

40. $V = C\left(1 - \dfrac{T}{N}\right)$ for N (Linear depreciation)

41. $I = kl(T - t)$ for T (Expansion allowance in highway construction)

42. $S = \dfrac{CRD}{12d}$ for d (Cutting speed of a circular saw)

43. $\dfrac{1}{R_n} = \dfrac{1}{R_1} + \dfrac{1}{R_2}$ for R_n (Resistance in parallel circuit design)

44. $f = \dfrac{1}{\dfrac{1}{a} + \dfrac{1}{b}}$ for b (Focal length of a camera lens)

For Problems 45–72, set up an equation and solve each problem.

45. The sum of two numbers is 98. If the larger is divided by the smaller, the quotient is 4 and the remainder is 13. Find the numbers.

46. One number is 100 larger than another number. If the larger number is divided by the smaller, the quotient is 15 and the remainder is 2. Find the numbers.

47. Working as a waiter, Tom made $157.50 in tips. Assuming that every customer tipped 15% of the cost of the meal, find the cost of all the meals Tom served.

48. A realtor who is paid 7% of the selling price in commission recently received $10,794 in commission on the sale of a property. What was the selling price of the property?

49. A sum of $2250 is to be divided between two people in the ratio of 2 to 3. How much does each person receive?

50. One type of motor requires a mixture of oil and gasoline in a ratio of 1 to 15 (that is, 1 part of oil to 15 parts of gasoline). How many liters of each are contained in a 20-liter mixture?

51. The ratio of students to teaching faculty in a certain high school is 20 to 1. If the total number of students and faculty is 777, find the number of each.

52. The ratio of the weight of sodium to that of chlorine in common table salt is 5 to 3. Find the amount of each element in a salt compound weighing 200 pounds.

53. Gary bought a coat at a 20% discount sale for $52. What was the original price of the coat?

54. Roya bought a pair of slacks at a 30% discount sale for $33.60. What was the original price of the slacks?

55. After a 7% increase in salary, Laurie makes $1016.50 per month. How much did she earn per month before the increase?

56. Russ bought a car for $11,025, including 5% sales tax. What was the selling price of the car without the tax?

57. A retailer has some shoes that cost $28 per pair. At what price should they be sold to obtain a profit of 15% of the cost?

58. If a head of lettuce costs a retailer $.40, at what price should it be sold to make a profit of 45% of the cost?

59. Karla sold a bicycle for $97.50. This selling price represented a 30% profit for her, based on what she had originally paid for the bike. Find Karla's original cost for the bicycle.

60. If a ring costs a jeweler $250, at what price should it be sold to make a profit of 60% of the selling price?

61. A retailer has some skirts that cost $18 each. She wants to sell them at a profit of 40% of the selling price. What price should she charge for the skirts?

62. Suppose that an item costs a retailer $50. How much more profit could be gained by fixing a 50% profit based on selling price rather than a 50% profit based on cost?

63. Derek has some nickels and dimes worth $3.60. The number of dimes is one more than twice the number of nickels. How many nickels and dimes does he have?

64. Robin has a collection of nickels, dimes, and quarters worth $38.50. She has 10 more dimes than nickels and twice as many quarters as dimes. How many coins of each kind does she have?

65. A collection of 70 coins consisting of dimes, quarters, and half-dollars has a value of $17.75. There are three times as many quarters as dimes. Find the number of each kind of coin.

66. A certain amount of money is invested at 8% per year, and $1500 more than that amount is invested at 9% per year. The annual interest from the 9% investment exceeds the annual interest from the 8% investment by $160. How much is invested at each rate?

67. A total of $5500 was invested, part of it at 9% per year and the remainder at 10% per year. If the total yearly interest amounted to $530, how much was invested at each rate?

68. A sum of $3500 is split between two investments, one paying 9% yearly interest and the other 11%. If the return on the 11% investment exceeds that on the 9% investment by $85 the first year, how much is invested at each rate?

69. Celia has invested $2500 at 11% yearly interest. How much must she invest at 12% so that the interest from both investments totals $695 after a year?

70. The length of a rectangle is 2 inches less than three times its width. If the perimeter of the rectangle is 108 inches, find its length and width.

71. The length of a rectangle is 4 centimeters more than its width. If the width is increased by 2 centimeters and the length is increased by 3 centimeters, a new rectangle is formed that has an area of 44 square centimeters more than the area of the original rectangle. Find the dimensions of the original rectangle.

72. The length of a picture without its border is 7 inches less than twice its width. If the border is 1 inch wide and its area is 62 square inches, what are the dimensions of the picture alone?

THOUGHTS INTO WORDS

73. Give a step-by-step description of how you would solve the formula $F = \dfrac{9}{5}C + 32$ for C.

74. What does the phrase "declare a variable" mean in the steps involved in solving a word problem?

75. Why must potential answers to word problems be checked back into the original statement of the problem?

76. From a consumer's viewpoint, would you prefer that retailers figure their profit on the basis of the cost or the selling price? Explain your answer.

77. Some people multiply by 2 and add 30 to estimate the change from a Celsius reading to a Fahrenheit reading. Why does this give an estimate? How good is the estimate?

Further Investigations

78. Is a 10% discount followed by a 20% discount equal to a 30% discount? Defend your answer.

79. Is a 10% discount followed by a 30% discount the same as a 30% discount followed by a 10% discount? Justify your answer.

80. A retailer buys an item for $90, resells it for $100, and claims that he is making only a 10% profit. Is his claim correct?

81. The following formula can be used to determine the selling price of an item when the profit is based on a percent of the selling price.

$$\dfrac{\text{Selling}}{\text{price}} = \dfrac{\text{Cost}}{100\% - \text{Percent of profit}}$$

Show how this formula is developed.

82. Use the formula from Problem 81 to determine the selling price of each of the following items. The given percent of profit is based on the selling price. Be sure to check each answer.

a. $.45 can of soup;   20% profit

b. $2.85 jar of coffee creamer;   25% profit

c. $.40 head of lettuce;   70% profit

d. $400 TV set;   45% profit

e. $18,000 car;   35% profit

1.3   QUADRATIC EQUATIONS

A **quadratic equation** in the variable x is defined as any equation that can be written in the form

$$ax^2 + bx + c = 0$$

where a, b, and c are real numbers and $a \neq 0$. The form $ax^2 + bx + c = 0$ is called the **standard form** of a quadratic equation. The choice of x for the variable is arbitrary. An equation such as $3t^2 + 5t - 4 = 0$ is a quadratic equation in the variable t.

Quadratic equations such as $x^2 + 2x - 15 = 0$, where the polynomial is factorable, can be solved by applying the following property: $ab = 0$ **if and only if** $a = 0$ **or** $b = 0$. Our work might take on the following format.

$$x^2 + 2x - 15 = 0$$
$$(x + 5)(x - 3) = 0$$

$$x + 5 = 0 \qquad \text{or} \qquad x - 3 = 0$$
$$x = -5 \qquad \text{or} \qquad x = 3$$

The solution set for this equation is $\{-5, 3\}$.

Let's consider another example of this type.

EXAMPLE 1

Solve the equation $n = -6n^2 + 12$.

Solution

$$n = -6n^2 + 12$$
$$6n^2 + n - 12 = 0$$
$$(3n - 4)(2n + 3) = 0$$
$$3n - 4 = 0 \qquad \text{or} \qquad 2n + 3 = 0$$
$$3n = 4 \qquad \text{or} \qquad 2n = -3$$
$$n = \frac{4}{3} \qquad \text{or} \qquad n = -\frac{3}{2}$$

The solution set is $\left\{ -\frac{3}{2}, \frac{4}{3} \right\}$. ■

Now suppose that we want to solve $x^2 = k$, where k is any real number. We can proceed as follows.

$$x^2 = k$$
$$x^2 - k = 0$$
$$\left(x + \sqrt{k} \right)\left(x - \sqrt{k} \right) = 0$$
$$x + \sqrt{k} = 0 \qquad \text{or} \qquad x - \sqrt{k} = 0$$
$$x = -\sqrt{k} \qquad \text{or} \qquad x = \sqrt{k}$$

Thus we can state the following property for any real number k.

PROPERTY 1.2

The solution set of $x^2 = k$ is $\left\{-\sqrt{k}, \sqrt{k}\right\}$, which can also be written $\left\{\pm\sqrt{k}\right\}$.

Property 1.2, along with our knowledge of the square root, makes it very easy to solve quadratic equations of the form $x^2 = k$.

EXAMPLE 2

Solve each of the following.

a. $x^2 = 72$ **b.** $(3n - 1)^2 = 26$ **c.** $(y + 2)^2 = -24$

Solutions

a. $x^2 = 72$

$\quad x = \pm\sqrt{72}$

$\quad x = \pm 6\sqrt{2}$

The solution set is $\left\{\pm 6\sqrt{2}\right\}$.

b. $(3n - 1)^2 = 26$

$\quad 3n - 1 = \pm\sqrt{26}$

$\quad 3n - 1 = \sqrt{26}$ or $3n - 1 = -\sqrt{26}$

$\quad\quad 3n = 1 + \sqrt{26}$ or $3n = 1 - \sqrt{26}$

$\quad\quad n = \dfrac{1 + \sqrt{26}}{3}$ or $n = \dfrac{1 - \sqrt{26}}{3}$

The solution set is $\left\{\dfrac{1 \pm \sqrt{26}}{3}\right\}$.

c. $(y + 2)^2 = -24$

$\quad y + 2 = \pm\sqrt{-24}$

$\quad y + 2 = \pm 2i\sqrt{6}$.   Remember that $\sqrt{-24} = i\sqrt{24} = i\sqrt{4}\sqrt{6} = 2i\sqrt{6}$.

$\quad y + 2 = 2i\sqrt{6}$ or $y + 2 = -2i\sqrt{6}$

$\quad\quad y = -2 + 2i\sqrt{6}$ or $y = -2 - 2i\sqrt{6}$

The solution set is $\left\{-2 \pm 2i\sqrt{6}\right\}$.   ■

Completing the Square

A factoring technique we reviewed in Chapter 0 relied on recognizing *perfect-square trinomials*. In each of the following examples, the perfect-square trinomial on the right side of the identity is the result of squaring the binomial on the left side.

$$(x + 5)^2 = x^2 + 10x + 25 \qquad (x - 7)^2 = x^2 - 14x + 49$$
$$(x + 9)^2 = x^2 + 18x + 81 \qquad (x - 12)^2 = x^2 - 24x + 144$$

Note that in each of the square trinomials, the constant term is equal to the square of one-half of the coefficient of the *x*-term. This relationship allows us to *form* a perfect-square trinomial by adding a proper constant term. For example, suppose that we want to form a perfect-square trinomial from $x^2 + 8x$. Because $\frac{1}{2}(8) = 4$ and $4^2 = 16$, the perfect-square trinomial is $x^2 + 8x + 16$. Now let's use this idea to solve a quadratic equation.

EXAMPLE 3

Solve $x^2 + 8x - 2 = 0$.

Solution

$$x^2 + 8x - 2 = 0$$
$$x^2 + 8x = 2$$
$$x^2 + 8x + 16 = 2 + 16 \qquad \text{We added 16 to the left side to form a perfect-square trinomial. Thus 16 has to be added to the right side.}$$

$$(x + 4)^2 = 18$$
$$x + 4 = \pm\sqrt{18}$$
$$x + 4 = \pm 3\sqrt{2}$$
$$x + 4 = 3\sqrt{2} \qquad \text{or} \qquad x + 4 = -3\sqrt{2}$$
$$x = -4 + 3\sqrt{2} \qquad \text{or} \qquad x = -4 - 3\sqrt{2}$$

The solution set is $\left\{-4 \pm 3\sqrt{2}\right\}$. ∎

We have been using a relationship for a perfect-square trinomial that states, *The constant term is equal to the square of one-half of the coefficient of the x-term.* This relationship holds only if the coefficient of x^2 is 1. Thus we need to make a slight adjustment when we are solving quadratic equations that have a coefficient of x^2 other than 1. The next example shows how to make this adjustment.

EXAMPLE 4

Solve $2x^2 + 6x - 3 = 0$.

Solution

$$2x^2 + 6x - 3 = 0$$
$$2x^2 + 6x = 3$$
$$x^2 + 3x = \frac{3}{2} \qquad \text{Multiply both sides by } \frac{1}{2}.$$
$$x^2 + 3x + \frac{9}{4} = \frac{3}{2} + \frac{9}{4} \qquad \text{Add } \frac{9}{4} \text{ to both sides.}$$

$$\left(x + \frac{3}{2}\right)^2 = \frac{15}{4}$$

$$x + \frac{3}{2} = \pm \frac{\sqrt{15}}{2}$$

$$x + \frac{3}{2} = \frac{\sqrt{15}}{2} \qquad \text{or} \qquad x + \frac{3}{2} = -\frac{\sqrt{15}}{2}$$

$$x = -\frac{3}{2} + \frac{\sqrt{15}}{2} \qquad \text{or} \qquad x = -\frac{3}{2} - \frac{\sqrt{15}}{2}$$

$$x = \frac{-3 + \sqrt{15}}{2} \qquad \text{or} \qquad x = \frac{-3 - \sqrt{15}}{2}$$

The solution set is $\left\{ \dfrac{-3 \pm \sqrt{15}}{2} \right\}$. ∎

Quadratic Formula

The process used in Examples 3 and 4 is called **completing the square**. It can be used to solve *any* quadratic equation. If we use this process of completing the square to solve the general quadratic equation $ax^2 + bx + c = 0$, we obtain a formula known as the **quadratic formula**. The details are as follows.

$$ax^2 + bx + c = 0, \qquad a \neq 0$$

$$ax^2 + bx = -c$$

$$x^2 + \frac{b}{a}x = -\frac{c}{a} \qquad \qquad \text{Multiply both sides by } \frac{1}{a}.$$

$$x^2 + \frac{b}{a}x + \frac{b^2}{4a^2} = -\frac{c}{a} + \frac{b^2}{4a^2} \qquad \qquad \begin{array}{l}\text{Complete the square by} \\ \text{adding } \dfrac{b^2}{4a^2} \text{ to both sides.}\end{array}$$

$$\left(x + \frac{b}{2a}\right)^2 = \frac{b^2 - 4ac}{4a^2} \qquad \qquad \begin{array}{l}\text{Combine the right side into} \\ \text{a single fraction.}\end{array}$$

$$x + \frac{b}{2a} = \pm \sqrt{\frac{b^2 - 4ac}{4a^2}}$$

$$x + \frac{b}{2a} = \pm \frac{\sqrt{b^2 - 4ac}}{\sqrt{4a^2}}$$

$$x + \frac{b}{2a} = \pm \frac{\sqrt{b^2 - 4ac}}{2a} \qquad \qquad \begin{array}{l}\sqrt{4a^2} = |2a| \text{ but } 2a \text{ can be used} \\ \text{because of the use of } \pm.\end{array}$$

$$x + \frac{b}{2a} = \frac{\sqrt{b^2 - 4ac}}{2a} \qquad \text{or} \qquad x + \frac{b}{2a} = -\frac{\sqrt{b^2 - 4ac}}{2a}$$

$$x = -\frac{b}{2a} + \frac{\sqrt{b^2 - 4ac}}{2a} \qquad \text{or} \qquad x = -\frac{b}{2a} - \frac{\sqrt{b^2 - 4ac}}{2a}$$

$$x = \frac{-b + \sqrt{b^2 - 4ac}}{2a} \qquad \text{or} \qquad x = \frac{-b - \sqrt{b^2 - 4ac}}{2a}$$

The quadratic formula can be stated as follows.

Quadratic Formula

If $a \neq 0$, then the solutions (roots) of the equation $ax^2 + bx + c = 0$ are given by

$$x = \frac{-b \pm \sqrt{b^2 - 4ac}}{2a}$$

This formula can be used to solve any quadratic equation by expressing the equation in the standard form, $ax^2 + bx + c = 0$, and substituting the values for a, b, and c into the formula. Let's consider some examples.

EXAMPLE 5

Solve each of the following by using the quadratic formula.

a. $3x^2 - x - 5 = 0$ **b.** $25n^2 - 30n = -9$ **c.** $t^2 - 2t + 4 = 0$

Solutions

a. We need to think of $3x^2 - x - 5 = 0$ as $3x^2 + (-x) + (-5) = 0$; thus $a = 3$, $b = -1$, and $c = -5$. We then substitute these values into the quadratic formula and simplify.

$$x = \frac{-b \pm \sqrt{b^2 - 4ac}}{2a}$$

$$x = \frac{-(-1) \pm \sqrt{(-1)^2 - 4(3)(-5)}}{2(3)}$$

$$= \frac{1 \pm \sqrt{61}}{6}$$

The solution set is $\left\{\dfrac{1 \pm \sqrt{61}}{6}\right\}$.

b. The quadratic formula is usually stated in terms of the variable x, but again the choice of variable is arbitrary. The given equation, $25n^2 - 30n = -9$, needs to be changed to standard form: $25n^2 - 30n + 9 = 0$. From this we obtain $a = 25$, $b = -30$, and $c = 9$. Now we use the formula.

$$n = \frac{-(-30) \pm \sqrt{(-30)^2 - 4(25)(9)}}{2(25)}$$

$$= \frac{30 \pm \sqrt{0}}{50}$$

$$= \frac{3}{5}$$

The solution set is $\left\{\dfrac{3}{5}\right\}$.

c. We substitute $a = 1$, $b = -2$, and $c = 4$ into the quadratic formula.

$$t = \frac{-(-2) \pm \sqrt{(-2)^2 - 4(1)(4)}}{2(1)}$$

$$= \frac{2 \pm \sqrt{-12}}{2}$$

$$= \frac{2 \pm 2i\sqrt{3}}{2}$$

$$= \frac{\cancel{2}(1 \pm i\sqrt{3})}{\cancel{2}}$$

The solution set is $\{1 \pm i\sqrt{3}\}$.   ■

From Example 5 we see that different kinds of solutions are obtained depending upon the radicand ($b^2 - 4ac$) inside the radical in the quadratic formula. For this reason, the number $b^2 - 4ac$ is called the **discriminant** of the quadratic equation. It can be used to determine the nature of the solutions as follows.

1. If $b^2 - 4ac > 0$, the equation has two unequal real solutions.

2. If $b^2 - 4ac = 0$, the equation has one real solution.

3. If $b^2 - 4ac < 0$, the equation has two complex but nonreal solutions.

The following examples illustrate each of these situations. (You may want to solve the equations completely to verify our conclusions.)

EQUATION	DISCRIMINANT	NATURE OF SOLUTIONS
$4x^2 - 7x - 1 = 0$	$\begin{aligned} b^2 - 4ac &= (-7)^2 - 4(4)(-1) \\ &= 49 + 16 \\ &= 65 \end{aligned}$	Two real solutions
$4x^2 + 12x + 9 = 0$	$\begin{aligned} b^2 - 4ac &= (12)^2 - 4(4)(9) \\ &= 144 - 144 \\ &= 0 \end{aligned}$	One real solution
$5x^2 + 2x + 1 = 0$	$\begin{aligned} b^2 - 4ac &= (2)^2 - 4(5)(1) \\ &= 4 - 20 \\ &= -16 \end{aligned}$	Two complex solutions

There is another useful relationship involving the solutions of a quadratic equation of the form $ax^2 + bx + c = 0$ and the numbers a, b, and c. Suppose that we let x_1 and x_2 be the two roots of the equation. (If $b^2 - 4ac = 0$, then $x_1 = x_2$ and the one-solution situation can be thought of as two equal solutions.) By the quadratic formula we have

$$x_1 = \frac{-b + \sqrt{b^2 - 4ac}}{2a} \quad \text{and} \quad x_2 = \frac{-b - \sqrt{b^2 - 4ac}}{2a}$$

Now let's consider both the sum and the product of the two roots.

Sum
$$x_1 + x_2 = \frac{-b + \sqrt{b^2 - 4ac}}{2a} + \frac{-b - \sqrt{b^2 - 4ac}}{2a}$$

$$= \frac{-2b}{2a} = -\frac{b}{a}$$

Product
$$(x_1)(x_2) = \left(\frac{-b + \sqrt{b^2 - 4ac}}{2a}\right)\left(\frac{-b - \sqrt{b^2 - 4ac}}{2a}\right)$$

$$= \frac{b^2 - (b^2 - 4ac)}{4a^2}$$

$$= \frac{b^2 - b^2 + 4ac}{4a^2}$$

$$= \frac{4ac}{4a^2} = \frac{c}{a}$$

These relationships provide another way of checking potential solutions when solving quadratic equations. We will illustrate this point in a moment.

Solving Quadratic Equations: Which Method?

Which method should be used to solve a particular quadratic equation? There is no definite answer to that question; it depends on the type of equation and perhaps on your personal preference. However, it is to your advantage to be able to use all three techniques and to know the strengths and weaknesses of each. In the next two examples, we will give our reasons for choosing a specific technique.

Solve $x^2 - 4x - 192 = 0$.

EXAMPLE 6

Solution

The size of the constant term makes the factoring approach a little cumbersome for this problem. However, because the coefficient of the x^2-term is 1 and the coefficient of the x-term is even, the method for completing the square should work effectively.

$$x^2 - 4x - 192 = 0$$
$$x^2 - 4x = 192$$
$$x^2 - 4x + 4 = 192 + 4$$
$$(x - 2)^2 = 196$$
$$x - 2 = \pm\sqrt{196}$$
$$x - 2 = \pm 14$$

$$x - 2 = 14 \quad \text{or} \quad x - 2 = -14$$
$$x = 16 \quad \text{or} \quad x = -12$$

Check

Sum of roots $16 + (-12) = 4$ and $-\dfrac{b}{a} = -\left(\dfrac{-4}{1}\right) = 4$

Product of roots $(16)(-12) = -192$ and $\dfrac{c}{a} = \dfrac{-192}{1} = -192.$

The solution set is $\{-12, 16\}$.   ■

EXAMPLE 7

Solve $2x^2 - x + 3 = 0$.

Solution

It would be reasonable first to try factoring the polynomial $2x^2 - x + 3$. Unfortunately, it is not factorable using integers; thus we must solve the equation by completing the square or by using the quadratic formula. The coefficient of the x^2 term is not 1, so let's avoid completing the square and use the formula instead.

$$x = \frac{-b \pm \sqrt{b^2 - 4ac}}{2a}$$
$$= \frac{-(-1) \pm \sqrt{(-1)^2 - 4(2)(3)}}{2(2)}$$
$$= \frac{1 \pm \sqrt{-23}}{4}$$
$$= \frac{1 \pm i\sqrt{23}}{4}$$

Check

Sum of roots $\dfrac{1 + i\sqrt{23}}{4} + \dfrac{1 - i\sqrt{23}}{4} = \dfrac{2}{4} = \dfrac{1}{2}$ and

$$-\frac{b}{a} = -\frac{-1}{2} = \frac{1}{2}$$

Product of roots $\left(\dfrac{1 + i\sqrt{23}}{4}\right)\left(\dfrac{1 - i\sqrt{23}}{4}\right) = \dfrac{1 - 23i^2}{16}$

$$= \frac{1 + 23}{16} = \frac{24}{16} = \frac{3}{2} \quad \text{and} \quad \frac{c}{a} = \frac{3}{2}$$

The solution set is $\left\{\dfrac{1 \pm i\sqrt{23}}{4}\right\}$.   ■

The ability to solve quadratic equations enables us to solve more word problems. Some of these problems involve geometric formulas and relationships. We have included a brief summary of some basic geometric formulas in the back sheets of this text.

PROBLEM 1

One leg of a right triangle is 7 meters longer than the other leg. If the length of the hypotenuse is 17 meters, find the length of each leg.

Solution

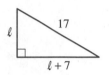

FIGURE 1.3

Look at Figure 1.3. Let l represent the length of one leg; then $l + 7$ represents the length of the other leg. Using the Pythagorean theorem as a guideline, we can set up and solve a quadratic equation.

$$l^2 + (l + 7)^2 = 17^2$$
$$l^2 + l^2 + 14l + 49 = 289$$
$$2l^2 + 14l - 240 = 0$$
$$l^2 + 7l - 120 = 0$$
$$(l + 15)(l - 8) = 0$$
$$l + 15 = 0 \quad \text{or} \quad l - 8 = 0$$
$$l = -15 \quad \text{or} \quad l = 8$$

The negative solution must be disregarded (because l is a length), so the length of one leg is 8 meters. The other leg, represented by $l + 7$, is $8 + 7 = 15$ meters long. ∎

PROBLEM SET 1.3

For Problems 1–16, solve each equation by factoring or by using the property, *If $x^2 = k$ then $x = \pm\sqrt{k}$.*

1. $x^2 - 3x - 28 = 0$

2. $x^2 - 4x - 12 = 0$

3. $3x^2 + 5x - 12 = 0$

4. $2x^2 - 13x + 6 = 0$

5. $2x^2 - 3x = 0$

6. $3n^2 = 3n$

7. $9y^2 = 12$

8. $(4n - 1)^2 = 16$

9. $(2n + 1)^2 = 20$

10. $3(4x - 1)^2 + 1 = 16$

11. $15n^2 + 19n - 10 = 0$

12. $6t^2 + 23t - 4 = 0$

13. $(x - 2)^2 = -4$

14. $24x^2 + 23x - 12 = 0$

15. $10y^2 + 33y - 7 = 0$

16. $(x - 3)^2 = -9$

For Problems 17–30, use the method of completing the square to solve each equation. Check your solutions by using the sum-and-product-of-roots relationships.

17. $x^2 - 10x + 24 = 0$

18. $x^2 + x - 20 = 0$

19. $n^2 + 10n - 2 = 0$

20. $n^2 + 6n - 1 = 0$

21. $y^2 - 3y = -1$

22. $y^2 + 5y = -2$

23. $x^2 + 4x + 6 = 0$

24. $x^2 - 6x + 21 = 0$

25. $2t^2 + 12t - 5 = 0$

26. $3p^2 + 12p - 2 = 0$

27. $x^2 - 2x - 288 = 0$

28. $x^2 + 4x - 221 = 0$

29. $3n^2 + 5n - 1 = 0$

30. $2n^2 + n - 4 = 0$

For Problems 31–44, use the quadratic formula to solve each equation. Check your solutions by using the sum-and-product-of-roots relationships.

31. $n^2 - 3n - 54 = 0$

32. $y^2 + 13y + 22 = 0$

33. $3x^2 + 16x = -5$

34. $10x^2 - 29x - 21 = 0$

35. $y^2 - 2y - 4 = 0$

36. $n^2 - 6n - 3 = 0$

37. $2a^2 - 6a + 1 = 0$

38. $2x^2 + 3x - 1 = 0$

39. $n^2 - 3n = -7$

40. $n^2 - 5n = -8$

41. $x^2 + 4 = 8x$

42. $x^2 + 31 = -14x$

43. $4x^2 - 4x + 1 = 0$

44. $x^2 + 24 = 0$

For Problems 45–60, solve each quadratic equation by using the method that seems most appropriate to you.

45. $8x^2 + 10x - 3 = 0$

46. $18x^2 - 39x + 20 = 0$

47. $x^2 + 2x = 168$

48. $x^2 + 28x = -187$

49. $2t^2 - 3t + 7 = 0$

50. $3n^2 - 2n + 5 = 0$

51. $(3n - 1)^2 + 2 = 18$

52. $20y^2 + 17y - 10 = 0$

53. $4y^2 + 4y - 1 = 0$

54. $(5n + 2)^2 + 1 = -27$

55. $x^2 - 16x + 14 = 0$

56. $x^2 - 18x + 15 = 0$

57. $t^2 + 20t = 25$

58. $n^2 - 18n = 9$

59. $5x^2 - 2x - 1 = 0$

60. $-x^2 + 11x - 18 = 0$

61. Find the discriminant of each of the following quadratic equations and determine whether the equation has (1) two complex but nonreal solutions, (2) one real solution, or (3) two unequal real solutions.

  a. $4x^2 + 20x + 25 = 0$

  b. $x^2 + 4x + 7 = 0$

  c. $x^2 - 18x + 81 = 0$

  d. $36x^2 - 31x + 3 = 0$

  e. $2x^2 + 5x + 7 = 0$

  f. $16x^2 = 40x - 25$

  g. $6x^2 - 4x - 7 = 0$

  h. $5x^2 - 2x - 4 = 0$

For Problems 62–77, set up a quadratic equation and solve each problem.

62. Find two consecutive even integers whose product is 528.

63. Find two consecutive whole numbers such that the sum of their squares is 265.

64. For a remodeling job, an architect suggested increasing the sides of a square patio by 3 feet per side. This made the area of the new patio 49 square feet. What was the area of the original patio?

65. A sailboat has a triangular sail with an area of 30 square feet. The height of the sail is 7 feet more than the length of the base of the sail. Find the height of the sail.

66. One leg of a right triangle is 4 inches longer than the other leg. If the length of the hypotenuse is 20 inches, find the length of each leg.

67. The sum of the lengths of the two legs of a right triangle is 34 meters. If the length of the hypotenuse is 26 meters, find the length of each leg.

68. The lengths of the three sides of a right triangle are consecutive even integers. Find the length of each side.

69. The perimeter of a rectangle is 44 inches and its area is 112 square inches. Find the length and width of the rectangle.

70. A page of a magazine contains 70 square inches of type. The height of the page is twice the width. If the margin around the type is 2 inches uniformly, what are the dimensions of the page?

71. The length of a rectangle is 4 meters more than twice its width. If the area of the rectangle is 126 square meters, find its length and width.

72. The length of one side of a triangle is 3 centimeters less than twice the length of the altitude to that side. If the area of the triangle is 52 square centimeters, find the length of the side and the length of the altitude to that side.

73. A rectangular plot of ground measuring 12 meters by 20 meters is surrounded by a sidewalk of uniform width. The area of the sidewalk is 68 square meters. Find the width of the walk.

74. A piece of wire 60 inches long is cut into two pieces and then each piece is bent into the shape of a square. If the sum of the areas of the two squares is 117 square inches, find the length of each piece of wire.

75. A rectangular piece of cardboard is 4 inches longer than it is wide. From each of its corners, a square piece 2 inches on a side is cut out. The flaps are then turned up to form an open box, which has a volume of 42 cubic inches. Find the length and width of the original piece of cardboard. See Figure 1.4.

76. The area of a rectangular region is 52 square feet. If the length of the rectangle is increased by 4 feet and the width by 2 feet, the area is increased by 50 square feet. Find the length and width of the original rectangular region.

77. The area of a circular region is numerically equal to four times the circumference of the circle. Find the length of a radius of the circle.

FIGURE 1.4

THOUGHTS INTO WORDS

78. Explain how you would solve $(x - 3)(x + 4) = 0$ and also how you would solve $(x - 3)(x + 4) = 8$.

79. Explain the process of completing the square to solve a quadratic equation.

80. Explain how to use the quadratic formula to solve $3x = x^2 - 2$.

81. Your friend states that the equation $-2x^2 + 4x - 1 = 0$ must be changed to $2x^2 - 4x + 1 = 0$ (by multiplying both sides by -1) before the quadratic formula can be applied. Is she right about this, and if not, how would you convince her?

Further Investigations

82. Solve each of the following equations for x.

a. $x^2 - 7kx = 0$ **b.** $x^2 = 25kx$

c. $x^2 - 3kx - 10k^2 = 0$ **d.** $6x^2 + kx - 2k^2 = 0$

e. $9x^2 - 6kx + k^2 = 0$ **f.** $k^2x^2 - kx - 6 = 0$

g. $x^2 + \sqrt{2}x - 3 = 0$ **h.** $x^2 - \sqrt{3}x + 5 = 0$

83. Solve each of the following for the indicated variable. (Assume that all letters represent positive numbers.)

a. $A = \pi r^2$ for r **b.** $E = c^2m - c^2m_0$ for c

c. $s = \frac{1}{2}gt^2$ for t **d.** $\frac{x^2}{a^2} + \frac{y^2}{b^2} = 1$ for x

e. $\frac{x^2}{a^2} - \frac{y^2}{b^2} = 1$ for y

f. $s = \frac{1}{2}gt^2 + V_0t$ for t

For Problems 84–86, use the discriminant to help solve each problem.

84. Determine k so that the solutions of $x^2 - 2x + k = 0$ are complex but nonreal.

85. Determine k so that $4x^2 - kx + 1 = 0$ has two equal real solutions.

86. Determine k so that $3x^2 - kx - 2 = 0$ has real solutions.

 87. The solution set for $x^2 - 4x - 37 = 0$ is $\{2 \pm \sqrt{41}\}$.

With a calculator, we found a rational approximation, to the nearest one-thousandth, for each of these solutions.

$$2 - \sqrt{41} = -4.403$$
$$2 + \sqrt{41} = 8.403$$

Thus the solution set is $\{-4.403, 8.403\}$, with answers rounded to the nearest one-thousandth.

Solve each of the following equations and express the solutions to the nearest one-thousandth.

a. $x^2 - 6x - 10 = 0$ *b.* $x^2 - 16x - 24 = 0$

c. $x^2 + 6x - 44 = 0$ *d.* $x^2 + 10x - 46 = 0$

e. $x^2 + 8x + 2 = 0$ *f.* $x^2 + 9x + 3 = 0$

g. $4x^2 - 6x + 1 = 0$ *h.* $5x^2 - 9x + 1 = 0$

i. $2x^2 - 11x - 5 = 0$ *j.* $3x^2 - 12x - 10 = 0$

1.4 APPLICATIONS OF LINEAR AND QUADRATIC EQUATIONS

Let's begin this section by considering two fractional equations, one that is equivalent to a linear equation and one that is equivalent to a quadratic equation.

EXAMPLE 1   Solve $\dfrac{3}{2x - 8} - \dfrac{x - 5}{x^2 - 2x - 8} = \dfrac{7}{x + 2}$.

Solution

$$\frac{3}{2x - 8} - \frac{x - 5}{x^2 - 2x - 8} = \frac{7}{x + 2}$$

$$\frac{3}{2(x - 4)} - \frac{x - 5}{(x - 4)(x + 2)} = \frac{7}{x + 2}, \qquad x \neq 4, x \neq -2$$

$$2(x - 4)(x + 2)\left(\frac{3}{2(x - 4)} - \frac{x - 5}{(x - 4)(x + 2)}\right) = 2(x - 4)(x + 2)\left(\frac{7}{x + 2}\right)$$

$$3(x + 2) - 2(x - 5) = 14(x - 4)$$

$$3x + 6 - 2x + 10 = 14x - 56$$

$$x + 16 = 14x - 56$$

$$72 = 13x$$

$$\frac{72}{13} = x$$

The solution set is $\left\{\dfrac{72}{13}\right\}$.   ■

In Example 1, notice that we did not indicate the restrictions until the denominators were expressed in factored form. It is usually easier to determine the necessary restrictions at that step.

EXAMPLE 2	Solve $\dfrac{3n}{n^2 + n - 6} + \dfrac{2}{n^2 + 4n + 3} = \dfrac{n}{n^2 - n - 2}$.

Solution

$$\frac{3n}{n^2 + n - 6} + \frac{2}{n^2 + 4n + 3} = \frac{n}{n^2 - n - 2}$$

$$\frac{3n}{(n + 3)(n - 2)} + \frac{2}{(n + 3)(n + 1)} = \frac{n}{(n - 2)(n + 1)}, \qquad n \neq -3, n \neq 2, n \neq -1$$

$$(n + 3)(n - 2)(n + 1)\left(\frac{3n}{(n + 3)(n - 2)} + \frac{2}{(n + 3)(n + 1)}\right) = (n + 3)(n - 2)(n + 1)\left(\frac{n}{(n - 2)(n + 1)}\right)$$

$$3n(n + 1) + 2(n - 2) = n(n + 3)$$

$$3n^2 + 3n + 2n - 4 = n^2 + 3n$$

$$3n^2 + 5n - 4 = n^2 + 3n$$

$$2n^2 + 2n - 4 = 0$$

$$n^2 + n - 2 = 0$$

$$(n + 2)(n - 1) = 0$$

$$n + 2 = 0 \qquad \text{or} \qquad n - 1 = 0$$

$$n = -2 \qquad \text{or} \qquad n = 1$$

The solution set is $\{-2, 1\}$. ■

More on Problem Solving

Before tackling a variety of applications of linear and quadratic equations, let's restate some suggestions made earlier in this chapter for solving word problems.

Suggestions for Solving Word Problems

1. Read the problem carefully, making certain that you understand the meanings of all the words. Be especially alert for any technical terms used in the statement of the problem.

2. Read the problem a second time (perhaps even a third time) to get an overview of the situation being described and to determine the known facts, as well as what is to be found.

3. Sketch any figure, diagram, or chart that might be helpful in analyzing the problem.

4. Choose a meaningful variable to represent an unknown quantity in the problem (for example, l for the length of a rectangle), and represent any other unknowns in terms of that variable.

5. Look for a *guideline* that can be used in setting up an equation. A guideline might be a formula, such as $A = lw$, or a relationship, such as *the fractional part of the job done by Bill plus the fractional part of the job done by Mary equals the total job.*

> **6.** Form an equation containing the variable to translate the conditions of the guideline from English to algebra.
> **7.** Solve the equation and use the solution to determine all the facts requested in the problem.
> **8.** Check all answers back into the *original statement of the problem.*

Suggestion 5 is a key part of the analysis of a problem. A formula to be used as a guideline may or may not be explicitly stated in the problem. Likewise, a relationship to be used as a guideline may not be actually stated in the problem but must be determined from what is stated. Let's consider some examples.

PROBLEM 1

A room contains 120 chairs. The number of chairs per row is one less than twice the number of rows. Find the number of rows and the number of chairs per row.

Solution

Let r represent the number of rows. Then $2r - 1$ represents the number of chairs per row. The statement of the problem implies a formation of chairs such that the total number of chairs is equal to the number of rows times the number of chairs per row. This gives us an equation.

Number of rows $\times$ Number of chairs per row = Total number of chairs

$$r \qquad \times \qquad (2r - 1) \qquad = \qquad 120$$

We solve this equation by the factorization method.

$$2r^2 - r = 120$$
$$2r^2 - r - 120 = 0$$
$$(2r + 15)(r - 8) = 0$$
$$2r + 15 = 0 \qquad \text{or} \qquad r - 8 = 0$$
$$2r = -15 \qquad \text{or} \qquad r = 8$$
$$r = -\frac{15}{2} \qquad \text{or} \qquad r = 8$$

The solution $-\dfrac{15}{2}$ must be disregarded, so there are 8 rows and $2(8) - 1 = 15$ chairs per row. ■

The basic relationship *distance equals rate times time* is used to help solve a variety of *uniform-motion problems.* This relationship may be expressed by any one of the following equations.

$$d = rt \qquad r = \frac{d}{t} \qquad t = \frac{d}{r}$$

| PROBLEM 2 | Domenica and Javier start from the same location at the same time and ride their bicycles in opposite directions for 4 hours, at which time they are 140 miles apart. If Domenica rides 3 miles per hour faster than Javier, find the rate of each rider. |

Solution

Let r represent Javier's rate; then $r + 3$ represents Domenica's rate. A sketch such as that in Figure 1.5 may help in our analysis. The fact that the total distance is 140 miles can be used as a guideline. We use the $d = rt$ equation.

Domenica riding
at $r + 3$ mph
for 4 hours

Javier riding
at r mph
for 4 hours

Total of 140 miles

F I G U R E   1 . 5

Distance Domenica rides + Distance Javier rides = 140

$$4(r + 3) \quad + \quad 4r \quad = 140$$

Solving this equation yields Javier's speed.

$$4r + 12 + 4r = 140$$
$$8r = 128$$
$$r = 16$$

Thus Javier rides at 16 miles per hour and Domenica at $16 + 3 = 19$ miles per hour.

■

Some people find it helpful to use a chart or a table to organize the known and unknown facts in uniform-motion problems. Let's illustrate this approach.

| PROBLEM 3 | Riding on a moped, Sue takes 2 hours less to travel 60 miles than Ann takes to travel 50 miles on a bicycle. Sue travels 10 miles per hour faster than Ann. Find the times and rates of both girls. |

Solution A

Let t represent Ann's time; then $t - 2$ represents Sue's time. We can record the information in a table as shown below. The fact that Sue travels 10 miles per hour faster than Ann can be used as a guideline.

	DISTANCE	TIME	$r = \dfrac{d}{t}$
Ann	50	t	$\dfrac{50}{t}$
Sue	60	$t - 2$	$\dfrac{60}{t - 2}$

$$\text{Sue's rate} = \text{Ann's rate} + 10$$

$$\frac{60}{t - 2} = \frac{50}{t} + 10$$

Solving this equation yields Ann's time.

$$t(t - 2)\left(\frac{60}{t - 2}\right) = t(t - 2)\left(\frac{50}{t} + 10\right), \qquad t \neq 0, t \neq 2$$

$$60t = 50(t - 2) + 10t(t - 2)$$
$$60t = 50t - 100 + 10t^2 - 20t$$
$$0 = 10t^2 - 30t - 100$$
$$0 = t^2 - 3t - 10$$
$$0 = (t - 5)(t + 2)$$
$$t - 5 = 0 \quad \text{or} \quad t + 2 = 0$$
$$t = 5 \quad \text{or} \quad t = -2$$

The solution -2 must be disregarded, because we're solving for time. Therefore, Ann rides for 5 hours at $\dfrac{50}{5} = 10$ miles per hour, and Sue rides for $5 - 2 = 3$ hours at $\dfrac{60}{3} = 20$ miles per hour.

Solution B

Let r represent Ann's rate; then $r + 10$ represents Sue's rate. Again, let's record the information in a table.

	DISTANCE	RATE	$t = \dfrac{d}{r}$
Ann	50	r	$\dfrac{50}{r}$
Sue	60	$r + 10$	$\dfrac{60}{r + 10}$

This time, let's use the fact that Sue's time is 2 hours less than Ann's time as a guideline.

Sue's time = Ann's time − 2

$$\frac{60}{r + 10} = \frac{50}{r} - 2$$

Solving this equation yields Ann's rate.

$$r(r + 10)\left(\frac{60}{r + 10}\right) = r(r + 10)\left(\frac{50}{r} - 2\right)$$
$$60r = (r + 10)(50) - 2r(r + 10)$$
$$60r = 50r + 500 - 2r^2 - 20r$$
$$2r^2 + 30r - 500 = 0$$
$$r^2 + 15r - 250 = 0$$
$$(r + 25)(r - 10) = 0$$
$$r + 25 = 0 \qquad \text{or} \qquad r - 10 = 0$$
$$r = -25 \qquad \text{or} \qquad r = 10$$

The solution -25 must be disregarded, because we are solving for a rate. Therefore, Ann rides at 10 miles per hour for $\dfrac{50}{10} = 5$ hours, and Sue rides at $10 + 10 = 20$ miles per hour for $\dfrac{60}{20} = 3$ hours. ■

Take a good look at both Solution A and Solution B for Problem 3. Both are reasonable approaches, but note that the approach in Solution A generates a quadratic equation that is a little easier to solve than the one generated in Solution B.

We might have expected this to happen, because the "times" in a motion problem are frequently smaller numbers than the "rates." Thus "thinking first" before "pushing the pencil" can make things a bit easier.

There are various applications commonly classified as *mixture problems.* Even though these problems arise in many different areas, essentially the same mathematical approach can be used to solve them. A general suggestion for solving mixture-type problems is to *work in terms of a pure substance.* We will illustrate what we mean by that statement.

PROBLEM 4

How many milliliters of pure acid must be added to 50 milliliters of a 40% acid solution to obtain a 50% acid solution?

Solution

Let a represent the number of milliliters of pure acid to be added. Thinking in terms of pure acid, we know that *the amount of pure acid to start with plus the amount of pure acid added equals the amount of pure acid in the final solution.* Let's use that as a guideline and set up an equation.

Pure acid to start with + Pure acid added = Pure acid in final solution

$$40\%(50) \quad + \quad a \quad = \quad 50\%(50 + a)$$

Solving this equation, we obtain the amount of acid we must add.

$$0.4(50) + a = 0.5(50 + a)$$
$$4(50) + 10a = 5(50 + a)$$
$$200 + 10a = 250 + 5a$$
$$5a = 50$$
$$a = 10$$

We need to add 10 milliliters of pure acid.   ■

There is another class of problems commonly referred to as *work problems,* or sometimes as *rate-time problems.* For example, if a certain machine produces 120 items in 10 minutes, then we say that it is working at a rate of $\frac{120}{10} = 12$ items per minute. Likewise, a person who can do a certain job in 5 hours is working at a rate of $\frac{1}{5}$ of the job per hour. In general, if Q is the quantity of something done in t units of time, then the rate, r, is given by $r = Q/t$. The rate is stated in terms of *so much quantity per unit of time.* The uniform-motion problems discussed earlier are a special kind of rate-time problem where the quantity is distance. Using tables to organize information (as we illustrated with the motion problems) is a convenient aid for rate-time problems in general. Let's consider some problems.

PROBLEM 5

Printing press A can produce 35 fliers per minute, and press B can produce 50 fliers per minute. Suppose that 2225 fliers are produced by first using press A alone for 15 minutes and then using presses A and B together until the job is done. How long does press B need to be used?

Solution

Let m represent the number of minutes that press B is used. Then $m + 15$ represents the number of minutes press A is used. The information in the problem can be organized in a table as shown below. Because the total quantity (the total number of fliers) is 2225, we can set up and solve the following equation.

	QUANTITY	TIME	RATE
A	$35(m + 15)$	$m + 15$	35
B	$50m$	m	50

$$35(m + 15) + 50m = 2225$$
$$35m + 525 + 50m = 2225$$
$$85m = 1700$$
$$m = 20$$

Therefore press B must be used for 20 minutes. ∎

PROBLEM 6

It takes Amy twice as long to deliver newspapers as it does Nancy. How long does it take each girl by herself if they can deliver the papers together in 40 minutes?

Solution

Let m represent the number of minutes that it takes Nancy by herself. Then $2m$ represents Amy's time by herself. Thus the information can be organized as shown below. (Note that the *quantity* is 1; there is one job to be done.)

	QUANTITY	TIME	RATE
Nancy	1	m	$\dfrac{1}{m}$
Amy	1	$2m$	$\dfrac{1}{2m}$

Because their combined rate is $\dfrac{1}{40}$, we can solve the following equation.

$$\frac{1}{m} + \frac{1}{2m} = \frac{1}{40}, \qquad m \neq 0$$

$$40m\left(\frac{1}{m} + \frac{1}{2m}\right) = 40m\left(\frac{1}{40}\right)$$

$$40 + 20 = m$$

$$60 = m$$

Therefore, Nancy can deliver the papers by herself in 60 minutes, and Amy can deliver them by herself in $2(60) = 120$ minutes. ∎

Our final examples of this section illustrate another approach that some people find works well for rate-time problems. The basic idea used in this approach involves representing the fractional parts of a job. For example, if a man can do a certain job in 7 hours, then at the end of 3 hours he has finished $\dfrac{3}{7}$ of the job. (Again, a constant rate of work is being assumed.) At the end of 5 hours he has finished $\dfrac{5}{7}$ of the job, and, in general, at the end of h hours he has finished $\dfrac{h}{7}$ of the job.

PROBLEM 7

Carlos can mow a lawn in 45 minutes and Felipe can mow the same lawn in 30 minutes. How long would it take the two of them working together to mow the lawn?

Solution

(Before you read any further, *estimate* an answer for this problem. Remember that Felipe can mow the lawn by himself in 30 minutes.) Let m represent the number of minutes that it takes them working together. Then we can set up the following equation.

Fractional part of the lawn that Carlos will mow		Fractional part of the lawn that Felipe will mow		The whole lawn
↓		↓		↓
$\dfrac{m}{45}$	$+$	$\dfrac{m}{30}$	$=$	1

Solving this equation yields the time that it will take when they work together.

$$90\left(\frac{m}{45} + \frac{m}{30}\right) = 90(1)$$
$$2m + 3m = 90$$
$$5m = 90$$
$$m = 18$$

It should take them 18 minutes to mow the lawn when they work together. ■

PROBLEM 8

Walt can mow a lawn in 50 minutes, and his son Mike can mow the same lawn in 40 minutes. One day Mike started to mow the lawn by himself and worked for 10 minutes. Then Walt joined him with another mower and they finished the lawn. How long did it take them to finish mowing the lawn after Walt started to help?

Solution

Let m represent the number of minutes that it takes for them to finish the mowing after Walt starts to help. Because Mike has been mowing for 10 minutes, he has done $\frac{10}{40}$, or $\frac{1}{4}$, of the lawn when Walt starts. Thus there is $\frac{3}{4}$ of the lawn yet to mow. The following guideline can be used to set up an equation.

Fractional part of the remaining $3/4$ of the lawn that Mike will mow in m minutes	+	Fractional part of the remaining $3/4$ of the lawn that Walt will mow in m minutes	=	$\dfrac{3}{4}$
↓		↓		↓
$\dfrac{m}{40}$	+	$\dfrac{m}{50}$	=	$\dfrac{3}{4}$

Solving this equation yields the time they mow the lawn together.

$$200\left(\frac{m}{40} + \frac{m}{50}\right) = 200\left(\frac{3}{4}\right)$$
$$5m + 4m = 150$$
$$9m = 150$$
$$m = \frac{150}{9} = \frac{50}{3}$$

They should finish the lawn in $16\frac{2}{3}$ minutes. ■

As you tackle word problems throughout this text, keep in mind that our primary objective is to expand your repertoire of problem-solving techniques. In the examples, we are sharing some of our ideas for solving problems, but don't hesitate to use your own ingenuity. Furthermore, don't become discouraged—all of us have difficulty with some problems. Give it your best shot.

PROBLEM SET 1.4

For Problems 1–20, solve each equation.

1. $\dfrac{x}{2x-8} + \dfrac{16}{x^2-16} = \dfrac{1}{2}$

2. $\dfrac{3}{n-5} - \dfrac{2}{2n+1} = \dfrac{n+3}{2n^2-9n-5}$

3. $\dfrac{5t}{2t+6} - \dfrac{4}{t^2-9} = \dfrac{5}{2}$

4. $\dfrac{x}{4x-4} + \dfrac{5}{x^2-1} = \dfrac{1}{4}$

5. $2 + \dfrac{4}{n-2} = \dfrac{8}{n^2-2n}$

6. $3 + \dfrac{6}{t-3} = \dfrac{6}{t^2-3t}$

7. $\dfrac{a}{a+2} + \dfrac{3}{a+4} = \dfrac{14}{a^2+6a+8}$

8. $\dfrac{3}{x+1} + \dfrac{2}{x+3} = 2$

9. $\dfrac{-2}{3x+2} + \dfrac{x-1}{9x^2-4} = \dfrac{3}{12x-8}$

10. $\dfrac{-1}{2x-5} + \dfrac{2x-4}{4x^2-25} = \dfrac{5}{6x+15}$

11. $\dfrac{n}{2n-3} + \dfrac{1}{n-3} = \dfrac{n^2-n-3}{2n^2-9n+9}$

12. $\dfrac{3y}{y^2+y-6} + \dfrac{2}{y^2+4y+3} = \dfrac{y}{y^2-y-2}$

13. $\dfrac{3y+1}{3y^2-4y-4} + \dfrac{9}{9y^2-4} = \dfrac{2y-2}{3y^2-8y+4}$

14. $\dfrac{4n+10}{2n^2-n-6} - \dfrac{3n+1}{2n^2-5n+2} = \dfrac{2}{4n^2+4n-3}$

15. $\dfrac{x+1}{2x^2+7x-4} - \dfrac{x}{2x^2-7x+3} = \dfrac{1}{x^2+x-12}$

16. $\dfrac{3}{x-2} + \dfrac{5}{x+3} = \dfrac{8x-1}{x^2+x-6}$

17. $\dfrac{7x+2}{12x^2+11x-15} - \dfrac{1}{3x+5} = \dfrac{2}{4x-3}$

18. $\dfrac{2n}{6n^2+7n-3} - \dfrac{n-3}{3n^2+11n-4} = \dfrac{5}{2n^2+11n+12}$

19. $\dfrac{3}{5x^2+18x-8} + \dfrac{x+1}{x^2-16} = \dfrac{5x}{5x^2-22x+8}$

20. $\dfrac{2}{4x^2+11x-3} - \dfrac{x+1}{3x^2+8x-3} = \dfrac{-4x}{12x^2-7x+1}$

For Problems 21–45, solve each problem.

21. An apple orchard contains 126 trees. The number of trees in each row is 4 less than twice the number of rows. Find the number of rows and the number of trees per row.

22. The sum of a number and its reciprocal is $\dfrac{10}{3}$. Find the number.

23. Jill starts at city A and travels toward city B at 50 miles per hour. At the same time, Russ starts at city B and travels on the same highway toward city A at 52 miles per hour. How long will it take before they meet if the two cities are 459 miles apart?

24. Two cars, which are 510 miles apart and whose speeds differ by 6 miles per hour, are moving toward each other. If they meet in 5 hours, find the speed of each car.

25. Rita rode her bicycle out into the country at a speed of 20 miles per hour and returned along the same route at 15 miles per hour. If the round trip took 5 hours and 50 minutes, how far out did she ride?

26. A jogger who can run an 8-minute mile starts a half-mile ahead of a jogger who can run a 6-minute mile. How long will it take the faster jogger to catch the slower jogger?

27. It takes a freight train 2 hours more to travel 300 miles than it takes an express train to travel 280 miles. The rate of the express train is 20 miles per hour greater than the rate of the freight train. Find the rates of both trains.

28. An airplane travels 2050 miles in the same time that a car travels 260 miles. If the rate of the plane is 358 miles per hour greater than the rate of the car, find the rate of the plane.

29. A container has 6 liters of a 40% alcohol solution in it. How much pure alcohol should be added to raise it to a 60% solution?

30. How many liters of a 60% acid solution must be added to 14 liters of a 10% acid solution to produce a 25% acid solution?

31. One solution contains 50% alcohol and another solution contains 80% alcohol. How many liters of each solution should be mixed to produce 10.5 liters of a 70% alcohol solution?

32. A contractor has a 24-pound mixture that is one-fourth cement and three-fourths sand. How much of a mixture that is half cement and half sand needs to be added to produce a mixture that is one-third cement?

33. A 10-quart radiator contains a 40% antifreeze solution. How much of the solution needs to be drained out and replaced with pure antifreeze in order to raise the solution to 70% antifreeze?

34. How much water must be evaporated from 20 gallons of a 10% salt solution in order to obtain a 20% salt solution?

35. One pipe can fill a tank in 4 hours and another pipe can fill the tank in 6 hours. How long will it take to fill the tank if both pipes are used?

36. Lolita and Doug working together can paint a shed in 3 hours and 20 minutes. If Doug can paint the shed by himself in 10 hours, how long would it take Lolita to paint the shed by herself?

37. An inlet pipe can fill a tank in 10 minutes. A drain can empty the tank in 12 minutes. If the tank is empty and both the pipe and drain are open, how long will it be before the tank overflows?

38. Mark can overhaul an engine in 20 hours and Phil can do the same job by himself in 30 hours. If they both work together for a time and then Mark finishes the job by himself in 5 hours, how long did they work together?

39. Pat and Mike working together can assemble a bookcase in 6 minutes. It takes Mike, working by himself, 9 minutes longer than it takes Pat working by himself to assemble the bookcase. How long does it take each, working alone, to do the job?

40. A printing company purchased a new copier that is twice as fast as the old copier. Using both copiers at the same time, it takes 5 hours to do a job. How long would it take the new copier working alone?

41. A professor can grade three tests in the time it takes a student assistant to grade one test. Working together, they can grade the tests for a class in 2 hours. How long would it take the student assistant working alone?

42. A car that averages 16 miles per gallon of gasoline for city driving and 22 miles per gallon for highway driving uses 14 gallons in 296 miles of driving. How much of the driving was city driving?

43. Angie bought some golf balls for $14. If each ball had cost $.25 less, she could have purchased one more ball for the same amount of money. How many golf balls did Angie buy?

44. A new labor contract provides for a wage increase of $1 per hour and a reduction of 5 hours in the workweek. A worker who received $320 per week under the old contract will receive $315 per week under the new contract. How long was the workweek under the old contract?

45. Todd contracted to paint a house for $480. It took him 4 hours longer than he had anticipated, so he earned $.50 per hour less than he originally calculated. How long had he anticipated it would take him to paint the house?

THOUGHTS INTO WORDS

46. One of our problem-solving suggestions is to "look for a guideline that can be used to help determine an equation." What does this suggestion mean to you?

47. Write a paragraph or two summarizing the various problem-solving ideas presented in this chapter.

| **1.5** | MISCELLANEOUS EQUATIONS |

Our previous work with solving linear and quadratic equations provides us with a basis for solving a variety of other types of equations. For example, the technique of factoring and applying the property

$$ab = 0 \quad \text{if and only if } a = 0 \text{ or } b = 0$$

can sometimes be used for equations other than quadratic equations.

EXAMPLE 1

Solve $x^3 + 2x^2 - 9x - 18 = 0$.

Solution

$$x^3 + 2x^2 - 9x - 18 = 0$$
$$x^2(x + 2) - 9(x + 2) = 0$$
$$(x + 2)(x^2 - 9) = 0$$
$$(x + 2)(x + 3)(x - 3) = 0$$
$$x + 2 = 0 \quad \text{or} \quad x + 3 = 0 \quad \text{or} \quad x - 3 = 0$$
$$x = -2 \quad \text{or} \quad x = -3 \quad \text{or} \quad x = 3$$

The solution set is $\{-3, -2, 3\}$. ∎

EXAMPLE 2

Solve $3x^5 + 5x^4 = 3x^3 + 5x^2$.

Solution

$$3x^5 + 5x^4 = 3x^3 + 5x^2$$
$$3x^5 + 5x^4 - 3x^3 - 5x^2 = 0$$
$$x^4(3x + 5) - x^2(3x + 5) = 0$$
$$(3x + 5)(x^4 - x^2) = 0$$
$$(3x + 5)(x^2)(x^2 - 1) = 0$$
$$(3x + 5)(x^2)(x + 1)(x - 1) = 0$$
$$3x + 5 = 0 \quad \text{or} \quad x^2 = 0 \quad \text{or} \quad x + 1 = 0 \quad \text{or} \quad x - 1 = 0$$
$$3x = -5$$
$$x = -\frac{5}{3} \quad \text{or} \quad x = 0 \quad \text{or} \quad x = -1 \quad \text{or} \quad x = 1$$

The solution set is $\left\{-\dfrac{5}{3}, 0, -1, 1\right\}$. ∎

Be careful with an equation like the one in Example 2. Don't be tempted to divide both sides of the equation by x^2. In so doing, you will lose the solution of zero. *In general, don't divide both sides of an equation by an expression containing the variable.*

Radical Equations

An equation such as

$$\sqrt{2x - 4} = x - 2$$

which contains a radical with the variable in the radicand, is often referred to as a **radical equation**. To solve radical equations, we need the following additional property of equality.

PROPERTY 1.3

Let a and b be real numbers and n a positive integer.

If $a = b$, then $a^n = b^n$.

Property 1.3 states that *we can raise both sides of an equation to a positive integral power*. However, we must be very careful when applying Property 1.3. Raising both sides of an equation to a positive integral power sometimes produces results that do not satisfy the original equation. Consider the following examples.

EXAMPLE 3

Solve $\sqrt{3x + 1} = 7$.

Solution

$$\sqrt{3x + 1} = 7$$
$$\left(\sqrt{3x + 1}\right)^2 = 7^2 \qquad \text{Square both sides.}$$
$$3x + 1 = 49$$
$$3x = 48$$
$$x = 16$$

Check $\sqrt{3x + 1} = 7$
$$\sqrt{3(16) + 1} \stackrel{?}{=} 7$$
$$\sqrt{49} \stackrel{?}{=} 7$$
$$7 = 7$$

The solution set is $\{16\}$. ∎

EXAMPLE 4

Solve $\sqrt{2x - 1} = -5$.

Solution

$$\sqrt{2x - 1} = -5$$
$$\left(\sqrt{2x - 1}\right)^2 = (-5)^2 \qquad \text{Square both sides.}$$
$$2x - 1 = 25$$
$$2x = 26$$
$$x = 13$$

Check $\sqrt{2x - 1} = -5$
$$\sqrt{2(13) - 1} \overset{?}{=} -5$$
$$\sqrt{25} \overset{?}{=} -5$$
$$5 \neq -5$$

Because 13 does not check, the equation has no solutions; the solution set is $\varnothing$. ■

REMARK It is true that the equation in Example 4 could be solved by inspection because the symbol $\sqrt{}$ refers to nonnegative numbers. However, we did want to demonstrate what happens if Property 1.3 is used.

Solve $\sqrt{x} + 6 = x$.

EXAMPLE 5

Solution

$$\sqrt{x} + 6 = x$$
$$\sqrt{x} = x - 6$$
$$\left(\sqrt{x}\right)^2 = (x - 6)^2 \qquad \text{Square both sides.}$$
$$x = x^2 - 12x + 36$$
$$0 = x^2 - 13x + 36$$
$$0 = (x - 4)(x - 9)$$
$$x - 4 = 0 \qquad \text{or} \qquad x - 9 = 0$$
$$x = 4 \qquad \text{or} \qquad x = 9$$

Check $\quad \sqrt{x} = x - 6 \qquad \sqrt{x} = x - 6$
$$\sqrt{4} \overset{?}{=} 4 - 6 \qquad \sqrt{9} \overset{?}{=} 9 - 6$$
$$2 \neq -2 \qquad\quad 3 = 3$$

The only solution is 9, so the solution set is $\{9\}$. ■

REMARK Notice what happens if we square both sides of the original equation. We obtain $x + 12\sqrt{x} + 36 = x^2$, an equation more complex than the original one and still containing a radical. Therefore, it is important first to isolate the term containing the radical on one side of the equation and then to square both sides of the equation.

In general, raising both sides of an equation to a positive integral power produces an equation that has all of the solutions of the original equation, *but* it may also have some extra solutions that will not satisfy the original equation. Such extra solutions are called extraneous solutions. Therefore, when using Property 1.3, you *must* check each potential solution in the original equation.

EXAMPLE 6

Solve $\sqrt[3]{2x + 3} = -3$.

Solution

$$\sqrt[3]{2x + 3} = -3$$
$$\left(\sqrt[3]{2x + 3}\right)^3 = (-3)^3 \qquad \text{Cube both sides.}$$
$$2x + 3 = -27$$
$$2x = -30$$
$$x = -15$$

 Check $\sqrt[3]{2x + 3} = -3$
$$\sqrt[3]{2(-15) + 3} \stackrel{?}{=} -3$$
$$\sqrt[3]{-27} \stackrel{?}{=} -3$$
$$-3 = -3$$

The solution set is $\{-15\}$. ∎

EXAMPLE 7

Solve $\sqrt{x + 4} = \sqrt{x - 1} + 1$.

Solution

$$\sqrt{x + 4} = \sqrt{x - 1} + 1$$
$$\left(\sqrt{x + 4}\right)^2 = \left(\sqrt{x - 1} + 1\right)^2 \qquad \text{Square both sides.}$$
$$x + 4 = x - 1 + 2\sqrt{x - 1} + 1 \qquad \begin{array}{l}\text{Remember the middle term when}\\ \text{squaring the binomial.}\end{array}$$

$$4 = 2\sqrt{x - 1}$$
$$2 = \sqrt{x - 1}$$
$$2^2 = \left(\sqrt{x - 1}\right)^2 \qquad \text{Square both sides.}$$
$$4 = x - 1$$
$$5 = x$$

Check $\sqrt{x + 4} = \sqrt{x - 1} + 1$
$$\sqrt{5 + 4} \stackrel{?}{=} \sqrt{5 - 1} + 1$$
$$\sqrt{9} \stackrel{?}{=} \sqrt{4} + 1$$
$$3 = 3$$

The solution set is $\{5\}$. ∎

Equations of Quadratic Form

An equation such as $x^4 + 5x^2 - 36 = 0$ is not a quadratic equation. However, if we let $u = x^2$, then we get $u^2 = x^4$. Substituting u for x^2 and u^2 for x^4 in $x^4 + 5x^2 - 36 = 0$ produces

$$u^2 + 5u - 36 = 0$$

which is a quadratic equation. In general, an equation in the variable x is said to be of **quadratic form** if it can be written in the form

$$au^2 + bu + c = 0$$

where $a \neq 0$ and u is some algebraic expression in x. We have two basic approaches to solving equations of quadratic form, as illustrated by the next two examples.

EXAMPLE 8

Solve $x^{2/3} + x^{1/3} - 6 = 0$.

Solution

Let $u = x^{1/3}$; then $u^2 = x^{2/3}$ and the given equation can be rewritten $u^2 + u - 6 = 0$. Solving this equation yields two solutions.

$$u^2 + u - 6 = 0$$
$$(u + 3)(u - 2) = 0$$
$$u + 3 = 0 \quad \text{or} \quad u - 2 = 0$$
$$u = -3 \quad \text{or} \quad u = 2$$

Now, substituting $x^{1/3}$ for u, we have

$$x^{1/3} = -3 \quad \text{or} \quad x^{1/3} = 2$$

from which we obtain

$$(x^{1/3})^3 = (-3)^3 \quad \text{or} \quad (x^{1/3})^3 = 2^3$$
$$x = -27 \quad \text{or} \quad x = 8$$

Check

$$x^{2/3} + x^{1/3} - 6 = 0 \qquad x^{2/3} + x^{1/3} - 6 = 0$$
$$(-27)^{2/3} + (-27)^{1/3} - 6 \overset{?}{=} 0 \qquad (8)^{2/3} + (8)^{1/3} - 6 \overset{?}{=} 0$$
$$9 + (-3) - 6 \overset{?}{=} 0 \qquad 4 + 2 - 6 \overset{?}{=} 0$$
$$0 = 0 \qquad 0 = 0$$

The solution set is $\{-27, 8\}$. ■

EXAMPLE 9

Solve $x^4 + 5x^2 - 36 = 0$.

Solution

$$x^4 + 5x^2 - 36 = 0$$
$$(x^2 + 9)(x^2 - 4) = 0$$

$$x^2 + 9 = 0 \quad \text{or} \quad x^2 - 4 = 0$$
$$x^2 = -9 \quad \text{or} \quad x^2 = 4$$
$$x = \pm 3i \quad \text{or} \quad x = \pm 2$$

The solution set is $\{\pm 3i, \pm 2\}$. ■

Notice in Example 8 that we made a substitution (u for $x^{1/3}$) to change the original equation to a quadratic equation in terms of the variable u. Then, after solving for u, we substituted $x^{1/3}$ for u to obtain the solutions of the original equation. However, in Example 9 we factored the given polynomial and proceeded without changing to a quadratic equation. Which approach you use may depend on the complexity of the given equation.

EXAMPLE 10

Solve $15x^{-2} - 11x^{-1} - 12 = 0$.

Solution

 Let $u = x^{-1}$; then $u^2 = x^{-2}$ and the given equation can be written and solved as follows.

$$15u^2 - 11u - 12 = 0$$
$$(5u + 3)(3u - 4) = 0$$

$$5u + 3 = 0 \quad \text{or} \quad 3u - 4 = 0$$
$$5u = -3 \quad \text{or} \quad 3u = 4$$
$$u = -\frac{3}{5} \quad \text{or} \quad u = \frac{4}{3}$$

Now, substituting x^{-1} back for u, we have

$$x^{-1} = -\frac{3}{5} \quad \text{or} \quad x^{-1} = \frac{4}{3}$$

from which we obtain

$$\frac{1}{x} = \frac{-3}{5} \quad \text{or} \quad \frac{1}{x} = \frac{4}{3}$$
$$-3x = 5 \quad \text{or} \quad 4x = 3$$
$$x = -\frac{5}{3} \quad \text{or} \quad x = \frac{3}{4}$$

The solution set is $\left\{-\frac{5}{3}, \frac{3}{4}\right\}$. ■

PROBLEM SET 1.5

For Problems 1–52, solve each equation. Don't forget that you *must* check potential solutions whenever Property 1.3 is applied.

1. $x^3 + x^2 - 4x - 4 = 0$

2. $x^3 - 5x^2 - x + 5 = 0$

3. $2x^3 - 3x^2 + 2x - 3 = 0$

4. $3x^3 + 5x^2 + 12x + 20 = 0$

5. $8x^5 + 10x^4 = 4x^3 + 5x^2$

6. $10x^5 + 15x^4 = 2x^3 + 3x^2$

7. $x^{3/2} = 4x$

8. $5x^4 = 6x^3$

9. $n^{-2} = n^{-3}$

10. $n^{4/3} = 4n$

11. $\sqrt{3x - 2} = 4$

12. $\sqrt{5x - 1} = -4$

13. $\sqrt{3x - 8} - \sqrt{x - 2} = 0$

14. $\sqrt{2x - 3} = 1$

15. $\sqrt{4x - 3} = -2$

16. $\sqrt{2x - 1} - \sqrt{x + 2} = 0$

17. $\sqrt[3]{2x + 3} + 3 = 0$

18. $\sqrt[3]{n^2 - 1} + 1 = 0$

19. $2\sqrt{n} + 3 = n$

20. $\sqrt{3t} - t = -6$

21. $\sqrt{3x - 2} = 3x - 2$

22. $5x - 4 = \sqrt{5x - 4}$

23. $\sqrt{2t - 1} + 2 = t$

24. $p = \sqrt{-4p + 17} + 3$

25. $\sqrt{x + 2} - 1 = \sqrt{x - 3}$

26. $\sqrt{x + 5} - 2 = \sqrt{x - 7}$

27. $\sqrt{7n + 23} - \sqrt{3n + 7} = 2$

28. $\sqrt{5t + 31} - \sqrt{t + 3} = 4$

29. $\sqrt{3x + 1} + \sqrt{2x + 4} = 3$

30. $\sqrt{2x - 1} - \sqrt{x + 3} = 1$

31. $\sqrt{x - 2} - \sqrt{2x - 11} = \sqrt{x - 5}$

32. $\sqrt{-2x - 7} + \sqrt{x + 9} = \sqrt{8 - x}$

33. $\sqrt{1 + 2\sqrt{x}} = \sqrt{x + 1}$

34. $\sqrt{7 + 3\sqrt{x}} = \sqrt{x + 1}$

35. $x^4 - 5x^2 + 4 = 0$

36. $x^4 - 25x^2 + 144 = 0$

37. $2n^4 - 9n^2 + 4 = 0$

38. $3n^4 - 4n^2 + 1 = 0$

39. $x^4 - 2x^2 - 35 = 0$

40. $2x^4 + 5x^2 - 12 = 0$

41. $x^4 - 4x^2 + 1 = 0$

42. $x^4 - 8x^2 + 11 = 0$

43. $x^{2/3} + 3x^{1/3} - 10 = 0$

44. $x^{2/3} + x^{1/3} - 2 = 0$

45. $6x^{2/3} - 5x^{1/3} - 6 = 0$

46. $3x^{2/3} - 11x^{1/3} - 4 = 0$

47. $x^{-2} + 4x^{-1} - 12 = 0$

48. $12t^{-2} - 17t^{-1} - 5 = 0$

49. $x - 11\sqrt{x} + 30 = 0$

50. $2x - 11\sqrt{x} + 12 = 0$

51. $x + 3\sqrt{x} - 10 = 0$

52. $6x - 19\sqrt{x} - 7 = 0$

For Problems 53–56, solve each problem.

53. The formula for the slant height of a right circular cone is $s = \sqrt{r^2 + h^2}$, where r is the length of a radius of the base and h is the altitude of the cone. Find the altitude of a cone whose slant height is 13 inches and whose radius is 5 inches.

54. A clockmaker wants to build a grandfather clock with a pendulum whose period will be 1.5 seconds. He knows the formula for the period is $T = 2\pi\sqrt{\dfrac{L}{32.144}}$, where T represents the period in seconds and L represents the length of the pendulum in feet. What length should the clockmaker use for the pendulum? Express your answer to the nearest hundredth of a foot.

55. Police sometimes use the formula $S = \sqrt{30Df}$ to correlate the speed of a car and the length of skid marks when the brakes have been applied. In this formula, S represents the speed of the car in miles per hour, D represents the length of skid marks measured in feet, and f represents a coefficient of friction. For a particular situation, the coefficient of friction is a constant that

depends on the type and condition of the road surface. Using .35 as a coefficient of friction, determine, to the nearest foot, how far a car will skid if the brakes are applied when the car is traveling at a speed of 58 miles per hour.

56. Using the formula given in Problem 55 and a coefficient of friction of .95, determine, to the nearest foot, how far a car will skid if the brakes are applied when the car is traveling at a speed of 65 miles per hour.

THOUGHTS INTO WORDS

57. Explain the concept of extraneous solutions.

58. What does it mean to say that an equation is of quadratic form?

59. Your friend attempts to solve the equation $3 + 2\sqrt{x} = x$ as follows:

$$\left(3 + 2\sqrt{x}\right)^2 = x^2$$
$$9 + 12\sqrt{x} + 4x = x^2$$

At this step, he stops and doesn't know how to proceed. What help would you give him?

Further Investigations

60. Verify that $x = a$ and $x^2 = a^2$ are *not* equivalent equations.

61. Solve the following equations, and express the solutions to the nearest hundredth.

    a. $x^4 - 3x^2 + 1 = 0$ **b.** $x^4 - 5x^2 + 2 = 0$

    c. $2x^4 - 7x^2 + 2 = 0$ **d.** $3x^4 - 9x^2 + 1 = 0$

    e. $x^4 - 100x^2 + 2304 = 0$

    f. $4x^4 - 373x^2 + 3969 = 0$

1.6 INEQUALITIES

Just as we use the symbol $=$ to represent *is equal to*, we also use the symbols $<$ and $>$ to represent *is less than* and *is greater than*, respectively. Thus various **statements of inequality** can be made:

    $a < b$ means a is less than b.

    $a \leq b$ means a is less than or equal to b.

    $a > b$ means a is greater than b.

    $a \geq b$ means a is greater than or equal to b.

The following are examples of **numerical statements of inequality**.

$7 + 8 > 10$	$-4 + (-6) \geq -10$
$-4 > -6$	$7 - 9 \leq -2$
$7 - 1 < 20$	$3 + 4 > 12$
$8(-3) < 5(-3)$	$7 - 1 < 0$

Notice that only $3 + 4 > 12$ and $7 - 1 < 0$ are *false*; the other six are *true* numerical statements.

Algebraic inequalities contain one or more variables. The following are examples of algebraic inequalities.

$$x + 4 > 8 \qquad\qquad 3x + 2y \leq 4$$
$$(x - 2)(x + 4) \geq 0 \qquad x^2 + y^2 + z^2 \leq 16$$

An algebraic inequality such as $x + 4 > 8$ is neither true nor false as it stands and is called an **open sentence**. For each numerical value substituted for x, the algebraic inequality $x + 4 > 8$ becomes a numerical statement of inequality that is true or false. For example, if $x = -3$, then $x + 4 > 8$ becomes $-3 + 4 > 8$, which is false. If $x = 5$, then $x + 4 > 8$ becomes $5 + 4 > 8$, which is true. **Solving an algebraic inequality** refers to the process of finding the numbers that make it a true numerical statement. Such numbers are called the **solutions** of the inequality and are said to **satisfy** it.

The general process for solving inequalities closely parallels that for solving equations. We repeatedly replace the given inequality with equivalent but simpler inequalities until the solution set is obvious. The following property provides the basis for producing equivalent inequalities.

PROPERTY 1.4

1. For all real numbers a, b, and c,

$$a > b \quad \text{if and only if } a + c > b + c$$

2. For all real numbers a, b, and c, with $c > 0$,

$$a > b \quad \text{if and only if } ac > bc$$

3. For all real numbers a, b, and c, with $c < 0$,

$$a > b \quad \text{if and only if } ac < bc$$

Similar properties exist if $>$ is replaced by $<$, $\leq$, or $\geq$. Part 1 of Property 1.4 is commonly called the **addition property of inequality**. Parts 2 and 3 together make up the **multiplication property of inequality**. Pay special attention to part 3. **If both sides of an inequality are multiplied by a negative number, the inequality symbol must be reversed.** For example, if both sides of $-3 < 5$ are multiplied by -2, the equivalent inequality $6 > -10$ is produced. Now let's consider using the addition and multiplication properties of inequality to help solve some inequalities.

EXAMPLE 1

Solve $3(2x - 1) < 8x - 7$.

Solution

$$3(2x - 1) < 8x - 7$$

$$6x - 3 < 8x - 7 \qquad \text{Apply distributive property to left side.}$$

$$-2x - 3 < -7 \qquad \text{Add } -8x \text{ to both sides.}$$

$$-2x < -4 \qquad \text{Add 3 to both sides.}$$

$$-\frac{1}{2}(-2x) > -\frac{1}{2}(-4) \qquad \text{Multiply both sides by } -\frac{1}{2}, \text{ which reverses the inequality.}$$

$$x > 2$$

The solution set is $\{x \mid x > 2\}$. ■

A graph of the solution set $\{x \mid x > 2\}$ in Example 1 is shown in Figure 1.6. The parenthesis indicates that 2 does not belong to the solution set.

FIGURE 1.6

Checking the solutions of an inequality presents a problem. Obviously, we cannot check all of the infinitely many solutions for a particular inequality. However, by checking at least one solution, especially when the multiplication property has been used, we might catch a mistake of forgetting to change the type of inequality. In Example 1 we are claiming that all numbers greater than 2 will satisfy the original inequality. Let's check the number 3.

$$3(2x - 1) < 8x - 7$$

$$3[2(3) - 1] \overset{?}{<} 8(3) - 7$$

$$3(5) \overset{?}{<} 17$$

$$15 < 17 \qquad \text{It checks!}$$

Interval Notation

It is also convenient to express solution sets of inequalities by using **interval notation**. For example, the symbol $(2, \infty)$ refers to the interval of all real numbers greater than 2. As on the graph in Figure 1.6, the left-hand parenthesis indicates that 2 is not to be included. The infinity symbol, ∞, along with the right-hand parenthesis, indicates that there is no right-hand endpoint. Following is a partial list of interval notations, along with the sets and graphs that they represent. Note the use of square brackets to *include* endpoints.

SET	GRAPH	INTERVAL NOTATION
$\{x \mid x > a\}$		(a, ∞)
$\{x \mid x \geq a\}$		$[a, \infty)$
$\{x \mid x < b\}$		$(-\infty, b)$
$\{x \mid x \leq b\}$		$(-\infty, b]$

EXAMPLE 2

Solve $\dfrac{-3x + 1}{2} > 4$.

Solution

$$\frac{-3x + 1}{2} > 4$$

$$2\left(\frac{-3x + 1}{2}\right) > 2(4) \qquad \text{Multiply both sides by 2.}$$

$$-3x + 1 > 8$$

$$-3x > 7$$

$$-\frac{1}{3}(-3x) < -\frac{1}{3}(7) \qquad \text{Multiply both sides by } -\frac{1}{3}, \text{ which reverses the inequality.}$$

$$x < -\frac{7}{3}$$

The solution set is $\left(-\infty, -\dfrac{7}{3}\right)$. ∎

EXAMPLE 3

Solve $\dfrac{x - 4}{6} - \dfrac{x - 2}{9} \leq \dfrac{5}{18}$.

Solution

$$\frac{x-4}{6} - \frac{x-2}{9} \le \frac{5}{18}$$

$$18\left(\frac{x-4}{6} - \frac{x-2}{9}\right) \le 18\left(\frac{5}{18}\right) \qquad \text{Multiply both sides by the LCD.}$$

$$18\left(\frac{x-4}{6}\right) - 18\left(\frac{x-2}{9}\right) \le 18\left(\frac{5}{18}\right)$$

$$3(x-4) - 2(x-2) \le 5$$

$$3x - 12 - 2x + 4 \le 5$$

$$x - 8 \le 5$$

$$x \le 13$$

The solution set is $(-\infty, 13]$.   ■

Compound Statements

We use the words <u>and</u> and <u>or</u> in mathematics to form **compound statements**. The following are examples of some compound numerical statements that use <u>and</u>. We call such statements **conjunctions**. We agree to call a conjunction true only if all of its component parts are true. Statements 1 and 2 below are true, but statements 3, 4, and 5 are false.

1. $3 + 4 = 7$ and $-4 < -3$ True
2. $-3 < -2$ and $-6 > -10$ True
3. $6 > 5$ and $-4 < -8$ False
4. $4 < 2$ and $0 < 10$ False
5. $-3 + 2 = 1$ and $5 + 4 = 8$ False

We call compound statements that use <u>or</u> **disjunctions**. The following are some examples of disjunctions that involve numerical statements.

6. $.14 > .13$ or $.235 < .237$ True

7. $\frac{3}{4} > \frac{1}{2}$ or $-4 + (-3) = 10$ True

8. $-\frac{2}{3} > \frac{1}{3}$ or $(.4)(.3) = .12$ True

9. $\frac{2}{5} < -\frac{2}{5}$ or $7 + (-9) = 16$ False

A disjunction is true if at least one of its component parts is true. In other words, disjunctions are false only if all of the component parts are false. In the statements above, 6, 7, and 8 are true, but 9 is false.

Now let's consider finding solutions for some compound statements that involve algebraic inequalities. Keep in mind that our previous agreements for labeling conjunctions and disjunctions true or false form the basis for our reasoning.

EXAMPLE 4

Graph the solution set for the conjunction $x > -1$ <u>and</u> $x < 3$.

Solution

The key word is <u>and</u>, so we need to satisfy both inequalities. Thus all numbers between -1 and 3 are solutions, and we can indicate this on a number line as in Figure 1.7.

FIGURE 1.7

Using interval notation, we can represent the interval enclosed in parentheses in Figure 1.7 by $(-1, 3)$. Using set-builder notation, we can express the same interval as $\{x \mid -1 < x < 3\}$, where the statement $-1 < x < 3$ is read "negative one is less than x and x is less than three." In other words, x is between -1 and 3. ∎

Example 4 represents another concept that pertains to sets. The set of all elements common to two sets is called the **intersection** of the two sets. Thus in Example 4 we found the intersection of the two sets $\{x \mid x > -1\}$ and $\{x \mid x < 3\}$ to be the set $\{x \mid -1 < x < 3\}$. In general, we define the intersection of two sets as follows.

DEFINITION 1.1

The **intersection** of two sets A and B (written $A \cap B$) is the set of all elements that are in both A and B. Using set-builder notation, we can write

$$A \cap B = \{x \mid x \in A \quad \underline{and} \quad x \in B\}$$

We can solve a conjunction such as $\dfrac{3x + 2}{2} > -2$ and $\dfrac{3x + 2}{2} < 7$, in which the same algebraic expression is contained in both inequalities, by using the compact form $-2 < \dfrac{3x + 2}{2} < 7$ as follows.

EXAMPLE 5

Solve $-2 < \dfrac{3x + 2}{2} < 7$.

Solution

$$-2 < \frac{3x + 2}{2} \quad < 7$$

$$2(-2) < 2\left(\frac{3x + 2}{2}\right) < 2(7) \qquad \text{Multiply through by 2.}$$

$$-4 < 3x + 2 \quad < 14$$

$$-6 < \quad 3x \quad < 12 \qquad \text{Add } -2 \text{ to all three quantities.}$$

$$-2 < \quad x \quad < 4 \qquad \text{Multiply through by } \tfrac{1}{3}.$$

The solution set is the interval $(-2, 4)$.   ■

The word <u>and</u> ties the concept of a conjunction to the set concept of intersection. In a like manner, the word <u>or</u> links the idea of a disjunction to the set concept of **union**. We define the union of two sets as follows.

> ## DEFINITION 1.2
>
> The **union** of two sets A and B (written $A \cup B$) is the set of all elements that are in A or in B or in both. Using set-builder notation, we can write
>
> $$A \cup B = \{x \mid x \in A \ \underline{\text{or}} \ x \in B\}.$$

EXAMPLE 6

Graph the solution set for the disjunction $x < -1$ <u>or</u> $x > 2$, and express it using interval notation.

Solution

The key word is <u>or</u>, so all numbers that satisfy either inequality (or both) are solutions. Thus all numbers less than -1, along with all numbers greater than 2, are the solutions. The graph of the solution set is shown in Figure 1.8.

FIGURE 1.8

Using interval notation and the set concept of union, we can express the solution set as $(-\infty, -1) \cup (2, \infty)$.   ■

Example 6 illustrates that in terms of set vocabulary, the solution set of a disjunction is the union of the solution sets of the component parts of the disjunction. Note that there is *no compact form* for writing $x < -1$ or $x > 2$ *or for any disjunction*.

The following agreements on the use of interval notation should be added to the list on page 137.

SET	GRAPH	INTERVAL NOTATION
$\{x \mid a < x < b\}$		(a, b)
$\{x \mid a \leq x < b\}$		$[a, b)$
$\{x \mid a < x \leq b\}$		$(a, b]$
$\{x \mid a \leq x \leq b\}$		$[a, b]$
$\{x \mid x$ is a real number$\}$		$(-\infty, \infty)$

Quadratic Inequalities

The equation $ax^2 + bx + c = 0$ has been referred to as the standard form of a quadratic equation in one variable. Similarly, the form $ax^2 + bx + c < 0$ is used to represent a **quadratic inequality**. (The symbol $<$ can be replaced by $>$, $\leq$, or $\geq$ to produce other forms of quadratic inequalities.)

The number line can be used to help solve quadratic inequalities where the quadratic polynomial is factorable. Let's consider two examples to illustrate this procedure.

EXAMPLE 7

Solve $x^2 + x - 6 < 0$.

Solution

First, let's factor the polynomial.

$$x^2 + x - 6 < 0$$
$$(x + 3)(x - 2) < 0$$

Second, let's locate the values where the product $(x + 3)(x - 2)$ is equal to zero. The numbers -3 and 2 divide the number line into three intervals (see Figure 1.9):

the numbers less than -3

the numbers between -3 and 2

the numbers greater than 2

FIGURE 1.9

We can choose a **test number** from each of these intervals and see how it affects the signs of the factors $x + 3$ and $x - 2$ and, consequently, the sign of the product of these factors. For example, if $x < -3$ (try $x = -4$), then $x + 3$ is negative, and $x - 2$ is negative; thus their product is positive. If $-3 < x < 2$ (try $x = 0$), then $x + 3$ is positive and $x - 2$ is negative; thus their product is negative. If $x > 2$ (try $x = 3$), then $x + 3$ is positive and $x - 2$ is positive; thus their product is positive. This information can be conveniently arranged by using a number line as in Figure 1.10.

FIGURE 1.10

Therefore, the given inequality, $x^2 + x - 6 < 0$, is satisfied by the numbers between -3 and 2. That is, the solution set is the open interval $(-3, 2)$.   ■

Numbers such as -3 and 2 in the preceding example, where the given polynomial or algebraic expression equals zero or is undefined, are referred to as **critical numbers**. Let's consider some additional examples where we make use of critical numbers and test numbers.

EXAMPLE 8

Solve $6x^2 + 17x - 14 \geq 0$.

Solution

First, we factor the polynomial.

$$6x^2 + 17x - 14 \geq 0$$
$$(2x + 7)(3x - 2) \geq 0$$

Second, we locate the values where the product $(2x + 7)(3x - 2)$ equals zero. We suggest putting dots at $-\dfrac{7}{2}$ and $\dfrac{2}{3}$ (see Figure 1.11) to remind ourselves that these

FIGURE 1.11

two numbers must be included in the solution set, because the given statement includes equality. Now let's choose a test number from each of the three intervals and observe the sign behavior of the factors, as in Figure 1.12.

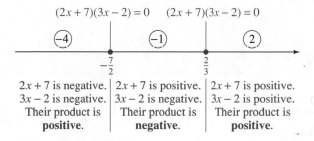

FIGURE 1.12

Using the concept of set union, we can write the solution set $\left(-\infty, -\dfrac{7}{2}\right] \cup \left[\dfrac{2}{3}, \infty\right)$.

■

> **REMARK** As you work with quadratic inequalities like those in Examples 7 and 8, you may be able to use a more abbreviated format than what we demonstrated. Basically, it is necessary to keep track of the sign of each factor in each of the intervals.

Let's conclude this section by considering a word problem that involves an inequality. All of the problem-solving techniques offered earlier continue to apply except that now we look for a guideline that can be used to generate an inequality rather than an equation.

PROBLEM 1

Lance has $500 to invest. If he invests $300 at 9%, at what rate must he invest the remaining $200 so that the total yearly interest from the two investments exceeds $47?

Solution

Let r represent the unknown rate of interest. The following guideline can be used to set up an inequality.

Interest from 9% investment + Interest from r percent investment > \$47

$$(9\%)(\$300) \qquad + \qquad r(\$200) \qquad > \$47$$

We solve this inequality using methods we have already acquired.

$$(0.9)(300) + 200r > 47$$
$$27 + 200r > 47$$
$$200r > 20$$
$$r > 0.1$$

The other \$200 must be invested at a rate greater than 10%. ■

PROBLEM SET 1.6

For Problems 1–12, express each solution set in interval notation and graph each solution set.

1. $x \le -2$

2. $x > -1$

3. $1 < x < 4$

4. $-1 < x \le 2$

5. $2 > x > 0$

6. $-3 \ge x$

7. $-2 \le x \le -1$

8. $1 \le x$

9. $x < 1$ or $x > 3$

10. $x > 2$ or $x < -1$

11. $x > -2$ or $x > 2$

12. $x > 2$ or $x < 4$

For Problems 13–20, solve each conjunction by using the compact form and express the solution sets in interval notation.

13. $-17 \le 3x - 2 \le 10$

14. $-25 \le 4x + 3 \le 19$

15. $2 > 2x - 1 > -3$

16. $4 > 3x + 1 > 1$

17. $-4 < \dfrac{x-1}{3} < 4$

18. $-1 \le \dfrac{x+2}{4} \le 1$

19. $-3 < 2 - x < 3$

20. $-4 < 3 - x < 4$

For Problems 21–62, solve each inequality and express the solution sets in interval notation.

21. $-2x + 1 > 5$

22. $6 - 3x < 12$

23. $-3n + 5n - 2 \ge 8n - 7 - 9n$

24. $3n - 5 > 8n + 5$

25. $6(2t - 5) - 2(4t - 1) \ge 0$

26. $3(2x + 1) - 2(2x + 5) < 5(3x - 2)$

27. $\dfrac{2}{3}x - \dfrac{3}{4} \le \dfrac{1}{4}x + \dfrac{2}{3}$

28. $\dfrac{3}{5} - \dfrac{x}{2} \ge \dfrac{1}{2} + \dfrac{x}{5}$

29. $\dfrac{n+2}{4} + \dfrac{n-3}{8} < 1$

30. $\dfrac{2n+1}{6} + \dfrac{3n-1}{5} > \dfrac{2}{15}$

31. $\dfrac{x}{2} - \dfrac{x-1}{5} \ge \dfrac{x+2}{10} - 4$

32. $\dfrac{4x-3}{6} - \dfrac{2x-1}{12} < -2$

33. $0.09x + 0.1(x + 200) > 77$

34. $0.06x + 0.08(250 - x) \ge 19$

35. $0 < \dfrac{5x-1}{3} < 2$

36. $-3 \le \dfrac{4x+3}{2} \le 1$

37. $3 \ge \dfrac{7-x}{2} \ge 1$

38. $-2 \le \dfrac{5-3x}{4} \le \dfrac{1}{2}$

39. $x^2 + 3x - 4 < 0$

40. $x^2 - 4 < 0$

41. $x^2 - 2x - 15 > 0$

42. $x^2 - 12x + 32 \ge 0$

43. $n^2 - n \le 2$

44. $n^2 + 5n \le 6$

45. $3t^2 + 11t - 4 > 0$

46. $2t^2 - 9t - 5 > 0$

47. $15x^2 - 26x + 8 \le 0$

48. $6x^2 + 25x + 14 \le 0$

49. $4x^2 - 4x + 1 > 0$ **50.** $9x^2 + 6x + 1 \le 0$

51. $(x + 1)(x - 3) > (x + 1)(2x - 1)$

52. $(x - 2)(2x + 5) > (x - 2)(x - 3)$

53. $(x + 1)(x - 2) \ge (x - 4)(x + 6)$

54. $(2x - 1)(x + 4) \ge (2x + 1)(x - 3)$

55. $(x - 1)(x - 2)(x + 4) > 0$

56. $(x + 1)(x - 3)(x + 7) \ge 0$

57. $(x + 2)(2x - 1)(x - 5) \le 0$

58. $(x - 3)(3x + 2)(x + 4) < 0$

59. $x^3 - 2x^2 - 24x \ge 0$ **60.** $x^3 + 2x^2 - 3x > 0$

61. $(x - 2)^2(x + 3) > 0$ **62.** $(x + 4)^2(x + 5) > 0$

For Problems 63–72, use inequalities to help solve each problem.

63. Felix has $1000 to invest. Suppose he invests $500 at 8% interest. At what rate must he invest the other $500 so that the two investments yield more than $100 of yearly interest?

64. Suppose that Annette invests $700 at 9%. How much must she invest at 11% so that the total yearly interest from the two investments exceeds $162?

65. Rhonda had scores of 94, 84, 86, and 88 on her first four history exams of the semester. What score must she obtain on the fifth exam to have an average of 90 or better for the five exams?

66. The average height of the two forwards and the center of a basketball team is 6 feet 8 inches. What must the average height of the two guards be so that the team average is at least 6 feet 4 inches?

67. At the food booths at a festival, the area of a pizza must be 150 square inches or more for the pizza to be classified as large. What must the diameter be for a pizza to be classified as large?

68. If the temperature for a 24-hour period ranged between 41°F and 59°F, inclusive, what was the range in Celsius degrees? $\left(F = \dfrac{9}{5}C + 32 \right)$

69. If the temperature for a 24-hour period ranged between −20°C and −5°C, inclusive, what was the range in Fahrenheit degrees? $\left(C = \dfrac{5}{9}(F - 32) \right)$

70. A person's intelligence quotient (IQ) is found by dividing mental age (M), as indicated by standard tests, by the chronological age (C), and then multiplying this ratio by 100. The formula $IQ = 100M/C$ can be used. If the IQ range of a group of 11-year-olds is given by $80 \le IQ \le 140$, find the mental-age range of this group.

71. A car can be rented from agency A at $75 per day plus $.10 a mile or from agency B at $50 a day plus $.20 a mile. If the car is driven m miles, for what values of m does it cost less to rent from agency A?

72. In statistics the formula for a z-score is $z = \dfrac{x - \bar{x}}{s}$, where x is a score, $\bar{x}$ is the mean, and s is the standard deviation. To give credibility to our results in a statistical claim, we want to determine the values of x that will produce a z-score greater than 2.5 when $\bar{x} = 8.7$ and $s = 1.2$. Find such values of x.

THOUGHTS INTO WORDS

73. Explain the difference between a conjunction and a disjunction. Give an example of each (outside the field of mathematics).

74. How do you know by inspection that the solution set of the inequality $x + 3 > x + 2$ is the entire set of real numbers?

75. Give a step-by-step description of how you would solve the inequality $-4 < 2(x - 1) - 3(x + 2)$.

76. Explain how you would solve the inequality $(x - 1)^2(x + 2)^2 > 0$.

77. Find the solution set for each of the following compound statements and explain your reasoning in each case.

   a. $x < 3$ and $5 > 2$

   b. $x < 3$ or $5 > 2$

   c. $x < 3$ and $6 < 4$

   d. $x < 3$ or $6 < 4$

🔍 **Further Investigations**

78. The product $(x - 2)(x + 3)$ is positive if both factors are negative *or* if both factors are positive. Therefore, we can solve $(x - 2)(x + 3) > 0$ as follows.

$(x - 2 < 0 \text{ and } x + 3 < 0)$ or $(x - 2 > 0 \text{ and } x + 3 > 0)$

$\quad (x < 2 \text{ and } x < -3)$ or $(x > 2 \text{ and } x > -3)$

$\qquad\qquad x < -3$ or $x > 2$

The solution set is $(-\infty, -3) \cup (2, \infty)$. Use this type of analysis to solve each of the following.

a. $(x - 1)(x + 5) > 0$

b. $(x + 2)(x - 4) \geq 0$

c. $(x + 4)(x - 3) < 0$

d. $(2x - 1)(x + 5) \leq 0$

e. $(x + 4)(x + 1)(x - 2) > 0$

f. $(x + 2)(x - 1)(x - 3) < 0$

79. If $a > b > 0$, verify that $1/a < 1/b$.

80. If $a > b$, is it always true that $1/a < 1/b$? Defend your answer.

1.7   INEQUALITIES INVOLVING QUOTIENTS AND ABSOLUTE VALUE

The same type of number-line analysis that we did in the previous section can be used for indicated quotients as well as for indicated products. In other words, inequalities such as

$$\frac{x - 2}{x + 3} > 0$$

can be solved very efficiently using the same basic approach that we used with quadratic inequalities in the previous section. Let's illustrate this procedure.

EXAMPLE 1

Solve $\dfrac{x - 2}{x + 3} > 0$.

Solution

First we find that at $x = 2$, the quotient $\dfrac{x - 2}{x + 3}$ equals zero and that at $x = -3$, the quotient is undefined. The critical numbers -3 and 2 divide the number line into three intervals. Then, using a test number from each interval (such as -4, 1, and 3), we can observe the sign behavior of the quotient, as in Figure 1.13.

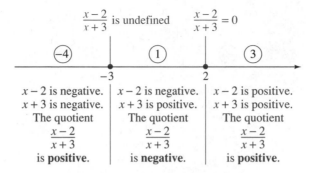

FIGURE 1.13

Therefore, the solution set for $\dfrac{x-2}{x+3} > 0$ is $(-\infty, -3) \cup (2, \infty)$.   ■

EXAMPLE 2

Solve $\dfrac{x+2}{x+4} \le 3$.

Solution

First, let's change the form of the given inequality.

$$\frac{x+2}{x+4} \le 3$$

$$\frac{x+2}{x+4} - 3 \le 0$$

$$\frac{x+2-3(x+4)}{x+4} \le 0$$

$$\frac{x+2-3x-12}{x+4} \le 0$$

$$\frac{-2x-10}{x+4} \le 0$$

Now we can proceed as before. If $x = -5$, then the quotient $\dfrac{-2x-10}{x+4}$ equals zero, and if $x = -4$, the quotient is undefined. Then, using test numbers such as $-6, -4\dfrac{1}{2}$, and -3, we are able to study the sign behavior of the quotient, as in Figure 1.14.

$$\frac{-2x - 10}{x + 4} = 0 \qquad \frac{-2x - 10}{x + 4} \text{ is undefined}$$

$$\begin{array}{c|c|c}
\boxed{-6} & \boxed{-4\tfrac{1}{2}} & \boxed{-3} \\
\hline
\underset{-5}{\bullet} & & \underset{-4}{\bullet} \\
\end{array}$$

$-2x - 10$ is positive.	$-2x - 10$ is negative.	$-2x - 10$ is negative.
$x + 4$ is negative.	$x + 4$ is negative.	$x + 4$ is positive.
The quotient	The quotient	The quotient
$\dfrac{-2x - 10}{x + 4}$	$\dfrac{-2x - 10}{x + 4}$	$\dfrac{-2x - 10}{x + 4}$
is **negative**.	is **positive**.	is **negative**.

FIGURE 1.14

Therefore, the solution set for $\dfrac{x + 2}{x + 4} \leq 3$ is $(-\infty, -5] \cup (-4, \infty)$. ∎ ■

Absolute Value

In Section 0.1 we defined the **absolute value** of a real number by

$$|a| = \begin{cases} a, & \text{if } a \geq 0 \\ -a, & \text{if } a < 0 \end{cases}$$

We also interpreted the absolute value of any real number to be the distance between the number and zero on the real number line. For example, $|6| = 6$ because the distance between 6 and 0 is six units. Likewise, $|-8| = 8$ because the distance between -8 and 0 is eight units.

Both the definition and the number-line interpretation of absolute value provide ways of analyzing a variety of equations and inequalities involving absolute value. For example, suppose that we need to solve the equation $|x| = 4$. Thinking in terms of distance on the number line, the equation $|x| = 4$ means that we are looking for numbers that are four units from zero. Thus x must be 4 or -4. From the definition viewpoint, we could proceed as follows.

If $x \geq 0$, then $|x| = x$ and the equation $|x| = 4$ becomes $x = 4$.

If $x < 0$, then $|x| = -x$ and the equation $|x| = 4$ becomes $-x = 4$, which is equivalent to $x = -4$.

Using either approach, we see that the solution set for $|x| = 4$ is $\{-4, 4\}$.

The following property should seem reasonable from the distance interpretation and can be verified using the definition of absolute value.

PROPERTY 1.5

For any real number $k > 0$,

if $|x| = k$, then $x = k$ or $x = -k$

To verify Property 1.5 using the definition of absolute value, we can reason as follows.

If $x \geq 0$, then $|x| = x$ and the equation $|x| = k$ becomes $x = k$.

If $x < 0$, then $|x| = -x$ and the equation $|x| = k$ becomes $-x = k$, which is equivalent to $x = -k$.

Therefore, the equation $|x| = k$ is equivalent to $x = k$ or $x = -k$. Now let's use Property 1.5 to solve an equation of the form $|ax + b| = k$.

EXAMPLE 3

Solve $|3x - 2| = 7$.

Solution

$$|3x - 2| = 7$$
$$3x - 2 = 7 \quad \text{or} \quad 3x - 2 = -7$$
$$3x = 9 \quad \text{or} \quad 3x = -5$$
$$x = 3 \quad \text{or} \quad x = -\frac{5}{3}$$

The solution set is $\left\{ -\frac{5}{3}, 3 \right\}$. ∎

The distance interpretation for absolute value also provides a good basis for solving some inequalities. For example, to solve $|x| < 4$, we know that the distance between x and 0 must be less than four units. In other words, x is to be less than four units away from zero. Thus $|x| < 4$ is equivalent to $-4 < x < 4$ and the solution set is the interval $(-4, 4)$. We will have you use the definition of absolute value and verify the following general property in the next set of exercises.

PROPERTY 1.6

For any real number $k > 0$,

$$\text{if } |x| < k, \quad \text{then } -k < x < k$$

Example 4 illustrates the use of Property 1.6.

EXAMPLE 4

Solve $|2x + 1| < 5$.

Solution

$$|2x + 1| < 5$$
$$-5 < 2x + 1 < 5$$
$$-6 < 2x < 4$$
$$-3 < x < 2$$

The solution set is the interval $(-3, 2)$. ∎

Property 1.6 can also be expanded to include the $\leq$ situation—that is, *if $|x| \leq k$, then $-k \leq x \leq k$.*

EXAMPLE 5

Solve $|-3x - 2| \leq 6$.

Solution

$$|-3x - 2| \leq 6$$
$$-6 \leq -3x - 2 \leq 6$$
$$-4 \leq -3x \leq 8$$
$$\frac{4}{3} \geq x \geq -\frac{8}{3} \qquad \text{Note that multiplying through by } -\frac{1}{3}$$
$$\text{reverses the inequalities}$$

The statement $\frac{4}{3} \geq x \geq -\frac{8}{3}$ is equivalent to $-\frac{8}{3} \leq x \leq \frac{4}{3}$. Therefore, the solution set is $\left[-\frac{8}{3}, \frac{4}{3}\right]$. ∎

Now suppose that we want to solve $|x| > 4$. The distance between x and 0 must be more than four units or, in other words, x is to be more than four units away from zero. Therefore, $|x| > 4$ is equivalent to $x < -4$ or $x > 4$, and the solution set is $(-\infty, -4) \cup (4, \infty)$. The following general property can be verified by using the definition of absolute value.

PROPERTY 1.7

For any real number $k > 0$,

$$\text{if } |x| > k, \quad \text{then } x < -k \text{ or } x > k$$

EXAMPLE 6

Solve $|4x - 3| > 9$.

Solution

$$|4x - 3| > 9$$
$$4x - 3 < -9 \qquad \text{or} \qquad 4x - 3 > 9$$
$$4x < -6 \qquad \text{or} \qquad 4x > 12$$
$$x < -\frac{6}{4} \qquad \text{or} \qquad x > 3$$
$$x < -\frac{3}{2} \qquad \text{or} \qquad x > 3$$

The solution set is $\left(-\infty, -\frac{3}{2}\right) \cup (3, \infty)$. ∎

Property 1.7 can also be expanded to include the $\geq$ situation—that is, *if $|x| \geq k$, then $x \leq -k$ or $x \geq k$.*

EXAMPLE 7

Solve $|-2 - x| \geq 9$.

Solution

$$|-2 - x| \geq 9$$

$$-2 - x \leq -9 \quad \text{or} \quad -2 - x \geq 9$$

$$-x \leq -7 \quad \text{or} \quad -x \geq 11$$

$$x \geq 7 \quad \text{or} \quad x \leq -11$$

The solution set is $(-\infty, -11] \cup [7, \infty)$. ∎

Properties 1.5, 1.6, and 1.7 provide a sound basis for solving many equations and inequalities involving absolute value. However, if at any time you become doubtful about which property applies, don't forget the definition and the distance interpretation for absolute value. Furthermore, there are some equations and inequalities where the properties do not apply. Let's consider one such example.

EXAMPLE 8

Solve the equation $|3x - 1| = |x + 4|$.

Solution

We could solve this equation by applying the definition of absolute value to both expressions; however, let's approach it in a less formal way. For the two numbers, $3x - 1$ and $x + 4$, to have the same absolute value, they must either be equal or be opposites of each other. Therefore, the equation $|3x - 1| = |x + 4|$ is equivalent to $3x - 1 = x + 4$ or $3x - 1 = -(x + 4)$, which can be solved as follows.

$$3x - 1 = x + 4 \quad \text{or} \quad 3x - 1 = -(x + 4)$$

$$2x = 5 \quad \text{or} \quad 3x - 1 = -x - 4$$

$$x = \frac{5}{2} \quad \text{or} \quad 4x = -3$$

$$x = \frac{5}{2} \quad \text{or} \quad x = -\frac{3}{4}$$

The solution set is $\left\{-\frac{3}{4}, \frac{5}{2}\right\}$. ∎

We should also note that in Properties 1.5 through 1.7, k is a positive number. This is not a serious restriction because problems where k is nonpositive are easily solved as follows.

$|x - 2| = 0$ The solution set is {2} because $x - 2$ has to equal zero.

$|3x - 7| = -4$ The solution set is ∅. For any real number, the absolute value of $3x - 7$ will always be nonnegative.

$|2x - 1| < -3$ The solution set is ∅. For any real number, the absolute value of $2x - 1$ will always be nonnegative.

$|5x + 2| > -4$ The solution set is $(-∞, ∞)$. The absolute value of $5x + 2$, regardless which real number is substituted for x, will always be greater than -4.

The number-line approach used in Examples 1 and 2 of this section, along with Properties 1.6 and 1.7, provide a systematic way of solving absolute value inequalities that have the variable in the denominator of a fraction. Let's analyze one such problem.

EXAMPLE 9   Solve $\left| \dfrac{x-2}{x+3} \right| < 4$.

Solution

By Property 1.6, $\left| \dfrac{x-2}{x+3} \right| < 4$ becomes $-4 < \dfrac{x-2}{x+3} < 4$, which can be written

$$\frac{x-2}{x+3} > -4 \quad \text{and} \quad \frac{x-2}{x+3} < 4$$

Each part of this *and* statement can be solved as we handled Example 2 earlier.

(a)		(b)
$\dfrac{x-2}{x+3} > -4$	and	$\dfrac{x-2}{x+3} < 4$
$\dfrac{x-2}{x+3} + 4 > 0$	and	$\dfrac{x-2}{x+3} - 4 < 0$
$\dfrac{x - 2 + 4(x+3)}{x+3} > 0$	and	$\dfrac{x - 2 - 4(x+3)}{x+3} < 0$
$\dfrac{x - 2 + 4x + 12}{x+3} > 0$	and	$\dfrac{x - 2 - 4x - 12}{x+3} < 0$
$\dfrac{5x + 10}{x+3} > 0$	and	$\dfrac{-3x - 14}{x+3} < 0$

This solution set is shown in Figure 1.15(a).   This solution set is shown in Figure 1.15(b).

(a) ⟵————————)——(————————⟶
                     -3 -2

(b) ⟵————————)——(————————⟶
                 $-\dfrac{14}{3}$ -3

FIGURE 1.15

The intersection of the two solution sets pictured is the set shown in Figure 1.16.

$-\frac{14}{3}$ -2

FIGURE 1.16

Therefore, the solution set of $\left|\dfrac{x-2}{x+3}\right| < 4$ is $\left(-\infty, -\dfrac{14}{3}\right) \cup (-2, \infty)$.   ■

Yes, Example 9 is a little messy, but it does illustrate the weaving together of previously used techniques to solve a more complicated problem. Don't be in a hurry when doing such problems. First analyze the general approach to be taken and then carry out the details in a neatly organized format to minimize your chances of making careless errors.

PROBLEM SET 1.7

For Problems 1–14, solve each inequality and express the solution set in interval notation.

1. $\dfrac{x+1}{x-5} > 0$

2. $\dfrac{x+2}{x+4} \le 0$

3. $\dfrac{2x-1}{x+2} < 0$

4. $\dfrac{3x+2}{x-1} > 0$

5. $\dfrac{-x+3}{3x-1} \ge 0$

6. $\dfrac{-n-2}{n+4} < 0$

7. $\dfrac{n}{n+2} \ge 3$

8. $\dfrac{x}{x-1} > 2$

9. $\dfrac{x-1}{x+2} < 2$

10. $\dfrac{t-1}{t-5} \le 2$

11. $\dfrac{t-3}{t+5} > 1$

12. $\dfrac{x+2}{x+7} < 1$

13. $\dfrac{1}{x-2} < \dfrac{1}{x+3}$

14. $\dfrac{2}{x+1} > \dfrac{3}{x-4}$

For Problems 15–34, solve each equation.

15. $|x-2| = 6$

16. $|x+3| = 4$

17. $\left|x + \dfrac{1}{4}\right| = \dfrac{2}{5}$

18. $\left|x - \dfrac{2}{3}\right| = \dfrac{3}{4}$

19. $|2n-1| = 7$

20. $|2n+1| = 11$

21. $|3x+4| = 5$

22. $|5x-3| = 10$

23. $|7x-1| = -4$

24. $|-2x-1| = 6$

25. $|-3x-2| = 8$

26. $|5x-4| = -3$

27. $\left|\dfrac{3}{k-1}\right| = 4$

28. $\left|\dfrac{-2}{n+3}\right| = 5$

29. $|3x-1| = |2x+3|$

30. $|2x+1| = |4x-3|$

31. $|-2n+1| = |-3n-1|$

32. $|-4n+5| = |-3n-5|$

33. $|x-2| = |x+4|$

34. $|2x-3| = |2x+5|$

For Problems 35–70, solve each inequality and express the solution set in interval notation.

35. $|x| < 6$

36. $|x| \ge 4$

37. $|x| > 8$ $(9, \infty)$ 8

38. $|x| \le 1$

39. $|x| \ge -4$

40. $|x| < -5$

41. $|t-3| > 5$

42. $|n+2| < 1$

43. $|2x-1| \le 7$

44. $|2x+1| \ge 3$

45. $|3n+2| > 9$

46. $|5n-2| < 2$

47. $|4x - 3| < -5$

48. $|2 - x| > 1$

49. $|3 - 2x| < 4$

50. $|4x + 5| > -3$

51. $|7x + 2| \geq -2$

52. $|-1 - x| \geq 8$

53. $|-2 - x| \leq 5$

54. $|x - 1| + 2 < 4$

55. $|x + 3| - 2 < 1$

56. $|x - 5| + 4 \leq 2$

57. $|x + 4| - 1 > 1$

58. $|x - 2| + 3 > 6$

59. $3|x - 2| \geq 6$

60. $2|x + 1| < 8$

61. $-2|x + 1| > -10$

62. $-|x - 4| \leq -4$

63. $\left|\dfrac{x + 1}{x - 2}\right| < 3$

64. $\left|\dfrac{x - 1}{x - 4}\right| < 2$

65. $\left|\dfrac{x - 1}{x + 3}\right| > 1$

66. $\left|\dfrac{x + 4}{x - 5}\right| \geq 3$

67. $\left|\dfrac{n + 2}{n}\right| \geq 4$

68. $\left|\dfrac{t + 6}{t - 2}\right| < 1$

69. $\left|\dfrac{k}{2k - 1}\right| \leq 2$

70. $\left|\dfrac{k}{k + 2}\right| > 4$

THOUGHTS INTO WORDS

71. Explain how you would solve the inequality
$$\frac{x - 2}{(x + 1)^2} > 0.$$

72. Explain how you would solve the inequality
$|3x - 7| > -2$.

73. Why is $\left\{\dfrac{3}{2}\right\}$ the solution set for $|2x - 3| \leq 0$?

74. Consider the following approach for solving the inequality in Example 2 of this section.

$$\frac{x + 2}{x + 4} \leq 3$$

$$(x + 4)\left(\frac{x + 2}{x + 4}\right) \leq 3(x + 4) \qquad \text{Multiply both sides by } x + 4.$$

$$x + 2 \leq 3x + 12$$
$$-2x \leq 10$$
$$x \geq -5$$

Obviously, the solution set that we obtain using this approach differs from what we obtained in the text. What is wrong with this approach? Can we make any adjustments so that this basic approach works?

Further Investigations

75. Use the definition of absolute value and prove Property 1.6.

76. Use the definition of absolute value and prove Property 1.7.

77. Solve each of the following inequalities by using the definition of absolute value. Do not use Properties 1.6 and 1.7.

a. $|x + 5| < 11$

b. $|x - 4| \leq 10$

c. $|2x - 1| > 7$

d. $|3x + 2| \geq 1$

e. $|2 - x| < 5$

f. $|3 - x| > 6$

CHAPTER I SUMMARY

This chapter covers three large topics: (1) solving equations, (2) solving inequalities, and (3) problem solving.

Solving Equations

The following properties are used extensively in the equation-solving process.

    1. $a = b$ if and only if $a + c = b + c$.   Addition property of equality

    2. $a = b$ if and only if $ac = bc$, $c \neq 0$.   Multiplication property of

    3. If $ab = 0$, then $a = 0$ or $b = 0$.   equality

    4. If $a = b$, then $a^n = b^n$, where n is a positive integer.

Remember that applying the fourth property may result in some extraneous solutions, so you *must check* all potential solutions.

The cross-multiplication property of proportions (if $a/b = c/d$ then $ad = bc$) can be used to solve some equations.

Quadratic equations can be solved by (1) factoring, (2) completing the square, or (3) using the quadratic formula, which can be stated as

$$x = \frac{-b \pm \sqrt{b^2 - 4ac}}{2a}$$

The discriminant of a quadratic equation, $b^2 - 4ac$, indicates the nature of the solutions of the equation.

    1. If $b^2 - 4ac > 0$, the equation has two unequal real solutions.

    2. If $b^2 - 4ac = 0$, the equation has one real solution.

    3. If $b^2 - 4ac < 0$, the equation has two complex but nonreal solutions.

If x_1 and x_2 are the solutions of a quadratic equation $ax^2 + bx + c = 0$, then (1) $x_1 + x_2 = -b/a$ and (2) $x_1 x_2 = c/a$. These relationships can be used to check potential solutions.

The property *If* $|x| = k$, *then* $x = k$ *or* $x = -k (k > 0)$ is often helpful for solving equations that involve absolute value.

Solving Inequalities

The following properties form a basis for solving inequalities.

    1. If $a > b$, then $a + c > b + c$.

    2. If $a > b$ and $c > 0$, then $ac > bc$.

    3. If $a > b$ and $c < 0$, then $ac < bc$.

To solve compound statements that involve inequalities, we proceed as follows.

1. Solve separately each inequality in the compound statement.
2. If it is a conjunction, the solution set is the intersection of the solution sets of the inequalities.
3. If it is a disjunction, the solution set is the union of the solution sets of the inequalities.

Quadratic inequalities such as $(x + 3)(x - 7) > 0$ can be solved by considering the *sign behavior* of the individual factors.

The following properties play an important role in solving inequalities that involve absolute value.

1. If $|x| < k$, where $k > 0$, then $-k < x < k$.
2. If $|x| > k$, where $k > 0$, then $x > k$ or $x < -k$.

Problem Solving

It would be helpful for you to reread the pages of this chapter that pertain to problem solving. Some key problem-solving ideas are illustrated on these pages.

CHAPTER 1 REVIEW PROBLEM SET

For Problems 1–22, solve each equation.

1. $2(3x - 1) - 3(x - 2) = 2(x - 5)$

2. $\dfrac{n - 1}{4} - \dfrac{2n + 3}{5} = 2$

3. $\dfrac{2}{x + 2} + \dfrac{5}{x - 4} = \dfrac{7}{2x - 8}$

4. $0.07x + 0.12(550 - x) = 56$

5. $(3x - 1)^2 = 16$

6. $4x^2 - 29x + 30 = 0$

7. $x^2 - 6x + 10 = 0$

8. $n^2 + 4n = 396$

9. $15x^3 + x^2 - 2x = 0$

10. $\dfrac{t + 3}{t - 1} - \dfrac{2t + 3}{t - 5} = \dfrac{3 - t^2}{t^2 - 6t + 5}$

11. $\dfrac{5 - x}{2 - x} - \dfrac{3 - 2x}{2x} = 1$

12. $x^4 + 4x^2 - 45 = 0$

13. $2n^{-4} - 11n^{-2} + 5 = 0$

14. $\left(x - \dfrac{2}{x}\right)^2 + 4\left(x - \dfrac{2}{x}\right) = 5$

15. $\sqrt{5 + 2x} = 1 + \sqrt{2x}$

16. $\sqrt{3 + 2n} + \sqrt{2 - 2n} = 3$

17. $\sqrt{3 - t} - \sqrt{3 + t} = \sqrt{t}$

18. $|5x - 1| = 7$

19. $|2x + 5| = |3x - 7|$

20. $\left|\dfrac{-3}{n - 1}\right| = 4$

21. $x^3 + x^2 - 2x - 2 = 0$

22. $2x^{2/3} + 5x^{1/3} - 12 = 0$

For Problems 23–40, solve each inequality. Express the solution sets using interval notation.

23. $3(2 - x) + 2(x - 4) > -2(x + 5)$

24. $\dfrac{3}{5}x - \dfrac{1}{3} \le \dfrac{2}{3}x + \dfrac{3}{4}$

25. $\dfrac{n-1}{3} - \dfrac{2n+1}{4} > \dfrac{1}{6}$

26. $0.08x + 0.09(700 - x) \geq 59$

27. $-16 \leq 7x - 2 \leq 5$

28. $5 > \dfrac{3y+4}{2} > 1$

29. $x^2 - 3x - 18 < 0$

30. $n^2 - 5n \geq 14$

31. $(x-1)(x-4)(x+2) < 0$

32. $\dfrac{x+4}{2x-3} \leq 0$

33. $\dfrac{5n-1}{n-2} > 0$

34. $\dfrac{x-1}{x+3} \geq 2$

35. $\dfrac{t+5}{t-4} < 1$

36. $|4x - 3| > 5$

37. $|3x + 5| \leq 14$

38. $|-3 - 2x| < 6$

39. $\left|\dfrac{x-1}{x}\right| > 2$

40. $\left|\dfrac{n+1}{n+2}\right| < 1$

For Problems 41–56, solve each problem.

41. The sum of three consecutive odd integers is 31 less than four times the largest integer. Find the integers.

42. The sum of two numbers is 74. If the larger is divided by the smaller, the quotient is 7 and the remainder is 2. Find the numbers.

43. The perimeter of a rectangle is 38 centimeters and its area is 84 square centimeters. Find the dimensions of the rectangle.

44. A sum of money amounting to $13.55 consists of nickels, dimes, and quarters. There are three times as many dimes as nickels and three fewer quarters than dimes. How many coins of each denomination are there?

45. A retailer has some shirts that cost him $14 each. He wants to sell them to make a profit of 30% of the selling price. What price should he charge for the shirts?

46. How many gallons of a solution of glycerine and water containing 55% glycerine should be added to 15 gallons of a 20% solution to give a 40% solution?

47. The sum of the present ages of Rosie and her mother is 47 years. In 5 years, Rosie will be one-half as old as her mother at that time. Find the present ages of both Rosie and her mother.

48. Kelly invested $800, part of it at 9% and the remainder at 12%. Her total yearly interest from the two investments was $85.50. How much did she invest at each rate?

49. Regina has scores of 93, 88, 89, and 95 on her first four math exams. What score must she get on the fifth exam to have an average of 92 or better for the five exams?

50. At how many minutes after 2 P.M. will the minute hand of a clock overtake the hour hand?

51. Russ started to mow the lawn, a task that usually takes him 40 minutes. After he had been working for 15 minutes, his friend Jay came along with his mower and began to help Russ. Working together, they finished the lawn in 10 minutes. How long would it have taken Jay to mow the lawn by himself?

52. Barry bought a number of shares of stock for $600. A week later the value of the stock increased $3 per share, and he sold all but 10 shares and regained his original investment of $600. How many shares did he sell and at what price per share?

53. Larry drove 156 miles in one hour more than it took Mike to drive 108 miles. Mike drove at an average rate of 2 miles per hour faster than Larry. How fast did each one travel?

54. It takes Bill 2 hours longer to do a certain job than it takes Cindy. They worked together for 2 hours; then Cindy left and Bill finished the job in 1 hour. How long would it take each of them to do the job alone?

55. One leg of a right triangle is 5 centimeters longer than the other leg. The hypotenuse is 25 centimeters long. Find the length of each leg.

56. The area of a rectangle is 35 square inches. If both the length and width are increased by 3 inches, the area is increased by 45 square inches. Find the dimensions of the original rectangle.

CHAPTER 1 TEST

For Problems 1–14, solve each equation.

1. $3(2x - 1) - 4(x + 2) = -7$

2. $10x^2 + 13x - 3 = 0$

3. $(5x + 2)^2 = 25$

4. $\dfrac{3n + 4}{4} - \dfrac{2n - 1}{10} = \dfrac{11}{20}$

5. $2x^2 - x + 4 = 0$

6. $(n - 2)(n + 7) = -18$

7. $0.06x + 0.08(1400 - x) = 100$

8. $|3x - 4| = 7$

9. $3x^2 - 2x - 2 = 0$

10. $3x^3 + 21x^2 - 54x = 0$

11. $\dfrac{x}{2x + 1} - 1 = \dfrac{-4}{7(x - 2)}$

12. $\sqrt{2x} = x - 4$

13. $\sqrt{x + 1} + 2 = \sqrt{x}$

14. $2n^{-2} + 5n^{-1} - 12 = 0$

For Problems 15–21, solve each inequality and express the solution set using interval notation.

15. $2(x - 1) - 3(3x + 1) \geq -6(x - 5)$

16. $\dfrac{x - 2}{6} - \dfrac{x + 3}{9} > -\dfrac{1}{2}$

17. $|6x - 4| < 10$

18. $|4x + 5| \geq 6$

19. $2x^2 - 9x - 5 \leq 0$

20. $\dfrac{3x - 1}{x + 2} > 0$

21. $\dfrac{x - 2}{x + 6} \geq 3$

For Problems 22–25, solve each problem.

22. How many cups of grapefruit juice must be added to 30 cups of a punch that contains 8% grapefruit juice to obtain a punch that is 10% grapefruit juice?

23. Lian can ride her bike 60 miles in one hour less time that it takes Tasya to ride 60 miles. Lian's rate is 3 miles per hour faster than Tasya's rate. Find Lian's rate.

24. Abdul bought a number of shares of stock for a total of $3000. Three months later the stock had increased in value by $5 per share, and he sold all but 50 shares and regained his original investment of $3000. How many shares did he sell?

25. The perimeter of a rectangle is 46 centimeters and its area is 126 square centimeters. Find the dimensions of the rectangle.

2

COORDINATE GEOMETRY AND GRAPHING TECHNIQUES

Bridges are often constructed with semi-elliptical arches.

*René Descartes, a French mathematician of the 17th century, was able to transform geometric problems into an algebraic setting so that he could use the tools of algebra to solve the problems. This merging of algebraic and geometric ideas is the foundation of a branch of mathematics called **analytic geometry**, today more commonly called **coordinate geometry**. Basically, there are two kinds of problems in coordinate geometry: given an algebraic equation, find its geometric graph; and given a set of conditions pertaining to a geometric graph, find its algebraic equation. We will discuss problems of both types in this chapter.*

2.1 COORDINATE GEOMETRY

Recall that the real number line (Figure 2.1) exhibits a one-to-one correspondence between the set of real numbers and the points on a line. That is, to each real number there corresponds one and only one point on the line, and to each point on the line there corresponds one and only one real number. The number that corresponds to a particular point on the line is called the **coordinate** of that point.

FIGURE 2.1

Suppose that on the number line we want to know the distance *from −2 to 6*. The *from–to* vocabulary implies a **directed distance**, which is $6 - (-2) = 8$ units. In other words, it is 8 units in a *positive direction* from −2 to 6. Likewise, the distance from 9 to −4 is $-4 - 9 = -13$; it is 13 units in a *negative direction*. In general, if x_1 and x_2 are the coordinates of two points on the number line, then the distance **from** x_1 **to** x_2 is given by $x_2 - x_1$, and the distance **from** x_2 **to** x_1 is given by $x_1 - x_2$.

Now suppose that we want to find the distance *between* −2 and 6. The *between* vocabulary implies **distance without regard to direction**. Thus the distance between −2 and 6 can be found by using either $|6 - (-2)| = 8$ or $|-2 - 6| = 8$. In general, if x_1 and x_2 are the coordinates of two points on the number line, then the distance between x_1 and x_2 can be found by using either $|x_2 - x_1|$ or $|x_1 - x_2|$.

Sometimes it is necessary to find the coordinate of a point located somewhere between the two given points. For example, in Figure 2.2 suppose that we want to find the coordinate (x) of the point located two-thirds of the distance *from 2 to 8*.

Because the total distance from 2 to 8 is $8 - 2 = 6$ units, we can start at 2 and move $\frac{2}{3}(6) = 4$ units toward 8. Thus

$$x = 2 + \frac{2}{3}(6) = 2 + 4 = 6$$

F I G U R E   2 . 2

The following examples further illustrate the process of finding the coordinate of a point somewhere between two given points (Figure 2.3).

Problem	**Solution**

a. Three-fourths of the distance from -2 to 10

$$x = -2 + \frac{3}{4}[10 - (-2)]$$
$$= -2 + \frac{3}{4}(12)$$
$$= 7$$

F I G U R E   2 . 3 a

b. Two-fifths of the distance from -1 to 7

$$x = -1 + \frac{2}{5}[7 - (-1)]$$
$$= -1 + \frac{2}{5}(8)$$
$$= \frac{11}{5}$$

F I G U R E   2 . 3 b

c. One-third of the distance from 9 to 1

$$x = 9 + \frac{1}{3}(1 - 9)$$
$$= 9 + \frac{1}{3}(-8)$$
$$= \frac{19}{3}$$

F I G U R E   2 . 3 c

d. a/b of the distance from x_1 to x_2

$$x = x_1 + \frac{a}{b}(x_2 - x_1)$$

F I G U R E   2 . 3 d

Problem (d) indicates that a general formula can be developed for this type of problem. However, it may be easier to remember the basic approach than it is to memorize the formula.

As we saw in Chapter 2, the real number line provides a geometric model for graphing solutions of algebraic equations and inequalities involving *one variable*. For example, the solutions of "$x > 2$ or $x \leq -1$" are graphed in Figure 2.4.

FIGURE 2.4

Rectangular Coordinate System

To expand our work with coordinate geometry, we now consider two number lines, one vertical and one horizontal, perpendicular to each other at the point associated with zero on both lines (Figure 2.5). We refer to these number lines as the **horizontal** and **vertical axes** and together as the **coordinate axes**. They partition the plane into four regions called **quadrants**. The quadrants are numbered counterclockwise from I through IV, as indicated in Figure 2.5. The point of intersection of the two axes is called the **origin**.

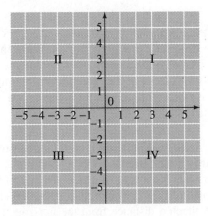

FIGURE 2.5

It is now possible to set up a one-to-one correspondence between *ordered pairs* of real numbers and the points in a plane. To each ordered pair of real numbers there corresponds a unique point in the plane, and to each point in the plane there corresponds a unique ordered pair of real numbers. Figure 2.6 shows examples of this correspondence. The ordered pair (3, 2) means that point *A* is located three units to the right and two units up from the origin. The ordered pair (−3, −5) means that point *D* is located three units to the left and five units down from the origin. The ordered pair (0, 0) is associated with the origin.

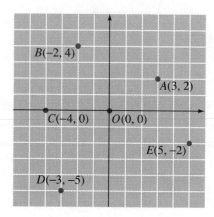

FIGURE 2.6

REMARK   We used the notation (−2, 4) in Chapter 1 to indicate an interval of the real number line. Now we are using the same notation to indicate an ordered pair of real numbers. This double meaning should not be confusing, because the context of the material will definitely indicate the meaning at a particular time. Throughout this chapter we will be using the ordered-pair interpretation.

In general, the real numbers *a* and *b* in the ordered pair (*a*, *b*) are associated with a point; they are referred to as the **coordinates of the point**. The first number, *a*, is called the **abscissa**; it is the directed distance of the point from the vertical axis, measured parallel to the horizontal axis. The second number, *b*, is called the **ordinate**; it is the directed distance from the horizontal axis, measured parallel to the vertical axis (Figure 2.7a). Thus, in the first quadrant, all points have a positive abscissa and a positive ordinate. In the second quadrant, all points have a negative abscissa and a positive ordinate. We have indicated the sign situations for all four quadrants in Figure 2.7(b). This system of associating points in a plane with pairs of real numbers is called the **rectangular coordinate** system or the **Cartesian coordinate system**.

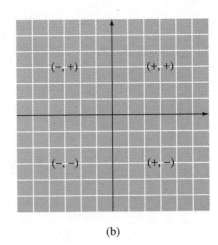

(a)   (b)

FIGURE 2.7

Distance Between Two Points

As we work with the rectangular coordinate system, it is sometimes necessary to express the length of certain line segments. In other words, we need to be able to find the *distance* between two points. Let's first consider two specific examples and then develop a general distance formula.

EXAMPLE 1

Find the distance between the points $A(2, 2)$ and $B(5, 2)$ and also between the points $C(-2, 5)$ and $D(-2, -4)$.

Solution

Let's plot the points and draw $\overline{AB}$ and $\overline{CD}$ as in Figure 2.8. (The symbol $\overline{AB}$ denotes the line segment with endpoints A and B.) Because $\overline{AB}$ is parallel to the horizontal

FIGURE 2.8

axis, its length can be expressed as $|5 - 2|$ or $|2 - 5|$. Thus the length of $\overline{AB}$ (we shall use the notation AB to represent the length of $\overline{AB}$) is $AB = 3$ units. Likewise, because $\overline{CD}$ is parallel to the vertical axis, we obtain $CD = |5 - (-4)| = 9$ units. ∎

EXAMPLE 2

Find the distance between the points $A(2, 3)$ and $B(5, 7)$.

Solution

Let's plot the points and form a right triangle using point D, as indicated in Figure 2.9. Note that the coordinates of point D are $(5, 3)$. Because $\overline{AD}$ is parallel to the

FIGURE 2.9

horizontal axis, as in Example 1, we have $AD = |5 - 2| = 3$ units. Likewise, $\overline{DB}$ is parallel to the vertical axis, and therefore $DB = |7 - 3| = 4$ units. Applying the Pythagorean theorem, we obtain

$$
\begin{aligned}
(AB)^2 &= (AD)^2 + (DB)^2 \\
&= 3^2 + 4^2 \\
&= 9 + 16 \\
&= 25
\end{aligned}
$$

Thus

$$AB = \sqrt{25} = 5 \text{ units}$$ ∎

Before we use the approach in Example 2 to develop a general distance formula, let's make another notational agreement. For most problems in coordinate geometry, it is customary to label the horizontal axis the x axis and the vertical axis the y axis. Then ordered pairs representing points in the xy plane are of the form (x, y); that is, x is the first coordinate and y is the second coordinate. Now let's develop a general distance formula.

Let $P_1(x_1, y_1)$ and $P_2(x_2, y_2)$ represent any two points in the xy plane. Form a right triangle using point R, as indicated in Figure 2.10. The coordinates of the vertex of the right angle, at point R, are (x_2, y_1). The length of $\overline{P_1R}$ is $|x_2 - x_1|$ and the length of $\overline{RP_2}$ is $|y_2 - y_1|$. Let d represent the length of $\overline{P_1P_2}$ and apply the Pythagorean theorem to obtain

$$d^2 = |x_2 - x_1|^2 + |y_2 - y_1|^2$$

F I G U R E   2 . 1 0

Because $|a|^2 = a^2$ for any real number a, the distance formula can be stated

$$d = \sqrt{(x_2 - x_1)^2 + (y_2 - y_1)^2}$$

It makes no difference which point you call P_1 and which you call P_2. Also, remember that if you forget the formula, there is no need to panic: Form a right triangle and apply the Pythagorean theorem as we did in Example 2.

Let's consider some examples that illustrate the use of the distance formula.

EXAMPLE 3

Find the distance between $(-2, 5)$ and $(1, -1)$.

Solution

Let $(-2, 5)$ be P_1 and $(1, -1)$ be P_2. Use the distance formula to obtain

$$\begin{aligned}
d &= \sqrt{(x_2 - x_1)^2 + (y_2 - y_1)^2} \\
&= \sqrt{[1 - (-2)]^2 + (-1 - 5)^2} \\
&= \sqrt{3^2 + (-6)^2} \\
&= \sqrt{9 + 36} \\
&= \sqrt{45} = 3\sqrt{5}
\end{aligned}$$

The distance between the two points is $3\sqrt{5}$ units.

In Example 3, note the simplicity of the approach when we use the distance formula. No diagram was needed; we merely plugged in the values and did the computation. However, many times a figure *is* helpful in the analysis of the problem, as we will see in the next example.

EXAMPLE 4

Verify that the points $(-3, 6)$, $(3, 4)$, and $(1, -2)$ are vertices of an isosceles triangle. (An isosceles triangle has two sides of the same length.)

Solution

Let's plot the points and draw the triangle (Figure 2.11). The lengths d_1, d_2, and d_3 can all be found by using the distance formula.

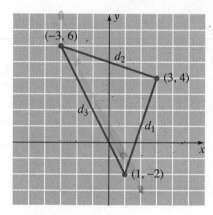

FIGURE 2.11

$$d_1 = \sqrt{(3 - 1)^2 + [4 - (-2)]^2}$$
$$= \sqrt{4 + 36}$$
$$= \sqrt{40} = 2\sqrt{10}$$
$$d_2 = \sqrt{(-3 - 3)^2 + (6 - 4)^2}$$
$$= \sqrt{36 + 4}$$
$$= \sqrt{40} = 2\sqrt{10}$$
$$d_3 = \sqrt{(-3 - 1)^2 + [6 - (-2)]^2}$$
$$= \sqrt{16 + 64}$$
$$= \sqrt{80} = 4\sqrt{5}$$

Because $d_1 = d_2$, it is an isosceles triangle.   ■

Points of Division of a Line Segment

Earlier in this section we discussed the process of finding the coordinate of a point on a number line, given that it is located somewhere between two other points on the line. This same type of problem can occur in the xy plane, and the approach we used earlier can be extended to handle it. Let's consider some examples.

EXAMPLE 5

Find the coordinates of the point P, which is two-thirds of the distance from $A(1, 2)$ to $B(7, 5)$.

Solution

In Figure 2.12 we plotted the given points A and B and completed a figure to help us analyze the problem. To find the coordinates of point P, we can proceed as follows. Point D is two-thirds of the distance from A to C because parallel lines cut off

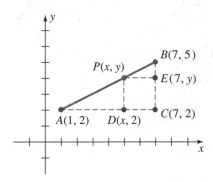

FIGURE 2.12

$$y = 2 + \frac{2}{3}(5\text{-}2)$$

FIGURE 2.13

proportional segments on every transversal that intersects the lines. Therefore, because $\overline{AC}$ is parallel to the x axis, it can be treated as a segment of the number line (see Figure 2.13). Thus we have

$$x = 1 + \frac{2}{3}(7 - 1) = 1 + \frac{2}{3}(6) = 5$$

Similarly, $\overline{CB}$ is parallel to the y axis, so it can also be treated as a segment of the number line (see Figure 2.14). Thus we obtain

$$y = 2 + \frac{2}{3}(5 - 2)$$

$$= 2 + \frac{2}{3}(3) = 4$$

$B \bullet\, 5$
$E \bullet\, y$

$C \bullet\, 2$

FIGURE 2.14

The point P has the coordinates $(5, 4)$. ■

EXAMPLE 6

Find the coordinates of the midpoint of the line segment determined by the points $P_1(x_1, y_1)$ and $P_2(x_2, y_2)$.

Solution

Figure 2.15 helps with the analysis of the problem. The line segment $\overline{P_1R}$ is parallel to the x axis, and $S(x, y_1)$ is the midpoint of $\overline{P_1R}$ (see Figure 2.16). Thus we can determine the x coordinate of S.

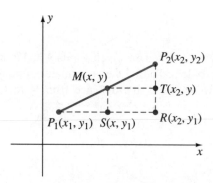

FIGURE 2.15

$P_1 \quad\quad S \quad\quad R$
$x_1 \quad\quad x \quad\quad x_2$

FIGURE 2.16

$$x = x_1 + \frac{1}{2}(x_2 - x_1)$$
$$= x_1 + \frac{1}{2}x_2 - \frac{1}{2}x_1$$
$$= \frac{1}{2}x_1 + \frac{1}{2}x_2 = \frac{x_1 + x_2}{2}$$

Similarly, $\overline{RP_2}$ is parallel to the y axis, and $T(x_2, y)$ is the midpoint of $\overline{RP_2}$ (see Figure 2.17). Therefore, we can calculate the y coordinate of T.

$P_2 \; \bullet \; y_2$
$T \; \bullet \; y$
$R \; \bullet \; y_1$

FIGURE 2.17

$$y = y_1 + \frac{1}{2}(y_2 - y_1)$$
$$= y_1 + \frac{1}{2}y_2 - \frac{1}{2}y_1$$
$$= \frac{1}{2}y_1 + \frac{1}{2}y_2 = \frac{y_1 + y_2}{2}$$

Thus the coordinates of the midpoint of a line segment determined by $P_1(x_1, y_1)$ and $P_2(x_2, y_2)$ are

$$\left(\frac{x_1 + x_2}{2}, \frac{y_1 + y_2}{2}\right)$$

∎

EXAMPLE 7

Find the coordinates of the midpoint of the line segment determined by the points $(-2, 4)$ and $(6, -1)$.

Solution

Using the midpoint formula, we obtain

$$\left(\frac{x_1 + x_2}{2}, \frac{y_1 + y_2}{2}\right) = \left(\frac{-2 + 6}{2}, \frac{4 + (-1)}{2}\right)$$

$$= \left(\frac{4}{2}, \frac{3}{2}\right)$$

$$= \left(2, \frac{3}{2}\right)$$ ■

We want to emphasize two ideas that emerge from Examples 5, 6, and 7. If we want to find a point of division of a line segment, then we use the same approach as in Example 5. However, for the special case of the midpoint, the formula developed in Example 6 is convenient to use.

PROBLEM SET 2.1

For Problems 1–8, find the indicated distances on a number line.

1. From -4 to 6

2. From 5 to -14

3. From -6 to -11

4. From -7 to 10

5. Between -2 and 4

6. Between -4 and -12

7. Between 5 and -10

8. Between -2 and 13

For Problems 9–14, find the coordinate of the indicated point on a number line.

9. Two-thirds of the distance from 1 to 10

10. Three-fourths of the distance from -2 to 14

11. One-third of the distance from -3 to 7

12. Two-fifths of the distance from -5 to 6

13. Three-fifths of the distance from -1 to -11

14. Five-sixths of the distance from 3 to -7

For Problems 15–22, find the length of $\overline{AB}$ and the midpoint of $\overline{AB}$.

15. $A(2, 1), B(10, 7)$

16. $A(-2, -1), B(7, 11)$

17. $A(1, -1), B(3, -4)$

18. $A(-5, 2), B(-1, 6)$

19. $A(6, -4), B(9, -7)$

20. $A(-3, 3), B(0, -3)$

21. $A\left(\frac{1}{2}, \frac{1}{3}\right), B\left(-\frac{1}{3}, \frac{3}{2}\right)$

22. $A\left(-\frac{3}{4}, 2\right), B\left(-1, -\frac{5}{4}\right)$

For Problems 23–28, find the coordinates of the indicated point in the *xy* plane.

23. One-third of the distance from $(2, 3)$ to $(5, 9)$

24. Two-thirds of the distance from $(1, 4)$ to $(7, 13)$

25. Two-fifths of the distance from $(-2, 1)$ to $(8, 11)$

26. Three-fifths of the distance from $(2, -3)$ to $(-3, 8)$

27. Five-eighths of the distance from $(-1, -2)$ to $(4, -10)$

28. Seven-eighths of the distance from $(-2, 3)$ to $(-1, -9)$

For Problems 29–44, solve each of the problems.

29. Find the coordinates of the point that is one-fourth of the distance from $(2, 4)$ to $(10, 13)$ by (a) using the midpoint formula twice, and (b) using the same approach as for Problems 23–28.

30. If one endpoint of a line segment is $(-6, 4)$ and the midpoint of the segment is $(-2, 7)$, find the other endpoint.

31. Use the distance formula to verify that the points $(-2, 7)$, $(2, 1)$, and $(4, -2)$ lie on a straight line.

32. Use the distance formula to verify that the points $(-3, 8)$, $(7, 4)$, and $(5, -1)$ are vertices of a right triangle.

33. Verify that the points $(0, 3)$, $(2, -3)$, and $(-4, -5)$ are vertices of an isosceles triangle.

34. Verify that the points $(7, 12)$ and $(11, 18)$ divide the line segment joining $(3, 6)$ and $(15, 24)$ into three segments of equal length.

35. Find the perimeter of the triangle whose vertices are $(-6, -4)$, $(0, 8)$, and $(6, 5)$.

36. Verify that $(-4, 9)$, $(8, 4)$, $(3, -8)$, and $(-9, -3)$ are vertices of a square.

37. Verify that the points $(4, -5)$, $(6, 7)$, and $(-8, -3)$ lie on a circle that has its center at $(-1, 2)$.

38. Suppose that $(-2, 5)$, $(6, 3)$ and $(-4, -1)$ are three vertices of a parallelogram. How many possibilities are there for the fourth vertex? Find the coordinates of each of these points. (*Hint:* The diagonals of a parallelogram bisect each other.)

39. Find x such that the line segment determined by $(x, -2)$ and $(-2, -14)$ is 13 units long.

40. Consider the triangle whose vertices are $(4, -6)$, $(2, 8)$, and $(-4, 2)$. Verify that the medians of this triangle intersect at a point that is two-thirds of the distance from a vertex to the midpoint of the opposite side. (A median of the triangle is the line segment determined by a vertex and the midpoint of the opposite side. Every triangle has three medians.)

41. Consider the line segment determined by $A(-1, 2)$ and $B(5, 11)$. Find the coordinates of a point P such that $AP/PB = 2/1$.

42. Verify that the midpoint of the hypotenuse of the right triangle formed by the points $A(4, 0)$, $B(0, 0)$, and $C(0, 6)$ is the same distance from all three vertices.

43. Consider the parallelogram determined by the points $A(1, 1)$, $B(5, 1)$, $C(6, 4)$, and $D(2, 4)$. Verify that the diagonals of this parallelogram bisect each other.

44. Consider the quadrilateral determined by the points $A(5, -3)$, $B(3, 4)$, $C(-2, 1)$, and $D(-1, -2)$. Verify that the line segments joining the midpoints of the opposite sides of this quadrilateral bisect each other.

THOUGHTS INTO WORDS

45. Consider the line segment determined by the two endpoints $A(2, 1)$ and $B(5, 10)$. Describe how you would find the coordinates of the point that is two-thirds of the distance from A to B. Then describe how you would find the point that is two-thirds of the distance from B to A.

46. How would you define the term *coordinate geometry* to a group of elementary algebra students?

Further Investigations

47. The tools of coordinate geometry can be used to prove various geometric properties. For example, consider the following way of proving that the diagonals of a rectangle are equal in length.

First we draw a rectangle and display it on coordinate axes by using a convenient position for the origin. Now we can use the distance formula to find the lengths of the diagonals $\overline{AC}$ and $\overline{BD}$. See Figure 2.18.

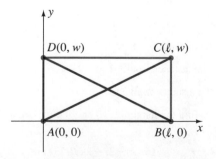

FIGURE 2.18

$$AC = \sqrt{(l - 0)^2 + (w - 0)^2} = \sqrt{l^2 + w^2}$$

$$BD = \sqrt{(0 - l)^2 + (w - 0)^2} = \sqrt{l^2 + w^2}$$

Thus $AC = BD$, and we have proved that the diagonals are equal in length. Prove each of the following.

a. The diagonals of an isosceles trapezoid are equal in length.

b. The line segment joining the midpoints of two sides of a triangle is equal in length to one-half of the third side.

c. The midpoint of the hypotenuse of a right triangle is equally distant from all three vertices.

d. The diagonals of a parallelogram bisect each other.

e. The line segments joining the midpoints of the opposite sides of a quadrilateral bisect each other.

f. The medians of a triangle intersect at a point that is two-thirds of the distance from a vertex to the midpoint of the opposite side. (See Problem 40.)

2.2 GRAPHING TECHNIQUES: LINEAR EQUATIONS AND INEQUALITIES

As you continue to study mathematics, you will find that the ability to sketch the graph of an equation quickly is important. Therefore, various curve-sketching techniques are discussed through precalculus and calculus courses. We will use a good portion of this chapter to expand your repertoire of graphing techniques.

First, let's briefly review some basic ideas by considering the solutions for the equation $y = x + 2$. A **solution** of an equation in two variables is an ordered pair of real numbers that satisfy the equation. When the variables are x and y, the ordered pairs are of the form (x, y). We see that $(1, 3)$ is a solution for $y = x + 2$ because replacing x by 1 and y by 3 yields a true numerical statement: $3 = 1 + 2$. Likewise, $(-2, 0)$ is a solution because $0 = -2 + 2$ is a true statement. An infinite number of pairs of real numbers that satisfy $y = x + 2$ can be found by arbitrarily choosing values for x and, for each value of x chosen, determining a corresponding value for y. Let's use a table to record some of the solutions for $y = x + 2$.

CHOOSE x	DETERMINE y FROM $y = x + 2$	SOLUTIONS FOR $y = x + 2$
0	2	$(0, 2)$
1	3	$(1, 3)$
3	5	$(3, 5)$
5	7	$(5, 7)$
-2	0	$(-2, 0)$
-4	-2	$(-4, -2)$
-6	-4	$(-6, -4)$

Plotting the points associated with the ordered pairs from the table produces Figure 2.19(a). The straight line containing the points (Figure 2.19b) is called the **graph of the equation** $y = x + 2$.

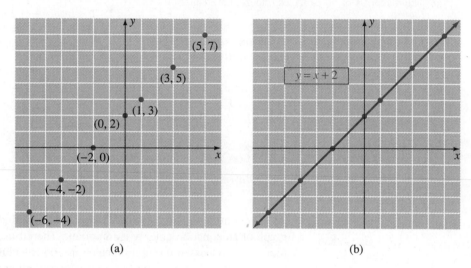

(a)   (b)

FIGURE 2.19

Graphing Linear Equations

Probably the most valuable graphing technique is the ability to recognize the kind of graph that is produced by a particular type of equation. For example, from previous mathematics courses you may remember that any equation of the form $Ax + By = C$, where A, B, and C are constants (A and B not both zero) and x and y are variables, is a **linear equation** and that its graph is a **straight line**. Two comments about this description of a linear equation should be made. First, the choice of x and y as variables is arbitrary; any two letters can be used to represent the variables. For example, an equation such as $3r + 2s = 9$ is also a linear equation in two variables. In order to avoid constantly changing the labeling of the coordinate axes when graphing equations, we will use the same two variables, x and y, in all equations. Second, the statement *any equation of the form* $Ax + By = C$ technically means any equation of that form or equivalent to that form. For example, the equation $y = 2x - 1$ is equivalent to $-2x + y = -1$ and therefore is linear and produces a straight-line graph.

Before we graph some linear equations, let's define in general the **intercepts** of a graph.

> The x coordinates of the points that a graph has in common with the x axis are called the x intercepts of the graph. (To compute the x intercepts, let $y = 0$ and solve for x.)
>
> The y coordinates of the points that a graph has in common with the y axis are called the y intercepts of the graph. (To compute the y intercepts, let $x = 0$ and solve for y.)

Once we know that any equation of the form $Ax + By = C$ produces a straight-line graph, along with the fact that two points determine a straight line, graphing linear equations becomes a simple process. We can find two points on the graph and draw the line determined by those two points. Usually the two points involving the intercepts are easy to find, and generally it's a good idea to plot a third point to serve as a check.

EXAMPLE 1

Graph $3x - 2y = 6$.

Solution

First let's find the intercepts. If $x = 0$, then

$$3(0) - 2y = 6$$
$$-2y = 6$$
$$y = -3$$

Therefore, the point $(0, -3)$ is on the line. If $y = 0$, then

$$3x - 2(0) = 6$$
$$3x = 6$$
$$x = 2$$

Thus the point $(2, 0)$ is also on the line. Now let's find a check point. If $x = -2$, then

$$3(-2) - 2y = 6$$
$$-6 - 2y = 6$$
$$-2y = 12$$
$$y = -6$$

Thus the point $(-2, -6)$ is also on the line. In Figure 2.20, the three points are plotted and the graph of $3x - 2y = 6$ is drawn.

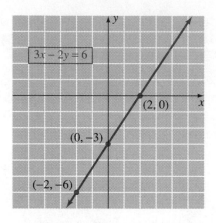

FIGURE 2.20   ■

Note in Example 1 that we did not solve the given equation for y in terms of x or for x in terms of y. Because we know the graph is a straight line, there is no need for an extensive table of values; thus there is no need to change the form of the original equation. Furthermore, the point $(-2, -6)$ served as a check point. If it had not been on the line determined by the two intercepts, then we would have known that we had made an error in finding the intercepts.

EXAMPLE 2

Graph $y = -2x$.

Solution

If $x = 0$, then $y = -2(0) = 0$, so the origin $(0, 0)$ is on the line. Because both intercepts are determined by the point $(0, 0)$, another point is necessary to determine the line. Then a third point should be found as a check point. The graph of $y = -2x$ is shown in Figure 2.21.

x	y
0	0
1	-2
-1	2

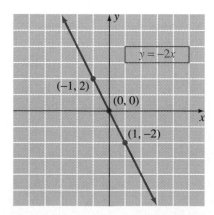

FIGURE 2.21   ■

Example 2 illustrates the general concept that for the form $Ax + By = C$, if $C = 0$ then the line contains the origin. Stated another way, **the graph of any equation of the form $y = kx$, where k is any real number, is a straight line containing the origin**

EXAMPLE 3

Graph $x = 2$.

Solution

Because we are considering linear equations *in two variables*, the equation $x = 2$ is equivalent to $x + 0(y) = 2$. Any value of y can be used, but the x value must always be 2. Therefore, some of the solutions are $(2, 0)$, $(2, 1)$, $(2, 2)$, $(2, -1)$, and $(2, -2)$. The graph of $x = 2$ is the vertical line shown in Figure 2.22.

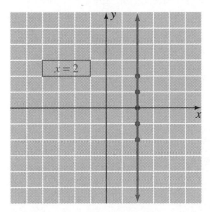

FIGURE 2.22

REMARK   It is important to realize that we are presently graphing **equations in two variables** (graphing in two-dimensional space). Thus, as shown in Example 3, the graph of $x = 2$ is a line. If we were graphing **equations in one variable** (graphing on a number line), then the graph of $x = 2$ would be a dot at 2. In subsequent mathematics courses, you may do some graphing of **equations in three variables** (graphing in three-dimensional space). At that time, the graph of $x = 2$ will be a plane.

In general, the graph of any equation of the form $Ax + By = C$, where $A = 0$ or $B = 0$ (not both), is a line parallel to one of the axes. More specifically, **any equation of the form $x = a$, where a is any nonzero real number, is a line parallel to the y axis** having an x intercept of a. Any equation of the form $y = b$, where b is a nonzero real number, is a line parallel to the x axis having a y intercept of b.

Graphing Linear Inequalities

Linear inequalities in two variables are of the form $Ax + By > C$ or $Ax + By < C$, where A, B, and C are real numbers. **(Combined linear equality and inequality**

statements are of the form $Ax + By \geq C$ or $Ax + By \leq C$.) Graphing linear inequalities is almost as easy as graphing linear equations. The following discussion will lead us to a simple, step-by-step process.

Let's consider the following equation and related inequalities.

$$x + y = 2 \qquad x + y > 2 \qquad x + y < 2$$

The straight line in Figure 2.23 is the graph of $x + y = 2$. The line divides the plane into two half-planes, one above the line and one below the line. For each point in the half-plane *above* the line, the ordered pair (x, y) associated with the point satisfies the inequality $x + y > 2$. For example, the ordered pair $(3, 4)$ produces the true statement $3 + 4 > 2$. Likewise, for each point in the half-plane *below* the line, the ordered pair (x, y) associated with the point satisfies the inequality $x + y < 2$. For example, $(-3, 1)$ produces the true statement $-3 + 1 < 2$.

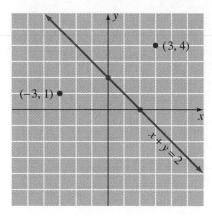

F I G U R E   2 . 2 3

Now let's use these ideas from the previous discussion to help graph some inequalities.

EXAMPLE 4

Graph $x - 2y > 4$.

Solution

First, graph $x - 2y = 4$ as a dashed line because equality is not included in $x - 2y > 4$ (Figure 2.24). Second, because *all* of the points in a specific half-plane satisfy either $x - 2y > 4$ or $x - 2y < 4$, try a *test point*. For example, try the origin.

$$x - 2y > 4 \quad \text{becomes } 0 - 2(0) > 4, \text{ which is a false statement.}$$

Because the ordered pairs in the half-plane containing the origin do not satisfy $x - 2y > 4$, the ordered pairs in the other half-plane must satisfy it. Therefore, the graph of $x - 2y > 4$ is the half-plane *below* the line, as indicated by the shaded portion in Figure 2.25.

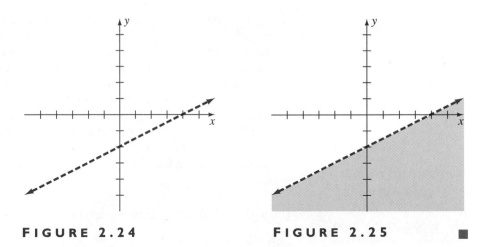

FIGURE 2.24   **FIGURE 2.25**   ∎

To graph a linear inequality, we suggest the following steps.

1. Graph the corresponding equality. Use a solid line if equality is included in the original statement and a dashed line if equality is not included.

2. Choose a *test point* not on the line and substitute its coordinates into the inequality. (The origin is a convenient point if it is not on the line.)

3. The graph of the original inequality is

   a. the half-plane containing the test point if the inequality is satisfied by that point.

   b. the half-plane not containing the test point if the inequality is not satisfied by the point.

EXAMPLE 5

Graph $2x + 3y \geq -6$.

Solution

STEP 1   Graph $2x + 3y = -6$ as a solid line (Figure 2.26).

STEP 2   Choose the origin as a test point.

$$2x + 3y \geq -6 \quad \text{becomes} \quad 2(0) + 3(0) \geq -6$$

which is true.

STEP 3   The test point satisfies the given inequality, so all points in the same half-plane as the test point satisfy it. The graph of $2x + 3y \geq -6$ is the line and the half-plane above the line (Figure 2.26).

∎

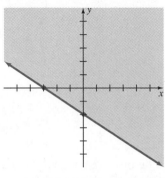

FIGURE 2.26

Graphing Utilities

The term **graphing utility** is used in current literature to refer to either a graphing calculator or a computer with a graphing software package. (We will frequently say,

"Use a graphing calculator to . . ." when we mean a graphing calculator or a computer with an appropriate software package.) These devices have a wide range of capabilities that enable the user not only to obtain a quick graph but also to study various characteristics of it—for example, the x intercepts, y intercepts, and turning points of a graph. We will introduce some of these features of graphing utilities as we need them in the text. Because so many different types of graphing utilities are available, we will use mostly generic terminology and let you consult your user's manual for specific key-punching instructions. We also suggest that you study the graphing utility examples in this text even if you do not have access to a graphing calculator or a computer. The examples were chosen to reinforce the concepts we are discussing.

EXAMPLE 6

Use a graphing utility to obtain a graph of the line $2.1x + 5.3y = 7.9$.

Solution

First, we need to solve the equation for y in terms of x. (If you are using a computer for this problem, you may not need to change the form of the given equation. Some software packages will allow you to graph two-variable equations without solving for y.)

$$2.1x + 5.3y = 7.9$$
$$5.3y = 7.9 - 2.1x$$
$$y = \frac{7.9 - 2.1x}{5.3}$$

Now we can enter the expression $\dfrac{7.9 - 2.1x}{5.3}$ for Y_1 and obtain the graph as shown in Figure 2.27.

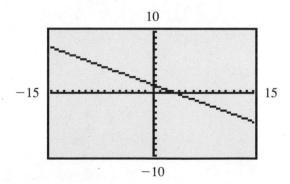

FIGURE 2.27

As indicated in Figure 2.27, the **viewing rectangle** of a graphing utility is a portion of the xy plane shown on the display of the utility. In this display the boundaries were set so that $-15 \leq x \leq 15$ and $-10 \leq y \leq 10$. These boundaries were set automatically; however, the fact that boundaries can be assigned as necessary is an important feature of graphing utilities.

PROBLEM SET 2.2

For Problems 1–16, graph each linear equation.

1. $x - 2y = 4$ **2.** $2x + y = -4$

3. $3x + 2y = 6$ **4.** $2x - 3y = 6$

5. $4x - 5y = 20$ **6.** $5x + 4y = 20$

7. $x - y = 3$ **8.** $-x + y = 4$

9. $y = 3x - 1$ **10.** $y = -2x + 3$

11. $y = -x$ **12.** $y = 4x$

13. $x = 0$ **14.** $y = -1$

15. $y = \dfrac{2}{3}x$ **16.** $y = -\dfrac{1}{2}x$

For Problems 17–30, graph each linear inequality.

17. $x + 2y > 4$ **18.** $2x - y < -4$

19. $3x - 2y < 6$ **20.** $2x + 3y < 6$

21. $2x + 5y \leq 10$ **22.** $4x + 5y \leq 20$

23. $y > -x - 1$ **24.** $y < 3x - 2$

25. $y \leq -x$ **26.** $y \geq x$

27. $x + 2y < 0$ **28.** $3x - y > 0$

29. $x > -1$ **30.** $y < 3$

THOUGHTS INTO WORDS

31. Explain how you would graph the inequality $-x + 2y > -4$.

32. What is the graph of the disjunction $x = 0$ **or** $y = 0$? What is the graph of the conjunction $x = 0$ **and** $y = 0$? Explain your answers.

Further Investigations

From our work with absolute value, we know that $|x + y| = 4$ is equivalent to $x + y = 4$ or $x + y = -4$. Therefore, the graph of $|x + y| = 4$ is the two lines $x + y = 4$ and $x + y = -4$. For Problems 33–38, graph each equation.

33. $|x - y| = 2$ **34.** $|2x + y| = 1$

35. $|x - 2y| \leq 4$ **36.** $|3x - 2y| \geq 6$

37. $|2x + 3y| > 6$ **38.** $|5x + 2y| < 10$

Using the definition of absolute value, the equation $y = |x| + 2$ becomes $y = x + 2$ for $x \geq 0$ and $y = -x + 2$ for $x < 0$. Therefore, the graph of $y = |x| + 2$ is as shown in Figure 2.28. For Problems 39–44, graph each equation.

39. $y = |x| - 1$ **40.** $y = |x - 2|$

41. $|y| = x$ **42.** $|y| = |x|$

43. $y = 2|x|$ **44.** $|x| + |y| = 4$

FIGURE 2.28

GRAPHING CALCULATOR ACTIVITIES

This is the first of many appearances of a group of problems called graphing calculator activities. These problems are specifically designed for those of you who have access to a graphing calculator or a computer with an appropriate software graphing package. Within the framework of these problems, you will be given the opportunity to reinforce concepts discussed in the text, lay groundwork for concepts to be introduced later in the text, predict shapes and locations of graphs on the basis of previous graphing experiences, solve problems that are unreasonable or perhaps impossible to solve without a graphing utility, and in general become familiar with the capabilities and limitations of your graphing utility.

The following problems are designed to get you started using your graphing utility and lay some groundwork for concepts we present in the next section. Set your boundaries so that the distance between the tic marks is the same on both axes.

45. a. Graph $y = 4x$, $y = 4x - 3$, $y = 4x + 2$, and $y = 4x + 5$ on the same set of axes. Do they appear to be parallel lines?

b. Graph $y = -2x + 1$, $y = -2x + 4$, $y = -2x - 2$, and $y = -2x - 5$ on the same set of axes. Do they appear to be parallel lines?

c. Graph $y = -\frac{1}{2}x + 3$, $y = -\frac{1}{2}x + 1$, $y = -\frac{1}{2}x - 1$, and $y = -\frac{1}{2}x - 4$ on the same set of axes. Do they appear to be parallel lines?

d. Graph $2x + 5y = 1$, $2x + 5y = -3$, $2x + 5y = 4$, and $2x + 5y = -5$ on the same set of axes. Do they appear to be parallel lines?

e. Graph $3x - 4y = 7$, $-3x + 4y = 8$, $3x - 4y = -2$, and $4x - 3y = 6$ on the same set of axes. Do they appear to be parallel lines?

f. On the basis of your results in parts (a) through (e), make a statement about how we can recognize parallel lines from their equations.

46. a. Graph $y = 4x$ and $y = -\frac{1}{4}x$ on the same set of axes. Do they appear to be perpendicular lines?

b. Graph $y = 3x$ and $y = \frac{1}{3}x$ on the same set of axes. Do they appear to be perpendicular lines?

c. Graph $y = \frac{2}{5}x - 1$ and $y = -\frac{5}{2}x + 2$ on the same set of axes. Do they appear to be perpendicular lines?

d. Graph $y = \frac{3}{4}x - 3$, $y = \frac{4}{3}x + 2$, and $y = -\frac{4}{3}x + 2$ on the same set of axes. Does there appear to be a pair of perpendicular lines?

e. On the basis of your results in parts (a) through (d), make a statement about how we can recognize perpendicular lines from their equations.

47. For each of the following pairs of equations, (1) predict whether they represent parallel lines, perpendicular lines, or lines that intersect but are not perpendicular, and (2) graph each pair of lines to check your prediction.

a. $5.2x + 3.3y = 9.4$ and $5.2x + 3.3y = 12.6$

b. $1.3x - 4.7y = 3.4$ and $1.3x - 4.7y = 11.6$

c. $2.7x + 3.9y = 1.4$ and $2.7x - 3.9y = 8.2$

d. $5x - 7y = 17$ and $7x + 5y = 19$

e. $9x + 2y = 14$ and $2x + 9y = 17$

f. $2.1x + 3.4y = 11.7$ and $3.4x - 2.1y = 17.3$

2.3   DETERMINING THE EQUATION OF A LINE

As we stated earlier, there are basically two types of problems in coordinate geometry: given an algebraic equation, find its geometric graph; and given a set of conditions pertaining to a geometric figure, find its algebraic equation. In the previous section, we considered some of the first type of problem; that is, we did some graphing. Now we want to consider some problems of the second type that deal specifically with straight lines; in other words, given certain facts about a line, we need to be able to determine its algebraic equation.

As we work with straight lines, it is often helpful to be able to refer to the *steepness* or *slant* of a particular line. The concept of *slope* is used as a measure of the slant of a line. The slope of a line is the ratio of the vertical change of distance to the horizontal change of distance as we move from one point on a line to another. Consider the line in Figure 2.29. From point A to point B there is a vertical change of two units and a horizontal change of three units; therefore, the slope of the line is $\frac{2}{3}$.

A precise definition for slope can be given by considering the coordinates of the points P_1, P_2, and R in Figure 2.30. The horizontal change of distance as we move from P_1 to P_2 is $x_2 - x_1$, and the vertical change is $y_2 - y_1$. Thus we have the following definition.

FIGURE 2.29

FIGURE 2.30

DEFINITION 2.1

If P_1 and P_2 are any two different points on a line, P_1 with coordinates (x_1, y_1) and P_2 with coordinates (x_2, y_2), then the slope of the line (denoted by m) is

$$m = \frac{y_2 - y_1}{x_2 - x_1}, \qquad x_2 \neq x_1$$

Because

$$\frac{y_2 - y_1}{x_2 - x_1} = \frac{y_1 - y_2}{x_1 - x_2}$$

how we designate P_1 and P_2 is not important. Let's use Definition 2.1 to find the slopes of some lines.

EXAMPLE 1

Find the slope of the line determined by each of the following pairs of points and graph each line.

a. $(-1, 1)$ and $(3, 2)$ *b.* $(4, -2)$ and $(-1, 5)$ *c.* $(2, -3)$ and $(-3, -3)$

Solutions

a. Let $(-1, 1)$ be P_1 and $(3, 2)$ be P_2 (Figure 2.31).

$$m = \frac{y_2 - y_1}{x_2 - x_1} = \frac{2 - 1}{3 - (-1)} = \frac{1}{4}$$

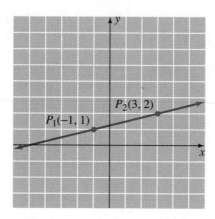

FIGURE 2.31

b. Let $(4, -2)$ be P_1 and $(-1, 5)$ be P_2 (Figure 2.32).

$$m = \frac{5 - (-2)}{-1 - 4} = \frac{7}{-5} = -\frac{7}{5}$$

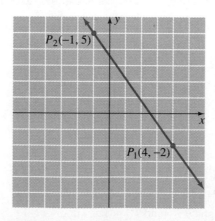

FIGURE 2.32

c. Let $(2, -3)$ be P_1 and $(-3, -3)$ be P_2 (Figure 2.33).

$$m = \frac{-3 - (-3)}{-3 - 2} = \frac{0}{-5} = 0$$

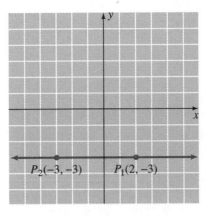

FIGURE 2.33

The three parts of Example 1 illustrate the three basic possibilities for slope; that is, the slope of a line can be positive, negative, or zero. A line that has a positive slope rises as we move from left to right, as in Figure 2.31. A line that has a negative slope falls as we move from left to right, as in Figure 2.32. A horizontal line, as in Figure 2.33, has a slope of zero. Finally, we need to realize that **the concept of slope is undefined for vertical lines.** This is due to the fact that for any vertical line, the horizontal change is zero as we move from one point on the line to another. Thus the ratio $(y_2 - y_1)/(x_2 - x_1)$ will have a denominator of zero and be undefined. Hence the restriction $x_2 \neq x_1$ is included in Definition 2.1.

Don't forget that **the slope of a line is a ratio**, the ratio of vertical change to horizontal change. For example, a slope of $\frac{2}{3}$ means that for every two units of vertical change, there must be a corresponding three units of horizontal change.

Applications of Slope

The concept of slope has many real-world applications even though the word *slope* is often not used. Technically, the concept of slope applies in most situations where the idea of an incline is used. Hospital beds are constructed so that both the head end and the foot end can be raised or lowered; that is, the slope of either end of the bed can be changed. Likewise, treadmills are designed so that the incline (slope) of the platform can be adjusted. A roofer, when making an estimate to replace a roof, is concerned not only about the total area to be covered but also about the pitch of the roof. (Contractors do not define pitch exactly in accordance with the mathematical definition of slope, but both concepts refer to "steepness.") In Figure 2.34, the two roofs might require the same amount of shingles, but the roof on the left will take longer to complete because the pitch is so great that scaffolding will be required.

FIGURE 2.34

The concept of slope is also used in the construction of stairways (Figure 2.35). The steepness (slope) of stairs can be expressed as the ratio of *rise* to *run*. In Figure 2.35, the stairs on the left, which have a ratio of $\frac{10}{11}$, are steeper than the stairs on the right, which have a ratio of $\frac{7}{11}$.

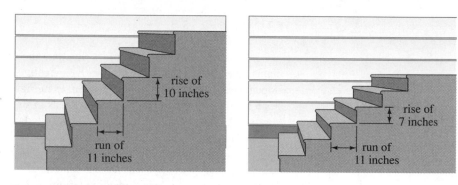

FIGURE 2.35

In highway construction, the word *grade* is used to describe the slope. For example, the highway in Figure 2.36 is said to have a grade of 17%. This means that for every horizontal distance of 100 feet, the highway rises or drops 17 feet. In other words, the slope of the highway is $\frac{17}{100}$.

FIGURE 2.36

EXAMPLE 2

A certain highway has a 3% grade. How many feet does it rise in a horizontal distance of 1 mile?

Solution

A 3% grade means a slope of $\dfrac{3}{100}$. Therefore, if we let y represent the unknown vertical distance and use the fact that 1 mile = 5280 feet, we can set up and solve the following proportion.

$$\frac{3}{100} = \frac{y}{5280}$$
$$100y = 3(5280) = 15{,}840$$
$$y = 158.4$$

The highway rises 158.4 feet in a horizontal distance of 1 mile. ■

Equations of Lines

Now let's consider some techniques for determining the equation of a line when given certain facts about the line.

EXAMPLE 3

Find the equation of the line that has a slope of $\dfrac{2}{5}$ and contains the point (3, 1).

Solution

First, let's draw the line and record the given information (Figure 2.37).

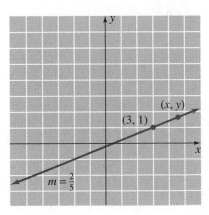

FIGURE 2.37

Then we choose a point (x, y) that represents any point on the line other than the given point $(3, 1)$. The slope determined by $(3, 1)$ and (x, y) is to be $\dfrac{2}{5}$. Thus

$$\frac{y - 1}{x - 3} = \frac{2}{5}$$
$$2(x - 3) = 5(y - 1)$$
$$2x - 6 = 5y - 5$$
$$2x - 5y = 1$$

■

EXAMPLE 4

Find the equation of the line determined by $(1, -2)$ and $(-3, 4)$.

Solution

First, let's draw the line determined by the two given points (Figure 2.38). These two points determine the slope of the line.

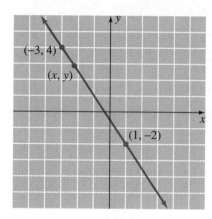

FIGURE 2.38

$$m = \frac{4 - (-2)}{-3 - 1} = \frac{6}{-4} = -\frac{3}{2}$$

Now we can use the same approach as in Example 3. Form an equation using one of the two given points, a point (x, y), and a slope of $-\dfrac{3}{2}$.

$$\frac{y + 2}{x - 1} = \frac{3}{-2}$$
$$3(x - 1) = -2(y + 2)$$
$$3x - 3 = -2y - 4$$
$$3x + 2y = -1$$

■

EXAMPLE 5

Find the equation of the line that has a slope of $\frac{1}{4}$ and a y intercept of 2.

Solution

A y intercept of 2 means that the point $(0, 2)$ is on the line (Figure 2.39). Choosing a point (x, y), we can proceed as in the previous examples.

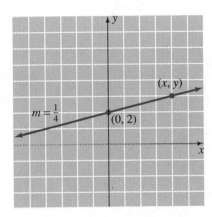

FIGURE 2.39

$$\frac{y - 2}{x - 0} = \frac{1}{4}$$
$$1(x - 0) = 4(y - 2)$$
$$x = 4y - 8$$
$$x - 4y = -8 \qquad\qquad ■$$

At this point you might pause for a moment and look back over Examples 3, 4, and 5. Note that we used the same basic approach in all three examples: We chose a point (x, y) and used it to determine the equation that satisfies the conditions stated in the problem. We will use this same approach later with figures other than straight lines. Furthermore, you should realize that this approach can be used to develop some general forms of equations of straight lines.

Point–Slope Form

EXAMPLE 6

Find the equation of the line that has a slope of m and contains the point (x_1, y_1).

Solution

Choosing (x, y) to represent another point on the line (Figure 2.40), the slope of the line is given by

$$m = \frac{y - y_1}{x - x_1}, \qquad x \neq x_1$$

from which we obtain

$$y - y_1 = m(x - x_1)$$

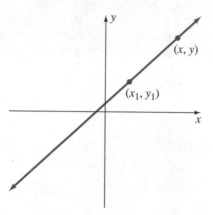

FIGURE 2.40

We refer to the equation

$$y - y_1 = m(x - x_1)$$

as the **point–slope form** of the equation of a straight line. Therefore, instead of using the approach of Example 3, we can substitute information into the point–slope form to write the equation of a line with a given slope that contains a given point. For example, the equation of the line that has a slope of $\frac{3}{5}$ and contains the point $(2, 4)$ can be determined this way. We substitute $(2, 4)$ for (x_1, y_1) and $\frac{3}{5}$ for m in the point–slope equation.

$$y - 4 = \frac{3}{5}(x - 2)$$
$$5(y - 4) = 3(x - 2)$$
$$5y - 20 = 3x - 6$$
$$-14 = 3x - 5y$$

Slope–Intercept Form

EXAMPLE 7

Find the equation of the line that has a slope of m and a y intercept of b.

Solution

A y intercept of b means that $(0, b)$ is on the line (Figure 2.41). Therefore, using the point–slope form with $(x_1, y_1) = (0, b)$, we obtain

$$y - y_1 = m(x - x_1)$$
$$y - b = m(x - 0)$$
$$y - b = mx$$
$$y = mx + b$$

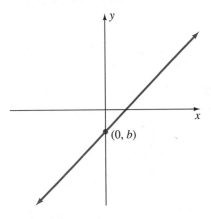

FIGURE 2.41 ∎

We refer to the equation

$$y = mx + b$$

as the **slope–intercept form** of the equation of a straight line. It can be used for two primary purposes, as the next two examples illustrate.

EXAMPLE 8

Find the equation of the line that has a slope of $\frac{1}{4}$ and a y intercept of 2.

Solution

This is a restatement of Example 5, but this time we will use the slope–intercept form ($y = mx + b$) of the equation of a line to write its equation. Because $m = \frac{1}{4}$ and $b = 2$, we obtain

$$y = mx + b$$
$$y = \frac{1}{4}x + 2$$
$$4y = x + 8$$
$$-8 = x - 4y \qquad \text{Same result as in Example 5}$$

∎

REMARK Sometimes we leave linear equations in slope–intercept form. We did not do so in Example 8 because we wanted to show that it was the same result as in Example 5.

EXAMPLE 9

Find the slope and y intercept of the line that has an equation $2x - 3y = 7$.

Solution

We can solve the equation for y in terms of x and then compare the result to the general slope–intercept form.

$$2x - 3y = 7$$
$$-3y = -2x + 7$$
$$y = \frac{2}{3}x - \frac{7}{3} \qquad y = mx + b$$

The slope of the line is $\frac{2}{3}$ and the y intercept is $-\frac{7}{3}$. ∎

In general, if the equation of a nonvertical line is written in slope–intercept form, the coefficient of x is the slope of the line and the constant term is the y intercept.

Parallel and Perpendicular Lines

Because the concept of slope is used to indicate the slant of a line, it seems reasonable to expect slope to be related to the concepts of parallelism and perpendicularity. Such is the case, and the following two properties summarize this link.

PROPERTY 2.1

If two nonvertical lines have slopes of m_1 and m_2, then

1. The two lines are parallel if and only if $m_1 = m_2$.
2. The two lines are perpendicular if and only if $m_1 m_2 = -1$.

We will test your ingenuity in devising proofs of these properties in the next problem set; here we will illustrate their use.

EXAMPLE 10

a. Verify that the graphs of $3x + 2y = 9$ and $6x + 4y = 19$ are parallel lines.
b. Verify that the graphs of $5x - 3y = 12$ and $3x + 5y = 27$ are perpendicular lines.

Solution

a. Let's change each equation to slope–intercept form.

$$3x + 2y = 9 \longrightarrow 2y = -3x + 9$$

$$y = -\frac{3}{2}x + \frac{9}{2}$$

$$6x + 4y = 19 \longrightarrow 4y = -6x + 19$$

$$y = -\frac{6}{4}x + \frac{19}{4}$$

$$y = -\frac{3}{2}x + \frac{19}{4}$$

The two lines have the same slope but different y intercepts. Therefore, they are parallel.

b. Change each equation to slope–intercept form.

$$5x - 3y = 12 \longrightarrow -3y = -5x + 12$$

$$y = \frac{5}{3}x - 4$$

$$3x + 5y = 27 \longrightarrow 5y = -3x + 27$$

$$y = -\frac{3}{5}x + \frac{27}{5}$$

Because $\left(\dfrac{5}{3}\right)\left(-\dfrac{3}{5}\right) = -1$, the product of the two slopes is -1 and the lines are perpendicular. ∎

> **REMARK**   The statement *The product of two slopes is* -1 is equivalent to saying that the two slopes are **negative reciprocals** of each other—that is, $m_1 = -1/m_2$.

EXAMPLE 11

Find the equation of the line that contains the point $(-1, 2)$ and is parallel to the line with the equation $2x - y = 4$.

Solution

First, we draw a figure to help in our analysis of the problem (Figure 2.42). Because the line through $(-1, 2)$ is to be parallel to the given line, it must have the same slope. Let's find the slope by changing $2x - y = 4$ to slope–intercept form.

$$2x - y = 4$$
$$-y = -2x + 4$$
$$y = 2x - 4$$

The slope of both lines is 2. Now, using the point–slope form with $(x_1, y_1) = (-1, 2)$, we obtain the equation of the line.

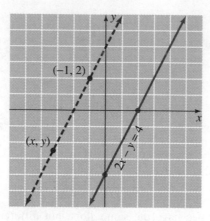

$$y - y_1 = m(x - x_1)$$
$$y - 2 = 2[x - (-1)]$$
$$y - 2 = 2(x + 1)$$
$$y - 2 = 2x + 2$$
$$-4 = 2x - y$$

■

EXAMPLE 12

Find the equation of the line that contains the point $(-1, -3)$ and is perpendicular to the line determined by $3x + 4y = 12$.

Solution

Again let's start by drawing a figure to help with our analysis (Figure 2.43). Because the line through $(-1, -3)$ is to be perpendicular to the given line, its slope must be the negative reciprocal of the slope of the line with the equation $3x + 4y = 12$. Let's find the slope of $3x + 4y = 12$ by changing to slope–intercept form.

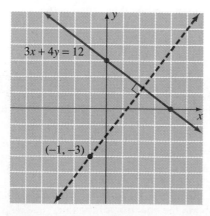

FIGURE 2.43

$$3x + 4y = 12$$
$$4y = -3x + 12$$
$$y = -\frac{3}{4}x + 3$$

The slope of the desired line is $\frac{4}{3}$ $\left(\text{the negative reciprocal of } -\frac{3}{4}\right)$, and we can proceed as before to obtain its equation.

$$y - y_1 = m(x - x_1)$$
$$y - (-3) = \frac{4}{3}[x - (-1)]$$
$$y + 3 = \frac{4}{3}(x + 1)$$
$$3y + 9 = 4x + 4$$
$$5 = 4x - 3y \qquad \blacksquare$$

Two forms of equations of straight lines are used extensively. They are the *standard form* and the *slope–intercept form*.

Standard Form $Ax + By = C$, where B and C are integers and A is a nonnegative integer (A and B are not both zero).

Slope–Intercept Form $y = mx + b$, where m is a real number representing the slope of the line and b is a real number representing the y intercept.

PROBLEM SET 2.3

For Problems 1–8, find the slope of the line determined by each pair of points.

1. $(3, 1)$ and $(7, 4)$

2. $(-1, 2)$ and $(5, -3)$

3. $(-2, -1)$ and $(-1, -6)$

4. $(-2, -4)$ and $(3, 7)$

5. $(-4, 2)$ and $(-2, 2)$

6. $(4, -5)$ and $(-1, -5)$

7. $(a, 0)$ and $(0, b)$

8. (a, b) and (c, d)

9. Find x if the line through $(-2, 4)$ and $(x, 6)$ has a slope of $\frac{2}{9}$.

10. Find y if the line through $(1, y)$ and $(4, 2)$ has a slope of $\frac{5}{3}$.

11. Find x if the line through $(x, 4)$ and $(2, -5)$ has a slope of $-\frac{9}{4}$.

12. Find y if the line through $(5, 2)$ and $(-3, y)$ has a slope of $-\frac{7}{8}$.

For each of the lines in Problems 13–18, you are given one point and the slope of the line. Find the coordinates of three other points on the line.

13. $(3, 2)$; $m = \frac{2}{3}$

14. $(-4, 4)$; $m = \frac{5}{6}$

15. $(-1, -4)$; $m = 4$

16. $(-5, -3)$; $m = 2$

17. $(2, -1)$; $m = -\frac{3}{5}$

18. $(5, -1)$; $m = -\frac{2}{3}$

For Problems 19–26, write the equation of the line that has the indicated slope and contains the indicated point. Express final equations in standard form.

19. $m = \frac{1}{3}$; $(2, 4)$

20. $m = \frac{3}{5}$; $(-1, 4)$

21. $m = 2$; $(-1, -2)$

22. $m = -3$; $(2, 5)$

23. $m = -\frac{2}{3}$; $(4, -3)$

24. $m = -\frac{1}{5}$; $(-3, 7)$

25. $m = 0$; $(5, -2)$

26. $m = \frac{4}{3}$; $(-4, -5)$

For Problems 27–34, write the equation of each line that contains the indicated pair of points. Express final equations in standard form.

27. $(2, 3)$ and $(9, 8)$

28. $(1, -4)$ and $(4, 4)$

29. $(-1, 7)$ and $(5, 2)$

30. $(-3, 1)$ and $(6, -2)$

31. $(4, 2)$ and $(-1, 3)$

32. $(2, 7)$ and $(2, 5)$

33. $(4, -3)$ and $(-7, -3)$

34. $(-4, 2)$ and $(2, -3)$

For Problems 35–42, write the equation of each line that has the indicated slope (m) and y-intercept (b). Express final equations in slope–intercept form.

35. $m = \frac{1}{2}$, $b = 3$

36. $m = \frac{5}{3}$, $b = -1$

37. $m = -\frac{3}{7}$, $b = 2$

38. $m = -3$, $b = -4$

39. $m = 4$, $b = \frac{3}{2}$

40. $m = \frac{2}{3}$, $b = \frac{3}{5}$

41. $m = -\frac{5}{6}$, $b = \frac{1}{4}$

42. $m = -\frac{4}{5}$, $b = 0$

For Problems 43–50, write the equation of each line that satisfies the given conditions. Express final equations in standard form.

43. The x intercept is 4 and the y intercept is -5.

44. Contains the point $(3, -1)$ and is parallel to the x axis

45. Contains the point $(-4, 3)$ and is parallel to the y axis

46. Contains the point $(1, 2)$ and is parallel to the line $3x - y = 5$

47. Contains the point $(4, -3)$ and is parallel to the line $5x + 2y = 1$

48. Contains the origin and is parallel to the line $5x - 2y = 10$

49. Contains the point $(-2, 6)$ and is perpendicular to the line $x - 4y = 7$

50. Contains the point $(-3, -5)$ and is perpendicular to the line $3x + 7y = 4$

For each pair of lines in Problems 51–58, determine whether they are parallel, perpendicular, or intersecting lines that are not perpendicular.

51. $y = \frac{5}{6}x + 2$

$y = \frac{5}{6}x - 4$

52. $y = 5x - 1$

$y = -\frac{1}{5}x + \frac{2}{3}$

53. $5x - 7y = 14$

$7x + 5y = 12$

54. $2x - y = 4$

$4x - 2y = 17$

55. $4x + 9y = 13$

$-4x + y = 11$

56. $y = 5x$

$y = -5x$

57. $x + y = 0$

$x - y = 0$

58. $2x - y = 14$

$3x - y = 17$

For Problems 59–66, find the slope and the y intercept of each line.

59. $2x - 3y = 4$

60. $3x + 4y = 7$

61. $x - 2y = 7$

62. $2x + y = 9$

63. $y = -3x$

64. $x - 5y = 0$

65. $7x - 5y = 12$

66. $-5x + 6y = 13$

67. The slope–intercept form of a line can also be used for graphing purposes. Suppose that we want to graph $y = \frac{2}{3}x + 1$. Because the y intercept is 1, the point $(0, 1)$ is on the line. Furthermore, because the slope is $\frac{2}{3}$, another point can be found by moving two units *up* and three units to the *right*. Thus the point $(3, 3)$ is also on the line. The two points $(0, 1)$ and $(3, 3)$ determine the line.

Use the slope–intercept form to help graph each of the following lines.

a. $y = \frac{3}{4}x + 2$

b. $y = \frac{1}{2}x - 4$

c. $y = -\frac{4}{5}x + 1$

d. $y = -\frac{2}{3}x - 6$

e. $y = -2x + \frac{5}{4}$

f. $y = x - \frac{3}{2}$

68. Use the concept of slope to verify that $(-4, 6)$, $(6, 10)$, $(10, 0)$, and $(0, -4)$ are the vertices of a square.

69. Use the concept of slope to verify that $(6, 6)$, $(2, -2)$, $(-8, -5)$, and $(-4, 3)$ are vertices of a parallelogram.

70. Use the concept of slope to verify that the triangle determined by $(4, 3)$, $(5, 1)$, and $(3, 0)$ is a right triangle.

71. Use the concept of slope to verify that the quadrilateral whose vertices are $(0, 7)$, $(-2, -1)$, $(2, -2)$, and $(4, 6)$ is a rectangle.

72. Use the concept of slope to verify that the points $(8, -3)$, $(2, 1)$, and $(-4, 5)$ lie on a straight line.

For Problems 73–80, solve the problem.

73. The midpoints of the sides of a triangle are $(-3, 4)$, $(1, -4)$, and $(7, 2)$. Find the equations of the lines that contain the sides of the triangle.

74. The vertices of a triangle are $(2, 6)$, $(5, 1)$, and $(1, -4)$. Find the equations of the lines that contain the three altitudes of the triangle. (An altitude of a triangle is the perpendicular line segment from a vertex to the opposite side.)

75. The vertices of a triangle are $(1, -6)$, $(3, 1)$, and $(-2, 2)$. Find the equations of the lines that contain the three medians of the triangle. (A median of a triangle is the line segment from a vertex to the midpoint of the opposite side.)

76. A certain highway has a 2% grade. How many feet does it rise in a horizontal distance of 1 mile? (1 mile = 5280 feet)

77. The grade of a highway up a hill is 30%. How much change in horizontal distance is there if the vertical height of the hill is 75 feet?

78. If the ratio of rise to run is to be $\dfrac{3}{5}$ for some stairs and the rise is 19 centimeters, find the measure of the run to the nearest centimeter.

79. If the ratio of rise to run is to be $\dfrac{2}{3}$ for some stairs and the run is 28 centimeters, find the rise to the nearest centimeter.

80. Suppose that a county ordinance requires a $2\dfrac{1}{4}\%$ fall for a sewage pipe from the house to the main pipe at the street. How much vertical drop must there be for a horizontal distance of 45 feet? Express the answer to the nearest tenth of a foot.

THOUGHTS INTO WORDS

81. How would you explain the concept of slope to someone who was absent from class the day it was discussed?

82. If one line has a slope of $\dfrac{2}{5}$ and another line has a slope of $\dfrac{3}{7}$, which line is steeper? Explain your answer.

83. What does it mean to say that two points *determine* a line? Do three points *determine* a line? Explain your answers.

84. Explain how you would find the slope of the line $y = 2$.

Further Investigations

85. The form

$$\frac{y - y_1}{x - x_1} = \frac{y_2 - y_1}{x_2 - x_1}$$

is called the **two-point form** of the equation of a straight line. (1) Using points (x_1, y_1) and (x_2, y_2), develop the two-point form for the equation of a line. (2) Use the two-point form to write the equation of each of the following lines, which contain the indicated pair of points. Express the final equations in standard form.

a. $(4, 3)$ and $(5, 6)$ **b.** $(-3, 5)$ and $(2, -1)$

c. $(0, 0)$ and $(-7, 2)$ **d.** $(-3, -4)$ and $(5, -1)$

86. The form $(x/a) + (y/b) = 1$ is called the intercept form of the equation of a straight line. (1) Using a to represent the x intercept and b to represent the y intercept, develop the intercept form. (2) Use the intercept form to write the equation of each of the following lines. Express the final equations in standard form.

a. $a = 2, b = 5$ **b.** $a = -3, b = 1$

c. $a = 6, b = -4$ **d.** $a = -1, b = -2$

87. Prove each of the following statements.

a. Two nonvertical parallel lines have the same slope.

b. Two lines with the same slope are parallel.

c. If two nonvertical lines are perpendicular, then their slopes are negative reciprocals of each other.

d. If the slopes of two lines are negative reciprocals of each other, then the lines are perpendicular.

88. Let $Ax + By = C$ and $A'x + B'y = C'$ represent two lines. Verify each of the following properties.

a. If $(A/A') = (B/B') \neq (C/C')$, then the lines are parallel.

b. If $AA' = -BB'$, then the lines are perpendicular.

89. The properties in Problem 88 give us another way to write the equation of a line parallel or perpendicular to a given line through a point not on the given line. For example, suppose that we want the equation of the line perpendicular to $3x + 4y = 6$ that contains the point $(1, 2)$. The form $4x - 3y = k$, where k is a constant, represents a family of lines perpendicular to $3x + 4y = 6$ because we have satisfied the condition $AA' = -BB'$. Therefore, to find the specific line of the family containing $(1, 2)$, we substitute 1 for x and 2 for y to determine k.

$$4x - 3y = k$$

$$4(1) - 3(2) = k$$

$$-2 = k$$

Thus the equation of the desired line is $4x - 3y = -2$. Use the properties from Problem 88 to help write the equation of each of the following lines.

a. Contains $(5, 6)$ and is parallel to the line $2x - y = 1$

b. Contains $(-3, 4)$ and is parallel to the line $3x + 7y = 2$

c. Contains $(2, -4)$ and is perpendicular to the line $2x - 5y = 9$

d. Contains $(-3, -5)$ and is perpendicular to the line $4x + 6y = 7$

90. Some real-world situations can be described by the use of linear equations in two variables. If two pairs of values are known, then the equation can be determined by using the approach we used in Example 4 of this section. For each of the following, assume that the relationship can be expressed as a linear equation in two variables, and use the given information to determine the equation. Express the equation in standard form.

a. A company produces 10 fiberglass shower stalls for $2015 and 15 stalls for $3015. Let y be the cost and x the number of stalls.

b. A company can produce 6 boxes of candy for $8 and 10 boxes of candy for $13. Let y represent the cost and x the number of boxes of candy.

c. Two banks on opposite corners of a town square have signs displaying the up-to-date temperature. One bank displays the temperature in Celsius degrees and the other in Fahrenheit. A temperature of 10°C was displayed at the same time as a temperature of 50°F. On another day, a temperature of -5°C was displayed at the same time as a temperature of 23°F. Let y represent the temperature in Fahrenheit and x the temperature in Celsius.

91. The relationships that tie slope to parallelism and perpendicularity are powerful tools for constructing coordinate geometry proofs. Prove each of the following using a coordinate geometry approach.

a. The diagonals of a square are perpendicular.

b. The line segment joining the midpoints of two sides of a triangle is parallel to the third side.

c. The line segments joining successive midpoints of the sides of a quadrilateral form a parallelogram.

d. The line segments joining successive midpoints of the sides of a rectangle form a rhombus. (A rhombus is a parallelogram with all sides of the same length.)

<table>
<tr><td>**2.4**</td><td>**MORE ON GRAPHING**</td></tr>
</table>

As we stated earlier, it is very helpful to recognize that a certain type of equation produces a particular kind of graph. In a later chapter, we will pursue that idea in much more detail. However, we also need to develop some general graphing techniques to use with equations where we do not recognize the graph. Let's begin with the following suggestions and then add to the list throughout the remainder of the text. (You may recognize some of the graphs in this section from previous graphing experiences, but keep in mind that the primary objective at this time is the development of some additional graphing techniques.)

1. Find the intercepts.
2. Solve the equation for y in terms of x or for x in terms of y if it is not already in such a form.
3. Set up a table of ordered pairs that satisfy the equation.
4. Plot the points associated with the ordered pairs and connect them with a smooth curve.

EXAMPLE 1

Graph $y = x^2 - 4$.

Solution

First, let's find the intercepts. If $x = 0$, then

$$y = 0^2 - 4$$
$$y = -4$$

This determines the point $(0, -4)$. If $y = 0$, then

$$0 = x^2 - 4$$
$$4 = x^2$$
$$\pm 2 = x$$

Thus the points $(2, 0)$ and $(-2, 0)$ are determined.

Second, because the given equation expresses y in terms of x, the form is convenient for setting up a table of ordered pairs. Plotting these points and connecting them with a smooth curve produces Figure 2.44.

x	y	
0	−4	
2	0	intercepts
−2	0	
1	−3	
−1	−3	other
3	5	points
−3	5	

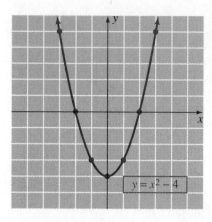

FIGURE 2.44   ■

The curve in Figure 2.44 is said to be **symmetric with respect to the** *y* **axis.** Stated another way, each half of the curve is a mirror image of the other half through the *y* axis. Note in the table of values that for each ordered pair (x, y), the ordered pair $(-x, y)$ is also a solution. Thus a general test for *y* axis symmetry can be stated as follows.

> *y* **Axis Symmetry**   The graph of an equation is symmetric with respect to the *y* axis if replacing *x* with $-x$ results in an equivalent equation.

Thus the equation $y = x^2 - 4$ exhibits *y* axis symmetry because replacing *x* with $-x$ produces $y = (-x)^2 - 4 = x^2 - 4$. Likewise, the equations $y = x^2 + 6$, $y = x^4$, and $y = x^4 + 2x^2$ exhibit *y* axis symmetry.

EXAMPLE 2

Graph $x - 1 = y^2$.

Solution

If $x = 0$, then

$$0 - 1 = y^2$$
$$-1 = y^2$$

The equation $y^2 = -1$ has no real number solutions; therefore, this graph has no points on the *y* axis. If $y = 0$, then

$$x - 1 = 0$$
$$x = 1$$

Thus the point $(1, 0)$ is determined.

Solving the original equation of x produces $x = y^2 + 1$, for which the table of values is easily determined. Plotting these points and connecting them with a smooth curve produces Figure 2.45.

x	y	
1	0	intercept
2	1	
2	−1	other
5	2	points
5	−2	

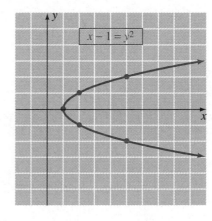

$x - 1 = y^2$

FIGURE 2.45 ■

The curve in Figure 2.45 is said to be **symmetric with respect to the x axis**. That is to say, each half of the curve is a mirror image of the other half through the x axis. Note in the table of values that for each ordered pair (x, y), the ordered pair $(x, -y)$ is also a solution. The following general test of x axis symmetry can be stated.

> x **Axis Symmetry**   The graph of an equation is symmetric with respect to the x axis if replacing y with $-y$ results in an equivalent equation.

Thus the equation $x - 1 = y^2$ exhibits x axis symmetry because replacing y with $-y$ produces $x - 1 = (-y)^2 = y^2$. Likewise, the equations $x = y^2$, $x = y^4 + 2$, and $x^3 = y^2$ exhibit x axis symmetry.

EXAMPLE 3

Graph $y = x^3$.

Solution

If $x = 0$, then

$$y = 0^3 = 0$$

Thus the origin $(0, 0)$ is on the graph.

The table of values is easily determined from the equation. Plotting these points and connecting them with a smooth curve produces Figure 2.46.

x	y	
0	0	intercept
1	1	
2	8	other
−1	−1	points
−2	−8	

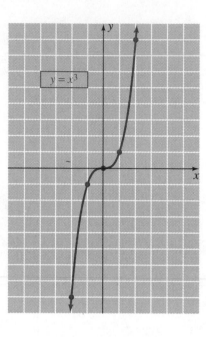

FIGURE 2.46 ■

The curve in Figure 2.46 is said to be **symmetric with respect to the origin.** Each half of the curve is a mirror image of the other half through the origin. In the table of values, we see that for each ordered pair (x, y), the ordered pair $(-x, -y)$ is also a solution. The following general test for origin symmetry can be stated.

> **Origin Symmetry** The graph of an equation is symmetric with respect to the origin if replacing x with $-x$ and y with $-y$ results in an equivalent equation.

The equation $y = x^3$ exhibits origin symmetry because replacing x with $-x$ and y with $-y$ produces $-y = -x^3$, which is equivalent to $y = x^3$. (Multiplying both sides of $-y = -x^3$ by -1 produces $y = x^3$.) Likewise, the equations $xy = 4$, $x^2 + y^2 = 10$, and $4x^2 - y^2 = 12$ exhibit origin symmetry.

> **REMARK** From the symmetry tests, we should observe that if a curve has both x-axis and y-axis symmetry, then it must have origin symmetry. However, it is possible for a curve to have origin symmetry and not be symmetric to either axis. Figure 2.46 is an example of such a curve.

Another graphing consideration is that of **restricting a variable** to ensure real number solutions. The following example illustrates this point.

EXAMPLE 4

Graph $y = \sqrt{x - 1}$.

Solution

The radicand, $x - 1$, must be nonnegative. Therefore,

$$x - 1 \geq 0$$
$$x \geq 1$$

The restriction $x \geq 1$ indicates that there is no y intercept. The x intercept can be found as follows: If $y = 0$, then

$$0 = \sqrt{x - 1}$$
$$0 = x - 1$$
$$1 = x$$

The point $(1, 0)$ is on the graph.

Now, keeping the restriction in mind, we can determine the table of values. Plotting these points and connecting them with a smooth curve produces Figure 2.47.

x	y	
1	0	intercept
2	1	
5	2	other points
10	3	

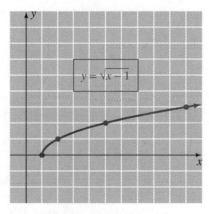

F I G U R E   2 . 4 7

Now let's restate and add the concepts of symmetry and restrictions to the list of graphing suggestions. The order of the suggestions also indicates the order in which we usually attack a graphing problem if it is a new graph—that is, one that we do not recognize from its equation.

1. Determine what type of symmetry the equation exhibits.

2. Find the intercepts.

3. Solve the equation for y in terms of x or for x in terms of y, if it is not already in such a form.

4. Determine the restrictions necessary to ensure real number solutions.

5. Set up a table of ordered pairs that satisfy the equation. The type of symmetry and the restrictions will affect your choice of values in the table.

6. Plot the points associated with the ordered pairs and connect them with a smooth curve. Then, if appropriate, reflect this curve according to the symmetry possessed by the graph.

The final two examples of this section should help you pull these ideas together and demonstrate the power of having these techniques at your fingertips.

EXAMPLE 5

Graph $x = -y^2 - 3$.

Solution

Symmetry   The graph is symmetric with respect to the x axis because replacing y with $-y$ produces $x = -(-y)^2 - 3$, which is equivalent to $x = -y^2 - 3$.

Intercepts   If $x = 0$, then

$$0 = -y^2 - 3$$
$$y^2 = -3$$

Therefore, the graph contains no points on the y axis. If $y = 0$, then

$$x = -0^2 - 3$$
$$x = -3$$

Thus the point $(-3, 0)$ is on the graph.

Restrictions   Because $x = -y^2 - 3$, y can take on any real number value, and for every value of y, x will be less than or equal to -3.

Table of Values   Because of the x-axis symmetry, let's choose only nonnegative values for y.

Plotting the Graph   Plotting the points determined by the table and connecting them with a smooth curve produces Figure 2.48(a). Then reflecting that portion of the curve across the x-axis produces the complete curve in Figure 2.48(b).

x	y
-3	0
$-\dfrac{13}{4}$	$\dfrac{1}{2}$
-4	1
$-\dfrac{21}{4}$	$\dfrac{3}{2}$
-7	2

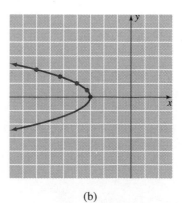

(a)   (b)

FIGURE 2.48

EXAMPLE 6

Graph $x^2 - y^2 = 4$.

Solution

Symmetry   The graph is symmetric with respect to both axes and the origin because replacing x with $-x$ and y with $-y$ produces $(-x)^2 - (-y)^2 = 4$, which is equivalent to $x^2 - y^2 = 4$.

Intercepts   If $x = 0$, then

$$0^2 - y^2 = 4$$
$$-y^2 = 4$$
$$y^2 = -4$$

Therefore, the graph contains no points on the y axis. If $y = 0$, then

$$x^2 - 0^2 = 4$$
$$x^2 = 4$$
$$x = \pm 2$$

Thus the points $(2, 0)$ and $(-2, 0)$ are on the graph.

Restrictions   Solving the given equation for y produces

$$x^2 - y^2 = 4$$
$$-y^2 = 4 - x^2$$
$$y^2 = x^2 - 4$$
$$y = \pm\sqrt{x^2 - 4}$$

Therefore, $x^2 - 4 \geq 0$, which is equivalent to $x \geq 2$ or $x \leq -2$.

Table of Values   Because of the restrictions and symmetries, we need only choose values corresponding to $x \geq 2$.

Plotting the Graph   Plotting the points in the table of values and connecting them with a smooth curve produces Figure 2.49(a). Because of the symmetry with respect to both axes and the origin, the portion of the curve in Figure 2.49(a) can be reflected across both axes and through the origin to produce the complete curve shown in Figure 2.49(b).

Even when you are using a graphing utility, it is often helpful to determine symmetry, intercepts, and restrictions before graphing the equations. This can serve as a partial check against using the utility incorrectly.

x	y
2	0
3	$\sqrt{5} \approx 2.2$
4	$2\sqrt{3} \approx 3.5$
5	$\sqrt{21} \approx 4.6$
6	$4\sqrt{2} \approx 5.7$

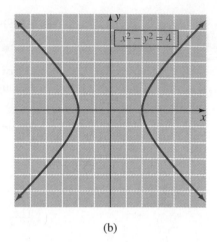

(a)   (b)

FIGURE 2.49   ■

| **EXAMPLE 7** | Use a graphing utility to obtain the graph of $y = \sqrt{x^2 - 49}$. |

Solution

Symmetry   The graph is symmetric with respect to the y axis because replacing x with $-x$ produces the same equation.

Intercepts   If $x = 0$, then $y = \sqrt{-49}$; thus the graph has no points on the y axis. If $y = 0$, then $x = \pm 7$; thus the points $(7, 0)$ and $(-7, 0)$ are on the graph.

Restrictions   Because $x^2 - 49$ has to be nonnegative, we know that $x \leq -7$ or $x \geq 7$. Now let's enter the expression $\sqrt{x^2 - 49}$ for Y_1 and obtain the graph in Figure 2.50. Note that the graph does exhibit the symmetry, intercepts, and restrictions that we determined earlier.

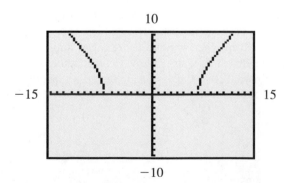

FIGURE 2.50   ■

PROBLEM SET 2.4

For Problems 1–6, determine the points that are symmetric to the given point with respect to the x axis, the y axis, and the origin.

1. $(4, 3)$ **2.** $(-2, 5)$ **3.** $(-6, -1)$

4. $(3, -7)$ **5.** $(0, 4)$ **6.** $(-5, 0)$

For Problems 7–20, determine the type of symmetry (x-axis, y-axis, origin) possessed by each graph. Do not sketch the graph.

7. $y = x^2 - 6$ **8.** $x = y^2 + 1$

9. $x^3 = y^2$ **10.** $x^2 y^2 = 4$

11. $x^2 + 2y^2 = 6$ **12.** $3x^2 - y^2 + 4x = 6$

13. $x^2 - 2x + y^2 - 3y - 4 = 0$

14. $xy = 4$ **15.** $y = x$

16. $2x - 3y = 15$ **17.** $y = x^3 + 2$

18. $y = x^4 + x^2$

19. $5x^2 - y^2 + 2y - 1 = 0$

20. $x^2 + y^2 - 2y - 4 = 0$

For Problems 21–48, use symmetry, intercepts, restrictions, and point plotting to help graph each equation.

21. $y = x^2$ **22.** $y = -x^2$

23. $y = x^2 + 2$ **24.** $y = -x^2 - 1$

25. $xy = 4$ **26.** $xy = -2$

27. $y = -x^3$ **28.** $y = x^3 + 2$

29. $y^2 = x^3$ **30.** $y^3 = x^2$

31. $y^2 - x^2 = 4$ **32.** $x^2 - 2y^2 = 8$

33. $y = -\sqrt{x}$ **34.** $y = \sqrt{x + 1}$

35. $x^2 y = 4$ **36.** $xy^2 = 4$

37. $x^2 + 2y^2 = 8$ **38.** $2x^2 + y^2 = 4$

39. $y = \dfrac{4}{x^2 + 1}$ **40.** $y = \dfrac{-2}{x^2 + 1}$

41. $y = \sqrt{x - 2}$ **42.** $y = 3 - x$

43. $-xy = 3$ **44.** $-x^2 y = 4$

45. $x = y^2 + 2$ **46.** $x = -y^2 + 4$

47. $x = -y^2 - 1$ **48.** $x = y^2 - 3$

THOUGHTS INTO WORDS

49. How does the concept of symmetry help when we are graphing equations?

50. Explain how you would go about graphing $x^2 y^2 = 4$.

 ### GRAPHING CALCULATOR ACTIVITIES

51. Graph $y = \dfrac{4}{x^2}$, $y = \dfrac{4}{(x - 2)^2}$, $y = \dfrac{4}{(x - 4)^2}$, and $y = \dfrac{4}{(x + 2)^2}$ on the same set of axes. Now predict the graph for $y = \dfrac{4}{(x - 6)^2}$. Check your prediction.

52. Graph $y = \sqrt{x}$, $y = \sqrt{x + 1}$, $y = \sqrt{x - 2}$, and $y = \sqrt{x - 4}$ on the same set of axes. Now predict the graph for $y = \sqrt{x + 3}$. Check your prediction.

53. Graph $y = \sqrt{x}$, $y = 2\sqrt{x}$, $y = 4\sqrt{x}$, and $y = 7\sqrt{x}$ on the same set of axes. How does the constant in front of the radical seem to affect the graph?

54. Graph $y = \dfrac{8}{x^2}$ and $y = -\dfrac{8}{x^2}$ on the same set of axes. How does the negative sign seem to affect the graph?

55. Graph $y = \sqrt{x}$ and $y = -\sqrt{x}$ on the same set of axes. How does the negative sign seem to affect the graph?

56. Graph $y = \sqrt{x}$, $y = \sqrt{x} + 2$, $y = \sqrt{x} + 4$, and $y = \sqrt{x} - 3$ on the same set of axes. How does the constant term seem to affect the graph?

57. Graph $y = \sqrt{x}$, $y = \sqrt{x+3}$, $y = \sqrt{x-1}$, and $y = \sqrt{x-5}$ on the same set of axes. How are the graphs related? Predict the location of $y = \sqrt{x+5}$. Check your prediction.

58. To graph $x = y^2$ we need first to solve for y in terms of x. This produces $y = \pm \sqrt{x}$. Now we can let $Y_1 = \sqrt{x}$ and $Y_2 = -\sqrt{x}$ and graph the two equations on the same set of axes. Then graph $x = y^2 + 4$ on this same set of axes.

How are the graphs related? Predict the location of the graph of $x = y^2 - 4$. Check your prediction.

59. To graph $x = y^2 + 2y$ we need first to solve for y in terms of x. Let's complete the square to do this.

$$y^2 + 2y = x$$
$$y^2 + 2y + 1 = x + 1$$
$$(y + 1)^2 = (\sqrt{x+1})^2$$
$$y + 1 = \sqrt{x+1} \qquad \text{or} \qquad y + 1 = -\sqrt{x+1}$$
$$y = -1 + \sqrt{x+1} \qquad \text{or} \qquad y = -1 - \sqrt{x+1}$$

Thus let's make the assignments $Y_1 = -1 + \sqrt{x+1}$ and $Y_2 = -1 - \sqrt{x+1}$ and graph them on the same set of axes to produce the graph of $x = y^2 + 2y$. Then graph $x = y^2 + 2y - 4$ on this same set of axes. Now predict the location of the graph of $x = y^2 + 2y + 4$. Check your prediction.

2.5   CIRCLES, ELLIPSES, AND HYPERBOLAS

When we apply the distance formula

$$d = \sqrt{(x_2 - x_1)^2 + (y_2 - y_1)^2}$$

(developed in Section 2.1) to the definition of a circle, we get what is known as the **standard form of the equation of a circle**. We start with a precise definition of a circle.

DEFINITION 2.2

A **circle** is the set of all points in a plane equidistant from a given fixed point called the **center**. A line segment determined by the center and any point on the circle is called a **radius**.

Now let's consider a circle that has a radius of length r and a center at (h, k) on a coordinate system (Figure 2.51). For any point P on the circle with coordinates (x, y), the length of a radius, denoted by r, can be expressed as

$$r = \sqrt{(x - h)^2 + (y - k)^2}$$

FIGURE 2.51

Squaring both sides of this equation, we obtain the standard form of the equation of a circle.

$$(x - h)^2 + (y - k)^2 = r^2$$

This form of the equation of a circle can be used to solve the two basic kinds of problems: (1) given the coordinates of the center of a circle and the length of a radius of a circle, find its equation; (2) given the equation of a circle, determine its graph. Let's illustrate each of these types of problems.

EXAMPLE 1

Find the equation of a circle that has its center at $(-3, 5)$ and has a radius of length four units.

Solution

Substitute -3 for h, 5 for k, and 4 for r in the standard equation and simplify to give us the equation of the circle.

$$(x - h)^2 + (y - k)^2 = r^2$$
$$[x - (-3)]^2 + (y - 5)^2 = 4^2$$
$$(x + 3)^2 + (y - 5)^2 = 4^2$$
$$x^2 + 6x + 9 + y^2 - 10y + 25 = 16$$
$$x^2 + y^2 + 6x - 10y + 18 = 0$$

■

Note that in Example 1 we simplified the equation to the form $x^2 + y^2 + Dx + Ey + F = 0$, where D, E, and F are constants. This is another form that we commonly use when working with circles.

EXAMPLE 2

Graph $x^2 + y^2 - 6x + 4y + 9 = 0$.

Solution

We can change the given equation into the standard form for a circle by completing the square on x and on y.

$$x^2 + y^2 - 6x + 4y + 9 = 0$$
$$(x^2 - 6x) + (y^2 + 4y) = -9$$
$$(x^2 - 6x + 9) + (y^2 + 4y + 4) = -9 + 9 + 4$$

Add 9 to complete the square on x.

Add 4 to complete the square on y.

Add 9 and 4 to compensate for the 4 and 9 added on the left side.

$$(x - 3)^2 + (y + 2)^2 = 2^2$$
$$(x - 3)^2 + [y - (-2)]^2 = 2^2$$

$\qquad\quad h \qquad\qquad\quad k \qquad\quad r$

The center is at $(3, -2)$ and the length of a radius is two units. The circle is drawn in Figure 2.52.

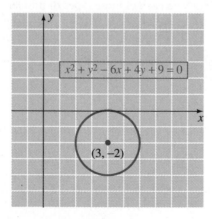

$x^2 + y^2 - 6x + 4y + 9 = 0$

$(3, -2)$

FIGURE 2.52

EXAMPLE 3

Find the center and length of a radius of the circle

$$4x^2 + 4x + 4y^2 - 12y - 26 = 0$$

Solution

$$4x^2 + 4x + 4y^2 - 12y - 26 = 0$$
$$4(x^2 + x + \_) + 4(y^2 - 3y + \_) = 26$$
$$4\left(x^2 + x + \frac{1}{4}\right) + 4\left(y^2 - 3y + \frac{9}{4}\right) = 26 + 1 + 9$$
$$4\left(x + \frac{1}{2}\right)^2 + 4\left(y - \frac{3}{2}\right)^2 = 36$$
$$\left(x + \frac{1}{2}\right)^2 + \left(y - \frac{3}{2}\right)^2 = 9$$
$$\left[x - \left(\frac{1}{2}\right)\right]^2 + \left(y - \frac{3}{2}\right)^2 = 3^2$$
$$\qquad\uparrow\qquad\qquad\uparrow\qquad\uparrow$$
$$\qquad h\qquad\qquad k\qquad r$$

Therefore, the center is at $\left(-\dfrac{1}{2}, \dfrac{3}{2}\right)$ and the length of a radius is 3 units. ■

Now suppose that we substitute 0 for h and 0 for k in the standard form of the equation of a circle.

$$(x - h)^2 + (y - k)^2 = r^2$$
$$(x - 0)^2 + (y - 0)^2 = r^2$$
$$x^2 + y^2 = r^2$$

The form $x^2 + y^2 = r^2$ is called the **standard form of the equation of a circle that has its center at the origin.** For example, by inspection we can recognize that $x^2 + y^2 = 9$ is a circle with its center at the origin and a radius of length three units. Likewise, the equation $5x^2 + 5y^2 = 10$ is equivalent to $x^2 + y^2 = 2$; therefore, its graph is a circle with its center at the origin and a radius of length $\sqrt{2}$ units. Furthermore, we can easily determine that the equation of the circle with its center at the origin and a radius of 8 units is $x^2 + y^2 = 64$.

Ellipses

Generally, it is true that any equation of the form $Ax^2 + By^2 = F$ (where $A = B$ and A, B, and F are nonzero constants that have the same sign) is a circle with its center at the origin. We can use the general equation $Ax^2 + By^2 = F$ to describe other geometric figures by changing the restrictions on A and B. For example, if A, B, and F are of the same sign but $A \neq B$, then the graph of the equation $Ax^2 + By^2 = F$ is an ellipse. Let's consider two examples.

EXAMPLE 4

Graph $4x^2 + 9y^2 = 36$.

Solution

Let's find the intercepts. if $x = 0$, then

$$4(0)^2 + 9y^2 = 36$$
$$9y^2 = 36$$
$$y^2 = 4$$
$$y = \pm 2$$

Thus the points $(0, 2)$ and $(0, -2)$ are on the graph. If $y = 0$, then

$$4x^2 + 9(0)^2 = 36$$
$$4x^2 = 36$$
$$x^2 = 9$$
$$x = \pm 3$$

Thus the points $(3, 0)$ and $(-3, 0)$ are on the graph.

Because we know that it is an ellipse, plotting the four points that we have gives us a pretty good sketch of the figure (Figure 2.53).

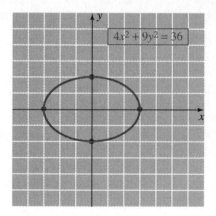

FIGURE 2.53   ■

In Figure 2.53, the line segment with endpoints at $(-3, 0)$ and $(3, 0)$ is called the **major axis** of the ellipse. The shorter segment with endpoints at $(0, -2)$ and $(0, 2)$ is called the **minor axis**. Establishing the endpoints of the major and minor axes provides a basis for sketching an ellipse. Also note that the equation $4x^2 + 9y^2 = 36$ exhibits symmetry with respect to both axes and the origin, as we see in Figure 2.53.

EXAMPLE 5

Graph $25x^2 + y^2 = 25$.

Solution

The endpoints of the major and minor axes can be found by finding the intercepts. If $x = 0$, then

$$25(0)^2 + y^2 = 25$$
$$y^2 = 25$$
$$y = \pm 5$$

The endpoints of the major axis are therefore at $(0, 5)$ and $(0, -5)$. If $y = 0$, then

$$25x^2 + (0)^2 = 25$$
$$25x^2 = 25$$
$$x^2 = 1$$
$$x = \pm 1$$

The endpoints of the minor axis are at $(1, 0)$ and $(-1, 0)$. The ellipse is sketched in Figure 2.54.

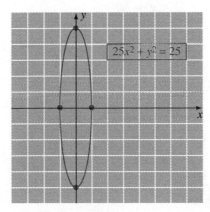

FIGURE 2.54

Hyperbolas

The graph of an equation of the form $Ax^2 + By^2 = F$, where A and B are of *unlike* signs, is a **hyperbola**. The next two examples illustrate the graphing of hyperbolas.

EXAMPLE 6

Graph $x^2 - 4y^2 = 4$.

Solution

If we let $y = 0$, then

$$x^2 - 4(0)^2 = 4$$
$$x^2 = 4$$
$$x = \pm 2$$

Thus the points $(2, 0)$ and $(-2, 0)$ are on the graph. If we let $x = 0$, then

$$0^2 - 4y^2 = 4$$
$$-4y^2 = 4$$
$$y^2 = -1$$

Because $y^2 = -1$ has no real number solutions, there are no points of the graph on the y axis.

Note that the equation $x^2 - 4y^2 = 4$ exhibits symmetry with respect to both axes and the origin. Now let's solve the given equation for y to get a more convenient form for finding other solutions.

$$x^2 - 4y^2 = 4$$
$$-4y^2 = 4 - x^2$$
$$4y^2 = x^2 - 4$$
$$y^2 = \frac{x^2 - 4}{4}$$
$$y = \frac{\pm\sqrt{x^2 - 4}}{2}$$

Because the radicand, $x^2 - 4$, must be nonnegative, the values chosen for x must be such that $x \geq 2$ or $x \leq -2$. Symmetry and the points determined by the table provide the basis for sketching Figure 2.55.

x	y	
2	0	intercepts
-2	0	
3	$\pm\frac{\sqrt{5}}{2}$	
4	$\pm\sqrt{3}$	other points
5	$\pm\frac{\sqrt{21}}{2}$	

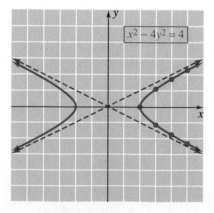

FIGURE 2.55

Note the dashed lines in Figure 2.55; they are called asymptotes. Each branch of the hyperbola approaches one of these lines but does not intersect it. Therefore, being able to sketch the asymptotes of a hyperbola is very helpful for graphing purposes. Fortunately, the equations of the asymptotes are easy to determine. They can be found by replacing the constant term in the given equation of the hyperbola with zero and then solving for y. (The reason why this works will be discussed in a later chapter.) For the hyperbola in Example 6, we obtain

$$x^2 - 4y^2 = 0$$
$$-4y^2 = -x^2$$
$$y^2 = \frac{1}{4}x^2$$
$$y = \pm\frac{1}{2}x$$

Thus the lines $y = \frac{1}{2}x$ and $y = -\frac{1}{2}x$ are the asymptotes indicated by the dashed lines in Figure 2.55.

EXAMPLE 7

Graph $4y^2 - 9x^2 = 36$.

Solution

If $x = 0$, then

$$4y^2 - 9(0)^2 = 36$$
$$4y^2 = 36$$
$$y^2 = 9$$
$$y = \pm 3$$

The points $(0, 3)$ and $(0, -3)$ are on the graph. If $y = 0$, then

$$4(0)^2 - 9x^2 = 36$$
$$-9x^2 = 36$$
$$x^2 = -4$$

Because $x^2 = -4$ has no real number solutions, we know that this hyperbola does not intersect the x axis. Solving the equation for y yields

$$4y^2 - 9x^2 = 36$$
$$4y^2 = 9x^2 + 36$$
$$y^2 = \frac{9x^2 + 36}{4}$$
$$y = \frac{\pm\sqrt{9x^2 + 36}}{2} = \frac{\pm\sqrt{9(x^2 + 4)}}{2} = \pm\frac{3\sqrt{x^2 + 4}}{2}$$

The table shows some additional solutions. The equations of the asymptotes are determined as follows.

$$4y^2 - 9x^2 = 0$$
$$4y^2 = 9x^2$$
$$y^2 = \frac{9}{4}x^2$$
$$y = \pm\frac{3}{2}x$$

Sketching the asymptotes, plotting the points from the table, and using symmetry, we determine the hyperbola in Figure 2.56.

x	y	
0	3	intercepts
0	-3	
1	$\pm\frac{3\sqrt{5}}{2}$	
2	$\pm3\sqrt{2}$	other points
3	$\pm\frac{3\sqrt{13}}{2}$	

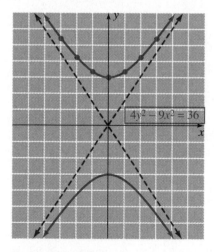

$4y^2 - 9x^2 = 36$

FIGURE 2.56

When using a graphing utility, we may find it necessary to change the boundaries on x or y (or both) to obtain a complete graph. Consider the following example.

EXAMPLE 8

Use a graphing utility to graph $x^2 - 40x + y^2 + 351 = 0$.

Solution

First we need to solve for y in terms of x.

$$x^2 - 40x + y^2 + 351 = 0$$
$$y^2 = -x^2 + 40x - 351$$
$$y = \pm\sqrt{-x^2 + 40x - 351}$$

Now we can make the following assignments.

$$Y_1 = \sqrt{-x^2 + 40x - 351}$$
$$Y_2 = -Y_1$$

(Note that we assigned Y_2 in terms of Y_1. By doing this, we avoid repetitive key strokes and reduce the chance for errors. You may need to consult your user's manual for instructions on how to key-stroke $-Y_1$.) Figure 2.57 shows the graph.

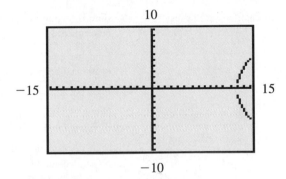

F I G U R E   2 . 5 7

We know from the original equation that this graph should be a circle, so we need to make some adjustments on the boundaries in order to get a complete graph. This can be done by completing the square on the original equation to change its form to $(x - 20)^2 + y^2 = 49$ or simply by a trial-and-error process. By changing the boundaries on x such that $-15 \leq x \leq 30$, we obtain Figure 2.58.

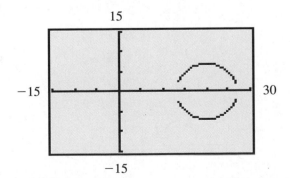

F I G U R E   2 . 5 8

In summarizing this section, we do want you to be aware of the continuity pattern used. We started by using the definition of a circle to generate the standard form of the equation of a circle. Then we discussed ellipses and hyperbolas, not from a definition viewpoint, but by considering variations of the general equation of a circle with its center at the origin ($Ax^2 + By^2 = F$, where A, B, and F are of the same sign and $A = B$). In Chapter 8, we will develop parabolas, ellipses, and hyperbolas from a definition viewpoint. In other words, we first define each of the concepts and then use those definitions to generate standard forms for their equations.

PROBLEM SET 2.5

For Problems 1–8, write the equation of each circle. Express the final equations in the form $x^2 + y^2 + Dx + Ey + F = 0$.

1. Center at (2, 3) and $r = 5$.

2. Center at (−3, 4) and $r = 2$.

3. Center at (−1, −5) and $r = 3$.

4. Center at (4, −2) and $r = 1$.

5. Center at (3, 0) and $r = 3$.

6. Center at (0, −4) and $r = 6$.

7. Center at the origin and $r = 7$.

8. Center at the origin and $r = 1$.

For Problems 9–18, find the center and length of a radius of each circle.

9. $x^2 + y^2 - 6x - 10y + 30 = 0$

10. $x^2 + y^2 + 8x - 12y + 43 = 0$

11. $x^2 + y^2 + 10x + 14y + 73 = 0$

12. $x^2 + y^2 + 6y - 7 = 0$

13. $x^2 + y^2 - 10x = 0$

14. $x^2 + y^2 - 4x + 2y = 0$

15. $x^2 + y^2 = 8$ **16.** $4x^2 + 4y^2 = 1$

17. $4x^2 + 4y^2 - 4x - 8y - 11 = 0$

18. $36x^2 + 36y^2 + 48x - 36y - 11 = 0$

19. Find the equation of the circle where the line segment determined by (−4, 9) and (10, −3) is a diameter.

20. Find the equation of the circle that passes through the origin and has its center at (−3, −4).

21. Find the equation of the circle that is tangent to both axes, has a radius of length seven units, and has its center in the fourth quadrant.

22. Find the equation of the circle that passes through the origin, has an x intercept of −6, and has a y intercept of 12. (The perpendicular bisector of a chord contains the center of the circle.)

23. Find the equations of the circles that are tangent to the x axis and have a radius of length five units. In each case, the abscissa of the center is −3. (There is more than one circle that satisfies these conditions.)

For Problems 24–40, graph each equation.

24. $4x^2 + 25y^2 = 100$ **25.** $9x^2 + 4y^2 = 36$

26. $x^2 - y^2 = 4$ **27.** $y^2 - x^2 = 9$

28. $x^2 + y^2 - 4x - 2y - 4 = 0$

29. $x^2 + y^2 - 4x = 0$

30. $4x^2 + y^2 = 4$ **31.** $x^2 + 9y^2 = 36$

32. $x^2 + y^2 + 2x - 6y - 6 = 0$

33. $y^2 - 3x^2 = 9$ **34.** $4x^2 - 9y^2 = 16$

35. $x^2 + y^2 + 4x + 6y - 12 = 0$

36. $2x^2 + 5y^2 = 50$ **37.** $4x^2 + 3y^2 = 12$

38. $x^2 + y^2 - 6x + 8y = 0$

39. $3x^2 - 2y^2 = 3$ **40.** $y^2 - 8x^2 = 9$

The graphs of equations of the form $xy = k$, where k is a nonzero constant, are also hyperbolas, sometimes referred to as **rectangular hyperbolas**. For Problems 41–44, graph each rectangular hyperbola.

41. $xy = 2$ **42.** $xy = 4$

43. $xy = -3$ **44.** $xy = -2$

THOUGHTS INTO WORDS

45. What is the graph of $xy = 0$? Explain your answer.

46. We have graphed various equations of the form $Ax^2 + By^2 = F$, where F is a nonzero constant. Describe the graph of each of the following and explain your answers.

a. $x^2 + y^2 = 0$ **b.** $2x^2 + 3y^2 = 0$

c. $x^2 - y^2 = 0$ **d.** $4x^2 - 9y^2 = 0$

Further Investigations

47. By expanding $(x - h)^2 + (y - k)^2 = r^2$, we obtain $x^2 - 2hx + h^2 + y^2 - 2ky + k^2 - r^2 = 0$. Comparing this result to the form $x^2 + y^2 + Dx + Ey + F = 0$, we see that $D = -2h$, $E = -2k$, and $F = h^2 + k^2 - r^2$. Therefore, the center and the length of a radius of a circle can be found by using $h = D/-2$, $k = E/-2$, and $r = \sqrt{h^2 + k^2 - F}$. Use these relationships to find the center and the length of a radius of each of the following circles.

a. $x^2 + y^2 - 2x - 8y + 8 = 0$

b. $x^2 + y^2 + 4x - 14y + 49 = 0$

c. $x^2 + y^2 + 12x + 8y - 12 = 0$

d. $x^2 + y^2 - 16x + 20y + 115 = 0$

e. $x^2 + y^2 - 12y - 45 = 0$

f. $x^2 + y^2 + 14x = 0$

48. Use a coordinate geometry approach to prove that an angle inscribed in a semicircle is a right angle (see Figure 2.59).

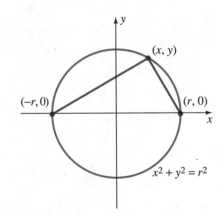

FIGURE 2.59

49. Use a coordinate geometry approach to prove that a line segment from the center of a circle bisecting a chord is perpendicular to the chord. [*Hint:* Let the ends of the chord be $(r, 0)$ and (a, b).]

GRAPHING CALCULATOR ACTIVITIES

50. For each of the following equations, (a) predict the type and location of the graph, and (b) use your graphics calculator to check your prediction.

a. $x^2 + y^2 = 9$ **b.** $2x^2 + y^2 = 4$

c. $x^2 - y^2 = 9$ **d.** $4x^2 - y^2 = 16$

e. $x^2 + 2x + y^2 - 4 = 0$ **f.** $x^2 + y^2 - 4y - 2 = 0$

g. $(x - 2)^2 + (y + 1)^2 = 4$

h. $(x + 3)^2 - (y - 4)^2 = 9$

i. $9y^2 - 4x^2 = 36$ **j.** $9y^2 + 4x^2 = 36$

CHAPTER 2 SUMMARY

We emphasized throughout this chapter that coordinate geometry contains two basic kinds of problems:

    1. Given an algebraic equation, determine its geometric graph.

    2. Given a set of conditions pertaining to a geometric figure, determine its algebraic equation.

Let's review this chapter in terms of those two kinds of problems.

Graphing

The following graphing techniques were discussed in this chapter.

    1. Recognize the type of graph that a certain kind of equation produces.

        a. $Ax + By = C$ produces a straight line.

        b. $x^2 + y^2 + Dx + Ey + F = 0$ produces a circle. The center and the length of a radius can be found by completing the square and comparing to the standard form of the equation of a circle:

$$(x - h)^2 + (y - k)^2 = r^2$$

        c. $Ax^2 + By^2 = F$, where A, B, and F have the same sign and $A = B$, produces a circle with the center at the origin.

        d. $Ax^2 + By^2 = F$, where A, B, and F are of the same sign but $A \neq B$, produces an ellipse.

        e. $Ax^2 + By^2 = F$, where A and B are of unlike signs, produces a hyperbola.

    2. Determine the symmetry that a graph possesses.

        a. The graph of an equation is symmetric with respect to the y axis if replacing x with $-x$ results in an equivalent equation.

b. The graph of an equation is symmetric with respect to the x axis if replacing y with $-y$ results in an equivalent equation.

c. The graph of an equation is symmetric with respect to the origin if replacing x with $-x$ and y with $-y$ results in an equivalent equation.

3. Find the intercepts. The x intercept is found by letting $y = 0$ and solving for x. The y intercept is found by letting $x = 0$ and solving for y.

4. Determine the restrictions necessary to ensure real number solutions.

5. Set up a table of ordered pairs that satisfy the equation. The type of symmetry and the restrictions will affect your choice of values in the table. Furthermore, it may be convenient to change the form of the original equation by solving for y in terms of x or for x in terms of y.

6. Plot the points associated with the ordered pairs in the table and connect them with a smooth curve. Then, if appropriate, reflect the curve according to any symmetries possessed by the graph.

Determining Equations When Given Certain Conditions

You should review Examples 3, 4, and 5 of Section 2.3 to be sure you are thoroughly familiar with the general approach of choosing a point (x, y) and using it to determine the equation that satisfies the conditions stated in the problem.

We developed some special forms that can be used to determine equations.

Point–slope form of a straight line:	$y - y_1 = m(x - x_1)$
Slope–intercept form of a straight line:	$y = mx + b$
Standard form of a circle:	$(x - h)^2 + (y - k)^2 = r^2$

The following formulas were used in different parts of the chapter.

Distance formula: $\qquad d = \sqrt{(x_2 - x_1)^2 + (y_2 - y_1)^2}$

Midpoint formula: $\qquad$ The coordinates of the midpoint of a line segment determined by (x_1, y_1) and (x_2, y_2) are

$$\left(\frac{x_1 + x_2}{2}, \frac{y_1 + y_2}{2} \right)$$

Slope formula: $\qquad m = \dfrac{y_2 - y_1}{x_2 - x_1}, \quad x_1 \neq x_2$

CHAPTER 2 REVIEW PROBLEM SET

1. On a number line, find the coordinate of the point located three-fifths of the distance from -4 to 11.

2. On a number line, find the coordinate of the point located four-ninths of the distance from 3 to -15.

3. On the xy-plane, find the coordinates of the point located five-sixths of the distance from $(-1, -3)$ to $(11, 1)$.

4. If one endpoint of a line segment is at $(8, 14)$ and the midpoint of the segment is $(3, 10)$, find the coordinates of the other endpoint.

5. Verify that the points $(2, 2)$, $(6, 4)$, and $(5, 6)$ are vertices of a right triangle.

6. Verify that the points $(-3, 1)$, $(1, 3)$, and $(9, 7)$ lie in a straight line.

For Problems 7–12, identify any symmetries (x-axis, y-axis, origin) that the equation exhibits.

7. $x = y^2 + 4$

8. $y = x^2 + 6x - 1$

9. $5x^2 - y^2 = 4$

10. $x^2 + y^2 - 2y - 4 = 0$

11. $y = -x$

12. $y = \dfrac{6}{x^2 + 4}$

For Problems 13–22, graph each of the following.

13. $x^2 + y^2 - 6x + 4y - 3 = 0$

14. $x^2 + 4y^2 = 16$

15. $x^2 - 4y^2 = 16$

16. $-2x + 3y = 6$

17. $2x - y < 4$

18. $x^2 y^2 = 4$

19. $4y^2 - 3x^2 = 8$

20. $x^2 + y^2 + 10y = 0$

21. $9x^2 + 2y^2 = 36$

22. $y \leq -2x - 3$

23. Find the slope of the line determined by $(-3, -4)$ and $(-5, 6)$.

24. Find the slope of the line with equation $5x - 7y = 12$.

For Problems 25–28, write the equation of the line that satisfies the stated conditions. Express final equations in standard form ($Ax + By = C$).

25. Contains the point $(7, 2)$ and has a slope of $-\dfrac{3}{4}$

26. Contains the points $(-3, -2)$ and $(1, 6)$

27. Contains the point $(2, -4)$ and is parallel to $4x + 3y = 17$

28. Contains the point $(-5, 4)$ and is perpendicular to $2x - y = 7$

For Problems 29–32, write the equation of the circle that satisfies the stated conditions. Express final equations in the form $x^2 + y^2 + Dx + Ey + F = 0$.

29. Center at $(5, -6)$ and $r = 1$

30. The endpoints of a diameter are $(-2, 4)$ and $(6, 2)$.

31. Center at $(-5, 12)$ and passes through the origin

32. Tangent to both axes, $r = 4$, and center in the third quadrant

CHAPTER 2 TEST

1. On a number line, find the coordinate of the point located two-thirds of the distance from -4 to 14.

2. On the xy plane, find the coordinates of the point located three-fourths of the distance from $(2, -3)$ to $(-6, 9)$.

3. If one endpoint of a line segment is at $(-2, -1)$ and the midpoint of the segment is at $(2, -\frac{5}{2})$, find the coordinates of the other endpoint.

4. Find the slope of the line determined by $(-4, -2)$ and $(5, -6)$.

5. Find the slope of the line determined by the equation $2x - 7y = -9$.

For Problems 6–10, determine the equation of the line that satisfies the stated conditions. Express final equations in standard form.

6. Has a slope of $-\frac{3}{4}$ and a y intercept of -3

7. Contains the points $(1, -4)$ and $(4, 7)$

8. Contains the point $(-1, 4)$ and is parallel to $x - 5y = 5$

9. Contains the point $(3, 5)$ and is perpendicular to $4x + 7y = 3$

10. Contains the point $(-2, -4)$ and is perpendicular to the x axis

For Problems 11–13, determine the equations of the circle that satisfies the stated conditions. Express final equations in the form $x^2 + y^2 + Dx + Ey + F = 0$.

11. Center at $(-3, -6)$ and a radius of length 4 units

12. The endpoints of a diameter are at $(-1, 3)$ and $(5, 5)$.

13. Center at $(4, -3)$ and passes through the origin

14. Find the center and the length of a radius of the circle $x^2 + 16x + y^2 - 10y + 80 = 0$.

15. Find the lengths of the three sides of the triangle determined by $(3, 2)$, $(5, -2)$, and $(-1, -1)$. Express the lengths in simplest radical form.

16. Find the x intercepts of the graph of the equation $x^2 - 6x + y^2 + 2y + 5 = 0$.

17. Find the y intercepts of the graph of the equation $5x^2 + 12y^2 = 36$.

18. Find the length of the major axis of the ellipse $9x^2 + 2y^2 = 18$.

19. Find the equations of the asymptotes for the hyperbola $9x^2 - 16y^2 = 48$.

20. Identify any symmetries (*x*-axis, *y*-axis, origin) that the equation exhibits.

    a. $x^2 + 2x + y^2 - 6 = 0$ ***b.*** $xy = -4$ ***c.*** $y = \dfrac{4}{x^2 + 1}$ ***d.*** $x^2y^2 = 5$

21. Graph the inequality $3x - y \le 6$.

For Problems 22–25 graph the equation.

22. $y^2 - 2x^2 = 9$ **23.** $x = y^2 - 4$

24. $3x^2 + 5y^2 = 45$ **25.** $x^2 + 4x + y^2 - 12 = 0$

CUMULATIVE REVIEW PROBLEM SET - CHAPTERS 0-2

For Problems 1–6, evaluate each expression.

1. 3^{-3}

2. -4^{-2}

3. $\left(\dfrac{2}{3}\right)^{-2}$

4. $-\sqrt[3]{\dfrac{8}{27}}$

5. $\left(\dfrac{1}{27}\right)^{-2/3}$

6. $\dfrac{1}{\left(\frac{3}{4}\right)^{-2}}$

For Problems 7–12, perform the indicated operations and simplify. Express final answers using positive exponents only.

7. $(5x^{-3}y^{-2})(4xy^{-1})$

8. $(-7a^{-3}b^2)(8a^4b^{-3})$

9. $\left(\dfrac{1}{2}x^{-2}y^{-1}\right)^{-2}$

10. $\dfrac{80x^{-3}y^{-4}}{16xy^{-6}}$

11. $\left(\dfrac{102x^{2/3}y^{3/4}}{6xy^{-1}}\right)^{-1}$

12. $\left(\dfrac{14a^3b^{-4}}{7a^{-1}b^3}\right)^2$

For Problems 13–20, express each in simplest radical form. All variables represent positive real numbers.

13. $-5\sqrt{72}$

14. $2\sqrt{27x^3y^2}$

15. $\sqrt[3]{56x^4y^7}$

16. $\dfrac{3\sqrt{18}}{5\sqrt{12}}$

17. $\sqrt{\dfrac{3x}{7y}}$

18. $\dfrac{5}{\sqrt{2}-3}$

19. $\dfrac{3\sqrt{7}}{2\sqrt{2}-\sqrt{6}}$

20. $\dfrac{4\sqrt{x}}{\sqrt{x}+3\sqrt{y}}$

For Problems 21–26, perform the indicated operations involving rational expressions. Express final answers in simplest form.

21. $\dfrac{12x^2y}{18x}\cdot\dfrac{9x^3y^3}{16xy^2}$

22. $\dfrac{-15ab^2}{14a^3b}\div\dfrac{20a}{7b^2}$

23. $\dfrac{3x^2+5x-2}{x^2-4}\cdot\dfrac{5x^2-9x-2}{3x^2-x}$

24. $\dfrac{2x-1}{4}+\dfrac{3x+2}{6}-\dfrac{x-1}{8}$

25. $\dfrac{5}{3n^2}-\dfrac{2}{n}+\dfrac{3}{2n}$

26. $\dfrac{5x}{x^2+6x-27}+\dfrac{3}{x^2-9}$

For Problems 27–38, solve each equation.

27. $3(-2x-1)-2(3x+4)=-4(2x-3)$

28. $(2x-1)(3x+4)=(x+2)(6x-5)$

29. $\dfrac{3x-1}{4}-\dfrac{2x-1}{5}=\dfrac{1}{10}$

30. $9x^2-4=0$

31. $5x^3+10x^2-40x=0$

32. $7t^2-31t+12=0$

33. $x^4+15x^2-16=0$

34. $|5x-2|=3$

35. $2x^2-3x-1=0$

36. $(3x-2)(x+4)=(2x-1)(x-1)$

37. $\sqrt{5-t}+1=\sqrt{7+2t}$

38. $(2x-1)^2+4=0$

For Problems 39–48, solve each inequality. Express the solution sets using interval notation.

39. $-2(x-1)+(3-2x)>4(x+1)$

40. $2n+1+\dfrac{3n-1}{4}\ge\dfrac{n-1}{2}$

41. $0.09x+0.12(450-x)\ge46.5$

42. $n^2+5n>24$

43. $6x^2+7x-3<0$

44. $(2x-1)(x+3)(x-4)>0$

45. $\dfrac{3x-2}{x+1}\le0$

46. $\dfrac{x+5}{x-1}\ge2$

47. $|3x-1|>5$

48. $|5x-3|<12$

For Problems 49–54, graph each equation.

49. $x^2+4y^2=36$

50. $4x^2-y^2=4$

51. $y=-x^3-1$

52. $y=-x+3$

53. $y^2-5x^2=9$

54. $y=-\dfrac{3}{4}x-1$

For Problems 55–58, solve each problem.

55. Find the center and the length of a radius of the circle with equation $x^2+y^2+14x-8y+56=0$.

56. Write the equation of the line that is parallel to $3x-4y=17$ and contains the point $(2, 8)$.

57. Find the coordinates of the point located one-fifth of the distance from $(-3, 4)$ to $(2, 14)$.

58. Write the equation of the perpendicular bisector of the line segment determined by $(-3, 4)$ and $(5, 10)$.

For Problems 59–65, set up an equation and solve the problem.

59. A retailer has some shirts that cost $22 per shirt. At what price should they be sold to obtain a profit of 30% of the cost? At what price should they be sold to obtain a profit of 30% of the selling price?

60. A total of $7500 was invested, part of it at 5% yearly interest and the remainder at 6%. The total yearly interest was $420. How much was invested at each rate?

61. The length of a rectangle is 1 inch less than twice the width. The area of the rectangular region is 36 square inches. Find the length and width of the rectangle.

62. The length of one side of a triangle is 4 centimeters less than three times the length of the altitude to that side. The area of the triangle is 80 square centimeters. Find the length of the side and the length of the altitude to that side.

63. How many milliliters of pure acid must be added to 40 milliliters of a 30% acid solution to obtain a 50% acid solution?

64. Amanda rode her bicycle out into the country at a speed of 15 miles per hour and returned along the same route at 10 miles per hour. The round trip took 5 hours. How far out did she ride?

65. If two inlet pipes are both open, they can fill a pool in 1 hour and 12 minutes. One of the pipes can fill the pool by itself in 2 hours. How long would it take the other pipe to fill the pool by itself?

3

FUNCTIONS

The volume of a right circular cylinder is a function of its height and the length of a radius of its base.

One of the fundamental concepts of mathematics is that of a function. Functions are used to unify different areas of mathematics, and they also serve as a meaningful way of applying mathematics to many real-world problems. They provide a means of studying quantities that vary with one another—that is, quantities such that a change in one produces a corresponding change in another. In this chapter, we will (1) introduce the basic ideas pertaining to functions, (2) use the idea of a function to unify some concepts from Chapter 2, and (3) discuss some applications in which functions are used.

3.1   CONCEPT OF A FUNCTION

The notion of correspondence is used in everyday situations and is central to the concept of a function. Consider the following correspondences.

1. To each person in a class, there corresponds an assigned seat.
2. To each day of a year, there corresponds an assigned integer that represents the average temperature for that day in a certain geographic location.
3. To each book in a library, there corresponds a whole number that represents the number of pages in the book.

Such correspondences can be depicted as in Figure 3.1. To each member in set *A* there corresponds *one and only one* member in set *B*. For example, in correspondence 1, set *A* would consist of the students in a class and set *B* would be the assigned seats. In the second example, set *A* would consist of the days of a year and set *B* would be a set of integers. Furthermore, the same integer might be assigned

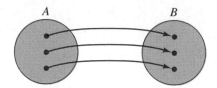

FIGURE 3.1

to more than one day of the year. (Different days might have the same average temperature.) The key idea is that *one and only one* integer is assigned to *each* day of the year. Likewise, in the third example, more than one book may have the same number of pages, but to each book there is assigned one and only one number of pages.

Mathematically, the general concept of a function can be defined as follows.

DEFINITION 3.1

A function *f* is a correspondence between two sets *X* and *Y* that assigns to each element *x* of set *X* one and only one element *y* of set *Y*. The element *y* being assigned is called the image of *x*. The set *X* is called the domain of the function, and the set of all images is called the range of the function.

In Definition 3.1, the image *y* is usually denoted by $f(x)$. Thus the symbol $f(x)$, which is read *f of x* or *the value of f at x*, represents the element in the range associated with the element *x* from the domain. Figure 3.2 depicts this situation. Again, we emphasize that each member of the domain has precisely one image in the range; however, different members in the domain, such as *a* and *b* in Figure 3.2, may have the same image.

FIGURE 3.2

In Definition 3.1 we named the function *f*. It is common to name functions by means of a single letter, and the letters *f*, *g*, and *h* are often used. We would suggest more meaningful choices when functions are used in real-world situations. For example, if a problem involves a profit function, then naming the function *p* or even *P* would seem natural. Be careful not to confuse *f* and $f(x)$. Remember that *f* is used to name a function, whereas $f(x)$ is an element of the range, namely the element assigned to *x* by *f*.

The assignments made by a function are often expressed as ordered pairs. For example, the assignments in Figure 3.2 could be expressed as $(a, f(a))$, $(b, f(b))$, $(c, f(c))$, and $(x, f(x))$, where the first components are from the domain and the second components are from the range. Thus a function can also be thought of as a set of ordered pairs where no two of the ordered pairs have the same first component.

REMARK In some texts, the concept of a relation is introduced first, and then functions are defined as special kinds of relations. A relation is defined as *a set of ordered pairs*, and a function is defined as *a relation in which no two ordered pairs have the same first element*.

The ordered pairs representing a function can be generated by various means, such as a graph or a chart. However, one of the most common ways of generating ordered pairs is by using equations. For example, the equation $f(x) = 2x + 3$ indicates that to each value of x in the domain, we assign $2x + 3$ from the range. For example,

$$f(1) = 2(1) + 3 = 5 \qquad \text{produces the ordered pair } (1, 5)$$
$$f(4) = 2(4) + 3 = 11 \qquad \text{produces the ordered pair } (4, 11)$$
$$f(-2) = 2(-2) + 3 = -1 \qquad \text{produces the ordered pair } (-2, -1)$$

It may be helpful for you to picture the concept of a function in terms of a *function machine*, as illustrated in Figure 3.3. Each time that a value of x is put into the machine, the equation $f(x) = 2x + 3$ is used to generate one and only one value for $f(x)$ to be ejected from the machine.

FIGURE 3.3

(a)

(b)

FIGURE 3.4

Using the ordered-pair interpretation of a function, we can define the **graph** of a function f to be the set of all points in a plane of the form $(x, f(x))$, where x is from the domain of f. In other words, the graph of f is the same as the graph of the equation $y = f(x)$. Furthermore, because $f(x)$, or y, takes on only one value for each value of x, we can easily tell whether a given graph represents a function. For example, in Figure 3.4(a), for any choice of x there is only one value for y. Geometrically, this means that no vertical line intersects the curve in more than one point. On the other hand, Figure 3.4(b) does not represent the graph of a function because certain values of x (all positive values) produce more than one value for y. In other words, some vertical lines intersect the curve in more than one point, as illustrated in Figure 3.4(b). A **vertical line test** for functions can be stated as follows.

Vertical Line Test If each vertical line intersects a graph in no more than one point, then the graph represents a function.

Let's consider some examples to help pull together some of these ideas about functions.

EXAMPLE 1

If $f(x) = x^2 - x + 4$ and $g(x) = x^3 - x^2$, find $f(3), f(-2), g(4)$, and $g(-3)$.

Solution

$$f(3) = 3^2 - 3 + 4 = 10 \qquad f(-2) = (-2)^2 - (-2) + 4 = 10$$
$$g(4) = 4^3 - 4^2 = 48 \qquad g(-3) = (-3)^3 - (-3)^2 = -36 \qquad \blacksquare$$

Note that in Example 1, we were working with two different functions in the same problem. That is why we used two different names, f and g. Sometimes the rule of assignment for a function may consist of more than one part. We often refer to such functions as **piecewise-defined functions**. Let's consider an example of such a function.

EXAMPLE 2

If $f(x) = \begin{cases} 2x + 1 & \text{for } x \geq 0 \\ 3x - 1 & \text{for } x < 0 \end{cases}$, find $f(2), f(4), f(-1)$, and $f(-3)$.

Solution

For $x \geq 0$, we use the assignment $f(x) = 2x + 1$.

$$f(2) = 2(2) + 1 = 5$$
$$f(4) = 2(4) + 1 = 9$$

For $x < 0$, we use the assignment $f(x) = 3x - 1$.

$$f(-1) = 3(-1) - 1 = -4$$
$$f(-3) = 3(-3) - 1 = -10 \qquad \blacksquare$$

The quotient $\dfrac{f(a + h) - f(a)}{h}$ is often called a **difference quotient**. We use it extensively with functions when studying the limit concept in calculus. The next examples illustrate finding the difference quotient for specific functions.

EXAMPLE 3

Find $\dfrac{f(a + h) - f(a)}{h}$ for each of the following functions.

a. $f(x) = x^2 + 6$ **b.** $f(x) = 2x^2 + 3x - 4$ **c.** $f(x) = \dfrac{1}{x}$

Solutions

a.
$$f(a) = a^2 + 6$$
$$f(a + h) = (a + h)^2 + 6 = a^2 + 2ah + h^2 + 6$$

Therefore,

$$f(a + h) - f(a) = (a^2 + 2ah + h^2 + 6) - (a^2 + 6)$$
$$= a^2 + 2ah + h^2 + 6 - a^2 - 6$$
$$= 2ah + h^2$$

and

$$\frac{f(a+h)-f(a)}{h} = \frac{2ah+h^2}{h} = \frac{\not{h}(2a+h)}{\not{h}} = 2a+h$$

b.
$$f(a) = 2a^2 + 3a - 4$$
$$f(a+h) = 2(a+h)^2 + 3(a+h) - 4$$
$$= 2(a^2 + 2ha + h^2) + 3a + 3h - 4$$
$$= 2a^2 + 4ha + 2h^2 + 3a + 3h - 4$$

Therefore,

$$f(a+h)-f(a) = (2a^2 + 4ha + 2h^2 + 3a + 3h - 4) - (2a^2 + 3a - 4)$$
$$= 2a^2 + 4ha + 2h^2 + 3a + 3h - 4 - 2a^2 - 3a + 4$$
$$= 4ha + 2h^2 + 3h$$

and

$$\frac{f(a+h)-f(a)}{h} = \frac{4ha + 2h^2 + 3h}{h}$$

$$= \frac{\not{h}(4a + 2h + 3)}{\not{h}}$$

$$= 4a + 2h + 3$$

c.
$$f(a) = \frac{1}{a}$$

$$f(a+h) = \frac{1}{a+h}$$

Therefore,

$$f(a+h)-f(a) = \frac{1}{a+h} - \frac{1}{a}$$

$$= \frac{a-(a+h)}{a(a+h)}$$

$$= \frac{a-a-h}{a(a+h)}$$

$$= \frac{-h}{a(a+h)}, \quad \text{or} \quad -\frac{h}{a(a+h)}$$

and

$$\frac{f(a+h)-f(a)}{h} = -\frac{\dfrac{h}{a(a+h)}}{h}$$

$$= -\frac{h}{a(a+h)} \cdot \frac{1}{h}$$

$$= -\frac{1}{a(a+h)}$$

For our purposes in this text, if the domain of a function is not specifically indicated or determined by a real-world application, then we will assume the domain to be *all real number* replacements for the variable, provided that they represent elements in the domain and produce *real number* functional values.

EXAMPLE 4

For the function $f(x) = \sqrt{x - 1}$, (a) specify the domain, (b) determine the range, and (c) evaluate $f(5), f(50)$, and $f(25)$.

Solutions

a. The radicand must be nonnegative, so $x - 1 \geq 0$ and thus $x \geq 1$. Therefore, the domain (D) is

$$D = \{x \mid x \geq 1\}$$

b. The symbol $\sqrt{}$ indicates the nonnegative square root; thus the range (R) is

$$R = \{f(x) \mid f(x) \geq 0\}$$

c.
$$f(5) = \sqrt{4} = 2$$
$$f(50) = \sqrt{49} = 7$$
$$f(25) = \sqrt{24} = 2\sqrt{6}$$

As we will see later, the range of a function is often easier to determine after we have graphed the function. However, our equation- and inequality-solving processes are frequently sufficient to determine the domain of a function. Let's consider some examples.

EXAMPLE 5

Determine the domain for each of the following functions.

a. $f(x) = \dfrac{3}{2x - 5}$ **b.** $g(x) = \dfrac{1}{x^2 - 9}$ **c.** $f(x) = \sqrt{x^2 + 4x - 12}$

Solutions

a. We can replace x with any real number except $\dfrac{5}{2}$, because $\dfrac{5}{2}$ makes the denominator zero. Thus the domain is

$$D = \left\{ x \mid x \neq \frac{5}{2} \right\}$$

b. We need to eliminate any values of x that will make the denominator zero. Therefore, let's solve the equation $x^2 - 9 = 0$.

$$x^2 - 9 = 0$$
$$x^2 = 9$$
$$x = \pm 3$$

The domain is thus the set

$$D = \{x \mid x \neq 3 \text{ and } x \neq -3\}$$

c. The radicand, $x^2 + 4x - 12$, must be nonnegative. Therefore, let's use a number line approach as we did in Chapter 2, to solve the inequality $x^2 + 4x = 12 \geq 0$ (see Figure 3.5).

$$x^2 + 4x - 12 \geq 0$$
$$(x + 6)(x - 2) \geq 0$$

FIGURE 3.5

The product $(x + 6)(x - 2)$ is nonnegative if $x \leq -6$ or $x \geq 2$. Using interval notation, we can express the domain as $(-\infty, -6] \cup [2, \infty)$. ■

Functions and function notation provide the basis for describing many real-world relationships. The next example illustrates this point.

EXAMPLE 6

Suppose a factory determines that the overhead for producing a quantity of a certain item is $500 and the cost for each item is $25. Express the total expenses as a function of the number of items produced and compute the expenses for producing 12, 25, 50, 75, and 100 items.

Solution

Let n represent the number of items produced. Then $25n + 500$ represents the total expenses. Using E to represent the *expense function*, we have

$$E(n) = 25n + 500, \qquad \text{where } n \text{ is a whole number}$$

Therefore, we obtain

$$E(12) = 25(12) + 500 = 800$$
$$E(25) = 25(25) + 500 = 1125$$
$$E(50) = 25(50) + 500 = 1750$$
$$E(75) = 25(75) + 500 = 2375$$
$$E(100) = 25(100) + 500 = 3000$$

Thus the total expenses for producing 12, 25, 50, 75, and 100 items are $800, $1125, $1750, $2375, and $3000, respectively. ■

As we stated before, an equation such as $f(x) = 5x - 7$ that is used to determine a function can also be written $y = 5x - 7$. In either form, we refer to x as the **independent variable** and to y (or $f(x)$) as the **dependent variable**. Many formulas in mathematics and other related areas also determine functions. For example, the area formula for a circular region, $A = \pi r^2$, assigns to each positive real value for r a unique value for A. This formula determines a function f, where $f(r) = \pi r^2$. The variable r is the independent variable, and A [or $f(r)$] is the dependent variable.

Many functions that we will study throughout this text can be classified as even or odd functions. A function f having the property that $f(-x) = f(x)$ for every x in the domain of f is called an **even function**. A function f having the property that $f(-x) = -f(x)$ for every x in the domain of f is called an **odd function**.

EXAMPLE 7

For each of the following, classify the function as even, odd, or neither even nor odd.

a. $f(x) = 2x^3 - 4x$ **b.** $f(x) = x^4 - 7x^2$ **c.** $f(x) = x^2 + 2x - 3$

Solution

a. The function $f(x) = 2x^3 - 4x$ is an odd function because $f(-x) = 2(-x)^3 - 4(-x) = -2x^3 + 4x$, which equals $-f(x)$.

b. The function $f(x) = x^4 - 7x^2$ is an even function because $f(-x) = (-x)^4 - 7(-x)^2 = x^4 - 7x^2$, which equals $f(x)$.

c. The function $f(x) = x^2 + 2x - 3$ is neither even nor odd because $f(-x) = (-x)^2 + 2(-x) - 3 = x^2 - 2x - 3$, which does not equal either $f(x)$ or $-f(x)$. ∎

PROBLEM SET 3.1

1. If $f(x) = -2x + 5$, find $f(3), f(5)$, and $f(-2)$.

2. If $f(x) = x^2 - 3x - 4$, find $f(2), f(4)$, and $f(-3)$.

3. If $g(x) = -2x^2 + x - 5$, find $g(3), g(-1)$, and $g(-4)$.

4. If $g(x) = -x^2 - 4x + 6$, find $g(0), g(5)$, and $g(-5)$.

5. If $h(x) = \frac{2}{3}x - \frac{3}{4}$, find $h(3), h(4)$, and $h\left(-\frac{1}{2}\right)$.

6. If $h(x) = -\frac{1}{2}x + \frac{2}{3}$, find $h(-2), h(6)$, and $h\left(-\frac{2}{3}\right)$.

7. If $f(x) = \sqrt{2x - 1}$, find $f(5), f\left(\frac{1}{2}\right)$, and $f(23)$.

8. If $f(x) = \sqrt{3x + 2}$, find $f\left(\frac{14}{3}\right), f(10)$, and $f\left(-\frac{1}{3}\right)$.

9. If $f(x) = \begin{cases} x & \text{for } x \geq 0 \\ x^2 & \text{for } x < 0 \end{cases}$, find $f(4), f(10), f(-3)$, and $f(-5)$.

10. If $f(x) = \begin{cases} 3x + 2 & \text{for } x \geq 0 \\ 5x - 1 & \text{for } x < 0 \end{cases}$, find $f(2), f(6), f(-1)$, and $f(-4)$.

11. If $f(x) = \begin{cases} 2x & \text{for } x \geq 0 \\ -2x & \text{for } x < 0 \end{cases}$, find $f(3), f(5), f(-3)$, and $f(-5)$.

12. If $f(x) = \begin{cases} 2 & \text{for } x < 0 \\ x^2 + 1 & \text{for } 0 \le x \le 4, \\ -1 & \text{for } x > 4 \end{cases}$ find $f(3), f(6),$

$f(0),$ and $f(-3)$.

13. If $f(x) = \begin{cases} 1 & \text{for } x > 0 \\ 0 & \text{for } -1 < x \le 0, \\ -1 & \text{for } x \le -1 \end{cases}$ find $f(2), f(0),$

$f\left(-\dfrac{1}{2}\right),$ and $f(-4)$.

For Problems 14–25, find $\dfrac{f(a + h - f(a)}{h}$.

14. $f(x) = 4x + 5$ **15.** $f(x) = -7x - 2$

16. $f(x) = x^2 - 3x$ **17.** $f(x) = -x^2 + 4x - 2$

18. $f(x) = 2x^2 + 7x - 4$ **19.** $f(x) = 3x^2 - x - 4$

20. $f(x) = x^3$ **21.** $f(x) = x^3 - x^2 + 2x - 1$

22. $f(x) = \dfrac{1}{x + 1}$ **23.** $f(x) = \dfrac{2}{x - 1}$

24. $f(x) = \dfrac{x}{x + 1}$ **25.** $f(x) = \dfrac{1}{x^2}$

For Problems 26–33 (Figures 3.6 through 3.13), determine whether the indicated graph represents a function of x.

26.
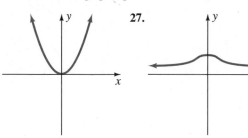

FIGURE 3.6   **FIGURE 3.7**

27.

28.

29.
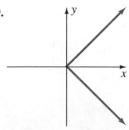

FIGURE 3.8   **FIGURE 3.9**

30.
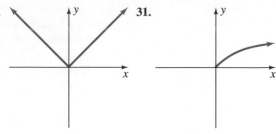

31.

FIGURE 3.10   **FIGURE 3.11**

32.
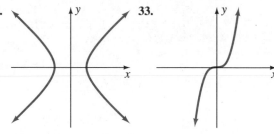

33.

FIGURE 3.12   **FIGURE 3.13**

For Problems 34–41, determine the domain and the range of the given function.

34. $f(x) = \sqrt{x}$ **35.** $f(x) = \sqrt{3x - 4}$

36. $f(x) = x^2 + 1$ **37.** $f(x) = x^2 - 2$

38. $f(x) = x^3$ **39.** $f(x) = |x|$

40. $f(x) = x^4$ **41.** $f(x) = -\sqrt{x}$

For Problems 42–51, determine the domain of the given function.

42. $f(x) = \dfrac{3}{x - 4}$ **43.** $f(x) = \dfrac{-4}{x + 2}$

44. $f(x) = \dfrac{2x}{(x - 2)(x + 3)}$ **45.** $f(x) = \dfrac{5}{(2x - 1)(x + 4)}$

46. $f(x) = \sqrt{5x + 1}$ **47.** $f(x) = \dfrac{1}{x^2 - 4}$

48. $g(x) = \dfrac{3}{x^2 + 5x + 6}$ **49.** $f(x) = \dfrac{4x}{x^2 - x - 12}$

50. $g(x) = \dfrac{5}{x^2 + 4x}$ **51.** $g(x) = \dfrac{x}{6x^2 + 13x - 5}$

For Problems 52–59, express the domain of the given function using interval notation.

52. $f(x) = \sqrt{x^2 - 1}$

53. $f(x) = \sqrt{x^2 - 16}$

54. $f(x) = \sqrt{x^2 + 4}$

55. $f(x) = \sqrt{x^2 + 1} - 4$

56. $f(x) = \sqrt{x^2 - 2x - 24}$

57. $f(x) = \sqrt{x^2 - 3x - 40}$

58. $f(x) = \sqrt{12x^2 + x - 6}$

59. $f(x) = -\sqrt{8x^2 + 6x - 35}$

For Problems 60–67, solve each problem.

60. Suppose that the profit function for selling n items is given by

$$P(n) = -n^2 + 500n - 61500$$

Evaluate $P(200)$, $P(230)$, $P(250)$, and $P(260)$.

61. The equation $A(r) = \pi r^2$ expresses the area of a circular region as a function of the length of a radius (r). Compute $A(2)$, $A(3)$, $A(12)$, and $A(17)$ and express your answers to the nearest hundredth.

62. In a physics experiment, it is found that the equation $V(t) = 1667t - 6940t^2$ expresses the velocity of an object as a function of time (t). Compute $V(0.1)$, $V(0.15)$, and $V(0.2)$.

63. The height of a projectile fired vertically into the air (neglecting air resistance) at an initial velocity of 64 feet per second is a function of the time (t) and is given by the equation $h(t) = 64t - 16t^2$. Compute $h(1)$, $h(2)$, $h(3)$, and $h(4)$.

64. A car rental agency charges $50 per day plus $.32 a mile. Therefore, the daily charge for renting a car is a function of the number of miles traveled (m) and can be expressed as $C(m) = 50 + 0.32 \ m$. Compute $C(75)$, $C(150)$, $C(225)$, and $C(650)$.

65. The equation $I(r) = 500r$ expresses the amount of simple interest earned by an investment of $500 for one year as a function of the rate of interest (r). Compute $I(0.11)$, $I(0.12)$, $I(0.135)$, and $I(0.15)$.

66. Suppose that the height of a semielliptical archway is given by the function $h(x) = \sqrt{64 - 4x^2}$, where x is the distance from the center line of the arch. Compute $h(0)$, $h(2)$, and $h(4)$.

67. The equation $A(r) = 2\pi r^2 + 16\pi r$ expresses the total surface area of a right circular cylinder of height 8 centimeters as a function of the length of a radius (r). Compute $A(2)$, $A(4)$, and $A(8)$ and express your answers to the nearest hundredth.

For Problems 68–77, determine whether f is even, odd, or neither even nor odd.

68. $f(x) = x^2$

69. $f(x) = x^3$

70. $f(x) = x^2 + 1$

71. $f(x) = 3x - 1$

72. $f(x) = x^2 + x$

73. $f(x) = x^3 + 1$

74. $f(x) = x^5$

75. $f(x) = x^4 + x^2 + 1$

76. $f(x) = -x^3$

77. $f(x) = x^5 + x^3 + x$

THOUGHTS INTO WORDS

78. Expand Definition 3.1 to include a definition for the concept of a relation.

79. What does it mean to say that the domain of a function may be restricted if the function represents a real-world situation? Give three examples of such functions.

80. Does $f(a + b) = f(a) + f(b)$ for all functions? Defend your answer.

81. Are there any functions for which $f(a + b) = f(a) + f(b)$? Defend your answer.

3.2   LINEAR AND QUADRATIC FUNCTIONS

As we use the function concept in our study of mathematics, it is helpful to classify certain types of functions and become familiar with their equations, characteristics, and graphs. In this section we will discuss two special types of functions: *linear* and *quadratic functions*. These functions are a natural extension of our earlier study of linear and quadratic equations.

Linear Functions

Any function that can be written in the form

$$f(x) = ax + b$$

where a and b are real numbers, is called a **linear function**. The following are examples of linear functions.

$$f(x) = -2x + 4 \qquad f(x) = 7x - 9 \qquad f(x) = \frac{2}{3}x + \frac{5}{6}$$

The equation $f(x) = ax + b$ can also be written $y = ax + b$. From our work with the slope–intercept form in Chapter 3, we know that $y = ax + b$ is the equation of a straight line having a slope of a and a y intercept of b. This information can be used to graph linear functions, as illustrated by the following example.

EXAMPLE 1

Graph $f(x) = -2x + 4$.

Solution

Because the y intercept is 4, the point (0, 4) is on the line. Furthermore, because the slope is -2, we can move two units down and one unit to the right of (0, 4) to determine the point (1, 2). The line determined by (0, 4) and (1, 2) is shown in Figure 3.14.

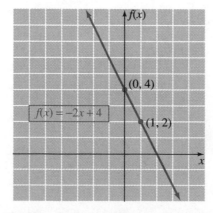

FIGURE 3.14

Note that in Figure 3.14 we labeled the vertical axis $f(x)$. It could also be labeled y, because $y = f(x)$. We will use the label $f(x)$ for most of our work with functions; however, we will continue to refer to y-axis symmetry instead of $f(x)$-axis symmetry.

Recall from Chapter 3 that we often graphed linear equations by finding the two intercepts. This same approach can be used with linear functions, as the next example illustrates.

EXAMPLE 2

Graph $f(x) = 3x - 6$.

Solution

First, we see that $f(0) = -6$; thus the point $(0, -6)$ is on the graph. Second, by setting $3x - 6$ equal to zero and solving for x, we obtain

$$3x - 6 = 0$$
$$3x = 6$$
$$x = 2$$

Therefore, $f(2) = 3(2) - 6 = 0$ and the point $(2, 0)$ is on the graph. The line determined by $(0, -6)$ and $(2, 0)$ is shown in Figure 3.15.

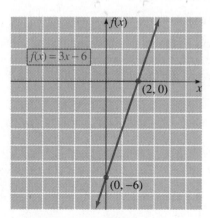

FIGURE 3.15

As you graph functions by using function notation, it is often helpful to think of the ordinate of every point on the graph as the value of the function at a specific value of x. Geometrically, **the functional value is the directed distance of the point from the x axis.** We have illustrated this idea in Figure 3.16 for the function $f(x) = x$ and in Figure 3.17 for the function $f(x) = 2$.

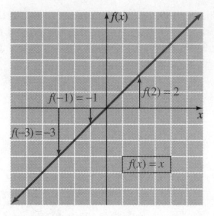

FIGURE 3.16

FIGURE 3.17

The linear function $f(x) = x$ is often called the **identity function**. Any linear function of the form $f(x) = ax + b$, where $a = 0$, is called a **constant function**, and its graph is a horizontal line.

Quadratic Functions

Any function that can be written in the form

$$f(x) = ax^2 + bx + c$$

where a, b, and c are real numbers and $a \neq 0$, is called a **quadratic function**. Furthermore, the graph of any quadratic function is a **parabola**. As we work with parabolas, we will use the vocabulary indicated in Figure 3.18.

FIGURE 3.18

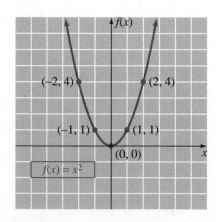

FIGURE 3.19

Graphing parabolas relies on finding the vertex, determining whether the parabola opens upward or downward, and locating two points on opposite sides of the axis of symmetry. It is also very helpful to compare the parabolas produced by various types of equations, such as $f(x) = x^2 + k$, $f(x) = ax^2$, $f(x) = (x - h)^2$, and $f(x) = a(x - h)^2 + k$. We are especially interested in how they compare to the **basic parabola** produced by the equation $f(x) = x^2$. The graph of $f(x) = x^2$ is shown in Figure 3.19. Note that the graph of $f(x) = x^2$ is symmetric with respect to the y, or $f(x)$, axis. Remember that an equation exhibits y-axis symmetry if replacing x with $-x$ produces an equivalent equation. Therefore, because $f(-x) = (-x)^2 = x^2$, the equation $f(x) = x^2$ exhibits y-axis symmetry.

Now let's consider an equation of the form $f(x) = x^2 + k$, where k is a constant. (Keep in mind that all such equations exhibit y-axis symmetry.)

EXAMPLE 3

Graph $f(x) = x^2 - 2$.

Solution

It should be observed that functional values for $f(x) = x^2 - 2$ are 2 less than corresponding functional values for $f(x) = x^2$. For example, $f(1) = -1$ for $f(x) = x^2 - 2$, but $f(1) = 1$ for $f(x) = x^2$. Thus the graph of $f(x) = x^2 - 2$ is the same as the graph of $f(x) = x^2$ except that it is *moved down 2 units* (Figure 3.20).

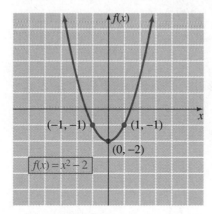

FIGURE 3.20 ■

In general, the graph of a quadratic function of the form $f(x) = x^2 + k$ is the same as the graph of $f(x) = x^2$ except that it is moved up or down $|k|$ units, depending on whether k is positive or negative. We say that the graph of $f(x) = x^2 + k$ is a **vertical translation** of the graph of $f(x) = x^2$.

Now let's consider some quadratic functions of the form $f(x) = ax^2$, where a is a nonzero constant. (The graphs of these equations also have y-axis symmetry.)

EXAMPLE 4

Graph $f(x) = 2x^2$.

Solution

Let's set up a table to make some comparisons of functional values. Note that in the table, the functional values for $f(x) = 2x^2$ are *twice* the corresponding functional values for $f(x) = x^2$. Thus the parabola associated with $f(x) = 2x^2$ has the same vertex (the origin) as the graph of $f(x) = x^2$, but it is *narrower*, as shown in Figure 3.21.

x	$f(x) = x^2$	$f(x) = 2x^2$
0	0	0
1	1	2
2	4	8
−1	1	2
−2	4	8

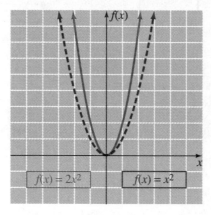

FIGURE 3.21

EXAMPLE 5

Graph $f(x) = \dfrac{1}{2}x^2$.

Solution

As we see from the table, the functional values for $f(x) = \dfrac{1}{2}x^2$ are *one-half* of the corresponding functional values for $f(x) = x^2$. Therefore, the parabola associated with $f(x) = \dfrac{1}{2}x^2$ is *wider* than the basic parabola, as shown in Figure 3.22.

x	$f(x) = x^2$	$f(x) = \frac{1}{2}x^2$
0	0	0
1	1	$\frac{1}{2}$
2	4	2
−1	1	$\frac{1}{2}$
−2	4	2

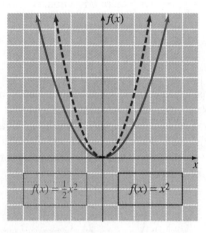

FIGURE 3.22

EXAMPLE 6

Graph $f(x) = -x^2$.

Solution

It should be evident that the functional values for $f(x) = -x^2$ are the *opposites* of the corresponding functional values for $f(x) = x^2$. Therefore, the graph of $f(x) = -x^2$ is a reflection across the x axis of the basic parabola (Figure 3.23).

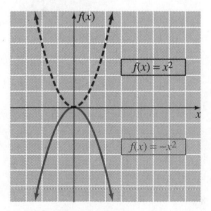

FIGURE 3.23

In general, the graph of a quadratic function of the form $f(x) = ax^2$ has its vertex at the origin and opens upward if a is positive and downward if a is negative. The parabola is narrower than the basic parabola if $|a| > 1$ and wider if $|a| < 1$.

Let's continue our investigation of quadratic functions by considering those of the form $f(x) = (x - h)^2$, where h is a nonzero constant.

EXAMPLE 7

Graph $f(x) = (x - 3)^2$.

Solution

A fairly extensive table of values illustrates a pattern. Note that $f(x) = (x - 3)^2$ and $f(x) = x^2$ take on the same functional values, but for different values of x. More specifically, if $f(x) = x^2$ achieves a certain functional value at a specific value of x, then $f(x) = (x - 3)^2$ achieves that same functional value at x *plus three*. In other words, the graph of $f(x) = (x - 3)^2$ is the graph of $f(x) = x^2$ *moved three units to the right* (Figure 3.24).

x	$f(x) = x^2$	$f(x) = (x-3)^2$
-1	1	16
0	0	9
1	1	4
2	4	1
3	9	0
4	16	1
5	25	4
6	36	9
7	49	16

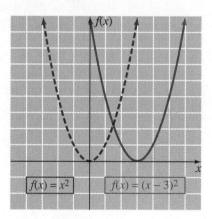

FIGURE 3.24

In general, the graph of a quadratic function of the form $f(x) = (x - h)^2$ is the same as the graph of $f(x) = x^2$ except that it is moved to the right h units if h is positive or moved to the left $|h|$ units if h is negative. We say that the graph of $f(x) = (x - h)^2$ is a **horizontal translation** of the graph of $f(x) = x^2$.

The following diagram summarizes our work thus far for graphing quadratic functions.

$f(x) = x^2 + ⓚ$	Moves the parabola up or down
$f(x) = ⓐx^2$	Affects the width and the way the parabola opens
$f(x) = (x - ⓗ)^2$	Moves the parabola right or left

$f(x) = x^2$ — basic parabola

Now let's consider two examples that combine these ideas.

EXAMPLE 8

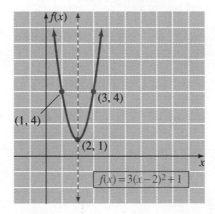

Graph $f(x) = 3(x - 2)^2 + 1$.

Solution

$f(x) = 3(x - 2)^2 + 1$

Narrows the parabola and opens it upward

Moves the parabola 2 units to the right

Moves the parabola 1 unit up

The vertex is (2, 1) and the line $x = 2$ is the axis of symmetry. If $x = 1$, then $f(1) = 3(1 - 2)^2 + 1 = 4$. Thus the point (1, 4) is on the graph, and so is its reflection, (3, 4), across the line of symmetry. The parabola is shown in Figure 3.25.

FIGURE 3.25

EXAMPLE 9

Graph $f(x) = -\dfrac{1}{2}(x + 1)^2 - 3$.

Solution

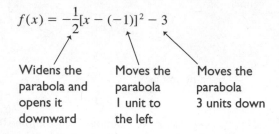

$$f(x) = -\frac{1}{2}[x - (-1)]^2 - 3$$

Widens the
parabola and
opens it
downward

Moves the
parabola
1 unit to
the left

Moves the
parabola
3 units down

The vertex is at $(-1, -3)$ and the line $x = -1$ is the axis of symmetry. If $x = 0$, then $f(0) = -\frac{1}{2}(0 + 1)^2 - 3 = -\frac{7}{2}$. Thus the point $\left(0, -\frac{7}{2}\right)$ is on the graph, and so is its reflection, $\left(-2, -\frac{7}{2}\right)$, across the line of symmetry. The parabola is shown in Figure 3.26.

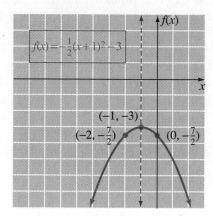

F I G U R E 3 . 2 6

Quadratic Functions of the Form $f(x) = ax^2 + bx + c$

We are now ready to graph quadratic functions of the form $f(x) = ax^2 + bx + c$. The general approach is to change from the form $f(x) = ax^2 + bx + c$ to the form $f(x) = a(x - h)^2 + k$ and then proceed as we did in Examples 8 and 9. The process of *completing the square* serves as the basis for making the change in form. Let's consider two examples to illustrate the details.

EXAMPLE 10

Graph $f(x) = x^2 - 4x + 3$.

Solution

$$f(x) = x^2 - 4x + 3$$
$$= (x^2 - 4x) + 3 \qquad \text{Add 4, which is the square of one-half of the coefficient of } x.$$
$$= (x^2 - 4x + 4) + 3 - 4 \leftarrow \text{Subtract 4 to compensate for the 4 that was added.}$$
$$= (x - 2)^2 - 1$$

The graph of $f(x) = (x - 2)^2 - 1$ is the basic parabola moved 2 units to the right and 1 unit down (Figure 3.27).

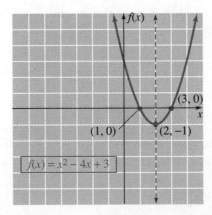

FIGURE 3.27

EXAMPLE 11

Graph $f(x) = -2x^2 - 4x + 1$.

Solution

$$f(x) = -2x^2 - 4x + 1$$
$$= -2(x^2 + 2x) + 1 \qquad \text{Factor } -2 \text{ from the first two terms.}$$
$$= -2(x^2 + 2x + 1) + 1 + 2 \qquad \text{Add 1 inside the parentheses to complete the square.}$$
$$= -2(x + 1)^2 + 3 \qquad \text{Add 2 to compensate for the 1 inside the parentheses times the factor } -2.$$

The graph of $f(x) = -2(x + 1)^2 + 3$ is shown in Figure 3.28.

FIGURE 3.28

Now let's graph a piecewise-defined function that involves both linear and quadratic rules of assignment.

EXAMPLE 12

Graph $f(x) = \begin{cases} 2x & \text{for } x \geq 0 \\ x^2 + 1 & \text{for } x < 0 \end{cases}$.

Solution

If $x \geq 0$, then $f(x) = 2x$. Thus for nonnegative values of x, we graph the linear function $f(x) = 2x$. If $x < 0$, then $f(x) = x^2 + 1$. Thus for negative values of x, we graph the quadratic function $f(x) = x^2 + 1$. The complete graph is shown in Figure 3.29.

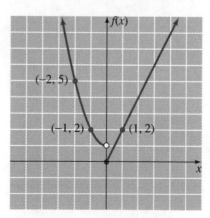

FIGURE 3.29

What we know about parabolas and the process of completing the square can be helpful when we are using a graphing utility to graph a quadratic function. Consider the following example.

EXAMPLE 13

Use a graphing utility to obtain the graph of the quadratic function

$$f(x) = -x^2 + 37x - 311$$

Solution

First, we know that the parabola opens downward and its width is the same as that of the basic parabola $f(x) = x^2$. Then we can start the process of completing the square to determine an approximate location of the vertex.

$$
\begin{aligned}
f(x) &= -x^2 + 37x - 311 \\
&= -(x^2 - 37x) - 311 \\
&= -\left(x^2 - 37x + \left(\frac{37}{2}\right)^2\right) - 311 + \left(\frac{37}{2}\right)^2 \\
&= -(x^2 - 37x + (18.5)^2) - 311 + 342.25
\end{aligned}
$$

Thus the vertex is near $x = 18$ and $y = 31$. Therefore, setting the boundaries of the viewing rectangle so that $-2 \le x \le 25$ and $-10 \le y \le 35$, we obtain the graph shown in Figure 3.30.

FIGURE 3.30

REMARK The graph in Figure 3.30 is sufficient for most purposes because it shows the vertex and the x intercepts of the parabola. Certainly, we could use other boundaries that would also give this information.

PROBLEM SET 3.2

For Problems 1–10, graph each linear function.

1. $f(x) = 2x - 4$

2. $f(x) = 3x + 6$

5. $f(x) = -2x$

6. $f(x) = 3x$

3. $f(x) = -x + 1$

4. $f(x) = -2x - 4$

7. $f(x) = \frac{1}{2}x - \frac{3}{4}$

8. $f(x) = -\frac{2}{3}x + \frac{1}{2}$

9. $f(x) = -1$ **10.** $f(x) = -3$

For Problems 11–34, graph each quadratic function.

11. $f(x) = x^2 + 1$ **12.** $f(x) = x^2 - 3$

13. $f(x) = 3x^2$ **14.** $f(x) = -2x^2$

15. $f(x) = -x^2 + 2$ **16.** $f(x) = -3x^2 - 1$

17. $f(x) = (x + 2)^2$ **18.** $f(x) = (x - 1)^2$

19. $f(x) = -2(x + 1)^2$ **20.** $f(x) = 3(x - 2)^2$

21. $f(x) = (x - 1)^2 + 2$ **22.** $f(x) = -(x + 2)^2 + 3$

23. $f(x) = \frac{1}{2}(x - 2)^2 - 3$ **24.** $f(x) = 2(x - 3)^2 - 1$

25. $f(x) = x^2 + 2x + 4$ **26.** $f(x) = x^2 - 4x + 2$

27. $f(x) = x^2 - 3x + 1$ **28.** $f(x) = x^2 + 5x + 5$

29. $f(x) = 2x^2 + 12x + 17$

30. $f(x) = 3x^2 - 6x$

31. $f(x) = -x^2 - 2x + 1$

32. $f(x) = -2x^2 + 12x - 16$

33. $f(x) = 2x^2 - 2x + 3$

34. $f(x) = 2x^2 + 3x - 1$

35. $f(x) = -2x^2 - 5x + 1$

36. $f(x) = -3x^2 + x - 2$

For Problems 37–44, graph each function.

37. $f(x) = \begin{cases} x & \text{for } x \geq 0 \\ 3x & \text{for } x < 0 \end{cases}$

38. $f(x) = \begin{cases} -x & \text{for } x \geq 0 \\ 4x & \text{for } x < 0 \end{cases}$

39. $f(x) = \begin{cases} 2x + 1 & \text{for } x \geq 0 \\ x^2 & \text{for } x < 0 \end{cases}$

40. $f(x) = \begin{cases} -x^2 & \text{for } x \geq 0 \\ 2x^2 & \text{for } x < 0 \end{cases}$

41. $f(x) = \begin{cases} 2 & \text{if } x \geq 0 \\ -1 & \text{if } x < 0 \end{cases}$

42. $f(x) = \begin{cases} 2 & \text{if } x > 2 \\ 1 & \text{if } 0 < x \leq 2 \\ -1 & \text{if } x \leq 0 \end{cases}$

43. $f(x) = \begin{cases} 1 & \text{if } 0 \leq x < 1 \\ 2 & \text{if } 1 \leq x < 2 \\ 3 & \text{if } 2 \leq x < 3 \\ 4 & \text{if } 3 \leq x < 4 \end{cases}$

44. $f(x) = \begin{cases} 2x + 3 & \text{if } x < 0 \\ x^2 & \text{if } 0 \leq x < 2 \\ 1 & \text{if } x \geq 2 \end{cases}$

45. The **greatest integer function** is defined by the equation $f(x) = [x]$, where $[x]$ refers to the largest integer less than or equal to x. For example, $[2.6] = 2$, $\left[\sqrt{2}\right] = 1$, $[4] = 4$, and $[-1.4] = -2$. Graph $f(x) = [x]$ for $-4 \leq x < 4$.

46. Explain the concept of a piecewise-defined function.

47. Suppose that Julian walks at a constant rate of 3 miles per hour. Explain what it means to say that the distance Julian walks is a *linear function* of the time he walks.

48. Is $f(x) = (3x - 2) - (2x + 1)$ a linear function? Explain your answer.

49. Give a step-by-step description of how you would use the ideas presented in this section to graph $f(x) = 5x^2 + 10x + 4$.

GRAPHING CALCULATOR ACTIVITIES

50. This problem is designed to reinforce ideas presented in this section. For each part, first predict the shapes and locations of the parabolas, and then use your graphing calculator to graph them on the same set of axes.

a. $f(x) = x^2$, $f(x) = x^2 - 4$, $f(x) = x^2 + 1$, $f(x) = x^2 + 5$

b. $f(x) = x^2, f(x) = (x - 5)^2, f(x) = (x + 5)^2, f(x) = (x - 3)^2$

c. $f(x) = x^2, f(x) = 5x^2, f(x) = \dfrac{1}{3}x^2, f(x) = -2x^2$

d. $f(x) = x^2, f(x) = (x - 7)^2 - 3, f(x) = -(x + 8)^2 + 4, f(x) = -3x^2 - 4$

e. $f(x) = x^2 - 4x - 2$, $f(x) = -x^2 + 4x + 2$, $f(x) = -x^2 - 16x - 58, f(x) = x^2 + 16x + 58$

51. a. Graph both $f(x) = x^2 - 14x + 51$ and $f(x) = x^2 + 14x + 51$ on the same set of axes. What relationship seems to exist between the two graphs?

b. Graph both $f(x) = x^2 + 12x + 34$ and $f(x) = x^2 - 12x + 34$ on the same set of axes. What relationship seems to exist between the two graphs?

c. Graph both $f(x) = -x^2 + 8x - 20$ and $f(x) = -x^2 - 8x - 20$ on the same set of axes. What relationship seems to exist between the two graphs?

d. Make a statement that generalizes your findings in parts (a) through (c).

52. Use your graphing calculator to graph the piecewise-defined functions in Problems 37–44. You may need to consult your user's manual for instructions on graphing these functions.

3.3   QUADRATIC FUNCTIONS AND PROBLEM SOLVING

In the previous section we used the process of completing the square to change a quadratic function such as $f(x) = x^2 - 4x + 3$ to the form $f(x) = (x - 2)^2 - 1$. From the form $f(x) = (x - 2)^2 - 1$, it is easy to identify the vertex $(2, -1)$ and the axis of symmetry $x = 2$ of the parabola. In general, if we complete the square on

$$f(x) = ax^2 + bx + c$$

we obtain

$$f(x) = a\left(x^2 + \frac{b}{a}x\right) + c$$

$$= a\left(x^2 + \frac{b}{a}x + \frac{b^2}{4a^2}\right) + c - \frac{b^2}{4a}$$

$$= a\left(x + \frac{b}{2a}\right)^2 + \frac{4ac - b^2}{4a}$$

Therefore, the parabola associated with the function $f(x) = ax^2 + bx + c$ has its vertex at

$$\left(-\frac{b}{2a}, \frac{4ac - b^2}{4a}\right)$$

and the equation of its axis of symmetry is $x = -b/2a$. These facts are illustrated in Figure 3.31.

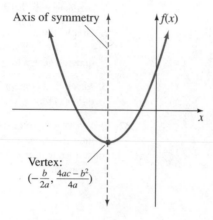

F I G U R E   3 . 3 1

By using the information from Figure 3.31, we now have another way of graphing quadratic functions of the form $f(x) = ax^2 + bx + c$, as indicated by the following steps.

1. Determine whether the parabola opens upward (if $a > 0$) or downward (if $a < 0$).

2. Find $-b/2a$, which is the x coordinate of the vertex.

3. Find $f(-b/2a)$, which is the y coordinate of the vertex, or find the y coordinate by evaluating

$$\frac{4ac - b^2}{4a}$$

4. Locate another point on the parabola and also locate its image across the axis of symmetry, which is the line with equation $x = -b/2a$.

The three points found in steps 2, 3, and 4 should determine the general shape of the parabola. Let's illustrate this procedure with two examples.

EXAMPLE 1

Graph $f(x) = 3x^2 - 6x + 5$.

Solution

STEP 1   Because $a > 0$, the parabola opens upward.

STEP 2 $-\dfrac{b}{2a} = -\dfrac{-6}{6} = 1$

STEP 3 $f\left(-\dfrac{b}{2a}\right) = f(1) = 3 - 6 + 5 = 2$. Thus the vertex is at $(1, 2)$.

STEP 4   Letting $x = 2$, we obtain $f(2) = 12 - 12 + 5 = 5$. Thus $(2, 5)$ is on the graph, and so is its reflection, $(0, 5)$, across the line of symmetry, $x = 1$.

The three points $(1, 2)$, $(2, 5)$, and $(0, 5)$ are used to graph the parabola in Figure 3.32.

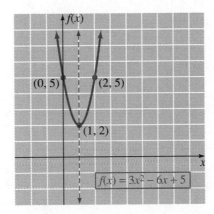

$f(x) = 3x^2 - 6x + 5$

FIGURE 3.32

Graph $f(x) = -x^2 - 4x - 7$.

EXAMPLE 2

Solution

STEP 1   Because $a < 0$, the parabola opens downward.

STEP 2 $-\dfrac{b}{2a} = -\dfrac{-4}{-2} = -2$

STEP 3 $f\left(-\dfrac{b}{2a}\right) = f(-2) = -(-2)^2 - 4(-2) - 7 = -3$. Thus the vertex is at $(-2, -3)$.

STEP 4   Letting $x = 0$, we obtain $f(0) = -7$. Thus $(0, -7)$ is on the graph, and so is its reflection, $(-4, -7)$, across the line of symmetry, $x = -2$.

The three points $(-2, -3)$, $(0, -7)$, and $(-4, -7)$ are used to draw the parabola in Figure 3.33.

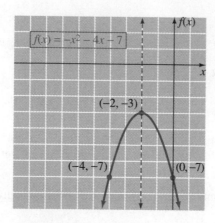

$$f(x) = -x^2 - 4x - 7$$

$(-2, -3)$

$(-4, -7)$ $(0, -7)$

FIGURE 3.33

In summary, we basically have two methods to graph a quadratic function.

1. We can express the function in the form $f(x) = a(x - h)^2 + k$ and use the values of a, h, and k to determine the parabola.

2. We can express the function in the form $f(x) = ax^2 + bx + c$ and use the approach demonstrated in Examples 1 and 2.

Parabolas possess various properties that make them very useful. For example, if a parabola is rotated about its axis, a parabolic surface is formed and such surfaces are used for light and sound reflectors. A projectile fired into the air will follow the curvature of a parabola. The *trend line* of profit and cost functions sometimes follows a parabolic curve. In most applications of the parabola, we are primarily interested in the x intercepts and the vertex. Let's consider some examples of finding the x intercepts and the vertex.

EXAMPLE 3

Find the x intercepts and the vertex for each of the following parabolas.

a. $f(x) = -x^2 + 11x - 18$ **b.** $f(x) = x^2 - 8x - 3$

c. $f(x) = 2x^2 - 12x + 23$

Solutions

a. To find the x intercepts, let $f(x) = 0$ and solve the resulting equation.

$$-x^2 + 11x - 18 = 0$$
$$x^2 - 11x + 18 = 0$$
$$(x - 2)(x - 9) = 0$$
$$x - 2 = 0 \quad \text{or} \quad x - 9 = 0$$
$$x = 2 \quad \text{or} \quad x = 9$$

Therefore, the x intercepts are 2 and 9. To find the vertex, let's determine the point $\left(-\dfrac{b}{2a}, f\left(-\dfrac{b}{2a}\right)\right)$.

$$f(x) = -x^2 + 11x - 18$$

$$-\frac{b}{2a} = -\frac{11}{2(-1)} = -\frac{11}{-2} = \frac{11}{2}$$

$$f\left(\frac{11}{2}\right) = -\left(\frac{11}{2}\right)^2 + 11\left(\frac{11}{2}\right) - 18$$

$$= -\frac{121}{4} + \frac{121}{2} - 18$$

$$= \frac{-121 + 242 - 72}{4}$$

$$= \frac{49}{4}$$

Therefore, the vertex is at $\left(\dfrac{11}{2}, \dfrac{49}{4}\right)$.

b. The find the x intercepts, let $f(x) = 0$, and solve the resulting equation.

$$x^2 - 8x - 3 = 0$$

$$x = \frac{-(-8) \pm \sqrt{(-8)^2 - 4(1)(-3)}}{2(1)}$$

$$= \frac{8 \pm \sqrt{76}}{2}$$

$$= \frac{8 \pm 2\sqrt{19}}{2}$$

$$= 4 \pm \sqrt{19}$$

Therefore, the x intercepts are $4 + \sqrt{19}$ and $4 - \sqrt{19}$. This time, to find the vertex, let's complete the square on x.

$$f(x) = x^2 - 8x - 3$$

$$= x^2 - 8x + 16 - 3 - 16$$

$$= (x - 4)^2 - 19$$

Therefore, the vertex is at $(4, -19)$.

c. To find the x intercepts, let $f(x) = 0$ and solve the resulting equation.

$$2x^2 - 12x + 23 = 0$$

$$x = \frac{-(-12) \pm \sqrt{(-12)^2 - 4(2)(23)}}{2(2)}$$

$$= \frac{12 \pm \sqrt{-40}}{4}$$

Because these solutions are nonreal complex numbers, there are no x intercepts. To find the vertex, let's determine the point $\left(-\dfrac{b}{2a}, f\left(-\dfrac{b}{2a}\right)\right)$.

$$f(x) = 2x^2 - 12x + 23$$
$$-\frac{b}{2a} = -\frac{-12}{2(2)}$$
$$= 3$$
$$f(3) = 2(3)^2 - 12(3) + 23$$
$$= 18 - 36 + 23$$
$$= 5$$

Therefore, the vertex is at $(3, 5)$. ∎

REMARK   Note that in parts (a) and (c) we used the general point

$$\left(-\frac{b}{2a}, f\left(-\frac{b}{2a}\right)\right)$$

to find the vertices. In part (b), however, we completed the square and used that form to determine the vertex. Which approach you use is up to you. We chose to complete the square in part (b) because the algebra involved was quite easy.

In Problem 60 of Problem Set 1.3 you were asked to solve the quadratic equation $-x^2 + 11x - 18 = 0$. You should have obtained a solution set of $\{2, 9\}$. Here, in part (a) of Example 3, we solved the same equation to determine that the x intercepts of the graph of the function $f(x) = -x^2 + 11x - 18$ are 2 and 9. The numbers 2 and 9 are also called the **real number zeros** of the function. In part (b), the real numbers $4 + \sqrt{19}$ and $4 - \sqrt{19}$ are the x intercepts of the graph of the function $f(x) = x^2 - 8x - 3$ and are the real number zeros of the function. In part (c), the nonreal complex numbers $\dfrac{12 \pm \sqrt{-40}}{4}$, which simplify to $\dfrac{6 \pm i\sqrt{10}}{2}$, indicate that the graph of the function $f(x) = 2x^2 - 12x + 23$ has no points on the x axis. The complex numbers are zeros of the function, but they have no physical significance for the graph other than indicating that the graph has no points on the x axis.

Figure 3.34 shows the result we got when we used a graphing calculator to graph the three functions of Example 3 on the same set of axes. This gives us a visual interpretation of the conclusions drawn regarding the x intercepts and vertices.

FIGURE 3.34

Back to Problem Solving

As we have seen, the vertex of the graph of a quadratic function is either the lowest or the highest point on the graph. Thus we often speak of the **minimum value** or **maximum value** of a function in applications of the parabola. The x value of the vertex indicates where the minimum or maximum occurs, and $f(x)$ yields the minimum or maximum value of the function. Let's consider some examples that illustrate these ideas.

PROBLEM 1

A farmer has 120 rods of fencing and wants to enclose a rectangular plot of land that requires fencing on only three sides because it is bounded by a river on one side. Find the length and width of the plot that will maximize the area.

Solution

Let x represent the width; then $120 - 2x$ represents the length, as indicated in Figure 3.35.

FIGURE 3.35

The function $A(x) = x(120 - 2x)$ represents the area of the plot in terms of the width x. Because

$$A(x) = x(120 - 2x)$$
$$= 120x - 2x^2$$
$$= -2x^2 + 120x$$

we have a quadratic function with $a = -2$, $b = 120$, and $c = 0$. Therefore, the *maximum* value ($a < 0$ so the parabola opens downward) of the function is obtained where the x value is

$$-\frac{b}{2a} = -\frac{120}{2(-2)} = 30$$

If $x = 30$, then $120 - 2x = 120 - 2(30) = 60$. Thus the farmer should make the plot 30 rods wide and 60 rods long to maximize the area at $(30)(60) = 1800$ square rods.   ■

PROBLEM 2

Find two numbers whose sum is 30, such that the sum of their squares is a minimum.

Solution

Let x represent one of the numbers; then $30 - x$ represents the other number. By expressing the sum of their squares as a function of x, we obtain

$$f(x) = x^2 + (30 - x)^2$$

which can be simplified to

$$f(x) = x^2 + 900 - 60x + x^2$$
$$= 2x^2 - 60x + 900$$

This is a quadratic function with $a = 2$, $b = -60$, and $c = 900$. Therefore, the x value where the *minimum* occurs is

$$-\frac{b}{2a} = -\frac{-60}{4}$$
$$= 15$$

If $x = 15$, then $30 - x = 30 - 15 = 15$. Thus the two numbers should both be 15.

■

PROBLEM 3

A golf pro shop operator finds that she can sell 30 sets of golf clubs at $500 per set in a year. Furthermore, she predicts that for each $25 decrease in price, she could sell three extra sets of golf clubs. At what price should she sell the clubs to maximize gross income?

Solution

In analyzing such a problem, it sometimes helps to start by setting up a table.

	NUMBER OF SETS	PRICE PER SET	INCOME
Three additional sets can be sold for ⟶ a $25 decrease in price.	30	$500	$15,000
	33	$475	$15,675
	36	$450	$16,200

Let x represent the number of $25 decreases in price. Then the income can be expressed as a function of x.

$$f(x) = (30 + 3x)(500 - 25x).$$

Number of sets   Price per set

Simplifying this, we obtain

$$f(x) = 15{,}000 - 750x + 1500x - 75x^2$$
$$= -75x^2 + 750x + 15{,}000$$

We complete the square in order to analyze the parabola.

$$f(x) = -75x^2 + 750x + 15{,}000$$
$$= -75(x^2 - 10x) + 15{,}000$$
$$= -75(x^2 - 10x + 25) + 15{,}000 + 1875$$
$$= -75(x - 5)^2 + 16{,}875$$

From this form we know that the vertex of the parabola is at $(5, 16875)$, and because $a = -75$, we know that a *maximum* occurs at the vertex. Thus five decreases of $25—that is, a $125 reduction in price—will give a maximum income of $16,875. The golf clubs should be sold at $375 per set.   ■

We have determined that the vertex of a parabola associated with $f(x) = ax^2 + bx + c$ is located at $\left(-\dfrac{b}{2a}, f\left(-\dfrac{b}{2a}\right)\right)$ and that the x intercepts of the graph can be found by solving the quadratic equation $ax^2 + bx + c = 0$. Therefore, a graphing utility does not provide us with much extra power when we are working with quadratic functions. However, as functions become more complex, a graphing utility becomes more helpful. Let's build our confidence in the use of a graphing utility at this time, while we have a way of checking our results.

| EXAMPLE 4 |

Use a graphing utility to graph $f(x) = x^2 - 8x - 3$ and find the x intercepts of the graph. [This is the parabola from part (b) of Example 3.]

Solution

A graph of the parabola is shown in Figure 3.36.

FIGURE 3.36

One x intercept appears to be between 0 and -1 and the other between 8 and 9. Let's zoom in on the x intercept between 8 and 9. This produces a graph like Figure 3.37.

FIGURE 3.37

Now we can use the trace function to determine that this x intercept is at approximately 8.4. (This agrees with the answer of $4 + \sqrt{19}$ that we got in Example 3.) In a similar fashion, we can determine that the other x intercept is at -0.4.

PROBLEM SET 3.3

For Problems 1–12, use the approach of Examples 1 and 2 of this section to graph each quadratic function.

1. $f(x) = x^2 - 8x + 15$

2. $f(x) = x^2 + 6x + 11$

3. $f(x) = 2x^2 + 20x + 52$

4. $f(x) = 3x^2 - 6x - 1$

5. $f(x) = -x^2 + 4x - 7$

6. $f(x) = -x^2 - 6x - 5$

7. $f(x) = -3x^2 + 6x - 5$

8. $f(x) = -2x^2 - 4x + 2$

9. $f(x) = x^2 + 3x - 1$

10. $f(x) = x^2 + 5x + 2$

11. $f(x) = -2x^2 + 5x + 1$

12. $f(x) = -3x^2 + 2x - 1$

For Problems 13–20, use the approach that you think is the most appropriate to graph each quadratic function.

13. $f(x) = -x^2 + 3$

14. $f(x) = (x + 1)^2 + 1$

15. $f(x) = x^2 + x - 1$

16. $f(x) = -x^2 + 3x - 4$

17. $f(x) = -2x^2 + 4x + 1$

18. $f(x) = 4x^2 - 8x + 5$

19. $f(x) = -\left(x + \dfrac{5}{2}\right)^2 + \dfrac{3}{2}$

20. $f(x) = x^2 - 4x$

For Problems 21–32, find the x intercepts and the vertex of each parabola.

21. $f(x) = x^2 - 8x + 15$

22. $f(x) = x^2 - 16x + 63$

23. $f(x) = 2x^2 - 28x + 96$

24. $f(x) = 3x^2 - 60x + 297$

25. $f(x) = -x^2 + 10x - 24$

26. $f(x) = -2x^2 + 36x - 160$

27. $f(x) = x^2 - 14x + 44$

28. $f(x) = x^2 - 18x + 68$

29. $f(x) = -x^2 + 9x - 21$

30. $f(x) = 2x^2 + 3x + 3$

31. $f(x) = -4x^2 + 4x + 4$

32. $f(x) = -2x^2 + 3x + 7$

For Problems 33–42, solve each problem.

33. Suppose that the equation $p(x) = -2x^2 + 280x - 1000$, where x represents the number of items sold, describes the profit function for a certain business. How many items should be sold to maximize the profit?

34. Suppose that the cost function for the production of a particular item is given by the equation $C(x) = 2x^2 - 320x + 12{,}920$, where x represents the number of items. How many items should be produced to minimize the cost?

35. The height of a projectile fired vertically into the air (neglecting air resistance) at an initial velocity of 96 feet per second is a function of time x and is given by the equation $f(x) = 96x - 16x^2$. Find the highest point reached by the projectile.

36. Find two numbers whose sum is 30, such that the sum of the square of one number plus ten times the other number is a minimum.

37. Find two numbers whose sum is 50 and whose product is a maximum.

38. Find two numbers whose difference is 40 and whose product is a minimum.

39. Two hundred and forty meters of fencing is available to enclose a rectangular playground. What should be the dimensions of the playground to maximize the area?

40. Motel managers advertise that they will provide dinner, dancing, and drinks for $50 per couple for a New Year's Eve party. They must have a guarantee of 30 couples. Furthermore, they will agree that for each couple in excess of 30, they will reduce the price per couple for all attending by $.50. How many couples will it take to maximize the motel's revenue?

41. A cable TV company has 1000 subscribers, each of whom pays $15 per month. On the basis of a survey, they believe that for each decrease of $.25 in the monthly rate, they could obtain 20 additional subscribers. At what rate will maximum revenue be obtained and how many subscribers will there be at that rate?

42. A manufacturer finds that for the first 500 units of its product that are produced and sold, the profit is $50 per unit. The profit on each of the units beyond 500 is decreased by $.10 times the number of additional units sold. What level of output will maximize profit?

THOUGHTS INTO WORDS

43. Suppose your friend was absent the day this section was discussed. How would you explain to her the ideas pertaining to x intercepts of the graph of a function, zeros of the function, and solutions of the equation $f(x) = 0$.

44. Give a step-by-step explanation of how to find the x intercepts of the graph of the function $f(x) = 2x^2 + 7x - 4$.

45. Give a step-by-step explanation of how to find the vertex of the parabola determined by the equation $f(x) = -x^2 - 6x - 5$.

Further Investigations

46. Suppose that an arch is shaped like a parabola. It is 20 feet wide at the base and 100 feet high. How wide is the arch 50 feet above the ground? (See Figure 3.38.)

FIGURE 3.38

47. A parabolic arch 27 feet high spans a parkway. The center section of the parkway is 50 feet wide. How wide is the arch if it has a minimum clearance of 15 feet above the center section?

48. A parabolic arch spans a stream 200 feet wide. How high must the arch be above the stream to give a minimum clearance of 40 feet over a 120-foot-wide channel in the center?

GRAPHING CALCULATOR ACTIVITIES

49. Suppose that the viewing window on your graphing calculator is set so that $-15 \leq x \leq 15$ and $-10 \leq y \leq 10$. Now try to graph the function $f(x) = x^2 - 8x + 28$. Nothing appears on the screen, so the parabola must be outside the viewing window. We could arbitrarily expand the window until the parabola appeared. However, let's be a little more systematic and use $\left(-\dfrac{b}{2a}, f\left(-\dfrac{b}{2a}\right)\right)$ to find the vertex. We find the vertex is at (4, 12), so let's change the y values of the window so that $0 \leq y \leq 25$. Now we get a good picture of the parabola.

Graph each of the following parabolas, and keep in mind that you may need to change the dimensions of the viewing window to obtain a good picture.

a. $f(x) = x^2 - 2x + 12$

b. $f(x) = -x^2 - 4x - 16$

c. $f(x) = x^2 + 12x + 44$

d. $f(x) = x^2 - 30x + 229$

e. $f(x) = -2x^2 + 8x - 19$

50. Use a graphing calculator to graph each of the following parabolas, and then use the trace function to help estimate the x intercepts and the vertex. Finally, use the approach of Example 3 to find the x intercepts and the vertex.

a. $f(x) = x^2 - 6x + 3$ **b.** $f(x) = x^2 - 18x + 66$

c. $f(x) = -x^2 + 8x - 3$

d. $f(x) = -x^2 + 24x - 129$

e. $f(x) = 14x^2 - 7x + 1$ **f.** $f(x) = -\dfrac{1}{2}x^2 + 5x - \dfrac{17}{2}$

51. In Problems 21–32, you were asked to find the x intercepts and the vertex of some parabolas. Now use a graphing calculator to graph each parabola and visually justify your answers.

3.4 TRANSFORMATIONS OF SOME BASIC CURVES

From our work in Section 3.2, we know that the graph of $f(x) = (x - 5)^2$ is the basic parabola $f(x) = x^2$ translated five units to the right. Likewise, we know that the graph of $f(x) = -x^2 - 2$ is the basic parabola reflected across the x axis and translated downward two units. Translations and reflections apply not only to parabolas but to curves in general. Therefore, if we know the shapes of a few basic curves, then it is easy to sketch numerous variations of these curves by using the concepts of translation and reflection.

Let's begin this section by establishing the graphs of four basic curves and then apply some transformations to these curves. First let's restate, in terms of function vocabulary, the graphing suggestions offered in Chapter 2. Pay special attention to suggestions 2 and 3, where we restate the concepts of intercepts and symmetry using function notation.

1. Determine the domain of the function.
2. Find the y intercept [we are labeling the y axis with $f(x)$] by evaluating $f(0)$. Find the x intercept by finding the value(s) of x such that $f(x) = 0$.

3. Determine any types of symmetry that the equation possesses. If $f(-x) = f(x)$, then the function exhibits y-axis symmetry. If $f(-x) = -f(x)$, then the function exhibits origin symmetry. (Note that the definition of a function rules out the possibility that the graph of a function has x-axis symmetry.)

4. Set up a table of ordered pairs that satisfy the equation. The type of symmetry and the domain will affect your choice of values of x in the table.

5. Plot the points associated with the ordered pairs and connect them with a smooth curve. Then, if appropriate, reflect this part of the curve according to any symmetries possessed by the graph.

EXAMPLE 1

Graph $f(x) = x^3$.

Solution

The domain is the set of real numbers. Because $f(0) = 0$, the origin is on the graph. Because $f(-x) = (-x)^3 = -x^3 = -f(x)$, the graph is symmetric with respect to the origin. Therefore, we can concentrate our table on the positive values of x. By connecting the points associated with the ordered pairs from the table with a smooth curve, and then reflecting it through the origin, we get the graph in Figure 3.39.

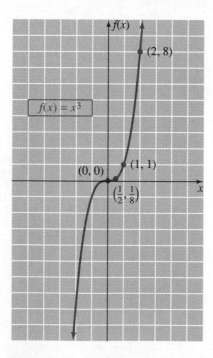

x	$f(x) = x^3$
0	0
1	1
2	8
$\frac{1}{2}$	$\frac{1}{8}$

FIGURE 3.39

EXAMPLE 2

Graph $f(x) = x^4$.

Solution

The domain is the set of real numbers. Because $f(0) = 0$, the origin is on the graph. Because $f(-x) = (-x)^4 = x^4 = f(x)$, the graph has y-axis symmetry, and we can concentrate our table of values on the positive values of x. If we connect the points associated with the ordered pairs from the table with a smooth curve, and then reflect across the vertical axis, we get the graph in Figure 3.40.

x	$f(x) = x^4$
0	0
1	1
2	16
$\frac{1}{2}$	$\frac{1}{16}$

FIGURE 3.40

REMARK   The curve in Figure 3.40 is not a parabola, even though it resembles one; this curve is flatter at the bottom and steeper.

EXAMPLE 3

Graph $f(x) = \sqrt{x}$.

Solution

The domain of the function is the set of nonnegative real numbers. Because $f(0) = 0$, the origin is on the graph. Because $f(-x) \neq f(x)$ and $f(-x) \neq -f(x)$, there is no symmetry, so let's set up a table of values using nonnegative values for x. Plotting the points determined by the table and connecting them with a smooth curve produces Figure 3.41.

x	$f(x) = \sqrt{x}$
0	0
1	1
4	2
9	3

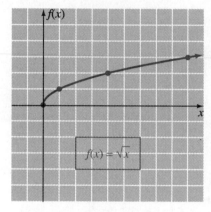

FIGURE 3.41

Sometimes a new function is defined in terms of old functions. In such cases, the definition plays an important role in the study of the new function. Consider the following example.

EXAMPLE 4

Graph $f(x) = |x|$.

Solution

The concept of absolute value is defined for all real numbers by

$$|x| = x \quad \text{if } x \geq 0$$
$$|x| = -x \quad \text{if } x < 0$$

Therefore, the absolute value function can be expressed as

$$f(x) = |x| = \begin{cases} x & \text{if } x \geq 0 \\ -x & \text{if } x < 0 \end{cases}$$

The graph of $f(x) = x$ for $x \geq 0$ is the ray in the first quadrant, and the graph of $f(x) = -x$ for $x < 0$ is the half-line (not including the origin) in the second quadrant, as indicated in Figure 3.42. Note that the graph has y-axis symmetry.

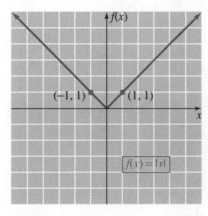

FIGURE 3.42

Translations of the Basic Curves

From our work in Section 3.2, we know that

1. the graph of $f(x) = x^2 + 3$ is the graph of $f(x) = x^2$ moved up three units.

2. the graph of $f(x) = x^2 - 2$ is the graph of $f(x) = x^2$ moved down two units.

Now let's describe in general the concept of a vertical translation.

Vertical Translation

The graph of $y = f(x) + k$ is the graph of $y = f(x)$ shifted k units upward if $k > 0$ or shifted $|k|$ units downward if $k < 0$.

In Figure 3.43, the graph of $f(x) = |x| + 2$ is obtained by shifting the graph of $f(x) = |x|$ upward two units, and the graph of $f(x) = |x| - 3$ is obtained by shifting the graph of $f(x) = |x|$ downward three units. [Remember that $f(x) = |x| - 3$ can be written as $f(x) = |x| + (-3)$.]

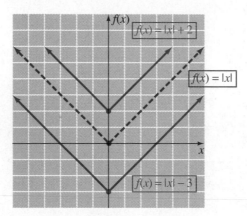

FIGURE 3.43

We also graphed horizontal translations of the basic parabola in Section 3.2. For example,

1.   the graph of $f(x) = (x - 4)^2$ is the graph of $f(x) = x^2$ shifted four units to the right.

2.   the graph of $f(x) = (x + 5)^2$ is the graph of $f(x) = x^2$ shifted five units to the left.

The general concept of a horizontal translation can be described as follows.

Horizontal Translation

The graph of $y = f(x - h)$ is the graph of $y = f(x)$ shifted h units to the right if $h > 0$ or shifted $|h|$ units to the left if $h < 0$.

In Figure 3.44, the graph of $f(x) = (x - 3)^3$ is obtained by shifting the graph of $f(x) = x^3$ three units to the right. Likewise, the graph of $f(x) = (x + 2)^3$ is obtained by shifting the graph of $f(x) = x^3$ two units to the left.

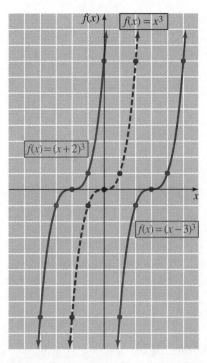

FIGURE 3.44

Reflections of the Basic Curves

From our work in Section 3.2, we know that the graph of $f(x) = -x^2$ is the graph of $f(x) = x^2$ reflected through the x axis. The general concept of an x-axis reflection can be described as follows.

---

x-axis Reflection

The graph of $y = -f(x)$ is the graph of $y = f(x)$ reflected through the x axis.

In Figure 3.45, the graph of $f(x) = -\sqrt{x}$ is obtained by reflecting the graph of $f(x) = \sqrt{x}$ through the x axis. Reflections are sometimes referred to as **mirror images**. Thus if we think of the x axis in Figure 3.45 as a mirror, then the graphs of $f(x) = \sqrt{x}$ and $f(x) = -\sqrt{x}$ are mirror images of each other.

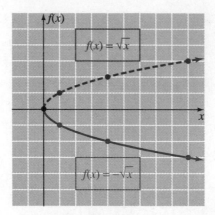

FIGURE 3.45

In Section 3.2, we did not consider a y-axis reflection of the basic parabola $f(x) = x^2$ because it is symmetric with respect to the y axis. In other words, a y-axis reflection of $f(x) = x^2$ produces the same figure. However, at this time let's describe the general concept of a y-axis reflection.

---

y-axis Reflection

The graph of $y = f(-x)$ is the graph of $y = f(x)$ reflected through the y axis.

Now suppose that we want to do a y-axis reflection of $f(x) = \sqrt{x}$. Because $f(x) = \sqrt{x}$ is defined for $x \geq 0$, the y-axis reflection $f(x) = \sqrt{-x}$ is defined for $-x \geq 0$, which is equivalent to $x \leq 0$. Figure 3.46 shows the y-axis reflection of $f(x) = \sqrt{x}$.

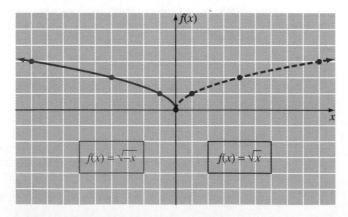

FIGURE 3.46

Vertical Stretching and Shrinking

Translations and reflections are called **rigid transformations** because the basic shape of the curve being transformed is not changed. In other words, only the positions of the graphs are changed. Now we want to consider some transformations that distort the shape of the original figure somewhat.

In Section 3.2, we graphed the equation $y = 2x^2$ by doubling the y coordinates of the ordered pairs that satisfy the equation $y = x^2$. We obtained a parabola with its vertex at the origin, symmetric to the y axis, but *narrower* than the basic parabola. Likewise, we graphed the equation $y = \frac{1}{2}x^2$ by halving the y coordinates of the ordered pairs that satisfy $y = x^2$. In this case, we obtained a parabola with its vertex at the origin, symmetric to the y axis, but *wider* than the basic parabola.

The concepts of *narrower* and *wider* can be used to describe parabolas, but they cannot be used to describe some other curves accurately. Instead, we use the more general concepts of vertical stretching and shrinking.

Vertical Stretching and Shrinking

The graph of $y = cf(x)$ is obtained from the graph of $y = f(x)$ by multiplying the y coordinates for $y = f(x)$ by c. If $c > 1$, the graph is said to be *stretched* by a factor of c, and if $0 < c < 1$, the graph is said to be *shrunk* by a factor of c.

In Figure 3.47, the graph of $f(x) = 2\sqrt{x}$ is obtained by doubling the y coordinates of points on the graph of $f(x) = \sqrt{x}$. Likewise, in Figure 3.47, the graph of $f(x) = \frac{1}{2}\sqrt{x}$ is obtained by halving the y coordinates of points on the graph of $f(x) = \sqrt{x}$.

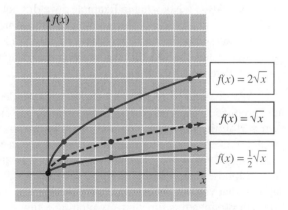

F I G U R E   3 . 4 7

Successive Transformations

Some curves are the result of performing more than one transformation on a basic curve. Let's consider the graph of a function that involves a stretching, a reflection, a horizontal translation, and a vertical translation of the basic absolute value function.

EXAMPLE 5

Graph $f(x) = -2|x - 3| + 1$.

Solution

This is the basic absolute value curve stretched by a factor of 2, reflected through the x axis, shifted three units to the right, and shifted one unit upward. To sketch the graph, we locate the point $(3, 1)$ and then determine a point on each of the rays. The graph is shown in Figure 3.48.

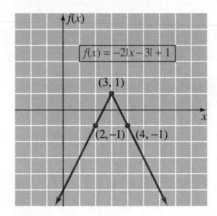

FIGURE 3.48

REMARK Note that in Example 5 we did not sketch the original basic curve $f(x) = |x|$ or any of the intermediate transformations. However, it is helpful to picture each transformation mentally. This locates the point $(3, 1)$ and establishes the fact that the two rays point downward. Then a point on each ray determines the final graph.

We do need to realize that changing the order of doing the transformations may produce an incorrect graph. In Example 5, performing the translations first, and then performing the stretching and x-axis reflection would locate the vertex of the graph at $(3, -1)$ instead of $(3, 1)$. **Unless parentheses indicate otherwise, stretchings, shrinkings, and reflections should be performed before translations.**

Suppose that you need to graph the function $f(x) = \sqrt{-3 - x}$. Furthermore, suppose that you are not certain which transformations of the basic square root function will produce this function. By plotting a few points and using your knowledge of the general shape of a square root curve, you should be able to sketch the curve as shown in Figure 3.49.

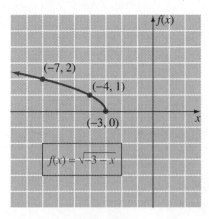

FIGURE 3.49

Now suppose that we want to graph a function such as

$$f(x) = \frac{2x^2}{x^2 + 4}$$

Because this is neither a basic function that we recognize nor a transformation of a basic function, we must revert to our previous graphing experiences. In other words, we need to find the domain, find the intercepts, check for symmetry, check for any restrictions, set up a table of values, plot the points, and sketch the curve. (If you want to do this at this time, you can check your result on page 408.) Furthermore, if the new function is defined in terms of an old function, we may be able to apply the definition of the old function and thereby simplify the new function for graphing purposes. For example, Problem 13 in Problem Set 3.4 asks you to graph the function $f(x) = |x| + x$. This function can be simplified by applying the definition of absolute value. We will leave that for you to do later.

Finally, let's use a graphing utility to give another illustration of the concept of stretching and shrinking a curve.

EXAMPLE 6

If $f(x) = \sqrt{25 - x^2}$, sketch a graph of $y = 2(f(x))$ and $y = \frac{1}{2}(f(x))$.

Solution

If $y = f(x) = \sqrt{25 - x^2}$, then

$$y = 2(f(x)) = 2\sqrt{25 - x^2} \qquad \text{and} \qquad y = \frac{1}{2}(f(x)) = \frac{1}{2}\sqrt{25 - x^2}$$

Graphing all three of these functions on the same set of axes produces Figure 3.50.

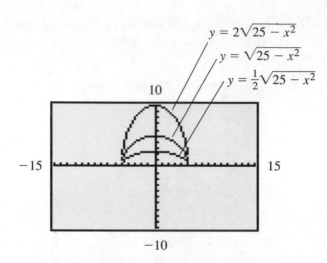

$$y = 2\sqrt{25 - x^2}$$

$$y = \sqrt{25 - x^2}$$

$$y = \tfrac{1}{2}\sqrt{25 - x^2}$$

FIGURE 3.50

PROBLEM SET 3.4

For Problems 1–30, graph each function.

1. $f(x) = x^4 + 2$

2. $f(x) = -x^4 - 1$

3. $f(x) = (x - 2)^4$

4. $f(x) = (x + 3)^4 + 1$

5. $f(x) = -x^3$

6. $f(x) = x^3 - 2$

7. $f(x) = (x + 2)^3$

8. $f(x) = (x - 3)^3 - 1$

9. $f(x) = |x - 1| + 2$

10. $f(x) = -|x + 2|$

11. $f(x) = |x + 1| - 3$

12. $f(x) = 2|x|$

13. $f(x) = x + |x|$

14. $f(x) = \dfrac{|x|}{x}$

15. $f(x) = -|x - 2| - 1$

16. $f(x) = 2|x + 1| - 4$

17. $f(x) = x - |x|$

18. $f(x) = |x| - x$

19. $f(x) = -2\sqrt{x}$

20. $f(x) = 2\sqrt{x - 1}$

21. $f(x) = \sqrt{x + 2} - 3$

22. $f(x) = -\sqrt{x + 2} + 2$

23. $f(x) = \sqrt{2 - x}$

24. $f(x) = \sqrt{-1 - x}$

25. $f(x) = -2x^4 + 1$

26. $f(x) = 2(x - 2)^4 - 4$

27. $f(x) = -2x^3$

28. $f(x) = 2x^3 + 3$

29. $f(x) = 3(x - 2)^3 - 1$

30. $f(x) = -2(x + 1)^3 + 2$

31. Suppose that the graph of $y = f(x)$ with a domain of $-2 \le x \le 2$ is shown in Figure 3.51.

FIGURE 3.51

Sketch the graph of each of the following transformations of $y = f(x)$.

a. $y = f(x) + 3$

b. $y = f(x - 2)$

c. $y = -f(x)$

d. $y = f(x + 3) - 4$

THOUGHTS INTO WORDS

32. Are the graphs of the two functions $f(x) = \sqrt{x - 2}$ and $g(x) = \sqrt{2 - x}$ y-axis reflections of each other? Defend your answer.

33. Are the graphs of $f(x) = 2\sqrt{x}$ and $g(x) = \sqrt{2x}$ identical? Defend your answer.

34. Are the graphs of $f(x) = \sqrt{x + 4}$ and $g(x) = \sqrt{-x + 4}$ y-axis reflections of each other? Defend your answer.

 GRAPHING CALCULATOR ACTIVITIES

35. Use your graphing calculator to check your graphs for Problems 13–30.

36. Graph $f(x) = \sqrt{x^2 + 8}$, $f(x) = \sqrt{x^2 + 4}$, and $f(x) = \sqrt{x^2 + 1}$ on the same set of axes. Look at these graphs and predict the graph of $f(x) = \sqrt{x^2 - 4}$. Now graph it with the calculator to test your prediction.

37. For each of the following, predict the general shape and location of the graph, and then use your calculator to graph the function to check your prediction.

   a. $f(x) = \sqrt{x^2}$ **b.** $f(x) = \sqrt{x^3}$

   c. $f(x) = |x^2|$ **d.** $f(x) = |x^3|$

38. Graph $f(x) = x^4 + x^3$. Now predict the graph for each of the following and check each prediction with your graphing calculator.

   a. $f(x) = x^4 + x^3 - 4$

   b. $f(x) = (x - 3)^4 + (x - 3)^3$

   c. $f(x) = -x^4 - x^3$ **d.** $f(x) = x^4 - x^3$

39. Graph $f(x) = \sqrt[3]{x}$. Now predict the graph for each of the following and check each prediction with your graphing calculator.

   a. $f(x) = 5 + \sqrt[3]{x}$ **b.** $f(x) = \sqrt[3]{x + 4}$

   c. $f(x) = -\sqrt[3]{x}$ **d.** $f(x) = \sqrt[3]{x - 3} - 5$

   e. $f(x) = \sqrt[3]{-x}$

3.5 COMBINING FUNCTIONS

In subsequent mathematics courses, it is common to encounter functions that are defined in terms of sums, differences, products, and quotients of simpler functions. For example, if $h(x) = x^2 + \sqrt{x - 1}$, then we may consider the function h as the sum of f and g, where $f(x) = x^2$ and $g(x) = \sqrt{x - 1}$. In general, *if f and g are functions and D is the intersection of their domains*, then the following definitions can be made.

Sum	$(f + g)(x) = f(x) + g(x)$
Difference	$(f - g)(x) = f(x) - g(x)$
Product	$(f \cdot g)(x) = f(x) \cdot g(x)$
Quotient	$\left(\dfrac{f}{g}\right)(x) = \dfrac{f(x)}{g(x)}, \quad g(x) \neq 0$

EXAMPLE 1

If $f(x) = 3x - 1$ and $g(x) = x^2 - x - 2$, find (a) $(f + g)(x)$, (b) $(f - g)(x)$, (c) $(f \cdot g)(x)$, and (d) $(f/g)(x)$. Determine the domain of each.

Solutions

a. $(f + g)(x) = f(x) + g(x) = (3x - 1) + (x^2 - x - 2) = x^2 + 2x - 3$

b. $(f - g)(x) = f(x) - g(x)$
$$= (3x - 1) - (x^2 - x - 2)$$
$$= 3x - 1 - x^2 + x + 2$$
$$= -x^2 + 4x + 1$$

c. $(f \cdot g)(x) = f(x) \cdot g(x)$
$$= (3x - 1)(x^2 - x - 2)$$
$$= 3x^3 - 3x^2 - 6x - x^2 + x + 2$$
$$= 3x^3 - 4x^2 - 5x + 2$$

d. $\left(\dfrac{f}{g}\right)(x) = \dfrac{f(x)}{g(x)} = \dfrac{3x - 1}{x^2 - x - 2}$

The domain of both f and g is the set of all real numbers. Therefore, the domain of $f + g, f - g,$ and $f \cdot g$ is the set of all real numbers. For f/g, the denominator $x^2 - x - 2$ cannot equal zero. Solving $x^2 - x - 2 = 0$ produces

$$(x - 2)(x + 1) = 0$$
$$x - 2 = 0 \quad \text{or} \quad x + 1 = 0$$
$$x = 2 \quad \text{or} \quad x = -1$$

Therefore, the domain for f/g is the set of all real numbers except 2 and -1. ∎

Composition of Functions

Besides adding, subtracting, multiplying, and dividing functions, there is another important operation called *composition*. The composition of two functions can be defined as follows.

DEFINITION 3.2

The composition of functions f and g is defined by

$$(f \circ g)(x) = f(g(x))$$

for all x in the domain of g such that $g(x)$ is in the domain of f.

The left side, $(f \circ g)(x)$, of the equation in Definition 3.2 is read *the composition of f and g*, and the right side is read *f of g of x*. It may also be helpful for you to have a mental picture of Definition 3.2 as two function machines hooked together to produce another function (called the **composite function**), as illustrated in Figure 3.52. Note that what comes out of the g function is substituted into the f function. Thus composition is sometimes called the **substitution of functions**.

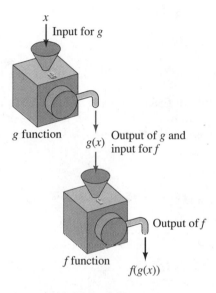

FIGURE 3.52

Figure 3.52 also illustrates the fact that $f \circ g$ is defined *for all x in the domain of g such that g(x) is in the domain of f.* In other words, what comes out of g must be capable of being fed into f. Let's consider some examples.

EXAMPLE 2

If $f(x) = x^2$ and $g(x) = 3x - 4$, find $(f \circ g)(x)$ and determine its domain.

Solution

Apply Definition 3.2 to obtain

$$(f \circ g)(x) = f(g(x))$$
$$= f(3x - 4)$$
$$= (3x - 4)^2$$
$$= 9x^2 - 24x + 16$$

Because g and f are both defined for all real numbers, so is $f \circ g$.   ∎

Definition 3.2, with f and g interchanged, defines the composition of g and f as $(g \circ f)(x) = g(f(x))$.

EXAMPLE 3

If $f(x) = x^2$ and $g(x) = 3x - 4$, find $(g \circ f)(x)$ and determine its domain.

Solution

$$
\begin{aligned}
(g \circ f)(x) &= g(f(x)) \\
&= g(x^2) \\
&= 3x^2 - 4
\end{aligned}
$$

Because f and g are defined for all real numbers, so is $g \circ f$. ∎

The results of Examples 2 and 3 demonstrate an important idea, that **the composition of functions is not a commutative operation.** In other words, $f \circ g \neq g \circ f$ for all functions f and g. However, as we will see in the next section, there is a special class of functions for which $f \circ g = g \circ f$.

EXAMPLE 4

If $f(x) = \sqrt{x}$ and $g(x) = 2x - 1$, find $(f \circ g)(x)$ and $(g \circ f)(x)$. Also determine the domain of each composite function.

Solution

$$
\begin{aligned}
(f \circ g)(x) &= f(g(x)) \\
&= f(2x - 1) \\
&= \sqrt{2x - 1}
\end{aligned}
$$

The domain and range of g are the set of all real numbers, but the domain of f is all *nonnegative* real numbers. Therefore $g(x)$, which is $2x - 1$, must be nonnegative.

$$
\begin{aligned}
2x - 1 &\geq 0 \\
2x &\geq 1 \\
x &\geq \frac{1}{2}
\end{aligned}
$$

Thus the domain of $f \circ g$ is $D = \left\{ x \mid x \geq \frac{1}{2} \right\}$.

$$
\begin{aligned}
(g \circ f)(x) &= g(f(x)) \\
&= g\!\left(\sqrt{x}\right) \\
&= 2\sqrt{x} - 1
\end{aligned}
$$

The domain and range of f are the set of nonnegative real numbers. The domain of g is the set of all real numbers. Therefore, the domain of $g \circ f$ is $D = \{ x \mid x \geq 0 \}$. ∎

EXAMPLE 5

If $f(x) = 2/(x - 1)$ and $g(x) = 1/x$, find $(f \circ g)(x)$ and $(g \circ f)(x)$. Determine the domain for each composite function.

Solution

$$(f \circ g)(x) = f(g(x))$$

$$= f\left(\frac{1}{x}\right)$$

$$= \frac{2}{\dfrac{1}{x} - 1} = \frac{2}{\dfrac{1 - x}{x}}$$

$$= \frac{2x}{1 - x}$$

The domain of g is all real numbers except 0, and the domain of f is all real numbers except 1. Because $g(x)$, which is $1/x$, cannot equal 1,

$$\frac{1}{x} \neq 1$$

$$x \neq 1$$

Therefore, the domain of $f \circ g$ is $D = \{x \mid x \neq 0 \text{ and } x \neq 1\}$.

$$(g \circ f)(x) = g(f(x))$$

$$= g\left(\frac{2}{x - 1}\right)$$

$$= \frac{1}{\dfrac{2}{x - 1}}$$

$$= \frac{x - 1}{2}$$

The domain of f is all real numbers except 1, and the domain of g is all real numbers except 0. Because $f(x)$, which is $2/(x - 1)$, will never equal 0, the domain of $g \circ f$ is $D = \{x \mid x \neq 1\}$. ∎

A graphing utility can be used to find the graph of a composite function without actually forming the function algebraically. Let's see how this works.

EXAMPLE 6

If $f(x) = x^3$ and $g(x) = x - 4$, use a graphing utility to obtain the graph of $y = (f \circ g)(x)$ and of $y = (g \circ f)(x)$.

Solution

To find the graph of $y = (f \circ g)(x)$, we can make the following assignments.

$$Y_1 = x - 4$$
$$Y_2 = (Y_1)^3$$

(Note that we have substituted Y_1 for x in $f(x)$ and assigned this expression to Y_2, much the same way as we would do it algebraically.) The graph of $y = (f \circ g)(x)$ is shown in Figure 3.53.

F I G U R E   3 . 5 3

To find the graph of $y = (g \circ f)(x)$, we can make the following assignments.

$$Y_1 = x^3$$
$$Y_2 = Y_1 - 4$$

The graph of $y = (g \circ f)(x)$ is shown in Figure 3.54.

F I G U R E   3 . 5 4   ■

Take another look at Figures 3.53 and 3.54. Note that in Figure 3.53, the graph of $y = (f \circ g)(x)$ is the basic cubic curve $f(x) = x^3$ shifted four units to the right. Likewise, in Figure 3.54, the graph of $y = (g \circ f)(x)$ is the basic cubic curve shifted four units downward. These are examples of a more general concept of using composite functions to represent various geometric transformations.

PROBLEM SET 3.5

For Problems 1–8, find $f + g, f - g, f \cdot g$, and $\dfrac{f}{g}$.

1. $f(x) = 3x - 4, \quad g(x) = 5x + 2$

2. $f(x) = -6x - 1, \quad g(x) = -8x + 7$

3. $f(x) = x^2 - 6x + 4, \quad g(x) = -x - 1$

4. $f(x) = 2x^2 - 3x + 5, \quad g(x) = x^2 - 4$

5. $f(x) = x^2 - x - 1, \quad g(x) = x^2 + 4x - 5$

6. $f(x) = x^2 - 2x - 24, \quad g(x) = x^2 - x - 30$

7. $f(x) = \sqrt{x - 1}, \quad g(x) = \sqrt{x}$

8. $f(x) = \sqrt{x + 2}, \quad g(x) = \sqrt{3x - 1}$

For Problems 9–26, find $(f \circ g)(x)$ and $(g \circ f)(x)$. Also specify the domain for each.

9. $f(x) = 2x, \quad g(x) = 3x - 1$

10. $f(x) = 4x + 1, \quad g(x) = 3x$

11. $f(x) = 5x - 3, \quad g(x) = 2x + 1$

12. $f(x) = 3 - 2x, \quad g(x) = -4x$

13. $f(x) = 3x + 4, \quad g(x) = x^2 + 1$

14. $f(x) = 3, \quad g(x) = -3x^2 - 1$

15. $f(x) = 3x - 4, \quad g(x) = x^2 + 3x - 4$

16. $f(x) = 2x^2 - x - 1, \quad g(x) = x + 4$

17. $f(x) = \dfrac{1}{x}, \quad g(x) = 2x + 7$

18. $f(x) = \dfrac{1}{x^2}, \quad g(x) = x$

19. $f(x) = \sqrt{x - 2}, \quad g(x) = 3x - 1$

20. $f(x) = \dfrac{1}{x}, \quad g(x) = \dfrac{1}{x^2}$

21. $f(x) = \dfrac{1}{x - 1}, \quad g(x) = \dfrac{2}{x}$

22. $f(x) = \dfrac{4}{x + 2}, \quad g(x) = \dfrac{3}{2x}$

23. $f(x) = 2x + 1, \quad g(x) = \sqrt{x - 1}$

24. $f(x) = \sqrt{x + 1}, \quad g(x) = 5x - 2$

25. $f(x) = \dfrac{1}{x - 1}, \quad g(x) = \dfrac{x + 1}{x}$

26. $f(x) = \dfrac{x - 1}{x + 2}, \quad g(x) = \dfrac{1}{x}$

For Problems 27–32, solve each problem.

27. If $f(x) = 3x - 2$ and $g(x) = x^2 + 1$, find $(f \circ g)(-1)$ and $(g \circ f)(3)$.

28. If $f(x) = x^2 - 2$ and $g(x) = x + 4$, find $(f \circ g)(2)$ and $(g \circ f)(-4)$.

29. If $f(x) = 2x - 3$ and $g(x) = x^2 - 3x - 4$, find $(f \circ g)(-2)$ and $(g \circ f)(1)$.

30. If $f(x) = 1/x$ and $g(x) = 2x + 1$, find $(f \circ g)(1)$ and $(g \circ f)(2)$.

31. If $f(x) = \sqrt{x}$ and $g(x) = 3x - 1$, find $(f \circ g)(4)$ and $(g \circ f)(4)$.

32. If $f(x) = x + 5$ and $g(x) = |x|$, find $(f \circ g)(-4)$ and $(g \circ f)(-4)$.

For Problems 33–38, show that $(f \circ g)(x) = x$ and that $(g \circ f)(x) = x$.

33. $f(x) = 2x, \quad g(x) = \dfrac{1}{2}x$

34. $f(x) = \dfrac{3}{4}x, \quad g(x) = \dfrac{4}{3}x$

35. $f(x) = x - 2, \quad g(x) = x + 2$

36. $f(x) = 2x + 1, \quad g(x) = \dfrac{x - 1}{2}$

37. $f(x) = 3x + 4, \quad g(x) = \dfrac{x - 4}{3}$

38. $f(x) = 4x - 3, \quad g(x) = \dfrac{x + 3}{4}$

39. Discuss whether addition, subtraction, multiplication, and division of functions are commutative operations.

40. Explain why the composition of two functions is not a commutative operation.

41. Explain how to find the domain of

$$\left(\frac{f}{g}\right)(x) \text{ if } f(x) = \frac{x-1}{x+2} \text{ and } g(x) = \frac{x+3}{x-5}.$$

Further Investigations

42. If $f(x) = 3x - 4$ and $g(x) = ax + b$, find conditions on a and b that will guarantee that $f \circ g = g \circ f$.

43. If $f(x) = x^2$ and $g(x) = \sqrt{x}$, with both having as domain the set of nonnegative real numbers, then show that $(f \circ g)(x) = x$ and $(g \circ f)(x) = x$.

44. If $f(x) = 3x^2 - 2x - 1$ and $g(x) = x$, find $f \circ g$ and $g \circ f$. (Recall that we have previously named $g(x) = x$ the *identity function*.)

45. In Section 3.1, we defined an *even function* to be a function such that $f(-x) = f(x)$ and an *odd function* to be one such that $f(-x) = -f(x)$. Verify that (a) the sum of two even functions is an even function, and (b) the sum of two odd functions is an odd function.

GRAPHING CALCULATOR ACTIVITIES

46. For each of the following, predict the general shape and location of the graph, and then use your calculator to graph the function to check your prediction. (Your knowledge of the graphs of the basic functions that are being added or subtracted should be helpful when you are making your predictions.)

a. $f(x) = x^4 + x^2$

b. $f(x) = x^3 + x^2$

c. $f(x) = x^4 - x^2$

d. $f(x) = x^2 - x^4$

e. $f(x) = x^2 - x^3$

f. $f(x) = x^3 - x^2$

g. $f(x) = |x| + \sqrt{x}$

h. $f(x) = |x| - \sqrt{x}$

47. For each of the following, find the graph of $y = (f \circ g)(x)$ and of $y = (g \circ f)(x)$.

a. $f(x) = x^2$ and $g(x) = x + 5$

b. $f(x) = x^3$ and $g(x) = x + 3$

c. $f(x) = x - 6$ and $g(x) = -x^3$

d. $f(x) = x^2 - 4$ and $g(x) = \sqrt{x}$

e. $f(x) = \sqrt{x}$ and $g(x) = x^2 + 4$

f. $f(x) = \sqrt[3]{x}$ and $g(x) = x^3 - 5$

3.6 DIRECT AND INVERSE VARIATION

The amount of simple interest earned by a fixed amount of money invested at a certain rate *varies directly* as the time.

At a constant temperature, the volume of an enclosed gas *varies inversely* as the pressure.

Such statements illustrate two basic types of functional relationships, **direct variation** and **inverse variation**, that are widely used, especially in the physical sciences. These relationships can be expressed by equations that determine functions. The purpose of this section is to investigate these special functions.

Direct Variation

The statement *y varies directly as x* means

$$y = kx$$

where k is a nonzero constant, called the **constant of variation**. The phrase *y is directly proportional to x* is also used to indicate direct variation; k is then referred to as the **constant of proportionality**.

> **REMARK**   Note that the equation $y = kx$ defines a function and can be written $f(x) = kx$. However, in this section it is more convenient not to use function notation but instead to use variables that are meaningful in terms of the physical entities involved in the particular problem.

Statements that indicate direct variation may also involve powers of a variable. For example, *y varies directly as the square of x* can be written $y = kx^2$. In general, *y varies directly as the nth power of x* $(n > 0)$ means

$$y = kx^n$$

There are basically three types of problems wherein we deal with direct variation: (1) translating an English statement into an equation expressing the direct variation, (2) finding the constant of variation from given values of the variables, and (3) finding additional values of the variables once the constant of variation has been determined. Let's consider an example of each of these types of problems.

EXAMPLE 1

Translate the statement *The tension on a spring varies directly as the distance it is stretched* into an equation, using k as the constant of variation.

Solution

Let t represent the tension and d the distance; the equation is

$$t = kd$$

EXAMPLE 2

If A varies directly as the square of e, and if $A = 96$ when $e = 4$, find the constant of variation.

Solution

Because A varies directly as the square of e, we have

$$A = ke^2$$

Substitute 96 for A and 4 for e to obtain

$$96 = k(4)^2$$
$$96 = 16k$$
$$6 = k$$

The constant of variation is 6. ■

EXAMPLE 3

If y is directly proportional to x, and if $y = 6$ when $x = 8$, find the value of y when $x = 24$.

Solution

The statement y *is directly proportional to* x translates into

$$y = kx$$

Let $y = 6$ and $x = 8$; the constant of variation becomes

$$6 = k(8)$$
$$\frac{6}{8} = k$$
$$\frac{3}{4} = k$$

Thus the specific equation is

$$y = \frac{3}{4}x$$

Now, let $x = 24$ to obtain

$$y = \frac{3}{4}(24) = 18$$ ■

Inverse Variation

The second basic type of variation, *inverse variation*, is defined as follows. The statement y *varies inversely as* x means

$$y = \frac{k}{x}$$

where k is a nonzero constant, which is again referred to as the constant of variation. The phrase *y is inversely proportional to x* is also used to express inverse variation. As with direct variation, statements indicating variation may involve powers of x. For example, *y varies inversely as the square of x* can be written $y = k/x^2$. In general, *y varies inversely as the nth power of x ($n > 0$)* means

$$y = \frac{k}{x^n}$$

The following examples illustrate the three basic kinds of problems that involve inverse variation.

EXAMPLE 4

Translate the statement *The length of a rectangle of fixed area varies inversely as the width* into an equation, using k as the constant of variation.

Solution

Let l represent the length and w the width; the equation is

$$l = \frac{k}{w}$$ ∎

EXAMPLE 5

If y is inversely proportional to x, and if $y = 14$ when $x = 4$, find the constant of variation.

Solution

Because y is inversely proportional to x, we have

$$y = \frac{k}{x}$$

Substitute 4 for x and 14 for y to obtain

$$14 = \frac{k}{4}$$

Solving this equation yields

$$k = 56$$

The constant of variation is 56. ∎

EXAMPLE 6

The time required for a car to travel a certain distance varies inversely as the rate at which it travels. If it takes 4 hours at 50 miles per hour to travel the distance, how long will it take at 40 miles per hour?

Solution

Let t represent time and r rate. The phrase *time required . . . varies inversely as the rate* translates into

$$t = \frac{k}{r}$$

Substitute 4 for t and 50 for r to find the constant of variation.

$$4 = \frac{k}{50}$$
$$k = 200$$

Thus the specific equation is

$$t = \frac{200}{r}$$

Now substitute 40 for r to produce

$$t = \frac{200}{40}$$
$$= 5$$

It will take 5 hours at 40 miles per hour.   ∎

The terms *direct* and *inverse*, as applied to variation, refer to the relative behavior of the variables involved in the equation. That is, in *direct variation* ($y = kx$), an assignment of **increasing absolute values** for x produces **increasing absolute values** for y. However, in *inverse variation* ($y = k/x$), an assignment of **increasing absolute values** for x produces **decreasing absolute values** for y.

Joint Variation

Variation may involve more than two variables. The table on page 287 illustrates some different types of variation statements and their equivalent algebraic equations that use k as the constant of variation. Statements 1, 2, and 3 illustrate the concept of *joint variation*. Statements 4 and 5 show that both direct and inverse variation may occur in the same problem. Statement 6 combines joint variation with inverse variation.

The final two examples of this section illustrate different kinds of problems involving some of these variation situations.

VARIATION STATEMENT	ALGEBRAIC EQUATION
1. *y varies jointly* as *x* and *z*	$y = kxz$
2. *y varies jointly* as *x*, *z*, and *w*	$y = kxzw$
3. *V varies jointly* as *h* and the square of *r*	$V = khr^2$
4. *h varies directly* as *V* and *inversely* as *w*	$h = \dfrac{kV}{w}$
5. *y* is *directly proportional* to *x* and *inversely proportional* to the square of *z*	$y = \dfrac{kx}{z^2}$
6. *y varies jointly* as *w* and *z* and *inversely* as *x*	$y = \dfrac{kwz}{x}$

EXAMPLE 7

The volume of a pyramid varies jointly as its altitude and the area of its base. If a pyramid with an altitude of 9 feet and a base with an area of 17 square feet has a volume of 51 cubic feet, find the volume of a pyramid with an altitude of 14 feet and a base with an area of 45 square feet.

Solution

Let's use the following variables.

V = volume h = altitude

B = area of base k = constant of variation

The fact that the volume varies jointly as the altitude and the area of the base can be represented by the equation

$V = kBh$

Substitute 51 for V, 17 for B, and 9 for h to obtain

$$51 = k(17)(9)$$
$$51 = 153k$$
$$\frac{51}{153} = k$$
$$\frac{1}{3} = k$$

Therefore, the specific equation is $V = \frac{1}{3}Bh$. Now substitute 45 for B and 14 for h to obtain

$$V = \frac{1}{3}(45)(14) = (15)(14) = 210$$

The volume is 210 cubic feet.   ■

EXAMPLE 8

Suppose that y varies jointly as x and z and inversely as w. If $y = 154$ when $x = 6$, $z = 11$, and $w = 3$, find y when $x = 8$, $z = 9$, and $w = 6$.

Solution

The statement *y varies jointly as x and z and inversely as w* translates into the equation

$$y = \frac{kxz}{w}$$

Substitute 154 for y, 6 for x, 11 for z, and 3 for w to produce

$$154 = \frac{(k)(6)(11)}{3}$$
$$154 = 22k$$
$$7 = k$$

Thus the specific equation is

$$y = \frac{7xz}{w}$$

Now substitute 8 for x, 9 for z, and 6 for w to obtain

$$y = \frac{7(8)(9)}{6} = 84$$ ■

PROBLEM SET 3.6

For Problems 1–8, translate each statement of variation into an equation; use k as the constant of variation.

1. y varies directly as the cube of x.

2. a varies inversely as the square of b.

3. A varies jointly as l and w.

4. s varies jointly as g and the square of t.

5. At a constant temperature, the volume (V) of a gas varies inversely as the pressure (P).

6. y varies directly as the square of x and inversely as the cube of w.

7. The volume (V) of a cone varies jointly as its height (h) and the square of a radius (r).

8. I is directly proportional to r and t.

For Problems 9–18, find the constant of variation for each stated condition.

9. y varies directly as x, and $y = 72$ when $x = 3$.

10. y varies inversely as the square of x, and $y = 4$ when $x = 2$.

11. A varies directly as the square of r, and $A = 154$ when $r = 7$.

12. V varies jointly as B and h, and $V = 104$ when $B = 24$ and $h = 13$.

13. A varies jointly as b and h, and $A = 81$ when $b = 9$ and $h = 18$.

14. s varies jointly as g and the square of t, and $s = -108$ when $g = 24$ and $t = 3$.

15. y varies jointly as x and z and inversely as w, and $y = 154$ when $x = 6$, $z = 11$, and $w = 3$.

16. V varies jointly as h and the square of r, and $V = 1100$ when $h = 14$ and $r = 5$.

17. y is directly proportional to the square of x and inversely proportional to the cube of w, and $y = 18$ when $x = 9$ and $w = 3$.

18. y is directly proportional to x and inversely proportional to the square root of w, and $y = \frac{1}{5}$ when $x = 9$ and $w = 10$.

For Problems 19–32, solve each problem.

19. If y is directly proportional to x, and $y = 5$ when $x = -15$, find the value of y when $x = -24$.

20. If y is inversely proportional to the square of x, and $y = \frac{1}{8}$ when $x = 4$, find y when $x = 8$.

21. If V varies jointly as B and h, and $V = 96$ when $B = 36$ and $h = 8$, find V when $b = 48$ and $h = 6$.

22. If A varies directly as the square of e, and $A = 150$ when $e = 5$, find A when $e = 10$.

23. The time required for a car to travel a certain distance varies inversely as the rate at which it travels. If it takes 3 hours to travel the distance at 50 miles per hour, how long will it take at 30 miles per hour?

24. The distance that a freely falling body falls varies directly as the square of the time it falls. If a body falls 144 feet in 3 seconds, how far will it fall in 5 seconds?

25. The period (the time required for one complete oscillation) of a simple pendulum varies directly as the square root of its length. If a pendulum 12 feet long has a period of 4 seconds, find the period of a pendulum of length 3 feet.

26. Suppose the number of days it takes to complete a construction job varies inversely as the number of people assigned to the job. If it takes 7 people 8 days to do the job, how long will it take 10 people to complete the job?

27. The number of days needed to assemble some machines varies directly as the number of machines and inversely as the number of people working. If it takes 4 people 32 days to assemble 16 machines, how many days will it take 8 people to assemble 24 machines?

28. The volume of a gas at a constant temperature varies inversely as the pressure. What is the volume of a gas under a pressure of 25 pounds if the gas occupies 15 cubic centimeters under a pressure of 20 pounds?

29. The volume (V) of a gas varies directly as the temperature (T) and inversely as the pressure (P). If $V = 48$ when $T = 320$ and $P = 20$, find V when $T = 280$ and $P = 30$.

30. The volume of a cylinder varies jointly as its altitude and the square of the radius of its base. If the volume of a cylinder is 1386 cubic centimeters when the radius of the base is 7 centimeters and its altitude is 9 centimeters, find the volume of a cylinder that has a base of radius 14 centimeters if the altitude of the cylinder is 5 centimeters.

31. The cost of labor varies jointly as the number of workers and the number of days that they work. If it costs $900 to have 15 people work for 5 days, how much will it cost to have 20 people work for 10 days?

32. The cost of publishing pamphlets varies directly as the number of pamphlets produced. If it costs $96 to publish 600 pamphlets, how much does it cost to publish 800 pamphlets?

THOUGHTS INTO WORDS

33. How would you explain the difference between direct variation and inverse variation?

34. Suppose that y varies directly as the square of x. Does doubling the value of x also double the value of y? Explain your answer.

35. Suppose that y varies inversely as x. Does doubling the value of x also double the value of y? Explain your answer.

Further Investigations

In the previous problems, we chose numbers to make computations reasonable without the use of a calculator. However, variation-type problems often involve messy computations and the calculator becomes a very useful tool. Use your calculator to help solve the following problems.

36. The simple interest earned by a certain amount of money varies jointly as the rate of interest and the time (in years) that the money is invested.

    a. If some money invested at 11% for 2 years earns $385, how much would the same amount earn at 12% for 1 year?

    b. If some money invested at 12% for 3 years earns $819, how much would the same amount earn at 14% for 2 years?

    c. If some money invested at 14% for 4 years earns $1960, how much would the same amount earn at 15% for 2 years?

37. The period (the time required for one complete oscillation) of a simple pendulum varies directly as the square root of its length. If a pendulum 9 inches long has a period of 2.4 seconds, find the period of a pendulum of length 12 inches. Express the answer to the nearest tenth of a second.

38. The volume of a cylinder varies jointly as its altitude and the square of the radius of its base. If the volume of a cylinder is 549.5 cubic meters when the radius of the base is 5 meters and its altitude is 7 meters, find the volume of a cylinder that has a base of radius 9 meters and an altitude of 14 meters.

39. If y is directly proportional to x and inversely proportional to the square of z, and if $y = 0.336$ when $x = 6$ and $z = 5$, find the constant of variation.

40. If y is inversely proportional to the square root of x, and $y = 0.08$ when $x = 225$, find y when $x = 625$.

CHAPTER 3 SUMMARY

The function concept serves as a thread to tie this chapter together.

Function Concept

DEFINITION 3.1

A function f is a correspondence between two sets X and Y that assigns to each element x of set X one and only one element y of set Y. The element y being assigned is called the image of x. The set X is called the domain of the function, and the set of all images is called the range of the function.

A function can also be thought of as a set of ordered pairs no two of which have the same first component. If each vertical line intersects a graph in no more than one point, then the graph represents a function.

Graphing Functions

Any function that can be written in the form

$$f(x) = ax + b$$

where a and b are real numbers, is a **linear function**. The graph of a linear function is a straight line.

Any function that can be written in the form

$$f(x) = ax^2 + bx + c$$

where a, b, and c are real numbers and $a \neq 0$, is a **quadratic function**. The graph of any quadratic function is a **parabola**, which can be drawn using either one of the following methods.

1. Express the function in the form $f(x) = a(x - h)^2 + k$ and use the values of a, h, and k to determine the parabola.

2. Express the function in the form $f(x) = ax^2 + bx + c$ and use the fact that the vertex is at

$$\left(-\frac{b}{2a}, f\left(-\frac{b}{2a} \right) \right)$$

and the axis of symmetry is

$$x = -\frac{b}{2a}$$

Another important skill in graphing is to be able to recognize equations of the transformations of basic curves. We worked with the following transformations in this chapter.

Vertical Translation   The graph of $y = f(x) + k$ is the graph of $y = f(x)$ shifted k units upward if $k > 0$ or shifted $|k|$ units downward if $k < 0$.

Horizontal Translation   The graph of $y = f(x - h)$ is the graph of $y = f(x)$ shifted h units to the right if $h > 0$ or shifted $|h|$ units to the left if $h < 0$.

x-axis Reflection   The graph of $y = -f(x)$ is the graph of $y = f(x)$ reflected through the x axis.

y-axis Reflection   The graph of $y = f(-x)$ is the graph of $y = f(x)$ reflected through the y axis.

Vertical Stretching and Shrinking   The graph of $y = cf(x)$ is obtained from the graph of $y = f(x)$ by multiplying the y coordinates of $y = f(x)$ by c. If $c > 1$, the graph is said to be **stretched** by a factor of c, and if $0 < c < 1$, the graph is said to be **shrunk** by a factor of c.

The following suggestions are helpful for graphing functions that are unfamiliar.

1. Determine the domain of the function.

2. Find the intercepts.

3. Determine what type of symmetry the equation exhibits.

4. Set up a table of values that satisfy the equation. The type of symmetry and the domain will affect your choice of values for x in the table.

5. Plot the points associated with the ordered pairs and connect them with a smooth curve. Then, if appropriate, reflect this part of the curve according to the symmetry possessed by the graph.

Operations on Functions

Sum of two functions $(f + g)(x) = f(x) + g(x)$

Difference of two functions $(f - g)(x) = f(x) - g(x)$

Product of two functions $(f \cdot g)(x) = f(x) \cdot g(x)$

Quotient of two functions $\left(\dfrac{f}{g}\right)(x) = \dfrac{f(x)}{g(x)}, \qquad g(x) \neq 0$

DEFINITION 3.2

The **composition** of functions f and g is defined by

$$(f \circ g)(x) = f(g(x))$$

for all x in the domain of g such that $g(x)$ is in the domain of f.

Remember that the composition of functions is *not a commutative operation*.

Applications of Functions

Quadratic functions produce parabolas that have either a **minimum** or a **maximum value**. Therefore, a real-world minimum- or maximum-value problem that can be described by a quadratic function can be solved using the techniques of this chapter.

Relationships that involve **direct** and **inverse variation** can be expressed by equations that determine functions. The statement *y varies directly as x* means

$$y = kx$$

where k is the **constant of variation**. The statement *y varies directly as the nth power of $x(n > 0)$* means.

$$y = kx^n$$

The statement *y varies inversely as x* means $y = \dfrac{k}{x}$.

The statement *y varies inversely as the nth power of x ($n > 0$)* means $y = \dfrac{k}{x^n}$.

The statement *y varies jointly as x and w* means $y = kxw$.

CHAPTER 3 REVIEW PROBLEM SET

1. If $f(x)$ is $3x^2 - 2x - 1$, find $f(2), f(-1)$, and $f(-3)$.

2. For each of the following functions, find
$$\frac{f(a + h) - f(a)}{h}$$

   a. $f(x) = -5x + 4$ **b.** $f(x) = 2x^2 - x + 4$

   c. $f(x) = -3x^2 + 2x - 5$

3. Determine the domain and range of the function $f(x) = x^2 + 5$.

4. Determine the domain of the function $f(x) = \dfrac{2}{2x^2 + 7x - 4}$.

5. Express the domain of $f(x) = \sqrt{x^2 - 7x + 10}$ using interval notation.

For Problems 6–15, graph each function.

6. $f(x) = -2x + 2$ 7. $f(x) = 2x^2 - 1$

8. $f(x) = -\sqrt{x - 2} + 1$ 9. $f(x) = x^2 - 8x + 17$

10. $f(x) = -x^3 + 2$ 11. $f(x) = 2|x - 1| + 3$

12. $f(x) = -2x^2 - 12x - 19$

13. $f(x) = -\dfrac{1}{3}x + 1$ 14. $f(x) = -\dfrac{2}{x^2}$

15. $f(x) = 2|x| - x$

16. If $f(x) = 2x + 3$ and $g(x) = x^2 - 4x - 3$, find $f + g$, $f - g, f \cdot g$, and f/g.

For Problems 17–20, find $(f \circ g)(x)$ and $(g \circ f)(x)$. Also specify the domain for each.

17. $f(x) = 3x - 9$ and $g(x) = -2x + 7$

18. $f(x) = x^2 - 5$ and $g(x) = 5x - 4$

19. $f(x) = \sqrt{x - 5}$ and $g(x) = x + 2$

20. $f(x) = \dfrac{1}{x - 3}$ and $g(x) = \dfrac{1}{x + 2}$

21. For each of the following, classify the function as even, odd, or neither even nor odd.

   a. $f(x) = 3x^2 - 4x + 6$ **b.** $f(x) = -x^3$

   c. $f(x) = -4x^2 + 6$ **d.** $f(x) = 2x^3 + x - 2$

22. If $f(x) = \begin{cases} x^2 - 2 & \text{for } x \geq 0 \\ -3x + 4 & \text{for } x < 0 \end{cases}$,

    find $f(5), f(0),$ and $f(-3)$.

23. If $f(x) = -x^2 - x + 4$ and $g(x) = \sqrt{x - 2}$, find $f(g(6))$ and $g(f(-2))$.

24. If $f(x) = |x|$ and $g(x) = x^2 - x - 1$, find $(f \circ g)(1)$ and $(g \circ f)(-3)$.

For Problems 25–30, solve each problem.

25. Find two numbers whose sum is 10, such that the sum of the square of one number plus four times the other number is a minimum.

26. A group of students is arranging a chartered flight to Europe. The charge per person is $496 if 100 students go on the flight. If more than 100 students go, the charge per student is reduced by an amount equal to $4 times the number of students above 100. How many students should the airline try to get in order to maximize its revenue?

27. If y varies directly as x and inversely as w, and if $y = 27$ when $x = 18$ and $w = 6$, find the constant of variation.

28. If y varies jointly as x and the square root of w, and if $y = 140$ when $x = 5$ and $w = 16$, find y when $x = 9$ and $w = 49$.

29. The weight of a body above the surface of the earth varies inversely as the square of its distance from the center of the earth. Assuming the radius of the earth to be 4000 miles, determine how much a man would weigh 1000 miles above the earth's surface if he weighs 200 pounds on the surface.

30. The number of hours needed to assemble some furniture varies directly as the number of pieces of furniture and inversely as the number of people working. If it takes 3 people 10 hours to assemble 20 pieces of furniture, how many hours will it take 4 people to assemble 40 pieces of furniture?

CHAPTER 3 TEST

1. If $f(x) = -\dfrac{1}{2}x + \dfrac{1}{3}$, find $f(-3)$.

2. If $f(x) = -x^2 - 6x + 3$, find $f(-2)$.

3. If $f(x) = 3x^2 + 2x - 5$, find $\dfrac{f(a+h) - f(a)}{h}$.

4. Determine the domain of the function $f(x) = \dfrac{-3}{2x^2 + 7x - 4}$.

5. Determine the domain of the function $f(x) = \sqrt{5 - 3x}$.

6. If $f(x) = 3x - 1$ and $g(x) = 2x^2 - x - 5$, find $f + g, f - g$, and $f \cdot g$.

7. If $f(x) = -3x + 4$ and $g(x) = 7x + 2$, find $(f \circ g)(x)$.

8. If $f(x) = 2x + 5$ and $g(x) = 2x^2 - x + 3$, find $(g \circ f)(x)$.

9. If $f(x) = \dfrac{3}{x - 2}$ and $g(x) = \dfrac{2}{x}$, find $(f \circ g)(x)$.

10. If $f(x) = x^2 - 2x - 3$ and $g(x) = |x - 3|$, find $f(g(-2))$ and $g(f(1))$.

11. Classify each of the following functions as even, odd, or neither even nor odd.

    a. $f(x) = 3x^2 - 10$ **b.** $f(x) = -x^5 + x^3$

    c. $f(x) = -x^2 + 6x - 4$ **d.** $f(x) = 2x^4 + x^2$

12. If $f(x) = \dfrac{3}{x}$ and $g(x) = \dfrac{2}{x - 1}$, determine the domain of $\left(\dfrac{f}{g}\right)(x)$.

13. If $f(x) = 2x^2 - x + 1$ and $g(x) = x^2 + 3$, find $(f + g)(-2)$, $(f - g)(4)$, and $(g - f)(-1)$.

14. If $f(x) = x^2 + 5x - 6$ and $g(x) = x - 1$, find $(f \cdot g)(x)$ and $\left(\dfrac{f}{g}\right)(x)$.

For Problems 15–18, solve the problem.

15. Find two numbers whose sum is 60, such that the sum of the square of one number plus twelve times the other number is a minimum.

16. If y varies jointly as x and z, and if $y = 18$ when $x = 8$ and $z = 9$, find y when $x = 5$ and $z = 12$.

17. If y varies inversely as x, and if $y = \dfrac{1}{2}$ when $x = -8$, find the constant of variation.

18. The simple interest earned by a certain amount of money varies jointly as the rate of interest and the time (in years) that the money is invested. If $140 is earned for the money invested at 7% for 5 years, how much is earned if the same amount is invested at 8% for 3 years?

For Problems 19–21, use the concepts of translation and/or reflection to describe how the second curve can be obtained from the first curve.

19. $f(x) = x^3, g(x) = (x - 6)^3 - 4$

20. $f(x) = |x|, g(x) = -|x| + 8$

21. $f(x) = \sqrt{x}, g(x) = -\sqrt{x + 5} + 7$

For Problems 22–25, graph each function.

22. $f(x) = -2x^2 - 12x - 14$

23. $f(x) = 3|x - 2| - 1$

24. $f(x) = \sqrt{-x + 2}$

25. $f(x) = -x - 1$

4

EXPONENTIAL AND LOGARITHMIC FUNCTIONS

Compound interest is a good illustration of exponential growth.

297

*In this chapter we will continue our study of exponents in several ways:
(1) we will extend the meaning of an exponent; (2) we will work with some
exponential functions; (3) we will introduce the concept of a logarithm; (4)
we will work some logarithmic functions; and (5) we will use the concepts of
exponent and logarithm to develop more problem-solving skills. Your calculator will be a valuable tool throughout this chapter.*

4.1 EXPONENTS AND EXPONENTIAL FUNCTIONS

In Chapter 0, the expression b^n was defined to mean n factors of b, where n is any
positive integer and b is any real number. For example,

$$2^3 = 2 \cdot 2 \cdot 2 = 8 \qquad \left(\frac{1}{3}\right)^4 = \left(\frac{1}{3}\right)\left(\frac{1}{3}\right)\left(\frac{1}{3}\right)\left(\frac{1}{3}\right) = \frac{1}{81}$$

$$(-4)^2 = (-4)(-4) = 16 \qquad -(0.5)^3 = -[(0.5)(0.5)(0.5)] = -0.125$$

Also in Chapter 0, by defining $b^0 = 1$ and $b^{-n} = 1/b^n$, where n is any positive integer and b is any nonzero real number, we extended the concept of an exponent to
include all integers. For example,

$$(0.76)^0 = 1 \qquad\qquad 2^{-3} = \frac{1}{2^3} = \frac{1}{8}$$

$$\left(\frac{2}{3}\right)^{-2} = \frac{1}{\left(\frac{2}{3}\right)^2} = \frac{1}{\frac{4}{9}} = \frac{9}{4} \qquad (0.4)^{-1} = \frac{1}{(0.4)^1} = \frac{1}{0.4} = 2.5$$

Finally, in Chapter 0 we provided for the use of any rational number as an exponent
by defining

$$b^{m/n} = \sqrt[n]{b^m} = \left(\sqrt[n]{b}\right)^m$$

where n is a positive integer greater than 1 and b is a real number such that $\sqrt[n]{b}$
exists. For example,

$$27^{2/3} = \left(\sqrt[3]{27}\right)^2 = 9 \qquad 16^{1/4} = \sqrt[4]{16^1} = 2$$

$$\left(\frac{1}{9}\right)^{1/2} = \sqrt{\frac{1}{9}} = \frac{1}{3} \qquad 32^{-1/5} = \frac{1}{32^{1/5}} = \frac{1}{\sqrt[5]{32}} = \frac{1}{2}$$

If we were to make a formal extension of the concept of an exponent to include
the use of irrational numbers, we would require some ideas from calculus, which is
beyond the scope of this text. However, we can give you a brief glimpse at the general idea involved. Consider the number $2^{\sqrt{3}}$. By using the nonterminating and nonrepeating decimal representation $1.73205\ldots$ for $\sqrt{3}$, we can form the sequence of
numbers $2^1, 2^{1.7}, 2^{1.73}, 2^{1.732}, 2^{1.7320}, 2^{1.73205}, \ldots$. It is a reasonable idea that each suc-

cessive power gets closer to $2^{\sqrt{3}}$. This is precisely what happens if n is irrational and b^n is properly defined by using the concept of a *limit*. Furthermore, this ensures that an expression such as 2^x will yield exactly one value for each value of x.

From now on, we can use any real number as an exponent, and the basic properties stated in Chapter 0 can be extended to include all real numbers as exponents. Let's restate those properties with the restriction that the bases a and b are to be positive numbers to avoid expressions such as $(-4)^{1/2}$, which do not represent real numbers.

PROPERTY 4.1

If a and b are positive real numbers and m and n are any real numbers, then the following properties hold.

1. $b^n \cdot b^m = b^{n+m}$ Product of two powers
2. $(b^n)^m = b^{mn}$ Power of a power
3. $(ab)^n = a^n b^n$ Power of a product
4. $\left(\dfrac{a}{b}\right)^n = \dfrac{a^n}{b^n}$ Power of a quotient
5. $\dfrac{b^n}{b^m} = b^{n-m}$ Quotient of two powers

Another property that can be used to solve certain types of equations involving exponents can be stated as follows.

PROPERTY 4.2

If $b > 0$ but $b \neq 1$, and if m and n are real numbers, then

$$b^n = b^m \qquad \text{if and only if } n = m$$

The following examples illustrate the use of Property 4.2.

EXAMPLE 1

Solve $2^x = 32$.

Solution

$$2^x = 32$$
$$2^x = 2^5 \qquad 32 = 2^5$$
$$x = 5 \qquad \text{Apply Property 4.2.}$$

The solution set is $\{5\}$.

EXAMPLE 2

Solve $2^{3x} = \dfrac{1}{64}$.

Solution

$$2^{3x} = \frac{1}{64}$$

$$2^{3x} = \frac{1}{2^6}$$

$$2^{3x} = 2^{-6}$$

$$3x = -6 \qquad \text{Apply Property 4.2.}$$

$$x = -2$$

The solution set is $\{-2\}$. ∎

EXAMPLE 3

Solve $\left(\dfrac{1}{5}\right)^{x-4} = \dfrac{1}{125}$.

Solution

$$\left(\frac{1}{5}\right)^{x-4} = \frac{1}{125}$$

$$\left(\frac{1}{5}\right)^{x-4} = \left(\frac{1}{5}\right)^3$$

$$x - 4 = 3 \qquad \text{Apply Property 4.2.}$$

$$x = 7$$

The solution set is $\{7\}$. ∎

EXAMPLE 4

Solve $9^x = 243$.

Solution

$$9^x = 243$$

$$(3^2)^x = 3^5$$

$$3^{2x} = 3^5$$

$$2x = 5 \qquad \text{Apply Property 4.2.}$$

$$x = \frac{5}{2}$$

The solution set is $\left\{\dfrac{5}{2}\right\}$. ∎

EXAMPLE 5

Solve $(8^{2x})(4^{2x-1}) = 16$.

Solution

$$(8^{2x})(4^{2x-1}) = 16$$

$$(2^3)^{2x}(2^2)^{2x-1} = 2^4$$

$$(2^{6x})(2^{4x-2}) = 2^4$$
$$2^{6x+4x-2} = 2^4$$
$$2^{10x-2} = 2^4$$
$$10x - 2 = 4 \qquad \text{Apply Property 4.2}$$
$$10x = 6$$
$$x = \frac{6}{10}$$
$$x = \frac{3}{5}$$

The solution set is $\left\{\dfrac{3}{5}\right\}$.

Exponential Functions

If b is any positive number, then the expression b^x designates exactly one real number for every real value of x. Therefore, the equation $f(x) = b^x$ defines a function whose domain is the set of real numbers. Furthermore, if we add the restriction $b \neq 1$, then any equation of the form $f(x) = b^x$ describes what we will call later a one-to-one function and is called an **exponential function**. This leads to the following definition.

DEFINITION 4.1

If $b > 0$ and $b \neq 1$, then the function f defined by

$$f(x) = b^x$$

where x is any real number, is called the **exponential function with base b**.

REMARK   The function $f(x) = 1^x$ is a constant function and therefore it is not a one-to-one function. Remember from Chapter 3 that one-to-one functions have inverses; this becomes a key issue in a later section.

Now let's graph some exponential functions.

EXAMPLE 6

Graph the function $f(x) = 2^x$.

Solution

Let's set up a table of values. Keep in mind that the domain is the set of real numbers and the equation $f(x) = 2^x$ exhibits no symmetry. Plot these points and connect them with a smooth curve to produce Figure 4.1.

x	2^x
-2	$\frac{1}{4}$
-1	$\frac{1}{2}$
0	1
1	2
2	4
3	8

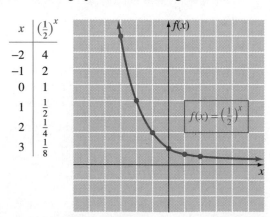

$f(x) = 2^x$

FIGURE 4.1

In the table for Example 6, we chose integral values for x to keep the computation simple. However, with a calculator, we could easily acquire functional values by using nonintegral exponents. Consider the following additional values for $f(x) = 2^x$.

$f(0.5) \approx 1.41$ $f(1.7) \approx 3.25$ $\approx$ means *is approximately*

$f(-0.5) \approx 0.71$ $f(-2.6) \approx 0.16$ *equal to*

Use your calculator to check these results. Also note that the points generated by these values fit the graph in Figure 4.1.

EXAMPLE 7

Graph $f(x) = \left(\dfrac{1}{2}\right)^x$.

Solution

Again, let's set up a table of values, plot the points, and connect them with a smooth curve. The graph is shown in Figure 4.2.

x	$\left(\frac{1}{2}\right)^x$
-2	4
-1	2
0	1
1	$\frac{1}{2}$
2	$\frac{1}{4}$
3	$\frac{1}{8}$

$f(x) = \left(\dfrac{1}{2}\right)^x$

FIGURE 4.2

REMARK   Because $\left(\dfrac{1}{2}\right)^x = 1/2^x = 2^{-x}$, the graphs of $f(x) = 2^x$ and $f(x) = \left(\dfrac{1}{2}\right)^x$ are reflections of each other across the y axis. Therefore, Figure 4.2 could have been drawn by reflecting Figure 4.1 across the y axis.

The graphs in Figures 4.1 and 4.2 illustrate a general behavior pattern of exponential functions. That is, if $b > 1$, then the graph of $f(x) = b^x$ goes up to the right, and the function is called an **increasing function**. If $0 < b < 1$, then the graph of $f(x) = b^x$ goes down to the right, and the function is called a **decreasing function**. These facts are illustrated in Figure 4.3. Notice that $b^0 = 1$ for any $b > 0$; thus, **all graphs of $f(x) = b^x$ contain the point (0, 1)**.

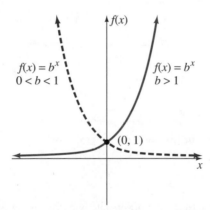

FIGURE 4.3

As you graph exponential functions, don't forget to use your previous graphing experience. For example, consider the following functions.

1. The graph of $f(x) = 2^x + 3$ is the graph of $f(x) = 2^x$ *moved up three units*.
2. The graph of $f(x) = 2^{x-4}$ is the graph of $f(x) = 2^x$ *moved to the right four units*.
3. The graph of $f(x) = -2^x$ is the graph of $f(x) = 2^x$ *reflected across the x axis*.
4. The graph of $f(x) = 2^x + 2^{-x}$ is symmetric with respect to the y axis because $f(-x) = 2^{-x} + 2^x = f(x)$.

Furthermore, if you are faced with an exponential function that is not of the form $f(x) = b^x$ or a variation thereof, don't forget the graphing suggestions offered in Chapter 2. Let's consider one such example.

EXAMPLE 8

Graph $f(x) = 2^{-x^2}$.

Solution

Because $f(-x) = 2^{-(-x)^2} = 2^{-x^2} = f(x)$, we know that this curve is symmetric with respect to the y axis. Therefore, let's set up a table of values using nonnegative values for x. Plot these points, connect them with a smooth curve, and reflect this portion of the curve across the y axis to produce the graph in Figure 4.4.

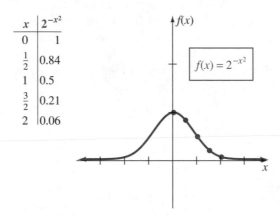

x	2^{-x^2}
0	1
$\frac{1}{2}$	0.84
1	0.5
$\frac{3}{2}$	0.21
2	0.06

$f(x) = 2^{-x^2}$

FIGURE 4.4

EXAMPLE 9

Use a graphing utility to obtain a graph of $f(x) = 50(2^x)$ and find an approximate value for x when $f(x) = 15,000$.

Solution

First, we must find an appropriate viewing rectangle. Because $50(2^{10}) = 51,200$, let's set the boundaries so that $0 \le x \le 10$ and $0 \le y \le 50,000$ with a scale of 10,000 on the y axis. (Certainly other boundaries could be used, but these will give us a graph that we can work with for this problem.) The graph of $f(x) = 50(2^x)$ is shown in Figure 4.5. Now we can use the trace and zoom-in features of the graphing utility to find that $x \approx 8.2$ at $y = 15,000$.

FIGURE 4.5

REMARK   In Example 9 we used a graphical approach to solve the equation $50(2^x) = 15,000$. In Section 4.5 we will use an algebraic approach to solve that same kind of equation.

PROBLEM SET 4.1

For Problems 1–20, solve each equation.

1. $3^x = 27$

2. $2^x = 64$

3. $\left(\dfrac{1}{2}\right)^x = \dfrac{1}{8}$

4. $\left(\dfrac{1}{2}\right)^n = 4$

5. $3^{-x} = \dfrac{1}{81}$

6. $3^{x+1} = 9$

7. $5^{2n-1} = 125$

8. $2^{3-n} = 8$

9. $\left(\dfrac{2}{3}\right)^t = \dfrac{9}{4}$

10. $\left(\dfrac{3}{4}\right)^n = \dfrac{64}{27}$

11. $4^{3x-1} = 256$

12. $16^x = 64$

13. $4^n = 8$

14. $27^{4x} = 9^{x+1}$

15. $32^x = 16^{1-x}$

16. $\left(\dfrac{1}{8}\right)^{-2t} = 2^{t+3}$

17. $(2^{2x-1})(2^{x+2}) = 32$

18. $(27)(3^x) = 9^x$

19. $(3^x)(3^{5x}) = 81$

20. $(4^x)(16^{3x-1}) = 8$

For Problems 21–40, graph each function.

21. $f(x) = 3^x$

22. $f(x) = \left(\dfrac{1}{3}\right)^x$

23. $f(x) = 4^x$

24. $f(x) = \left(\dfrac{1}{4}\right)^x$

25. $f(x) = \left(\dfrac{2}{3}\right)^x$

26. $f(x) = \left(\dfrac{3}{2}\right)^x$

27. $f(x) = 2^x + 1$

28. $f(x) = 2^x - 3$

29. $f(x) = 2^{x-1}$

30. $f(x) = 2^{x+2}$

31. $f(x) = -3^x$

32. $f(x) = -2^x$

33. $f(x) = 2^{-x+1}$

34. $f(x) = 2^{-x-2}$

35. $f(x) = 2^x + 2^{-x}$

36. $f(x) = 2^{x^2}$

37. $f(x) = 3^{1-x^2}$

38. $f(x) = 2^{|x|}$

39. $f(x) = 2^{-|x|}$

40. $f(x) = 2^x - 2^{-x}$

41. Graph $f(x) = -\left(\dfrac{1}{2}\right)^x$. Then, on the same set of axes, graph $f(x) = -\left(\dfrac{1}{2}\right)^x + 2$, $f(x) = -\left(\dfrac{1}{2}\right)^{x+3}$, and $f(x) = -\left(\dfrac{1}{2}\right)^{-x}$.

THOUGHTS INTO WORDS

42. Why is the base of an exponential function restricted to positive numbers not including 1?

43. How would you go about graphing the function $f(x) = -\left(\dfrac{1}{3}\right)^x$?

44. Explain how you would solve the equation
$$(4^{x-1})(8^{2x+3}) = 128$$

GRAPHING CALCULATOR ACTIVITIES

45. Use your graphing calculator to check your graphs for Problems 36–40.

46. Graph $f(x) = 4^x$. Where should the graphs of $f(x) = 4^{x-2}$, $f(x) = 4^{x-4}$, and $f(x) = 4^{x+3}$ be located? Graph all three functions on the same set of axes with $f(x) = 4^x$.

47. Graph $f(x) = \left(\dfrac{1}{4}\right)^x$. Where should the graphs of $f(x) = \left(\dfrac{1}{4}\right)^x - 2$, $f(x) = \left(\dfrac{1}{4}\right)^x + 3$, and $f(x) = \left(\dfrac{1}{4}\right)^x - 4$ be located? Graph all three functions on the same set of axes with $f(x) = \left(\dfrac{1}{4}\right)^x$.

48. Graph $f(x) = \left(\dfrac{3}{4}\right)^x$. Now predict the graphs for $f(x) = -\left(\dfrac{3}{4}\right)^x$, $f(x) = \left(\dfrac{3}{4}\right)^{-x}$, and $f(x) = -\left(\dfrac{3}{4}\right)^{-x}$. Graph all three functions on the same set of axes with $f(x) = \left(\dfrac{3}{4}\right)^x$.

49. Graph $f(x) = (-2)^x$. Explain your result.

50. What is the solution for $3^x = 5$? Do you agree that it is between 1 and 2 because $3^1 = 3$ and $3^2 = 9$? Now graph $f(x) = 3^x - 5$ and use the zoom-in and trace features of your graphing calculator to find an approximation, to the nearest hundredth, for the x intercept. You should get the answer 1.46. This should be an approximation for the solution for $3^x = 5$. Try it; raise 3 to the 1.46 power.

Find an approximate solution, to the nearest hundredth, for each of the following equations by graphing the appropriate function and finding the x intercept.

a. $2^x = 19$ **b.** $3^x = 50$ **c.** $4^x = 47$

d. $5^x = 120$ **e.** $2^x = 1500$ **f.** $3^{x-1} = 34$

4.2 APPLICATIONS OF EXPONENTIAL FUNCTIONS

Many real-world situations that exhibit growth or decay can be represented by equations that describe exponential functions. For example, suppose that an economist predicts an annual inflation rate of 5% per year for the next 10 years. This means that an item that presently costs $8 will cost $8(105%) = $8(1.05) = $8.40 a year from now. The same item will cost [$8(105%)] × (105%) = $8(1.05)^2 = $8.82 in 2 years. In general, the equation

$$P = P_0(1.05)^t$$

yields the predicted price P of an item in t years if the present cost is P_0 and the annual inflation rate is 5%. By using this equation, we can look at some future prices based on the prediction of a 5% inflation rate.

A $0.79 jar of mustard will cost $0.79(1.05)^3 = $0.91 in 3 years.

A $2.69 bag of potato chips will cost $2.69(1.05)^5 = $3.43 in 5 years.

A $6.69 can of coffee will cost $6.69(1.05)^7 = $9.41 in 7 years.

Compound Interest

Compound interest provides another illustration of exponential growth. Suppose that $500 (called the **principal**) is invested at an interest rate of 8% **compounded**

annually. The interest earned the first year is $500(0.08) = $40, and this amount is added to the original $500 to form a new principal of $540 for the second year. The interest earned during the second year is $540(0.08) = $43.20, and this amount is added to $540 to form a new principal of $583.20 for the third year. Each year a new principal is formed by reinvesting the interest earned during that year.

In general, suppose that a sum of money P (called the principal) is invested at an interest rate of r percent compounded annually. The interest earned the first year is Pr, and the new principal for the second year is $P + Pr$ or $P(1 + r)$. Note that the new principal for the second year can be found by multiplying the original principal P by $(1 + r)$. In like fashion, we can find the new principal for the third year by multiplying the previous principal, $P(1 + r)$, by $1 + r$, thus obtaining $P(1 + r)^2$. If this process is continued, then after t years the total amount of money accumulated, A, is given by

$$A = P(1 + r)^t$$

Consider the following examples of investments made at a certain rate of interest compounded annually.

1. $750 invested for 5 years at 9% compounded annually produces

$$A = \$750(1.09)^5 = \$1153.97$$

2. $1000 invested for 10 years at 11% compounded annually produces

$$A = \$1000(1.11)^{10} = \$2839.42$$

3. $5000 invested for 20 years at 12% compounded annually produces

$$A = \$5000(1.12)^{20} = \$48,231.47$$

The compound interest formula can be used to determine what rate of interest is needed to accumulate a certain amount of money based on a given initial investment. The next example illustrates this idea.

PROBLEM 1

What rate of interest is needed for an investment of $1000 to yield $4000 in 10 years if the interest is compounded annually?

Solution

Let's substitute $1000 for P, $4000 for A, and 10 years for t in the compound interest formula and solve for r.

$$A = P(1 + r)^t$$
$$4000 = 1000(1 + r)^{10}$$
$$4 = (1 + r)^{10}$$
$$4^{0.1} = [(1 + r)^{10}]^{0.1} \qquad \text{Raise both sides to the 0.1 power.}$$
$$1.148698355 \approx 1 + r$$
$$0.148698355 \approx r$$
$$r = 14.9\% \quad \text{to the nearest tenth of a percent}$$

Therefore, a rate of interest of approximately 14.9% is needed. (Perhaps you should check this answer.)   ■

If money invested at a certain rate of interest is to be compounded more than once a year, then the basic formula $A = P(1 + r)^t$ can be adjusted according to the number of compounding periods in a year. For example, for **compounding semiannually**, the formula becomes

$$A = P\left(1 + \frac{r}{2}\right)^{2t}$$

and for **compounding quarterly**, the formula becomes

$$A = P\left(1 + \frac{r}{4}\right)^{4t}$$

In general, if n represents the number of **compounding periods** in a year, then the formula becomes

$$A = P\left(1 + \frac{r}{n}\right)^{nt}$$

The following examples illustrate the use of the formula.

1. $750 invested for 5 years at 9% compounded semiannually produces

$$A = \$750\left(1 + \frac{0.09}{2}\right)^{2(5)} = \$750(1.045)^{10} = \$1164.73$$

2. $1000 invested for 10 years at 11% compounded quarterly produces

$$A = \$1000\left(1 + \frac{0.11}{4}\right)^{4(10)} = \$1000(1.0275)^{40} = \$2959.87$$

3. $5000 invested for 20 years at 12% compounded monthly produces

$$A = \$5000\left(1 + \frac{0.12}{12}\right)^{12(20)} = \$5000(1.01)^{240} = \$54,462.77$$

You may find it interesting to compare these results with those obtained earlier for compounding annually.

Exponential Decay

Suppose that it is estimated that the value of a car depreciates 15% per year for the first 5 years. Therefore, a car that costs $9500 will be worth $9500 × (100% − 15%) = $9500(85%) = $9500(0.85) = $8075 in 1 year. In 2 years the value of the car will have depreciated to $9500(0.85)^2 = 6864 (to the nearest dollar). The equation

$$V = V_0(0.85)^t$$

yields the value V of a car in t years if the initial cost is V_0 and it depreciates 15% per year. Therefore, we can estimate some car values to the nearest dollar.

A \$13,000 car will be worth $\$13{,}000(0.85)^3 = \$7984$ in 3 years.

A \$17,000 car will be worth $\$17{,}000(0.85)^5 = \$7543$ in 5 years.

A \$25,000 car will be worth $\$25{,}000(0.85)^4 = \$13{,}050$ in 4 years.

Another example of exponential decay is associated with radioactive substances. The rate of decay can be described exponentially and is based on the half-life of a substance. The **half-life** of a radioactive substance is the amount of time that it takes for one-half of an initial amount of the substance to disappear as the result of decay. For example, suppose that we have 200 grams of a certain substance that has a half-life of 5 days. After 5 days, $200\left(\dfrac{1}{2}\right) = 100$ grams remain. After 10 days, $200\left(\dfrac{1}{2}\right)^2 = 50$ grams remain. After 15 days, $200\left(\dfrac{1}{2}\right)^3 = 25$ grams remain. In general, after t days, $200\left(\dfrac{1}{2}\right)^{t/5}$ grams remain.

The previous discussion leads into the following half-life formula. Suppose there is an initial amount, Q_0, of a radioactive substance with a half-life of h. The amount of substance remaining, Q, after a time period of t, is given by the formula

$$Q = Q_0\left(\frac{1}{2}\right)^{t/h}.$$

The units of measure for t and h must be the same.

PROBLEM 2

Barium-140 has a half-life of 13 days. If there are 500 milligrams of barium initially, how many milligrams remain after 26 days? After 100 days?

Solution

Using $Q_0 = 500$ and $h = 13$, the half-life formula becomes

$$Q = 500\left(\frac{1}{2}\right)^{t/13}$$

If $t = 26$, then

$$Q = 500\left(\frac{1}{2}\right)^{26/13}$$

$$= 500\left(\frac{1}{2}\right)^2$$

$$= 500\left(\frac{1}{4}\right)$$

$$= 125$$

Thus 125 milligrams remain after 26 days. If $t = 100$, then

$$Q = 500\left(\frac{1}{2}\right)^{100/13}$$

$$= 500(0.5)^{100/13}$$

$$= 2.4 \quad \text{to the nearest tenth of a milligram}$$

Approximately 2.4 milligrams remain after 100 days.   ■

REMARK   The solution to Example 2 clearly demonstrates one facet of the role of the calculator in the application of mathematics. We solved the first part of the problem easily without the calculator, but the calculator certainly was helpful for the second part of the problem.

Number e

An interesting situation occurs if we consider the compound interest formula for $P = \$1$, $r = 100\%$, and $t = 1$ year. The formula becomes $A = 1\left(1 + \dfrac{1}{n}\right)^n$. The following table shows some values, rounded to eight decimal places, of $\left(1 + \dfrac{1}{n}\right)^n$ for different values of n.

n	$\left(1 + \dfrac{1}{n}\right)^n$
1	2.00000000
10	2.59374246
100	2.70481383
1000	2.71692393
10,000	2.71814593
100,000	2.71826824
1,000,000	2.71828047
10,000,000	2.71828169
100,000,000	2.71828181
1,000,000,000	2.71828183

The table suggests that as n increases, the value of $\left(1 + \dfrac{1}{n}\right)^n$ gets closer and closer to some fixed number. This does happen, and the fixed number is called e. To five decimal places, $e = 2.71828$.

The function defined by the equation $f(x) = e^x$ is the **natural exponential function**. It has a great many real-world applications, some of which we will look at in a moment. First, however, let's get a picture of the natural exponential function. Because $2 < e < 3$, the graph of $f(x) = e^x$ must fall between the graphs of $f(x) = 2^x$ and $f(x) = 3^x$. To be more specific, let's use our calculator to determine a table of values. Use the $\boxed{e^x}$ key, and round the results to the nearest tenth to obtain the table. Plot the points determined by this table and connect them with a smooth curve to produce Figure 4.6.

x	$f(x) = e^x$
0	1.0
1	2.7
2	7.4
−1	0.4
−2	0.1

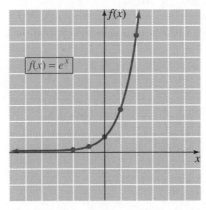

FIGURE 4.6

Back to Compound Interest

Let's return to the concept of compound interest. If the number of compounding periods in a year is increased indefinitely, we arrive at the concept of **compounding continuously**. Mathematically, this can be accomplished by applying the limit concept to the expression

$$P\left(1 + \frac{r}{n}\right)^{nt}$$

We will not show the details here, but the following result is obtained. The formula

$$A = Pe^{rt}$$

yields the accumulated value (A) of a sum of money (P) that has been invested for t years at a rate of r percent compounded continuously. The following examples illustrate the use of this formula.

1. $750 invested for 5 years at 9% compounded continuously produces

$$A = 750e^{(0.09)(5)} = 750e^{0.45} = \$1176.23$$

2. $1000 invested for 10 years at 11% compounded continuously produces

$$A = 1000e^{(0.11)(10)} = 1000e^{1.1} = \$3004.17$$

3. $5000 invested for 20 years at 12% compounded continuously produces

$$A = 5000e^{(0.12)(20)} = 5000e^{2.4} = \$55,115.88$$

Again, you may find it interesting to compare these results with those you obtained earlier using a different number of compounding periods.

Is it better to invest at 6% compounded quarterly or at 5.75% compounded continuously? To answer such a question, we can use the concept of **effective yield** (sometimes called *effective annual rate of interest*). The effective yield of an investment is the simple interest rate that would yield the same amount in 1 year. Thus, for the *6% compounded quarterly* investment, we can calculate the effective yield as follows.

$$P(1 + r) = P\left(1 + \frac{0.06}{4}\right)^4$$

$$1 + r = \left(1 + \frac{0.06}{4}\right)^4 \qquad \text{Multiply both sides by } \frac{1}{P}.$$

$$1 + r = (1.015)^4$$

$$r = (1.015)^4 - 1$$

$$r \approx 0.0613635506$$

$$r = 6.14\% \qquad \text{to the nearest hundredth of a percent}$$

Likewise, for the 5.75% *compounded continuously* investment we can calculate the effective yield as follows.

$$P(1 + r) = Pe^{0.0575}$$

$$1 + r = e^{0.0575}$$

$$r = e^{0.0575} - 1$$

$$r \approx 0.0591852707$$

$$r = 5.92\% \qquad \text{to the nearest hundredth of a percent}$$

Therefore, comparing the two effective yields, we see that it is better to invest at 6% compounded quarterly than to invest at 5.75% compounded continuously.

Law of Exponential Growth

The ideas behind compounded continuously carry over to other growth situations. The law of exponential growth,

$$Q(t) = Q_0 e^{kt}$$

is used as a mathematical model for numerous growth-and-decay applications. In this equation, $Q(t)$ represents the quantity of a given substance at any time t; Q_0 is the initial amount of the substance (when $t = 0$); and k is a constant that depends on the particular application. If $k < 0$, then $Q(t)$ decreases as t increases, and we refer to the model as the **law of decay**.

Let's consider some growth-and-decay applications.

PROBLEM 3

Suppose that in a certain culture, the equation $Q(t) = 15000e^{0.3t}$ expresses the number of bacteria present as a function of the time t, where t is expressed in hours. Find (a) the initial number of bacteria, and (b) the number of bacteria after 3 hours.

Solution

a. The initial number of bacteria is produced when $t = 0$.

$$Q(0) = 15000e^{0.3(0)}$$
$$= 15000e^0$$
$$= 15000 \qquad e^0 = 1$$

b. $Q(3) = 15000e^{0.3(3)}$
$$= 15000e^{0.9} = 36894 \qquad \text{to the nearest whole number}$$

There should be approximately 36,894 bacteria present after 3 hours. ■

PROBLEM 4

Suppose the number of bacteria present in a certain culture after t minutes is given by the equation $Q(t) = Q_0 e^{0.05t}$, where Q_0 represents the initial number of bacteria. If 5000 bacteria are present after 20 minutes, how many bacteria were present initially?

Solution

If 5000 bacteria are present after 20 minutes, then $Q(20) = 5000$.

$$5000 = Q_0 e^{0.05(20)}$$
$$5000 = Q_0 e^1$$
$$\frac{5000}{e} = Q_0$$
$$1839 = Q_0 \qquad \text{to the nearest whole number}$$

Thus approximately 1839 bacteria were present initially. ■

PROBLEM 5

The number of grams of a certain radioactive substance present after t seconds is given by the equation $Q(t) = 200e^{-0.3t}$. How many grams remain after 7 seconds?

Solution

Use $Q(t) = 200e^{-0.3t}$ to obtain

$$Q(7) = 200e^{(-0.3)(7)}$$
$$= 200e^{-2.1} = 24.5 \qquad \text{to the nearest tenth}$$

Thus approximately 24.5 grams remain after 7 seconds. ■

Finally, let's use the graphical approach to solve two problems.

PROBLEM 6

Suppose that $1000 is invested at 6.5% interest compounded continuously. How long will it take for the money to double?

Solution

Substitute $1000 for P and 0.065 for r in the formula $A = Pe^{rt}$ to produce $A = 1000e^{0.065t}$. If we let $y = A$ and $x = t$, we can graph the equation $y = 1000e^{0.065x}$. By letting $x = 20$, we obtain $y = 1000e^{0.065(20)} = 1000e^{1.3} \approx 3670$. Therefore, let's set the boundaries of the viewing rectangle so that $0 \leq x \leq 20$ and $0 \leq y \leq 3700$ with a y scale of 1000. Then we obtain the graph in Figure 4.7. Now we want to find the value of x so that $y = 2000$. (The money is to double.) Using the zoom and trace features of the graphing utility, we can determine that an x value of approximately 10.7 will produce a y value of 2000. Thus it will take approximately 10.7 years for the $1000 investment to double.

FIGURE 4.7

EXAMPLE 1

Graph the function $y = \dfrac{1}{\sqrt{2\pi}} e^{-x^2/2}$ and find its maximum value.

Solution

If $x = 0$, then

$$y = \frac{1}{\sqrt{2\pi}} e^0 = \frac{1}{\sqrt{2\pi}} \approx 0.4.$$

Let's set the boundaries of the viewing rectangle so that $-5 \leq x \leq 5$ and $0 \leq y \leq 1$ with a y scale of 0.1; the graph of the function is shown in Figure 4.8. From the graph, we see that the maximum value of the function occurs at $x = 0$, which we have already determined to be approximately 0.4.

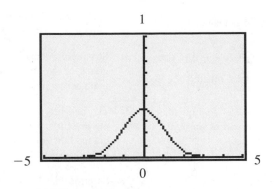

FIGURE 4.8

REMARK   The curve in Figure 4.8 is called the **normal distribution curve**. You may want to ask your instructor to explain what it means to assign grades on the basis of the normal distribution curve.

PROBLEM SET 4.2

1. Assuming that the rate of inflation is 7% per year, the equation $P = P_0(1.07)^t$ yields the predicted price P of an item in t years if it presently costs P_0. Find the predicted price of each of the following items for the indicated years ahead.

a. $.55 can of soup in 3 years

b. $3.43 container of cocoa mix in 5 years

c. $1.76 jar of coffee creamer in 4 years

d. $.44 can of beans and bacon in 10 years

e. $9000 car in 5 years (to the nearest dollar)

f. $50,000 house in 8 years (to the nearest dollar)

g. $500 TV set in 7 years (to the nearest dollar)

2. Suppose that it is estimated that the value of a car declines, or the car depreciates, 20% per year for the first 5 years. The equation $A = P_0(0.8)^t$ yields the value (A) of a car after t years if the original price is P_0. Find the value (to the nearest dollar) of each of the following cars after the indicated time.

a. $9000 car after 4 years

b. $14,000 car after 2 years

c. $18,000 car after 5 years

d. $25,000 car after 3 years

For Problems 3–14, use the formula

$$A = P\left(1 + \frac{r}{n}\right)^{nt}$$

to find the total amount of money accumulated at the end of the indicated time period for each of the following investments. Estimate to the nearest cent.

3. $250 for 5 years at 9% compounded annually

4. $350 for 7 years at 11% compounded annually

5. $300 for 6 years at 8% compounded semiannually

6. $450 for 10 years at 10% compounded semiannually

7. $600 for 12 years at 12% compounded quarterly

8. $750 for 15 years at 9% compounded quarterly

9. $1000 for 5 years at 12% compounded monthly

10. $1250 for 8 years at 9% compounded monthly

11. $600 for 10 years at $8\frac{1}{2}$% compounded annually

12. $1500 for 15 years at $9\frac{1}{4}$% compounded semiannually

13. $8000 for 10 years at 10.5% compounded quarterly

14. $10,000 for 25 years at 9.25% compounded monthly

For Problems 15–23, use the formula $A = Pe^{rt}$ to find the total amount of money accumulated at the end of the indicated time period by continuous compounding.

15. $400 for 5 years at 7%

16. $500 for 7 years at 6%

17. $750 for 8 years at 8%

18. $1000 for 10 years at 9%

19. $2000 for 15 years at 10%

20. $5000 for 20 years at 11%

21. $7500 for 10 years at 8.5%

22. $10,000 for 25 years at 9.25%

23. $15,000 for 10 years at 7.75%

For Problems 24–39, solve the problem.

24. What rate of interest, to the nearest tenth of a percent, compounded annually is needed for an investment of $200 to grow to $350 in 5 years?

25. What rate of interest, to the nearest tenth of a percent, compounded quarterly is needed for an investment of $1500 to grow to $2700 in 10 years?

26. Find the effective yield, to the nearest tenth of a percent, of an investment at 7.5% compounded monthly.

27. Find the effective yield, to the nearest hundredth of a percent, of an investment at 7.75% compounded continuously.

28. Which investment yields the greater return: 7% compounded monthly or 6.85% compounded continuously?

29. Which investment yields the greater return: 8.25% compounded quarterly or 8.3% compounded semiannually?

30. Suppose that a certain radioactive substance has a half-life of 20 years. If there are presently 2500 milligrams of

the substance, how much, to the nearest milligram, will remain after 40 years? After 50 years?

31. Strontium-90 has a half-life of 29 years. If there are 400 grams of strontium initially, how much, to the nearest gram, will remain after 87 years? After 100 years?

32. The half-life of radium is approximately 1600 years. If the present amount of radium in a certain location is 500 grams, how much will remain after 800 years? Express your answer to the nearest gram.

33. Suppose that in a certain culture, the equation $Q(t) = 1000e^{0.4t}$ expresses the number of bacteria present as a function of the time t, where t is expressed in hours. How many bacteria are present at the end of 2 hours? 3 hours? 5 hours?

34. The number of bacteria present at a given time under certain conditions is given by the equation $Q = 5000e^{0.05t}$, where t is expressed in minutes. How many bacteria are present at the end of 10 minutes? 30 minutes? 1 hour?

35. The number of bacteria present in a certain culture after t hours is given by the equation $Q = Q_0e^{0.3t}$, where Q_0 represents the initial number of bacteria. If 6640 bacteria are present after 4 hours, how many bacteria were present initially?

36. The number of grams Q of a certain radioactive substance present after t seconds is given by the equation $Q = 1500e^{-0.4t}$. How many grams remain after 5 seconds? 10 seconds? 20 seconds?

37. The atmospheric pressure, measured in pounds per square inch, is a function of the altitude above sea level. The equation $P(a) = 14.7e^{-0.21a}$, where a is the altitude measured in miles, can be used to approximate atmospheric pressure. Find the atmospheric pressure at each of the following locations.

 a. Mount McKinley in Alaska—altitude of 3.85 miles

 b. Denver, Colorado—the mile-high city

 c. Asheville, North Carolina—altitude of 1985 feet

 d. Phoenix, Arizona—altitude of 1090 feet

38. Suppose that the present population of a city is 75,000. Using the equation $P(t) = 75,000e^{0.01t}$ to estimate future growth, estimate the population (a) 10 years from now, (b) 15 years from now, and (c) 25 years from now.

39. The brightness of a star viewed from Earth is measured in magnitudes. A star of any given magnitude is 2.512 times as bright as a star of the next higher magnitude. Therefore, to determine how many times brighter one star is than another, we can use the exponential function $f(x) = 2.512^x$, where x is the higher magnitude minus the lower magnitude.

   a. How many times brighter is a star of magnitude 1 than a star of magnitude 6?

   b. The star Altair has a magnitude of 0.9, and the Kapteyn's star has a magnitude of 8.8. How many times brighter than Kapteyn's star is Altair?

   c. The sun has a magnitude of -26.7, and Sirius has a magnitude of -1.6. How many times brighter is the sun than Sirius?

For Problems 40–45, graph each exponential function.

40. $f(x) = e^x + 1$ **41.** $f(x) = e^x - 2$

42. $f(x) = 2e^x$ **43.** $f(x) = -e^x$

44. $f(x) = e^{2x}$ **45.** $f(x) = e^{-x}$

THOUGHTS INTO WORDS

46. Explain the difference between simple interest and compound interest.

47. How would you explain the concept of effective yield to someone who missed class the day it was discussed?

48. How would you explain the half-life formula to someone who missed class the day it was discussed?

Further Investigations

49. Complete the following chart that illustrates what happens to $1000 invested at various rates of interest for different lengths of time but always compounded continuously. Round your answers to the nearest dollar.

$1000 Compounded continuously

	8%	10%	12%	14%
5 years				
10 years				
15 years				
20 years				
25 years				

50. Complete the following chart that illustrates what happens to $1000 invested at 12% for different lengths of time and different numbers of compounding periods. Round all of your answers to the nearest dollar.

$1000 at 12%

	1 year	5 years	10 years	20 years
Compounded annually				
Compounded semiannually				
Compounded quarterly				
Compounded monthly				
Compounded continuously				

51. Complete the following chart that illustrates what happens to $1000 in 10 years based on different rates of interest and different numbers of compounding periods. Round your answers to the nearest dollar.

$1000 for 10 years

	8%	10%	12%	14%
Compounded annually				
Compounded semiannually				
Compounded quarterly				
Compounded monthly				
Compounded continuously				

For Problems 52–56, graph each function.

52. $f(x) = x(2^x)$

53. $f(x) = \dfrac{e^x + e^{-x}}{2}$

54. $f(x) = \dfrac{2}{e^x + e^{-x}}$

55. $f(x) = \dfrac{e^x - e^{-x}}{2}$

56. $f(x) = \dfrac{2}{e^x - e^{-x}}$

 GRAPHING CALCULATOR ACTIVITIES

57. Use your graphing calculator to check your graphs for Problems 40–45 and 52–56.

58. How should the graphs of $f(x) = 2^x$, $f(x) = e^x$, and $f(x) = 3^x$ compare? Graph them on the same set of axes.

59. Graph $f(x) = e^x$. Where should the graphs of $f(x) = e^{x-2}$, $f(x) = e^{x+4}$, and $f(x) = e^{x-6}$ be located? Graph all three functions on the same set of axes.

60. Graph $f(x) = e^x$ again. Now predict the graphs for $f(x) = -e^x$, $f(x) = e^{-x}$, and $f(x) = -e^{-x}$. Graph these three functions on the same set of axes.

61. How do you think the graphs of $f(x) = e^x$, $f(x) = e^{2x}$, and $f(x) = 2e^x$ will compare? Graph them on the same set of axes to see whether you were right.

62. Find an approximate solution, to the nearest hundredth, for each of the following equations by graphing the appropriate function and finding the x intercept.

a. $e^x = 7$ **b.** $e^x = 21$ **c.** $e^x = 53$

d. $2e^x = 60$ **e.** $e^{x+1} = 150$ **f.** $e^{x-2} = 300$

63. Use a graphing approach to argue that it is better to invest money at 6% compounded quarterly than at 5.75% compounded continuously.

64. How long will it take $500 to be worth $1500 if it is invested at 7.5% interest compounded semiannually?

65. How long will it take $5000 to triple if it is invested at 6.75% interest compounded quarterly?

4.3 INVERSE FUNCTIONS

Recall the *vertical line test*: If each vertical line intersects a graph in no more than one point, then the graph represents a function. There is also a useful distinction between two basic types of functions. Consider the graphs of the two functions in Figure 4.9: (a) $f(x) = 2x - 1$ and (b) $g(x) = x^2$. In part (a), any *horizontal line* will intersect the graph in no more than one point. Therefore, every value of $f(x)$ has only one value of x associated with it. Any function that has this property of having exactly one value of x associated with each value of $f(x)$ is called a **one-to-one function**. The function $g(x) = x^2$ is not a one-to-one function because the horizontal line in Figure 4.9(b) intersects the parabola in two points.

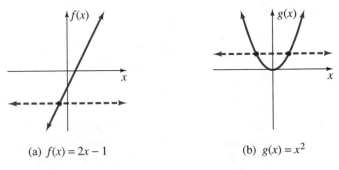

(a) $f(x) = 2x - 1$ (b) $g(x) = x^2$

FIGURE 4.9

Stated another way, a function f is said to be one-to-one if $x_1 \neq x_2$ implies that $f(x_1) \neq f(x_2)$. In other words, different values for x always result in different values for $f(x)$. Thus, without a graph, we can show that $f(x) = 2x - 1$ is a one-to-one function as follows: If $x_1 \neq x_2$, then $2x_1 \neq 2x_2$ and therefore $2x_1 - 1 \neq 2x_2 - 1$. Furthermore, we can show that $f(x) = x^2$ is not a one-to-one function because $f(2) = 4$ and $f(-2) = 4$; that is, different values for x produce the same value for $f(x)$.

Now let's consider a one-to-one function f that assigns the value $f(x)$ in its range R to each x in its domain D (Figure 4.10a). We can define a new function g that goes from R to D; it assigns $f(x)$ in R back to x in D, as indicated in Figure 4.10(b). The functions f and g are called *inverse functions* of one another. The following definition precisely states this concept.

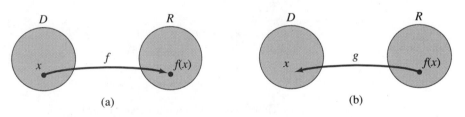

(a)   (b)

FIGURE 4.10

DEFINITION 4.2

Let f be a one-to-one function with a domain of X and a range of Y. A function g with a domain of Y and a range of X is called the inverse function of f if

$$(f \circ g)(x) = x \qquad \text{for every } x \text{ in } Y$$

and

$$(g \circ f)(x) = x \qquad \text{for every } x \text{ in } X$$

In Definition 4.2, note that for f and g to be inverses of each other, the domain of f must equal the range of g, and the range of f must equal the domain of g. Furthermore, g must reverse the correspondences given by f, and f must reverse the correspondences given by g. In other words, inverse functions *undo* each other. Let's use Definition 4.2 to verify that two specific functions are inverses of each other.

EXAMPLE I

Verify that $f(x) = 4x - 5$ and $g(x) = \dfrac{x + 5}{4}$ are inverse functions.

Solution

Because the set of real numbers is the domain and range of both functions, we know that the domain of f equals the range of g and that the range of f equals the domain of g. Furthermore,

$$(f \circ g)(x) = f(g(x))$$
$$= f\left(\frac{x + 5}{4}\right)$$
$$= 4\left(\frac{x + 5}{4}\right) - 5 = x$$

and

$$(g \circ f)(x) = g(f(x))$$
$$= g(4x - 5)$$
$$= \frac{4x - 5 + 5}{4} = x$$

Therefore, f and g are inverses of each other. ∎

EXAMPLE 2

Verify that $f(x) = x^2 + 1$ for $x \geq 0$ and $g(x) = \sqrt{x - 1}$ for $x \geq 1$ are inverse functions.

Solution

First, note that the domain of f equals the range of g—namely, the set of nonnegative real numbers. Also, the range of f equals the domain of g—namely, the set of real numbers greater than or equal to 1. Furthermore,

$$(f \circ g)(x) = f(g(x))$$
$$= f\left(\sqrt{x-1}\right)$$
$$= \left(\sqrt{x - 1}\right)^2 + 1$$
$$= x - 1 + 1 = x$$

and

$$(g \circ f)(x) = g(f(x))$$
$$= g(x^2 + 1)$$
$$= \sqrt{x^2 + 1 - 1} = \sqrt{x^2} = x \qquad \sqrt{x^2} = x \text{ because } x \geq 1$$

Therefore, f and g are inverses of each other.   ■

The inverse of a function f is commonly denoted by f^{-1} (read *f inverse* or *the inverse of f*). Do not confuse the -1 in f^{-1} with a negative exponent. The symbol f^{-1} *does not* mean $1/f^1$ but refers to the inverse function of function f.

Remember that a function can also be thought of as a set of ordered pairs no two of which have the same first element. Along those lines, a one-to-one function further requires that no two of the ordered pairs have the same second element. Then, if the components of each ordered pair of a given one-to-one function are interchanged, the resulting function and the given function are inverses of each other. Thus, if

$$f = \{(1, 4), (2, 7), (5, 9)\}$$

then

$$f^{-1} = \{(4, 1), (7, 2), (9, 5)\}$$

Graphically, two functions that are inverses of each other are **mirror images with reference to the line** $y = x$. This is due to the fact that ordered pairs (a, b) and (b, a) are reflections of each other with respect to the line $y = x$, as illustrated in Figure 4.11. (You will verify this in the next set of exercises.) Therefore, if the graph of a function f is known, as in Figure 4.12(a), then the graph of f^{-1} can be determined by reflecting f across the line $y = x$ (Figure 4.12b).

FIGURE 4.11

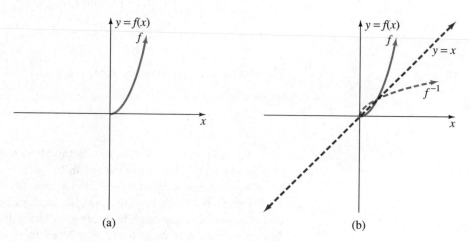

(a)   (b)

FIGURE 4.12

Finding Inverse Functions

The idea of inverse functions *undoing each other* provides the basis for an informal approach to finding the inverse of a function. Consider the function

$$f(x) = 2x + 1$$

To each x this function assigns twice x plus 1. To undo this function, we can subtract 1 and divide by 2. Hence the inverse is

$$f^{-1}(x) = \frac{x - 1}{2}$$

Now let's verify that f and f^{-1} are indeed inverses of each other.

$$(f \circ f^{-1})(x) = f(f^{-1}(x)) \qquad\qquad (f^{-1} \circ f)(x) = f^{-1}(f(x))$$
$$= f\left(\frac{x-1}{2}\right) \qquad\qquad\qquad = f^{-1}(2x+1)$$
$$= 2\left(\frac{x-1}{2}\right) + 1 \qquad\qquad\qquad = \frac{2x+1-1}{2}$$
$$= x - 1 + 1 = x \qquad\qquad\qquad = \frac{2x}{2} = x$$

Thus the inverse of $f(x) = 2x + 1$ is $f^{-1}(x) = \dfrac{x-1}{2}$.

This informal approach may not work very well with more complex functions, but it does emphasize how inverse functions are related to each other. A more formal and systematic technique for finding the inverse of a function can be described as follows.

1. Replace the symbol $f(x)$ with y.

2. Interchange x and y.

3. Solve the equation for y in terms of x.

4. Replace y with the symbol $f^{-1}(x)$.

The following examples illustrate this technique.

EXAMPLE 3   Find the inverse of $f(x) = \dfrac{2}{3}x + \dfrac{3}{5}$.

Solution

When we replace $f(x)$ with y, the equation becomes $y = \dfrac{2}{3}x + \dfrac{3}{5}$. Interchanging x and y produces $x = \dfrac{2}{3}y + \dfrac{3}{5}$.

Now, solving for y, we obtain

$$x = \frac{2}{3}y + \frac{3}{5}$$
$$15(x) = 15\left(\frac{2}{3}y + \frac{3}{5}\right)$$
$$15x = 10y + 9$$
$$15x - 9 = 10y$$
$$\frac{15x - 9}{10} = y$$

Finally, by replacing y with $f^{-1}(x)$, we can express the inverse function as

$$f^{-1}(x) = \frac{15x - 9}{10}$$

The domain of f is equal to the range of f^{-1} (both are the set of real numbers) and the range of f equals the domain of f^{-1} (both are the set of real numbers). Furthermore, we could show that $(f \circ f^{-1})(x) = x$ and $(f^{-1} \circ f)(x) = x$. We leave this for you to complete.   ∎

Does $f(x) = x^2 - 2$ have an inverse function? Sometimes a graph of the function helps to answer such a question. In Figure 4.13(a), it should be evident that f is not a one-to-one function and therefore cannot have an inverse. However, it should also be apparent from the graph that if we restrict the domain of f to be the nonnegative real numbers, then it is a one-to-one function and should have an inverse (Figure 4.13b). The next example illustrates how to find the inverse function.

(a)

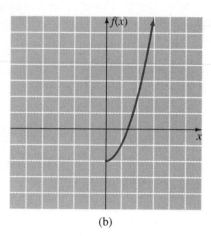

(b)

FIGURE 4.13

EXAMPLE 4

Find the inverse of $f(x) = x^2 - 2$, where $x \geq 0$.

Solution

When we replace $f(x)$ with y, the equation becomes

$$y = x^2 - 2, \qquad x \geq 0$$

Interchanging x and y produces

$$x = y^2 - 2, \qquad y \geq 0$$

Now let's solve for y; keep in mind that y is to be nonnegative.

$$x = y^2 - 2$$
$$x + 2 = y^2$$
$$\sqrt{x + 2} = y, \qquad x \geq -2$$

Finally, by replacing y with $f^{-1}(x)$, we can express the inverse function as

$$f^{-1}(x) = \sqrt{x + 2}, \qquad x \geq -2$$

The domain of f equals the range of f^{-1} (both are the nonnegative real numbers), and the range of f equals the domain of f^{-1} (both are the real numbers greater than or equal to -2). It can also be shown that $(f \circ f^{-1})(x) = x$ and $(f^{-1} \circ f)(x) = x$. Again, we leave this for you to complete.   ∎

Increasing and Decreasing Functions

Some general ideas can be formulated that were specifically illustrated in Example 4. In Figure 4.14 the function f is said to be *increasing* on the intervals $(-\infty, x_1]$ and $[x_2, \infty)$, and f is said to be *decreasing* on the interval $[x_1, x_2]$.

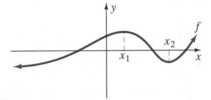

FIGURE 4.14

More specifically, increasing and decreasing functions are defined as follows.

DEFINITION 4.3

Let f be a function, with the interval I a subset of the domain of f. Let x_1 and x_2 be in I. Then

1. f is *increasing on I* if $f(x_1) < f(x_2)$ whenever $x_1 < x_2$,
2. f is *decreasing on I* if $f(x_1) > f(x_2)$ whenever $x_1 < x_2$, and
3. f is *constant on I* if $f(x_1) = f(x_2)$ for every x_1 and x_2.

Apply Definition 4.3 and you will see that the quadratic function $f(x) = x^2$ shown in Figure 4.15 is decreasing on $(-\infty, 0]$ and increasing on $[0, \infty)$. Likewise, the linear function $f(x) = 2x$ in Figure 4.16 is increasing throughout its domain of real numbers, so we say it is increasing on $(-\infty, \infty)$. The function $f(x) = -2x$ in Figure 4.17 is decreasing on $(-\infty, \infty)$. For our purposes in this text, we will rely on our knowledge of the graphs of the functions to determine where functions are increasing and decreasing. More formal techniques for determining where functions increase and decrease will be developed in calculus.

FIGURE 4.15

FIGURE 4.16

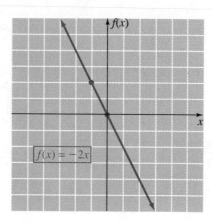

FIGURE 4.17

A function that is always increasing (or always decreasing) over its entire domain is one-to-one and so has an inverse. Furthermore, as illustrated by Example 4, even if a function is not one-to-one over its entire domain, it may be so over some subset of the domain. It then has an inverse over this restricted domain.

As functions become more complex, a graphing utility can be used to help with the problems we have discussed in this section. For example, suppose that we want to know whether the function $f(x) = \dfrac{3x + 1}{x - 4}$ is a one-to-one function and therefore has an inverse. Using a graphing utility, we can quickly get a sketch of the graph (see Figure 4.18). Then, by applying the horizontal line test to the graph, we can be fairly certain that the function is one-to-one. (Later we will develop some concepts that will allow us to be absolutely certain of this conclusion.)

FIGURE 4.18

A graphing utility can also be used to help determine intervals on which a function is increasing or decreasing. For example, to determine such intervals for the function $f(x) = \sqrt{x^2 + 4}$, let's use a graphing utility to get a sketch of the curve (Figure 4.19). From this graph we see that the function is decreasing on the interval $(-\infty, 0]$ and increasing on the interval $[0, \infty)$.

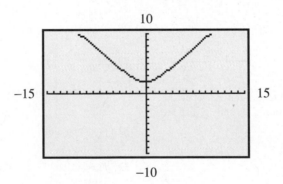

FIGURE 4.19

PROBLEM SET 4.3

For Problems 1–6, determine whether the graph represents a one-to-one function.

1.   **2.**   **3.**   **4.**

5. 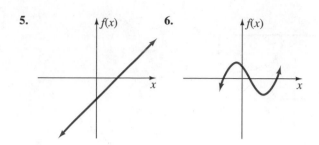   **6.**

For Problems 7–14, determine whether the function f is one-to-one.

7. $f(x) = 5x + 4$

8. $f(x) = -3x + 4$

9. $f(x) = x^3$

10. $f(x) = x^5 + 1$

11. $f(x) = |x| + 1$

11. $f(x) = -|x| - 2$

13. $f(x) = -x^4$

14. $f(x) = x^4 + 1$

For Problems 15–18, (a) list the domain and range of the function, (b) form the inverse function f^{-1}, and (c) list the domain and range of f^{-1}.

15. $f = \{(1, 5), (2, 9), (5, 21)\}$

16. $f = \{(1, 1), (4, 2), (9, 3), (16, 4)\}$

17. $f = \{(0, 0), (2, 8), (-1, -1), (-2, -8)\}$

18. $f = \{(-1, 1), (-2, 4), (-3, 9), (-4, 16)\}$

For Problems 19–26, verify that the two given functions are inverses of each other.

19. $f(x) = 5x - 9$ and $g(x) = \dfrac{x + 9}{5}$

20. $f(x) = -3x + 4$ and $g(x) = \dfrac{4 - x}{3}$

21. $f(x) = -\dfrac{1}{2}x + \dfrac{5}{6}$ and $g(x) = -2x + \dfrac{5}{3}$

22. $f(x) = x^3 + 1$ and $g(x) = \sqrt[3]{x - 1}$

23. $f(x) = \dfrac{1}{x - 1}$ for $x > 1$ and

$g(x) = \dfrac{x + 1}{x}$ for $x > 0$

24. $f(x) = x^2 + 2$ for $x \geq 0$ and

$g(x) = \sqrt{x - 2}$ for $x \geq 2$

25. $f(x) = \sqrt{2x - 4}$ for $x \geq 2$ and

$g(x) = \dfrac{x^2 + 4}{2}$ for $x \geq 0$

26. $f(x) = x^2 - 4$ for $x \geq 0$ and

$g(x) = \sqrt{x + 4}$ for $x \geq -4$

For Problems 27–36, determine whether f and g are inverse functions.

27. $f(x) = 3x$ and $g(x) = -\dfrac{1}{3}x$

28. $f(x) = \dfrac{3}{4}x - 2$ and $g(x) = \dfrac{4}{3}x + \dfrac{8}{3}$

29. $f(x) = x^3$ and $g(x) = \sqrt[3]{x}$

30. $f(x) = \dfrac{1}{x + 1}$ and $g(x) = \dfrac{1 - x}{x}$

31. $f(x) = x$ and $g(x) = \dfrac{1}{x}$

32. $f(x) = \dfrac{3}{5}x + \dfrac{1}{3}$ and $g(x) = \dfrac{5}{3}x - 3$

33. $f(x) = x^2 - 3$ for $x \geq 0$ and

$g(x) = \sqrt{x + 3}$ for $x \geq -3$

34. $f(x) = |x - 1|$ for $x \geq 1$ and

$g(x) = |x + 1|$ for $x \geq 0$

35. $f(x) = \sqrt{x + 1}$ and $g(x) = x^2 - 1$ for $x \geq 0$

36. $f(x) = \sqrt{2x - 2}$ and $g(x) = \dfrac{1}{2}x^2 + 1$

For Problems 37–50, (a) find f^{-1} and (b) verify that $(f \circ f^{-1})(x) = x$ and $(f^{-1} \circ f)(x) = x$.

37. $f(x) = x - 4$

38. $f(x) = 2x - 1$

39. $f(x) = -3x - 4$

40. $f(x) = -5x + 6$

41. $f(x) = \dfrac{3}{4}x - \dfrac{5}{6}$

42. $f(x) = \dfrac{2}{3}x - \dfrac{1}{4}$

43. $f(x) = -\dfrac{2}{3}x$

44. $f(x) = \dfrac{4}{3}x$

45. $f(x) = \sqrt{x}$ for $x \geq 0$

46. $f(x) = \dfrac{1}{x}$ for $x \neq 0$

47. $f(x) = x^2 + 4$ for $x \geq 0$

48. $f(x) = x^2 + 1$ for $x \leq 0$

49. $f(x) = 1 + \dfrac{1}{x}$ for $x > 0$

50. $f(x) = \dfrac{x}{x + 1}$ for $x > -1$

For Problems 51–58, (a) find f^{-1} and (b) graph f and f^{-1} on the same set of axes.

51. $f(x) = 3x$

52. $f(x) = -x$

53. $f(x) = 2x + 1$

54. $f(x) = -3x - 3$

55. $f(x) = \dfrac{2}{x - 1}$ for $x > 1$

56. $f(x) = \dfrac{-1}{x - 2}$ for $x > 2$

57. $f(x) = x^2 - 4$ for $x \geq 0$

58. $f(x) = \sqrt{x - 3}$ for $x \geq 3$

For Problems 59–66, find the intervals on which the given function is increasing and the intervals on which it is decreasing.

59. $f(x) = x^2 + 1$

60. $f(x) = x^3$

61. $f(x) = -3x + 1$

62. $f(x) = (x - 3)^2 + 1$

63. $f(x) = -(x + 2)^2 - 1$

64. $f(x) = x^2 - 2x + 6$

65. $f(x) = -2x^2 - 16x - 35$

66. $f(x) = x^2 + 3x - 1$

THOUGHTS INTO WORDS

67. Does the function $f(x) = 4$ have an inverse? Explain your answer.

68. Explain why every nonconstant linear function has an inverse.

69. Are the functions $f(x) = x^4$ and $g(x) = \sqrt[4]{x}$ inverses of each other? Explain your answer.

70. What does it mean to say that 2 and -2 are additive inverses of each other? What does it mean to say that 2 and $\dfrac{1}{2}$ are multiplicative inverses of each other? What does it mean to say that the functions $f(x) = x - 2$ and $f(x) = x + 2$ are inverses of each other? Do you think that the concept of "inverse" is being used in a consistent manner? Explain your answer.

Further Investigations

71. The function notation and the operation of composition can be used to find inverses as follows: To find the inverse of $f(x) = 5x + 3$, we know that $f(f^{-1}(x))$ must produce x. Therefore,

$$f(f^{-1}(x)) = 5[f^{-1}(x)] + 3 = x$$
$$5[f^{-1}(x)] = x - 3$$
$$f^{-1}(x) = \frac{x - 3}{5}$$

Use this approach to find the inverse of each of the following functions.

a. $f(x) = 3x - 9$ **b.** $f(x) = -2x + 6$

c. $f(x) = -x + 1$ **d.** $f(x) = 2x$

e. $f(x) = -5x$ **f.** $f(x) = x^2 + 6$ for $x \geq 0$

72. If $f(x) = 2x + 3$ and $g(x) = 3x - 5$, find

a. $(f \circ g)^{-1}(x)$ **b.** $(f^{-1} \circ g^{-1})(x)$

c. $(g^{-1} \circ f^{-1})(x)$

GRAPHING CALCULATOR ACTIVITIES

73. For Problems 37–44, graph the given function, the inverse function that you found, and $f(x) = x$ on the same set of axes. In each case, the given function and its inverse should produce graphs that are reflections of each other through the line $f(x) = x$.

74. There is another way in which we can use the graphing calculator to help show that two functions are inverses of each other. Suppose we want to show that $f(x) = x^2 - 2$ for $x \geq 0$ and $g(x) = \sqrt{x + 2}$ for $x \geq -2$ are inverses of each other. Let's make the following assignments for our graphing calculator.

$$
\begin{aligned}
f: & \quad Y_1 = x^2 - 2 \\
g: & \quad Y_2 = \sqrt{x + 2} \\
f \circ g: & \quad Y_3 = (Y_2)^2 - 2 \\
g \circ f: & \quad Y_4 = \sqrt{Y_1 + 2}
\end{aligned}
$$

Now we can proceed as follows:

1. Graph $Y_1 = x^2 - 2$ and note that for $x > 0$, the range is greater than or equal to -2.

2. Graph $Y_2 = \sqrt{x + 2}$ and note that for $x \geq -2$, the range is greater than or equal to 0.

> Thus the domain of f equals the range of g and the range of f equals the domain of g.

3. Graph $Y_3 = (Y_2)^2 - 2$ for $x \geq -2$ and observe the line $y = x$ for $x \geq -2$.

4. Graph $Y_4 = \sqrt{Y_1 + 2}$ for $x \geq 0$ and observe the line $y = x$ for $x \geq 0$.

> Thus $(f \circ g)(x) = x$ and $(g \circ f)(x) = x$, and the two functions are inverses of each other.

Use this approach to check your answers for Problems 45–50.

75. Use the technique demonstrated in Problem 74 to show that

$$f(x) = \frac{x}{\sqrt{x^2 + 1}}$$

and

$$g(x) = \frac{x}{\sqrt{1 - x^2}} \quad \text{for } -1 < x < 1$$

are inverses of each other.

4.4 LOGARITHMS

In Sections 4.1 and 4.2, we gave meaning to exponential expressions of the form b^n, where b is any positive real number and n is any real number; we next used exponential expressions of the form b^n to define exponential functions; and then we used exponential functions to help solve problems. In the next three sections, we will follow the same basic pattern with respect to a new concept, that of a *logarithm*. Let's begin with the following definition.

DEFINITION 4.4

If r is any positive real number, then the unique exponent t such that $b^t = r$ is called the logarithm of r with base b and is denoted by $\log_b r$.

According to Definition 4.4, the logarithm of 16 base 2 is the exponent t such that $2^t = 16$; thus we can write $\log_2 16 = 4$. Likewise, we can write $\log_{10} 1000 = 3$ because $10^3 = 1000$. In general, Definition 4.4 can be remembered in terms of the statement

$$\log_b r = t \quad \text{is equivalent to } b^t = r$$

Therefore, we can easily switch back and forth between exponential and logarithmic forms of equations, as the next examples illustrate.

$$\log_2 8 = 3 \qquad \text{is equivalent to } 2^3 = 8$$
$$\log_{10} 100 = 2 \qquad \text{is equivalent to } 10^2 = 100$$
$$\log_3 81 = 4 \qquad \text{is equivalent to } 3^4 = 81$$
$$\log_{10} 0.001 = -3 \qquad \text{is equivalent to } 10^{-3} = 0.001$$
$$2^7 = 128 \qquad \text{is equivalent to } \log_2 128 = 7$$
$$5^3 = 125 \qquad \text{is equivalent to } \log_5 125 = 3$$
$$\left(\frac{1}{2}\right)^4 = \frac{1}{16} \qquad \text{is equivalent to } \log_{1/2} \frac{1}{16} = 4$$
$$10^{-2} = 0.01 \qquad \text{is equivalent to } \log_{10} 0.01 = -2$$

Some logarithms can be determined by changing to exponential form and using the properties of exponents, as in the next two examples.

EXAMPLE 1

Evaluate $\log_{10} 0.0001$.

Solution

Let $\log_{10} 0.0001 = x$. Changing to exponential form yields $10^x = 0.0001$, which can be solved as follows.

$$10^x = 0.0001$$

$$10^x = 10^{-4} \qquad 0.0001 = \frac{1}{10,000} = \frac{1}{10^4} = 10^{-4}$$

$$x = -4$$

Thus we have $\log_{10} 0.0001 = -4$. ∎

EXAMPLE 2

Evaluate $\log_9\left(\sqrt[5]{27}/3\right)$.

Solution

Let $\log_9\left(\sqrt[5]{27}/3\right) = x$. Changing to exponential form yields $9^x = \sqrt[5]{27}/3$, which can be solved as follows.

$$9^x = \frac{(27)^{1/5}}{3}$$

$$(3^2)^x = \frac{(3^3)^{1/5}}{3}$$

$$3^{2x} = \frac{3^{3/5}}{3}$$

$$3^{2x} = 3^{-2/5}$$

$$2x = -\frac{2}{5}$$

$$x = -\frac{1}{5}$$

Therefore, we have $\log_9 \dfrac{\sqrt[5]{27}}{3} = -\dfrac{1}{5}$. ∎

Some equations that involve logarithms can also be solved by changing to exponential form and using our knowledge of exponents.

EXAMPLE 3

Solve $\log_8 x = \dfrac{2}{3}$.

Solution

Change $\log_8 x = \dfrac{2}{3}$ to exponential form to obtain

$$8^{2/3} = x$$

Therefore,

$$x = \left(\sqrt[3]{8}\right)^2 = 2^2 = 4$$

The solution set is {4}.   ■

EXAMPLE 4

Solve $\log_b \dfrac{27}{64} = 3$.

Solution

Change $\log_b \dfrac{27}{64} = 3$ to exponential form to obtain

$$b^3 = \frac{27}{64}$$

Therefore,

$$b = \sqrt[3]{\frac{27}{64}} = \frac{3}{4}$$

The solution set is $\left\{\dfrac{3}{4}\right\}$.   ■

Properties of Logarithms

There are some properties of logarithms that are a direct consequence of Definition 4.4 and the properties of exponents. For example, by writing the exponential equations $b^1 = b$ and $b^0 = 1$ in logarithmic form, we obtain the following property.

PROPERTY 4.3

For $b > 0$ and $b \neq 1$,

$$\log_b b = 1 \qquad \text{and} \qquad \log_b 1 = 0$$

Therefore, according to Property 4.3, we can write

$$\log_{10} 10 = 1 \qquad \log_4 4 = 1$$
$$\log_{10} 1 = 0 \qquad \log_5 1 = 0$$

Also, from Definition 4.4 we know that $\log_b r$ is the exponent t such that $b^t = r$. Therefore, raising b to the $\log_b r$ power must produce r. This fact is stated in Property 4.4.

> ### PROPERTY 4.4
>
> For $b > 0$, $b \neq 1$, and $r > 0$,
>
> $$b^{\log_b r} = r$$

Therefore, according to Property 4.4, we can write

$$10^{\log_{10} 72} = 72 \qquad 3^{\log_3 85} = 85 \qquad e^{\log_e 7} = 7$$

Because a logarithm is by definition an exponent, it is reasonable to predict that logarithms will have some properties that correspond to the basic exponential properties. This is an accurate prediction; these properties provide a basis for computational work with logarithms. Let's state the first of these properties and show how we can verify it by using our knowledge of exponents.

> ### PROPERTY 4.5
>
> For positive numbers b, r, and s, where $b \neq 1$,
>
> $$\log_b rs = \log_b r + \log_b s$$

To verify Property 4.5, we can proceed as follows. Let $m = \log_b r$ and $n = \log_b s$. Change each of these equations to exponential form.

$$m = \log_b r \qquad \text{becomes} \quad r = b^m$$
$$n = \log_b s \qquad \text{becomes} \quad s = b^n$$

Thus the product rs becomes

$$rs = b^m \cdot b^n = b^{m+n}$$

Now, by changing $rs = b^{m+n}$ back to logarithmic form, we obtain

$$\log_b rs = m + n$$

Replacing m with $\log_b r$ and n with $\log_b s$ yields

$$\log_b rs = \log_b r + \log_b s$$

The following two examples demonstrate a use of Property 4.5.

EXAMPLE 5

If $\log_2 5 = 2.3219$ and $\log_2 3 = 1.5850$, evaluate $\log_2 15$.

Solution

Because $15 = 5 \cdot 3$, we can apply Property 4.5 as follows.

$$\log_2 15 = \log_2(5 \cdot 3)$$
$$= \log_2 5 + \log_2 3$$
$$= 2.3219 + 1.5850 = 3.9069 \qquad \blacksquare$$

EXAMPLE 6

If $\log_{10} 178 = 2.2504$ and $\log_{10} 89 = 1.9494$, evaluate $\log_{10}(178 \cdot 89)$.

Solution

$$\log_{10}(178 \cdot 89) = \log_{10} 178 + \log_{10} 89$$
$$= 2.2504 + 1.9494 = 4.1998 \qquad \blacksquare$$

Because $b^m/b^n = b^{m-n}$, we would expect a corresponding property that pertains to logarithms. Property 4.6 is that property. It can be verified by using an approach similar to the one we used for Property 4.5. This verification is left for you to do as an exercise in the next problem set.

PROPERTY 4.6

For positive numbers b, r, and s, where $b \neq 1$,
$$\log_b\left(\frac{r}{s}\right) = \log_b r - \log_b s$$

Property 4.6 can be used to change a division problem into an equivalent subtraction problem, as the next two examples illustrate.

EXAMPLE 7

If $\log_5 36 = 2.2266$ and $\log_5 4 = 0.8614$, evaluate $\log_5 9$.

Solution

Because $9 = \dfrac{36}{4}$, we can use Property 4.6 as follows.

$$\log_5 9 = \log_5\left(\frac{36}{4}\right)$$
$$= \log_5 36 - \log_5 4$$
$$= 2.2266 - 0.8614 = 1.3652 \qquad \blacksquare$$

EXAMPLE 8

Evaluate $\log_{10}\left(\dfrac{379}{86}\right)$, given that $\log_{10} 379 = 2.5786$ and $\log_{10} 86 = 1.9345$.

Solution

$$\log_{10}\left(\frac{379}{86}\right) = \log_{10} 379 - \log_{10} 86$$
$$= 2.5786 - 1.9345 = 0.6441 \qquad \blacksquare$$

Another property of exponents states that $(b^n)^m = b^{mn}$. The corresponding property of logarithms is stated in Property 4.7. Again, we leave the verification of this property as an exercise for you to do in the next set of problems.

PROPERTY 4.7

If r is a positive real number, b is a positive real number other than 1, and p is any real number, then

$$\log_b r^p = p(\log_b r)$$

The next two examples demonstrate a use of Property 4.7.

EXAMPLE 9

Evaluate $\log_2 22^{1/3}$, given that $\log_2 22 = 4.4594$.

Solution

$$\log_2 22^{1/3} = \frac{1}{3}\log_2 22 \qquad \text{Property 4.7}$$

$$= \frac{1}{3}(4.4594) = 1.4865 \qquad \blacksquare$$

EXAMPLE 10

Evaluate $\log_{10}(8540)^{3/5}$, given that $\log_{10} 8540 = 3.9315$.

Solution

$$\log_{10}(8540)^{3/5} = \frac{3}{5}\log_{10} 8540$$

$$= \frac{3}{5}(3.9315) = 2.3589 \qquad \blacksquare$$

The properties of logarithms can be used to change the forms of various logarithmic expressions, as we will see in the next two examples.

EXAMPLE 11

Express $\log_b \sqrt{xy/z}$ in terms of the logarithms of x, y, and z.

Solution

$$\log_b \sqrt{\frac{xy}{z}} = \log_b\left(\frac{xy}{z}\right)^{1/2}$$

$$= \frac{1}{2}\log_b\left(\frac{xy}{z}\right) \qquad \text{Property 4.7}$$

$$= \frac{1}{2}(\log_b xy - \log_b z) \qquad \text{Property 4.6}$$

$$= \frac{1}{2}(\log_b x + \log_b y - \log_b z) \qquad \text{Property 4.5} \qquad \blacksquare$$

EXAMPLE 12

Express $2 \log_b x + 3 \log_b y - 4 \log_b z$ as one logarithm.

Solution

$$
\begin{aligned}
2 \log_b x + 3 \log_b y - 4 \log_b z &= \log_b x^2 + \log_b y^3 - \log_b z^4 \\
&= \log_b x^2 y^3 - \log_b z^4 \\
&= \log_b \left(\frac{x^2 y^3}{z^4} \right)
\end{aligned}
$$

∎

Sometimes we need to change from an indicated sum or difference of logarithmic quantities to an indicated product or quotient. This is especially helpful when we are solving certain kinds of equations that involve logarithms. Note in these next two examples how we can use the properties, along with the process of changing from logarithmic form to exponential form, to solve some equations.

EXAMPLE 13

Solve $\log_{10} x + \log_{10}(x + 9) = 1$.

Solution

$$
\begin{aligned}
\log_{10} x + \log_{10}(x + 9) &= 1 \\
\log_{10}[x(x + 9)] &= 1 \qquad \text{Property 4.5} \\
x(x + 9) &= 10^1 \qquad \text{Change to exponential form.} \\
x^2 + 9x &= 10 \\
x^2 + 9x - 10 &= 0 \\
(x + 10)(x - 1) &= 0
\end{aligned}
$$

$$
x + 10 = 0 \qquad \text{or} \qquad x - 1 = 0
$$
$$
x = -10 \qquad \text{or} \qquad x = 1
$$

Because the left-hand side of the original equation is meaningful only if $x > 0$ and $x + 9 > 0$, the solution -10 must be discarded. Thus the solution set is $\{1\}$.   ∎

EXAMPLE 14

Solve $\log_5(x + 4) - \log_5 x = 2$.

Solution

$$
\begin{aligned}
\log_5(x + 4) - \log_5 x &= 2 \\
\log_5 \left(\frac{x + 4}{x} \right) &= 2 \qquad \text{Property 4.6} \\
5^2 &= \frac{x + 4}{x} \qquad \text{Change to exponential form.} \\
25 &= \frac{x + 4}{x} \\
25x &= x + 4 \\
24x &= 4 \\
x &= \frac{4}{24} = \frac{1}{6} \qquad \text{The solution set is } \left\{ \frac{1}{6} \right\}.
\end{aligned}
$$

∎

PROBLEM SET 4.4

For Problems 1–8, write each equation in logarithmic form. For example, $2^4 = 16$ becomes $\log_2 16 = 4$.

1. $3^2 = 9$ **2.** $2^5 = 32$

3. $5^3 = 125$ **4.** $10^1 = 10$

5. $2^{-4} = \dfrac{1}{16}$ **6.** $\left(\dfrac{2}{3}\right)^{-3} = \dfrac{27}{8}$

7. $10^{-2} = 0.01$ **8.** $10^5 = 100{,}000$

For Problems 9–14, write each equation in exponential form. For example, $\log_2 8 = 3$ becomes $2^3 = 8$.

9. $\log_2 64 = 6$ **10.** $\log_3 27 = 3$

11. $\log_{10} 0.1 = -1$ **12.** $\log_5\left(\dfrac{1}{25}\right) = -2$

13. $\log_2\left(\dfrac{1}{16}\right) = -4$ **14.** $\log_{10} 0.00001 = -5$

For Problems 15–30, evaluate each expression.

15. $\log_6 36$ **16.** $\log_3 243$

17. $\log_5\left(\dfrac{1}{5}\right)$ **18.** $\log_4\left(\dfrac{1}{64}\right)$

19. $\log_{10} 10$ **20.** $\log_{10} 1$

21. $\log_3 \sqrt{3}$ **22.** $\log_5 \sqrt[3]{25}$

23. $\log_3\left(\dfrac{\sqrt{27}}{3}\right)$ **24.** $\log_{1/2}\left(\dfrac{\sqrt[4]{8}}{2}\right)$

25. $\log_{1/4}\left(\dfrac{\sqrt[4]{32}}{2}\right)$ **26.** $\log_2\left(\dfrac{\sqrt[3]{16}}{4}\right)$

27. $10^{\log_{10} 7}$ **28.** $5^{\log_5 13}$

29. $\log_2(\log_5 5)$ **30.** $\log_6(\log_2 64)$

For Problems 31–38, solve each equation.

31. $\log_5 x = 2$ **32.** $\log_{10} x = 3$

33. $\log_8 t = \dfrac{5}{3}$ **34.** $\log_4 m = \dfrac{3}{2}$

35. $\log_b 3 = \dfrac{1}{2}$ **36.** $\log_b 2 = \dfrac{1}{2}$

37. $\log_{10} x = 0$ **38.** $\log_{10} x = 1$

For Problems 39–46, given that $\log_2 5 = 2.3219$ and $\log_2 7 = 2.8074$, evaluate each expression by using Properties 4.5–4.7.

39. $\log_2 35$ **40.** $\log_2\left(\dfrac{7}{5}\right)$

41. $\log_2 125$ **42.** $\log_2 49$

43. $\log_2 \sqrt{7}$ **44.** $\log_2 \sqrt[3]{5}$

45. $\log_2 175$ **46.** $\log_2 56$

47. $\log_2 80$

For Problems 48–56, given that $\log_8 5 = 0.7740$ and $\log_8 11 = 1.1531$, evaluate each expression using Properties 4.5–4.7.

48. $\log_8 55$ **49.** $\log_8\left(\dfrac{5}{11}\right)$

50. $\log_8 25$ **51.** $\log_8 \sqrt{11}$

52. $\log_8(5)^{2/3}$ **53.** $\log_8 88$

54. $\log_8 320$ **55.** $\log_8\left(\dfrac{25}{11}\right)$

56. $\log_8\left(\dfrac{121}{25}\right)$

For Problems 57–64, express each as the sum or difference of simpler logarithmic quantities. (Assume that all variables represent positive real numbers.) For example,

$$\log_b\left(\dfrac{x^3}{y^2}\right) = \log_b x^3 - \log_b y^2$$
$$= 3\log_b x - 2\log_b y$$

57. $\log_b xyz$ **58.** $\log_b\left(\dfrac{x^2}{y}\right)$

59. $\log_b x^2 y^3$ **60.** $\log_b x^{2/3} y^{3/4}$

61. $\log_b \sqrt{xy}$ **62.** $\log_b \sqrt[3]{x^2 z}$

63. $\log_b \sqrt{\dfrac{x}{y}}$ **64.** $\log_b\left[x\left(\sqrt{\dfrac{x}{y}}\right)\right]$

For Problems 65–72, express each as a single logarithm. (Assume that all variables represent positive numbers.) For example,

$$3 \log_b x + 5 \log_b y = \log_b x^3 y^5$$

65. $\log_b x + \log_b y - \log_b z$

66. $2 \log_b x - 4 \log_b y$

67. $(\log_b x - \log_b y) - \log_b z$

68. $\log_b x - (\log_b y - \log_b z)$

69. $\log_b x + \dfrac{1}{2} \log_b y$

70. $2 \log_b x + 4 \log_b y - 3 \log_b z$

71. $2 \log_b x + \dfrac{1}{2} \log_b(x - 1) - 4 \log_b(2x + 5)$

72. $\dfrac{1}{2} \log_b x - 3 \log_b x + 4 \log_b y$

For Problems 73–84, solve each equation.

73. $\log_3 x + \log_3 4 = 2$

74. $\log_7 5 + \log_7 x = 1$

75. $\log_{10} x + \log_{10}(x - 21) = 2$

76. $\log_{10} x + \log_{10}(x - 3) = 1$

77. $\log_2 x + \log_2(x - 3) = 2$

78. $\log_3 x + \log_3(x - 2) = 1$

79. $\log_{10}(2x - 1) - \log_{10}(x - 2) = 1$

80. $\log_{10}(9x - 2) = 1 + \log_{10}(x - 4)$

81. $\log_5(3x - 2) = 1 + \log_5(x - 4)$

82. $\log_6 x + \log_6(x + 5) = 2$

83. $\log_8(x + 7) + \log_8 x = 1$

84. $\log_6(x + 1) + \log_6(x - 4) = 2$

85. Verify Property 4.6.

86. Verify Property 4.7.

THOUGHTS INTO WORDS

87. How would you explain the concept of a logarithm to someone who has never studied algebra?

88. Explain, without using Property 4.4, why $4^{\log_4 9}$ equals 9.

89. In the next section we are going to show that the logarithmic function $f(x) = \log_2 x$ is the inverse of the exponential function $f(x) = 2^x$. From that information, how could you sketch a graph of $f(x) = \log_2 x$?

4.5 LOGARITHMIC FUNCTIONS

The concept of a logarithm can now be used to define a logarithmic function.

DEFINITION 4.5

If $b > 0$ and $b \neq 1$, then the function defined by

$$f(x) = \log_b x$$

where x is any positive real number, is called the **logarithmic function with base b**.

We can obtain the graph of a specific logarithmic function in various ways. For example, we can change the equation $y = \log_2 x$ to the exponential equation $2^y = x$, from which we can determine a table of values. We will instruct you to use this approach to graph some logarithmic functions in the next set of exercises.

We can also obtain the graph of a logarithmic function by setting up a table of values directly from the logarithmic equation. We will demonstrate this approach.

EXAMPLE 1

Graph $f(x) = \log_2 x$.

Solution

Let's choose some values for x where the corresponding values for $\log_2 x$ are easily determined. (Remember that logarithms are defined only for the positive real numbers.) Plot the points determined by the table and connect them with a smooth curve to produce Figure 4.20.

x	$f(x)$
$\frac{1}{8}$	-3
$\frac{1}{4}$	-2
$\frac{1}{2}$	-1
1	0
2	1
4	2
8	3

$\log_2 \frac{1}{8} = -3$ because $2^{-3} = \frac{1}{2^3} = \frac{1}{8}$

$\log_2 1 = 0$ because $2^0 = 1$

[handwritten: $\log_2 2 = \varnothing$ $\log_2 4 = \varnothing$ $2^x = 4$ $2^x = 2^2$]

FIGURE 4.20

Now suppose that we consider two functions f and g as follows.

$f(x) = b^x$ Domain: all real numbers
  Range: positive real numbers

$g(x) = \log_b x$ Domain: positive real numbers
  Range: all real numbers

Furthermore, suppose that we consider the composition of f and g and the composition of g and f.

$$(f \circ g)(x) = f(g(x)) = f(\log_b x) = b^{\log_b x} = x$$

$$(g \circ f)(x) = g(f(x)) = g(b^x) = \log_b b^x = x \log_b b = x(1) = x$$

Therefore, because the domain of f is the range of g and the range of f is the domain of g, and because $f(g(x)) = x$ and $g(f(x)) = x$, the two functions f and g are *inverses of each other*.

Remember that the graphs of a function and its inverse are reflections of each other through the line $y = x$. Thus the graph of a logarithmic function can be determined by reflecting the graph of its inverse exponential function through the line $y = x$. This idea is illustrated in Figure 4.21, where the graph of $y = 2^x$ has been reflected across the line $y = x$ to produce the graph of $y = \log_2 x$.

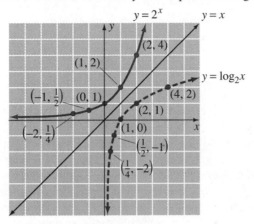

F I G U R E   4 . 2 1

Figure 4.3 illustrated the general behavior patterns of exponential functions with two graphs. We can now reflect each of these graphs through the line $y = x$ and observe the general behavior patterns of logarithmic functions, as shown in Figure 4.22.

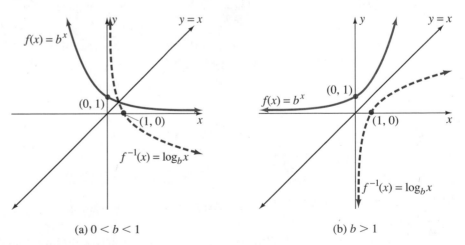

(a) $0 < b < 1$ (b) $b > 1$

F I G U R E   4 . 2 2

Finally, when graphing logarithmic functions, don't forget about transformations of the basic curves.

1. The graph of $f(x) = 3 + \log_2 x$ is the graph of $f(x) = \log_2 x$ *moved up three units*. (Because $\log_2 x + 3$ is apt to be confused with $\log_2(x + 3)$, we commonly write $3 + \log_2 x$.)

2. The graph of $f(x) = \log_2(x - 4)$ is the graph of $f(x) = \log_2 x$ *moved four units to the right*.

3. The graph of $f(x) = -\log_2 x$ is the graph of $f(x) = \log_2 x$ *reflected across the x axis*.

Common Logarithms: Base 10

The properties of logarithms that we discussed in Section 4.4 are true for any valid base. However, because the Hindu–Arabic numeration system that we use is a base-10 system, logarithms to base 10 have historically been used for computational purposes. Base-10 logarithms are called **common logarithms**.

Originally, common logarithms were developed to aid in complicated numerical calculations that involve products, quotients, and powers of real numbers. Today they are seldom used for that purpose because the calculator and computer can much more effectively handle the messy computational problems. However, common logarithms do still occur in applications, so they deserve our attention.

> **REMARK** In Appendix A we have included a short discussion about the computational aspects of common logarithms. You may find it interesting to browse through this material. It probably will enhance your appreciation of the calculator.

As we know from earlier work, the definition of a logarithm allows us to evaluate $\log_{10} x$ for values of x that are integral powers of 10. Consider the following examples.

$$\log_{10} 1000 = 3 \qquad \text{because } 10^3 = 1000$$
$$\log_{10} 100 = 2 \qquad \text{because } 10^2 = 100$$
$$\log_{10} 10 = 1 \qquad \text{because } 10^1 = 10$$
$$\log_{10} 1 = 0 \qquad \text{because } 10^0 = 1$$
$$\log_{10} 0.1 = -1 \qquad \text{because } 10^{-1} = \frac{1}{10} = 0.1$$
$$\log_{10} 0.01 = -2 \qquad \text{because } 10^{-2} = \frac{1}{10^2} = 0.01$$
$$\log_{10} 0.001 = -3 \qquad \text{because } 10^{-3} = \frac{1}{10^3} = 0.001$$

When working exclusively with base-10 logarithms, it is customary to omit writing the numeral 10 to designate the base. Thus the expression $\log_{10} x$ is written as $\log x$, and a statement such as $\log_{10} 1000 = 3$ becomes $\log 1000 = 3$. We will follow this practice from now on in this chapter, but don't forget that the base is understood to be 10.

$$\log_{10} x = \log x$$

To find the common logarithm of a positive number that is not an integral power of 10, we can use an appropriately equipped calculator or a table such as the one that appears in Appendix A. We used a calculator equipped with a common logarithmic function (ordinarily a key labeled [log] is used) to obtain the following results rounded to four decimal places.

$\log 1.75 = 0.2430$

$\log 23.8 = 1.3766$

$\log 134 = 2.1271$ **Be sure that you can use a calculator and obtain these results.**

$\log 0.192 = -0.7167$

$\log 0.0246 = -1.6091$

In order to use logarithms to solve problems, we sometimes need to be able to determine a number when the logarithm of the number is known. That is, we may need to determine x when $\log x$ is known. Let's consider an example.

EXAMPLE 2

Find x if $\log x = 0.2430$.

Solution

If $\log x = 0.2430$, then changing to exponential form yields $10^{0.2430} = x$; use the [10^x] key to find x.

$$x = 10^{0.2430} \approx 1.749846689$$

Therefore, $x = 1.7498$ rounded to five significant digits.   ∎

Be sure that you can use your calculator and obtain the following results. We have rounded the values for x to five significant digits.

If $\log x = 0.7629$, then $x = 10^{0.7629} = 5.7930$.

If $\log x = 1.4825$, then $x = 10^{1.4825} = 30.374$.

If $\log x = 4.0214$, then $x = 10^{4.0214} = 10505$.

If $\log x = -1.5162$, then $x = 10^{-1.5162} = 0.030465$.

If $\log x = -3.8921$, then $x = 10^{-3.8921} = 0.00012820$.

The **common logarithmic function** is defined by the equation $f(x) = \log x$. It should now be a simple matter to set up a table of values and sketch the function. You will do this in the next set of exercises. Remember that $f(x) = 10^x$ and $g(x) = \log x$ are inverses of each other. Therefore, we could also get the graph of $g(x) = \log x$ by reflecting the exponential curve $f(x) = 10^x$ across the line $y = x$.

Natural Logarithms—Base e

In many practical applications of logarithms, the number e (remember $e \approx 2.71828$) is used as a base. Logarithms with a base of e are called **natural logarithms** and the symbol $\ln x$ is commonly used instead of $\log_e x$.

$$\log_e x = \ln x$$

Natural logarithms can be found with an appropriately equipped calculator or with a table of natural logarithms. (A table of natural logarithms is provided in Appendix B.) Use a calculator with a natural logarithm function (ordinarily a key labeled $\boxed{\ln x}$) to obtain the following results rounded to four decimal places.

$\ln 3.21 = 1.1663$

$\ln 47.28 = 3.8561$

$\ln 842 = 6.7358$

$\ln 0.21 = -1.5606$

$\ln 0.0046 = -5.3817$

$\ln 10 = 2.3026$

Be sure that you can use your calculator to obtain these results. Keep in mind the significance of a statement such as $\ln 3.21 = 1.1663$. By changing to exponential form, we are claiming that e raised to the 1.1663 power is approximately 3.21. Using a calculator, we obtain $e^{1.1663} = 3.210093293$.

Let's do a few more problems to find x when given $\ln x$. Be sure that you agree with these results.

If $\ln x = 2.4156$, then $x = e^{2.4156} = 11.196$.

If $\ln x = 0.9847$, then $x = e^{0.9847} = 2.6770$.

If $\ln x = 4.1482$, then $x = e^{4.1482} = 63.320$.

If $\ln x = -1.7654$, then $x = e^{-1.7654} = 0.17112$.

The **natural logarithmic function** is defined by the equation $f(x) = \ln x$. It is the inverse of the natural exponential function $f(x) = e^x$. Thus one way to graph $f(x) = \ln x$ is to reflect the graph of $f(x) = e^x$ across the line $y = x$. We will ask you to do this in the next set of problems.

In Figure 4.23 we have used a graphing utility to sketch the graph of $f(x) = e^x$. Now on the basis of our previous work with trnsformations, we should be able to make the statements that follow.

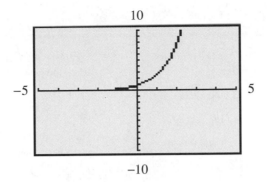

FIGURE 4.23

1. The graph of $f(x) = -e^x$ is the graph of $f(x) = e^x$ reflected through the x axis.

2. The graph of $f(x) = e^{-x}$ is the graph of $f(x) = e^x$ reflected through the y axis.

3. The graph of $f(x) = e^x + 4$ is the graph of $f(x) = e^x$ shifted upward four units.

4. The graph of $f(x) = e^{x+2}$ is the graph of $f(x) = e^x$ shifted two units to the left.

These statements are confirmed in Figure 4.24, which shows the result of graphing these four functions on the same set of axes using a graphing utility.

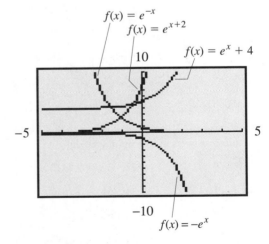

FIGURE 4.24

REMARK   So far, we have used a graphing utility to graph only common logarithmic and natural logarithmic functions. In the next section, we will see how logarithms with bases other than 10 or e are related to common and natural logarithms. This will provide a way of using a graphing utility to graph a logarithmic function with any valid base.

PROBLEM SET 4.5

For Problems 1–10, use a calculator to find each common logarithm. Express answers to four decimal places.

1. log 7.24   **2.** log 2.05   **3.** log 52.23

4. log 825.8   **5.** log 3214.1   **6.** log 14.189

7. log 0.729   **8.** log 0.04376   **9.** log 0.00034

10. log 0.000069

For Problems 11–20, use your calculator to find x when given log x. Express answers to five significant digits.

11. log $x = 2.6143$ **12.** log $x = 1.5263$

13. log $x = 4.9547$ **14.** log $x = 3.9335$

15. log $x = 1.9006$ **16.** log $x = 0.5517$

17. log $x = -1.3148$ **18.** log $x = -0.1452$

19. log $x = -2.1928$ **20.** log $x = -2.6542$

For Problems 21–30, use your calculator to find each natural logarithm. Express answers to four decimal places.

21. ln 5   **22.** ln 18   **23.** ln 32.6

24. ln 79.5   **25.** ln 430   **26.** ln 371.8

27. ln 0.46   **28.** ln 0.524   **29.** ln 0.0314

30. ln 0.008142

For Problems 31–40, use your calculator to find x when given ln x. Express answers to five significant digits.

31. ln $x = 0.4721$ **32.** ln $x = 0.9413$

33. ln $x = 1.1425$ **34.** ln $x = 2.7619$

35. ln $x = 4.6873$ **36.** ln $x = 3.0259$

37. ln $x = -0.7284$ **38.** ln $x = -1.6246$

39. ln $x = -3.3244$ **40.** ln $x = -2.3745$

41. a. Complete the following table and then graph $f(x) = $ log x. (Express the values for log x to the nearest tenth.)

x	0.1	0.5	1	2	4	8	10
log x							

b. Complete the following table and express values for 10^x to the nearest tenth.

x	-1	-0.3	0	0.3	0.6	0.9	1
10^x							

Then graph $f(x) = 10^x$ and reflect it across the line $y = x$ to produce the graph for $f(x) = $ log x.

42. a. Complete the following table and then graph $f(x) = $ ln x. (Express the values for ln x to the nearest tenth.)

x	0.1	0.5	1	2	4	8	10
ln x							

b. Complete the following table and express values for e^x to the nearest tenth.

x	-2.3	-0.7	0	0.7	1.4	2.1	2.3
e^x							

Then graph $f(x) = e^x$ and reflect it across the line $y = x$ to produce the graph for $f(x) = $ ln x.

43. Graph $y = \log_{1/2} x$ by graphing $\left(\dfrac{1}{2}\right)^y = x$.

44. Graph $y = \log_2 x$ by graphing $2^y = x$.

45. Graph $f(x) = \log_3 x$ by reflecting the graph of $g(x) = 3^x$ across the line $y = x$.

46. Graph $f(x) = \log_4 x$ by reflecting the graph of $g(x) = 4^x$ across the line $y = x$.

For Problems 47–53, graph each function. Remember that the graph of $f(x) = \log_2 x$ is given in Figure 4.20.

47. $f(x) = 3 + \log_2 x$ **48.** $f(x) = -2 + \log_2 x$

49. $f(x) = \log_2 (x + 3)$ **50.** $f(x) = \log_2 (x - 2)$

51. $f(x) = \log_2 2x$ **52.** $f(x) = -\log_2 x$

53. $f(x) = 2 \log_2 x$

54. In chemistry the term pH, meaning "hydrogen power," is defined as the negative base-10 logarithm of the concentration, in moles per liter, of H^+ ions. In other words, pH is a function of the number of H^+ ions and can be expressed as

$$f(x) = -\log x$$

where x is the number of H^+ ions in moles per liter of the solution. A solution with a pH below 7 is called an *acid solution*, and a solution with a pH above 7 is called a *basic solution*.

Find, to the nearest tenth, the pH of each of the following solutions with the given H^+ concentrations and identify each as an acid or basic solution.

a. $2(10)^{-9}$ **b.** $7.1(10)^{-4}$

c. $8(10)^{-2}$ **d.** $6(10)^{-7}$

e. $1.8(10)^{-11}$

THOUGHTS INTO WORDS

55. Describe three ways in which the graph of $f(x) = \log_3 x$ can be obtained.

56. How do we know that $\log_2 6$ is between 2 and 3?

Further Investigations

57. Graph the function $f(x) = \log_2 x^2$.

58. Graph the function $f(x) = 2\log_2 x$.

59. According to Property 4.7, $\log_2 x^2 = 2\log_2 x$. Why are the graphs for Problems 57 and 58 different?

 ## GRAPHING CALCULATOR ACTIVITIES

60. Graph $f(x) = x$, $f(x) = e^x$, and $f(x) = \ln x$ on the same set of axes.

61. Graph $f(x) = x$, $f(x) = 10^x$, and $f(x) = \log x$ on the same set of axes.

62. Graph $f(x) = \ln x$. How should the graphs of $f(x) = 2\ln x$, $f(x) = 4\ln x$, and $f(x) = 6\ln x$ compare to this basic curve? Graph the three functions on the same set of axes with the graph of $f(x) = \ln x$.

63. Graph $f(x) = \log x$. Now predict the graphs for $f(x) = 3 + \log x$, $f(x) = -2 + \log x$, and $f(x) = -4 + \log x$. Graph them on the same set of axes with the graph of $f(x) = \log x$.

For each of the following, (a) predict the general shape and location of the graph, and (b) use your graphing calculator to graph the function to check your prediction.

a. $f(x) = \log x + \ln x$ **b.** $f(x) = \log x - \ln x$

c. $f(x) = \ln x - \log x$ **d.** $f(x) = \ln x^2$

| **4.6** | EXPONENTIAL AND LOGARITHMIC EQUATIONS; PROBLEM SOLVING |

In Section 4.1 we solved exponential equations such as $3^x = 81$ by expressing both sides of the equation as a power of 3 and then applying the property *If $b^n = b^m$, then $n = m$.* However, if we try this same approach with an equation such as $3^x = 5$, we face the difficulty of expressing 5 as a power of 3. We can solve this type of problem by using the properties of logarithms and the following property of equality.

PROPERTY 4.8

If $x > 0$, $y > 0$, $b > 0$, and $b \neq 1$, then

$$x = y \qquad \text{if and only if } \log_b x = \log_b y$$

EXAMPLE 1

Property 4.8 is stated in terms of any valid base b; however, for most applications we use either common logarithms or natural logarithms. Let's consider some examples.

Solve $3^x = 5$ to the nearest hundredth.

Solution

By using common logarithms, we can proceed as follows.

$$3^x = 5$$
$$\log 3^x = \log 5 \qquad \text{Property 4.8}$$
$$x \log 3 = \log 5 \qquad \log r^p = p \log r$$
$$x = \frac{\log 5}{\log 3}$$
$$x = 1.46 \qquad \text{nearest hundredth}$$

 Check   Because $3^{1.46} \approx 4.972754647$, we say that, to the nearest hundredth, the solution set for $3^x = 5$ is $\{1.46\}$. ∎

A WORD OF CAUTION!   The expression $\dfrac{\log 5}{\log 3}$ means that we must *divide*, not subtract, the logarithms. That is, $\dfrac{\log 5}{\log 3}$ *does not* mean $\log\left(\dfrac{5}{3}\right)$. Remember that $\log\left(\dfrac{5}{3}\right) = \log 5 - \log 3$.

EXAMPLE 2

Solve $e^{x+1} = 5$ to the nearest hundredth.

Solution

Because base e is used in the exponential expression, let's use natural logarithms to help solve this equation.

$$e^{x+1} = 5$$
$$\ln e^{x+1} = \ln 5 \qquad \text{Property 4.8}$$
$$(x + 1)\ln e = \ln 5 \qquad \ln r^p = p \ln r$$
$$(x + 1)(1) = \ln 5 \qquad \ln e = 1$$
$$x = \ln 5 - 1$$
$$x = 0.61 \qquad \text{nearest hundredth}$$

The solution set is $\{0.61\}$. Check it! ∎

EXAMPLE 3

Solve $2^{3x-2} = 3^{2x+1}$ to the nearest hundredth.

Solution

$$2^{3x-2} = 3^{2x+1}$$
$$\log 2^{3x-2} = \log 3^{2x+1}$$
$$(3x - 2)\log 2 = (2x + 1)\log 3$$
$$3x \log 2 - 2 \log 2 = 2x \log 3 + \log 3$$
$$3x \log 2 - 2x \log 3 = \log 3 + 2 \log 2$$
$$x(3 \log 2 - 2 \log 3) = \log 3 + 2 \log 2$$
$$x = \frac{\log 3 + 2 \log 2}{3 \log 2 - 2 \log 3}.$$
$$x = -21.10 \qquad \text{nearest hundredth}$$

The solution set is $\{-21.10\}$. Check it! ∎

Logarithmic Equations

In Example 13 of Section 4.4, we solved the logarithmic equation

$$\log_{10} x + \log_{10}(x + 9) = 1$$

by simplifying the left side of the equation to $\log_{10}[x(x + 9)]$ and then changing the equation to exponential form to complete the solution. At this time, we can use Property 4.8 to solve this type of logarithmic equation another way, and we can also expand our equation-solving capabilities. Let's consider some examples.

EXAMPLE 4

Solve $\log x + \log(x - 15) = 2$.

Solution

Because $\log 100 = 2$, the given equation becomes

$$\log x + \log(x - 15) = \log 100$$

Now simplify the left side, apply Property 4.8, and proceed as follows.

$$\log[(x)(x - 15)] = \log 100$$
$$x(x - 15) = 100$$
$$x^2 - 15x - 100 = 0$$
$$(x - 20)(x + 5) = 0$$
$$x - 20 = 0 \qquad \text{or} \qquad x + 5 = 0$$
$$x = 20 \qquad \text{or} \qquad x = -5$$

The domain of a logarithmic function must contain only positive numbers, so x and $x - 15$ must be positive in this problem. Therefore, we discard the solution of -5, and the solution set is $\{20\}$. ∎

EXAMPLE 5

Solve $\ln(x + 2) = \ln(x - 4) + \ln 3$.

Solution

$$\ln(x + 2) = \ln(x - 4) + \ln 3$$
$$\ln(x + 2) = \ln[3(x - 4)]$$
$$x + 2 = 3(x - 4)$$
$$x + 2 = 3x - 12$$
$$14 = 2x$$
$$7 = x$$

The solution set is $\{7\}$.

■

EXAMPLE 6

Solve $\log_b(x + 2) + \log_b(2x - 1) = \log_b x$.

Solution

$$\log_b(x + 2) + \log_b(2x - 1) = \log_b x$$
$$\log_b[(x + 2)(2x - 1)] = \log_b x$$
$$(x + 2)(2x - 1) = x$$
$$2x^2 + 3x - 2 = x$$
$$2x^2 + 2x - 2 = 0$$
$$x^2 + x - 1 = 0$$

Use the quadratic formula to obtain

$$x = \frac{-1 \pm \sqrt{1 + 4}}{2}$$

$$= \frac{-1 \pm \sqrt{5}}{2}$$

Because $x + 2$, $2x - 1$, and x all have to be positive, the solution of $\left(-1 - \sqrt{5}\right)/2$ has to be discarded, and the solution set is

$$\left\{\frac{-1 + \sqrt{5}}{2}\right\}$$

■

Problem Solving

In Section 4.2 we used the compound interest formula

$$A = P\left(1 + \frac{r}{n}\right)^{nt}$$

to determine the amount of money (A) accumulated at the end of t years if P dollars is invested at rate r of interest compounded n times per year. Now let's use this formula to solve other types of problems that deal with compound interest.

PROBLEM 1

How long will it take $500 to double if it is invested at 12% compounded quarterly?

Solution

To *double* $500 means that the $500 will grow into $1000. We want to find out how long it will take; that is, what is t? Thus

$$1000 = 500\left(1 + \frac{0.12}{4}\right)^{4t}$$

$$= 500(1 + 0.03)^{4t}$$

$$= 500(1.03)^{4t}$$

Multiply both sides of $1000 = 500(1.03)^{4t}$ by $\frac{1}{500}$ to yield

$$2 = (1.03)^{4t}$$

(handwritten: $\log 2 = 4t \log 1.03$ *)*

Therefore,

$$\log 2 = \log(1.03)^{4t} \qquad \text{Property 4.8}$$
$$= 4t \log 1.03 \qquad \log r^p = p \log r$$

(handwritten: $\frac{\log 2}{\log 1.03 \cdot 4} = t$ *)*

Now let's solve for t.

$$4t \log 1.03 = \log 2$$

$$t = \frac{\log 2}{4 \log 1.03}$$

$$t = 5.9 \qquad \text{nearest tenth}$$

Therefore, we are claiming that $500 invested at 12% interest compounded quarterly will double in approximately 5.9 years.

Check $500 invested at 12% compounded quarterly for 5.9 years will produce

$$A = \$500\left(1 + \frac{0.12}{4}\right)^{4(5.9)}$$

$$= \$500(1.03)^{23.6}$$

$$= \$1004.45$$

∎

PROBLEM 2

Suppose that the number of bacteria present in a certain culture after t minutes is given by the equation $Q(t) = Q_0 e^{0.04t}$, where Q_0 represents the initial number of bacteria. How long would it take for the bacteria count to grow from 500 to 2000?

Solution

Substituting into $Q(t) = Q_0 e^{0.04t}$ and solving for t, we obtain

$$2000 = 500e^{0.04t}$$

$$4 = e^{0.04t}$$

$$\ln 4 = \ln e^{0.04t}$$

$$\ln 4 = 0.04t \ln e$$

$$\ln 4 = 0.04t \qquad \ln e = 1$$

$$\frac{\ln 4}{0.04} = t$$

$$34.7 = t \qquad \text{nearest tenth}$$

It should take approximately 34.7 minutes.

Richter Numbers

Seismologists use the Richter scale to measure and report the magnitude of earthquakes. The equation

$$R = \log \frac{I}{I_0} \qquad \text{R is called a Richter number.}$$

compares the intensity I of an earthquake to a minimal or reference intensity I_0. The reference intensity is the smallest earth movement that can be recorded on a seismograph. Suppose that the intensity of an earthquake was determined to be 50,000 times the reference intensity. In this case, $I = 50{,}000 \, I_0$ and the Richter number would be calculated as follows.

$$R = \log \frac{50{,}000 \, I_0}{I_0}$$

$$R = \log 50{,}000$$

$$R \approx 4.698970004$$

Thus a Richter number of 4.7 would be reported. Let's consider two more examples that involve Richter numbers.

PROBLEM 3

An earthquake that occurred in San Francisco in 1989 was reported to have a Richter number of 6.9. How did its intensity compare to the reference intensity?

Solution

$$6.9 = \log \frac{I}{I_0}$$

$$10^{6.9} = \frac{I}{I_0}$$

$$I = (10^{6.9})(I_0)$$

$$I \approx 7{,}943{,}282 \, I_0$$

Thus its intensity was a little less than 8 million times the reference intensity.

PROBLEM 4

An earthquake in Iran in 1990 had a Richter number of 7.7. Compare the intensity of this earthquake to that of the one in San Francisco (Problem 3).

Solution

From Problem 3 we have $I = (10^{6.9})(I_0)$ for the earthquake in San Francisco. Then, using a Richter number of 7.7, we obtain $I = (10^{7.7})(I_0)$ for the earthquake in Iran. Therefore, by comparison,

$$\frac{(10^{7.7})(I_0)}{(10^{6.9})(I_0)} = 10^{7.7-6.9} = 10^{0.8} \approx 6.3$$

The earthquake in Iran was about 6 times as intense as the one in San Francisco. ∎

Logarithms with Base Other Than 10 or e

The basic approach of applying Property 4.8 and using either common or natural logarithms can also be used to evaluate a logarithm to some base other than 10 or e. The next example illustrates this idea.

EXAMPLE 7

Evaluate $\log_3 41$.

Solution

Let $x = \log_3 41$. Changing to exponential form, we obtain

$$3^x = 41$$

Now we can apply Property 4.8.

$$\log 3^x = \log 41$$
$$x \log 3 = \log 41$$
$$x = \frac{\log 41}{\log 3}$$
$$x = 3.3802 \qquad \text{rounded to four decimal places}$$

[handwritten: $x\log 3 = \log 41$; $x = \dfrac{\log 41}{\log 3}$]

Therefore, we are claiming that 3 raised to the 3.3802 power will produce approximately 41. Check it! ∎

Using the method of Example 7 to evaluate $\log_a r$ produces the following formula, which is often referred to as the **change-of-base formula** for logarithms.

PROPERTY 4.9

If a, b, and r are positive numbers, with $a \neq 1$ and $b \neq 1$, then

$$\log_a r = \frac{\log_b r}{\log_b a}$$

Property 4.9 provides us with a convenient way to express logarithms with bases other than 10 or e in terms of common or natural logarithms. For example, $\log_3 41$ is of the form $\log_a r$ with $r = 41$ and $a = 3$. Therefore, in terms of common logarithms (base 10), we have

$$\log_3 41 = \frac{\log_{10} 41}{\log_{10} 3}$$

Using the abbreviated notation for base-10 logarithms, we have

$$\log_3 41 = \frac{\log 41}{\log 3}$$

Thus the following format could be used to evaluate $\log_3 41$.

$$\log_3 41 = \frac{\log 41}{\log 3}$$
$$= 3.3802 \qquad \text{rounded to four decimal places}$$

In a similar fashion, we can use natural logarithms to evaluate expressions such as $\log_3 41$.

$$\log_3 41 = \frac{\ln 41}{\ln 3}$$
$$= 3.3802 \qquad \text{rounded to four decimal places}$$

Property 4.9 also provides us with another way of solving equations such as $3^x = 5$.

$$3^x = 5$$
$$x = \log_3 5 \qquad \text{Changed to logarithmic form}$$
$$x = \frac{\log 5}{\log 3} \qquad \text{Applied Property 4.9}$$
$$x = 1.46 \qquad \text{to the nearest hundredth}$$

Finally, by using Property 4.9, we can obtain a relationship between common and natural logarithms by letting $a = 10$ and $b = e$. Then

$$\log_a r = \frac{\log_b r}{\log_b a}$$

becomes

$$\log_{10} r = \frac{\log_e r}{\log_e 10}$$
$$\log_e r = (\log_e 10)(\log_{10} r)$$
$$\log_e r = (2.3026)(\log_{10} r)$$

Thus the natural logarithm of any positive number is approximately equal to 2.3026 times the common logarithm of the number.

Now we can use a graphing utility to graph logarithmic functions such as $f(x) = \log_2 x$. Using the change-of-base formula, we can express this function as $f(x) = \dfrac{\log x}{\log 2}$ or as $f(x) = \dfrac{\ln x}{\ln 2}$. The graph of $f(x) = \log_2 x$ is shown in Figure 4.25.

FIGURE 4.25   ■

Finally, let's use a graphical approach to solve an equation that is cumbersome to solve with an algebraic approach.

EXAMPLE 8

Solve the equation $(5^x - 5^{-x})/2 = 3$.

Solution

First, we need to recognize that the solutions for the equation $(5^x - 5^{-x})/2 = 3$ are the x intercepts of the graph of the equation $y = (5^x - 5^{-x})/2 - 3$. Thus let's use a graphing utility to obtain the graph of this equation as shown in Figure 4.26. Use the zoom and trace features to determine that the graph crosses the x axis at approximately 1.13. Thus the solution set of the original equation is $\{1.13\}$.

FIGURE 4.26   ■

PROBLEM SET 4.6

For Problems 1–18, solve each exponential equation. Express approximate solutions to the nearest hundredth.

1. $2^x = 9$ **2.** $3^x = 20$ **3.** $5^t = 123$

4. $4^t = 12$ **5.** $2^{x+1} = 7$ **6.** $3^{x-2} = 11$

7. $7^{2t-1} = 35$ **8.** $5^{3t+1} = 9$ **9.** $e^x = 4.1$

10. $e^x = 30$ **11.** $e^{x-1} = 8.2$ **12.** $e^{x-2} = 13.1$

13. $2e^x = 12.4$ **14.** $3e^x - 1 = 17$

15. $3^{x-1} = 2^{x+3}$ **16.** $5^{2x+1} = 7^{x+3}$

17. $5^{x-1} = 2^{2x+1}$ **18.** $3^{2x+1} = 2^{3x+2}$

For Problems 19–30, solve each logarithmic equation. Express irrational solutions in simplest radical form.

19. $\log x + \log(x + 3) = 1$

20. $\log x + \log(x + 21) = 2$

21. $\log(2x - 1) - \log(x - 3) = 1$

22. $\log(3x - 1) = 1 + \log(5x - 2)$

23. $\log(x - 2) = 1 - \log(x + 3)$

24. $\log(x + 1) = \log 3 - \log(2x - 1)$

25. $\log(x + 1) - \log(x + 2) = \log \dfrac{1}{x}$

26. $\log(x + 2) - \log(2x + 1) = \log x$

27. $\ln(3t - 4) - \ln(t + 1) = \ln 2$

28. $\ln(2t + 5) = \ln 3 + \ln(t - 1)$

29. $\log(x^2) = (\log x)^2$ **30.** $\log \sqrt{x} = \sqrt{\log x}$

For Problems 31–38, evaluate each logarithm to three decimal places.

31. $\log_3 14$ **32.** $\log_4 94$ **33.** $\log_5 2.1$

34. $\log_6 0.345$ **35.** $\log_7 176$ **36.** $\log_8 296$

37. $\log_9 14.32$ **38.** $\log_7 0.024$

For Problems 39–57, solve each problem.

39. How long will it take $1000 to double if it is invested at 9% interest compounded semiannually?

40. How long will it take $750 to be worth $1000 if it is invested at 12% interest compounded quarterly?

41. How long will it take $500 to triple if it is invested at 9% interest compounded continuously?

42. How long will it take $2000 to double if it is invested at 13% interest compounded continuously?

43. At what rate of interest (to the nearest tenth of a percent) compounded annually will an investment of $200 grow to $350 in 5 years?

44. At what rate of interest (to the nearest tenth of a percent) compounded continuously will an investment of $500 grow to $900 in 10 years?

45. A piece of machinery valued at $30,000 depreciates at a rate of 10% yearly. How long will it take until the machinery has a value of $15,000?

46. For a certain strain of bacteria, the number present after t hours is given by the equation $Q = Q_0 e^{0.34t}$, where Q_0 represents the initial number of bacteria. How long will it take 400 bacteria to increase to 4000 bacteria?

47. The number of grams of a certain radioactive substance present after t hours is given by the equation $Q = Q_0 e^{-0.45t}$, where Q_0 represents the initial number of grams. How long will it take 2500 grams to be reduced to 1250 grams?

48. The atmospheric pressure in pounds per square inch is expressed by the equation $P(a) = 14.7 e^{-0.21a}$, where a is the altitude above sea level measured in miles. If the atmospheric pressure in Cheyenne, Wyoming, is approximately 11.53 pounds per square inch, find its altitude above sea level. Express your answer to the nearest hundred feet.

49. Suppose you are given the equation $P(t) = P_0 e^{0.02t}$ to predict population growth, where P_0 represents an initial population and t is the time in years. How long does this equation predict it will take a city of 50,000 to double in population?

50. In a certain bacterial culture, the equation $Q(t) = Q_0 e^{0.4t}$ yields the number of bacteria as a function of the time, where Q_0 is an initial number of bacteria and t is time measured in hours. How long will it take 500 bacteria to increase to 2000?

51. An earthquake in Los Angeles in 1971 had an intensity of approximately five million times the reference intensity. What was the Richter number associated with that earthquake?

52. An earthquake in San Francisco in 1906 was reported to have a Richter number of 8.3. How did its intensity compare to the reference intensity?

53. Calculate how many times more intense an earthquake with a Richter number of 7.3 is than an earthquake with a Richter number of 6.4.

54. Calculate how many times more intense an earthquake with a Richter number of 8.9 is than an earthquake with a Richter number of 6.2.

55. In Problem 39 of Problem Set 4.2, we used the function $f(x) = 2.512^x$, where x is the higher magnitude minus the lower magnitude, to compare the relative brightness of stars. Suppose star A is 212 times brighter than star B. Find the difference, to the nearest tenth, of their magnitudes.

56. See Problem 55. If star C has a magnitude of 7 and is 100 times brighter than star D, find the magnitude of star D. Express your answer to the nearest whole number.

57. See Problem 55. If star E is 10,000 times brighter than star F, and if star F has a magnitude of 20, find the magnitude of star E. Express your answer to the nearest whole number.

THOUGHTS INTO WORDS

58. Explain the concept of a Richter number.

59. Explain how you would solve the equation $7^x = 134$.

60. Explain how you would evaluate $\log_4 79$.

61. How do logarithms with a base of 9 compare to logarithms with a base of 3?

Further Investigations

62. Use the approach of Example 7 to develop Property 4.9.

63. Let $r = b$ in Property 4.9 and verify that $\log_a b = \dfrac{1}{\log_b a}$.

64. To solve the equation $(5^x - 5^{-x})/2 = 3$, let's begin as follows.

$$\frac{5^x - 5^{-x}}{2} = 3$$

$$5^x - 5^{-x} = 6$$

$$
\begin{array}{ll}
5^x(5^x - 5^{-x}) = 6(5^x) & \text{Multiply both} \\
5^{2x} - 1 = 6(5^x) & \text{sides by } 5^x. \\
5^{2x} - 6(5^x) - 1 = 0 &
\end{array}
$$

This final equation is of quadratic form. Finish the solution and check your answer against the answer in Example 8.

65. Solve the equation $y = (10^x + 10^{-x})/2$ for x in terms of y.

66. Solve the equation $y = (e^x - e^{-x})/2$ for x in terms of y.

GRAPHING CALCULATOR ACTIVITIES

67. Check your answers for Problems 15–18 by graphing the appropriate function and finding the x intercept.

68. Graph $f(x) = \log_2 x$. Then predict the graphs for $f(x) = \log_3 x$, $f(x) = \log_4 x$, and $\log_8 x$. Now graph these three functions on the same set of axes with the graph of $f(x) = \log_2 x$.

69. Graph $f(x) = x$, $f(x) = 2^x$, and $f(x) = \log_2 x$ on the same set of axes.

70. Graph $f(x) = x$, $f(x) = \left(\dfrac{1}{2}\right)^x$, and $f(x) = \log_{1/2} x$ on the same set of axes.

71. Use both a graphical and an algebraic approach to solve the equation $(2^x - 2^{-x})/3 = 4$.

CHAPTER 4 SUMMARY

This chapter can be summarized in terms of four main topics: (1) exponents and exponential functions, (2) inverse functions, (3) logarithms and logarithmic functions, and (4) applications of exponential and logarithmic functions.

Exponents and Exponential Functions

If a and b are positive numbers, and m and n are real numbers, then the following properties hold.

1. $b^n \cdot b^m = b^{n+m}$ Product of two powers

2. $(b^n)^m = b^{mn}$ Power of a power

3. $(ab)^n = a^n b^n$ Power of a product

4. $\left(\dfrac{a}{b}\right)^n = \dfrac{a^n}{b^n}$ Power of a quotient

5. $\dfrac{b^n}{b^m} = b^{n-m}$ Quotient of two powers

A function defined by an equation of the form

$$f(x) = b^x \qquad b > 0 \text{ and } b \neq 1$$

is called an **exponential function**. Figure 4.27 illustrates the general behavior of the graph of an exponential function of the form $f(x) = b^x$.

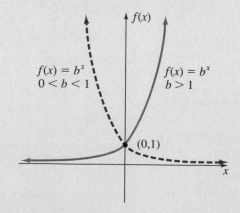

FIGURE 4.27

Inverse Functions

DEFINITION 4.2

Let f be a one-to-one function with a domain of X and a range of Y. A function g, with a domain of Y and a range of X, is called the **inverse function** of f if

$$(f \circ g)(x) = x \qquad \text{for every } x \text{ in } Y$$

and

$$(g \circ f)(x) = x \qquad \text{for every } x \text{ in } X$$

The inverse of a function f is denoted by f^{-1}. Graphically, two functions that are inverses of each other are mirror images with reference to the line $y = x$.

A systematic technique for finding the inverse of a function can be described as follows.

1. Let $y = f(x)$.

2. Interchange x and y.

3. Solve the equation for y in terms of x.

4. The inverse function $f^{-1}(x)$ is determined by the equation in step 3.

Don't forget that the domain of f must equal the range of f^{-1}, and the domain of f^{-1} must equal the range of f.

Increasing and decreasing functions are defined as follows.

DEFINITION 4.3

Let f be a function, with the interval I a subset of the domain of f. Let x_1 and x_2 be in I. Then

1. f is *increasing on I* if $f(x_1) < f(x_2)$ whenever $x_1 < x_2$.

2. f is *decreasing on I* if $f(x_1) > f(x_2)$ whenever $x_1 < x_2$, and

3. f is *constant on I* if $f(x_1) = f(x_2)$ for every x_1 and x_2.

A function that is always increasing or always decreasing over its entire domain is a one-to-one function and therefore has an inverse.

Logarithms and Logarithmic Functions

If r is any positive real number, then the unique exponent t such that $b^t = r$ is called the **logarithm of r with base b**; it is denoted by $\log_b r$.

The following properties of logarithms are used frequently.

1. $\log_b b = 1$

2. $\log_b 1 = 0$

3. $b^{\log_b r} = r$

4. $\log_b rs = \log_b r + \log_b s$

5. $\log_b \left(\dfrac{r}{s}\right) = \log_b r - \log_b s$

6. $\log_b(r^p) = p \log_b r$

Logarithms with a base of 10 are called **common logarithms**. The expression $\log_{10} x$ is usually written $\log x$.

Many calculators are equipped with a common logarithm function. Often a key labeled $\boxed{\log}$ is used to find common logarithms.

Natural logarithms are logarithms that have a base of e, where e is an irrational number whose decimal approximation to eight digits is 2.7182818. Natural logarithms are denoted by $\log_e x$ or $\ln x$.

Many calculators are also equipped with a natural logarithmic function. Often a key labeled $\boxed{\ln x}$ is used for this purpose.

A function defined by an equation of the form

$$f(x) = \log_b x \qquad b > 0 \text{ and } b \neq 1$$

is called a **logarithmic function**.

The graph of a logarithmic function (such as $y = \log_2 x$) can be determined by changing the equation to exponential form ($2^y = x$) and plotting points, or by reflecting the graph of the inverse function ($y = 2^x$) across the line $y = x$. This last approach is based on the fact that exponential and logarithmic functions are inverses of each other.

Figure 4.28 illustrates the general behavior of the graph of a logarithmic function of the form $f(x) = \log_b x$.

FIGURE 4.28

Applications

We use the following properties of equality frequently when solving exponential and logarithmic equations.

> **1.** If $b > 0$, $b \neq 1$, and m and n are real numbers, then
>
> $$b^n = b^m \qquad \text{if and only if } n = m$$
>
> **2.** If $x > 0$, $y > 0$, $b > 0$, and $b \neq 1$, then
>
> $$x = y \qquad \text{if and only if } \log_b x = \log_b y$$

A general formula for any principal (P) that is compounded n times per year for any number (t) of years at a given rate (r) is

$$A = P\left(1 + \frac{r}{n}\right)^{nt}$$

where A represents the total amount of money accumulated at the end of the t years.

As n gets infinitely large, the value of $[1 + (1/n)]^n$ approaches the number e, where e equals 2.71828 to five decimal places.

The formula

$$A = Pe^{rt}$$

yields the accumulated value (A) of a sum of money (P) that has been invested for t years at a rate of r percent **compounded continuously**.

The formula

$$Q = Q_0\left(\frac{1}{2}\right)^{t/h}$$

is referred to as the **half-life formula**.

The equation

$$Q(t) = Q_0 e^{kt}$$

is used as a mathematical model for exponential growth and decay problems.

The formula

$$R = \log \frac{I}{I_0}$$

yields the Richter number associated with an earthquake.

The formula

$$\log_a r = \frac{\log_b r}{\log_b a}$$

is often called the **change-of-base formula**.

CHAPTER 4 REVIEW PROBLEM SET

For Problems 1–10, evaluate each expression.

1. $8^{5/3}$

2. $-25^{3/2}$

3. $(-27)^{4/3}$

4. $\log_6 216$

5. $\log_7 \left(\dfrac{1}{49} \right)$

6. $\log_2 \sqrt[3]{2}$

7. $\log_2 \left(\dfrac{\sqrt[4]{32}}{2} \right)$

8. $\log_{10} 0.00001$

9. $\ln e$

10. $7^{\log_7 12}$

For Problems 11–24, solve each equation. Express approximate solutions to the nearest hundredth.

11. $\log_{10} 2 + \log_{10} x = 1$

12. $\log_3 x = -2$

13. $4^x = 128$

14. $3^t = 42$

15. $\log_2 x = 3$

16. $\left(\dfrac{1}{27} \right)^{3x} = 3^{2x-1}$

17. $2e^x = 14$

18. $2^{2x+1} = 3^{x+1}$

19. $\ln(x + 4) - \ln(x + 2) = \ln x$

20. $\log x + \log(x - 15) = 2$

21. $\log(\log x) = 2$

22. $\log(7x - 4) - \log(x - 1) = 1$

23. $\ln(2t - 1) = \ln 4 + \ln(t - 3)$

24. $64^{2t+1} = 8^{-t+2}$

For Problems 25–28, if $\log 3 = 0.4771$ and $\log 7 = 0.8451$, evaluate each of the following.

25. $\log \left(\dfrac{7}{3} \right)$

26. $\log 21$

27. $\log 27$

28. $\log 7^{2/3}$

29. Express each of the following as the sum or difference of simpler logarithmic quantities. Assume that all variables represent positive real numbers.

a. $\log_b \left(\dfrac{x}{y^2} \right)$

b. $\log_b \sqrt[4]{xy^2}$

c. $\log_b \left(\dfrac{\sqrt{x}}{y^3} \right)$

30. Express each of the following as a single logarithm. Assume that all variables represent positive real numbers.

a. $3 \log_b x + 2 \log_b y$ **b.** $\dfrac{1}{2} \log_b y - 4 \log_b x$

c. $\dfrac{1}{2}(\log_b x + \log_b y) - 2 \log_b z$

For Problems 31–34, approximate each of the logarithms to three decimal places.

31. $\log_2 3$ **32.** $\log_3 2$

33. $\log_4 191$ **34.** $\log_2 0.23$

For Problems 35–42, graph each function.

35. $f(x) = \left(\dfrac{3}{4}\right)^x$ **36.** $f(x) = 2^{x+2}$

37. $f(x) = e^{x-1}$ **38.** $f(x) = -1 + \log x$

39. $f(x) = 3^x - 3^{-x}$ **40.** $f(x) = e^{-x^2/2}$

41. $f(x) = \log_2(x - 3)$ **42.** $f(x) = 3 \log_3 x$

For Problems 43–45, use the compound interest formula

$$A = P\left(1 + \frac{r}{n}\right)^{nt}$$

to find the total amount of money accumulated at the end of the indicated time period for each of the investments.

43. $750 for 10 years at 11% compounded quarterly

44. $1250 for 15 years at 9% compounded monthly

45. $2500 for 20 years at 9.5% compounded semiannually

For Problems 46–49, determine whether f and g are inverse functions.

46. $f(x) = 7x - 1$ and $g(x) = \dfrac{x + 1}{7}$

47. $f(x) = -\dfrac{2}{3}x$ and $g(x) = \dfrac{3}{2}x$

48. $f(x) = x^2 - 6$ for $x \geq 0$ and $g(x) = \sqrt{x + 6}$ for $x \geq -6$

49. $f(x) = 2 - x^2$ for $x \geq 0$ and $g(x) = \sqrt{2 - x}$ for $x \leq 2$

For Problems 50–53, (a) find f^{-1} and (b) verify that $(f \circ f^{-1})(x) = x$ and $(f^{-1} \circ f)(x) = x$.

50. $f(x) = 4x + 5$ **51.** $f(x) = -3x - 7$

52. $f(x) = \dfrac{5}{6}x - \dfrac{1}{3}$

53. $f(x) = -2 - x^2$ for $x \geq 0$

For Problems 54 and 55, find the intervals on which the function is increasing and the intervals on which it is decreasing.

54. $f(x) = -2x^2 + 16x - 35$

55. $f(x) = 2\sqrt{x - 3}$

For Problems 56–62, solve each problem.

56. How long will it take $100 to double if it is invested at 14% interest compounded annually?

57. How long will it take $1000 to be worth $3500 if it is invested at 10.5% interest compounded quarterly?

58. At what rate of interest (to the nearest tenth of a percent) compounded continuously will an investment of $500 grow to $1000 in 8 years?

59. Suppose that the present population of a city is 50,000 and suppose that the equation $P(t) = P_0 e^{0.02t}$, where P_0 represents an initial population, can be used to estimate future populations. Estimate the population of that city in 10 years, 15 years, and 20 years.

60. The number of bacteria present in a certain culture after t hours is given by the equation $Q = Q_0 e^{0.29t}$, where Q_0 represents the initial number of bacteria. How long will it take 500 bacteria to increase to 2000 bacteria?

61. Suppose that a certain radioactive substance has a half-life of 40 days. If there are presently 750 grams of the substance, how much, to the nearest gram, will remain after 100 days?

62. An earthquake occurred in Mexico City in 1985 that had an intensity level about 125,000,000 times the reference intensity. Find the Richter number for that earthquake.

CHAPTER 4 TEST

For Problems 1–4, evaluate each expression.

1. $\log_3 \sqrt{3}$ **2.** $\log_2 (\log_2 4)$ **3.** $-2 + \ln e^3$ **4.** $\log_2 (0.5)$

For Problems 5–10, solve each equation.

5. $4^x = \dfrac{1}{64}$ **6.** $9^x = \dfrac{1}{27}$ **7.** $2^{3x-1} = 128$ **8.** $\log_9 x = \dfrac{5}{2}$

9. $\log x + \log (x + 48) = 2$ **10.** $\ln x = \ln 2 + \ln(3x - 1)$

For Problems 11–13, given that $\log_3 4 = 1.2619$ and $\log_3 5 = 1.4650$, evaluate each expression.

11. $\log_3 100$ **12.** $\log_3 1.25$ **13.** $\log_3 \sqrt{5}$

14. Solve $e^x = 176$ to the nearest hundredth.

15. Solve $2^{x-2} = 314$ to the nearest hundredth.

16. Determine $\log_5 632$ to four decimal places.

17. Find the inverse of the function $f(x) = -3x - 6$.

18. Find the inverse of the function $f(x) = \dfrac{2}{3}x - \dfrac{3}{5}$.

19. Are $f(x) = \dfrac{1}{2}x + 3$ and $g(x) = 2x - 6$ inverses of each other?

For Problems 20–23, solve each problem.

20. If \$3500 is invested at 7.5% interest compounded quarterly, how much money has accumulated at the end of 8 years?

21. How long will it take \$5000 to be worth \$12,500 if it is invested at 7% compounded annually? Express your answer to the nearest tenth of a year.

22. The number of bacteria present in a certain culture after t hours is given by $Q(t) = Q_0 e^{0.23t}$, where Q_0 represents the initial number of bacteria. How long will it take 400 bacteria to increase to 2400 bacteria? Express your answer to the nearest tenth of an hour.

23. Suppose that a certain radioactive substance has a half-life of 50 years. If there are presently 7500 grams of the substance, how much will remain after 32 years? Express your answer to the nearest gram.

For Problems 24 and 25, graph each of the functions.

24. $f(x) = e^x - 2$ **25.** $f(x) = \log_2 (x - 2)$

5

POLYNOMIAL AND RATIONAL FUNCTIONS

Many problems that involve maximum and minimum values can be solved by using polynomial and rational functions.

In earlier chapters we solved linear and quadratic equations and graphed linear and quadratic functions. In this chapter we will expand our equation-solving processes and graphing techniques to include more general polynomial equations and functions. Our knowledge of polynomial functions will then allow us to work with rational functions; *the function concept will unify the chapter. To facilitate our study in this chapter, we will first review the concept of dividing polynomials; then we will introduce a special division technique called* synthetic division.

5.1 DIVIDING POLYNOMIALS

In Chapter 0 we used the properties

$$\frac{a+b}{c} = \frac{a}{c} + \frac{b}{c} \quad \text{and} \quad \frac{a-b}{c} = \frac{a}{c} - \frac{b}{c}$$

as a basis for dividing a polynomial by a monomial. For example,

$$\frac{18x^3 + 24x^2}{6x} = \frac{18x^3}{6x} + \frac{24x^2}{6x} = 3x^2 + 4x$$

and

$$\frac{35x^2y^3 - 55x^3y^4}{5xy^2} = \frac{35x^2y^3}{5xy^2} - \frac{55x^3y^4}{5xy^2} = 7xy = 11x^2y^2$$

You may recall from a previous algebra course that the format used to divide a polynomial by a binomial resembles the long-division format in arithmetic. Let's work through an example step by step.

STEP 1 Use the conventional long-division format and arrange both the dividend and the divisor in descending powers of the variable.

$$3x + 1 \overline{\smash{)}3x^3 - 5x^2 + 10x + 1}$$

STEP 2 Find the first term of the quotient by dividing the first term of the dividend by the first term of the divisor.

$$\begin{array}{r} x^2 \\ 3x + 1 \overline{\smash{)}3x^3 - 5x^2 + 10x + 1} \end{array}$$

STEP 3 Multiply the entire divisor by the quotient term in step 2 and place this product in position to be subtracted from the dividend.

$$\begin{array}{r} x^2 \\ 3x + 1 \overline{)3x^3 - 5x^2 + 10x + 1} \\ \underline{3x^3 + x^2} \end{array}$$

STEP 4 Subtract.

$$\begin{array}{r} x^2 \\ 3x + 1 \overline{)3x^3 - 5x^2 + 10x + 1} \\ \underline{3x^3 + x^2} \\ -6x^2 + 10x + 1 \end{array}$$

STEP 5 Repeat steps 2, 3, and 4 and use $-6x^2 + 10x + 1$ as a new dividend.

$$\begin{array}{r} x^2 - 2x \\ 3x + 1 \overline{)3x^3 - 5x^2 + 10x + 1} \\ \underline{3x^3 + x^2} \\ -6x^2 + 10x + 1 \\ \underline{-6x^2 - 2x} \\ 12x + 1 \end{array}$$

STEP 6 Repeat steps 2, 3, and 4 and use $12x + 1$ as a new dividend.

$$\begin{array}{r} x^2 - 2x + 4 \\ 3x + 1 \overline{)3x^3 - 5x^2 + 10x + 1} \\ \underline{3x^3 + x^2} \\ -6x^2 + 10x + 1 \\ \underline{-6x^2 - 2x} \\ 12x + 1 \\ \underline{12x + 4} \\ -3 \end{array}$$

Therefore, $3x^3 - 5x^2 + 10x + 1 = (3x + 1)(x^2 - 2x + 4) + (-3)$, which is of the familiar form

Dividend = (divisor)(quotient) + remainder

This result is commonly called the **division algorithm for polynomials**, which can be stated in general terms as follows.

Division Algorithm for Polynomials

If $f(x)$ and $g(x)$ are polynomials and $g(x) \neq 0$, then unique polynomials $q(x)$ and $r(x)$ exist such that

$$f(x) = g(x)q(x) + r(x)$$

Dividend   Divisor   Quotient   Remainder

where $r(x) = 0$ or the degree of $r(x)$ is less than the degree of $g(x)$.

Let's consider one more example to illustrate this division process further.

EXAMPLE I

Divide $t^2 - 3t + 2t^4 - 1$ by $t^2 + 4t$.

Solution

Don't forget to arrange both the dividend and the divisor in descending powers of the variable.

$$
\begin{array}{r}
2t^2 - 8t + 33 \\
t^2 + 4t{\overline{\smash{\big)}\,2t^4 + 0t^3 + t^2 - 3t - 1}} \\
\underline{2t^4 + 8t^3} \\
-8t^3 + t^2 - 3t - 1 \\
\underline{-8t^3 - 32t^2} \\
33t^2 - 3t - 1 \\
\underline{33t^2 + 132t} \\
-135t - 1
\end{array}
$$

⟵ Notice the insertion of a t^3 term with a zero coefficient.

The division process is completed when the degree of the remainder is less than the degree of the divisor. ∎

Synthetic Division

If the divisor is of the form $x - c$, where c is a constant, then the typical long-division algorithm can be simplified to a process called **synthetic division**. First, let's consider another division problem and use the regular-division algorithm. Then, in a step-by-step fashion, we will demonstrate some shortcuts that will lead us into the synthetic-division procedure. Consider the division problem $(2x^4 + x^3 - 17x^2 + 13x + 2) \div (x - 2)$.

$$
\begin{array}{r}
2x^3 + 5x^2 - 7x - 1 \\
x - 2{\overline{\smash{\big)}\,2x^4 + x^3 - 17x^2 + 13x + 2}} \\
\underline{2x^4 - 4x^3} \\
5x^3 - 17x^2 \\
\underline{5x^3 - 10x^2} \\
-7x^2 + 13x \\
\underline{-7x^2 + 14x} \\
-x + 2 \\
\underline{-x + 2}
\end{array}
$$

Because the dividend is written in descending powers of x, the quotient is produced in descending powers of x. In other words, the numerical coefficients are the *key issues*, so let's rewrite the problem in terms of its coefficients.

$$
\begin{array}{r}
\ 2 \quad\ 5 \quad -7 \quad -1 \\
\hline
1-2)\,2 \quad\ \ 1 \quad -17 \quad\ 13 \quad 2 \\
\textcircled{2} \quad -4 \\
\hline
5 \quad \textcircled{-17} \\
5 \quad -10 \\
\hline
-7 \quad \textcircled{13} \\
\textcircled{-7} \quad 14 \\
\hline
-1 \ \textcircled{2} \\
\textcircled{-1}\ 2
\end{array}
$$

Now observe that the circled numbers are simply repetitions of the numbers directly above them in the format. Thus the circled numbers can be omitted, and the format will be as follows (disregard the arrows for the moment).

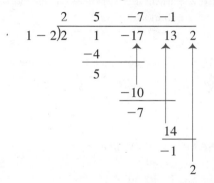

Next, by moving some numbers up (indicated by the arrows) and by not writing the 1 that is the coefficient of x in the divisor, we obtain the following more compact form.

$$
\begin{array}{ll}
\ \ \ \ \ 2 \quad\ 5 \quad -7 \quad -1 & \textbf{(1)} \\
\hline
-2)\,2 \quad\ 1 \quad -17 \quad 13 \quad 2 & \textbf{(2)} \\
\ \ \ \ \ \ \ -4 \quad -10 \quad 14 \quad 2 & \textbf{(3)} \\
\hline
\ \ \ \ \ \ \ \ \ 5 \quad -7 \quad -1 & \textbf{(4)}
\end{array}
$$

Note that line 4 reveals all of the coefficients of the quotient (line 1) except for the first coefficient, 2. Thus we can omit line 1, begin line 4 with the first coefficient, and then use the following form.

$$
\begin{array}{ll}
-2|\,2 \quad\ 1 \quad -17 \quad 13 \quad 2 & \textbf{(5)} \\
\ \ \ \ \ \ \ -4 \quad -10 \quad 14 \quad 2 & \textbf{(6)} \\
\hline
\ \ \ 2 \quad\ 5 \quad -7 \quad -1 \quad 0 & \textbf{(7)}
\end{array}
$$

Line 7 contains the coefficients of the quotient, where the zero indicates the remainder. Finally, by changing the constant in the divisor to 2 (instead of -2), which changes the signs of the numbers in line 6, we can *add* the corresponding entries in lines 5 and 6 rather than subtract. Thus the final synthetic division form for this problem is

$$2 \underline{|2 \quad 1 \quad -17 \quad 13 \quad 2}$$
$$\quad \quad 4 \quad 10 \quad -14 \quad -2$$
$$\overline{2 \quad 5 \quad -7 \quad -1 \quad 0}$$

Now we will consider another problem and indicate a step-by-step procedure for setting up and carrying out the synthetic-division process. Suppose that we want to do the following division problem.

$$x + 4\overline{)2x^3 + 5x^2 - 13x - 2}$$

STEP 1   Write the coefficients of the dividend as follows.

$$\overline{)2 \quad 5 \quad -13 \quad -2}$$

STEP 2   In the divisor, use -4 instead of 4 so that later we can add rather than subtract.

$$-4\overline{)2 \quad 5 \quad -13 \quad -2}$$

STEP 3   Bring down the first coefficient of the dividend.

$$-4\overline{)2 \quad 5 \quad -13 \quad -2}$$
$$\overline{2}$$

STEP 4   Multiply that first coefficient times the divisor, which yields $2(-4) = -8$; add this result to the second coefficient of the dividend.

$$-4\overline{)2 \quad 5 \quad -13 \quad -2}$$
$$\quad \quad 2 \quad -8$$
$$\overline{2 \quad -3}$$

STEP 5   Multiply $(-3)(-4)$, which yields 12; add this result to the third coefficient of the dividend.

$$-4\overline{)2 \quad 5 \quad -13 \quad -2}$$
$$\quad \quad \quad -8 \quad 12$$
$$\overline{2 \quad -3 \quad -1}$$

STEP 6   Multiply $(-1)(-4)$, which yields 4; add this result to the last term of the dividend.

$$-4\overline{)2 \quad 5 \quad -13 \quad -2}$$
$$\quad \quad \quad -8 \quad 12 \quad 4$$
$$\overline{2 \quad -3 \quad -1 \quad 2}$$

The last row indicates a quotient of $2x^2 - 3x - 1$ and a remainder of 2.

Now let's consider some examples in which we show only the final compact form of synthetic division.

E X A M P L E 2

Find the quotient and remainder for $(x^3 + 8x^2 + 13x - 6) \div (x + 3)$.

Solution

$$
\begin{array}{r|rrrr}
-3) & 1 & 8 & 13 & -6 \\
    &   & -3 & -15 & 6 \\
\hline
    & 1 & 5 & -2 & 0
\end{array}
$$

Thus the quotient is $x^2 + 5x - 2$ and the remainder is zero. ■

E X A M P L E 3

Find the quotient and the remainder for $(3x^4 + 5x^3 - 29x^2 - 45x + 14) \div (x - 3)$.

Solution

$$
\begin{array}{r|rrrrr}
3) & 3 & 5 & -29 & -45 & 14 \\
   &   & 9 & 42 & 39 & -18 \\
\hline
   & 3 & 14 & 13 & -6 & -4
\end{array}
$$

Thus the quotient is $3x^3 + 14x^2 + 13x - 6$ and the remainder is -4. ■

E X A M P L E 4

Find the quotient and the remainder for $(4x^4 - 2x^3 + 6x - 1) \div (x - 1)$.

Solution

$$
\begin{array}{r|rrrrr}
1) & 4 & -2 & 0 & 6 & -1 \\
   &   & 4 & 2 & 2 & 8 \\
\hline
   & 4 & 2 & 2 & 8 & 7
\end{array}
$$

Note that a zero has been inserted as the coefficient of the missing x^2 term.

Thus the quotient is $4x^3 + 2x^2 + 2x + 8$ and the remainder is 7. ■

E X A M P L E 5

Find the quotient and the remainder for $(x^4 + 16) \div (x + 2)$.

Solution

$$
\begin{array}{r|rrrrr}
-2) & 1 & 0 & 0 & 0 & 16 \\
    &   & -2 & 4 & -8 & 16 \\
\hline
    & 1 & -2 & 4 & -8 & 32
\end{array}
$$

Note that zeros have been inserted as coefficients of the missing terms in the dividend.

Thus the quotient is $x^3 - 2x^2 + 4x - 8$ and the remainder is 32. ■

PROBLEM SET 5.1

For Problems 1–14, find the quotient and remainder for each division problem.

1. $(12x^2 + 7x - 10) \div (3x - 2)$

2. $(20x^2 - 39x + 18) \div (5x - 6)$

3. $(3t^3 + 7t^2 - 10t - 4) \div (3t + 1)$

4. $(4t^3 - 17t^2 + 7t + 10) \div (4t - 5)$

5. $(6x^2 + 19x + 11) \div (3x + 2)$

6. $(20x^2 + 3x - 1) \div (5x + 2)$

7. $(3x^3 + 2x^2 - 5x - 1) \div (x^2 + 2x)$

8. $(4x^3 - 5x^2 + 2x - 6) \div (x^2 - 3x)$

9. $(5y^3 - 6y^2 - 7y - 2) \div (y^2 - y)$

10. $(8y^3 - y^2 - y + 5) \div (y^2 + y)$

11. $(4a^3 - 2a^2 + 7a - 1) \div (a^2 - 2a + 3)$

12. $(5a^3 + 7a^2 - 2a - 9) \div (a^2 + 3a - 4)$

13. $(3x^2 - 2xy - 8y^2) \div (x - 2y)$

14. $(4a^2 - 8ab + 4b^2) \div (a - b)$

For Problems 15–38, use *synthetic division* to determine the quotient and remainder for each division problem.

15. $(3x^2 + x - 4) \div (x - 1)$

16. $(2x^2 - 5x - 3) \div (x - 3)$

17. $(x^2 + 2x - 10) \div (x - 4)$

18. $(x^2 - 10x + 15) \div (x - 8)$

19. $(4x^2 + 5x - 4) \div (x + 2)$

20. $(5x^2 + 18x - 8) \div (x + 4)$

21. $(x^3 - 2x^2 - x + 2) \div (x - 2)$

22. $(x^3 - 5x^2 + 2x + 8) \div (x + 1)$

23. $(3x^4 - x^3 + 2x^2 - 7x - 1) \div (x + 1)$

24. $(2x^3 - 5x^2 - 4x + 6) \div (x - 2)$

25. $(x^3 - 7x - 6) \div (x + 2)$

26. $(x^3 + 6x^2 - 5x - 1) \div (x - 1)$

27. $(x^4 + 4x^3 - 7x - 1) \div (x - 3)$

28. $(2x^4 + 3x^2 + 3) \div (x + 2)$

29. $(x^3 + 6x^2 + 11x + 6) \div (x + 3)$

30. $(x^3 - 4x^2 - 11x + 30) \div (x - 5)$

31. $(x^5 - 1) \div (x - 1)$

32. $(x^5 - 1) \div (x + 1)$

33. $(x^5 + 1) \div (x - 1)$

34. $(x^5 + 1) \div (x + 1)$

35. $(2x^3 + 3x^2 - 2x + 3) \div \left(x + \dfrac{1}{2}\right)$

36. $(9x^3 - 6x^2 + 3x - 4) \div \left(x - \dfrac{1}{3}\right)$

37. $(4x^4 - 5x^2 + 1) \div \left(x - \dfrac{1}{2}\right)$

38. $(3x^4 - 2x^3 + 5x^2 - x - 1) \div \left(x + \dfrac{1}{3}\right)$

THOUGHTS INTO WORDS

39. How would you describe synthetic division to someone who had just completed an elementary algebra course?

40. Why is synthetic division restricted to situations where the divisor is of the form $x - c$?

5.2   REMAINDER AND FACTOR THEOREMS

Let's consider the division algorithm (stated in the previous section) when the dividend, $f(x)$, is divided by a *linear polynomial* of the form $x - c$. Then the division algorithm,

$$f(x) = g(x)q(x) + r(x)$$

Dividend   Divisor   Quotient   Remainder

becomes

$$f(x) = (x - c)q(x) + r(x)$$

Because the degree of the remainder, $r(x)$, must be less than the degree of the divisor, $x - c$, the remainder is a constant. Therefore, letting R represent the remainder, we have

$$f(x) = (x - c)q(x) + R$$

If we evaluate f at c, we obtain

$$\begin{aligned}
f(c) &= (c - c)q(c) + R \\
&= 0 \cdot q(c) + R \\
&= R
\end{aligned}$$

In other words, if a polynomial is divided by a linear polynomial of the form $x - c$, then the remainder is the value of the polynomial at c. Let's state this more formally as the **remainder theorem**.

PROPERTY 5.1   *Remainder Theorem*

If a polynomial $f(x)$ is divided by $x - c$, then the remainder is equal to $f(c)$.

EXAMPLE 1

If $f(x) = x^3 + 2x^2 - 5x - 1$, find $f(2)$ first (a) by using synthetic division and the remainder theorem and then (b) by evaluating $f(2)$ directly.

Solutions

a.
$$\begin{array}{r|rrrr}
2 & 1 & 2 & -5 & -1 \\
  &   & 2 & 8 & 6 \\
\hline
  & 1 & 4 & 3 & \boxed{5}
\end{array}$$
$\longleftarrow R = f(2)$

b. $f(2) = 2^3 + 2(2)^2 - 5(2) - 1 = 8 + 8 - 10 - 1 = 5$

EXAMPLE 2

If $f(x) = x^4 + 7x^3 + 8x^2 + 11x + 5$, find $f(-6)$ first (a) by using synthetic division and the remainder theorem and then (b) by evaluating $f(-6)$ directly.

Solutions

a.

$$
\begin{array}{r|rrrrr}
-6) & 1 & 7 & 8 & 11 & 5 \\
    &   & -6 & -6 & -12 & 6 \\
\hline
    & 1 & 1 & 2 & -1 & \text{⑪} \leftarrow R = f(-6)
\end{array}
$$

b. $f(-6) = (-6)^4 + 7(-6)^3 + 8(-6)^2 + 11(-6) + 5$
$$= 1296 - 1512 + 288 - 66 + 5 = 11$$ ∎

In Example 2, note that finding $f(-6)$ by synthetic division and the remainder theorem involves easier computation than evaluating $f(-6)$ directly. This is often the case.

EXAMPLE 3

Find the remainder when $x^3 + 3x^2 - 13x - 15$ is divided by $x + 1$.

Solution

Let $f(x) = x^3 + 3x^2 - 13x - 15$ and write $x + 1$ as $x - (-1)$ so that we can apply the remainder theorem.

$$f(-1) = (-1)^3 + 3(-1)^2 - 13(-1) - 15 = 0$$

Thus the remainder is zero. ∎

Example 3 illustrates an important special case of the remainder theorem in which the remainder is *zero*. In this case, we say that $x + 1$ is a **factor** of $x^3 + 3x^2 - 13x - 15$.

Factor Theorem

A general *factor theorem* can be formulated by considering the equation

$$f(x) = (x - c)q(x) + R$$

If $x - c$ is a factor of $f(x)$, then the remainder R, which is also $f(c)$, must be zero. Conversely, if $R = f(c) = 0$, then $f(x) = (x - c)q(x)$; in other words, $x - c$ is a factor of $f(x)$. The **factor theorem** can be stated as follows.

PROPERTY 5.2   *Factor Theorem*

A polynomial $f(x)$ has a factor $x - c$ if and only if $f(c) = 0$.

EXAMPLE 4

Is $x - 1$ a factor of $x^3 + 5x^2 + 2x - 8$?

Solution

Let $f(x) = x^3 + 5x^2 + 2x - 8$ and compute $f(1)$ to obtain

$$f(1) = 1^3 + 5(1)^2 + 2(1) - 8 = 0$$

Therefore, by the factor theorem, $x - 1$ is a factor of $f(x)$.   ■

EXAMPLE 5

Is $x + 3$ a factor of $2x^3 + 5x^2 - 6x - 7$?

Solution

Using synthetic division, we obtain

$$-3\overline{)\begin{array}{rrrr} 2 & 5 & -6 & -7 \\ & -6 & 3 & 9 \\ \hline 2 & -1 & -3 & ②\end{array}} \longleftarrow R = f(-3)$$

Because $f(-3) \neq 0$, we know that $x + 3$ is not a factor of the given polynomial.   ■

In Examples 4 and 5, we were concerned only with determining whether a linear polynomial of the form $x - c$ was a factor of another polynomial. For such problems, it is reasonable to compute $f(c)$ either directly or by synthetic division, whichever way seems easier. However, if more information is required, such as complete factorization of the given polynomial, then using synthetic division becomes appropriate, as in the next two examples.

EXAMPLE 6

Show that $x - 1$ is a factor of $x^3 - 2x^2 - 11x + 12$ and find the other linear factors of the polynomial.

Solution

Let's use synthetic division to divide $x^3 - 2x^2 - 11x + 12$ by $x - 1$.

$$1\overline{)\begin{array}{rrrr} 1 & -2 & -11 & 12 \\ & 1 & -1 & -12 \\ \hline 1 & -1 & -12 & 0 \end{array}}$$

The last line indicates a quotient of $x^2 - x - 12$ and a remainder of zero. The zero remainder means that $x - 1$ is a factor. Furthermore, we can write

$$x^3 - 2x^2 - 11x + 12 = (x - 1)(x^2 - x - 12)$$

We can factor the quadratic polynomial $x^2 - x - 12$ as $(x - 4)(x + 3)$ by using our conventional factoring techniques. Thus we obtain

$$x^3 - 2x^2 - 11x + 12 = (x - 1)(x - 4)(x + 3)$$ ■

EXAMPLE 7

Show that $x + 4$ is a factor of $f(x) = x^3 - 5x^2 - 22x + 56$ and complete the factorization of $f(x)$.

Solution

We use synthetic division to divide $x^3 - 5x^2 - 22x + 56$ by $x + 4$.

$$
\begin{array}{r|rrrr}
-4) & 1 & -5 & -22 & 56 \\
   &   & -4 & 36 & -56 \\
\hline
   & 1 & -9 & 14 & 0 \\
\end{array}
$$

The last line indicates a quotient of $x^2 - 9x + 14$ and a remainder of zero. The zero remainder means that $x + 4$ is a factor. Furthermore, we can write

$$x^3 - 5x^2 - 22x + 56 = (x + 4)(x^2 - 9x + 14)$$

and then complete the factoring to obtain

$$f(x) = x^3 - 5x^2 - 22x + 56 = (x + 4)(x - 7)(x - 2)$$ ∎

The factor theorem also plays a significant role in determining some general factorization ideas, such as the last example of this section illustrates.

EXAMPLE 8

Verify that $x + 1$ is a factor of $x^n + 1$ whenever n is an odd positive integer.

Solution

Let $f(x) = x^n + 1$ and compute $f(-1)$ to obtain

$$
\begin{aligned}
f(-1) &= (-1)^n + 1 \\
      &= -1 + 1 \qquad \text{Any odd power of } -1 \text{ is } -1. \\
      &= 0
\end{aligned}
$$

Because $f(-1) = 0$, we know that $x + 1$ is a factor of $f(x)$. ∎

PROBLEM SET 5.2

For Problems 1–10, find $f(c)$ (a) by using synthetic division and the remainder theorem and (b) by evaluating $f(c)$ directly.

1. $f(x) = x^2 + x - 8$ and $c = 2$

2. $f(x) = x^3 + x^2 - 2x - 4$ and $c = -1$

3. $f(x) = 3x^3 + 4x^2 - 5x + 3$ and $c = -4$

4. $f(x) = 2x^4 + x^2 + 6$ and $c = 1$

5. $f(x) = x^4 - 2x^3 - 3x^2 + 5x - 1$ and $c = -2$

6. $f(x) = 2x^4 + x^3 - 4x^2 - x + 1$ and $c = 2$

7. $f(t) = 6t^3 - 35t^2 + 8t - 10$ and $c = 6$

8. $f(t) = 2t^5 - 1$ and $c = -2$

9. $f(n) = 3n^4 - 2n^3 + 4n - 1$ and $c = 3$

10. $f(n) = -2n^4 + 4n - 5$ and $c = -3$

For Problems 11–18, find $f(c)$ *either* by using synthetic division and the remainder theorem or by evaluating $f(c)$ directly.

11. $f(x) = 5x^6 - x^3 - 1$ and $c = -1$

12. $f(x) = 2x^3 - 3x^2 - 5x + 4$ and $c = 4$

13. $f(x) = x^4 - 8x^3 + 9x^2 - 15x + 2$ and $c = 7$

14. $f(t) = 5t^3 - 8t^2 + 9t - 4$ and $c = -5$

15. $f(n) = -2n^4 + 2n^2 - n - 5$ and $c = -2$

16. $f(x) = 4x^7 + 3$ and $c = 3$

17. $f(x) = 2x^3 - 5x^2 + 4x - 3$ and $c = \dfrac{1}{2}$

18. $f(x) = 3x^3 + 4x^2 - 5x - 7$ and $c = -\dfrac{1}{3}$

For Problems 19–28, use the factor theorem to help answer each question about factors.

19. Is $x - 2$ a factor of $3x^2 - 4x - 4$?

20. Is $x + 3$ a factor of $6x^2 + 13x - 15$?

21. Is $x + 2$ a factor of $x^3 + x^2 - 7x - 10$?

22. Is $x - 3$ a factor of $2x^3 - 3x^2 - 10x + 3$?

23. Is $x - 1$ a factor of $3x^3 + 5x^2 - x - 2$?

24. Is $x + 4$ a factor of $x^3 - 4x^2 + 2x - 8$?

25. Is $x - 2$ a factor of $x^3 - 8$?

26. Is $x + 2$ a factor of $x^3 + 8$?

27. Is $x - 3$ a factor of $x^4 - 81$?

28. Is $x + 3$ a factor of $x^4 - 81$?

For Problems 29–34, use synthetic division to show that $g(x)$ is a factor of $f(x)$ and complete the factorization of $f(x)$.

29. $g(x) = x + 2$; $f(x) = x^3 + 7x^2 + 4x - 12$

30. $g(x) = x - 1$; $f(x) = 3x^3 + 19x^2 - 38x + 16$

31. $g(x) = x - 3$; $f(x) = 6x^3 - 17x^2 - 5x + 6$

32. $g(x) = x + 2$; $f(x) = 12x^3 + 29x^2 + 8x - 4$

33. $g(x) = x + 1$; $f(x) = x^3 - 2x^2 - 7x - 4$

34. $g(x) = x - 5$; $f(x) = 2x^3 + x^2 - 61x + 30$

For Problems 35–38, find the value(s) of k that make(s) the second polynomial a factor of the first.

35. $x^3 - kx^2 + 5x + k$; $x - 2$

36. $k^2x^4 + 3kx^2 - 4$; $x - 1$

37. $x^3 + 4x^2 - 11x + k$; $x + 2$

38. $kx^3 + 19x^2 + x - 6$; $x + 3$

39. Show that $x + 2$ is a factor of $x^{12} - 4096$.

40. Argue that $f(x) = 2x^4 + x^2 + 3$ has no factor of the form $x - c$, where c is a real number.

41. Verify that $x - 1$ is a factor of $x^n - 1$ for all positive integral values of n.

42. Verify that $x + 1$ is a factor of $x^n - 1$ for all even positive integral values of n.

43. a. Verify that $x - y$ is a factor of $x^n - y^n$ whenever n is a positive integer.

   b. Verify that $x + y$ is a factor of $x^n - y^n$ whenever n is an even positive integer.

   c. Verify that $x + y$ is a factor of $x^n + y^n$ whenever n is an odd positive integer.

THOUGHTS INTO WORDS

44. In your own words, explain how the remainder theorem is used to prove the factor theorem.

45. It is sometimes said that the factor theorem is a special case of the remainder theorem. What does this statement mean?

Further Investigations

The remainder and factor theorems are true for any complex value of c. Therefore, for Problems 46–48, find $f(c)$ (a) by using synthetic division and the remainder theorem and (b) by evaluating $f(c)$ directly.

46. $f(x) = x^3 - 5x^2 + 2x + 1$ and $c = i$

47. $f(x) = x^2 + 4x - 2$ and $c = 1 + i$

48. $f(x) = x^3 + 2x^2 + x - 2$ and $c = 2 - 3i$

For Problems 49 and 50, solve each problem.

49. Show that $x - 2i$ is a factor of $f(x) = x^4 + 6x^2 + 8$.

50. Show that $x + 3i$ is a factor of $f(x) = x^4 + 14x^2 + 45$.

51. Consider changing the form of the polynomial $f(x) = x^3 + 4x^2 - 3x + 2$ as follows.

$$f(x) = x^3 + 4x^2 - x + 2$$
$$= (x^2 + 4x - 3)x + 2$$
$$= [x(x + 4) - 3]x + 2$$

The final form, $f(x) = [x(x + 4) - 3]x + 2$, is called the **nested form** of the polynomial. It is particularly well suited to evaluating functional values of f either by hand or with a calculator.

For each of the following, find the indicated functional values, using the nested form of the given polynomial.

a. $f(4)$, $f(-5)$, and $f(7)$ for $f(x) = x^3 + 5x^2 - 2x + 1$

b. $f(3)$, $f(6)$, and $f(-7)$ for $f(x) = 2x^3 - 4x^2 - 3x + 2$

c. $f(4)$, $f(5)$, and $f(-3)$ for $f(x) = -2x^3 + 5x^2 - 6x - 7$

d. $f(5)$, $f(6)$, and $f(-3)$ for $f(x) = x^4 + 3x^3 - 2x^2 + 5x - 1$

5.3   POLYNOMIAL EQUATIONS

In Chapter 1 we solved a large variety of *linear equations* of the form $ax + b = 0$ and *quadratic equations* of the form $ax^2 + bx + c = 0$. Linear and quadratic equations are special cases of a general class of equations we refer to as **polynomial equations**. The equation

$$a_n x^n + a_{n-1} x^{n-1} + \cdots + a_1 x + a_0 = 0$$

where the coefficients $a_0, a_1, \ldots a_n$ are real numbers and n is a positive integer, is called a **polynomial equation of degree** n. The following are examples of polynomial equations.

$$\sqrt{2}x - 6 = 0 \qquad \text{Degree 1}$$

$$\frac{3}{4}x^2 - \frac{2}{3}x + 5 = 0 \qquad \text{Degree 2}$$

$$4x^3 - 3x^2 - 7x - 9 = 0 \qquad \text{Degree 3}$$

$$5x^4 - x + 6 = 0 \qquad \text{Degree 4}$$

REMARK   The most general polynomial equation allows complex numbers as coefficients. However, for our purposes in this text, we will restrict the coefficients to real numbers. We refer to such equations as **polynomial equations over the reals**.

In general, solving polynomial equations of degree greater than 2 can be very difficult and often requires mathematics beyond the scope of this text. However, there are some general methods for solving polynomial equations that you should know, because there are certain types of polynomial equations that we *can* solve with the techniques available to us at this time.

Let's begin by listing some (previously encountered) polynomial equations and their solution sets.

Equation	Solution Set
$3x + 4 = 7$	$\{1\}$
$x^2 + x - 6 = 0$	$\{-3, 2\}$
$2x^3 - 3x^2 - 2x + 3 = 0$	$\left\{-1, 1, \dfrac{3}{2}\right\}$
$x^4 - 16 = 0$	$\{-2, 2, -2i, 2i\}$

Note that in each of these examples, the number of solutions corresponds to the degree of the equation. The first-degree equation has one solution, the second-degree equation has two solutions, the third-degree equation has three solutions, and the fourth-degree equation has four solutions. Now consider the equation

$$(x - 4)^2(x + 5)^3 = 0$$

It can be written

$$(x - 4)(x - 4)(x + 5)(x + 5)(x + 5) = 0$$

which implies that

$$x - 4 = 0 \quad \text{or} \quad x - 4 = 0 \quad \text{or} \quad x + 5 = 0 \quad \text{or}$$
$$x + 5 = 0 \quad \text{or} \quad x + 5 = 0$$

Therefore,

$$x = 4 \quad \text{or} \quad x = 4 \quad \text{or} \quad x = -5 \quad \text{or} \quad x = -5 \quad \text{or}$$
$$x = -5$$

We say that the solution set of the original equation is $\{-5, 4\}$, but we also say that the equation has a solution of 4 with a **multiplicity of two** and a solution of -5 with a **multiplicity of three**. Furthermore, note that the sum of the multiplicities is 5, which agrees with the degree of the equation.

We can state the following general property.

PROPERTY 5.3

A polynomial equation of degree n has n solutions, where any solution of multiplicity p is counted p times.

Finding Rational Solutions

As we stated earlier, solving polynomial equations of degree greater than 2 can be very difficult. However, *rational solutions* of polynomial equations with integral coefficients can be found by using techniques from this chapter. The following property restricts the possible rational solutions of such an equation.

PROPERTY 5.4 *Rational Root Theorem*

Consider the polynomial equation

$$a_n x^n + a_{n-1} x^{n-1} + \cdots + a_1 x + a_0 = 0$$

where the coefficients $a_0, a_1, \ldots, a_n$ are integers. If the rational number c/d, reduced to lowest terms, is a solution of the equation, then c is a factor of the constant term a_0, and d is a factor of the leading coefficient a_n.

The *why* behind the rational root theorem is based on some simple factoring ideas, as indicated by the following outline of a proof for the theorem.

Outline of Proof

If c/d is to be a solution, then

$$a_n\left(\frac{c}{d}\right)^n + a_{n-1}\left(\frac{c}{d}\right)^{n-1} + \cdots + a_1\left(\frac{c}{d}\right) + a_0 = 0$$

Multiply both sides of this equation by d^n and then add $-a_0 d^n$ to both sides.

$$a_n c^n + a_{n-1} c^{n-1} d + \cdots + a_1 c d^{n-1} = -a_0 d^n$$

Because c is a factor of the left side of this equation, c must also be a factor of $-a_0 d^n$. Furthermore, because c/d is in reduced form, c and d have no common factors other than -1 or 1. Thus c must be a factor of a_0. In the same way, from the equation

$$a_{n-1} c^{n-1} d + \cdots + a_1 c d^{n-1} + a_0 d^n = -a_n c^n$$

we can conclude that d is a factor of the left side and that therefore d is also a factor of a_n. ∎

The rational root theorem, synthetic division, the factor theorem, and some previous knowledge about solving linear and quadratic equations all merge to form a basis for finding rational solutions. Let's consider some examples.

EXAMPLE 1

Find all rational solutions of $3x^3 + 8x^2 - 15x + 4 = 0$.

Solution

If c/d is a rational solution, then c must be a factor of 4 and d must be a factor of 3. Therefore, the possible values for c and d are as follows.

For c	$\pm 1, \pm 2, \pm 4$
For d	$\pm 1, \pm 3$

Thus the possible values for c/d are

$$\pm 1, \pm\frac{1}{3}, \pm 2, \pm\frac{2}{3}, \pm 4, \pm\frac{4}{3}$$

By using synthetic division, we can test $x - 1$.

$$
\begin{array}{r|rrrr}
1) & 3 & 8 & -15 & 4 \\
   &   & 3 & 11 & -4 \\
\hline
   & 3 & 11 & -4 & 0
\end{array}
$$

This shows that $x - 1$ is a factor of the given polynomial; therefore, 1 is a rational solution of the equation. Furthermore, the synthetic-division result also indicates how to factor the given polynomial.

$$3x^3 + 8x^2 - 15x + 4 = 0$$
$$(x - 1)(3x^2 + 11x - 4) = 0$$

The quadratic factor can be further factored by using techniques we are familiar with.

$$(x - 1)(3x^2 + 11x - 4) = 0$$
$$(x - 1)(3x - 1)(x + 4) = 0$$

$$x - 1 = 0 \quad \text{or} \quad 3x - 1 = 0 \quad \text{or} \quad x + 4 = 0$$

$$x = 1 \quad \text{or} \quad x = \frac{1}{3} \quad \text{or} \quad x = -4$$

Thus the entire solution set consists of rational numbers and can be listed as $\left\{-4, \frac{1}{3}, 1\right\}$. ∎

In Example 1, we were fortunate that the first time we used synthetic division, we got a rational solution. But this often does not happen, and then we need to conduct a little organized search, as the next example illustrates.

EXAMPLE 2

Find all rational solutions of $3x^3 + 7x^2 - 22x - 8 = 0$.

Solution

If c/d is a rational solution, then c must be a factor of -8 and d must be a factor of 3. Therefore, the possible values for c and d are as follows.

For c	$\pm 1, \pm 2, \pm 4, \pm 8$
For d	$\pm 1, \pm 3$

Thus the possible values for c/d are

$$\pm 1, \ \pm \frac{1}{3}, \ \pm 2, \ \pm \frac{2}{3}, \ \pm 4, \ \pm \frac{4}{3}, \ \pm 8, \ \pm \frac{8}{3}$$

Let's begin our search for rational solutions by trying the integers first.

$$
\begin{array}{r|rrrr}
1) & 3 & 7 & -22 & -8 \\
  &   & 3 & 10 & -12 \\
\hline
  & 3 & 10 & -12 & \boxed{-20} \\
\end{array}
$$

This indicates that $x - 1$ is not a factor and thus 1 is not a solution.

$$
\begin{array}{r|rrrr}
-1) & 3 & 7 & -22 & -8 \\
   &   & -3 & -4 & 26 \\
\hline
   & 3 & 4 & -26 & \boxed{18} \\
\end{array}
$$

This indicates that -1 is not a solution.

$$
\begin{array}{r|rrrr}
2) & 3 & 7 & -22 & -8 \\
  &   & 6 & 26 & 8 \\
\hline
  & 3 & 13 & 4 & 0 \\
\end{array}
$$

Now we know that $x - 2$ is a factor, and we can proceed as follows.

$$3x^3 + 7x^2 - 22x - 8 = 0$$
$$(x - 2)(3x^2 + 13x + 4) = 0$$
$$(x - 2)(3x + 1)(x + 4) = 0$$

$$
\begin{array}{cccccc}
x - 2 = 0 & \text{or} & 3x + 1 = 0 & \text{or} & x + 4 = 0 \\
x = 2 & \text{or} & 3x = -1 & \text{or} & x = -4 \\
x = 2 & \text{or} & x = -\dfrac{1}{3} & \text{or} & x = -4 \\
\end{array}
$$

The solution set is $\left\{ -4, \ -\dfrac{1}{3}, \ 2 \right\}$. ■

In Examples 1 and 2, we were solving third-degree equations. Therefore, once we found one linear factor by synthetic division, we were able to factor the remaining quadratic factor in the usual way. However, if the given equation is of degree 4 or more, then we may need to find more than one linear factor by synthetic division, as in the next example.

EXAMPLE 3

Solve $x^4 - 6x^3 + 22x^2 - 30x + 13 = 0$.

Solution

The possible values for c/d are ± 1 and ± 13. By synthetic division we test 1.

$$
\begin{array}{r|rrrrr}
1) & 1 & -6 & 22 & -30 & 13 \\
  &   & 1 & -5 & 17 & -13 \\
\hline
  & 1 & -5 & 17 & -13 & 0 \\
\end{array}
$$

This indicates that $x - 1$ is a factor of the given polynomial. The bottom line of the synthetic division indicates that the given polynomial can now be factored as follows.

$$x^4 - 6x^3 + 22x^2 - 30x + 13 = 0$$
$$(x - 1)(x^3 - 5x^2 + 17x - 13) = 0$$

Therefore,

$$x - 1 = 0 \quad \text{or} \quad x^3 - 5x^2 + 17x - 13 = 0$$

Now we can use the same approach to look for rational solutions of $x^3 - 5x^2 + 17x - 13 = 0$. The possible values for c/d are, again, ± 1 and ± 13. By synthetic division we test 1 again.

$$
\begin{array}{r|rrrr}
1 & 1 & -5 & 17 & -13 \\
  &   & 1 & -4 & 13 \\
\hline
  & 1 & -4 & 13 & 0
\end{array}
$$

This indicates that $x - 1$ is also a factor of $x^3 - 5x^2 + 17x - 13$, and the other factor is $x^2 - 4x + 13$. Now we can solve the original equation.

$$x^4 - 6x^3 + 22x^2 - 30x + 13 = 0$$
$$(x - 1)(x^3 - 5x^2 + 17x - 13) = 0$$
$$(x - 1)(x - 1)(x^2 - 4x + 13) = 0$$

$$x - 1 = 0 \quad \text{or} \quad x - 1 = 0 \quad \text{or} \quad x^2 - 4x + 13 = 0$$
$$x = 1 \quad \text{or} \quad x = 1 \quad \text{or} \quad x^2 - 4x + 13 = 0$$

Use the quadratic formula on $x^2 - 4x + 13 = 0$ to produce

$$x = \frac{4 \pm \sqrt{16 - 52}}{2} = \frac{4 \pm \sqrt{-36}}{2} = \frac{4 \pm 6i}{2} = 2 \pm 3i$$

Thus the original equation has a rational solution of 1 with a multiplicity of two and two complex solutions, $2 + 3i$ and $2 - 3i$. We list the solution set as $\{1, 2 \pm 3i\}$.

■

Example 3 illustrates two general properties. First, note that the coefficient of x^4 is 1, which forces the possible rational solutions to be integers. In general, the **possible rational solutions of $x^n + a_{n-1}x^{n-1} + \cdots + a_1x + a_0 = 0$ are the integral factors of a_0.** Second, note that the complex solutions of Example 3 are conjugates of each other. The following general property can be stated.

PROPERTY 5.5

If a polynomial equation with real coefficients has any nonreal complex solutions, they must occur in conjugate pairs.

REMARK The justification for Property 5.5 is based on some properties of conjugates that were presented in Problem 79 of Problem Set 0.8. We will not show the details of such a proof at this time.

Each of Properties 5.3, 5.4, and 5.5 yields some information about the solutions of a polynomial equation. Before we state one more property that will give us some additional information, we need to illustrate two ideas.

In a polynomial that is arranged in descending powers of x, if two successive terms differ in sign, we say that there is a **variation in sign**. Terms with zero coefficients are disregarded when counting sign variations. For example, the polynomial

$$+3x^3 - 2x^2 + 4x + 7$$

has *two* sign variations, whereas the polynomial

$$+x^5 - 4x^3 + x - 5$$

has *three* variations.

Another idea that we need to understand is the fact that the solutions of

$$a_n(-x)^n + a_{n-1}(-x)^{n-1} + \cdots + a_1(-x) + a_0 = 0$$

are the opposites of the solutions of

$$a_n x^n + a_{n-1}x^{n-1} + \cdots + a_1 x + a_0 = 0$$

In other words, if a new equation is formed by replacing x with $-x$ in a given equation, then the solutions of the new equation are the opposites of the solutions of the original equation. For example, the solution set of $x^2 + 7x + 12 = 0$ is $\{-4, -3\}$; the solution set of $(-x)^2 + 7(-x) + 12 = 0$, which simplifies to $x^2 - 7x + 12 = 0$, is $\{3, 4\}$.

Now we can state a property that can help us to determine the nature of the solutions of a polynomial equation without actually solving the equation.

PROPERTY 5.6 *Descartes' Rule of Signs*

Let $a_n x^n + a_{n-1}x^{n-1} + \cdots + a_1 x + a_0 = 0$ be a polynomial equation with real coefficients.

1. The number of **positive real solutions** of the given equation is either equal to the number of variations in sign of the polynomial or less than that number of variations by a positive even integer.

2. The number of **negative real solutions** of the given equation is either equal to the number of variations in sign of the polynomial $a_n(-x)^n + a_{n-1}(-x)^{n-1} + \cdots + a_1(-x) + a_0$ or less than that number of variations by a positive even integer.

Property 5.6, along with Properties 5.3 and 5.5, allows us to acquire some information about the solutions of a polynomial equation without actually solving the equation. Let's consider some equations and indicate how much we know about their solutions without solving them.

$$x^3 + 3x^2 + 5x + 4 = 0$$

1. No variations of sign in $x^3 + 3x^2 + 5x + 4$ means that there are *no positive solutions*.
2. Replace x with $-x$ in the given polynomial to produce $(-x)^3 + 3(-x)^2 + 5(-x) + 4$, which simplifies to $-x^3 + 3x^2 - 5x + 4$. This polynomial contains three variations of sign; thus there are *three or one negative solution(s)*.

Conclusion The given equation has either three negative real solutions or one negative real solution and two nonreal complex solutions.

$$2x^4 + 3x^2 - x - 1 = 0$$

1. There is one variation of sign in the given polynomial; thus the equation has *one positive solution*.
2. Replace x with $-x$ to produce $2(-x)^4 + 3(-x)^2 - (-x) - 1$, which simplifies to $2x^4 + 3x^2 + x - 1$ and contains one variation of sign. Thus the given equation has *one negative solution*.

Conclusion The given equation has one positive, one negative, and two nonreal complex solutions.

$$3x^4 + 2x^2 + 5 = 0$$

1. No variations of sign in the given polynomial means that there are *no positive solutions*.
2. Replace x with $-x$ to produce $3(-x)^4 + 2(-x)^2 + 5$, which simplifies to $3x^4 + 2x^2 + 5$ and still contains no variations of sign. Thus there are *no negative solutions*.

Conclusion The given equation contains four nonreal complex solutions. We also know that these solutions will appear in conjugate pairs.

$$2x^5 - 4x^3 + 2x - 5 = 0$$

1. Three variations of sign in the given polynomial imply that *the number of positive solutions is three or one.*

2. Replace x with $-x$ to produce $2(-x)^5 - 4(-x)^3 + 2(-x) - 5 = -2x^5 + 4x^3 - 2x - 5$, which contains two variations of sign. Thus *the number of negative solutions is two or zero.*

Conclusion   The given equation has

three positive and two negative solutions, or

three positive and two nonreal complex solutions, or

one positive, two negative, and two nonreal complex solutions, or

one positive and four nonreal complex solutions

It should be evident from the previous discussions that sometimes we can truly pinpoint the nature of the solutions of a polynomial equation. However, for some equations (such as the last example), if we use the properties discussed in this section, the best that we can do is to restrict the nature of the solutions to a few possibilities.

Finally, we need to realize that some of the properties presented in these last two sections help us to determine polynomial equations with specified roots. Let's consider some examples.

EXAMPLE 4

Find a polynomial equation with integral coefficients that has the given numbers as solutions and the indicated degree.

a. $1, \dfrac{1}{2}, -2$; degree 3   **b.** 2 of multiplicity four; degree 4

c. $1 + i, -3i$; degree 4

Solution

a. If $1, \dfrac{1}{2}$, and -2 are solutions, then $(x - 1), \left(x - \dfrac{1}{2}\right)$, and $(x + 2)$ are factors of the polynomial. Thus the following third-degree polynomial equation can be formed.

$$(x - 1)\left(x - \frac{1}{2}\right)(x + 2) = 0$$
$$(x - 1)(2x - 1)(x + 2) = 0$$
$$2x^3 + x^2 - 5x + 2 = 0$$

b. If 2 is to be a solution with multiplicity four, then the equation $(x - 2)^4 = 0$ can be formed. Using the binomial expansion pattern, we can express the equation as follows.

$$(x - 2)^4 = 0$$
$$x^4 - 8x^3 + 24x^2 - 32x + 16 = 0$$

c. By Property 5.5, if $1 + i$ is a solution, then so is $1 - i$. Likewise, because $-3i$ is a solution, so is $3i$. Therefore, we can form the following equation.

$$[x - (1 + i)][x - (1 - i)](x + 3i)(x - 3i) = 0$$
$$[(x - 1) - i][(x - 1) + i](x^2 + 9) = 0$$
$$[(x - 1)^2 - i^2](x^2 + 9) = 0$$
$$(x^2 - 2x + 1 + 1)(x^2 + 9) = 0$$
$$x^4 - 2x^3 + 11x^2 - 18x + 18 = 0 \qquad \blacksquare$$

A graphing utility can be very helpful when solving polynomial equations, especially if they are of degree greater than 2. Even the search for possible rational solutions can be simplified by looking at a graph. To find the rational solutions of $3x^3 + 8x^2 - 15x + 4 = 0$ (Example 1), we could begin by graphing the equation $y = 3x^3 + 8x^2 - 15x + 4$. This graph is shown in Figure 5.1. From the graph it looks as if 1 and -4 are two of the x intercepts and therefore solutions of the original equation. Let's check them in the equation.

$$3(1)^3 + 8(1)^2 - 15(1) + 4 = 3 + 8 - 15 + 4 = 0$$
$$3(-4)^3 + 8(-4)^2 - 15(-4) + 4 = -192 + 128 + 60 + 4 = 0$$

FIGURE 5.1

Thus $x - 1$ and $x + 4$ are factors of $3x^3 + 8x^2 - 15x + 4$ and the remaining factor could be found by division. We could then determine the solution set as we did in Example 1. Let's consider an example where we use a graphing utility to approximate the real number solutions of a polynomial equation.

Find the real number solutions of the equation $x^4 - 2x^3 - 5 = 0$.

EXAMPLE 5

Solution

Let's use a graphing utility to get a sketch of the graph of $y = x^4 - 2x^3 - 5$ (Figure 5.2). From this graph we see that one x intercept is between -1 and -2 and another is between 2 and 3. We can use the zoom and trace features to approximate these values at -1.2 and 2.4, to the nearest tenth. Thus the real solutions for the equation $x^4 - 2x^3 - 5 = 0$ are approximately -1.2 and 2.4. (The other two solutions must be conjugate complex numbers.)

FIGURE 5.2

PROBLEM SET 5.3

For Problems 1–20, use the rational root theorem and the factor theorem to help solve each equation. Be sure that the number of solutions for each equation agrees with Property 5.3; take into account the multiplicity of solutions.

1. $x^3 + x^2 - 4x - 4 = 0$

2. $x^3 - 2x^2 - 11x + 12 = 0$

3. $6x^3 + x^2 - 10x + 3 = 0$

4. $8x^3 - 2x^2 - 41x - 10 = 0$

5. $3x^3 + 13x^2 - 52x + 28 = 0$

6. $15x^3 + 14x^2 - 3x - 2 = 0$

7. $x^3 - 2x^2 - 7x - 4 = 0$

8. $x^3 - x^2 - 8x + 12 = 0$

9. $x^4 - 4x^3 - 7x^2 + 34x - 24 = 0$

10. $x^4 + 4x^3 - x^2 - 16x - 12 = 0$

11. $x^3 - 10x - 12 = 0$ 12. $x^3 - 4x^2 + 8 = 0$

13. $3x^4 - x^3 - 8x^2 + 2x + 4 = 0$

14. $2x^4 + 3x^3 - 11x^2 - 9x + 15 = 0$

15. $6x^4 - 13x^3 - 19x^2 + 12x = 0$

16. $x^3 - x^2 + x - 1 = 0$

17. $x^4 - 3x^3 + 2x^2 + 2x - 4 = 0$

18. $x^4 + x^3 - 3x^2 - 17x - 30 = 0$

19. $2x^5 - 5x^4 + x^3 + x^2 - x + 6 = 0$

20. $4x^4 + 12x^3 + x^2 - 12x + 4 = 0$

For Problems 21–26, verify that each equation has no rational solutions.

21. $x^4 - x^3 - 8x^2 - 3x + 1 = 0$

22. $x^4 + 3x - 2 = 0$

23. $2x^4 - 3x^3 + 6x^2 - 24x + 5 = 0$

24. $3x^4 - 4x^3 - 10x^2 + 3x - 4 = 0$

25. $x^5 - 2x^4 + 3x^3 + 4x^2 + 7x - 1 = 0$

26. $x^5 + 2x^4 - 2x^3 + 5x^2 - 2x - 3 = 0$

27. The rational root theorem pertains to polynomial equations with integral coefficients. However, if the coefficients are nonintegral rational numbers, we can first apply the *multiplication property of equality* to produce an equivalent equation with integral coefficients. Use this method to solve each of the following equations.

a. $\frac{1}{10}x^3 + \frac{1}{2}x^2 + \frac{1}{5}x - \frac{4}{5} = 0$ $x^3 + 5x^2 + 2x - 8 = 0$

b. $\frac{1}{10}x^3 + \frac{1}{5}x^2 - \frac{1}{2}x - \frac{3}{5} = 0$

c. $x^3 + \frac{9}{2}x^2 - x - 12 = 0$

d. $x^3 - \frac{5}{6}x^2 - \frac{22}{3}x + \frac{5}{2} = 0$

For Problems 28–37, use Descartes' rule of signs (Property 5.6) to determine the possibilities for the nature of the solutions for each of the equations. *Do not solve the equations.*

28. $6x^2 + 7x - 20 = 0$

29. $8x^2 - 14x + 3 = 0$

30. $2x^3 + x - 3 = 0$

31. $4x^3 + 3x + 7 = 0$

32. $3x^3 - 2x^2 + 6x + 5 = 0$

33. $4x^3 + 5x^2 - 6x - 2 = 0$

34. $x^5 - 3x^4 + 5x^3 - x^2 + 2x - 1 = 0$

35. $2x^5 + 3x^3 - x + 1 = 0$

36. $x^5 + 32 = 0$

37. $2x^6 + 3x^4 - 2x^2 - 1 = 0$

For Problems 38–47, find a polynomial equation with integral coefficients that has the given numbers as solutions and the indicated degree.

38. $2, 4, -3$;   degree 3

39. $1, -1, 2, -4$;   degree 4

40. $-2, \frac{1}{2}, \frac{2}{3}$;   degree 3

41. $3, -\frac{2}{3}, \frac{3}{4}$;   degree 3

42. 1 of multiplicity 5;   degree 5

43. -3 of multiplicity 4;   degree 4

44. $3, 2 + 3i$;   degree 3

45. $-2, 1 - 4i$;   degree 3

46. $1 - i, 2i$;   degree 4

47. $-2 + 3i, -i$;   degree 4

THOUGHTS INTO WORDS

48. Explain the concept of *multiplicity of roots* of an equation.

49. How would you defend the statement that the equation $2x^4 + 3x^3 + x^2 + 5 = 0$ has no positive solutions? Does it have any negative solutions? Defend your answer.

50. How do we know by inspection that the equation $2x^4 + 3x^2 + 6 = 0$ has no real number solutions?

Further Investigations

51. Use the rational root theorem to argue that $\sqrt{2}$ is not a rational number. [*Hint:* The solutions of $x^2 - 2 = 0$ are $\pm\sqrt{2}$.]

52. Use the rational root theorem to argue that $\sqrt{12}$ is not a rational number.

53. Defend the following statement. *Every polynomial equation of odd degree with real coefficients has at least one real number solution.*

54. The following synthetic division shows that 2 is a solution of $x^4 + x^3 + x^2 - 9x - 10 = 0$.

$$
\begin{array}{r|rrrrr}
2 & 1 & 1 & 1 & -9 & -10 \\
  &   & 2 & 6 & 14 & 10 \\
\hline
  & 1 & 3 & 7 & 5 & 0 \longleftarrow
\end{array}
$$

Note that the new quotient row (indicated by the arrow) consists entirely of nonnegative numbers. This indicates that searching for solutions greater than 2 would be a waste of time, because larger divisors would continue to increase each of the numbers (except the 1 on the far left) in the new quotient row. (Try 3 as a divisor!) Thus we say that 2 is an **upper bound** for the real number solutions of the given equation.

Now consider the following synthetic division, which shows that -1 is also a solution of $x^4 + x^3 + x^2 - 9x - 10 = 0$.

$$
\begin{array}{r|rrrrr}
-1) & 1 & 1 & 1 & -9 & -10 \\
    &   & -1 & 0 & -1 & 10 \\
\hline
    & 1 & 0 & 1 & -10 & 0 \longleftarrow
\end{array}
$$

The new quotient row (indicated by the arrow) shows that there is no need to look for solutions less than -1, because any divisor less than -1 would increase the size (in absolute value) of each number in the new quotient row (except the 1 on the far left). (Try -2 as a divisor!) Thus we say that -1 is a **lower bound** for the real number solutions of the given equation.

The following general property can be stated: If $a_n + a_{n-1}x^{n-1} + \cdots + a_1x + a_0 = 0$ is a polynomial equation with real coefficients, where $a_n > 0$, and if the polynomial is divided synthetically by $x - c$, then:

1. If $c > 0$ and all numbers in the new quotient row of the synthetic division are nonnegative, then c is an upper bound for the real number solutions of the given equation.

2. If $c < 0$ and the numbers in the new quotient row alternate in sign (with 0 considered either positive or negative, as needed), then c is a lower bound for the real number solutions of the given equation.

Find the smallest positive integer and the largest negative integer that are upper and lower bounds, respectively, for the real number solutions of each of the following equations. Keep in mind that the integers that serve as bounds do not necessarily have to be solutions of the equation.

a. $x^3 - 3x^2 + 25x - 75 = 0$

b. $x^3 + x^2 - 4x - 4 = 0$

c. $x^4 + 4x^3 - 7x^2 - 22x + 24 = 0$

d. $3x^3 + 7x^2 - 22x - 8 = 0$

e. $x^4 - 2x^3 - 9x^2 + 2x + 8 = 0$

 GRAPHING CALCULATOR ACTIVITIES

55. Suppose that we want to solve the equation $x^3 + 2x^2 - 14x - 40 = 0$. Let's graph the function $f(x) = x^3 + 2x^2 - 14x - 40$. The graph has only one x intercept, so the equation must have one real number solution and two nonreal complex solutions. The graph also indicates that the real solution is approximately 4. We can determine that 4 is a solution, and then we can proceed to solve the equation using the ideas of this section.

Solve each of the following equations, using a graphing calculator whenever it seems to be helpful. Express all irrational solutions in lowest radical form.

a. $x^3 + 2x^2 - 14x - 40 = 0$

b. $x^3 + x^2 - 7x + 65 = 0$

c. $x^4 - 6x^3 - 6x^2 + 32x + 24 = 0$

d. $x^4 + 3x^3 - 39x^2 + 11x + 24 = 0$

e. $x^3 - 14x^2 + 26x - 24 = 0$

f. $x^4 + 2x^3 - 3x^2 - 4x + 4 = 0$

56. Use a graphing calculator to help determine the nature of the solutions for each of the following equations. You may also need to use the property stated in Problem 54.

a. $2x^3 - 3x^2 - 3x + 2 = 0$

b. $3x^3 + 7x^2 + 8x + 2 = 0$

c. $2x^4 + 3x^2 + 1 = 0$

d. $4x^5 - 8x^4 - 5x^3 + 10x^2 + x - 2 = 0$

e. $x^4 - x^3 + 2x^2 - x - 1 = 0$

f. $x^5 - x^4 + x^3 - x^2 + x - 3 = 0$

g. $x^4 - 14x^3 + 23x^2 + 14x - 24 = 0$

h. $x^3 + 13x^2 - 28x + 30 = 0$

57. Find approximations, to the nearest hundredth, of the real number solutions of each of the following equations.

a. $x^2 - 4x + 1 = 0$

b. $3x^3 - 2x^2 + 12x - 8 = 0$

c. $x^4 - 8x^3 + 14x^2 - 8x + 13 = 0$

d. $x^4 + 6x^3 - 10x^2 - 22x + 161 = 0$

e. $7x^5 - 5x^4 + 35x^3 - 25x^2 + 28x - 20 = 0$

5.4 GRAPHING POLYNOMIAL FUNCTIONS

Just as we have a vocabulary to deal with linear, quadratic, and polynomial equations, we also have terms that classify functions. In Chapter 3 we defined a **linear function** by means of the equation

$$f(x) = ax + b$$

and a **quadratic function** by means of the equation

$$f(x) = ax^2 + bx + c$$

Both of these are special cases of a general class of functions called **polynomial functions**. Any function of the form

$$f(x) = a_n x^n + a_{n-1} x^{n-1} + \cdots + a_1 x + a_0$$

is called a **polynomial function of degree** n, where a_n is a nonzero real number; $a_{n-1}, \ldots, a_1, a_0$ are real numbers; and n is a nonnegative integer. The following are examples of polynomial functions.

$f(x) = 5x^3 - 2x^2 + x - 4$	Degree 3
$f(x) = -2x^4 - 5x^3 + 3x^2 + 4x - 1$	Degree 4
$f(x) = 3x^5 + 2x^2 - 3$	Degree 5

REMARK Our previous work with polynomial equations is sometimes presented as *finding zeros of polynomial functions*. The *solutions*, or *roots*, of a polynomial equation are also called the **zeros** of the polynomial function. For example, -2 and 2 are solutions of $x^2 - 4 = 0$ and they are zeros of $f(x) = x^2 - 4$. That is, $f(-2) = 0$ and $f(2) = 0$.

For a complete discussion of graphing polynomial functions, we would need some tools from calculus. However, the graphing techniques that we have discussed so far enable us to graph certain kinds of polynomial functions. For example, polynomial functions of the form

$$f(x) = ax^n$$

are quite easy to graph. We know from our previous work that if $n = 1$, then functions such as $f(x) = 2x$, $f(x) = -3x$, and $f(x) = \frac{1}{2}x$ are lines through the origin that have slopes 2, -3, and $\frac{1}{2}$, respectively.

Furthermore, if $n = 2$, then we know that the graphs of functions of the form $f(x) = ax^2$ are parabolas that are symmetric with respect to the y axis and have vertices at the origin.

We have also previously graphed the special case of $f(x) = ax^n$, where $a = 1$ and $n = 3$—namely, the function $f(x) = x^3$. This graph is shown in Figure 5.3.

From our work with transformations of graphs in Section 3.4, we know that the graphs of functions of the form $f(x) = ax^3$, where $a > 1$, are vertical stretchings of $f(x) = x^3$ and can be easily determined by plotting a few points. Likewise, if $0 < a < 1$, the graph of $f(x) = ax^3$ is a shrinking of $f(x) = x^3$. Furthermore, we know that $f(x) = -x^3$ is an x-axis (and also a y-axis) reflection of $f(x) = x^3$. Figure 5.4 shows graphs of $f(x) = \dfrac{1}{2}x^3$ and $f(x) = -x^3$.

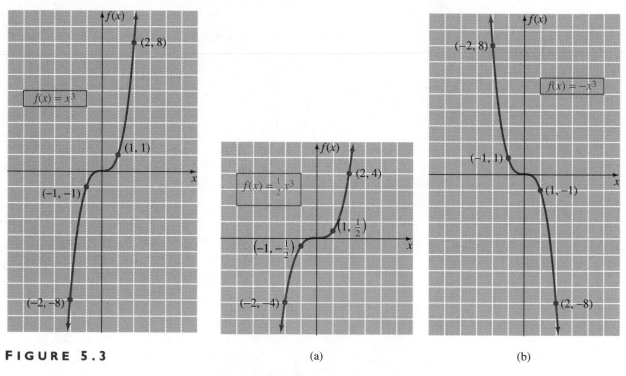

FIGURE 5.3

(a)

(b)

FIGURE 5.4

Two general patterns emerge from studying functions of the form $f(x) = x^n$. If n is odd and greater than 3, then the graph of $f(x) = x^n$ closely resembles Figure 5.3. For example, the graph of $f(x) = x^5$ is shown in Figure 5.5. Note that it *flattens out* a little more rapidly around the origin than the graph of $f(x) = x^3$ and that it increases and decreases more rapidly because of the larger exponent. If n is even and greater than 2, then the graphs of $f(x) = x^n$ are not parabolas; they resemble the basic parabola, but they are flatter at the bottom and steeper. Figure 5.6 shows the graph of $f(x) = x^4$.

FIGURE 5.5

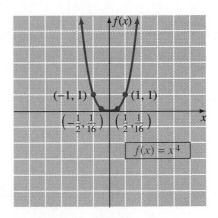

FIGURE 5.6

Graphs of functions of the form $f(x) = ax^n$, where n is an integer greater than 2 and $a \neq 1$, are variations of those shown in Figures 5.3 and 5.6. If n is odd, the curve is symmetric about the origin; if n is even, the graph is symmetric about the y axis.

Transformations of these basic curves are easy to sketch. For example, in Figure 5.7 we translated the graph of $f(x) = x^3$ upward two units to produce the graph of $f(x) = x^3 + 2$. In Figure 5.8 we obtained the graph of $f(x) = (x - 1)^5$ by translating the graph of $f(x) = x^5$ one unit to the right. In Figure 5.9 we sketched the graph of $f(x) = -x^4$ as the x-axis reflection of $f(x) = x^4$.

FIGURE 5.7

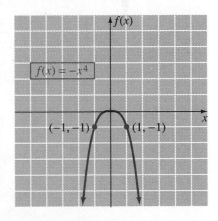

FIGURE 5.8

FIGURE 5.9

Graphing Polynomial Functions in Factored Form

As we mentioned earlier, a complete discussion of graphing polynomials of degree greater than 2 requires some tools from calculus. In fact, as the degree increases, the graphs often become more complicated. We do know that polynomial functions produce smooth continuous curves with a number of turning points, as illustrated in Figures 5.10 and 5.11. Figure 5.10 shows some graphs of polynomial functions of odd degree. As suggested by the graphs, every polynomial function of odd degree has at least *one real zero*—that is, at least one real number c such that $f(c) = 0$. Geometrically, the zeros of the function are the x intercepts of the graph. Figure 5.11 shows some graphs of polynomial functions of even degree.

Degree 3 with one real zero

(a)

Degree 3 with three real zeros

(b)

Degree 5 with five real zeros

(c)

FIGURE 5.10

Degree 4 with no real zeros

(a)

Degree 4 with four real zeros

(b)

Degree 6 with two real zeros

(c)

FIGURE 5.11

As indicated by the graphs in Figures 5.10 and 5.11, polynomial functions usually have **turning points** where the function either changes from increasing to decreasing or from decreasing to increasing. In calculus we are able to verify that *a polynomial function of degree n has at most n − 1 turning points*. Now let's illustrate how this information, along with some other techniques, can be used to graph polynomial functions that are expressed in factored form.

EXAMPLE 1

Graph $f(x) = (x + 2)(x - 1)(x - 3)$.

Solution

First, let's find the x intercepts (zeros of the function) by setting each factor equal to zero and solving for x.

$$x + 2 = 0 \qquad \text{or} \qquad x - 1 = 0 \qquad \text{or} \qquad x - 3 = 0$$
$$x = -2 \qquad \text{or} \qquad x = 1 \qquad \text{or} \qquad x = 3$$

Thus the points $(-2, 0)$, $(1, 0)$, and $(3, 0)$ are on the graph. Second, the points associated with the x intercepts divide the x axis into four intervals, as we see in Figure 5.12.

FIGURE 5.12

In each of these intervals, $f(x)$ is either always positive or always negative. That is, the graph is either completely above or completely below the x axis. The sign can be determined by selecting a *test value* for x in each of the intervals. Any additional points that are easily obtained improve the accuracy of the graph. The following table summarizes these results.

INTERVAL	TEST VALUE	SIGN OF $f(x)$	LOCATION OF GRAPH
$x < -2$	$f(-3) = -24$	Negative	Below x axis
$-2 < x < 1$	$f(0) = 6$	Positive	Above x axis
$1 < x < 3$	$f(2) = -4$	Negative	Below x axis
$x > 3$	$f(4) = 18$	Positive	Above x axis

Additional values: $f(-1) = 8$

Making use of the x intercepts and the information in the table, we sketched the graph in Figure 5.13. (The points $(-3, -24)$ and $(4, 18)$ are not shown, but they are used to indicate a rapid decrease and increase of the curve in those regions.)

> **REMARK**   In Figure 5.13, we indicated turning points of the graph at $(2, -4)$ and $(-1, 8)$. Keep in mind that these are only approximations; again, the tools of calculus are needed to find the exact turning points.

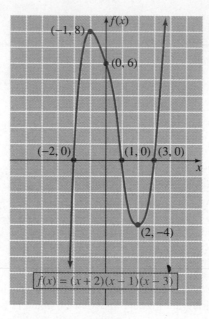

$$f(x) = (x + 2)(x - 1)(x - 3)$$

FIGURE 5.13

EXAMPLE 2

Graph $f(x) = -x^4 + 3x^3 - 2x^2$.

Solution

The polynomial can be factored as follows.

$$f(x) = -x^4 + 3x^3 - 2x^2$$
$$= -x^2(x^2 - 3x + 2)$$
$$= -x^2(x - 1)(x - 2)$$

Now we can find the x intercepts.

$$-x^2 = 0 \quad \text{or} \quad x - 1 = 0 \quad \text{or} \quad x - 2 = 0$$
$$x = 0 \quad \text{or} \quad x = 1 \quad \text{or} \quad x = 2$$

Thus the points $(0, 0)$, $(1, 0)$, and $(2, 0)$ are on the graph and divide the x axis into four intervals (see Figure 5.14).

$x < 0$	$0 < x < 1$	$1 < x < 2$	$x > 2$
0	1	2	

FIGURE 5.14

The following table determines some points and summarizes the sign behavior of $f(x)$.

INTERVAL	TEST VALUE	SIGN OF $f(x)$	LOCATION OF GRAPH
$x < 0$	$f(-1) = -6$	Negative	Below x axis
$0 < x < 1$	$f\left(\dfrac{1}{2}\right) = -\dfrac{3}{16}$	Negative	Below x axis
$1 < x < 2$	$f\left(\dfrac{3}{2}\right) = \dfrac{9}{16}$	Positive	Above x axis
$x > 2$	$f(3) = -18$	Negative	Below x axis

Make use of the table and the x intercepts to graph Figure 5.15.

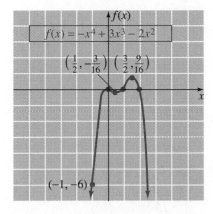

FIGURE 5.15

EXAMPLE 3

Graph $f(x) = x^3 + 3x^2 - 4$.

Solution

By using the rational root theorem, synthetic division, and the factor theorem, we can factor the given polynomial as follows.

$$f(x) = x^3 + 3x^2 - 4$$
$$= (x - 1)(x^2 + 4x + 4)$$
$$= (x - 1)(x + 2)^2$$

Now we can find the x intercepts.

$$x - 1 = 0 \quad \text{or} \quad (x + 2)^2 = 0$$
$$x = 1 \quad \text{or} \quad x = -2$$

Thus the points $(-2, 0)$ and $(1, 0)$ are on the graph and divide the x axis into three intervals (see Figure 5.16).

FIGURE 5.16

The following table determines some points and summarizes the sign behavior of $f(x)$.

INTERVAL	TEST VALUE	SIGN OF $f(x)$	LOCATION OF GRAPH
$x < -2$	$f(-3) = -4$	Negative	Below x axis
$-2 < x < 1$	$f(0) = -4$	Negative	Below x axis
$x > 1$	$f(2) = 16$	Positive	Above x axis

Additional values: $f(-1) = -2; f(-4) = -20$

With the results of the table and the x intercepts, we sketched the graph in Figure 5.17.

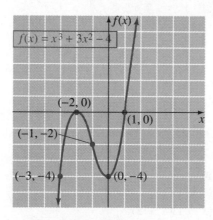

FIGURE 5.17   ▪

Finally, let's use a graphical approach to solve a problem involving a polynomial function.

PROBLEM I

Suppose that we have a rectangular piece of cardboard that measures 20 inches by 14 inches. From each corner a square piece is cut out, and then the flaps are turned up to form an open box (see Figure 5.18). Determine the length of a side of the square pieces to be cut out so that the volume of the box is as large as possible.

(a)   (b)

FIGURE 5.18

Solution

Let x represent the length of a side of the square to be cut from each corner. Then $20 - 2x$ represents the length of the open box and $14 - 2x$ represents the width. The volume of a rectangular box is given by the formula $V = lwh$, so the volume of this box can be represented by $V = x(20 - 2x)(14 - 2x)$. Now let $y = V$ and graph the function $y = x(20 - 2x)(14 - 2x)$ as shown in Figure 5.19. For this problem we are interested only in the part of the graph between $x = 0$ and $x = 7$, because the length of a side of the square has to be less than 7 inches for a box to be formed. Figure 5.20 gives us a view of that part of the graph. Now we can use the zoom and trace features to determine that as x equals approximately 2.7, the value of y is a maximum of approximately 339.0. Thus square pieces of length approximately 2.7 inches on a side should be cut from each corner of the rectangular piece of cardboard. The open box formed will have a volume of approximately 339.0 cubic inches.

FIGURE 5.19

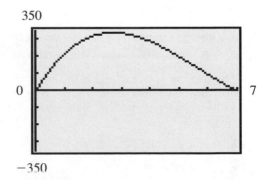

FIGURE 5.20

PROBLEM SET 5.4

For Problems 1–22, graph each polynomial function.

1. $f(x) = x^3 - 3$

2. $f(x) = (x + 1)^3$

3. $f(x) = (x - 2)^3 + 1$

4. $f(x) = -(x - 3)^3$

5. $f(x) = x^4 - 2$

6. $f(x) = (x + 3)^4$

7. $f(x) = (x + 1)^4 + 3$

8. $f(x) = -x^5$

9. $f(x) = (x - 1)^5 + 2$

10. $f(x) = -(x - 2)^4$

11. $f(x) = (x - 1)(x + 1)(x - 3)$

12. $f(x) = (x - 2)(x + 1)(x + 3)$

13. $f(x) = (x + 4)(x + 1)(1 - x)$

14. $f(x) = x(x + 2)(2 - x)$

15. $f(x) = -x(x + 3)(x - 2)$

16. $f(x) = -x^2(x - 1)(x + 1)$

17. $f(x) = (x + 3)(x + 1)(x - 1)(x - 2)$

18. $f(x) = (2x - 1)(x - 2)(x - 3)$

19. $f(x) = (x - 1)^2(x + 2)$

20. $f(x) = (x + 2)^3(x - 4)$

21. $f(x) = (x + 1)^2(x - 1)^2$

22. $f(x) = x(x - 2)^2(x + 1)$

For Problems 23–34, graph each polynomial function by first factoring the given polynomial. You may need to use some factoring techniques from Chapter 0, as well as the rational root theorem and the factor theorem.

23. $f(x) = x^3 + x^2 - 2x$

24. $f(x) = -x^3 - x^2 + 6x$

25. $f(x) = -x^4 - 3x^3 - 2x^2$

26. $f(x) = x^4 - 6x^3 + 8x^2$

27. $f(x) = x^3 - x^2 - 4x + 4$

28. $f(x) = x^3 + 2x^2 - x - 2$

29. $f(x) = x^3 - 13x + 12$

30. $f(x) = x^3 - x^2 - 9x + 9$

31. $f(x) = x^3 - 2x^2 - 11x + 12$

32. $f(x) = 2x^3 - 3x^2 - 3x + 2$

33. $f(x) = -x^3 + 6x^2 - 11x + 6$

34. $f(x) = x^4 - 5x^2 + 4$

For Problems 35–41, find (a) the y intercepts, (b) the x intercepts, and (c) the intervals of x where $f(x) > 0$ and where $f(x) < 0$. *Do not* sketch the graph.

35. $f(x) = (x - 5)(x + 4)(x - 3)$

36. $f(x) = (x + 3)(x - 6)(8 - x)$

37. $f(x) = (x - 4)^2(x + 3)^3$

38. $f(x) = (x + 3)^4(x - 1)^3$

39. $f(x) = (x + 2)^2(x - 1)^3(x - 2)$

40. $f(x) = x(x - 6)^2(x + 4)$

41. $f(x) = (x + 2)^5(x - 4)^2$

THOUGHTS INTO WORDS

42. The graph of $f(x) = x^3 - 3$ is the graph of $f(x) = x^3$ translated three units downward. Describe each of the following as transformations of the basic cubic function $f(x) = x^3$.

   a. $f(x) = (x + 4)^3$ *b.* $f(x) = -3x^3$

   c. $f(x) = (x - 2)^3 + 6$

   d. $f(x) = 2(x + 1)^3 - 4$

43. How would you go about graphing $f(x) = -(x - 1)(x + 2)^3$?

44. Give a general description of how to graph polynomial functions that are in factored form.

⌕ **Further Investigations**

45. A polynomial function with real coefficients is **continuous everywhere;** that is, its graph has no holes or breaks. This is the basis for the following property: **If $f(x)$ is a polynomial with real coefficients, and if $f(a)$ and $f(b)$ are of opposite sign, then there is at least one real zero between a and b.** This property, along with what we already know about polynomial functions, provides the basis for locating and approximating irrational solutions of a polynomial equation.

Consider the equation $x^3 + 2x - 4 = 0$. Apply Descartes' rule of signs to determine that this equation has one positive real solution and two nonreal complex solutions. (You may want to confirm this!) The rational root theorem indicates that the only possible positive *rational* solutions are 1, 2, and 4. Use a little more compact format for synthetic division to obtain the following results when testing for 1 and 2 as possible solutions.

	1	0	2	−4
1	1	1	3	−1
2	1	2	6	8

Because $f(1) = -1$ (negative) and $f(2) = 8$ (positive), there must be an *irrational* solution between 1 and 2. Furthermore, because -1 is closer to 0 than 8, our guess is that the solution is closer to 1 than to 2. Let's start looking at 1.0, 1.1, 1.2, and so on, until we can clamp the solution between two numbers.

	1	0	2	−4
1.0	1	1	3	−1
1.1	1	1.1	3.21	−0.469
1.2	1	1.2	3.44	0.128

A calculator is very helpful at this time.

Because $f(1.1) = -0.469$ and $f(1.2) = 0.128$, the irrational solution must be between 1.1 and 1.2. Furthermore, because 0.128 is closer to 0 than -0.469, our guess is that the solution is closer to 1.2 than to 1.1. Let's start looking at 1.15, 1.16, and so on.

	1	0	2	−4
1.15	1	1.15	3.3225	−0.179
1.16	1	1.16	3.3456	−0.119
1.17	1	1.17	3.3689	−0.058
1.18	1	1.18	3.3924	0.003

Because $f(1.17) = -0.058$ and $f(1.18) = 0.003$, the irrational solution must be between 1.17 and 1.18. Therefore, we can use 1.2 as a rational approximation to the nearest tenth.

For each of the following equations, verify that the equation has exactly one irrational solution, and find an approximation, to the nearest tenth, of that solution.

a. $x^3 + x - 6 = 0$

b. $x^3 - 6x - 4 = 0$

c. $x^3 - 27x + 18 = 0$

d. $x^3 - x^2 - x - 1 = 0$

e. $x^3 - 24x - 32 = 0$

f. $x^3 - 5x^2 + 3 = 0$

▦ **GRAPHING CALCULATOR ACTIVITIES**

46. Graph $f(x) = x^3$. Now predict the graphs for $f(x) = x^3 + 2$, $f(x) = -x^3 + 2$, and $f(x) = -x^3 - 2$. Graph these three functions on the same set of axes with the graph of $f(x) = x^3$.

47. Draw a rough sketch of the graphs of the functions $f(x) = x^3 - x^2$, $f(x) = -x^3 + x^2$, and $f(x) =$ $-x^3 - x^2$. Now graph these three functions to check your sketches.

48. Graph $f(x) = x^4 + x^3 + x^2$. What should the graphs of $f(x) = x^4 - x^3 + x^2$ and $f(x) = -x^4 - x^3 - x^2$ look like? Graph them to see whether you were right.

49. How should the graphs of $f(x) = x^3$, $f(x) = x^5$, and $f(x) = x^7$ compare? Graph these three functions on the same set of axes.

50. How should the graphs of $f(x) = x^2$, $f(x) = x^4$, and $f(x) = x^6$ compare? Graph these three functions on the same set of axes.

51. For each of the following functions, find the x intercepts and find the intervals of x where $f(x) > 0$ and those where $f(x) < 0$.

a. $f(x) = x^3 - 3x^2 - 6x + 8$

b. $f(x) = x^3 - 8x^2 - x + 8$

c. $f(x) = x^3 - 7x^2 + 16x - 12$

d. $f(x) = x^3 - 19x^2 + 90x - 72$

e. $f(x) = x^4 + 3x^3 - 3x^2 - 11x - 6$

f. $f(x) = x^4 + 12x^2 - 64$

52. Find the coordinates of the turning points of each of the following graphs. Express x and y values to the nearest integer.

a. $f(x) = 2x^3 - 3x^2 - 12x + 40$

b. $f(x) = 2x^3 - 33x^2 + 60x + 1050$

c. $f(x) = -2x^3 - 9x^2 + 24x + 100$

d. $f(x) = x^4 - 4x^3 - 2x^2 + 12x + 3$

e. $f(x) = x^3 - 30x^2 + 288x - 900$

f. $f(x) = x^5 - 2x^4 - 3x^3 - 2x^2 + x - 1$

53. For each of the following functions, find the x intercepts and find the turning points. Express your answers to the nearest tenth.

a. $f(x) = x^3 + 2x^2 - 3x + 4$

b. $f(x) = 42x^3 - x^2 - 246x - 35$

c. $f(x) = x^4 - 4x^2 - 4$

54. A rectangular piece of cardboard is 13 inches long and 9 inches wide. From each corner a square piece is cut out, and then the flaps are turned up to form an open box. Determine the length of a side of the square pieces so that the volume of the box is as large as possible.

55. A company determines that its weekly profit from manufacturing and selling x units of a certain item is given by $P(x) = -x^3 + 3x^2 + 2880x - 500$. What weekly production rate will maximize the profit?

5.5   GRAPHING RATIONAL FUNCTIONS

A function of the form

$$f(x) = \frac{p(x)}{q(x)}, \qquad q(x) \neq 0$$

where $p(x)$ and $q(x)$ are both polynomial functions, is called a **rational function** The following are examples of rational functions.

$$f(x) = \frac{2}{x - 1} \qquad\qquad f(x) = \frac{x}{x - 2}$$

$$f(x) = \frac{x^2}{x^2 - x - 6} \qquad f(x) = \frac{x^3 - 8}{x + 4}$$

In each example, the domain of the rational function is the set of all real numbers except those that make the denominator zero. For example, the domain of $f(x) = \frac{2}{x - 1}$ is the set of all real numbers except 1. As you will see, these exclusions from the domain are important numbers from a graphing standpoint. They represent breaks in an otherwise continuous curve.

Let's set the stage for graphing rational functions by considering in detail the function $f(x) = 1/x$. First, note that at $x = 0$, the function is undefined. Second, let's consider an extensive table of values to show some number trends and to build a basis for defining the concept of an *asymptote*.

x	$f(x) = \dfrac{1}{x}$	
1	1	
2	0.5	These values indicate that the value of $f(x)$ is positive and approaches zero from above as x gets larger and larger.
10	0.1	
100	0.01	
1000	0.001	
0.5	2	
0.1	10	These values indicate that $f(x)$ is positive and is getting larger and larger as x approaches zero from the right.
0.01	100	
0.001	1000	
0.0001	10000	
−0.5	−2	
−0.1	−10	These values indicate that $f(x)$ is negative and is getting smaller and smaller as x approaches zero from the left.
−0.01	−100	
−0.001	−1000	
−0.0001	−10000	
−1	−1	
−2	−0.5	These values indicate that $f(x)$ is negative and approaches zero from below as x gets smaller and smaller.
−10	−0.1	
−100	−0.01	
−1000	−0.001	

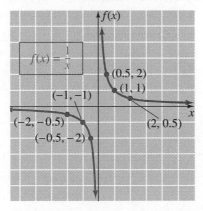

FIGURE 5.21

Using a few points from this table and the patterns we discussed, we have sketched $f(x) = 1/x$ in Figure 5.21. Note that the graph approaches both axes but does not touch either axis. We say that the y axis [or $f(x)$ axis] is a *vertical asymptote* and the x axis is a *horizontal asymptote*. In general, the following definitions can be given.

REMARK   Observe that the equation $f(x) = 1/x$ exhibits origin symmetry because $f(-x) = -f(x)$. Thus we could have drawn the graph in Figure 5.21 by first determining the part of the curve in the first quadrant and then reflecting that through the origin.

Vertical Asymptote   A line $x = a$ is a **vertical asymptote** for the graph of a function f if it satisfies either of the following two properties.

1. $f(x)$ either increases or decreases without bound as x approaches the number a from the right, as in Figure 5.22, *or*

2. $f(x)$ either increases or decreases without bound as x approaches the number a from the left, as in Figure 5.23.

F I G U R E   5 . 2 2

F I G U R E   5 . 2 3

Horizontal Asymptote   A line $y = b$ [or $f(x) = b$] is a **horizontal asymptote** for the graph of a function f if it satisfies either of the following two properties.

1. $f(x)$ approaches the number b from above or below as x gets infinitely small, as in Figure 5.24, *or*

2. $f(x)$ approaches the number b from above or below as x gets infinitely large, as in Figure 5.25.

F I G U R E   5 . 2 4

F I G U R E   5 . 2 5

The following suggestions will help you graph rational functions of the type we are considering in this section.

1. Check for y-axis symmetry and origin symmetry.
2. Find any vertical asymptote(s) by setting the denominator equal to zero and solving for x.
3. Find any horizontal asymptote(s) by studying the behavior of $f(x)$ as x gets infinitely large or as x gets infinitely small.
4. Study the behavior of the graph when it is close to the asymptotes.
5. Plot as many points as necessary to determine the shape of the graph. This may be affected by whether the graph has any symmetry.

Keep these suggestions in mind as you study the following examples.

EXAMPLE 1

Graph $f(x) = \dfrac{-2}{x-1}$.

Solution

Because $x = 1$ makes the denominator zero, the line $x = 1$ is a vertical asymptote; we have indicated this with a dashed line in Figure 5.26. Now let's look for a horizontal asymptote by checking some large and small values of x in the tables that accompany Figure 5.26.

x	$f(x)$	
10	$-\dfrac{2}{9}$	
100	$-\dfrac{2}{99}$	This portion of the table shows that as x gets very large, the value of $f(x)$ approaches 0 from below.
1000	$-\dfrac{2}{999}$	
−10	$\dfrac{2}{11}$	
−100	$\dfrac{2}{101}$	This portion shows that as x gets very small, the value of $f(x)$ approaches 0 from above.
−1000	$\dfrac{2}{1001}$	

Therefore, the x axis is a horizontal asymptote. Finally, let's check the behavior of the graph near the vertical asymptote.

x	$f(x)$
2	-2
1.5	-4
1.1	-20
1.01	-200
1.001	-2000
0	2
0.5	4
0.9	20
0.99	200
0.999	2000

As x approaches 1 from the right side, the value of $f(x)$ gets smaller and smaller.

As x approaches 1 from the left side, the value of $f(x)$ gets larger and larger.

The graph of $f(x) = \dfrac{-2}{x-1}$ is shown in Figure 5.26.

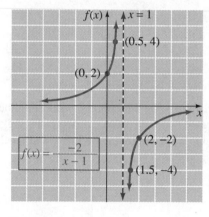

FIGURE 5.26

EXAMPLE 2   Graph $f(x) = \dfrac{x}{x+2}$.

Solution

Because $x = -2$ makes the denominator zero, the line $x = -2$ is a vertical asymptote. To study the behavior of $f(x)$ as x gets very large or very small, let's change the form of the rational expression by dividing both the numerator and the denominator by x.

$$f(x) = \frac{x}{x+2} = \frac{\dfrac{x}{x}}{\dfrac{x+2}{x}} = \frac{1}{\dfrac{x}{x} + \dfrac{2}{x}} = \frac{1}{1 + \dfrac{2}{x}}$$

Now we can see that (1) as x gets larger and larger, the value of $f(x)$ approaches 1 from below, and (2) as x gets smaller and smaller, the value of $f(x)$ approaches 1 from above. (Perhaps you should check these claims by plugging in some values for x.) Thus the line $f(x) = 1$ is a horizontal asymptote. Drawing the asymptotes (dashed lines) and plotting a few points enable us to complete the graph shown in Figure 5.27.

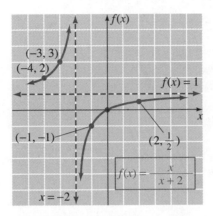

FIGURE 5.27

In the next two examples, pay special attention to the role of symmetry. It will allow us to focus on the portion of a curve in quadrants I and IV and then to reflect that portion of the curve across the vertical axis to complete the graph.

EXAMPLE 3

Graph $f(x) = \dfrac{2x^2}{x^2 + 4}$.

Solution

First, note that $f(-x) = f(x)$; therefore, this graph is symmetric with respect to the y axis. Second, the denominator $x^2 + 4$ cannot equal zero for any real number x. Thus there is no vertical asymptote. Third, dividing both the numerator and the denominator of the rational expression by x^2 produces

$$f(x) = \frac{2x^2}{x^2 + 4} = \frac{\dfrac{2x^2}{x^2}}{\dfrac{x^2 + 4}{x^2}} = \frac{2}{\dfrac{x^2}{x^2} + \dfrac{4}{x^2}} = \frac{2}{1 + \dfrac{4}{x^2}}$$

Now we can see that as x gets larger and larger, the value of $f(x)$ approaches 2 from below. Therefore, the line $f(x) = 2$ is a horizontal asymptote. We can plot a few points using positive values for x, sketch this part of the curve, and then reflect across the $f(x)$ axis to obtain the complete graph shown in Figure 5.28.

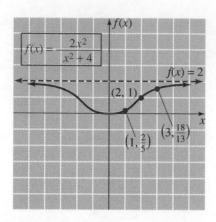

FIGURE 5.28

EXAMPLE 4

Graph $f(x) = \dfrac{3}{x^2 - 4}$.

Solution

First, note that $f(-x) = f(x)$; therefore, this graph is symmetric about the $f(x)$ axis. Second, by setting the denominator equal to zero and solving for x, we obtain

$$x^2 - 4 = 0$$
$$x^2 = 4$$
$$x = \pm 2$$

The lines $x = 2$ and $x = -2$ are vertical asymptotes. Next, we can see that $\dfrac{3}{x^2 - 4}$ approaches zero from above as x gets larger and larger. Finally, we can plot a few points using positive values for x (not 2), sketch this part of the curve, and then reflect it across the $f(x)$ axis to obtain the complete graph shown in Figure 5.29.

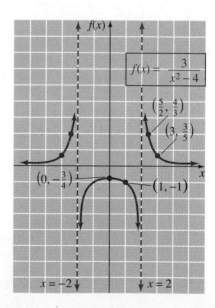

FIGURE 5.29

Now suppose that we are going to use a graphing utility to obtain a graph of the function $f(x) = \dfrac{4x^2}{x^4 - 16}$. Before we enter this function into a graphing utility, let's analyze what we know about the graph.

1. Because $f(0) = 0$, the origin is a point on the graph.

2. Because $f(-x) = f(x)$, the graph is symmetric with respect to the y axis.

3. By setting the denominator equal to zero and solving for x, we can determine the vertical asymptotes.

$$x^4 - 16 = 0$$
$$(x^2 + 4)(x^2 - 4) = 0$$

$$x^2 + 4 = 0 \quad \text{or} \quad x^2 - 4 = 0$$
$$x^2 = -4 \quad \text{or} \quad x^2 = 4$$
$$x = \pm 2i \quad \text{or} \quad x = \pm 2$$

Remember that we are working with ordered pairs of real numbers. Thus the lines $x = -2$ and $x = 2$ are vertical asymptotes.

4. Divide both the numerator and the denominator of the rational expression by x^4 to produce

$$\frac{4x^2}{x^4 - 16} = \frac{\dfrac{4x^2}{x^4}}{\dfrac{x^4 - 16}{x^4}} = \frac{\dfrac{4}{x^2}}{1 - \dfrac{16}{x^4}}$$

From the last expression, we see that as $|x|$ gets larger and larger, the value of $f(x)$ approaches zero from above. Therefore, the x axis is a horizontal asymptote.

Let's enter the function in a graphing utility and obtain the graph shown in Figure 5.30. Note that the graph is consistent with all of the information we determined before we used the graphing utility. In other words, our knowledge of graphing techniques enhances our use of a graphing utility.

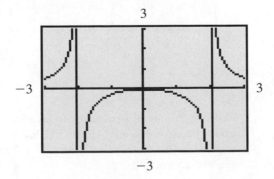

FIGURE 5.30

REMARK   In Figure 5.30, the origin is a point of the graph that is on the horizontal asymptote. More will be said about such situations in the next section.

Back in Problem Set 1.4, you were asked to solve the following problem: How much pure alcohol should be added to 6 liters of a 40% alcohol solution to raise it to a 60% alcohol solution? The answer of 3 liters can be found by solving the following equation, where x represents the amount of pure alcohol to be added.

$$0.40(6) \quad + \quad x \quad = 0.60(6 + x)$$

Now let's consider this problem in a more general setting. Again, x represents the amount of pure alcohol to be added, and the rational expression $\dfrac{2.4 + x}{6 + x}$ represents the concentration of pure alcohol in the final solution. Let's graph the rational function $y = \dfrac{2.4 + x}{6 + x}$ as shown in Figure 5.31. For this particular problem, x is nonnegative, so we are interested only in the part of the graph that is in the first quadrant. Change the boundaries of the viewing rectangle so that $0 \le x \le 15$ and $0 \le y \le 2$ to obtain Figure 5.32. Now we are ready to answer questions about this situation.

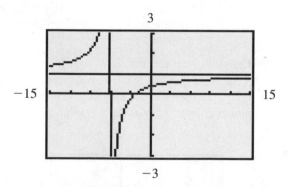

F I G U R E   5 . 3 1

1. How much pure alcohol needs to be added to raise the 40% solution to a 60% alcohol solution? (*Answer:* Using the trace feature of the graphing utility, we find that $y = 0.6$ when $x = 3$. Therefore, 3 liters of pure alcohol must be added.)

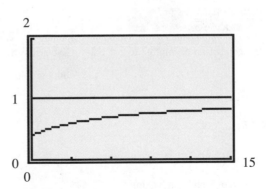

FIGURE 5.32

2. How much pure alcohol needs to be added to raise the 40% solution to a 70% alcohol solution? (*Answer:* Using the trace feature, we find that $y = 0.7$ when $x = 6$. Therefore, 6 liters of pure alcohol must be added.)

3. What percent of alcohol do we have if we add 9 liters of pure alcohol to the 6 liters of a 40% solution? (*Answer:* Using the trace feature, we find that $y = 0.76$ when $x = 9$. Therefore, adding 9 liters of pure alcohol will give us a 76% alcohol solution.)

PROBLEM SET 5.5

For Problems 1–22, graph each rational function.

1. $f(x) = \dfrac{-1}{x}$

2. $f(x) = \dfrac{1}{x^2}$

3. $f(x) = \dfrac{3}{x + 1}$

4. $f(x) = \dfrac{-1}{x - 3}$

5. $f(x) = \dfrac{2}{(x - 1)^2}$

6. $f(x) = \dfrac{-3}{(x + 2)^2}$

7. $f(x) = \dfrac{x}{x - 3}$

8. $f(x) = \dfrac{2x}{x - 1}$

9. $f(x) = \dfrac{-3x}{x + 2}$

10. $f(x) = \dfrac{-x}{x + 1}$

11. $f(x) = \dfrac{1}{x^2 - 1}$

12. $f(x) = \dfrac{-2}{x^2 - 4}$

13. $f(x) = \dfrac{-2}{(x + 1)(x - 2)}$

14. $f(x) = \dfrac{3}{(x + 2)(x - 4)}$

15. $f(x) = \dfrac{2}{x^2 + x - 2}$

16. $f(x) = \dfrac{-1}{x^2 + x - 6}$

17. $f(x) = \dfrac{x + 2}{x}$

18. $f(x) = \dfrac{2x - 1}{x}$

19. $f(x) = \dfrac{4}{x^2 + 2}$

20. $f(x) = \dfrac{4x^2}{x^2 + 1}$

21. $f(x) = \dfrac{2x^4}{x^4 + 1}$

22. $f(x) = \dfrac{x^2 - 4}{x^2}$

THOUGHTS INTO WORDS

23. How would you explain the concept of an asymptote to an elementary algebra student?

24. Give a step-by-step description of how you would go about graphing $f(x) = \dfrac{-2}{x^2 - 9}$.

Further Investigations

25. The rational function $f(x) = \dfrac{(x - 2)(x + 3)}{x - 2}$ has a domain of all the real numbers except 2 and can be simplified to $f(x) = x + 3$. Thus its graph is a straight line with a hole at (2, 5). Graph each of the following functions.

a. $f(x) = \dfrac{(x + 4)(x - 1)}{x + 4}$

b. $f(x) = \dfrac{x^2 - 5x + 6}{x - 2}$

c. $f(x) = \dfrac{x - 1}{x^2 - 1}$

d. $f(x) = \dfrac{x + 2}{x^2 + 6x + 8}$

GRAPHING CALCULATOR ACTIVITIES

26. Use a graphing calculator to check your graphs for Problem 25. What feature of the graph does not show up on the calculator?

27. Each of the following graphs is a transformation of $f(x) = 1/x$. First predict the general shape and location of the graph, and then check your prediction with a graphing calculator.

a. $f(x) = \dfrac{1}{x} - 2$

b. $f(x) = \dfrac{1}{x + 3}$

c. $f(x) = -\dfrac{1}{x}$

d. $f(x) = \dfrac{1}{x - 2} + 3$

e. $f(x) = \dfrac{2x + 1}{x}$

28. Graph $f(x) = \dfrac{1}{x^2}$. How should the graphs of $f(x) = \dfrac{1}{(x - 4)^2}, f(x) = \dfrac{1 + 3x^2}{x^2}$, and $f(x) = -\dfrac{1}{x^2}$ compare to the graph of $f(x) = \dfrac{1}{x^2}$? Graph the three functions on the same set of axes with the graph of $f(x) = \dfrac{1}{x^2}$.

29. Graph $f(x) = \dfrac{1}{x^3}$. How should the graphs of $f(x) = \dfrac{2x^3 + 1}{x^3}, f(x) = \dfrac{1}{(x + 2)^3}$, and $f(x) = -\dfrac{1}{x^3}$ compare to the graph of $f(x) = \dfrac{1}{x^3}$? Graph the three functions on the same set of axes with the graph of $f(x) = \dfrac{1}{x^3}$.

30. Use a graphing calculator to check your graphs for Problems 19–22.

31. Graph each of the following functions. Be sure that you get a complete graph for each one. Sketch each graph on a sheet of paper and keep them all handy as you study the next section.

a. $f(x) = \dfrac{x^2}{x^2 - x - 2}$

b. $f(x) = \dfrac{x}{x^2 - 4}$

c. $f(x) = \dfrac{3x}{x^2 + 1}$

d. $f(x) = \dfrac{x^2 - 1}{x - 2}$

32. Suppose that x ounces of pure acid has been added to 14 ounces of a 15% acid solution.

a. Set up the rational expression that represents the concentration of pure acid in the final solution.

b. Graph the rational function that displays the level of concentration.

c. How many ounces of pure acid must be added to the 14 ounces of a 15% solution to raise it to a 40.5% solution? Check your answer.

d. How many ounces of pure acid must be added to the 14 ounces of a 15% solution to raise it to a 50% solution? Check your answer.

e. What percent of acid do we obtain if we add 12 ounces of pure acid to the 14 ounces of a 15% solution? Check your answer.

33. Solve the following problem both algebraically and graphically: One solution contains 50% alcohol, and another solution contains 80% alcohol. How many liters of each solution should be mixed to produce 10.5 liters of a 70% alcohol solution? Check your answer.

5.6   MORE ON GRAPHING RATIONAL FUNCTIONS

The rational functions that we studied in the previous section "behaved rather well." In fact, once we established the vertical and horizontal asymptotes, a little bit of point plotting usually determined the graph rather easily. Such is not always the case with rational functions. In this section, we want to investigate some rational functions that behave a little differently.

Vertical asymptotes occur at values of x where the denominator is zero, so there can be no points of a graph on a vertical asymptote. However, recall that horizontal asymptotes are created by the behavior of $f(x)$ as x gets infinitely large or infinitely small. This does not restrict the possibility that for some values of x, there will be points of the graph on the horizontal asymptote. Let's consider some examples.

EXAMPLE 1

Graph $f(x) = \dfrac{x^2}{x^2 - x - 2}$.

Solution

First, let's identify the vertical asymptotes by setting the denominator equal to zero and solving for x.

$$x^2 - x - 2 = 0$$
$$(x - 2)(x + 1) = 0$$
$$x - 2 = 0 \quad \text{or} \quad x + 1 = 0$$
$$x = 2 \quad \text{or} \quad x = -1$$

Thus the lines $x = 2$ and $x = -1$ are vertical asymptotes. Next, we can divide both the numerator and the denominator of the rational expression by x^2.

$$f(x) = \frac{x^2}{x^2 - x - 2} = \frac{\dfrac{x^2}{x^2}}{\dfrac{x^2 - x - 2}{x^2}} = \frac{1}{1 - \dfrac{1}{x} - \dfrac{2}{x^2}}$$

Now we can see that as x gets larger and larger, the value of $f(x)$ approaches 1 from above. Thus the line $f(x) = 1$ is a horizontal asymptote. To determine whether any points of the graph are *on* the horizontal asymptote, we can see whether the equation

$$\frac{x^2}{x^2 - x - 2} = 1$$

has any solutions.

$$\frac{x^2}{x^2 - x - 2} = 1$$
$$x^2 = x^2 - x - 2$$
$$0 = -x - 2$$
$$x = -2$$

Therefore, the point $(-2, 1)$ is on the graph. Now, by drawing the asymptotes, plotting a few points [including $(-2, 1)$], and studying the behavior of the function close to the asymptotes, we can sketch the curve shown in Figure 5.33.

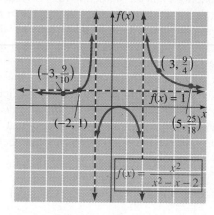

FIGURE 5.33

EXAMPLE 2

Graph $f(x) = \dfrac{x}{x^2 - 4}$.

Solution

First, note that $f(-x) = -f(x)$; therefore, this graph has origin symmetry. Second, let's identify the vertical asymptotes.

$$x^2 - 4 = 0$$
$$x^2 = 4$$
$$x = \pm 2$$

Thus the lines $x = -2$ and $x = 2$ are vertical asymptotes. Next, by dividing the numerator and the denominator of the rational expression by x^2, we obtain

$$f(x) = \frac{x}{x^2 - 4} = \frac{\dfrac{x}{x^2}}{\dfrac{x^2 - 4}{x^2}} = \frac{\dfrac{1}{x}}{1 - \dfrac{4}{x^2}}$$

From this form, we can see that as x gets larger and larger, the value of $f(x)$ approaches zero from above. Therefore, the x axis is a horizontal asymptote. Because $f(0) = 0$, we know that the origin is a point of the graph. Finally, by concentrating our point plotting on positive values of x, we can sketch the portion of the curve to the right of the vertical axis and then use the fact that the graph is symmetric with respect to the origin to complete the graph. Figure 5.34 shows the completed graph.

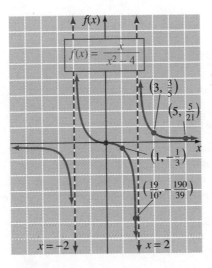

FIGURE 5.34

EXAMPLE 3   Graph $f(x) = \dfrac{3x}{x^2 + 1}$.

Solution

First, observe that $f(-x) = -f(x)$; therefore, this graph is symmetric with respect to the origin. Second, because $x^2 + 1$ is a positive number for all real number values of x, there are no vertical asymptotes for this graph. Next, by dividing the numerator and denominator of the rational expression by x^2, we obtain

$$f(x) = \frac{3x}{x^2 + 1} = \frac{\dfrac{3x}{x^2}}{\dfrac{x^2 + 1}{x^2}} = \frac{\dfrac{3}{x}}{1 + \dfrac{1}{x^2}}$$

From this form, we see that as x gets larger and larger, the value of $f(x)$ approaches zero from above. Thus the x axis is a horizontal asymptote. Because $f(0) = 0$, the origin is a point of the graph. Finally, by concentrating our point plotting on positive values of x, we can sketch the portion of the curve to the right of the vertical axis and then use origin symmetry to complete the graph, as shown in Figure 5.35.

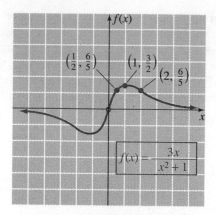

FIGURE 5.35

Oblique Asymptotes

Thus far we have restricted our study of rational functions to those where the degree of the numerator is less than or equal to the degree of the denominator. As our final examples of graphing rational functions, we will consider functions where the degree of the numerator is one greater than the degree of the denominator.

EXAMPLE 4   Graph $f(x) = \dfrac{x^2 - 1}{x - 2}$.

Solution

First, let's observe that $x = 2$ is a vertical asymptote. Second, because the degree of the numerator is greater than the degree of the denominator, we can change the form of the rational expression by division. We use synthetic division.

$$
\begin{array}{r}
2)\overline{1 \quad\; 0 \quad -1} \\
\underline{ 2 \quad\;\; 4} \\
1 \quad\; 2 \quad\;\; 3
\end{array}
$$

Therefore, the original function can be rewritten

$$
f(x) = \frac{x^2 - 1}{x - 2} = x + 2 + \frac{3}{x - 2}
$$

Now, for very large values of $|x|$, the fraction $\dfrac{3}{x - 2}$ is close to zero. Therefore, as $|x|$ gets larger and larger, the graph of $f(x) = x + 2 + \dfrac{3}{x - 2}$ gets closer and closer to the line $f(x) = x + 2$. We call this line an **oblique asymptote** and indicate it with a dashed line in Figure 5.36. Finally, because this is a new situation, it may be necessary to plot a large number of points on both sides of the vertical asymptote, so let's make an extensive table of values. The graph of the function is shown in Figure 5.36.

x	$f(x) = \dfrac{x^2 - 1}{x - 2}$
2.1	34.1
2.5	10.5
3	8
4	7.5
5	8
6	8.75
10	12.375

These values indicate the behavior of $f(x)$ to the right of the vertical asymptote $x = 2$.

x	$f(x)$
1.9	−26.1
1.5	−2.5
1	0
0	0.5
−1	0
−3	−1.6
−5	−3.4
−10	−8.25

These values indicate the behavior of $f(x)$ to the left of the vertical asymptote $x = 2$.

F I G U R E   5 . 3 6

If the degree of the numerator of a rational function is *exactly one more* than the degree of its denominator, then the graph of the function has an oblique asymptote. (If the graph is a line, as is the case with $f(x) = \dfrac{(x - 2)(x + 1)}{x - 2}$, then we consider it to be its own asymptote.) As in Example 4, we find the equation of the oblique asymptote by changing the form of the function using long division. Let's consider another example.

EXAMPLE 5

Graph $f(x) = \dfrac{x^2 - x - 2}{x - 1}$.

Solution

From the given form of the function, we see that $x = 1$ is a vertical asymptote. Then, by factoring the numerator, we can change the form to

$$f(x) = \frac{(x - 2)(x + 1)}{(x - 1)}$$

which indicates x intercepts of 2 and -1. Then, by long division, we can change the original form of the function to

$$f(x) = x - \frac{2}{x - 1}$$

which indicates an oblique asymptote $f(x) = x$. Finally, by plotting a few additional points, we can determine the graph as shown in Figure 5.37.

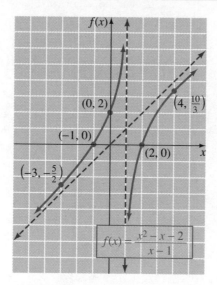

FIGURE 5.37

Finally, let's combine our knowledge of rational functions with the use of a graphing utility to obtain the graph of a fairly complex rational function.

EXAMPLE 6

Graph the rational function $f(x) = \dfrac{x^3 - 2x^2 - x - 1}{x^2 - 36}$.

Solution

Before entering this function into a graphing utility, let's analyze what we know about the graph.

1. Because $f(0) - \dfrac{1}{36}$, the point $\left(0, \dfrac{1}{36}\right)$ is on the graph.

2. Because $f(-x) \neq f(x)$ and $f(-x) \neq -f(x)$, there is no symmetry with respect to the origin or the y axis.

3. The denominator is zero at $x = \pm 6$. Thus the lines $x = 6$ and $x = -6$ are vertical asymptotes.

4. Let's change the form of the rational expression by division.

$$
\begin{array}{r}
x - 2 \\
x^2 - 36 \overline{)\, x^3 - 2x^2 - x - 1} \\
\underline{x^3 - 36x } \\
-2x^2 + 35x - 1 \\
\underline{-2x^2 + 72} \\
35x - 73
\end{array}
$$

Thus the original function can be rewritten as

$$
f(x) = x - 2 + \frac{35x - 73}{x^2 - 36}
$$

Therefore, the line $y = x - 2$ is an oblique asymptote. Now let $Y_1 = x - 2$ and $Y_2 = \dfrac{x^3 - 2x^2 - x - 1}{x^2 - 36}$ and use a viewing rectangle where $-15 \leq x \leq 15$ and $-30 \leq y \leq 30$ (Figure 5.38).

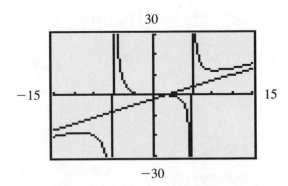

FIGURE 5.38   ■

Note that the graph in Figure 5.38 is consistent with the information we had before we used the graphing utility. (The graph may appear to have origin symmetry, but remember that the point $\left(0, \dfrac{1}{36}\right)$ is on the graph whereas the point $\left(0, -\dfrac{1}{36}\right)$ is not.) Also note that the curve does intersect the oblique asymptote. We can use the zoom and trace features of the graphing utility to find that point of intersection, or we can do it algebraically as follows: Because $y = \dfrac{x^3 - 2x^2 - x - 1}{x^2 - 36}$ and $y = x - 2$, we can equate the two expressions for y and solve the resulting equation for x.

$$\frac{x^3 - 2x^2 - x - 1}{x^2 - 36} = x - 2$$

$$x^3 - 2x^2 - x - 1 = (x - 2)(x^2 - 36)$$

$$x^3 - 2x^2 - x - 1 = x^3 - 2x^2 - 36x + 72$$

$$35x = 73$$

$$x = \frac{73}{35}$$

If $x = \dfrac{73}{35}$, then $y = x - 2 = \dfrac{73}{35} - 2 = \dfrac{3}{35}$. The point of intersection of the curve and the oblique asymptote is $\left(\dfrac{73}{35}, \dfrac{3}{35} \right)$.

PROBLEM SET 5.6

For Problems 1–20, graph each rational function. Check first for symmetry and identify the asymptotes.

1. $f(x) = \dfrac{x^2}{x^2 + x - 2}$

2. $f(x) = \dfrac{x^2}{x^2 + 2x - 3}$

3. $f(x) = \dfrac{2x^2}{x^2 - 2x - 8}$

4. $f(x) = \dfrac{-x^2}{x^2 + 3x - 4}$

5. $f(x) = \dfrac{-x}{x^2 - 1}$

6. $f(x) = \dfrac{2x}{x^2 - 9}$

7. $f(x) = \dfrac{x}{x^2 + x - 6}$

8. $f(x) = \dfrac{-x}{x^2 - 2x - 8}$

9. $f(x) = \dfrac{x^2}{x^2 - 4x + 3}$

10. $f(x) = \dfrac{1}{x^3 + x^2 - 6x}$

11. $f(x) = \dfrac{x}{x^2 + 2}$

12. $f(x) = \dfrac{6x}{x^2 + 1}$

13. $f(x) = \dfrac{-4x}{x^2 + 1}$

14. $f(x) = \dfrac{-5x}{x^2 + 2}$

15. $f(x) = \dfrac{x^2 + 2}{x - 1}$

16. $f(x) = \dfrac{x^2 - 3}{x + 1}$

17. $f(x) = \dfrac{x^2 - x - 6}{x + 1}$

18. $f(x) = \dfrac{x^2 + 4}{x + 2}$

19. $f(x) = \dfrac{x^2 + 1}{1 - x}$

20. $f(x) = \dfrac{x^3 + 8}{x^2}$

THOUGHTS INTO WORDS

21. Explain the concept of an oblique asymptote.

22. Explain why it is possible for curves to intersect horizontal and oblique asymptotes but not to intersect vertical asymptotes.

23. Give a step-by-step description of how you would go about graphing $f(x) = \dfrac{x^2 - x - 12}{x - 2}$.

24. Your friend is having difficulty finding the point of intersection of a curve and the oblique asymptote. How would you help?

GRAPHING CALCULATOR ACTIVITIES

25. First check for symmetry and identify the asymptotes for the graphs of the following rational functions. Then use your graphing utility to graph each function.

a. $f(x) = \dfrac{4x^2}{x^2 + x - 2}$

b. $f(x) = \dfrac{-2x}{x^2 - 5x - 6}$

c. $f(x) = \dfrac{x^2}{x^2 - 9}$

d. $f(x) = \dfrac{x^2 - 4}{x^2 - 9}$

e. $f(x) = \dfrac{x^2 - 9}{x^2 - 4}$

f. $f(x) = \dfrac{x^2 + 2x + 1}{x^2 - 5x + 6}$

c. $f(x) = \dfrac{2x^2 + x + 1}{x + 1}$

d. $f(x) = \dfrac{x^2 + 4}{x - 3}$

e. $f(x) = \dfrac{3x^2 - x - 2}{x - 2}$

f. $f(x) = \dfrac{4x^2 + x + 1}{x + 1}$

g. $f(x) = \dfrac{x^3 + x^2 - x - 1}{x^2 + 2x + 3}$

h. $f(x) = \dfrac{x^3 + 2x^2 + x - 3}{x^2 - 4}$

26. For each of the following rational functions, first determine and graph any oblique asymptotes. Then, on the same set of axes, graph the function.

a. $f(x) = \dfrac{x^2 - 1}{x - 2}$

b. $f(x) = \dfrac{x^2 + 1}{x + 2}$

5.7 PARTIAL FRACTIONS

In Chapter 0 we reviewed the process of adding rational expressions. For example,

$$\frac{3}{x - 2} + \frac{2}{x + 3} = \frac{3(x + 3) + 2(x - 2)}{(x - 2)(x + 3)} = \frac{3x + 9 + 2x - 4}{(x - 2)(x + 3)} = \frac{5x + 5}{(x - 2)(x + 3)}$$

Now suppose that we want to reverse the process. That is, suppose we are given the rational expression

$$\frac{5x + 5}{(x - 2)(x + 3)}$$

and we want to express it as the sum of two simpler rational expressions called **partial fractions**. This process, called **partial fraction decomposition**, has several applications in calculus and differential equations. The following property provides the basis for partial fraction decomposition.

PROPERTY 5.7

Let $f(x)$ and $g(x)$ be polynomials with real coefficients, such that the degree of $f(x)$ is less than the degree of $g(x)$. The indicated quotient $f(x)/g(x)$ can be decomposed into partial fractions as follows.

1. If $g(x)$ has a linear factor of the form $ax + b$, then the partial fraction decomposition will contain a term of the form

$$\frac{A}{ax + b},\qquad \text{where } A \text{ is a constant}$$

2. If $g(x)$ has a linear factor of the form $ax + b$ raised to the kth power, then the partial fraction decomposition will contain terms of the form

$$\frac{A_1}{ax + b} + \frac{A_2}{(ax + b)^2} + \cdots + \frac{A_k}{(ax + b)^k}$$

where $A_1, A_2, \ldots, A_k$ are constants.

3. If $g(x)$ has a quadratic factor of the form $ax^2 + bx + c$, where $b^2 - 4ac < 0$, then the partial fraction decomposition will contain a term of the form

$$\frac{Ax + B}{ax^2 + bx + c},\qquad \text{where } A \text{ and } B \text{ are constants}$$

4. If $g(x)$ has a quadratic factor of the form $ax^2 + bx + c$ raised to the kth power, where $b^2 - 4ac < 0$, then the partial fraction decomposition will contain terms of the form

$$\frac{A_1x + B_1}{ax^2 + bx + c} + \frac{A_2x + B_2}{(ax^2 + bx + c)^2} + \cdots + \frac{A_kx + B_kx}{(ax^2 + bx + c)^k}$$

where $A_1, A_2, \ldots, A_k$, and $B_1, B_2, \ldots, B_k$ are constants.

Note that Property 5.7 applies only to **proper fractions**—that is, fractions where the degree of the numerator is less than the degree of the denominator. If the numerator is not of lower degree, we can divide and then apply Property 5.7 to the remainder, which will be a proper fraction. For example,

$$\frac{x^3 - 3x^2 - 3x - 5}{x^2 - 4} = x - 3 + \frac{x - 17}{x^2 - 4}$$

and the proper fraction $\dfrac{x - 17}{x^2 - 4}$ can be decomposed into partial fractions by applying Property 5.7. Now let's consider some examples to illustrate the four cases in Property 5.7.

EXAMPLE 1

Find the partial fraction decomposition of $\dfrac{11x + 2}{2x^2 + x - 1}$.

Solution

The denominator can be expressed as $(x + 1)(2x - 1)$. Therefore, according to part 1 of Property 5.7, each of the linear factors produces a partial fraction of the form *constant over linear factor*. In other words, we can write

$$\frac{11x + 2}{(x + 1)(2x - 1)} = \frac{A}{x + 1} + \frac{B}{2x - 1} \tag{1}$$

for some constants A and B. To find A and B, we multiply both sides of equation (1) by the least common denominator $(x + 1)(2x - 1)$:

$$11x + 2 = A(2x - 1) + B(x + 1) \tag{2}$$

Equation (2) is an **identity**: *It is true for all values of x*. Therefore, let's choose some convenient values for x that will determine the values for A and B. If we let $x = -1$, then equation (2) becomes an equation only in A.

$$11(-1) + 2 = A[2(-1) - 1] + B(-1 + 1)$$
$$-9 = -3A$$
$$3 = A$$

If we let $x = \dfrac{1}{2}$, then equation (2) becomes an equation only in B.

$$11\left(\frac{1}{2}\right) + 2 = A\left[2\left(\frac{1}{2}\right) - 1\right] + B\left(\frac{1}{2} + 1\right)$$
$$\frac{15}{2} = \frac{3}{2}B$$
$$5 = B$$

Therefore, the given rational expression can now be written

$$\frac{11x + 2}{2x^2 + x - 1} = \frac{3}{x + 1} + \frac{5}{2x - 1} \qquad ■$$

The key idea in Example 1 is the statement that equation (2) is true for all values of x. If we had chosen *any* two values for x, we still would have been able to determine the values for A and B. For example, letting $x = 1$ and then $x = 2$ produces the equations $13 = A + 2B$ and $24 = 3A + 3B$. Solving this system of two equations in two unknowns produces $A = 3$ and $B = 5$. In Example 1, our choices of letting $x = -1$ and then $x = \dfrac{1}{2}$ simply eliminated the need for solving a system of equations to find A and B.

EXAMPLE 2

Find the partial fraction decomposition of

$$\frac{-2x^2 + 7x + 2}{x(x-1)^2}$$

Solution

Apply part 1 of Property 5.7 to determine that there is a partial fraction of the form A/x corresponding to the factor of x. Next, applying part 2 of Property 5.7 and the squared factor $(x-1)^2$ gives rise to a sum of partial fractions of the form

$$\frac{B}{x-1} + \frac{C}{(x-1)^2}$$

Therefore, the complete partial fraction decomposition is of the form

$$\frac{-2x^2 + 7x + 2}{x(x-1)^2} = \frac{A}{x} + \frac{B}{x-1} + \frac{C}{(x-1)^2} \tag{1}$$

Multiply both sides of equation (1) by $x(x-1)^2$ to produce

$$-2x^2 + 7x + 2 = A(x-1)^2 + Bx(x-1) + Cx \tag{2}$$

which is true for all values of x. If we let $x = 1$, then equation (2) becomes an equation only in C.

$$-2(1)^2 + 7(1) + 2 = A(1-1)^2 + B(1)(1-1) + C(1)$$
$$7 = C$$

If we let $x = 0$, then equation (2) becomes an equation just in A.

$$-2(0)^2 + 7(0) + 2 = A(0-1)^2 + B(0)(0-1) + C(0)$$
$$2 = A$$

If we let $x = 2$, then equation (2) becomes an equation in A, B, and C.

$$-2(2)^2 + 7(2) + 2 = A(2-1)^2 + B(2)(2-1) + C(2)$$
$$8 = A + 2B + 2C$$

But we already know that $A = 2$ and $C = 7$, so we can easily determine B.

$$8 = 2 + 2B + 14$$
$$-8 = 2B$$
$$-4 = B$$

Therefore, the original rational expression can be written

$$\frac{-2x^2 + 7x + 2}{x(x-1)^2} = \frac{2}{x} - \frac{4}{x-1} + \frac{7}{(x-1)^2}$$ ■

EXAMPLE 3

Find the partial fraction decomposition of

$$\frac{4x^2 + 6x - 10}{(x + 3)(x^2 + x + 2)}$$

Solution

Apply part 1 of Property 5.7 to determine that there is a partial fraction of the form $A/(x + 3)$ that corresponds to the factor $x + 3$. Apply part 3 of Property 5.7 to determine that there is also a partial fraction of the form

$$\frac{Bx + C}{x^2 + x + 2}$$

Thus the complete partial fraction decomposition is of the form

$$\frac{4x^2 + 6x - 10}{(x + 3)(x^2 + x + 2)} = \frac{A}{x + 3} + \frac{Bx + C}{x^2 + x + 2} \tag{1}$$

Multiply both sides of equation (1) by $(x + 3)(x^2 + x + 2)$ to produce

$$4x^2 + 6x - 10 = A(x^2 + x + 2) + (Bx + C)(x + 3) \tag{2}$$

which is true for all values of x. If we let $x = -3$, then equation (2) becomes an equation in A alone.

$$4(-3)^2 + 6(-3) - 10 = A[(-3)^2 + (-3) + 2] + [B(-3) + C][(-3) + 3]$$
$$8 = 8A$$
$$1 = A$$

If we let $x = 0$, then equation (2) becomes an equation in A and C.

$$4(0)^2 + 6(0) - 10 = A(0^2 + 0 + 2) + [B(0) + C](0 + 3)$$
$$-10 = 2A + 3C$$

Because $A = 1$, we obtain the value of C.

$$-10 = 2 + 3C$$
$$-12 = 3C$$
$$-4 = C$$

If we let $x = 1$, then equation (2) becomes an equation in A, B, and C.

$$4(1)^2 + 6(1) - 10 = A(1^2 + 1 + 2) + [B(1) + C](1 + 3)$$
$$0 = 4A + 4B + 4C$$
$$0 = A + B + C$$

But because $A = 1$ and $C = -4$, we obtain the value of B.

$$0 = A + B + C$$
$$0 = 1 + B + (-4)$$
$$3 = B$$

Therefore, the original rational expression can now be written

$$\frac{4x^2 + 6x - 10}{(x + 3)(x^2 + x + 2)} = \frac{1}{x + 3} + \frac{3x - 4}{x^2 + x + 2}$$ ∎

EXAMPLE 4

Find the partial fraction decomposition of

$$\frac{x^3 + x^2 + x + 3}{(x^2 + 1)^2}$$

Solution

Apply part 4 of Property 5.7 to determine that the partial fraction decomposition of this fraction is of the form

$$\frac{x^3 + x^2 + x + 3}{(x^2 + 1)^2} = \frac{Ax + B}{x^2 + 1} + \frac{Cx + D}{(x^2 + 1)^2}$$ **(1)**

Multiply both sides of equation (1) by $(x^2 + 1)^2$ to produce

$$x^3 + x^2 + x + 3 = (Ax + B)(x^2 + 1) + Cx + D$$ **(2)**

which is true for all values of x. Equation (2) is an identity, so we know that the coefficients of similar terms on both sides of the equation must be equal. Therefore, let's collect similar terms on the right side of equation (2).

$$x^3 + x^2 + x + 3 = Ax^3 + Ax + Bx^2 + B + Cx + D$$
$$= Ax^3 + Bx^2 + (A + C)x + B + D$$

Now we can equate coefficients from both sides:

$$1 = A \qquad 1 = B \qquad 1 = A + C \qquad \text{and} \qquad 3 = B + D$$

From these equations we can determine that $A = 1$, $B = 1$, $C = 0$, and $D = 2$. Therefore, the original rational expression can be written

$$\frac{x^3 + x^2 + x + 3}{(x^2 + 1)^2} = \frac{x + 1}{x^2 + 1} + \frac{2}{(x^2 + 1)^2}$$ ∎

PROBLEM SET 5.7

For Problems 1–22, find the partial fraction decomposition for each rational expression.

1. $\dfrac{11x - 10}{(x - 2)(x + 1)}$

2. $\dfrac{11x - 2}{(x + 3)(x - 4)}$

3. $\dfrac{-2x - 8}{x^2 - 1}$

4. $\dfrac{-2x + 32}{x^2 - 4}$

5. $\dfrac{20x - 3}{6x^2 + 7x - 3}$

6. $\dfrac{-2x - 8}{10x^2 - x - 2}$

7. $\dfrac{x^2 - 18x + 5}{(x - 1)(x + 2)(x - 3)}$

8. $\dfrac{-9x^2 + 7x - 4}{x^3 - 3x^2 - 4x}$

9. $\dfrac{-6x^2 + 7x + 1}{x(2x - 1)(4x + 1)}$

10. $\dfrac{15x^2 + 20x + 30}{(x + 3)(3x + 2)(2x + 3)}$

11. $\dfrac{2x + 1}{(x - 2)^2}$ **12.** $\dfrac{-3x + 1}{(x + 1)^2}$

13. $\dfrac{-6x^2 + 19x + 21}{x^2(x + 3)}$ **14.** $\dfrac{10x^2 - 73x + 144}{x(x - 4)^2}$

15. $\dfrac{-2x^2 - 3x + 10}{(x^2 + 1)(x - 4)}$ **16.** $\dfrac{8x^2 + 15x + 12}{(x^2 + 4)(3x - 4)}$

17. $\dfrac{3x^2 + 10x + 9}{(x + 2)^3}$ **18.** $\dfrac{2x^3 + 8x^2 + 2x + 4}{(x + 1)^2(x^2 + 3)}$

19. $\dfrac{5x^2 + 3x + 6}{x(x^2 - x + 3)}$ **20.** $\dfrac{x^3 + x^2 + 2}{(x^2 + 2)^2}$

21. $\dfrac{2x^3 + x + 3}{(x^2 + 1)^2}$ **22.** $\dfrac{4x^2 + 3x + 14}{x^3 - 8}$

THOUGHTS INTO WORDS

23. Give a general description of partial fraction decomposition for someone who missed class the day it was discussed.

24. Give a step-by-step explanation of how to find the partial fraction decomposition of $\dfrac{11x + 5}{2x^2 + 5x - 3}$.

CHAPTER 5 SUMMARY

Two themes unify this chapter: (1) solving polynomial equations and (2) graphing polynomial and rational functions.

Solving Polynomial Equations

The following concepts and properties provide the basis for solving polynomial equations.

1. Synthetic division.

2. The factor theorem: A polynomial $f(x)$ has a factor $x - c$ if and only if $f(c) = 0$.

3. Property 5.3: A polynomial equation of degree n has n solutions, where any solution of multiplicity p is counted p times.

4. The rational root theorem: Consider the polynomial equation

$$a_n x^n + a_{n-1} x^{n-1} + \cdots + a_1 x + a_0 = 0$$

where *the coefficients are integers*. If the rational number c/d, reduced to lowest terms, is a solution of the equation, then c is a factor of the constant term, a_0, and d is a factor of the leading coefficient, a_n.

5. Property 5.5: If a polynomial equation with real coefficients has any non-real complex solutions, they must occur in conjugate pairs.

6. Descartes' rule of signs: Let $a_n x^n + a_{n-1}x^{n-1} + \cdots + a_1 x + a_0 = 0$ be a polynomial equation with real coefficients.

a. The number of *positive real solutions* either is equal to the number of sign variations in the given polynomial or is less than the number of sign variations by a positive even integer.

b. The number of *negative real solutions* either is equal to the number of sign variations in

$$a_n(-x)^n + a_{n-1}(-x)^{n-1} + \cdots + a_1(-x) + a_0$$

or is less than that number of sign variations by a positive even integer.

Graphing Polynomial and Rational Functions

Graphs of polynomial functions of the form $f(x) = ax^n$, where n is an integer greater than 2 and $a \neq 1$, are variations of the graphs shown in Figures 5.3 and 5.6. If n is odd, the curve is symmetric about the origin, and if n is even, the graph is symmetric about the vertical axis.

Graphs of polynomial functions of the form $f(x) = ax^n$ can be translated horizontally and vertically and reflected across the x axis. For example:

1. The graph of $f(x) = 2(x-4)^3$ is the graph of $f(x) = 2x^3$ moved four units to the right.

2. The graph of $f(x) = 3x^4 + 4$ is the graph of $f(x) = 3x^4$ moved up four units.

3. The graph of $f(x) = -x^5$ is the graph of $f(x) = x^5$ reflected across the x axis.

To graph a polynomial function that is expressed in factored form, the following steps are helpful.

1. Find the x intercepts, which are also called the *zeros* of the polynomial.

2. Use a test value in each of the intervals determined by the x intercepts to find out whether the function is positive or negative over that interval.

3. Plot any additional points that are needed to determine the graph.

To graph a rational function, the following steps are useful.

1. Check for vertical-axis and origin symmetry.

2. Find any vertical asymptotes by setting the denominator equal to zero and solving it for x.

3. Find any horizontal asymptotes by studying the behavior of $f(x)$ as x gets very large or very small. This may require changing the form of the original rational expression.

4. If the degree of the numerator is one larger than the degree of the denominator, determine the equation of the oblique asymptote.

5. Study the behavior of the graph when it is close to the asymptotic lines.

6. Plot as many points as necessary to determine the graph. This may be affected by whether the graph has any symmetries.

Be sure that you understand the process of partial fraction decomposition outlined in Property 5.7.

CHAPTER 5 REVIEW PROBLEM SET

For Problems 1 and 2, find the quotient and remainder of each division problem.

1. $(6x^3 + 11x^2 - 27x + 32) \div (2x + 7)$

2. $(2a^3 - 3a^2 + 13a - 1) \div (a^2 - a + 6)$

For Problems 3–6, use synthetic division to determine the quotient and remainder.

3. $(3x^3 - 4x^2 + 6x - 2) \div (x - 1)$

4. $(5x^3 + 7x^2 - 9x + 10) \div (x + 2)$

5. $(-2x^4 + x^3 - 2x^2 - x - 1) \div (x + 4)$

6. $(-3x^4 - 5x^2 + 9) \div (x + 3)$

For Problems 7–10, find $f(c)$ either by using synthetic division and the remainder theorem or by evaluating $f(c)$ directly.

7. $f(x) = 4x^5 - 3x^3 + x^2 - 1$ and $c = 1$

8. $f(x) = 4x^3 - 7x^2 + 6x - 8$ and $c = -3$

9. $f(x) = -x^4 + 9x^2 - x - 2$ and $c = -2$

10. $f(x) = x^4 - 9x^3 + 9x^2 - 10x + 16$ and $c = 8$

For Problems 11–14, use the factor theorem to help answer some questions about factors.

11. Is $x + 2$ a factor of $2x^3 + x^2 - 7x - 2$?

12. Is $x - 3$ a factor of $x^4 + 5x^3 - 7x^2 - x + 3$?

13. Is $x - 4$ a factor of $x^5 - 1024$?

14. Is $x + 1$ a factor of $x^5 + 1$?

For Problems 15–18, use the rational root theorem and the factor theorem to help solve each equation.

15. $x^3 - 3x^2 - 13x + 15 = 0$

16. $8x^3 + 26x^2 - 17x - 35 = 0$

17. $x^4 - 5x^3 + 34x^2 - 82x + 52 = 0$

18. $x^3 - 4x^2 - 10x + 4 = 0$

For Problems 19 and 20, use Descartes' rule of signs (Property 5.6) to list the possibilities for the nature of the solutions. *Do not* solve the equations.

19. $4x^4 - 3x^3 + 2x^2 + x + 4 = 0$

20. $x^5 + 3x^3 + x + 7 = 0$

For Problems 21–24, graph each polynomial function.

21. $f(x) = -(x - 2)^3 + 3$

22. $f(x) = (x + 3)(x - 1)(3 - x)$

23. $f(x) = x^4 - 4x^2$

24. $f(x) = x^3 - 4x^2 + x + 6$

For Problems 25–28, graph each rational function. Be sure to identify the asymptotes.

25. $f(x) = \dfrac{2x}{x-3}$

26. $f(x) = \dfrac{-3}{x^2+1}$

27. $f(x) = \dfrac{-x^2}{x^2-x-6}$

28. $f(x) = \dfrac{x^2+3}{x+1}$

For Problems 29 and 30, find the partial fraction decomposition.

29. $\dfrac{5x^2-4}{x^2(x+2)}$

30. $\dfrac{x^2-x-21}{(x^2+4)(2x-1)}$

CHAPTER 5 TEST

1. Find the quotient and remainder for the division problem $(6x^3 - 19x^2 + 3x + 20) \div (3x - 5)$.

2. Find the quotient and remainder for the division problem $(3x^4 + 8x^3 - 5x^2 - 12x - 15) \div (x + 3)$.

3. Find the quotient and remainder for the division problem $(4x^4 - 7x^2 + 4) \div (x - 2)$.

4. If $f(x) = x^5 - 8x^4 + 9x^3 - 13x^2 - 9x - 10$, find $f(7)$.

5. If $f(x) = 3x^4 + 20x^3 - 6x^2 + 9x + 19$, find $f(-7)$.

6. If $f(x) = x^5 - 35x^3 - 32x + 15$, find $f(6)$.

7. Is $x - 5$ a factor of $3x^3 - 11x^2 - 22x - 20$?

8. Is $x + 2$ a factor of $5x^3 + 9x^2 - 9x - 17$?

9. Is $x + 3$ a factor of $x^4 - 16x^2 - 17x + 12$?

10. Is $x - 6$ a factor of $x^4 - 2x^2 + 3x - 12$?

11. Use Descartes' rule of signs to determine the nature of the roots of $5x^4 + 3x^3 - x^2 - 9 = 0$.

12. Find the x intercepts of the graph of the function $f(x) = 3x^3 + 19x^2 - 14x$.

13. Find the equation of the vertical asymptote for the graph of the function $f(x) = \dfrac{5x}{x + 3}$.

14. Find the equation of the horizontal asymptote for the graph of the function $f(x) = \dfrac{5x^2}{x^2 - 4}$.

15. What type of symmetry does the equation $f(x) = \dfrac{x^2}{x^2 + 2}$ exhibit?

16. What type of symmetry does the equation $f(x) = \dfrac{-3x}{x^2 + 1}$ exhibit?

17. Find the equation of the oblique asymptote for the graph of the function $f(x) = \dfrac{4x^2 + x + 1}{x + 1}$.

For Problems 18–20, find the partial fraction decomposition.

18. $\dfrac{11x - 22}{(2x - 1)(x - 6)}$

19. $\dfrac{x^2 - 2x - 4}{x(x + 2)^2}$

20. $\dfrac{3x^2 - x + 1}{(x + 1)(x^2 - x + 3)}$

For Problems 21–25, graph each of the functions. Be sure to identify the asymptotes for the rational functions.

21. $f(x) = (2 - x)(x - 1)(x + 1)$ **22.** $f(x) = \dfrac{-x}{x - 3}$

23. $f(x) = (x + 2)^2(x - 1)$ **24.** $f(x) = \dfrac{-2}{x^2 - 4}$

25. $f(x) = \dfrac{4x^2 + x + 1}{x + 1}$

CUMULATIVE REVIEW PROBLEM SET · CHAPTERS 0–5

For Problems 1–10, evaluate each numerical expression.

1. $\left(\dfrac{3}{4}\right)^{-3}$

2. $\sqrt[3]{-\dfrac{8}{27}}$

3. -5^{-2}

4. $8^{4/3}$

5. $9^{-(3/2)}$

6. $\log_4 64$

7. $\log_{10} 0.0001$

8. $\log_2\left(\dfrac{1}{32}\right)$

9. $(-64)^{2/3}$

10. $\ln e^3$

For Problems 11–33, solve each problem.

11. Express the domain of the function $f(x) = \sqrt{2x^2 + 11x - 6}$ using interval notation.

12. If $f(x) = 3x - 1$ and $g(x) = x^2 - x + 3$, find $(f \circ g)(-2)$ and $(g \circ f)(3)$.

13. If $f(x) = -\dfrac{2}{x}$ and $g(x) = \dfrac{1}{x - 4}$, find $(f \circ g)(x)$ and $(g \circ f)(x)$. Also indicate the domain of each composite function.

14. If $f(x) = -2x + 7$, find the inverse of f.

15. If $f(x) = x^2 + 7x - 2$, find $\dfrac{f(a + h) - f(a)}{h}$.

16. If $f(x) = 2x^4 - 17x^3 - 10x^2 + 11x + 15$, find $f(9)$.

17. Find the quotient for $(3x^5 - 25x^3 - 7x^2 + x + 6) \div (x - 3)$.

18. Is $x + 2$ a factor of $2x^4 + 3x^3 + x^2 + 2x - 16$?

19. Evaluate $\log_2 50$ to the nearest hundredth.

20. Find the center and the length of a radius of the circle $x^2 + y^2 + 6x - 4y + 4 = 0$.

21. Write the equation of the line that contains the points $(-4, 2)$ and $(5, -1)$.

22. Write the equation of the perpendicular bisector of the line segment determined by $(-2, -4)$ and $(6, 2)$.

23. Find the length of the major axis of the ellipse $16x^2 + y^2 = 64$.

24. Find the equations of the asymptotes of the hyperbola $x^2 - 9y^2 = 18$.

25. If y varies directly as x, and if $y = 3$ when $x = 4$, find y when $x = 16$.

26. If y varies inversely as the square root of x, and if $y = \dfrac{2}{5}$ when $x = 25$, find y when $x = 49$.

27. Find the total amount of money accumulated at the end of 8 years if $450 is invested at 7% compounded quarterly.

28. How long will it take $500 to double if it is invested at 8% interest compounded continuously?

29. Sandy has a collection of 57 coins worth $10. They consist of nickels, dimes, and quarters, and the number of quarters is 2 more than three times the number of nickels. How many coins of each kind does she have?

30. A retailer bought a dress for $75 and wants to sell it at a profit of 40% of the selling price. What price should she ask for the dress?

31. A container has 8 quarts of a 30% alcohol solution. How much pure alcohol should be added to raise it to a 40% solution?

32. Claire rode her bicycle out into the country at a speed of 15 miles per hour and returned along the same route at 10 miles per hour. If the entire trip took $7\dfrac{1}{2}$ hours, how far out did she ride?

33. Adam can do a job in 2 hours less time than it takes Carl to do the same job. Working together, they can do the job in 2 hours and 24 minutes. How long would it take Adam to do the job by himself?

For Problems 34–45, solve each equation.

34. $(2x - 5)(6x + 1) = (3x + 2)(4x - 7)$

35. $(2x + 1)(x - 2) = (3x - 2)(x + 4)$

36. $4x^3 + 20x^2 - 56x = 0$

37. $6x^3 + 17x^2 + x - 10 = 0$

38. $|4x - 3| = 7$

39. $\dfrac{2x - 1}{3} - \dfrac{3x + 2}{4} = -\dfrac{5}{6}$

40. $3^{x-2} = 27^x$

41. $\ln(t + 2) = \ln t + \ln 4$

42. $\log 5 + \log(x - 1) = 1$

43. $x^4 + 3x^2 - 54 = 0$

44. $(2x - 1)(x + 3) = 49$

45. $x^4 - 2x^3 + 2x^2 - 7x + 6 = 0$

For Problems 46–53, solve each inequality and express the solution set using interval notation.

46. $3(x - 1) - 5(x + 2) > 3(x + 4)$

47. $\dfrac{x - 1}{2} + \dfrac{2x + 1}{5} \geq \dfrac{x - 2}{3}$

48. $x^2 - 3x < 18$

49. $(x - 1)(x + 3)(2 - x) \leq 0$

50. $|2x - 1| > 6$

51. $|3x + 2| \leq 8$

52. $\dfrac{4x - 3}{x - 2} \geq 0$

53. $\dfrac{x + 3}{x - 4} < 3$

For Problems 54–64, graph each function.

54. $f(x) = -2x + 4$

55. $f(x) = 2x^2 - 3$

56. $f(x) = 2^x - 3$

57. $f(x) = \log_2(x - 1)$

58. $f(x) = \dfrac{2x}{x + 1}$

59. $f(x) = -|x - 2| + 1$

60. $f(x) = 2\sqrt{x} + 1$

61. $f(x) = 3x^2 + 12x + 9$

62. $f(x) = -(x - 3)^3 + 1$

63. $f(x) = (x + 1)(x - 2)(x - 4)$

64. $f(x) = x^4 - x^2$

SYSTEMS OF EQUATIONS

A system of two linear equations in two variables can be used to approximate the effect of the jet stream on airline schedules.

435

In this chapter we will begin by reviewing some techniques for solving sys-tems of linear equations that involve two or three variables. Then, because many applications of mathematics require the use of large numbers of vari-ables and equations, we will introduce some additional techniques for solv-ing such extensive systems. These new techniques also form a basis for solving systems by using a computer.

6.1   SYSTEMS OF TWO LINEAR EQUATIONS IN TWO VARIABLES

In Chapter 2 we stated that any equation of the form $Ax + By = C$, where A, B, and C are real numbers (A and B not both zero), is a **linear equation** in the two variables x and y, and its graph is a straight line. Two linear equations in two variables consid-ered together form a **system of two linear equations in two variables**, as illustrated by the following examples.

$$\begin{pmatrix} x + y = 6 \\ x - y = 2 \end{pmatrix} \qquad \begin{pmatrix} 3x + 2y = 1 \\ 5x - 2y = 23 \end{pmatrix} \qquad \begin{pmatrix} 4x - 5y = 21 \\ -3x + y = -7 \end{pmatrix}$$

To solve a system, (such as any of these three examples) means to find all of the ordered pairs that simultaneously satisfy both equations in the system. For example, if we graph the two equations $x + y = 6$ and $x - y = 2$ on the same set of axes, as in Figure 6.1, then the ordered pair associated with the point of intersection of the two

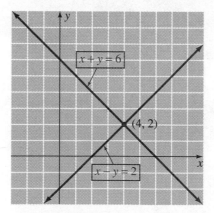

FIGURE 6.1

lines is the **solution of the system.** Thus we say that $\{(4, 2)\}$ is the solution set of the system

$$\begin{pmatrix} x + y = 6 \\ x - y = 2 \end{pmatrix}$$

To check the solution, we substitute 4 for x and 2 for y in the two equations.

$x + y = 6$ becomes $4 + 2 = 6$, a true statement

$x - y = 2$ becomes $4 - 2 = 2$, a true statement

Because the graph of a linear equation in two variables is a straight line, there are three possible situations that can occur when we are solving a system of two linear equations in two variables. Each situation is shown in Figure 6.2.

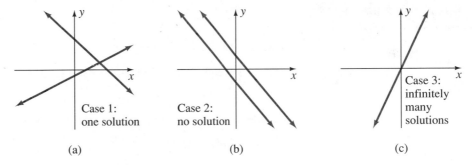

(a)   (b)   (c)

F I G U R E   6 . 2

CASE I	The graphs of the two equations are two lines intersecting in *one* point. There is exactly one solution, and the system is called a **consistent system**.
CASE 2	The graphs of the two equations are parallel lines. There is *no solution*, and the system is called an **inconsistent system**.
CASE 3	The graphs of the two equations are the same line, and there are *infinitely many solutions* of the system. Any pair of real numbers that satisfies one of the equations also satisfies the other equation, and we say that the equations are dependent.

Thus, as we solve a system of two linear equations in two variables, we can expect one of three outcomes: The system will have *no* solutions, *one* ordered pair as a solution, or *infinitely many* ordered pairs as solutions.

The Substitution Method

Solving specific systems of equations by graphing requires accurate graphs. However, unless the solutions are integers, it is difficult to obtain exact solutions from a graph. Therefore, we will consider some other techniques for solving systems of equations.

The **substitution method**, which works especially well with systems of two equations in two unknowns, can be described as follows.

STEP 1   Solve one of the equations for one variable in terms of the other. (If possible, make a choice that will avoid fractions.)

STEP 2   Substitute the expression obtained in step 1 into the other equation, producing an equation in one variable.

STEP 3   Solve the equation obtained in step 2.

STEP 4   Use the solution obtained in step 3, along with the expression obtained in step 1, to determine the solution of the system.

EXAMPLE 1

Solve the system $\begin{pmatrix} x - 3y = -25 \\ 4x + 5y = 19 \end{pmatrix}$.

Solution

Solve the first equation for x in terms of y to produce

$$x = 3y - 25$$

Substitute $3y - 25$ for x in the second equation and solve for y.

$$\begin{aligned}
4x + 5y &= 19 \\
4(3y - 25) + 5y &= 19 \\
12y - 100 + 5y &= 19 \\
17y &= 119 \\
y &= 7
\end{aligned}$$

Next, substitute 7 for y in the equation $x = 3y - 25$ to obtain

$$x = 3(7) - 25 = -4$$

The solution set of the given system is $\{(-4, 7)\}$. (You should check this solution in both of the original equations.) ■

EXAMPLE 2

Solve the system $\begin{pmatrix} 5x + 9y = -2 \\ 2x + 4y = -1 \end{pmatrix}$.

Solution

A glance at the system should tell us that solving either equation for either variable will produce a fractional form, so let's just use the first equation and solve for x in terms of y.

$$\begin{aligned}
5x + 9y &= -2 \\
5x &= -9y - 2 \\
x &= \frac{-9y - 2}{5}
\end{aligned}$$

Now we can substitute this value for x into the second equation and solve for y.

$$2x + 4y = -1$$

$$2\left(\frac{-9y - 2}{5}\right) + 4y = -1$$

$$2(-9y - 2) + 20y = -5 \qquad \text{Multiplied both sides by 5.}$$

$$-18y - 4 + 20y = -5$$

$$2y - 4 = -5$$

$$2y = -1$$

$$y = -\frac{1}{2}$$

Now we can substitute $-\dfrac{1}{2}$ for y in $x = \dfrac{-9y - 2}{5}$.

$$x = \frac{-9\left(-\dfrac{1}{2}\right) - 2}{5} = \frac{\dfrac{9}{2} - 2}{5} = \frac{1}{2}$$

The solution set is $\left\{\left(\dfrac{1}{2}, -\dfrac{1}{2}\right)\right\}$.

EXAMPLE 3

Solve the system

$$\left(\begin{array}{c} 6x - 4y = 18 \\ y = \dfrac{3}{2}x - \dfrac{9}{2} \end{array}\right)$$

Solution

The second equation is given in appropriate form for us to begin the substitution process. Substitute $\dfrac{3}{2}x - \dfrac{9}{2}$ for y in the first equation to yield

$$6x - 4y = 18$$

$$6x - 4\left(\frac{3}{2}x - \frac{9}{2}\right) = 18$$

$$6x - 6x + 18 = 18$$

$$18 = 18$$

Our obtaining a true numerical statement ($18 = 18$) indicates that the system has infinitely many solutions. Any ordered pair that satisfies one of the equations will also satisfy the other equation. Thus, in the second equation of the original system, if we let $x = k$, then $y = \dfrac{3}{2}k - \dfrac{9}{2}$. Therefore, the solution set can be expressed $\left\{\left(k, \dfrac{3}{2}k - \dfrac{9}{2}\right) \mid k \text{ is a real number}\right\}$. If some specific solutions are needed, they can be generated by the ordered pair $\left(k, \dfrac{3}{2}k - \dfrac{9}{2}\right)$. For example, if we let $k = 1$, then we get

$\frac{3}{2}(1) - \frac{9}{2} = -\frac{6}{2} = -3$. Thus the ordered pair $(1, -3)$ is a member of the solution set of the given system. ∎

The Elimination-by-Addition Method

Now let's consider the **elimination-by-addition method** for solving a system of equations. This is a very important method because it is the basis for developing other techniques for solving systems that contain many equations and variables. The method involves replacing systems of equations with *simpler equivalent systems* until we obtain a system where the solutions are obvious. **Equivalent systems of equations are systems that have exactly the same solution set.** The following operations or transformations can be applied to a system of equations to produce an equivalent system.

    1. Any two equations of the system can be interchanged.

    2. Both sides of any equation of the system can be multiplied by any nonzero real number.

    3. Any equation of the system can be replaced by the sum of that equation and a nonzero multiple of another equation.

EXAMPLE 4

Solve the system $\begin{pmatrix} 3x + 5y = -9 \\ 2x - 3y = 13 \end{pmatrix}$.

        (1)
        (2)

Solution

We can replace the given system with an equivalent system by multiplying equation (2) by -3.

$$\begin{pmatrix} 3x + 5y = -9 \\ -6x + 9y = -39 \end{pmatrix}$$

        (3)
        (4)

Now let's replace equation (4) with an equation formed by multiplying equation (3) by 2 and adding this result to equation (4).

$$\begin{pmatrix} 3x + 5y = -9 \\ 19y = -57 \end{pmatrix}$$

        (5)
        (6)

From equation (6) we can easily determine that $y = -3$. Then, substituting -3 for y in equation (5) produces

$$3x + 5(-3) = -9$$
$$3x - 15 = -9$$
$$3x = 6$$
$$x = 2$$

The solution set for the given system is $\{(2, -3)\}$. ∎

    REMARK We are using a format for the elimination-by-addition method that highlights the use of equivalent systems. In Section 6.3, this format will lead naturally to an approach using matrices. Thus it is beneficial to stress the use of equivalent systems at this time.

EXAMPLE 5

Solve the system

$$\left(\begin{array}{l} \dfrac{1}{2}x + \dfrac{2}{3}y = -4 \\[2mm] \dfrac{1}{4}x - \dfrac{3}{2}y = 20 \end{array}\right)$$

(7)

(8)

Solution

The given system can be replaced with an equivalent system by multiplying equation (7) by 6 and equation (8) by 4.

$$\left(\begin{array}{l} 3x + 4y = -24 \\ x - 6y = 80 \end{array}\right)$$

(9)

(10)

Now let's exchange equations (9) and (10).

$$\left(\begin{array}{l} x - 6y = 80 \\ 3x + 4y = -24 \end{array}\right)$$

(11)

(12)

We can replace equation (12) with an equation formed by multiplying equation (11) by -3 and adding this result to equation (12).

$$\left(\begin{array}{l} x - 6y = 80 \\ 22y = -264 \end{array}\right)$$

(13)

(14)

From equation (14) we can determine that $y = -12$. Then, substituting -12 for y in equation (13) produces

$$x - 6(-12) = 80$$
$$x + 72 = 80$$
$$x = 8$$

The solution set of the given system is $\{(8, -12)\}$. (Check this!)   ■

EXAMPLE 6

Solve the system $\left(\begin{array}{l} x - 4y = 9 \\ x - 4y = 3 \end{array}\right)$.

(15)

(16)

Solution

We can replace equation (16) with an equation formed by multiplying equation (15) by -1 and adding this result to equation (16).

$$\left(\begin{array}{l} x - 4y = 9 \\ 0 = -6 \end{array}\right)$$

(17)

(18)

The statement $0 = -6$ is a contradiction, and therefore the original system is *inconsistent*; it has no solution. The solution set is $\varnothing$.   ■

Both the elimination-by-addition and substitution methods can be used to obtain exact solutions for any system of two linear equations in two unknowns. Sometimes the issue is one of deciding which method to use on a particular system. Some systems lend themselves to one or the other of the methods by virtue of the original format of the equations. We will illustrate this idea in a moment when we solve some word problems.

Using Systems to Solve Problems

Many word problems that we solved earlier in this text with one variable and one equation can also be solved by using a system of two linear equations in two variables. In fact, in many of these problems you may find it more natural to use two variables and two equations.

The two-variable expression $10t + u$ can be used to represent any two-digit whole number. The t represents the tens digit, and the u represents the units digit. For example, if $t = 4$ and $u = 8$, then $10t + u$ becomes $10(4) + 8 = 48$. Now let's use this general representation for a two-digit number to help solve a problem.

PROBLEM 1

The units digit of a two-digit number is one more than twice the tens digit. The number with the digits reversed is 45 larger than the original number. Find the original number.

Solution

Let u represent the units digit of the original number, and let t represent the tens digit. Then $10t + u$ represents the original number, and $10u + t$ represents the new number with the digits reversed. The problem translates into the following system.

$$\begin{pmatrix} u = 2t + 1 \\ 10u + t = 10t + u + 45 \end{pmatrix}$$

The units digit is one more than twice the tens digit.

The number with the digits reversed is 45 larger than the original number.

Simplify the second equation, and the system becomes

$$\begin{pmatrix} u = 2t + 1 \\ u - t = 5 \end{pmatrix}$$

Because of the form of the first equation, this system lends itself to solution by the substitution method. Substitute $2t + 1$ for u in the second equation to produce

$$(2t + 1) - t = 5$$
$$t + 1 = 5$$
$$t = 4$$

Now substitute 4 for t in the equation $u = 2t + 1$ to get

$$u = 2(4) + 1 = 9$$

The tens digit is 4 and the units digit is 9, so the number is 49. ∎

PROBLEM 2

Lucinda invested $950, part of it at 11% interest and the remainder at 12%. Her total yearly income from the two investments was $111.50. How much did she invest at each rate?

  Solution

Let x represent the amount invested at 11% and y the amount invested at 12%. The problem translates into the following system.

$$\begin{pmatrix} x + y = 950 \\ 0.11x + 0.12y = 111.50 \end{pmatrix}$$

— The two investments total $950.

The yearly interest from the two investments totals $111.50.

Multiply the second equation by 100 to produce an equivalent system.

$$\begin{pmatrix} x + y = 950 \\ 11x + 12y = 11150 \end{pmatrix}$$

Because neither equation is solved for one variable in terms of the other, let's use the elimination-by-addition method to solve the system. The second equation can be replaced by an equation formed by multiplying the first equation by -11 and adding this result to the second equation.

$$\begin{pmatrix} x + y = 950 \\ y = 700 \end{pmatrix}$$

Now we substitute 700 for y in the equation $x + y = 950$.

$$x + 700 = 950$$
$$x = 250$$

Therefore, Lucinda must have invested $250 at 11% and $700 at 12%.   ∎

In our final example of this section, we will use a graphing utility to help solve a system of equations.

EXAMPLE 7

Solve the system $\begin{pmatrix} 1.14x + 2.35y = -7.12 \\ 3.26x - 5.05y = 26.72 \end{pmatrix}$

Solution

First, we need to solve each equation for y in terms of x. Thus the system becomes

$$\begin{pmatrix} y = \dfrac{-7.12 - 1.14x}{2.35} \\ y = \dfrac{3.26x - 26.72}{5.05} \end{pmatrix}$$

Now we can enter both of these equations into a graphing utility and obtain Figure 6.3. From this figure it appears that the point of intersection is at approximately $x = 2$ and $y = -4$. By direct substitution into the given equations, we can verify that the point of intersection is exactly $(2, -4)$.

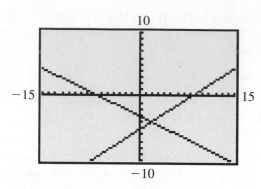

FIGURE 6.3

PROBLEM SET 6.1

For Problems 1–18, solve each system by using the substitution method.

1. $\begin{pmatrix} x + y = 16 \\ y = x + 2 \end{pmatrix}$

2. $\begin{pmatrix} 2x + 3y = -5 \\ y = 2x + 9 \end{pmatrix}$

3. $\begin{pmatrix} x = 3y - 25 \\ 4x + 5y = 19 \end{pmatrix}$

4. $\begin{pmatrix} 3x - 5y = 25 \\ x = y + 7 \end{pmatrix}$

5. $\begin{pmatrix} y = \dfrac{2}{3}x - 1 \\ 5x - 7y = 9 \end{pmatrix}$

6. $\begin{pmatrix} y = \dfrac{3}{4}x + 5 \\ 4x - 3y = -1 \end{pmatrix}$

7. $\begin{pmatrix} a = 4b + 13 \\ 3a + 6b = -33 \end{pmatrix}$

8. $\begin{pmatrix} 9a - 2b = 28 \\ b = -3a + 1 \end{pmatrix}$

9. $\begin{pmatrix} 2x - 3y = 4 \\ y = \dfrac{2}{3}x - \dfrac{4}{3} \end{pmatrix}$

10. $\begin{pmatrix} t + u = 11 \\ t = u + 7 \end{pmatrix}$

11. $\begin{pmatrix} u = t - 2 \\ t + u = 12 \end{pmatrix}$

12. $\begin{pmatrix} y = 5x - 9 \\ 5x - y = 9 \end{pmatrix}$

13. $\begin{pmatrix} 4x + 3y = -7 \\ 3x - 2y = 16 \end{pmatrix}$

14. $\begin{pmatrix} 5x - 3y = -34 \\ 2x + 7y = -30 \end{pmatrix}$

15. $\begin{pmatrix} 5x - y = 4 \\ y = 5x + 9 \end{pmatrix}$

16. $\begin{pmatrix} 2x + 3y = 3 \\ 4x - 9y = -4 \end{pmatrix}$

17. $\begin{pmatrix} 4x - 5y = 3 \\ 8x + 15y = -24 \end{pmatrix}$

18. $\begin{pmatrix} 4x + y = 9 \\ y = 15 - 4x \end{pmatrix}$

For Problems 19–34, solve each system by using the elimination-by-addition method.

19. $\begin{pmatrix} 3x + 2y = 1 \\ 5x - 2y = 23 \end{pmatrix}$

20. $\begin{pmatrix} 4x + 3y = -22 \\ 4x - 5y = 26 \end{pmatrix}$

21. $\begin{pmatrix} x - 3y = -22 \\ 2x + 7y = 60 \end{pmatrix}$

22. $\begin{pmatrix} 6x - y = 3 \\ 5x + 3y = -9 \end{pmatrix}$

23. $\begin{pmatrix} 4x - 5y = 21 \\ 3x + 7y = -38 \end{pmatrix}$

24. $\begin{pmatrix} 5x - 3y = -34 \\ 2x + 7y = -30 \end{pmatrix}$

25. $\begin{pmatrix} 5x - 2y = 19 \\ 5x - 2y = 7 \end{pmatrix}$

26. $\begin{pmatrix} 3a - 2b = 5 \\ 2a + 7b = 9 \end{pmatrix}$

27. $\begin{pmatrix} 6a - 3b = 4 \\ 5a + 2b = -1 \end{pmatrix}$

28. $\begin{pmatrix} 7x + 2y = 11 \\ 7x + 2y = -4 \end{pmatrix}$

29. $\begin{pmatrix} \dfrac{2}{3}s + \dfrac{1}{4}t = -1 \\ \dfrac{1}{2}s - \dfrac{1}{3}t = -7 \end{pmatrix}$

30. $\begin{pmatrix} \dfrac{1}{4}s - \dfrac{2}{3}t = -3 \\ \dfrac{1}{3}s + \dfrac{1}{3}t = 7 \end{pmatrix}$

31. $\begin{pmatrix} \dfrac{x}{2} - \dfrac{2y}{5} = \dfrac{-23}{60} \\ \dfrac{2x}{3} + \dfrac{y}{4} = \dfrac{-1}{4} \end{pmatrix}$

32. $\begin{pmatrix} \dfrac{2x}{3} - \dfrac{y}{2} = \dfrac{3}{5} \\ \dfrac{x}{4} + \dfrac{y}{2} = \dfrac{7}{80} \end{pmatrix}$

33. $\begin{pmatrix} \dfrac{4x}{5} - \dfrac{3y}{2} = \dfrac{1}{5} \\ -2x + y = -1 \end{pmatrix}$

34. $\begin{pmatrix} \dfrac{3x}{2} - \dfrac{2y}{7} = -1 \\ 4x + y = 2 \end{pmatrix}$

For Problems 35–50, solve each system by either the substitution method or the elimination-by-addition method, whichever seems more appropriate.

35. $\begin{pmatrix} 5x - y = -22 \\ 2x + 3y = -2 \end{pmatrix}$ **36.** $\begin{pmatrix} 4x + 5y = -41 \\ 3x - 2y = 21 \end{pmatrix}$

37. $\begin{pmatrix} x = 3y - 10 \\ x = -2y + 15 \end{pmatrix}$ **38.** $\begin{pmatrix} y = 4x - 24 \\ 7x + y = 42 \end{pmatrix}$

39. $\begin{pmatrix} 3x - 5y = 9 \\ 6x - 10y = -1 \end{pmatrix}$ **40.** $\begin{pmatrix} y = \dfrac{2}{5}x - 3 \\ 4x - 7y = 33 \end{pmatrix}$

41. $\begin{pmatrix} \dfrac{1}{2}x - \dfrac{2}{3}y = 22 \\ \dfrac{1}{2}x + \dfrac{1}{4}y = 0 \end{pmatrix}$ **42.** $\begin{pmatrix} \dfrac{2}{5}x - \dfrac{1}{3}y = -9 \\ \dfrac{3}{4}x + \dfrac{1}{3}y = -14 \end{pmatrix}$

43. $\begin{pmatrix} t = 2u + 2 \\ 9u - 9t = -45 \end{pmatrix}$ **44.** $\begin{pmatrix} 9u - 9t = 36 \\ u = 2t + 1 \end{pmatrix}$

45. $\begin{pmatrix} x + y = 1000 \\ 0.12x + 0.14y = 136 \end{pmatrix}$ **46.** $\begin{pmatrix} x + y = 10 \\ 0.3x + 0.7y = 4 \end{pmatrix}$

47. $\begin{pmatrix} y = 2x \\ 0.09x + 0.12y = 132 \end{pmatrix}$ **48.** $\begin{pmatrix} y = 3x \\ 0.1x + 0.11y = 64.5 \end{pmatrix}$

49. $\begin{pmatrix} x + y = 10.5 \\ 0.5x + 0.8y = 7.35 \end{pmatrix}$ **50.** $\begin{pmatrix} 2x + y = 7.75 \\ 3x + 2y = 12.5 \end{pmatrix}$

For Problems 51–70, solve each problem by using a system of equations.

51. The sum of two numbers is 53 and their difference is 19. Find the numbers.

52. The sum of two numbers is -3 and their difference is 25. Find the numbers.

53. The measure of the larger of two complementary angles is $15°$ more than four times the measure of the smaller angle. Find the measures of both angles.

54. Assume that a plane is flying at a constant speed under unvarying wind conditions. Traveling against a head wind, it takes the plane 4 hours to travel 1540 miles. Traveling with a tail wind, the plane flies 1365 miles in 3 hours. Find the speed of the plane and the speed of the wind.

55. The tens digit of a two-digit number is 1 more than three times the units digit. If the sum of the digits is 9, find the number.

56. The units digit of a two-digit number is 1 less than twice the tens digit. The sum of the digits is 8. Find the number.

57. The sum of the digits of a two-digit number is 7. If the digits are reversed, the newly formed number is 9 larger than the original number. Find the original number.

58. The units digit of a two-digit number is 1 less than twice the tens digit. If the digits are reversed, the newly formed number is 27 larger than the original number. Find the original number.

59. A motel rents double rooms at $32 per day and single rooms at $26 per day. If 23 rooms were rented one day for a total of $688, how many rooms of each kind were rented?

60. An apartment complex rents one-bedroom apartments for $325 per month and two-bedroom apartments for $375 per month. One month the number of one-bedroom apartments was twice the number of two-bedroom apartments. If the total income for that month was $12,300, how many apartments of each kind were rented?

61. The income from a student production was $10,000. The price of a student ticket was $3, and nonstudent tickets were sold at $5 each. Three thousand tickets were sold. How many tickets of each kind were sold?

62. Michelle can enter a small business as a full partner and receive a salary of $10,000 a year and 15% of the year's profit, or she can be sales manager for a salary of $25,000 plus 5% of the year's profit. What must the year's profit be for her total earnings to be the same whether she is a full partner or a sales manager?

63. Melinda invested three times as much money at 11% yearly interest as she did at 9%. Her total yearly interest from the two investments was $210. How much did she invest at each rate?

64. Sam invested $1950, part of it at 10% and the rest at 12% yearly interest. The yearly income on the 12% investment was $6 less than twice the income from the 10% investment. How much did he invest at each rate?

65. One day last summer, Jim went kayaking on the Little Susitna River in Alaska. Paddling upstream against the current, he traveled 20 miles in 4 hours. Then he turned around and paddled twice as fast downstream and, with the help of the current, traveled 19 miles in 1 hour. Find the rate of the current.

66. One solution contains 30% alcohol and a second solution contains 70% alcohol. How many liters of each solution should be mixed to make 10 liters containing 40% alcohol?

67. Bill bought 4 tennis balls and 3 golf balls for a total of $10.25. Bret went into the same store and bought 2 tennis balls and 5 golf balls for $11.25. What was the price for a tennis ball and the price for a golf ball?

68. Six cans of pop and 2 bags of potato chips cost $5.12. At the same prices, 8 cans of pop and 5 bags of potato chips cost $9.86. Find the price per can of pop and the price per bag of potato chips.

69. A cash drawer contains only five- and ten-dollar bills. There are 12 more five-dollar bills than ten-dollar bills. If the drawer contains $330, find the number of each kind of bill.

70. Brad has a collection of dimes and quarters totaling $47.50. The number of quarters is ten more than twice the number of dimes. How many coins of each kind does he have?

THOUGHTS INTO WORDS

71. Give a general description of how to use the substitution method to solve a system of two linear equations in two variables.

72. Give a general description of how to use the elimination-by-addition method to solve a system of two linear equations in two variables.

73. Which method would you use to solve the system $\begin{pmatrix} 9x + 4y = 7 \\ 3x + 2y = 6 \end{pmatrix}$? Why?

74. Which method would you use to solve the system $\begin{pmatrix} 5x + 3y = 12 \\ 3x - y = 10 \end{pmatrix}$? Why?

Further Investigations

A system such a

$$\begin{pmatrix} \dfrac{2}{x} + \dfrac{3}{y} = \dfrac{19}{15} \\ -\dfrac{2}{x} + \dfrac{1}{y} = -\dfrac{7}{15} \end{pmatrix}$$

is not a linear system, but it can be solved using the elimination-by-addition method as follows. Add the first equation to the second to produce the equivalent system

$$\begin{pmatrix} \dfrac{2}{x} + \dfrac{3}{y} = \dfrac{19}{15} \\ \dfrac{4}{y} = \dfrac{12}{15} \end{pmatrix}$$

Now solve $\dfrac{4}{y} = \dfrac{12}{15}$ to produce $y = 5$. Substitute 5 for y in the first equation and solve for x to produce

$$\frac{2}{x} + \frac{3}{5} = \frac{19}{15}$$

$$\frac{2}{x} = \frac{10}{15}$$

$$10x = 30$$

$$x = 3$$

The solution set of the original system is $\{(3, 5)\}$.

For Problems 75–80, solve each system.

75. $\begin{pmatrix} \dfrac{1}{x} + \dfrac{2}{y} = \dfrac{7}{12} \\ \dfrac{3}{x} - \dfrac{2}{y} = \dfrac{5}{12} \end{pmatrix}$

76. $\begin{pmatrix} \dfrac{3}{x} + \dfrac{2}{y} = 2 \\ \dfrac{2}{x} - \dfrac{3}{y} = \dfrac{1}{4} \end{pmatrix}$

77. $\left(\begin{array}{l} \dfrac{3}{x} - \dfrac{2}{y} = \dfrac{13}{6} \\ \dfrac{2}{x} + \dfrac{3}{y} = 0 \end{array} \right)$

78. $\left(\begin{array}{l} \dfrac{4}{x} + \dfrac{1}{y} = 11 \\ \dfrac{3}{x} - \dfrac{5}{y} = -9 \end{array} \right)$

79. $\left(\begin{array}{l} \dfrac{5}{x} - \dfrac{2}{y} = 23 \\ \dfrac{4}{x} + \dfrac{3}{y} = \dfrac{23}{2} \end{array} \right)$

80. $\left(\begin{array}{l} \dfrac{2}{x} - \dfrac{7}{y} = \dfrac{9}{10} \\ \dfrac{5}{x} + \dfrac{4}{y} = -\dfrac{41}{20} \end{array} \right)$

81. Consider the linear system $\left(\begin{array}{l} a_1 x + b_1 y = c_1 \\ a_2 x + b_2 y = c_2 \end{array} \right)$.

   a. Prove that this system has exactly one solution if and only if $\dfrac{a_1}{a_2} \neq \dfrac{b_1}{b_2}$.

   b. Prove that this system has no solutions if and only if
   $$\dfrac{a_1}{a_2} = \dfrac{b_1}{b_2} \neq \dfrac{c_1}{c_2}.$$

   c. Prove that this system has infinitely many solutions if and only if $\dfrac{a_1}{a_2} = \dfrac{b_1}{b_2} = \dfrac{c_1}{c_2}$.

82. For each of the following systems, use the results from Problem 81 to determine whether the system is consistent or inconsistent or the equations are dependent.

   a. $\left(\begin{array}{l} 5x + y = 9 \\ x - 5y = 4 \end{array} \right)$ **b.** $\left(\begin{array}{l} 3x - 2y = 14 \\ 2x + 3y = 9 \end{array} \right)$

   c. $\left(\begin{array}{l} x - 7y = 4 \\ x - 7y = 9 \end{array} \right)$ **d.** $\left(\begin{array}{l} 3x - 5y = 10 \\ 6x - 10y = 1 \end{array} \right)$

   e. $\left(\begin{array}{l} 3x + 6y = 2 \\ \dfrac{3}{5}x + \dfrac{6}{5}y = \dfrac{2}{5} \end{array} \right)$ **f.** $\left(\begin{array}{l} \dfrac{2}{3}x - \dfrac{3}{4}y = 2 \\ \dfrac{1}{2}x + \dfrac{2}{5}y = 9 \end{array} \right)$

   g. $\left(\begin{array}{l} 7x + 9y = 14 \\ 8x - 3y = 12 \end{array} \right)$ **h.** $\left(\begin{array}{l} 4x - 5y = 3 \\ 12x - 15y = 9 \end{array} \right)$

GRAPHING CALCULATOR ACTIVITIES

83. For each of the systems of equations in Problem 82, use your graphing calculator to help determine whether the system is consistent or inconsistent or the equations are dependent.

84. Use your graphing calculator to help determine the solution set for each of the following systems. Be sure to check your answers.

   a. $\left(\begin{array}{l} y = 3x - 1 \\ y = 9 - 2x \end{array} \right)$ **b.** $\left(\begin{array}{l} 5x + y = -9 \\ 3x - 2y = 5 \end{array} \right)$

   c. $\left(\begin{array}{l} 4x - 3y = 18 \\ 5x + 6y = 3 \end{array} \right)$ **d.** $\left(\begin{array}{l} 2x - y = 20 \\ 7x + y = 79 \end{array} \right)$

   e. $\left(\begin{array}{l} 13x - 12y = 37 \\ 15x + 13y = -11 \end{array} \right)$ **f.** $\left(\begin{array}{l} 1.98x + 2.49y = 13.92 \\ 1.19x + 3.45y = 16.18 \end{array} \right)$

6.2   SYSTEMS OF THREE LINEAR EQUATIONS IN THREE VARIABLES

Consider a linear equation in three variables x, y and z, such as $3x - 2y + z = 7$. Any **ordered triple** (x, y, z) that makes the equation a true numerical statement is said to be a **solution** of the equation. For example, the ordered triple $(2, 1, 3)$ is a solution because $3(2) - 2(1) + 3 = 7$. However, the ordered triple $(5, 2, 4)$ is not a solution because $3(5) - 2(2) + 4 \neq 7$. There are infinitely many solutions in the solution set.

REMARK   The idea of a linear equation is generalized to include equations of more than two variables. Thus an equation such as $5x - 2y + 9z = 8$ is called a *linear equation in three variables*; the equation $5x - 7y + 2z - 11w = 1$ is called a *linear equation in four variables*, and so on.

To *solve* a system of three linear equations in three variables, such as

$$\left. \begin{array}{l} 3x - y + 2z = 13 \\ 4x + 2y + 5z = 30 \\ 5x - 3y - z = 3 \end{array} \right)$$

means to find all of the ordered triples that satisfy all three equaions. In other words, the solution set of the system is the intersection of the solution sets of all three equations in the system.

The graph of a linear equation in three variables is a *plane*, not a line. In fact, graphing equations in three variables requires the use of a three-dimensional coordinate system. Thus using a graphing approach to solve systems of three linear equations in three variables is not at all practical. However, a simple graphical analysis does provide us with some direction as to what we can expect as we begin solving such systems.

In general, because each linear equation in three variables produces a plane, a system of three such equations produces three planes. There are various ways in which three planes can be related. For example, they may be mutually parallel, or two of the planes may be parallel with the third intersecting the other two. (You may want to analyze all of the other possibilities for the three planes!) However, for our purposes at this time, we need to realize that from a solution set viewpoint, a system of three linear equations in three variables produces one of the following possibilities.

1. There is *one ordered triple* that satisfies all three equations. The three planes have a common point of intersection, as indicated in Figure 6.4.

FIGURE 6.4

2. There are *infinitely many ordered triples* in the solution set, all of which are coordinates of *points on a line* common to the three planes. This can happen if the three planes have a common line of intersection (Figure 6.5(a)), or if two of the planes coincide and the third plane intersects them (Figure 6.5(b)).

(a)   (b)

FIGURE 6.5

3. There are *infinitely many ordered triples* in the solution set, all of which are coordinates of *points on a plane*. This can happen if the three planes coincide, as illustrated in Figure 6.6.

FIGURE 6.6

4. The solution set is *empty*; thus we write $\varnothing$. This can happen in various ways, as illustrated in Figure 6.7. Notice that in each situation there are no points common to all three planes.

(a) Three parallel planes

(b) Two planes coincide and the third one is parallel to the coinciding planes.

(c) Two planes are parallel and the third intersects them in parallel lines.

(d) No two planes are parallel, but two of them intersect in a line that is parallel to the third plane.

FIGURE 6.7

Now that we know what possibilities exist, let's consider finding the solution sets for some systems. Our approach will be the elimination-by-addition method, whereby systems are replaced with equivalent systems until a system is obtained where we can easily determine the solution set. The details of this approach will become apparent as we work a few examples.

EXAMPLE 1

Solve the system

$$\begin{pmatrix} 4x - 3y - 2z = 5 \\ 5y + z = -11 \\ 3z = 12 \end{pmatrix}$$

(1)
(2)
(3)

Solution

The form of this system makes it easy to solve. From equation (3) we obtain $z = 4$. Then, substituting 4 for z in equation (2), we get

$$5y + 4 = -11$$
$$5y = -15$$
$$y = -3$$

Finally, substituting 4 for z and -3 for y in equation (1) yields

$$4x - 3(-3) - 2(4) = 5$$
$$4x + 1 = 5$$
$$4x = 4$$
$$x = 1$$

Thus the solution set of the given system is $\{(1, -3, 4)\}$. ∎

EXAMPLE 2

Solve the system

$$\begin{pmatrix} x - 2y + 3z = 22 \\ 2x - 3y - z = 5 \\ 3x + y - 5z = -32 \end{pmatrix}$$

(4)
(5)
(6)

Solution

Equation (5) can be replaced with the equation formed by multiplying equation (4) by -2 and adding this result to equation (5). Equation (6) can be replaced with the equation formed by multiplying equation (4) by -3 and adding this result to equation (6). The following equivalent system is produced, in which equations (8) and (9) contain only the two variables y and z.

$$\begin{pmatrix} x - 2y + 3z = 22 \\ y - 7z = -39 \\ 7y - 14z = -98 \end{pmatrix}$$

(7)
(8)
(9)

Equation (9) can be replaced with the equation formed by multiplying equation (8) by -7 and adding this result to equation (9). This produces the following equivalent system.

$$\begin{cases} x - 2y + 3z = 22 \\ \qquad y - 7z = -39 \\ \qquad\qquad 35z = 175 \end{cases}$$

(10)
(11)
(12)

From equation (12) we obtain $z = 5$. Then, substituting 5 for z in equation (11), we obtain

$$y - 7(5) = -39$$
$$y - 35 = -39$$
$$y = -4$$

Finally, substituting -4 for y and 5 for z in equation (10) produces

$$x - 2(-4) + 3(5) = 22$$
$$x + 8 + 15 = 22$$
$$x + 23 = 22$$
$$x = -1$$

The solution set of the original system is $\{(-1, -4, 5)\}$. (Perhaps you should check this ordered triple in all three of the original equations.) ■

EXAMPLE 3

Solve the system

$$\begin{cases} 3x - y + 2z = 13 \\ 5x - 3y - z = 3 \\ 4x + 2y + 5z = 30 \end{cases}$$

(13)
(14)
(15)

Solution

Equation (14) can be replaced with the equation formed by multiplying equation (13) by -3 and adding this result to equation (14). Equation (15) can be replaced with the equation formed by multiplying equation (13) by 2 and adding this result to equation (15). Thus we produce the following equivalent system, in which equations (17) and (18) contain only the two variables x and z.

$$\begin{cases} 3x - y + 2z = 13 \\ -4x - 7z = -36 \\ 10x + 9z = 56 \end{cases}$$

(16)
(17)
(18)

Now if we multiply equation (17) by 5 and equation (18) by 2, we get the following equivalent system.

$$\begin{pmatrix} 3x - y + 2z = 13 \\ -20x - 35z = -180 \\ 20x + 18z = 112 \end{pmatrix}$$

(19)
(20)
(21)

Equation (21) can be replaced with the equation formed by adding equation (20) to equation (21).

$$\begin{pmatrix} 3x - y + 2z = 13 \\ -20x - 35z = -180 \\ -17z = -68 \end{pmatrix}$$

(22)
(23)
(24)

From equation (24), we obtain $z = 4$. Then we can substitute 4 for z in equation (23).

$$-20x - 35(4) = -180$$
$$-20x - 140 = -180$$
$$-20x = -40$$
$$x = 2$$

Now we can substitute 2 for x and 4 for z in equation (22).

$$3(2) - y + 2(4) = 13$$
$$6 - y + 8 = 13$$
$$-y + 14 = 13$$
$$-y = -1$$
$$y = 1$$

The solution set of the original system is $\{(2, 1, 4)\}$. ∎

EXAMPLE 4

Solve the system

$$\begin{pmatrix} 2x + 3y + z = 14 \\ 3x - 4y - 2z = -30 \\ 5x + 7y + 3z = 32 \end{pmatrix}$$

(25)
(26)
(27)

Solution

Equation (26) can be replaced with the equation formed by multiplying equation (25) by 2 and adding this result to equation (26). Equation (27) can be replaced with the equation formed by multiplying equation (25) by -3 and adding this result to equation (27). The following equivalent system is produced, in which equations (29) and (30) contain only the two variables x and y.

$$\begin{pmatrix} 2x + 3y + z = 14 \\ 7x + 2y = -2 \\ -x - 2y = -10 \end{pmatrix}$$

(28)
(29)
(30)

Now equation (30) can be replaced with the equation formed by adding equation (29) to equation (30).

$$\left(\begin{array}{l} 2x + 3y + z = 14 \\ \quad\ \ 7x + 2y = -2 \\ \quad\qquad 6x = -12 \end{array}\right)$$

(31)
(32)
(33)

From equation (33) we obtain $x = -2$. Then, substituting -2 for x in equation (32), we obtain

$$7(-2) + 2y = -2$$
$$2y = 12$$
$$y = 6$$

Finally, substituting 6 for y and -2 for x in equation (31) yields

$$2(-2) + 3(6) + z = 14$$
$$14 + z = 14$$
$$z = 0$$

The solution set of the original system is $\{(-2, 6, 0)\}$. ■

The ability to solve systems of three linear equations in three unknowns enhances our problem-solving capabilities. Let's conclude this section with a problem that we can solve using such a system.

PROBLEM I

A small company that manufactures sporting equipment produces three different styles of golf shirts. Each style of shirt requires the services of three departments, as indicated by the following table.

	STYLE A	STYLE B	STYLE C
Cutting department	0.1 hr	0.1 hr	0.3 hr
Sewing department	0.3 hr	0.2 hr	0.4 hr
Packaging department	0.1 hr	0.2 hr	0.1 hr

The cutting, sewing, and packaging departments have available a maximum of 340, 580, and 255 work-hours per week, respectively. How many of each style of golf shirt should be produced each week so that the company is operating at full capacity?

Solution

Let a represent the number of shirts of style A produced per week, b the number of style B per week, and c the number of style C per week. Then the problem translates into the following system of equations.

$$\begin{pmatrix} 0.1a + 0.1b + 0.3c = 340 \\ 0.3a + 0.2b + 0.4c = 580 \\ 0.1a + 0.2b + 0.1c = 255 \end{pmatrix} \begin{array}{l} \longleftarrow \text{Cutting department} \\ \longleftarrow \text{Sewing department} \\ \longleftarrow \text{Packaging department} \end{array}$$

Solving this system (we will leave the details for you to carry out) produces $a = 500$, $b = 650$, and $c = 750$. Thus the company should produce 500 golf shirts of style A, 650 of style B, and 750 of style C per week. ■

PROBLEM SET 6.2

For Problems 1–20, solve each system.

1. $\begin{pmatrix} 2x - 3y + 4z = 10 \\ 5y - 2z = -16 \\ 3z = 9 \end{pmatrix}$

2. $\begin{pmatrix} -3x + 2y + z = -9 \\ 4x \qquad -3z = 18 \\ 4z = -8 \end{pmatrix}$

3. $\begin{pmatrix} x + 2y - 3z = 2 \\ 3y - z = 13 \\ 3y + 5z = 25 \end{pmatrix}$

4. $\begin{pmatrix} 2x + 3y - 4z = -10 \\ 2y + 3z = 16 \\ 2y - 5z = -16 \end{pmatrix}$

5. $\begin{pmatrix} 3x + 2y - 2z = 14 \\ x \qquad - 6z = 16 \\ 2x \qquad + 5z = -2 \end{pmatrix}$

6. $\begin{pmatrix} 3x + 2y - z = -11 \\ 2x - 3y = -1 \\ 4x + 5y = -13 \end{pmatrix}$

7. $\begin{pmatrix} x - 2y + 3z = 7 \\ 2x + y + 5z = 17 \\ 3x - 4y - 2z = 1 \end{pmatrix}$

8. $\begin{pmatrix} x - 2y + z = -4 \\ 2x + 4y - 3z = -1 \\ -3x - 6y + 7z = 4 \end{pmatrix}$

9. $\begin{pmatrix} 2x - y + z = 0 \\ 3x - 2y + 4z = 11 \\ 5x + y - 6z = -32 \end{pmatrix}$

10. $\begin{pmatrix} 2x - y + 3z = -14 \\ 4x + 2y - z = 12 \\ 6x - 3y + 4z = -22 \end{pmatrix}$

11. $\begin{pmatrix} 3x + 2y - z = -11 \\ 2x - 3y + 4z = 11 \\ 5x + y - 2z = -17 \end{pmatrix}$

12. $\begin{pmatrix} 9x + 4y - z = 0 \\ 3x - 2y + 4z = 6 \\ 6x - 8y - 3z = 3 \end{pmatrix}$

13. $\begin{pmatrix} 2x + 3y - 4z = -10 \\ 4x - 5y + 3z = 2 \\ 2y + z = 8 \end{pmatrix}$

14. $\begin{pmatrix} x + 2y - 3z = 2 \\ 3x - z = -8 \\ 2x - 3y + 5z = -9 \end{pmatrix}$

15. $\begin{pmatrix} 3x + 2y - 2z = 14 \\ 2x - 5y + 3z = 7 \\ 4x - 3y + 7z = 5 \end{pmatrix}$

16. $\begin{pmatrix} 4x + 3y - 2z = -11 \\ 3x - 7y + 3z = 10 \\ 9x - 8y + 5z = 9 \end{pmatrix}$

17. $\begin{pmatrix} 2x - 3y + 4z = -12 \\ 4x + 2y - 3z = -13 \\ 6x - 5y + 7z = -31 \end{pmatrix}$

18. $\begin{pmatrix} 3x + 5y - 2z = -27 \\ 5x - 2y + 4z = 27 \\ 7x + 3y - 6z = -55 \end{pmatrix}$

19. $\begin{pmatrix} 5x - 3y - 6z = 22 \\ x - y + z = -3 \\ -3x + 7y - 5z = 23 \end{pmatrix}$

20. $\begin{pmatrix} 4x + 3y - 5z = -29 \\ 3x - 7y - z = -19 \\ 2x + 5y + 2z = -10 \end{pmatrix}$

For Problems 21–31, solve each problem by setting up and solving a system of three linear equations in three variables.

21. A gift store is making a mixture of almonds, pecans, and peanuts, which cost \$3.50 per pound, \$4 per pound, and \$2 per pound, respectively. The storekeeper wants to make 20 pounds of the mix to sell at \$2.70 per pound. The number of pounds of peanuts is to be three times the number of pounds of pecans. Find the number of pounds of each to be used in the mixture.

22. The organizer for a church picnic ordered coleslaw, potato salad, and beans amounting to 50 pounds. There was to be three times as much potato salad as coleslaw. The number of pounds of beans was to be six less than the number of pounds of potato salad. Find the number of pounds of each.

23. A box contains \$7.15 in nickels, dimes, and quarters. There are 42 coins in all, and the sum of the numbers of nickels and dimes is two less than the number of quarters. How many coins of each kind are there?

24. A handful of 65 coins consists of pennies, nickels, and dimes. The number of nickels is four less than twice the number of pennies, and there are 13 more dimes than nickels. How many coins of each kind are there?

25. The measure of the largest angle of a triangle is twice the smallest angle. The sum of the smallest angle and the largest angle is twice the other angle. Find the measure of each angle.

26. The perimeter of a triangle is 45 centimeters. The longest side is 4 centimeters less than twice the shortest side. The sum of the lengths of the shortest and longest sides is 7 centimeters less than three times the length of the remaining side. Find the lengths of all three sides of the triangle.

27. Part of $3000 is invested at 12%, another part at 13%, and the remainder at 14% yearly interest. The total yearly income from the three investments is $400. The sum of the amounts invested at 12% and 13% equals the amount invested at 14%. How much is invested at each rate?

28. Different amounts are invested at 10%, 11%, and 12% yearly interest. The amount invested at 11% is $300 more than what is invested at 10%, and the total yearly income from all three investments is $324. A total of $2900 is invested. Find the amount invested at each rate.

29. A small company makes three different types of bird houses. Each type requires the services of three different departments, as indicated by the following table.

	TYPE A	TYPE B	TYPE C
Cutting department	0.1 hr	0.2 hr	0.1 hr
Finishing department	0.4 hr	0.4 hr	0.3 hr
Assembly department	0.2 hr	0.1 hr	0.3 hr

The cutting, finishing, and assembly departments have available a maximum of 35, 95, and 62.5 work-hours per week, respectively. How many bird houses of each type should be made per week so that the company is operating at full capacity?

30. A certain diet consists of dishes A, B, and C. Each serving of A has 1 gram of fat, 2 grams of carbohydrate, and 4 grams of protein. Each serving of B has 2 grams of fat, 1 gram of carbohydrate, and 3 grams of protein. Each serving of C has 2 grams of fat, 4 grams of carbohydrate, and 3 grams of protein. The diet allows 15 grams of fat, 24 grams of carbohydrate, and 30 grams of protein. How many servings of each dish can be eaten?

31. Recall that one form of the equation of a circle is $x^2 + y^2 + Dx + Ey + F = 0$. Find the equation of the circle that passes through the points $(-3, 1)$, $(7, 1)$, and $(-7, 5)$.

THOUGHTS INTO WORDS

32. Give a general description of how to solve a system of three linear equations in three variables.

33. Give a step-by-step description of how to solve the system

$$\begin{pmatrix} x - 2y + 3z = -23 \\ 5y - 2z = 32 \\ 4z = -24 \end{pmatrix}$$

34. Give a step-by-step description of how to solve the system

$$\begin{pmatrix} 3x - 2y + 7z = 9 \\ x - 3z = 4 \\ 2x + z = 9 \end{pmatrix}$$

6.3   MATRIX APPROACH TO SOLVING SYSTEMS

In the first two sections of this chapter, we found that the techniques of substitution and elimination-by-addition worked effectively with two equations and two unknowns, but they started to get a bit cumbersome with three equations and three unknowns. Therefore, we shall now begin to analyze some techniques that lend themselves to use with larger systems of equations. Furthermore, some of these techniques form the basis for using a computer to solve systems. Even though these techniques are primarily designed for large systems of equations, we shall study them in the context of small systems so that we won't get bogged down with the computational aspects of the techniques.

Matrices

A **matrix** is an array of numbers arranged in horizontal rows and vertical columns and enclosed in brackets. For example, the matrix

has 2 rows and 3 columns and is called a 2×3 (read *two by three*) matrix. Each number in a matrix is called an **element** of the matrix. Some additional examples of matrices (*matrices* is the plural of matrix) follow.

$$
\begin{array}{cccc}
3 \times 2 & 2 \times 2 & 1 \times 2 & 4 \times 1 \\
\begin{bmatrix} 2 & 1 \\ 1 & -4 \\ \dfrac{1}{2} & \dfrac{2}{3} \end{bmatrix} &
\begin{bmatrix} 17 & 18 \\ -14 & 16 \end{bmatrix} &
\begin{bmatrix} 7 & 14 \end{bmatrix} &
\begin{bmatrix} 3 \\ -2 \\ 1 \\ 19 \end{bmatrix}
\end{array}
$$

In general, a matrix of m rows and n columns is called a matrix of **dimension $m \times n$** or **order $m \times n$.**

With every system of linear equations we can associate a matrix that consists of the coefficients and constant terms. For example, with the system

$$
\begin{pmatrix}
a_1 x + b_1 y + c_1 z = d_1 \\
a_2 x + b_2 y + c_2 z = d_2 \\
a_3 x + b_3 y + c_3 z = d_3
\end{pmatrix}
$$

we can associate the matrix

$$
\begin{bmatrix}
a_1 & b_1 & c_1 & \vdots & d_1 \\
a_2 & b_2 & c_2 & \vdots & d_2 \\
a_3 & b_3 & c_3 & \vdots & d_3
\end{bmatrix}
$$

which is commonly called the **augmented matrix** of the system of equations. The dashed line simply separates the coefficients from the constant terms and reminds us that we are working with an augmented matrix.

On page 440 we listed the operations or transformations that can be applied to a system of equations to produce an equivalent system. Because augmented matrices are essentially abbreviated forms of systems of linear equations, there are analogous transformations that can be applied to augmented matrices. These transformations are usually referred to as **elementary row operations** and can be stated as follows.

> For any augmented matrix of a system of linear equations, the follow-ing elementary row operations will produce a matrix of an equivalent system.
>
> **1.** Any two rows of the matrix can be interchanged.
>
> **2.** Any row of the matrix can be multiplied by a nonzero real number.
>
> **3.** Any row of the matrix can be replaced by the sum of a nonzero multiple of another row plus that row.

Let's illustrate the use of augmented matrices and elementary row operations to solve a system of two linear equations in two variables.

EXAMPLE I

Solve the system

$$\begin{pmatrix} x - 3y = -17 \\ 2x + 7y = 31 \end{pmatrix}$$

Solution

The augmented matrix of the system is

$$\begin{bmatrix} 1 & -3 & \vdots & -17 \\ 2 & 7 & \vdots & 31 \end{bmatrix}$$

We would like to change this matrix to one of the form

$$\begin{bmatrix} 1 & 0 & \vdots & a \\ 0 & 1 & \vdots & b \end{bmatrix}$$

where we can easily determine that the solution is $x = a$ and $y = b$. Let's begin by adding -2 times row 1 to row 2 to produce a new row 2.

$$\begin{bmatrix} 1 & -3 & \vdots & -17 \\ 0 & 13 & \vdots & 65 \end{bmatrix}$$

Now we can multiply row 2 by $\dfrac{1}{13}$.

$$\begin{bmatrix} 1 & -3 & \vdots & -17 \\ 0 & 1 & \vdots & 5 \end{bmatrix}$$

Finally, we can add 3 times row 2 to row 1 to produce a new row 1.

$$\begin{bmatrix} 1 & 0 & \vdots & -2 \\ 0 & 1 & \vdots & 5 \end{bmatrix}$$

From this last matrix we see that $x = -2$ and $y = 5$. In other words, the solution set of the original system is $\{(-2, 5)\}$. ∎

It may seem that the matrix approach does not provide us with much extra power for solving systems of two linear equations in two unknowns. However, as the systems get larger, the compactness of the matrix approach becomes more convenient. Let's consider a system of three equations in three variables.

EXAMPLE 2

Solve the system

$$\begin{pmatrix} x + 2y - 3z = 15 \\ -2x - 3y + z = -15 \\ 4x + 9y - 4z = 49 \end{pmatrix}$$

Solution

The augmented matrix of this system is

$$\begin{bmatrix} 1 & 2 & -3 & \vdots & 15 \\ -2 & -3 & 1 & \vdots & -15 \\ 4 & 9 & -4 & \vdots & 49 \end{bmatrix}$$

If the system has a unique solution, then we will be able to change the augmented matrix to the form

$$\begin{bmatrix} 1 & 0 & 0 & \vdots & a \\ 0 & 1 & 0 & \vdots & b \\ 0 & 0 & 1 & \vdots & c \end{bmatrix}$$

where we will be able to read the solution $x = a$, $y = b$, and $z = c$.

Add 2 times row 1 to row 2 to produce a new row 2. Likewise, add -4 times row 1 to row 3 to produce a new row 3.

$$\begin{bmatrix} 1 & 2 & -3 & \vdots & 15 \\ 0 & 1 & -5 & \vdots & 15 \\ 0 & 1 & 8 & \vdots & -11 \end{bmatrix}$$

Now add -2 times row 2 to row 1 to produce a new row 1. Also, add -1 times row 2 to row 3 to produce a new row 3.

$$\begin{bmatrix} 1 & 0 & 7 & \vdots & -15 \\ 0 & 1 & -5 & \vdots & 15 \\ 0 & 0 & 13 & \vdots & -26 \end{bmatrix}$$

Now let's multiply row 3 by $\dfrac{1}{13}$.

$$\begin{bmatrix} 1 & 0 & 7 & \vdots & -15 \\ 0 & 1 & -5 & \vdots & 15 \\ 0 & 0 & 1 & \vdots & -2 \end{bmatrix}$$

Finally, we can add -7 times row 3 to row 1 to produce a new row 1, and we can add 5 times row 3 to row 2 for a new row 2.

$$\begin{bmatrix} 1 & 0 & 0 & \vdots & -1 \\ 0 & 1 & 0 & \vdots & 5 \\ 0 & 0 & 1 & \vdots & -2 \end{bmatrix}$$

From this last matrix, we can see that the solution set of the original system is $\{(-1, 5, -2)\}$. ∎

The final matrices of Examples 1 and 2,

$$\begin{bmatrix} 1 & 0 & \vdots & -2 \\ 0 & 1 & \vdots & 5 \end{bmatrix} \quad \text{and} \quad \begin{bmatrix} 1 & 0 & 0 & \vdots & -1 \\ 0 & 1 & 0 & \vdots & 5 \\ 0 & 0 & 1 & \vdots & -2 \end{bmatrix}$$

are said to be in **reduced echelon form**. In general a matrix is in reduced echelon form if the following conditions are satisfied.

1. Reading from left to right, the first nonzero entry of each row is 1.

2. In the *column* containing the leftmost 1 of a row, all the remaining entries are zeros.

3. The leftmost 1 of any row is to the right of the leftmost 1 of the preceding row.

4. Rows containing only zeros are below all the rows containing nonzero entries.

Like the final matrices of Examples 1 and 2, the following are in reduced echelon form.

$$\begin{bmatrix} 1 & 2 & \vdots & -3 \\ 0 & 0 & \vdots & 0 \end{bmatrix} \quad \begin{bmatrix} 1 & 0 & -2 & \vdots & 5 \\ 0 & 1 & 4 & \vdots & 7 \\ 0 & 0 & 0 & \vdots & 0 \end{bmatrix} \quad \begin{bmatrix} 1 & 0 & 0 & 0 & \vdots & 8 \\ 0 & 1 & 0 & 0 & \vdots & -9 \\ 0 & 0 & 1 & 0 & \vdots & -2 \\ 0 & 0 & 0 & 1 & \vdots & 12 \end{bmatrix}$$

In contrast, the following matrices are *not* in reduced echelon form for the reason indicated below each matrix.

$$\begin{bmatrix} 1 & 0 & 0 & \vdots & 11 \\ 0 & 3 & 0 & \vdots & -1 \\ 0 & 0 & 1 & \vdots & -2 \end{bmatrix} \qquad \begin{bmatrix} 1 & 2 & -3 & \vdots & 5 \\ 0 & 1 & 7 & \vdots & 9 \\ 0 & 0 & 1 & \vdots & -6 \end{bmatrix}$$

Violates conditon 1   Violates conditon 2

$$\begin{bmatrix} 1 & 0 & 0 & \vdots & 7 \\ 0 & 0 & 1 & \vdots & -8 \\ 0 & 1 & 0 & \vdots & 14 \end{bmatrix} \qquad \begin{bmatrix} 1 & 0 & 0 & 0 & \vdots & -1 \\ 0 & 0 & 0 & 0 & \vdots & 0 \\ 0 & 0 & 1 & 0 & \vdots & 7 \\ 0 & 0 & 0 & 0 & \vdots & 0 \end{bmatrix}$$

Violates conditon 3   Violates conditon 4

Once we have an augmented matrix in reduced echelon form, it is easy to determine the solution set of the system. Furthermore, the procedure for changing a given augmented matrix to reduced echelon form can be described in a very systematic way. For example, if an augmented matrix of a system of three linear equations in three unknowns has a unique solution, then it can be changed to reduced echelon form as follows.

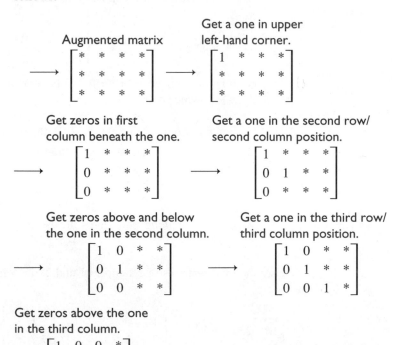

Augmented matrix

Get a one in upper left-hand corner.

Get zeros in first column beneath the one.

Get a one in the second row/second column position.

Get zeros above and below the one in the second column.

Get a one in the third row/third column position.

Get zeros above the one in the third column.

We can identify inconsistent and dependent systems while we are changing a matrix to reduced echelon form. We will show some examples of such cases in a moment, but first let's consider another example of a system of three linear equations in three unknowns where there is a unique solution.

EXAMPLE 3

Solve the system

$$\begin{pmatrix} 2x + 4y - 5z = 37 \\ x + 3y - 4z = 29 \\ 5x - y + 3z = -20 \end{pmatrix}$$

Solution

The augmented matrix

$$\begin{bmatrix} 2 & 4 & -5 & \vdots & 37 \\ 1 & 3 & -4 & \vdots & 29 \\ 5 & -1 & 3 & \vdots & -20 \end{bmatrix}$$

does not have a one in the upper left-hand corner, but this can be remedied by exchanging rows 1 and 2.

$$\begin{bmatrix} 1 & 3 & -4 & \vdots & 29 \\ 2 & 4 & -5 & \vdots & 37 \\ 5 & -1 & 3 & \vdots & -20 \end{bmatrix}$$

Now we can get zeros in the first column beneath the one by adding -2 times row 1 to row 2 and by adding -5 times row 1 to row 3.

$$\begin{bmatrix} 1 & 3 & -4 & \vdots & 29 \\ 0 & -2 & 3 & \vdots & -21 \\ 0 & -16 & 23 & \vdots & -165 \end{bmatrix}$$

Next, we can get a one for the first nonzero entry of the second row by multiplying the second row by $-\dfrac{1}{2}$.

$$\begin{bmatrix} 1 & 3 & -4 & \vdots & 29 \\ 0 & 1 & -\dfrac{3}{2} & \vdots & \dfrac{21}{2} \\ 0 & -16 & 23 & \vdots & -165 \end{bmatrix}$$

Now we can get zeros above and below the one in the second column by adding -3 times row 2 to row 1 and by adding 16 times row 2 to row 3.

$$\begin{bmatrix} 1 & 0 & \dfrac{1}{2} & \vdots & -\dfrac{5}{2} \\ 0 & 1 & -\dfrac{3}{2} & \vdots & \dfrac{21}{2} \\ 0 & 0 & -1 & \vdots & 3 \end{bmatrix}$$

Next, we can get a one in the first nonzero entry of the third row by multiplying the third row by -1.

$$\begin{bmatrix} 1 & 0 & \frac{1}{2} & \vdots & -\frac{5}{2} \\ 0 & 1 & -\frac{3}{2} & \vdots & \frac{21}{2} \\ 0 & 0 & 1 & \vdots & -3 \end{bmatrix}$$

Finally, we can get zeros above the one in the third colum by adding $-\frac{1}{2}$ times row 3 to row 1 and by adding $\frac{3}{2}$ times row 3 to row 2.

$$\begin{bmatrix} 1 & 0 & 0 & \vdots & -1 \\ 0 & 1 & 0 & \vdots & 6 \\ 0 & 0 & 1 & \vdots & -3 \end{bmatrix}$$

From this last matrix, we see that the solution set of the original system is $\{(-1, 6, -3)\}$. ■

Example 3 illustrates that even though the process of changing to reduced echelon form can be systematically described, it can involve some rather messy calculations. However, with the aid of a computer, such calculations are not troublesome. For our purposes in this text, the examples and problems involve systems that minimize messy calculations. This will allow us to concentrate on the procedures.

We want to call your attention to another issue in the solution of Example 3. Consider the matrix

$$\begin{bmatrix} 1 & 3 & -4 & \vdots & 29 \\ 0 & 1 & -\frac{3}{2} & \vdots & \frac{21}{2} \\ 0 & -16 & 23 & \vdots & -165 \end{bmatrix}$$

which is obtained about halfway through the solution. At this step it seems evident that the calculations are getting a little messy. Therefore, instead of continuing toward the reduced echelon form, let's add 16 times row 2 to row 3 to produce a new row 3.

$$\begin{bmatrix} 1 & 3 & -4 & \vdots & 29 \\ 0 & 1 & -\frac{3}{2} & \vdots & \frac{21}{2} \\ 0 & 0 & -1 & \vdots & 3 \end{bmatrix}$$

The system represented by this matrix is

$$\begin{pmatrix} x + 3y - 4z = 29 \\ y - \frac{3}{2}z = \frac{21}{2} \\ -z = 3 \end{pmatrix}$$

and it is said to be in **triangular form**. The last equation determines the value for z; then we can use the process of back-substitution to determine the values for y and x.

Finally, let's consider two examples to illustrate what happens when we use the matrix approach on inconsistent and dependent systems.

EXAMPLE 4

Solve the sytem

$$\begin{pmatrix} x - 2y + 3z = 3 \\ 5x - 9y + 4z = 2 \\ 2x - 4y + 6z = -1 \end{pmatrix}$$

Solution

The augmented matrix of the system is

$$\begin{bmatrix} 1 & -2 & 3 & \vdots & 3 \\ 5 & -9 & 4 & \vdots & 2 \\ 2 & -4 & 6 & \vdots & -1 \end{bmatrix}$$

We can get zeros below the one in the first column by adding -5 times row 1 to row 2 and by adding -2 times row 1 to row 3.

$$\begin{bmatrix} 1 & -2 & 3 & \vdots & 3 \\ 0 & 1 & -11 & \vdots & -13 \\ 0 & 0 & 0 & \vdots & -7 \end{bmatrix}$$

At this step we can stop, because the bottom row of the matrix represents the statement $0(x) + 0(y) + 0(z) = -7$, which is obviously false for all values of x, y, and z. Thus the original system is inconsistent; its solution set is $\varnothing$. ∎

EXAMPLE 5

Solve the system

$$\begin{pmatrix} x + 2y + 2z = 9 \\ x + 3y - 4z = 5 \\ 2x + 5y - 2z = 14 \end{pmatrix}$$

Solution

The augmented matrix of the system is

$$\begin{bmatrix} 1 & 2 & 2 & \vdots & 9 \\ 1 & 3 & -4 & \vdots & 5 \\ 2 & 5 & -2 & \vdots & 14 \end{bmatrix}$$

We can get zeros in the first column below the one in the upper left-hand corner by adding -1 times row 1 to row 2 and adding -2 times row 1 to row 3.

$$\begin{bmatrix} 1 & 2 & 2 & \vdots & 9 \\ 0 & 1 & -6 & \vdots & -4 \\ 0 & 1 & -6 & \vdots & -4 \end{bmatrix}$$

Now we can get zeros in the second column above and below the one in the second row by adding -2 times row 2 to row 1 and adding -1 times row 2 to row 3.

$$\begin{bmatrix} 1 & 0 & 14 & | & 17 \\ 0 & 1 & -6 & | & -4 \\ 0 & 0 & 0 & | & 0 \end{bmatrix}$$

The bottom row of zeros represents the statement $0(x) + 0(y) + 0(z) = 0$, which is true for all values of x, y, and z. The second row represents the statement $y - 6z = -4$, which can be rewritten $y = 6z - 4$. The top row represents the statement $x + 14z = 17$, which can be rewritten $x = -14z + 17$. Therefore, if we let $z = k$, where k is any real number, the solution set of infinitely many ordered triples can be represented by $\{(-14k + 17, 6k - 4, k) \,|\, k \text{ is a real number}\}$. Specific solutions can be generated by letting k take on a value. For example, if $k = 2$, then $6k - 4$ becomes $6(2) - 4 = 8$ and $-14k + 17$ becomes $-14(2) + 17 = -11$. Thus the ordered triple $(-11, 8, 2)$ is a member of the solution set.   ■

PROBLEM SET 6.3

For Problems 1–10, indicate whether each matrix is in reduced echelon form.

1. $\begin{bmatrix} 1 & 0 & | & -4 \\ 0 & 1 & | & 14 \end{bmatrix}$

2. $\begin{bmatrix} 1 & 2 & | & 8 \\ 0 & 0 & | & 0 \end{bmatrix}$

3. $\begin{bmatrix} 1 & 0 & 2 & | & 5 \\ 0 & 1 & 3 & | & 7 \\ 0 & 0 & 0 & | & 0 \end{bmatrix}$

4. $\begin{bmatrix} 1 & 0 & 0 & | & 5 \\ 0 & 3 & 0 & | & 8 \\ 0 & 0 & 1 & | & -11 \end{bmatrix}$

5. $\begin{bmatrix} 1 & 0 & 0 & | & 17 \\ 0 & 0 & 0 & | & 0 \\ 0 & 1 & 0 & | & -14 \end{bmatrix}$

6. $\begin{bmatrix} 1 & 0 & 0 & | & -7 \\ 0 & 1 & 0 & | & 0 \\ 0 & 0 & 1 & | & 9 \end{bmatrix}$

7. $\begin{bmatrix} 1 & 1 & 0 & | & -3 \\ 0 & 1 & 2 & | & 5 \\ 0 & 0 & 1 & | & 7 \end{bmatrix}$

8. $\begin{bmatrix} 1 & 0 & 3 & | & 8 \\ 0 & 1 & 2 & | & -6 \\ 0 & 0 & 0 & | & 0 \end{bmatrix}$

9. $\begin{bmatrix} 1 & 0 & 0 & 3 & | & 4 \\ 0 & 1 & 0 & 5 & | & -3 \\ 0 & 0 & 1 & -1 & | & 7 \\ 0 & 0 & 0 & 0 & | & 0 \end{bmatrix}$

10. $\begin{bmatrix} 1 & 0 & 0 & 0 & | & 2 \\ 0 & 0 & 1 & 0 & | & 4 \\ 0 & 1 & 0 & 0 & | & -3 \\ 0 & 0 & 0 & 1 & | & 9 \end{bmatrix}$

For Problems 11–30, use a matrix approach to solve each system.

11. $\left(\begin{array}{l} x - 3y = 14 \\ 3x + 2y = -13 \end{array} \right)$

12. $\left(\begin{array}{r} x + 5y = -18 \\ -2x + 3y = -16 \end{array} \right)$

13. $\left(\begin{array}{l} 3x - 4y = 33 \\ x + 7y = -39 \end{array} \right)$

14. $\left(\begin{array}{r} 2x + 7y = -55 \\ x - 4y = 25 \end{array} \right)$

15. $\left(\begin{array}{l} x - 6y = -2 \\ 2x - 12y = 5 \end{array} \right)$

16. $\left(\begin{array}{l} 2x - 3y = -12 \\ 3x + 2y = 8 \end{array} \right)$

17. $\left(\begin{array}{l} 3x - 5y = 39 \\ 2x + 7y = -67 \end{array} \right)$

18. $\left(\begin{array}{l} 3x + 9y = -1 \\ x + 3y = 10 \end{array} \right)$

19. $\left(\begin{array}{l} x - 2y - 3z = -6 \\ 3x - 5y - z = 4 \\ 2x + y + 2z = 2 \end{array} \right)$

20. $\left(\begin{array}{l} x + 3y - 4z = 13 \\ 2x + 7y - 3z = 11 \\ -2x - y + 2z = -8 \end{array} \right)$

21. $\left(\begin{array}{r} -2x - 5y + 3z = 11 \\ x + 3y - 3z = -12 \\ 3x - 2y + 5z = 31 \end{array} \right)$

22. $\begin{pmatrix} -3x + 2y + z = 17 \\ x - y + 5z = -2 \\ 4x - 5y - 3z = -36 \end{pmatrix}$

23. $\begin{pmatrix} x - 3y - z = 2 \\ 3x + y - 4z = -18 \\ -2x + 5y + 3z = 2 \end{pmatrix}$

24. $\begin{pmatrix} x - 4y + 3z = 16 \\ 2x + 3y - 4z = -22 \\ -3x + 11y - z = -36 \end{pmatrix}$

25. $\begin{pmatrix} x - y + 2z = 1 \\ -3x + 4y - z = 4 \\ -x + 2y + 3z = 6 \end{pmatrix}$
    26. $\begin{pmatrix} x + 2y - 5z = -1 \\ 2x + 3y - 2z = 2 \\ 3x + 5y - 7z = 4 \end{pmatrix}$

27. $\begin{pmatrix} -2x + y + 5z = -5 \\ 3x + 8y - z = -34 \\ x + 2y + z = -12 \end{pmatrix}$

28. $\begin{pmatrix} 4x - 10y + 3z = -19 \\ 2x + 5y - z = -7 \\ x - 3y - 2z = -2 \end{pmatrix}$

29. $\begin{pmatrix} 2x + 3y - z = 7 \\ 3x + 4y + 5z = -2 \\ 5x + y + 3z = 13 \end{pmatrix}$
    30. $\begin{pmatrix} 4x + 3y - z = 0 \\ 3x + 2y + 5z = 6 \\ 5x - y - 3z = 3 \end{pmatrix}$

Subscript notation is frequently used for working with larger systems of equations. For Problems 31–34, use a matrix approach to solve each system. Express the solutions as 4-tuples of the form (x_1, x_2, x_3, x_4).

31. $\begin{pmatrix} x_1 - 3x_2 - 2x_3 + x_4 = -3 \\ -2x_1 + 7x_2 + x_3 - 2x_4 = -1 \\ 3x_1 - 7x_2 - 3x_3 + 3x_4 = -5 \\ 5x_1 + x_2 + 4x_3 - 2x_4 = 18 \end{pmatrix}$

32. $\begin{pmatrix} x_1 - 2x_2 + 2x_3 - x_4 = -2 \\ -3x_1 + 5x_2 - x_3 - 3x_4 = 2 \\ 2x_1 + 3x_2 + 3x_3 + 5x_4 = -9 \\ 4x_1 - x_2 - x_3 - 2x_4 = 8 \end{pmatrix}$

33. $\begin{pmatrix} x_1 + 3x_2 - x_3 + 2x_4 = -2 \\ 2x_1 + 7x_2 + 2x_3 - x_4 = 19 \\ -3x_1 - 8x_2 + 3x_3 + x_4 = -7 \\ 4x_1 + 11x_2 - 2x_3 - 3x_4 = 19 \end{pmatrix}$

34. $\begin{pmatrix} x_1 + 2x_2 - 3x_3 + x_4 = -2 \\ -2x_1 - 3x_2 + x_3 - x_4 = 5 \\ 4x_1 + 9x_2 - 2x_3 - 2x_4 = -28 \\ -5x_1 - 9x_2 + 2x_3 - 3x_4 = 14 \end{pmatrix}$

In Problems 35–42, each matrix is the reduced echelon matrix for a system with variables x_1, x_2, x_3 and x_4. Find the solution set of each system.

35. $\begin{bmatrix} 1 & 0 & 0 & 0 & | & -2 \\ 0 & 1 & 0 & 0 & | & 4 \\ 0 & 0 & 1 & 0 & | & -3 \\ 0 & 0 & 0 & 1 & | & 0 \end{bmatrix}$

36. $\begin{bmatrix} 1 & 0 & 0 & 0 & | & 0 \\ 0 & 1 & 0 & 0 & | & -5 \\ 0 & 0 & 1 & 0 & | & 0 \\ 0 & 0 & 0 & 1 & | & 4 \end{bmatrix}$

37. $\begin{bmatrix} 1 & 0 & 0 & 0 & | & -8 \\ 0 & 1 & 0 & 0 & | & 5 \\ 0 & 0 & 1 & 0 & | & -2 \\ 0 & 0 & 0 & 0 & | & 1 \end{bmatrix}$

38. $\begin{bmatrix} 1 & 0 & 0 & 0 & | & 2 \\ 0 & 1 & 0 & 2 & | & -3 \\ 0 & 0 & 1 & 3 & | & 4 \\ 0 & 0 & 0 & 0 & | & 0 \end{bmatrix}$

39. $\begin{bmatrix} 1 & 0 & 0 & 3 & | & 5 \\ 0 & 1 & 0 & 0 & | & -1 \\ 0 & 0 & 1 & 4 & | & 2 \\ 0 & 0 & 0 & 0 & | & 0 \end{bmatrix}$

40. $\begin{bmatrix} 1 & 3 & 0 & | & 0 \\ 0 & 0 & 1 & | & 0 \\ 0 & 0 & 0 & | & 1 \\ 0 & 0 & 0 & | & 0 \end{bmatrix}$

41. $\begin{bmatrix} 1 & 3 & 0 & 0 & | & 9 \\ 0 & 0 & 1 & 0 & | & 2 \\ 0 & 0 & 0 & 1 & | & -3 \\ 0 & 0 & 0 & 0 & | & 0 \end{bmatrix}$

42. $\begin{bmatrix} 1 & 0 & 0 & 0 & | & 7 \\ 0 & 1 & 0 & 0 & | & -3 \\ 0 & 0 & 1 & -2 & | & 5 \\ 0 & 0 & 0 & 0 & | & 0 \end{bmatrix}$

THOUGHTS INTO WORDS

---

43. What is a matrix? What is an augmented matrix of a system of linear equations?

44. Describe how to use matrices to solve the system $\begin{pmatrix} x - 2y = 5 \\ 2x + 7y = 9 \end{pmatrix}$.

Further Investigations

For Problems 45–50, change each augmented matrix of the system to reduced echelon form and then indicate the solutions of the system.

45. $\begin{pmatrix} x - 2y + 3z = 4 \\ 3x - 5y - z = 7 \end{pmatrix}$ **46.** $\begin{pmatrix} x + 3y - 2z = -1 \\ 2x - 5y + 7z = 4 \end{pmatrix}$

47. $\begin{pmatrix} 2x - 4y + 3z = 8 \\ 3x + 5y - z = 7 \end{pmatrix}$ **48.** $\begin{pmatrix} 3x + 6y - z = 9 \\ 2x - 3y + 4z = 1 \end{pmatrix}$

49. $\begin{pmatrix} x - 2y + 4z = 9 \\ 2x - 4y + 8z = 3 \end{pmatrix}$

50. $\begin{pmatrix} x + y - 2z = -1 \\ 3x + 3y - 6z = -3 \end{pmatrix}$

GRAPHING CALCULATOR ACTIVITIES

---

51. If your graphing calculator has the capability of manipulating matrices, this is a good time to become familiar with those operations. You may need to refer to your user's manual for the key-punching instructions. To begin the familiarization process, load your calculator with the three augmented matrices in Examples 1, 2, and 3. Then, for each one, carry out the row operations as described in the text.

6.4 DETERMINANTS

Before we introduce the concept of a determinant, let's agree on some convenient new notation. A **general** $m \times n$ (*m*-by-*n*) **matrix** can be represented by

$$A = \begin{bmatrix} a_{11} & a_{12} & a_{13} & \cdots & a_{1n} \\ a_{21} & a_{22} & a_{23} & \cdots & a_{2n} \\ \vdots & \vdots & \vdots & & \vdots \\ a_{m1} & a_{m2} & a_{m3} & \cdots & a_{mn} \end{bmatrix}$$

where the double subscripts are used to identify the number of the row and the number of the column, in that order. For example, a_{23} is the entry at the intersection of the second row and the third column. In general, the entry at the intersection of row i and column j is denoted by a_{ij}.

A **square matrix** is one that has the same number of rows as columns. Each square matrix A with real number entries can be associated with a real number called the **determinant** of the matrix, denoted by $|A|$. We will first define $|A|$ for a 2×2 matrix.

DEFINITION 6.1

If $A = \begin{bmatrix} a_{11} & a_{12} \\ a_{21} & a_{22} \end{bmatrix}$, then

$$|A| = \begin{vmatrix} a_{11} & a_{12} \\ a_{21} & a_{22} \end{vmatrix} = a_{11}a_{22} - a_{12}a_{21}$$

EXAMPLE 1

If $A = \begin{bmatrix} 3 & -2 \\ 5 & 8 \end{bmatrix}$, find $|A|$.

Solution

Use Definition 6.1 to obtain

$$|A| = \begin{vmatrix} 3 & -2 \\ 5 & 8 \end{vmatrix} = 3(8) - (-2)(5)$$
$$= 24 + 10$$
$$= 34 \qquad \blacksquare$$

Finding the determinant of a square matrix is commonly called **evaluating the determinant**, and the matrix notation is often omitted.

EXAMPLE 2

Evaluate $\begin{vmatrix} -3 & 6 \\ 2 & 8 \end{vmatrix}$.

Solution

$$\begin{vmatrix} -3 & 6 \\ 2 & 8 \end{vmatrix} = (-3)(8) - (6)(2)$$
$$= -24 - 12$$
$$= -36 \qquad \blacksquare$$

To find the determinants of 3×3 and larger square matrices, it is convenient to introduce some additional terminology.

DEFINITION 6.2

If A is a 3×3 matrix, then the **minor** (denoted by M_{ij}) of the a_{ij} element is the determinant of the 2×2 matrix obtained by deleting row i and column j of A.

EXAMPLE 3

If $A = \begin{bmatrix} 2 & 1 & 4 \\ -6 & 3 & -2 \\ 4 & 2 & 5 \end{bmatrix}$, find (a) M_{11} and b) M_{23}.

Solution

a. To find M_{11} we first delete row 1 and column 1 of matrix A.

$$\begin{bmatrix} 2 & 1 & 4 \\ -6 & 3 & -2 \\ 4 & 2 & 5 \end{bmatrix}$$

Thus

$$M_{11} = \begin{vmatrix} 3 & -2 \\ 2 & 5 \end{vmatrix} = 3(5) - (-2)(2) = 19$$

b. To find M_{23} we first delete row 2 and column 3 of matrix A.

$$\begin{bmatrix} 2 & 1 & 4 \\ 6 & 3 & 2 \\ 4 & 2 & 5 \end{bmatrix}$$

Thus

$$M_{23} = \begin{bmatrix} 2 & 1 \\ 4 & 2 \end{bmatrix} = 2(2) - (1)(4) = 0 \quad \blacksquare$$

The following definition will also be used.

DEFINITION 6.3

If A is a 3×3 matrix, then the **cofactor** (denoted by C_{ij}) of the element a_{ij} is defined by

$$C_{ij} = (-1)^{i+j} M_{ij}$$

According to Definition 6.3, to find the cofactor of any element a_{ij} of a square matrix A, we find the minor of a_{ij} and multiply it by 1 if $i + j$ is even, or multiply it by -1 if $i + j$ is odd.

EXAMPLE 4

If $A = \begin{vmatrix} 3 & 2 & -4 \\ 1 & 5 & 4 \\ 2 & -3 & 1 \end{vmatrix}$, find C_{32}.

Solution

First, let's find M_{32} by deleting row 3 and column 2 of matrix A.

$$\begin{bmatrix} 3 & 2 & -4 \\ 1 & 5 & 4 \\ 2 & -3 & 1 \end{bmatrix}$$

Thus

$$M_{32} = \begin{vmatrix} 3 & -4 \\ 1 & 4 \end{vmatrix} = 3(4) - (-4)(1) = 16$$

Therefore,

$$C_{32} = (-1)^{3+2}M_{32} = (-1)^5(16) = -16 \qquad \blacksquare$$

The concept of a cofactor can be used to define the determinant of a 3×3 matrix as follows.

DEFINITION 6.4

If $A = \begin{bmatrix} a_{11} & a_{12} & a_{13} \\ a_{21} & a_{22} & a_{23} \\ a_{31} & a_{32} & a_{33} \end{bmatrix}$, then

$$|A| = a_{11}C_{11} + a_{21}C_{21} + a_{31}C_{31}$$

Definition 6.4 simply states that the determinant of a 3×3 matrix can be found by multiplying each element of the first column by its corresponding cofactor and then adding the three results. Let's illustrate this procedure.

EXAMPLE 5

Find $|A|$ if $A = \begin{bmatrix} -2 & 1 & 4 \\ 3 & 0 & 5 \\ 1 & -4 & -6 \end{bmatrix}$.

Solution

$$|A| = a_{11}C_{11} + a_{21}C_{21} + a_{31}C_{31}$$

$$= (-2)(-1)^{1+1}\begin{vmatrix} 0 & 5 \\ -4 & -6 \end{vmatrix} + (3)(-1)^{2+1}\begin{vmatrix} 1 & 4 \\ -4 & -6 \end{vmatrix} + (1)(-1)^{3+1}\begin{vmatrix} 1 & 4 \\ 0 & 5 \end{vmatrix}$$

$$= (-2)(1)(20) + (3)(-1)(10) + (1)(1)(5)$$

$$= -40 - 30 + 5$$

$$= -65 \qquad \blacksquare$$

When we use Definition 6.4, we often say that *the determinant is being expanded about the first column*. It can also be shown that **any row or column can be used to expand a determinant.** For example, for matrix A in Example 5, the expansion of the determinant about the *second row* is as follows.

$$\begin{vmatrix} -2 & 1 & 4 \\ 3 & 0 & 5 \\ 1 & -4 & -6 \end{vmatrix} = (3)(-1)^{2+1}\begin{vmatrix} 1 & 4 \\ -4 & -6 \end{vmatrix} + (0)(-1)^{2+2}\begin{vmatrix} -2 & 4 \\ 1 & -6 \end{vmatrix} + (5)(-1)^{2+3}\begin{vmatrix} -2 & 1 \\ 1 & -4 \end{vmatrix}$$

$$= (3)(-1)(10) + (0)(1)(8) + (5)(-1)(7)$$
$$= -30 + 0 - 35$$
$$= -65$$

Note that when we expanded about the second row, the computation was simplified by the presence of a zero. In general, it is helpful to expand about the row or column that contains the most zeros.

The concepts of minor and cofactor have been defined in terms of 3×3 matrices. Analogous definitions can be given for any square matrix (that is, any $n \times n$ matrix with $n \geq 2$), and the determinant can then be expanded about any row or column. Certainly as the matrices become larger than 3×3, the computations get more tedious. We will concentrate most of our efforts in this text on 2×2 and 3×3 matrices.

Properties of Determinants

Determinants have several interesting properties, some of which are important primarily from a theoretical standpoint. But some of the properties are also very useful when evaluating determinants. We will state these properties for square matrices in general, but we will use 2×2 or 3×3 matrices as examples. We can demonstrate some of the proofs of these properties by evaluating the determinants involved, and some of the proofs for 3×3 matrices will be left for you to verify in the next problem set.

> ### PROPERTY 6.1
>
> If any row (or column) of a square matrix A contains only zeros, then $|A| = 0$.

If every element of a row (or column) of a square matrix A is 0, then it should be evident that expanding the determinant about that row (or column) of zeros will produce 0.

> ### PROPERTY 6.2
>
> If square matrix B is obtained from square matrix A by interchanging two rows (or two columns), then $|B| = -|A|$.

Property 6.2 states that **interchanging two rows (or columns) changes the sign of the determinant.** As an example of this property, suppose that

$$A = \begin{bmatrix} 2 & 5 \\ -1 & 6 \end{bmatrix}$$

and that rows 1 and 2 are interchanged to form

$$B = \begin{bmatrix} -1 & 6 \\ 2 & 5 \end{bmatrix}$$

Calculating $|A|$ and $|B|$ yields

$$|A| = \begin{vmatrix} 2 & 5 \\ -1 & 6 \end{vmatrix} = 2(6) - (5)(-1) = 17$$

and

$$|B| = \begin{vmatrix} -1 & 6 \\ 2 & 5 \end{vmatrix} = (-1)(5) - (6)(2) = -17$$

PROPERTY 6.3

If square matrix B is obtained from square matrix A by multiplying each element of any row (or column) of A by some real number k, then $|B| = k|A|$.

Property 6.3 states that **multiplying any row (or column) by a factor of k affects the value of the determinant by a factor of k.** As an example of this property, suppose that

$$A = \begin{bmatrix} 1 & -2 & 8 \\ 2 & 1 & 12 \\ 3 & 2 & -16 \end{bmatrix}$$

and that B is formed by multiplying each element of the third column by $\frac{1}{4}$.

$$B = \begin{bmatrix} 1 & -2 & 2 \\ 2 & 1 & 3 \\ 3 & 2 & -4 \end{bmatrix}$$

Now let's calculate $|A|$ and $|B|$ by expanding about the third column in each case.

$$|A| = \begin{vmatrix} 1 & -2 & 8 \\ 2 & 1 & 12 \\ 3 & 2 & -16 \end{vmatrix} = (8)(-1)^{1+3} \begin{vmatrix} 2 & 1 \\ 3 & 2 \end{vmatrix} + (12)(-1)^{2+3} \begin{vmatrix} 1 & -2 \\ 3 & 2 \end{vmatrix} + (-16)(-1)^{3+3} \begin{vmatrix} 1 & -2 \\ 2 & 1 \end{vmatrix}$$

$$= (8)(1)(1) + (12)(-1)(8) + (-16)(1)(5)$$

$$= -168$$

$$|B| = \begin{vmatrix} 1 & -2 & 2 \\ 2 & 1 & 3 \\ 3 & 2 & 4 \end{vmatrix} = (2)(-1)^{1+3} \begin{vmatrix} 2 & 1 \\ 3 & 2 \end{vmatrix} + (3)(-1)^{2+3} \begin{vmatrix} 1 & -2 \\ 3 & 2 \end{vmatrix} + (-4)(-1)^{3+3} \begin{vmatrix} 1 & -2 \\ 2 & 1 \end{vmatrix}$$

$$= (2)(1)(1) + (3)(-1)(8) + (-4)(1)(5)$$

$$= -42$$

We see that $|B| = \dfrac{1}{4}|A|$. This example also illustrates the usual computational use of Property 6.3: We can factor out a common factor from a row or column and then adjust the value of the determinant by that factor. For example,

$$\begin{vmatrix} 2 & 6 & 8 \\ -1 & 2 & 7 \\ 5 & 2 & 1 \end{vmatrix} = 2 \begin{vmatrix} 1 & 3 & 4 \\ -1 & 2 & 7 \\ 5 & 2 & 1 \end{vmatrix}$$

Factor a 2 from the top row.

PROPERTY 6.4

If square matrix B is obtained from square matrix A by adding k times a row (or column) of A to another row (or column) of A, then $|B| = |A|$.

Property 6.4 states that **adding the product of k times a row (or column) to another row (or column) does not affect the value of the determinant.** As an example of this property, suppose that

$$A = \begin{bmatrix} 1 & 2 & 4 \\ 2 & 4 & 7 \\ -1 & 3 & 5 \end{bmatrix}$$

Now let's form B by replacing row 2 with the result of adding -2 times row 1 to row 2.

$$B = \begin{bmatrix} 1 & 2 & 4 \\ 0 & 0 & -1 \\ -1 & 3 & 5 \end{bmatrix}$$

Next, let's evaluate $|A|$ and $|B|$ by expanding about the second row in each case.

$$|A| = \begin{vmatrix} 1 & 2 & 4 \\ 2 & 4 & 7 \\ -1 & 3 & 5 \end{vmatrix} = (2)(-1)^{2+1} \begin{vmatrix} 2 & 4 \\ 3 & 5 \end{vmatrix} + (4)(-1)^{2+2} \begin{vmatrix} 1 & 4 \\ -1 & 5 \end{vmatrix} + (7)(-1)^{2+3} \begin{vmatrix} 1 & 2 \\ -1 & 3 \end{vmatrix}$$

$$= 2(-1)(-2) + (4)(1)(9) + (7)(-1)(5)$$

$$= 5$$

$$|B| = \begin{vmatrix} 1 & 2 & 4 \\ 0 & 0 & -1 \\ -1 & 3 & 5 \end{vmatrix} = (0)(-1)^{2+1} \begin{vmatrix} 2 & 4 \\ 3 & 5 \end{vmatrix} + (0)(-1)^{2+2} \begin{vmatrix} 1 & 4 \\ -1 & 5 \end{vmatrix} + (-1)(-1)^{2+3} \begin{vmatrix} 1 & 2 \\ -1 & 3 \end{vmatrix}$$

$$= 0 + 0 + (-1)(-1)(5)$$
$$= 5$$

Note that $|B| = |A|$. Furthermore, note that because of the zeros in the second row, evaluating $|B|$ is much easier than evaluating $|A|$. Property 6.4 can often be used to obtain some zeros before evaluating a determinant.

A word of caution is in order at this time. Be careful not to confuse Properties 6.2, 6.3, and 6.4 with the three elementary row transformations of augmented matrices that were used in Section 6.3. The statements of the two sets of properties do resemble each other, but the properties pertain to *two different concepts*, so be sure you understand the distinction between them.

One final property of determinants should be mentioned.

PROPERTY 6.5

If two rows (or columns) of a square matrix A are identical, then $|A| = 0$.

Property 6.5 is a direct consequence of Property 6.2. Suppose that A is a square matrix (any size) with two identical rows. Square matrix B can be formed from A by interchanging the two identical rows. Because identical rows were interchanged, $|B| = |A|$. *But* by Property 6.2, $|B| = -|A|$. For both of these statements to hold, $|A| = 0$.

Let's conclude this section by evaluating a 4×4 determinant, using Properties 6.3 and 6.4 to facilitate the computation.

EXAMPLE 6

Evaluate $\begin{vmatrix} 6 & 2 & 1 & -2 \\ 9 & -1 & 4 & 1 \\ 12 & -2 & 3 & -1 \\ 0 & 0 & 9 & 3 \end{vmatrix}$.

Solution

First, let's add -3 times the fourth column to the third column.

$$\begin{vmatrix} 6 & 2 & 7 & -2 \\ 9 & -1 & 1 & 1 \\ 12 & -2 & 6 & -1 \\ 0 & 0 & 0 & 3 \end{vmatrix}$$

Now if we expand about the fourth row, we get only one nonzero product.

$$(3)(-1)^{4+4}\begin{vmatrix} 6 & 2 & 7 \\ 9 & -1 & 1 \\ 12 & -2 & 6 \end{vmatrix}$$

Factoring a 3 out of the first column of the 3 × 3 determinant, we obtain

$$(3)(-1)^8(3)\begin{vmatrix} 2 & 2 & 7 \\ 3 & -1 & 1 \\ 4 & -2 & 6 \end{vmatrix}$$

Now, working with the 3 × 3 determinant, we can first add column 3 to column 2 and then add −3 times column 3 to column 1.

$$(3)(-1)^8(3)\begin{vmatrix} -19 & 9 & 7 \\ 0 & 0 & 1 \\ -14 & 4 & 6 \end{vmatrix}$$

Finally, by expanding this 3 × 3 determinant about the second row, we obtain

$$(3)(-1)^8(3)(1)(-1)^{2+3}\begin{vmatrix} -19 & 9 \\ -14 & 4 \end{vmatrix}$$

Our final result is

$$(3)(-1)^8(3)(1)(-1)^5(50) = -450 \qquad \blacksquare$$

PROBLEM SET 6.4

For Problems 1–12, evaluate each 2 × 2 determinant by using Definition 6.1.

1. $\begin{vmatrix} 4 & 3 \\ 2 & 7 \end{vmatrix}$

2. $\begin{vmatrix} 3 & 5 \\ 6 & 4 \end{vmatrix}$

3. $\begin{vmatrix} -3 & 2 \\ 7 & 5 \end{vmatrix}$

4. $\begin{vmatrix} 5 & 3 \\ 6 & -1 \end{vmatrix}$

5. $\begin{vmatrix} 2 & -3 \\ 8 & -2 \end{vmatrix}$

6. $\begin{vmatrix} -5 & 5 \\ -6 & 2 \end{vmatrix}$

7. $\begin{vmatrix} -2 & -3 \\ -1 & -4 \end{vmatrix}$

8. $\begin{vmatrix} -4 & -3 \\ -5 & -7 \end{vmatrix}$

9. $\begin{vmatrix} \dfrac{1}{2} & \dfrac{1}{3} \\ -3 & -6 \end{vmatrix}$

10. $\begin{vmatrix} \dfrac{2}{3} & 3 \\ 8 & 6 \end{vmatrix}$

11. $\begin{vmatrix} \dfrac{1}{2} & \dfrac{2}{3} \\ \dfrac{3}{4} & -\dfrac{1}{3} \end{vmatrix}$

12. $\begin{vmatrix} \dfrac{2}{3} & \dfrac{1}{5} \\ -\dfrac{1}{4} & \dfrac{3}{2} \end{vmatrix}$

For Problems 13–28, evaluate each 3 × 3 determinant. Use the properties of determinants to your advantage.

13. $\begin{vmatrix} 1 & 2 & -1 \\ 3 & 1 & 2 \\ 2 & 4 & 3 \end{vmatrix}$

14. $\begin{vmatrix} 1 & -2 & 1 \\ 2 & 1 & -1 \\ 3 & 2 & 4 \end{vmatrix}$

15. $\begin{vmatrix} 1 & -4 & 1 \\ 2 & 5 & -1 \\ 3 & 3 & 4 \end{vmatrix}$

16. $\begin{vmatrix} 3 & -2 & 1 \\ 2 & 1 & 4 \\ -1 & 3 & 5 \end{vmatrix}$

17. $\begin{vmatrix} 6 & 12 & 3 \\ -1 & 5 & 1 \\ -3 & 6 & 2 \end{vmatrix}$

18. $\begin{vmatrix} 2 & 35 & 5 \\ 1 & -5 & 1 \\ -4 & 15 & 2 \end{vmatrix}$

19. $\begin{vmatrix} 2 & -1 & 3 \\ 0 & 3 & 1 \\ 1 & -2 & -1 \end{vmatrix}$

20. $\begin{vmatrix} 2 & -17 & 3 \\ 0 & 5 & 1 \\ 1 & -3 & -1 \end{vmatrix}$

21. $\begin{vmatrix} -3 & -2 & 1 \\ 5 & 0 & 6 \\ 2 & 1 & -4 \end{vmatrix}$ **22.** $\begin{vmatrix} -5 & 1 & -1 \\ 3 & 4 & 2 \\ 0 & 2 & -3 \end{vmatrix}$

23. $\begin{vmatrix} 3 & -4 & -2 \\ 5 & -2 & 1 \\ 1 & 0 & 0 \end{vmatrix}$ **24.** $\begin{vmatrix} -6 & 5 & 3 \\ 2 & 0 & -1 \\ 4 & 0 & 7 \end{vmatrix}$

25. $\begin{vmatrix} 24 & -1 & 4 \\ 40 & 2 & 0 \\ -16 & 6 & 0 \end{vmatrix}$ **26.** $\begin{vmatrix} 2 & -1 & 3 \\ 0 & 3 & 1 \\ 4 & -8 & -4 \end{vmatrix}$

27. $\begin{vmatrix} 2 & 3 & -4 \\ 4 & 6 & -1 \\ -6 & 1 & -2 \end{vmatrix}$ **28.** $\begin{vmatrix} 1 & 2 & -3 \\ -3 & -1 & 1 \\ 4 & 5 & 4 \end{vmatrix}$

For Problems 29–32, evaluate each 4×4 determinant. Use the properties of determinants to your advantage.

29. $\begin{vmatrix} 1 & -2 & 3 & 2 \\ 2 & -1 & 0 & 4 \\ -3 & 4 & 0 & -2 \\ -1 & 1 & 1 & 5 \end{vmatrix}$

30. $\begin{vmatrix} 1 & 2 & 5 & 7 \\ -6 & 3 & 0 & 9 \\ -3 & 5 & 2 & 7 \\ 2 & 1 & 4 & 3 \end{vmatrix}$

31. $\begin{vmatrix} 3 & -1 & 2 & 3 \\ 1 & 0 & 2 & 1 \\ 2 & 3 & 0 & 1 \\ 5 & 2 & 4 & -5 \end{vmatrix}$ **32.** $\begin{vmatrix} 1 & 2 & 0 & 0 \\ 3 & -1 & 4 & 5 \\ -2 & 4 & 1 & 6 \\ 2 & -1 & -2 & -3 \end{vmatrix}$

For Problems 33–42, use the appropriate property of determinants from this section to justify each true statement. *Do not* evaluate the determinants.

33. $(-4)\begin{vmatrix} 2 & 1 & -1 \\ 3 & 2 & 1 \\ 2 & 1 & 3 \end{vmatrix} = \begin{vmatrix} 2 & -4 & -1 \\ 3 & -8 & 1 \\ 2 & -4 & 3 \end{vmatrix}$

34. $\begin{vmatrix} 1 & -2 & 3 \\ 4 & -6 & -8 \\ 0 & 2 & 7 \end{vmatrix} = (-2)\begin{vmatrix} 1 & -2 & 3 \\ -2 & 3 & 4 \\ 0 & 2 & 7 \end{vmatrix}$

35. $\begin{vmatrix} 4 & 7 & 9 \\ 6 & -8 & 2 \\ 4 & 3 & -1 \end{vmatrix} = -\begin{vmatrix} 4 & 9 & 7 \\ 6 & 2 & -8 \\ 4 & -1 & 3 \end{vmatrix}$

36. $\begin{vmatrix} 3 & -1 & 4 \\ 5 & 2 & 7 \\ 3 & -1 & 4 \end{vmatrix} = 0$

37. $\begin{vmatrix} 1 & 3 & 4 \\ -2 & 5 & 7 \\ -3 & -1 & 2 \end{vmatrix} = \begin{vmatrix} 1 & 3 & 4 \\ -2 & 5 & 7 \\ 0 & 8 & 14 \end{vmatrix}$

38. $\begin{vmatrix} 3 & 2 & 0 \\ 1 & 4 & 1 \\ -4 & 9 & 2 \end{vmatrix} = \begin{vmatrix} 3 & 2 & -3 \\ 1 & 4 & 0 \\ -4 & 9 & 6 \end{vmatrix}$

39. $\begin{vmatrix} 6 & 2 & 2 \\ 3 & -1 & 4 \\ 9 & -3 & 6 \end{vmatrix} = 6\begin{vmatrix} 2 & 2 & 1 \\ 1 & -1 & 2 \\ 3 & -3 & 3 \end{vmatrix} = 18\begin{vmatrix} 2 & 2 & 1 \\ 1 & -1 & 2 \\ 1 & -1 & 1 \end{vmatrix}$

40. $\begin{vmatrix} 2 & 1 & -3 \\ 0 & 2 & -4 \\ -5 & 1 & 3 \end{vmatrix} = -\begin{vmatrix} 2 & 1 & -3 \\ -5 & 1 & 3 \\ 0 & 2 & -4 \end{vmatrix}$

41. $\begin{vmatrix} 2 & -3 & 2 \\ 1 & -4 & 1 \\ 7 & 8 & 7 \end{vmatrix} = 0$

42. $\begin{vmatrix} 3 & 1 & 2 \\ -4 & 5 & -1 \\ 2 & -2 & -4 \end{vmatrix} = \begin{vmatrix} 3 & 1 & 0 \\ -4 & 5 & -11 \\ 2 & -2 & 0 \end{vmatrix}$

THOUGHTS INTO WORDS

43. Explain the difference between a matrix and a determinant.

44. Explain the concept of a cofactor and how it is used to help expand a determinant.

45. What does it mean to say that any row or column can be used to expand a determinant?

46. Give a step-by-step explanation of how to evaluate the determinant

$$\begin{vmatrix} 3 & 0 & 2 \\ 1 & -2 & 5 \\ 6 & 0 & 9 \end{vmatrix}$$

Further Investigations

For Problems 47–50, use

$$A = \begin{bmatrix} a_{11} & a_{12} & a_{13} \\ a_{21} & a_{22} & a_{23} \\ a_{31} & a_{32} & a_{33} \end{bmatrix}$$

as a general representation for any 3×3 matrix.

47. Verify Property 6.2 for 3×3 matrices.

48. Verify Property 6.3 for 3×3 matrices.

49. Verify Property 6.4 for 3×3 matrices.

50. Show that $|A| = a_{11}a_{22}a_{33}a_{44}$ if

$$A = \begin{bmatrix} a_{11} & a_{12} & a_{13} & a_{14} \\ 0 & a_{22} & a_{23} & a_{24} \\ 0 & 0 & a_{33} & a_{34} \\ 0 & 0 & 0 & a_{44} \end{bmatrix}$$

GRAPHING CALCULATOR ACTIVITIES

51. Use a calculator to check your answers for Problems 29–32.

52. Consider the following matrix.

$$A = \begin{bmatrix} 2 & 5 & 7 & 9 \\ -4 & 6 & 2 & 4 \\ 6 & 9 & 12 & 3 \\ 5 & 4 & -2 & 8 \end{bmatrix}$$

Form matrix B by interchanging rows 1 and 3 of matrix A. Now use your calculator to show that $|B| = -|A|$.

53. Consider the following matrix.

$$A = \begin{bmatrix} 2 & 1 & 7 & 6 & 8 \\ 3 & -2 & 4 & 5 & -1 \\ 6 & 7 & 9 & 12 & 13 \\ -4 & -7 & 6 & 2 & 1 \\ 9 & 8 & 12 & 14 & 17 \end{bmatrix}$$

Form matrix B by multiplying each element of the second row of matrix A by 3. Now use your calculator to show that $|B| = 3|A|$.

54. Consider the following matrix.

$$A = \begin{bmatrix} 4 & 3 & 2 & 1 & 5 & -3 \\ 5 & 2 & 7 & 8 & 6 & 3 \\ 0 & 9 & 1 & 4 & 7 & 2 \\ 4 & 3 & 2 & 1 & 5 & -3 \\ -4 & -6 & 7 & 12 & 11 & 9 \\ 5 & 8 & 6 & -3 & 2 & -1 \end{bmatrix}$$

Use your calculator to show that $|A| = 0$.

6.5   CRAMER'S RULE

Determinants provide the basis for another method of solving linear systems. Consider the following linear system of two equations and two unknowns.

$$\begin{pmatrix} a_1x + b_1y = c_1 \\ a_2x + b_2y = c_2 \end{pmatrix}$$

The augmented matrix of this system is

$$\begin{bmatrix} a_1 & b_1 & \vdots & c_1 \\ a_2 & b_2 & \vdots & c_2 \end{bmatrix}$$

Using the elementary row transformations of augmented matrices, we can change this matrix to the following reduced echelon form. (The details of this are left for you to do as an exercise.)

$$\begin{bmatrix} 1 & 0 & \vdots & \dfrac{c_1b_2 - c_2b_1}{a_1b_2 - a_2b_1} \\ 0 & 1 & \vdots & \dfrac{a_1c_2 - a_2c_1}{a_1b_2 - a_2b_1} \end{bmatrix}, \qquad a_1b_2 - a_2b_1 \neq 0$$

The solution for x and y can be expressed in determinant form as follows.

$$x = \frac{c_1b_2 - c_2b_1}{a_1b_2 - a_2b_1} = \frac{\begin{vmatrix} c_1 & b_1 \\ c_2 & b_2 \end{vmatrix}}{\begin{vmatrix} a_1 & b_1 \\ a_2 & b_2 \end{vmatrix}}$$

$$y = \frac{a_1c_2 - a_2c_1}{a_1b_2 - a_2b_1} = \frac{\begin{vmatrix} a_1 & c_1 \\ a_2 & c_2 \end{vmatrix}}{\begin{vmatrix} a_1 & b_1 \\ a_2 & b_2 \end{vmatrix}}$$

This method of using determinants to solve a system of two linear equations in two variables is called **Cramer's rule** and can be stated as follows.

Cramer's Rule (2 × 2 case)

Given the system

$$\begin{pmatrix} a_1x + b_1y = c_1 \\ a_2x + b_2y = c_2 \end{pmatrix}$$

with

$$D = \begin{vmatrix} a_1 & b_1 \\ a_2 & b_2 \end{vmatrix} \neq 0$$

$$D_x = \begin{vmatrix} c_1 & b_1 \\ c_2 & b_2 \end{vmatrix} \qquad \text{and} \qquad D_y = \begin{vmatrix} a_1 & c_1 \\ a_2 & c_2 \end{vmatrix}$$

then the solution for this system is given by

$$x = \frac{D_x}{D} \qquad \text{and} \qquad y = \frac{D_y}{D}$$

Note that the elements of D are the coefficients of the variables in the given system. In D_x the coefficients of x are replaced by the corresponding constants, and in D_y the coefficients of y are replaced by the corresponding constants. Let's illustrate the use of Cramer's rule to solve some systems.

EXAMPLE 1

Solve the system $\begin{pmatrix} 6x + 3y = 2 \\ 3x + 2y = -4 \end{pmatrix}$.

Solution

The system is in the proper form for us to apply Cramer's rule, so let's determine D, D_x, and D_y.

$$D = \begin{vmatrix} 6 & 3 \\ 3 & 2 \end{vmatrix} = 12 - 9 = 3$$

$$D_x = \begin{vmatrix} 2 & 3 \\ -4 & 2 \end{vmatrix} = 4 + 12 = 16$$

$$D_y = \begin{vmatrix} 6 & 2 \\ 3 & -4 \end{vmatrix} = -24 - 6 = -30$$

Therefore,

$$x = \frac{D_x}{D} = \frac{16}{3}$$

and

$$y = \frac{D_y}{D} = \frac{-30}{3} = -10$$

The solution set is $\left\{ \left(\frac{16}{3}, -10 \right) \right\}$.

■

EXAMPLE 2

Solve the system $\begin{pmatrix} y = -2x - 2 \\ 4x - 5y = 17 \end{pmatrix}$.

Solution

To begin, we must change the form of the first equation so that the system fits the form given in Cramer's rule. The equation $y = -2x - 2$ can be rewritten $2x + y = -2$. The system now becomes

$$\begin{pmatrix} 2x + y = -2 \\ 4x - 5y = 17 \end{pmatrix}$$

and we can proceed to determine D, D_x, and D_y.

$$D = \begin{vmatrix} 2 & 1 \\ 4 & -5 \end{vmatrix} = -10 - 4 = -14$$

$$D_x = \begin{vmatrix} -2 & 1 \\ 17 & -5 \end{vmatrix} = 10 - 17 = -7$$

$$D_y = \begin{vmatrix} 2 & -2 \\ 4 & 17 \end{vmatrix} = 34 - (-8) = 42$$

Thus

$$x = \frac{D_x}{D} = \frac{-7}{-14} = \frac{1}{2} \quad \text{and} \quad y = \frac{D_y}{D} = \frac{42}{-14} = -3$$

The solution set is $\left\{ \left(\frac{1}{2}, -3 \right) \right\}$, which can be verified, as always, by substituting back into the original equations.   ■

EXAMPLE 3

Solve the system

$$\begin{pmatrix} \dfrac{1}{2}x + \dfrac{2}{3}y = -4 \\ \dfrac{1}{4}x - \dfrac{3}{2}y = 20 \end{pmatrix}$$

Solution

With such a system, either we can first produce an equivalent system with integral coefficients and then apply Cramer's rule, or we can apply the rule immediately. Let's avoid some work with fractions by multiplying the first equation by 6 and the second equation by 4 to produce the following equivalent system.

$$\begin{pmatrix} 3x + 4y = -24 \\ x - 6y = 80 \end{pmatrix}$$

Now we can proceed as before.

$$D = \begin{vmatrix} 3 & 4 \\ 1 & -6 \end{vmatrix} = -18 - 4 = -22$$

$$D_x = \begin{vmatrix} -24 & 4 \\ 80 & -6 \end{vmatrix} = 144 - 320 = -176$$

$$D_y = \begin{vmatrix} 3 & -24 \\ 1 & 80 \end{vmatrix} = 240 - (-24) = 264$$

Therefore,

$$x = \frac{D_x}{D} = \frac{-176}{-22} = 8 \quad \text{and} \quad y = \frac{D_y}{D} = \frac{264}{-22} = -12$$

The solution set is $\{(8, -12)\}$.   ■

In the statement of Cramer's rule, the condition that $D \neq 0$ was imposed. If $D = 0$ and either D_x or D_y (or both) is nonzero, then the system is inconsistent and has no solution. If $D = 0$, $D_x = 0$, and $D_y = 0$, then the equations are dependent and there are infinitely many solutions.

Cramer's Rule Extended

Without showing the details, we will simply state that Cramer's rule also applies to solving systems of three linear equations in three variables. It can be stated as follows.

Cramer's Rule (3 × 3 case)

Given the system

$$\begin{pmatrix} a_1x + b_1y + c_1z = d_1 \\ a_2x + b_2y + c_2z = d_2 \\ a_3x + b_3y + c_3z = d_3 \end{pmatrix}$$

with

$$D = \begin{vmatrix} a_1 & b_1 & c_1 \\ a_2 & b_2 & c_2 \\ a_3 & b_3 & c_3 \end{vmatrix} \neq 0 \qquad D_x = \begin{vmatrix} d_1 & b_1 & c_1 \\ d_2 & b_2 & c_2 \\ d_3 & b_3 & c_3 \end{vmatrix}$$

$$D_y = \begin{vmatrix} a_1 & d_1 & c_1 \\ a_2 & d_2 & c_2 \\ a_3 & d_3 & c_3 \end{vmatrix} \qquad D_z = \begin{vmatrix} a_1 & b_1 & d_1 \\ a_2 & b_2 & d_2 \\ a_3 & b_3 & d_3 \end{vmatrix}$$

then

$$x = \frac{D_x}{D} \qquad y = \frac{D_y}{D} \qquad \text{and} \qquad z = \frac{D_z}{D}$$

Again, note the restriction that $D \neq 0$. If $D = 0$ and at least one of D_x, D_y, and D_z is not zero, then the system is inconsistent. If D, D_x, D_y, and D_z are all zero, then the equations are dependent and there are infinitely many solutions.

EXAMPLE 4

Solve the system

$$\begin{pmatrix} x - 2y + z = -4 \\ 2x + y - z = 5 \\ 3x + 2y + 4z = 3 \end{pmatrix}$$

Solution

We will simply indicate the values of D, D_x, D_y, and D_z and leave the computations for you to check.

$$D = \begin{vmatrix} 1 & -2 & 1 \\ 2 & 1 & -1 \\ 3 & 2 & 4 \end{vmatrix} = 29 \qquad D_x = \begin{vmatrix} -4 & -2 & 1 \\ 5 & 1 & -1 \\ 3 & 2 & 4 \end{vmatrix} = 29$$

$$D_y = \begin{vmatrix} 1 & -4 & 1 \\ 2 & 5 & -1 \\ 3 & 3 & 4 \end{vmatrix} = 58 \qquad D_z = \begin{vmatrix} 1 & -2 & -4 \\ 2 & 1 & 5 \\ 3 & 2 & 3 \end{vmatrix} = -29$$

Therefore,

$$x = \frac{D_x}{D} = \frac{29}{29} = 1$$

$$y = \frac{D_y}{D} = \frac{58}{29} = 2$$

and

$$z = \frac{D_z}{D} = \frac{-29}{29} = -1$$

The solution set is $\{(1, 2, -1)\}$. (Be sure to check it!)

EXAMPLE 5

Solve the system

$$\begin{pmatrix} x + 3y - z = 4 \\ 3x - 2y + z = 7 \\ 2x + 6y - 2z = 1 \end{pmatrix}$$

Solution

$$D = \begin{vmatrix} 1 & 3 & -1 \\ 3 & -2 & 1 \\ 2 & 6 & -2 \end{vmatrix} = 2 \begin{vmatrix} 1 & 3 & -1 \\ 3 & -2 & 1 \\ 1 & 3 & -1 \end{vmatrix} = 2(0) = 0$$

$$D_x = \begin{vmatrix} 4 & 3 & -1 \\ 7 & -2 & 1 \\ 1 & 6 & -2 \end{vmatrix} = -7$$

Therefore, because $D = 0$ and at least one of D_x, D_y, and D_z is not zero, the system is inconsistent. The solution set is $\varnothing$.

Example 5 illustrates why D should be determined first. Once we found that $D = 0$ and $D_x \neq 0$, we knew that the system was inconsistent and there was no need to find D_y and D_z.

Finally, it should be noted that Cramer's rule can be extended to systems of n linear equations in n variables; however, that method is not considered to be a very efficient way of solving a large system of linear equations.

PROBLEM SET 6.5

For Problems 1–32, use Cramer's rule to find the solution set for each system. If the equations are dependent, simply indicate that there are infinitely many solutions.

1. $\begin{pmatrix} 2x - y = -2 \\ 3x + 2y = 11 \end{pmatrix}$

2. $\begin{pmatrix} 3x + y = -9 \\ 4x - 3y = 1 \end{pmatrix}$

3. $\begin{pmatrix} 5x + 2y = 5 \\ 3x - 4y = 29 \end{pmatrix}$

4. $\begin{pmatrix} 4x - 7y = -23 \\ 2x + 5y = -3 \end{pmatrix}$

5. $\begin{pmatrix} 5x - 4y = 14 \\ -x + 2y = -4 \end{pmatrix}$

6. $\begin{pmatrix} -x + 2y = 10 \\ 3x - y = -10 \end{pmatrix}$

7. $\begin{pmatrix} y = 2x - 4 \\ 6x - 3y = 1 \end{pmatrix}$

8. $\begin{pmatrix} -3x - 4y = 14 \\ -2x + 3y = -19 \end{pmatrix}$

9. $\begin{pmatrix} -4x + 3y = 3 \\ 4x - 6y = -5 \end{pmatrix}$

10. $\begin{pmatrix} x = 4y - 1 \\ 2x - 8y = -2 \end{pmatrix}$

11. $\begin{pmatrix} 9x - y = -2 \\ 8x + y = 4 \end{pmatrix}$

12. $\begin{pmatrix} 6x - 5y = 1 \\ 4x - 7y = 2 \end{pmatrix}$

13. $\begin{pmatrix} -\dfrac{2}{3}x + \dfrac{1}{2}y = -7 \\ \dfrac{1}{3}x - \dfrac{3}{2}y = 6 \end{pmatrix}$

14. $\begin{pmatrix} \dfrac{1}{2}x + \dfrac{2}{3}y = -6 \\ \dfrac{1}{4}x - \dfrac{1}{3}y = -1 \end{pmatrix}$

15. $\begin{pmatrix} 2x + 7y = -1 \\ x = 2 \end{pmatrix}$

16. $\begin{pmatrix} 5x - 3y = 2 \\ y = 4 \end{pmatrix}$

17. $\begin{pmatrix} x - y + 2z = -8 \\ 2x + 3y - 4z = 18 \\ -x + 2y - z = 7 \end{pmatrix}$

18. $\begin{pmatrix} x - 2y + z = 3 \\ 3x + 2y + z = -3 \\ 2x - 3y - 3z = -5 \end{pmatrix}$

19. $\begin{pmatrix} 2x - 3y + z = -7 \\ -3x + y - z = -7 \\ x - 2y - 5z = -45 \end{pmatrix}$

20. $\begin{pmatrix} 3x - y - z = 18 \\ 4x + 3y - 2z = 10 \\ -5x - 2y + 3z = -22 \end{pmatrix}$

21. $\begin{pmatrix} 4x + 5y - 2z = -14 \\ 7x - y + 2z = 42 \\ 3x + y + 4z = 28 \end{pmatrix}$

22. $\begin{pmatrix} -5x + 6y + 4z = -4 \\ -7x - 8y + 2z = -2 \\ 2x + 9y - z = 1 \end{pmatrix}$

23. $\begin{pmatrix} 2x - y + 3z = -17 \\ 3y + z = 5 \\ x - 2y - z = -3 \end{pmatrix}$

24. $\begin{pmatrix} 2x - y + 3z = -5 \\ 3x + 4y - 2z = -25 \\ -x + z = 6 \end{pmatrix}$

25. $\begin{pmatrix} x + 3y - 4z = -1 \\ 2x - y + z = 2 \\ 4x + 5y - 7z = 0 \end{pmatrix}$

26. $\begin{pmatrix} x - 2y + z = 1 \\ 3x + y - z = 2 \\ 2x - 4y + 2z = -1 \end{pmatrix}$

27. $\begin{pmatrix} 3x - 2y - 3z = -5 \\ x + 2y + 3z = -3 \\ -x + 4y - 6z = 8 \end{pmatrix}$

28. $\begin{pmatrix} 3x - 2y + z = 11 \\ 5x + 3y = 17 \\ x + y - 2z = 6 \end{pmatrix}$

29. $\begin{pmatrix} x - 2y + 3z = 1 \\ -2x + 4y - 3z = -3 \\ 5x - 6y + 6z = 10 \end{pmatrix}$

30. $\begin{pmatrix} 2x - y + 2z = -1 \\ 4x + 3y - 4z = 2 \\ x + 5y - z = 9 \end{pmatrix}$

31. $\begin{pmatrix} -x - y + 3z = -2 \\ -2x + y + 7z = 14 \\ 3x + 4y - 5z = 12 \end{pmatrix}$

32. $\begin{pmatrix} -2x + y - 3z = -4 \\ x + 5y - 4z = 13 \\ 7x - 2y - z = 37 \end{pmatrix}$

THOUGHTS INTO WORDS

33. Give a step-by-step description of how you would solve the system

$$\begin{pmatrix} 2x - y + 3z = 31 \\ x - 2y - z = 8 \\ 3x + 5y + 8z = 35 \end{pmatrix}$$

34. Give a step-by-step description of how you would find the value of x in the solution for the system

$$\begin{pmatrix} x + 5y - z = -9 \\ 2x - y + z = 11 \\ -3x - 2y + 4z = 20 \end{pmatrix}$$

Further Investigations

35. A linear system in which the constant terms are all zero is called a homogeneous system.

   a. Verify that for a 3×3 homogeneous system, if $D \neq 0$, then $(0, 0, 0)$ is the only solution for the system.

   b. Verify that for a 3×3 homogeneous system, if $D = 0$, then the equations are dependent.

For Problems 36–39, solve each of the homogeneous systems (see problem 35). If the equations are dependent, indicate that the system has infinitely many solutions.

36. $\begin{pmatrix} x - 2y + 5z = 0 \\ 3x + y - 2z = 0 \\ 4x - y + 3z = 0 \end{pmatrix}$

37. $\begin{pmatrix} 2x - y + z = 0 \\ 3x + 2y + 5z = 0 \\ 4x - 7y + z = 0 \end{pmatrix}$

38. $\begin{pmatrix} 3x + y - z = 0 \\ x - y + 2z = 0 \\ 4x - 5y - 2z = 0 \end{pmatrix}$

39. $\begin{pmatrix} 2x - y + 2z = 0 \\ x + 2y + z = 0 \\ x - 3y + z = 0 \end{pmatrix}$

 ## GRAPHING CALCULATOR ACTIVITIES

40. Use determinants and your calculator to solve each of the following systems.

   a. $\begin{pmatrix} 4x - 3y + z = 10 \\ 8x + 5y - 2z = -6 \\ -12x - 2y + 3z = -2 \end{pmatrix}$

   b. $\begin{pmatrix} 2x + y - z + w = -4 \\ x + 2y + 2z - 3w = 6 \\ 3x - y - z + 2w = 0 \\ 2x + 3y + z + 4w = -5 \end{pmatrix}$

   c. $\begin{pmatrix} x - 2y + z - 3w = 4 \\ 2x + 3y - z - 2w = -4 \\ 3x - 4y + 2z - 4w = 12 \\ 2x - y - 3z + 2w = -2 \end{pmatrix}$

   d. $\begin{pmatrix} 1.98x + 2.49y + 3.45z = 80.10 \\ 2.15x + 3.20y + 4.19z = 97.16 \\ 1.49x + 4.49y + 2.79z = 83.92 \end{pmatrix}$

CHAPTER 6 SUMMARY

The primary focus of this entire chapter is the development of different techniques for solving systems of linear equations.

Substitution Method

With the aid of an example, we can describe the substitution method as follows. Suppose we want to solve the system

$$\begin{pmatrix} x - 2y = 22 \\ 3x + 4y = -24 \end{pmatrix}$$

STEP 1 Solve the first equation for x in terms of y.

$$x - 2y = 22$$
$$x = 2y + 22$$

STEP 2 Substitute $2y + 22$ for x in the second equation.

$$3(2y + 22) + 4y = -24$$

STEP 3 Solve the equation obtained in step 2.

$$6y + 66 + 4y = -24$$
$$10y + 66 = -24$$
$$10y = -90$$
$$y = -9$$

STEP 4 Substitute -9 for y in the equation of step 1.

$$x = 2(-9) + 22 = 4$$

The solution set is $\{(4, -9)\}$.

Elimination-by-Addition Method

This method allows us to replace systems of equations with *simpler equivalent systems* until we obtain a system where we can easily determine the solution. The following operations produce equivalent systems.

1. Any two equations of a system can be interchanged.
2. Both sides of any equation of the system can be multiplied by any nonzero real number.
3. Any equation of the system can be replaced by the sum of a nonzero multiple of another equation plus that equation.

For example, through a sequence of operations, we can transform the system

$$\begin{pmatrix} 5x + 3y = -28 \\ \dfrac{1}{2}x - y = -8 \end{pmatrix}$$

to the equivalent system

$$\begin{pmatrix} x - 2y = -16 \\ 13y = 52 \end{pmatrix}$$

where we can easily determine the solution set $\{(-8, 4)\}$.

Matrix Approach

We can change the augmented matrix of a system to reduced echelon form by applying the following elementary row operations.

1. Any two rows of the matrix can be interchanged.

2. Any row of the matrix can be multiplied by a nonzero real number.

3. Any row of the matrix can be replaced by the sum of a nonzero multiple of another row plus that row.

For example, the augmented matrix of the system

$$\begin{pmatrix} x - 2y + 3z = 4 \\ 2x + y - 4z = 3 \\ -3x + 4y - z = -2 \end{pmatrix}$$

is

$$\begin{bmatrix} 1 & -2 & 3 & 4 \\ 2 & 1 & -4 & 3 \\ -3 & 4 & -1 & -2 \end{bmatrix}$$

We can change this matrix to the reduced echelon form

$$\begin{bmatrix} 1 & 0 & 0 & 4 \\ 0 & 1 & 0 & 3 \\ 0 & 0 & 1 & 2 \end{bmatrix}$$

where the solution set $\{(4, 3, 2)\}$ is obvious.

Cramer's Rule

Cramer's rule for solving systems of linear equations involves the use of determinants. It is stated for the 2×2 case on page 477 and for the 3×3 case on page 480. For example the solution set of the system

$$\begin{pmatrix} 3x - y - z = 2 \\ 2x + y + 3z = 9 \\ -x + 5y - 6z = -29 \end{pmatrix}$$

is determined by

$$x = \frac{\begin{vmatrix} 2 & -1 & -1 \\ 9 & 1 & 3 \\ -29 & 5 & -6 \end{vmatrix}}{\begin{vmatrix} 3 & -1 & -1 \\ 2 & 1 & 3 \\ -1 & 5 & -6 \end{vmatrix}} = \frac{-83}{-83} = 1$$

$$y = \frac{\begin{vmatrix} 3 & 2 & -1 \\ 2 & 9 & 3 \\ -1 & -29 & -6 \end{vmatrix}}{\begin{vmatrix} 3 & -1 & -1 \\ 2 & 1 & 3 \\ -1 & 5 & -6 \end{vmatrix}} = \frac{166}{-83} = -2$$

and

$$z = \frac{\begin{vmatrix} 3 & -1 & 2 \\ 2 & 1 & 9 \\ -1 & 5 & -29 \end{vmatrix}}{\begin{vmatrix} 3 & -1 & -1 \\ 2 & 1 & 3 \\ -1 & 5 & -6 \end{vmatrix}} = \frac{-249}{-83} = 3$$

CHAPTER 6 REVIEW PROBLEM SET

For Problems 1–4, solve each system by using the *substitution* method.

1. $\begin{pmatrix} 3x - y = 16 \\ 5x + 7y = -34 \end{pmatrix}$

2. $\begin{pmatrix} 6x + 5y = -21 \\ x - 4y = 11 \end{pmatrix}$

3. $\begin{pmatrix} 2x - 3y = 12 \\ 3x + 5y = -20 \end{pmatrix}$

4. $\begin{pmatrix} 5x + 8y = 1 \\ 4x + 7y = -2 \end{pmatrix}$

For Problems 5–8, solve each system by using the *elimination-by-addition* method.

5. $\begin{pmatrix} 4x - 3y = 34 \\ 3x + 2y = 0 \end{pmatrix}$

6. $\begin{pmatrix} \frac{1}{2}x - \frac{2}{3}y = 1 \\ \frac{3}{4}x + \frac{1}{6}y = -1 \end{pmatrix}$

7. $\begin{pmatrix} 2x - y + 3z = -19 \\ 3x + 2y - 4z = 21 \\ 5x - 4y - z = -8 \end{pmatrix}$

8. $\begin{pmatrix} 3x + 2y - 4z = 4 \\ 5x + 3y - z = 2 \\ 4x - 2y + 3z = 11 \end{pmatrix}$

For Problems 9–12, solve each system by *changing the augmented matrix to reduced echelon form.*

9. $\begin{pmatrix} x - 3y = 17 \\ -3x + 2y = -23 \end{pmatrix}$

10. $\begin{pmatrix} 2x + 3y = 25 \\ 3x - 5y = -29 \end{pmatrix}$

11. $\begin{pmatrix} x - 2y + z = -7 \\ 2x - 3y + 4z = -14 \\ -3x + y - 2z = 10 \end{pmatrix}$

12. $\begin{pmatrix} -2x - 7y + z = 9 \\ x + 3y - 4z = -11 \\ 4x + 5y - 3z = -11 \end{pmatrix}$

For Problems 13–16, solve each system by using *Cramer's rule.*

13. $\begin{pmatrix} 5x + 3y = -18 \\ 4x - 9y = -3 \end{pmatrix}$

14. $\begin{pmatrix} 0.2x + 0.3y = 2.6 \\ 0.5x - 0.1y = 1.4 \end{pmatrix}$

15. $\begin{pmatrix} 2x - 3y - 3z = 25 \\ 3x + y + 2z = -5 \\ 5x - 2y - 4z = 32 \end{pmatrix}$

16. $\begin{pmatrix} 3x - y + z = -10 \\ 6x - 2y + 5z = -35 \\ 7x + 3y - 4z = 19 \end{pmatrix}$

For Problems 17–24, solve each system by using the method you think is most appropriate.

17. $\begin{pmatrix} 4x + 7y = -15 \\ 3x - 2y = 25 \end{pmatrix}$

18. $\begin{pmatrix} \dfrac{3}{4}x - \dfrac{1}{2}y = -15 \\ \dfrac{2}{3}x + \dfrac{1}{4}y = -5 \end{pmatrix}$

19. $\begin{pmatrix} x + 4y = 3 \\ 3x - 2y = 1 \end{pmatrix}$

20. $\begin{pmatrix} 7x - 3y = -49 \\ y = \dfrac{3}{5}x - 1 \end{pmatrix}$

21. $\begin{pmatrix} x - y - z = 4 \\ -3x + 2y + 5z = -21 \\ 5x - 3y - 7z = 30 \end{pmatrix}$

22. $\begin{pmatrix} 2x - y + z = -7 \\ -5x + 2y - 3z = 17 \\ 3x + y + 7z = -5 \end{pmatrix}$

23. $\begin{pmatrix} 3x - 2y - 5z = 2 \\ -4x + 3y + 11z = 3 \\ 2x - y + z = -1 \end{pmatrix}$

24. $\begin{pmatrix} 7x - y + z = -4 \\ -2x + 9y - 3z = -50 \\ x - 5y + 4z = 42 \end{pmatrix}$

Problems 25–30, evaluate each determinant.

25. $\begin{vmatrix} -2 & 6 \\ 3 & 8 \end{vmatrix}$

26. $\begin{vmatrix} 5 & -4 \\ 7 & -3 \end{vmatrix}$

27. $\begin{vmatrix} 2 & 3 & -1 \\ 3 & 4 & -5 \\ 6 & 4 & 2 \end{vmatrix}$

28. $\begin{vmatrix} 3 & -2 & 4 \\ 1 & 0 & 6 \\ 3 & -3 & 5 \end{vmatrix}$

29. $\begin{vmatrix} 5 & 4 & 3 \\ 2 & -7 & 0 \\ 3 & -2 & 0 \end{vmatrix}$

30. $\begin{vmatrix} 5 & -4 & 2 & 1 \\ 3 & 7 & 6 & 2 \\ 2 & 1 & -5 & 0 \\ 3 & -2 & 4 & 0 \end{vmatrix}$

For Problems 31–34, solve each problem by setting up and solving a system of linear equations.

31. The sum of the digits of a two-digit number is 9. If the digits are reversed, the newly formed number is 45 less than the original number. Find the original number.

32. Sara invested $2500, part of it at 10% and the rest at 12% yearly interest. The yearly income on the 12% investment was $102 more than the income on the 10% investment. How much money did she invest at each rate?

33. A box contains $17.70 in nickels, dimes, and quarters. The number of dimes is eight less than twice the number of nickels. The number of quarters is two more than the sum of the numbers of nickels and dimes. How many coins of each kind are there in the box?

34. The measure of the largest angle of a triangle is 10° more than four times the smallest angle. The sum of the smallest and largest angles is three times the measure of the other angle. Find the measure of each angle of the triangle.

CHAPTER 6 TEST

For Problems 1–4, refer to the following systems of equations.

I. $\begin{pmatrix} 3x - 2y = 4 \\ 9x - 6y = 12 \end{pmatrix}$ **II.** $\begin{pmatrix} 5x - y = 4 \\ 3x + 7y = 9 \end{pmatrix}$ **III.** $\begin{pmatrix} 2x - y = 4 \\ 2x - y = -6 \end{pmatrix}$

1. For which system are the graphs parallel lines?

2. For which system are the equations dependent?

3. For which system is the solution set $\varnothing$?

4. Which system is consistent?

For Problems 5–8, evaluate each determinant.

5. $\begin{vmatrix} -2 & 4 \\ -5 & 6 \end{vmatrix}$ 6. $\begin{vmatrix} \dfrac{1}{2} & \dfrac{1}{3} \\ \dfrac{3}{4} & -\dfrac{2}{3} \end{vmatrix}$ 7. $\begin{vmatrix} -1 & 2 & 1 \\ 3 & 1 & -2 \\ 2 & -1 & 1 \end{vmatrix}$ 8. $\begin{vmatrix} 2 & 4 & -5 \\ -4 & 3 & 0 \\ -2 & 6 & 1 \end{vmatrix}$

9. How many ordered pairs of real numbers are in the solution set for the system $\begin{pmatrix} y = 3x - 4 \\ 9x - 3y = 12 \end{pmatrix}$?

10. Solve the system $\begin{pmatrix} 3x - 2y = -14 \\ 7x + 2y = -6 \end{pmatrix}$.

11. Solve the system $\begin{pmatrix} 4x - 5y = 17 \\ y = -3x + 8 \end{pmatrix}$.

12. Find the value of x in the solution for the system

$$\begin{pmatrix} \dfrac{3}{4}x - \dfrac{1}{2}y = -21 \\ \dfrac{2}{3}x + \dfrac{1}{6}y = -4 \end{pmatrix}$$

13. Find the value of y in the solution for the system $\begin{pmatrix} 4x - y = 7 \\ 3x + 2y = 2 \end{pmatrix}$.

14. Is $(1, -1, 4)$ a solution of the following system?

$$\begin{pmatrix} 2x - y + z = 7 \\ 3x - 2y + 2z = 13 \\ x - 4y + 5z = 17 \end{pmatrix}$$

15. Suppose that the augmented matrix of a system of three linear equations in the three variables x, y, and z can be changed to the matrix

$$\begin{bmatrix} 1 & 1 & -4 & \vdots & 3 \\ 0 & 1 & 4 & \vdots & 5 \\ 0 & 0 & 3 & \vdots & 6 \end{bmatrix}$$

Find the value of x in the solution for the system.

16. Suppose that the augmented matrix of a system of three linear equations in the three variables x, y, and z can be changed to the matrix

$$\begin{bmatrix} 1 & 2 & -3 & \vert & 4 \\ 0 & 1 & 2 & \vert & 5 \\ 0 & 0 & 2 & \vert & -8 \end{bmatrix}$$

Find the value of y in the solution for the system.

17. How many ordered triples are there in the solution set for the following system?

$$\left(\begin{array}{l} x + 3y - z = 5 \\ 2x - y - z = 7 \\ 5x + 8y - 4z = 22 \end{array} \right)$$

18. How many ordered triples are there in the solution set for the following system?

$$\left(\begin{array}{l} 3x - y - 2z = 1 \\ 4x + 2y + z = 5 \\ 6x - 2y - 4z = 9 \end{array} \right)$$

19. Solve the following system.

$$\left(\begin{array}{l} 5x - 3y - 2z = -1 \\ 4y + 7z = 3 \\ 4z = -12 \end{array} \right)$$

20. Solve the following system.

$$\left(\begin{array}{l} x - 2y + z = 0 \\ y - 3z = -1 \\ 2y + 5z = -2 \end{array} \right)$$

21. Find the value of x in the solution for the system

$$\left(\begin{array}{l} x - 4y + z = 12 \\ -2x + 3y - z = -11 \\ 5x - 3y + 2z = 17 \end{array} \right)$$

22. Find the value of y in the solution for the system

$$\left(\begin{array}{l} x - 3y + z = -13 \\ 3x + 5y - z = 17 \\ 5x - 2y + 2z = -13 \end{array} \right)$$

23. The tens digit of a two-digit number is 1 more than twice the units digit. The number formed by reversing the digits is 27 times smaller than the original number. Find the original number.

24. One solution is 30% alcohol and another solution is 70% alcohol. Some of each of the two solutions is mixed to produce 8 liters of a 40% solution. How many liters of the 70% solution should be used?

25. A box contains $7.25 in nickels, dimes, and quarters. There are 43 coins, and the number of quarters is 1 more than three times the number of nickels. Find the number of quarters in the box.

ALGEBRA OF MATRICES

Matrices can be used to organize and manipulate data that determine population trends over a certain time interval.

In Section 6.3, we used matrices strictly as a device to help solve systems of linear equations. Our primary objective was the development of techniques for solving systems of equations, not the study of matrices. However, matrices can be studied from an algebraic viewpoint, much as we study the set of real numbers. That is, we can define certain operations on matrices and verify properties of those operations. This algebraic approach to matrices is the focal point of this chapter. In order to get a simplified view of the algebra of matrices, we will begin by studying 2 × 2 matrices, and then later we will enlarge our discussion to include m × n matrices. As a bonus, another technique for solving systems of equations will emerge from our study.

7.1 ALGEBRA OF 2 × 2 MATRICES

Throughout these next two sections, we will be working primarily with 2 × 2 matrices; therefore, any reference to matrices means 2 × 2 matrices unless stated otherwise. The following 2 × 2 matrix notation will be used frequently.

$$A = \begin{bmatrix} a_{11} & a_{12} \\ a_{21} & a_{22} \end{bmatrix} \quad B = \begin{bmatrix} b_{11} & b_{12} \\ b_{21} & b_{22} \end{bmatrix} \quad C = \begin{bmatrix} c_{11} & c_{12} \\ c_{21} & c_{22} \end{bmatrix}$$

Two matrices are **equal** if and only if all elements in corresponding positions are equal. Thus $A = B$ if and only if $a_{11} = b_{11}$, $a_{12} = b_{12}$, $a_{21} = b_{21}$, and $a_{22} = b_{22}$.

Addition of Matrices

To add two matrices, we add the elements that appear in corresponding positions. Therefore, the sum of matrix A and matrix B is defined as follows.

> **DEFINITION 7.1**
>
> $$A + B = \begin{bmatrix} a_{11} & a_{12} \\ a_{21} & a_{22} \end{bmatrix} + \begin{bmatrix} b_{11} & b_{12} \\ b_{21} & b_{22} \end{bmatrix}$$
> $$= \begin{bmatrix} a_{11} + b_{11} & a_{12} + b_{12} \\ a_{21} + b_{21} & a_{22} + b_{22} \end{bmatrix}$$

For example,

$$\begin{bmatrix} 2 & -1 \\ -3 & 4 \end{bmatrix} + \begin{bmatrix} -5 & 4 \\ -1 & 7 \end{bmatrix} = \begin{bmatrix} -3 & 3 \\ -4 & 11 \end{bmatrix}$$

It is not difficult to show that **the commutative and associative properties are valid for the addition of matrices.** Thus we can state that

$$A + B = B + A \qquad \text{and} \qquad (A + B) + C = A + (B + C)$$

Because

$$\begin{bmatrix} a_{11} & a_{12} \\ a_{21} & a_{22} \end{bmatrix} + \begin{bmatrix} 0 & 0 \\ 0 & 0 \end{bmatrix} = \begin{bmatrix} a_{11} & a_{12} \\ a_{21} & a_{22} \end{bmatrix}$$

we see that $\begin{bmatrix} 0 & 0 \\ 0 & 0 \end{bmatrix}$, which is called the **zero matrix** and represented by O, is the **additive identity element.** Thus we can state that

$$A + O = O + A = A$$

Because every real number has an additive inverse, it follows that any matrix A has an **additive inverse**, $-A$, that is formed by taking the additive inverse of each element of A. For example, if

$$A = \begin{bmatrix} 4 & -2 \\ -1 & 0 \end{bmatrix} \qquad \text{then} \qquad -A = \begin{bmatrix} -4 & 2 \\ 1 & 0 \end{bmatrix}$$

and

$$A + (-A) = \begin{bmatrix} 4 & -2 \\ -1 & 0 \end{bmatrix} + \begin{bmatrix} -4 & 2 \\ 1 & 0 \end{bmatrix} = \begin{bmatrix} 0 & 0 \\ 0 & 0 \end{bmatrix}$$

In general, we can state that **every matrix A has an additive inverse $-A$ such that**

$$A + (-A) = (-A) + A = O$$

Subtraction of Matrices

Again, paralleling the algebra of real numbers, *subtraction* of matrices can be defined in terms of *adding the additive inverse*. Therefore, we can define subtraction as follows.

DEFINITION 7.2

$$A - B = A + (-B)$$

For example,

$$\begin{bmatrix} 2 & -7 \\ -6 & 5 \end{bmatrix} - \begin{bmatrix} 3 & 4 \\ -2 & -1 \end{bmatrix} = \begin{bmatrix} 2 & -7 \\ -6 & 5 \end{bmatrix} + \begin{bmatrix} -3 & -4 \\ 2 & 1 \end{bmatrix}$$

$$= \begin{bmatrix} -1 & -11 \\ -4 & 6 \end{bmatrix}$$

Scalar Multiplication

When we work with matrices, we commonly refer to a single real number as a **scalar** to distinguish it from a matrix. Then taking the **product** of a scalar and a matrix (often referred to as a **scalar multiplication**) can be accomplished by multiplying each element of the matrix by the scalar. For example,

$$3\begin{bmatrix} -4 & -6 \\ 1 & -2 \end{bmatrix} = \begin{bmatrix} 3(-4) & 3(-6) \\ 3(1) & 3(-2) \end{bmatrix} = \begin{bmatrix} -12 & -18 \\ 3 & -6 \end{bmatrix}$$

In general, scalar multiplication can be defined as follows.

DEFINITION 7.3

$$kA = k\begin{bmatrix} a_{11} & a_{12} \\ a_{21} & a_{22} \end{bmatrix} = \begin{bmatrix} ka_{11} & ka_{12} \\ ka_{21} & ka_{32} \end{bmatrix}$$

where k is any real number.

EXAMPLE 1

If $A = \begin{bmatrix} -4 & 3 \\ 2 & -5 \end{bmatrix}$ and $B = \begin{bmatrix} 2 & -3 \\ 7 & -6 \end{bmatrix}$, find

a. $-2A$ **b.** $3A + 2B$ **c.** $A - 4B$

Solutions

a. $-2A = -2\begin{bmatrix} -4 & 3 \\ 2 & -5 \end{bmatrix} = \begin{bmatrix} 8 & -6 \\ -4 & 10 \end{bmatrix}$

b. $3A + 2B = 3\begin{bmatrix} -4 & 3 \\ 2 & -5 \end{bmatrix} + 2\begin{bmatrix} 2 & -3 \\ 7 & -6 \end{bmatrix}$

$$= \begin{bmatrix} -12 & 9 \\ 6 & -15 \end{bmatrix} + \begin{bmatrix} 4 & -6 \\ 14 & -12 \end{bmatrix}$$

$$= \begin{bmatrix} -8 & 3 \\ 20 & -27 \end{bmatrix}$$

c. $A - 4B = \begin{bmatrix} -4 & 3 \\ 2 & -5 \end{bmatrix} - 4\begin{bmatrix} 2 & -3 \\ 7 & -6 \end{bmatrix}$

$ = \begin{bmatrix} -4 & 3 \\ 2 & -5 \end{bmatrix} - \begin{bmatrix} 8 & -12 \\ 28 & -24 \end{bmatrix}$

$ = \begin{bmatrix} -4 & 3 \\ 2 & -5 \end{bmatrix} + \begin{bmatrix} -8 & 12 \\ -28 & 24 \end{bmatrix}$

$ = \begin{bmatrix} -12 & 15 \\ -26 & 19 \end{bmatrix}$ ■

The following properties, which are easy to check, pertain to scalar multiplication and matrix addition (where k and l represent any real numbers).

$$k(A + B) = kA + kB$$
$$(k + l)A = kA + lA$$
$$(kl)A = k(lA)$$

Multiplication of Matrices

At this time, it probably would seem quite natural to define matrix multiplication by multiplying corresponding elements of two matrices. However, it turns out that such a definition does not have many worthwhile applications. Therefore, we use a special type of **matrix multiplication**, sometimes referred to as a *row-by-column* multiplication. We will state the definition, paraphrase what it says, and then give some examples.

DEFINITION 7.4

$$AB = \begin{bmatrix} a_{11} & a_{12} \\ a_{21} & a_{22} \end{bmatrix}\begin{bmatrix} b_{11} & b_{12} \\ b_{21} & b_{22} \end{bmatrix}$$

$$= \begin{bmatrix} a_{11}b_{11} + a_{12}b_{21} & a_{11}b_{12} + a_{12}b_{22} \\ a_{21}b_{11} + a_{22}b_{21} & a_{21}b_{12} + a_{22}b_{22} \end{bmatrix}$$

Notice the row-by-column pattern of Definition 7.4. We multiply the rows of A times the columns of B in a pairwise entry fashion, adding the results. For example, the element in the first row and second column of the product is obtained by multiplying the elements of the first row of A times the elements of the second column of B and adding the results.

$$\begin{bmatrix} \boxed{a_{11}} & a_{12} \\ a_{21} & a_{22} \end{bmatrix} \begin{bmatrix} b_{11} & \boxed{b_{12}} \\ b_{21} & \boxed{b_{22}} \end{bmatrix} = \begin{bmatrix} & a_{11}b_{12} + a_{12}b_{22} \end{bmatrix}$$

Now let's look at some specific examples.

EXAMPLE 2

If $A = \begin{bmatrix} -2 & 1 \\ 4 & 5 \end{bmatrix}$ and $B = \begin{bmatrix} 3 & -2 \\ -1 & 7 \end{bmatrix}$, find (a) AB and (b) BA.

Solutions

a. $AB = \begin{bmatrix} -2 & 1 \\ 4 & 5 \end{bmatrix} \begin{bmatrix} 3 & -2 \\ -1 & 7 \end{bmatrix}$

$= \begin{bmatrix} (-2)(3) + (1)(-1) & (-2)(-2) + (1)(7) \\ (4)(3) + (5)(-1) & (4)(-2) + (5)(7) \end{bmatrix}$

$= \begin{bmatrix} -7 & 11 \\ 7 & 27 \end{bmatrix}$

b. $BA = \begin{bmatrix} 3 & -2 \\ -1 & 7 \end{bmatrix} \begin{bmatrix} -2 & 1 \\ 4 & 5 \end{bmatrix}$

$= \begin{bmatrix} (3)(-2) + (-2)(4) & (3)(1) + (-2)(5) \\ (-1)(-2) + (7)(4) & (-1)(1) + (7)(5) \end{bmatrix} = \begin{bmatrix} -14 & -7 \\ 30 & 34 \end{bmatrix}$ ∎

Example 2 makes it immediately apparent that **matrix multiplication is not a commutative operation.**

EXAMPLE 3

If $A = \begin{bmatrix} 2 & -6 \\ -3 & 9 \end{bmatrix}$ and $B = \begin{bmatrix} -3 & 6 \\ -1 & 2 \end{bmatrix}$, find AB.

Solution

Once you feel comfortable with Definition 7.4, you can do the addition mentally.

$$AB = \begin{bmatrix} 2 & -6 \\ -3 & 9 \end{bmatrix} \begin{bmatrix} -3 & 6 \\ -1 & 2 \end{bmatrix} = \begin{bmatrix} 0 & 0 \\ 0 & 0 \end{bmatrix}$$ ∎

Example 3 illustrates that the product of two matrices can be the zero matrix even though neither of the two matrices is the zero matrix. This is different from the property of real numbers that states $ab = 0$ *if and only if* $a = 0$ *or* $b = 0$.

As we illustrated and stated earlier, matrix multiplication is *not* a commutative operation. However, it is an **associative operation** and it does abide by two **distributive properties.** These properties can be stated as follows.

$$(AB)C = A(BC)$$
$$A(B + C) = AB + AC$$
$$(B + C)A = BA + CA$$

We will ask you to verify these properties in the next set of problems.

PROBLEM SET 7.1

For Problems 1–12, compute the indicated matrix by using the following matrices:

$$A = \begin{bmatrix} 1 & -2 \\ 3 & 4 \end{bmatrix} \qquad B = \begin{bmatrix} 2 & -3 \\ 5 & -1 \end{bmatrix}$$

$$C = \begin{bmatrix} 0 & 6 \\ -4 & 2 \end{bmatrix} \qquad D = \begin{bmatrix} -2 & 3 \\ 5 & -4 \end{bmatrix}$$

$$E = \begin{bmatrix} 2 & 5 \\ 7 & 3 \end{bmatrix}$$

1. $A + B$

2. $B - C$

3. $3C + D$

4. $2D - E$

5. $4A - 3B$

6. $2B + 3D$

7. $(A - B) - C$

8. $B - (D - E)$

9. $2D - 4E$

10. $3A - 4E$

11. $B - (D + E)$

12. $A - (B + C)$

For Problems 13–26, compute AB and BA.

13. $A = \begin{bmatrix} 1 & -1 \\ 2 & -2 \end{bmatrix}, \qquad B = \begin{bmatrix} 3 & -4 \\ -1 & 2 \end{bmatrix}$

14. $A = \begin{bmatrix} -3 & 4 \\ 2 & 1 \end{bmatrix}, \qquad B = \begin{bmatrix} -2 & 5 \\ 6 & -1 \end{bmatrix}$

15. $A = \begin{bmatrix} 1 & -3 \\ -4 & 6 \end{bmatrix}, \qquad B = \begin{bmatrix} 7 & -3 \\ 4 & 5 \end{bmatrix}$

16. $A = \begin{bmatrix} 5 & 0 \\ -2 & 3 \end{bmatrix}, \qquad B = \begin{bmatrix} -3 & 6 \\ 4 & 1 \end{bmatrix}$

17. $A = \begin{bmatrix} 2 & -4 \\ 1 & -2 \end{bmatrix}, \qquad B = \begin{bmatrix} 1 & -2 \\ -3 & 6 \end{bmatrix}$

18. $A = \begin{bmatrix} 1 & 2 \\ 1 & 2 \end{bmatrix}, \qquad B = \begin{bmatrix} 2 & 2 \\ -1 & -1 \end{bmatrix}$

19. $A = \begin{bmatrix} -3 & -2 \\ -4 & -1 \end{bmatrix}, \qquad B = \begin{bmatrix} 2 & -1 \\ 4 & 5 \end{bmatrix}$

20. $A = \begin{bmatrix} -2 & 3 \\ -1 & 7 \end{bmatrix}, \qquad B = \begin{bmatrix} -1 & -3 \\ -5 & -7 \end{bmatrix}$

21. $A = \begin{bmatrix} 2 & -1 \\ -5 & 3 \end{bmatrix}, \qquad B = \begin{bmatrix} 3 & 1 \\ 5 & 2 \end{bmatrix}$

22. $A = \begin{bmatrix} -8 & -5 \\ 3 & 2 \end{bmatrix}, \qquad B = \begin{bmatrix} -2 & -5 \\ 3 & 8 \end{bmatrix}$

23. $A = \begin{bmatrix} \dfrac{1}{2} & -\dfrac{1}{3} \\ \dfrac{1}{3} & \dfrac{1}{4} \end{bmatrix}, \qquad B = \begin{bmatrix} 4 & -6 \\ 6 & -4 \end{bmatrix}$

24. $A = \begin{bmatrix} \dfrac{1}{3} & -\dfrac{1}{2} \\ \dfrac{3}{2} & -\dfrac{2}{3} \end{bmatrix}, \qquad B = \begin{bmatrix} -6 & -18 \\ 12 & -12 \end{bmatrix}$

25. $A = \begin{bmatrix} 5 & 6 \\ 2 & 3 \end{bmatrix}, \qquad B = \begin{bmatrix} 1 & -2 \\ -\dfrac{2}{3} & \dfrac{5}{3} \end{bmatrix}$

26. $A = \begin{bmatrix} -3 & -5 \\ 2 & 4 \end{bmatrix}, \qquad B = \begin{bmatrix} -2 & -\dfrac{5}{2} \\ 1 & \dfrac{3}{2} \end{bmatrix}$

For Problems 27–30, use the following matrices.

$$A = \begin{bmatrix} -2 & 3 \\ 5 & 4 \end{bmatrix} \qquad B = \begin{bmatrix} 0 & 1 \\ 1 & 0 \end{bmatrix}$$

$$C = \begin{bmatrix} 1 & 0 \\ 1 & 0 \end{bmatrix} \qquad D = \begin{bmatrix} 1 & 1 \\ 1 & 1 \end{bmatrix}$$

$$I = \begin{bmatrix} 1 & 0 \\ 0 & 1 \end{bmatrix}$$

27. Compute AB and BA.   **28.** Compute AC and CA.

29. Compute AD and DA.   **30.** Compute AI and IA.

For Problems 31–34, use the following matrices.

$$A = \begin{bmatrix} 2 & 4 \\ 5 & -3 \end{bmatrix} \qquad B = \begin{bmatrix} -2 & 3 \\ -1 & 2 \end{bmatrix}$$

$$C = \begin{bmatrix} 2 & 1 \\ 3 & 7 \end{bmatrix}$$

31. Show that $(AB)C = A(BC)$.

32. Show that $A(B + C) = AB + AC$.

33. Show that $(A + B)C = AC + BC$.

34. Show that $(3 + 2)A = 3A + 2A$.

For Problems 35–43, use the following matrices.

$$A = \begin{bmatrix} a_{11} & a_{12} \\ a_{21} & a_{22} \end{bmatrix} \qquad B = \begin{bmatrix} b_{11} & b_{12} \\ b_{21} & b_{22} \end{bmatrix}$$

$$C = \begin{bmatrix} c_{11} & c_{12} \\ c_{21} & c_{22} \end{bmatrix} \qquad O = \begin{bmatrix} 0 & 0 \\ 0 & 0 \end{bmatrix}$$

35. Show that $A + B = B + A$.

36. Show that $(A + B) + C = A + (B + C)$.

37. Show that $A + (-A) = O$.

38. Show that $k(A + B) = kA + kB$ for any real number k.

39. Show that $(k + l)A = kA + lA$ for any real numbers k and l.

40. Show that $(kl)A = k(lA)$ for any real numbers k and l.

41. Show that $(AB)C = A(BC)$.

42. Show that $A(B + C) = AB + AC$.

43. Show that $(A + B)C = AC + BC$.

THOUGHTS INTO WORDS

44. How would you show that addition of 2×2 matrices is a commutative operation?

45. How would you show that subtraction of 2×2 matrices is not a commutative operation?

46. How would you explain matrix multiplication to someone who missed class the day it was discussed?

47. Your friend says that because multiplication of real numbers is a commutative operation, it seems reasonable that multiplication of matrices should also be a commutative operation. How would you react to that statement?

Further Investigations

48. If $A = \begin{bmatrix} 2 & 0 \\ 0 & 3 \end{bmatrix}$, calculate A^2 and A^3, where A^2 means AA and A^3 means AAA.

49. If $A = \begin{bmatrix} 1 & -1 \\ 2 & 3 \end{bmatrix}$, calculate A^2 and A^3.

50. Does $(A + B)(A - B) = A^2 - B^2$ for all 2×2 matrices? Defend your answer.

GRAPHING CALCULATOR ACTIVITIES

51. Use your calculator to check the answers to all three parts of Example 1.

52. Use a calculator to check your answers for Problems 21–26.

53. Use the following matrices.

$$A = \begin{bmatrix} 7 & -4 \\ 6 & 9 \end{bmatrix} \qquad B = \begin{bmatrix} -3 & 8 \\ -5 & 7 \end{bmatrix}$$

$$C = \begin{bmatrix} 8 & -2 \\ 4 & -7 \end{bmatrix}$$

a. Show that $(AB)C = A(BC)$.

b. Show that $A(B + C) = AB + AC$.

c. Show that $(B + C)A = BA + CA$.

7.2 MULTIPLICATIVE INVERSES

We know that 1 is a multiplicative identity element for the set of real numbers. That is, $a(1) = 1(a) = a$ for any real number a. Is there a multiplicative identity element for 2×2 matrices? Yes. The matrix

$$I = \begin{bmatrix} 1 & 0 \\ 0 & 1 \end{bmatrix}$$

is the **multiplicative identity element**, because

$$\begin{bmatrix} 1 & 0 \\ 0 & 1 \end{bmatrix} \begin{bmatrix} a_{11} & a_{12} \\ a_{21} & a_{22} \end{bmatrix} = \begin{bmatrix} a_{11} & a_{12} \\ a_{21} & a_{22} \end{bmatrix}$$

and

$$\begin{bmatrix} a_{11} & a_{12} \\ a_{21} & a_{22} \end{bmatrix} \begin{bmatrix} 1 & 0 \\ 0 & 1 \end{bmatrix} = \begin{bmatrix} a_{11} & a_{12} \\ a_{21} & a_{22} \end{bmatrix}$$

Therefore, we can state that

$$AI = IA = A$$

for all 2×2 matrices.

Again, refer to the real numbers, where every nonzero real number a has a multiplicative inverse $1/a$ such that $a(1/a) = (1/a)a = 1$. Does every 2×2 matrix have a multiplicative inverse? To help answer this question, let's think about finding the multiplicative inverse (if one exists) for a specific matrix. This should give us some clues about a general approach.

EXAMPLE I   Find the multiplicative inverse of $A = \begin{bmatrix} 3 & 5 \\ 2 & 4 \end{bmatrix}$.

Solution

We are looking for a matrix A^{-1} such that $AA^{-1} = A^{-1}A = I$. In other words, we want to solve the following matrix equation

$$\begin{bmatrix} 3 & 5 \\ 2 & 4 \end{bmatrix}\begin{bmatrix} x & y \\ z & w \end{bmatrix} = \begin{bmatrix} 1 & 0 \\ 0 & 1 \end{bmatrix}$$

We need to multiply the two matrices on the left side of this equation and then set the elements of the product matrix equal to the corresponding elements of the identity matrix. We obtain the following system of equations.

$$\begin{cases} 3x + 5z = 1 \\ 3y + 5w = 0 \\ 2x + 4z = 0 \\ 2y + 4w = 1 \end{cases}$$

    (1)
    (2)
    (3)
    (4)

Solving equations (1) and (3) simultaneously produces values for x and z as follows.

$$x = \frac{\begin{vmatrix} 1 & 5 \\ 0 & 4 \end{vmatrix}}{\begin{vmatrix} 3 & 5 \\ 2 & 4 \end{vmatrix}} = \frac{1(4) - 5(0)}{3(4) - 5(2)} = \frac{4}{2} = 2$$

$$z = \frac{\begin{vmatrix} 3 & 1 \\ 2 & 0 \end{vmatrix}}{\begin{vmatrix} 3 & 5 \\ 2 & 4 \end{vmatrix}} = \frac{3(0) - 1(2)}{3(4) - 5(2)} = \frac{-2}{2} = -1$$

Likewise, solving equations (2) and (4) simultaneously produces values for y and w.

$$y = \frac{\begin{vmatrix} 0 & 5 \\ 1 & 4 \end{vmatrix}}{\begin{vmatrix} 3 & 5 \\ 2 & 4 \end{vmatrix}} = \frac{0(4) - 5(1)}{3(4) - 5(2)} = \frac{-5}{2} = -\frac{5}{2}$$

$$w = \frac{\begin{vmatrix} 3 & 0 \\ 2 & 1 \end{vmatrix}}{\begin{vmatrix} 3 & 5 \\ 2 & 4 \end{vmatrix}} = \frac{3(1) - 0(2)}{3(4) - 5(2)} = \frac{3}{2}$$

Therefore,

$$A^{-1} = \begin{bmatrix} x & y \\ z & w \end{bmatrix} = \begin{bmatrix} 2 & -\dfrac{5}{2} \\ -1 & \dfrac{3}{2} \end{bmatrix}$$

To check this, we perform the following multiplication.

$$\begin{bmatrix} 3 & 5 \\ 2 & 4 \end{bmatrix} \begin{bmatrix} 2 & -\dfrac{5}{2} \\ -1 & \dfrac{3}{2} \end{bmatrix} = \begin{bmatrix} 2 & -\dfrac{5}{2} \\ -1 & \dfrac{3}{2} \end{bmatrix} \begin{bmatrix} 3 & 5 \\ 2 & 4 \end{bmatrix} = \begin{bmatrix} 1 & 0 \\ 0 & 1 \end{bmatrix}$$

Now let's use the approach in Example 1 on the general matrix

$$A = \begin{bmatrix} a_{11} & a_{12} \\ a_{21} & a_{22} \end{bmatrix}$$

We want to find

$$A^{-1} = \begin{bmatrix} x & y \\ z & w \end{bmatrix}$$

such that $AA^{-1} = I$. Therefore, we need to solve the matrix equation

$$\begin{bmatrix} a_{11} & a_{12} \\ a_{21} & a_{22} \end{bmatrix} \begin{bmatrix} x & y \\ z & w \end{bmatrix} = \begin{bmatrix} 1 & 0 \\ 0 & 1 \end{bmatrix}$$

for x, y, z, and w. Once again, we multiply the two matrices on the left side of the equation and set the elements of this product matrix equal to the corresponding elements of the identity matrix. We then obtain the following system of equations.

$$\begin{cases} a_{11}x + a_{12}z = 1 \\ a_{11}y + a_{12}w = 0 \\ a_{21}x + a_{22}z = 0 \\ a_{21}y + a_{22}w = 1 \end{cases}$$

Solving this system produces

$$x = \frac{a_{22}}{a_{11}a_{22} - a_{12}a_{21}} \qquad y = \frac{-a_{12}}{a_{11}a_{22} - a_{12}a_{21}}$$

$$z = \frac{-a_{21}}{a_{11}a_{22} - a_{12}a_{21}} \qquad w = \frac{a_{11}}{a_{11}a_{22} - a_{12}a_{21}}$$

Note that the number in each denominator, $a_{11}a_{22} - a_{12}a_{21}$, is the determinant of the matrix A. Thus, if $|A| \neq 0$, then

$$A^{-1} = \frac{1}{|A|} \begin{bmatrix} a_{22} & -a_{12} \\ -a_{21} & a_{11} \end{bmatrix}$$

Matrix multiplication will show that $AA^{-1} = A^{-1}A = I$. If $|A| = 0$, then the matrix A has no multiplicative inverse.

EXAMPLE 2

Find A^{-1} if $A = \begin{bmatrix} 3 & 5 \\ -2 & -4 \end{bmatrix}$.

Solution

First, let's find $|A|$.

$$|A| = (3)(-4) - (5)(-2) = -2$$

Therefore,

$$A^{-1} = \frac{1}{-2} \begin{bmatrix} -4 & -5 \\ 2 & 3 \end{bmatrix} = -\frac{1}{2} \begin{bmatrix} -4 & -5 \\ 2 & 3 \end{bmatrix} = \begin{bmatrix} 2 & \frac{5}{2} \\ -1 & -\frac{3}{2} \end{bmatrix}$$

It is easy to check that $AA^{-1} = A^{-1}A = I$.   ∎

EXAMPLE 3

Find A^{-1} if $A = \begin{bmatrix} 8 & -2 \\ -12 & 3 \end{bmatrix}$.

Solution

$$|A| = (8)(3) - (-2)(-12) = 0$$

Therefore, A has no multiplicative inverse.   ∎

More About Multiplication of Matrices

Thus far we have found the product only of 2×2 matrices. The *row-by-column* multiplication pattern can be applied to many different kinds of matrices, which we shall see in the next section. For now let's find the product of a 2×2 matrix and a 2×1 matrix, with the 2×2 matrix on the left, as follows.

$$\begin{bmatrix} a_{11} & a_{12} \\ a_{21} & a_{22} \end{bmatrix} \begin{bmatrix} b_{11} \\ b_{21} \end{bmatrix} = \begin{bmatrix} a_{11}b_{11} + a_{12}b_{21} \\ a_{21}b_{11} + a_{22}b_{21} \end{bmatrix}$$

Note that the product matrix is a 2×1 matrix. The following example illustrates this pattern.

$$\begin{bmatrix} -2 & 3 \\ 1 & -4 \end{bmatrix} \begin{bmatrix} 5 \\ 7 \end{bmatrix} = \begin{bmatrix} (-2)(5) + (3)(7) \\ (1)(5) + (-4)(7) \end{bmatrix} = \begin{bmatrix} 11 \\ -23 \end{bmatrix}$$

Back to Solving Systems of Equations

The linear system of equations

$$\begin{pmatrix} a_{11}x + a_{12}y = d_1 \\ a_{21}x + a_{22}y = d_2 \end{pmatrix}$$

can be represented by the matrix equation

$$\begin{bmatrix} a_{11} & a_{12} \\ a_{21} & a_{22} \end{bmatrix} \begin{bmatrix} x \\ y \end{bmatrix} = \begin{bmatrix} d_1 \\ d_2 \end{bmatrix}$$

If we let

$$A = \begin{bmatrix} a_{11} & a_{12} \\ a_{21} & a_{22} \end{bmatrix} \qquad X = \begin{bmatrix} x \\ y \end{bmatrix} \qquad \text{and} \qquad B = \begin{bmatrix} d_1 \\ d_2 \end{bmatrix}$$

then the previous matrix equation can be written $AX = B$.

If A^{-1} exists, then we can multiply both sides of $AX = B$ by A^{-1} (on the left) and simplify as follows.

$$AX = B$$
$$A^{-1}(AX) = A^{-1}(B)$$
$$(A^{-1}A)X = A^{-1}B$$
$$IX = A^{-1}B$$
$$X = A^{-1}B$$

Therefore, the product $A^{-1}B$ is the solution of the system.

EXAMPLE 4

Solve the system $\begin{pmatrix} 5x + 4y = 10 \\ 6x + 5y = 13 \end{pmatrix}$.

Solution

If we let

$$A = \begin{bmatrix} 5 & 4 \\ 6 & 5 \end{bmatrix} \qquad X = \begin{bmatrix} x \\ y \end{bmatrix} \qquad \text{and} \qquad B = \begin{bmatrix} 10 \\ 13 \end{bmatrix}$$

then the given system can be represented by the matrix equation $AX = B$. From our previous discussion, we know that the solution of this equation is $X = A^{-1}B$, so we need to find A^{-1} and the product $A^{-1}B$.

$$A^{-1} = \frac{1}{|A|} \begin{bmatrix} 5 & -4 \\ -6 & 5 \end{bmatrix} = \frac{1}{1} \begin{bmatrix} 5 & -4 \\ -6 & 5 \end{bmatrix} = \begin{bmatrix} 5 & -4 \\ -6 & 5 \end{bmatrix}$$

Therefore,

$$A^{-1}B = \begin{bmatrix} 5 & -4 \\ -6 & 5 \end{bmatrix} \begin{bmatrix} 10 \\ 13 \end{bmatrix} = \begin{bmatrix} -2 \\ 5 \end{bmatrix}$$

The solution set of the given system is $\{(-2. 5)\}$.

EXAMPLE 5

Solve the system $\begin{pmatrix} 3x - 2y = 9 \\ 4x + 7y = -17 \end{pmatrix}$.

Solution

If we let

$$A = \begin{bmatrix} 3 & -2 \\ 4 & 7 \end{bmatrix} \qquad X = \begin{bmatrix} x \\ y \end{bmatrix} \qquad \text{and} \qquad B = \begin{bmatrix} 9 \\ -17 \end{bmatrix}$$

then the system is represented by $AX = B$, where $X = A^{-1}B$ and

$$A^{-1} = \frac{1}{|A|} \begin{bmatrix} 7 & 2 \\ -4 & 3 \end{bmatrix} = \frac{1}{29} \begin{bmatrix} 7 & 2 \\ -4 & 3 \end{bmatrix} = \begin{bmatrix} \dfrac{7}{29} & \dfrac{2}{29} \\ -\dfrac{4}{29} & \dfrac{3}{29} \end{bmatrix}$$

Therefore,

$$A^{-1}B = \begin{bmatrix} \dfrac{7}{29} & \dfrac{2}{29} \\ -\dfrac{4}{29} & \dfrac{3}{29} \end{bmatrix} \begin{bmatrix} 9 \\ -17 \end{bmatrix} = \begin{bmatrix} 1 \\ -3 \end{bmatrix}$$

The solution set of the given system is $\{(1, -3)\}$. ■

This technique of using matrix inverses to solve systems of linear equations is especially useful when there are many systems to be solved that have the same coefficients but different constant terms.

PROBLEM SET 7.2

For Problems 1–18, find the multiplicative inverse (if one exists) of each matrix.

1. $\begin{bmatrix} 5 & 7 \\ 2 & 3 \end{bmatrix}$ **2.** $\begin{bmatrix} 3 & 4 \\ 2 & 3 \end{bmatrix}$

3. $\begin{bmatrix} 3 & 8 \\ 2 & 5 \end{bmatrix}$ **4.** $\begin{bmatrix} 2 & 9 \\ 3 & 13 \end{bmatrix}$

5. $\begin{bmatrix} -1 & 2 \\ 3 & 4 \end{bmatrix}$ **6.** $\begin{bmatrix} 1 & -2 \\ 4 & -3 \end{bmatrix}$

7. $\begin{bmatrix} -2 & -3 \\ 4 & 6 \end{bmatrix}$ **8.** $\begin{bmatrix} 5 & -1 \\ 3 & 4 \end{bmatrix}$

9. $\begin{bmatrix} -3 & 2 \\ -4 & 5 \end{bmatrix}$ **10.** $\begin{bmatrix} 3 & -4 \\ 6 & -8 \end{bmatrix}$

11. $\begin{bmatrix} 0 & 1 \\ 5 & 3 \end{bmatrix}$ **12.** $\begin{bmatrix} -2 & 0 \\ -3 & 5 \end{bmatrix}$

13. $\begin{bmatrix} -2 & -3 \\ -1 & -4 \end{bmatrix}$ **14.** $\begin{bmatrix} -2 & -5 \\ -3 & -6 \end{bmatrix}$

15. $\begin{bmatrix} -2 & 5 \\ -3 & 6 \end{bmatrix}$ **16.** $\begin{bmatrix} -3 & 4 \\ 1 & -2 \end{bmatrix}$

17. $\begin{bmatrix} 1 & 1 \\ 1 & -1 \end{bmatrix}$ **18.** $\begin{bmatrix} 1 & -1 \\ 1 & 1 \end{bmatrix}$

For Problems 19–26, compute AB.

19. $A = \begin{bmatrix} 4 & 3 \\ 2 & 5 \end{bmatrix}$, $\quad B = \begin{bmatrix} 3 \\ 6 \end{bmatrix}$

20. $A = \begin{bmatrix} 5 & -2 \\ 3 & 1 \end{bmatrix}$, $\quad B = \begin{bmatrix} 5 \\ 8 \end{bmatrix}$

21. $A = \begin{bmatrix} -3 & -4 \\ 2 & 1 \end{bmatrix}$, $\quad B = \begin{bmatrix} 4 \\ -3 \end{bmatrix}$

22. $A = \begin{bmatrix} 5 & 2 \\ -1 & -3 \end{bmatrix}$, $\quad B = \begin{bmatrix} 3 \\ -5 \end{bmatrix}$

23. $A = \begin{bmatrix} -4 & 2 \\ 7 & -5 \end{bmatrix}$, $\quad B = \begin{bmatrix} -1 \\ -4 \end{bmatrix}$

24. $A = \begin{bmatrix} 0 & -3 \\ 2 & 9 \end{bmatrix}$, $\quad B = \begin{bmatrix} -3 \\ -6 \end{bmatrix}$

25. $A = \begin{bmatrix} -2 & -3 \\ -5 & -6 \end{bmatrix}$, $\quad B = \begin{bmatrix} 5 \\ -2 \end{bmatrix}$

26. $A = \begin{bmatrix} -3 & -5 \\ 4 & -7 \end{bmatrix}$, $\quad B = \begin{bmatrix} -3 \\ -10 \end{bmatrix}$

For Problems 27–40, use the method of matrix inverses to solve each system.

27. $\begin{pmatrix} 2x + 3y = 13 \\ x + 2y = 8 \end{pmatrix}$

28. $\begin{pmatrix} 3x + 2y = 10 \\ 7x + 5y = 23 \end{pmatrix}$

29. $\begin{pmatrix} 4x - 3y = -23 \\ -3x + 2y = 16 \end{pmatrix}$

30. $\begin{pmatrix} 6x - y = -14 \\ 3x + 2y = -17 \end{pmatrix}$

31. $\begin{pmatrix} x - 7y = 7 \\ 6x + 5y = -5 \end{pmatrix}$

32. $\begin{pmatrix} x + 9y = -5 \\ 4x - 7y = -20 \end{pmatrix}$

33. $\begin{pmatrix} 3x - 5y = 2 \\ 4x - 3y = -1 \end{pmatrix}$

34. $\begin{pmatrix} 5x - 2y = 6 \\ 7x - 3y = 8 \end{pmatrix}$

35. $\begin{pmatrix} y = 19 - 3x \\ 9x - 5y = 1 \end{pmatrix}$

36. $\begin{pmatrix} 4x + 3y = 31 \\ x = 5y + 2 \end{pmatrix}$

37. $\begin{pmatrix} 3x + 2y = 0 \\ 30x - 18y = -19 \end{pmatrix}$

38. $\begin{pmatrix} 12x + 30y = 23 \\ 12x - 24y = -13 \end{pmatrix}$

39. $\begin{pmatrix} \dfrac{1}{3}x + \dfrac{3}{4}y = -18 \\ \dfrac{2}{3}x + \dfrac{1}{5}y = -2 \end{pmatrix}$

40. $\begin{pmatrix} \dfrac{3}{2}x + \dfrac{1}{6}y = 11 \\ \dfrac{2}{3}x - \dfrac{1}{4}y = 1 \end{pmatrix}$

THOUGHTS INTO WORDS

41. Describe how to solve the system $\begin{pmatrix} x - 2y = -10 \\ 3x + 5y = 14 \end{pmatrix}$ using each of the following techniques.

   a. substitution method

   b. elimination-by-addition method

   c. reduced echelon form of the augmented matrix

   d. determinants

   e. the method of matrix inverses

GRAPHING CALCULATOR ACTIVITIES

42. Use your calculator to find the multiplicative inverse (if it exists) of each of the following matrices. Be sure to check your answers by showing that $A^{-1}A = I$.

   a. $\begin{bmatrix} 7 & 6 \\ 8 & 7 \end{bmatrix}$

   b. $\begin{bmatrix} -12 & 5 \\ -19 & 8 \end{bmatrix}$

   c. $\begin{bmatrix} -7 & 9 \\ 6 & -8 \end{bmatrix}$

   d. $\begin{bmatrix} -6 & -11 \\ -4 & -8 \end{bmatrix}$

   e. $\begin{bmatrix} 13 & 12 \\ 4 & 4 \end{bmatrix}$

   f. $\begin{bmatrix} 15 & -8 \\ -9 & 5 \end{bmatrix}$

   g. $\begin{bmatrix} 9 & 36 \\ 3 & 12 \end{bmatrix}$

   h. $\begin{bmatrix} 1.2 & 1.5 \\ 7.6 & 4.5 \end{bmatrix}$

43. Use your calculator to find the multiplicative inverse of $\begin{bmatrix} \dfrac{1}{2} & \dfrac{2}{5} \\ \dfrac{3}{4} & \dfrac{1}{4} \end{bmatrix}$. What difficulty did you encounter?

44. Use your calculator and the method of matrix inverses to solve each of the following systems. Be sure to check your solutions.

a. $\begin{pmatrix} 5x + 7y = 82 \\ 7x + 10y = 116 \end{pmatrix}$ **b.** $\begin{pmatrix} 9x - 8y = -150 \\ -10x + 9y = 168 \end{pmatrix}$

d. $\begin{pmatrix} 15x - 8y = -15 \\ -9x + 5y = 12 \end{pmatrix}$ **d.** $\begin{pmatrix} 1.2x + 1.5y = 5.85 \\ 7.6x + 4.5y = 19.55 \end{pmatrix}$

e. $\begin{pmatrix} 12x - 7y = -34.5 \\ 8x + 9y = 79.5 \end{pmatrix}$ **f.** $\begin{pmatrix} \dfrac{3x}{2} + \dfrac{y}{6} = 11 \\ \dfrac{2x}{3} - \dfrac{y}{4} = 1 \end{pmatrix}$

g. $\begin{pmatrix} 114x + 129y = 2832 \\ 127x + 214y = 4139 \end{pmatrix}$ **h.** $\begin{pmatrix} \dfrac{x}{2} + \dfrac{2y}{5} = 14 \\ \dfrac{3x}{4} + \dfrac{y}{4} = 14 \end{pmatrix}$

7.3 $m \times n$ MATRICES

Now let's see how much of the algebra of 2×2 matrices extends to $m \times n$ matrices, that is, to matrices of any dimension. In Section 7.4 we represented a general $m \times n$ matrix by

$$A = \begin{bmatrix} a_{11} & a_{12} & a_{13} & \cdots & a_{1n} \\ a_{21} & a_{22} & a_{23} & \cdots & a_{2n} \\ \vdots & \vdots & \vdots & & \vdots \\ a_{m1} & a_{m2} & a_{m3} & \cdots & a_{mn} \end{bmatrix}$$

We denote the element at the intersection of row i and column j by a_{ij}. It is also customary to denote a matrix A with the abbreviated notation (a_{ij}).

Addition of matrices can be extended to matrices of any dimensions by the following definition.

DEFINITION 7.5

Let $A = (a_{ij})$ and $B = (b_{ij})$ be two matrices *of the same dimension.* Then

$$A + B = (a_{ij}) + (b_{ij}) = (a_{ij} + b_{ij})$$

Definition 7.5 states that to add two matrices, we add the elements that appear in corresponding positions in the matrices. For this to work, the matrices must be of the same dimension. An example of the sum of two 3×2 matrices is

$$\begin{bmatrix} 3 & 2 \\ 4 & -1 \\ -3 & 8 \end{bmatrix} + \begin{bmatrix} -2 & 1 \\ -3 & -7 \\ 5 & 9 \end{bmatrix} = \begin{bmatrix} 1 & 3 \\ 1 & -8 \\ 2 & 17 \end{bmatrix}$$

The **commutative** and **associative properties** hold for any matrices that can be added. The *m* × *n* **zero matrix**, denoted by *O*, is the matrix that contains all zeros. It is the **identity element for addition**. For example,

$$\begin{bmatrix} 2 & 3 & -1 & -5 \\ -7 & 6 & 2 & 8 \end{bmatrix} + \begin{bmatrix} 0 & 0 & 0 & 0 \\ 0 & 0 & 0 & 0 \end{bmatrix} = \begin{bmatrix} 2 & 3 & -1 & -5 \\ -7 & 6 & 2 & 8 \end{bmatrix}$$

Every matrix *A* has an **additive inverse**, −*A*, that can be found by changing the sign of each element of *A*. For example, if

$$A = [2 \quad -3 \quad 0 \quad 4 \quad -7]$$

then

$$-A = [-2 \quad 3 \quad 0 \quad -4 \quad 7]$$

Furthermore, $A + (-A) = O$ for all matrices.

The definition we gave earlier for subtraction, $A - B = A + (-B)$, can be extended to any two matrices of the same dimension. For example,

$$[-4 \quad 3 \quad -5] - [7 \quad -4 \quad -1] = [-4 \quad 3 \quad -5] + [-7 \quad 4 \quad 1]$$
$$= [-11 \quad 7 \quad -4]$$

The **scalar product** of any real number *k* and any *m* × *n* matrix $A = (a_{ij})$ is defined by

$$kA = (ka_{ij})$$

In other words, to find *kA*, we simply multiply each element of *A* by *k*. For example,

$$(-4)\begin{bmatrix} 1 & -1 \\ -2 & 3 \\ 4 & 5 \\ 0 & -8 \end{bmatrix} = \begin{bmatrix} -4 & 4 \\ 8 & -12 \\ -16 & -20 \\ 0 & 32 \end{bmatrix}$$

The properties $k(A + B) = kA + kB$, $(k + l)A = kA + lA$, and $(kl)A = k(lA)$ hold for all matrices. The matrices *A* and *B* must be of the same dimension to be added.

The *row-by-column* definition for multiplying two matrices can be extended, but we must take care. In order for us to define the product *AB* of two matrices *A* and *B*, **the number of columns of *A* must equal the number of rows of *B*.** Therefore, suppose $A = (a_{ij})$ is *m* × *n* and $B = (b_{ij})$ is *n* × *p*. Then

$$AB = \begin{bmatrix} a_{11} & a_{12} & \cdots & a_{1n} \\ & \cdot & & \cdot \\ & \cdot & \cdots & \cdot \\ \boxed{a_{i1}} & a_{i2} & & a_{in} \\ & \cdot & & \cdot \\ & \cdot & & \cdot \\ & \cdot & \cdots & \\ a_{m1} & a_{m2} & & a_{mn} \end{bmatrix} \begin{bmatrix} b_{11} & \cdots & b_{1j} & \cdots & b_{1p} \\ b_{21} & \cdots & b_{2j} & \cdots & b_{2p} \\ \cdot & & \cdot & & \cdot \\ \cdot & & \cdot & & \cdot \\ b_{n1} & \cdots & b_{nj} & \cdots & b_{np} \end{bmatrix} = C$$

The product matrix C is of dimension $m \times p$, and the general element, c_{ij}, is determined as follows.

$$c_{ij} = a_{i1}b_{ij} + a_{i2}b_{2j} + \cdots + a_{in}b_{nj}$$

A specific element of the product matrix, such as c_{23}, is the result of multiplying the elements in row 2 of matrix A times the elements in column 3 of matrix B and adding the results. Therefore,

$$c_{23} = a_{21}b_{13} + a_{22}b_{23} + \cdots + a_{2n}b_{n3}$$

The following example illustrates the product of a 2×3 matrix and a 3×2 matrix.

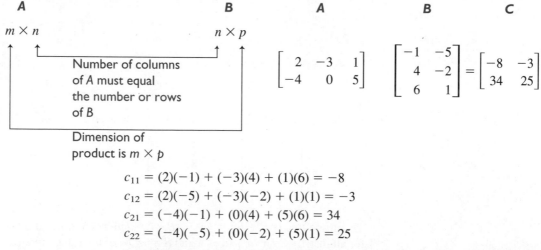

$$c_{11} = (2)(-1) + (-3)(4) + (1)(6) = -8$$
$$c_{12} = (2)(-5) + (-3)(-2) + (1)(1) = -3$$
$$c_{21} = (-4)(-1) + (0)(4) + (5)(6) = 34$$
$$c_{22} = (-4)(-5) + (0)(-2) + (5)(1) = 25$$

Recall that **matrix multiplication is not commutative.** In fact, it may be that AB is defined and BA is not defined. For example, if A is a 2×3 matrix and B is a 3×4 matrix, then the product AB is a 2×4 matrix, but the product BA is not defined because the number of columns of B does not equal the number of rows of A.

The **associative property for multiplication** and the two **distributive properties** hold if the matrices have the proper number of rows and columns for the operations to be defined. In that case, we have $(AB)C = A(BC)$, $A(B + C) = AB + AC$, and $(A + B)C = AC + BC$.

Square Matrices

Now let's extend some of the algebra of 2×2 matrices to all square matrices (where the number of rows equals the number of columns). For example, the general **multiplicative identity element** for square matrices contains 1's in the main diagonal from the upper left-hand corner to the lower right-hand corner and 0s elsewhere. Therefore, for 3×3 and 4×4 matrices, the multiplicative identity elements are as follows.

$$I_3 = \begin{bmatrix} 1 & 0 & 0 \\ 0 & 1 & 0 \\ 0 & 0 & 1 \end{bmatrix} \quad I_4 = \begin{bmatrix} 1 & 0 & 0 & 0 \\ 0 & 1 & 0 & 0 \\ 0 & 0 & 1 & 0 \\ 0 & 0 & 0 & 1 \end{bmatrix}$$

We saw in Section 7.2 that some, but not all, 2×2 matrices have multiplicative inverses. In general, some, but not all, square matrices of a particular dimension have multiplicative inverses. If an $n \times n$ square matrix A does have a multiplicative inverse A^{-1}, then

$$AA^{-1} = A^{-1}A = I_n$$

The technique used in Section 7.2 for finding multiplicative inverses of 2×2 matrices does generalize, but it becomes quite complicated. Therefore, we shall now describe another technique that works for all square matrices. Given an $n \times n$ matrix A, we begin by forming the $n \times 2n$ matrix

$$\begin{bmatrix} a_{11} & a_{12} & \cdots & a_{1n} & 1 & 0 & 0 & \cdots & 0 \\ a_{21} & a_{22} & \cdots & a_{2n} & 0 & 1 & 0 & \cdots & 0 \\ \vdots & \vdots & & \vdots & \vdots & \vdots & \vdots & & \vdots \\ a_{n1} & a_{n2} & \cdots & a_{nn} & 0 & 0 & 0 & \cdots & 1 \end{bmatrix}$$

where the identity matrix I_n appears to the right of A. Now we apply a succession of elementary row transformations to this double matrix until we obtain a matrix of the form

$$\begin{bmatrix} 1 & 0 & 0 & \cdots & 0 & b_{11} & b_{12} & \cdots & b_{1n} \\ 0 & 1 & 0 & \cdots & 0 & b_{21} & b_{22} & \cdots & b_{2n} \\ \vdots & \vdots & \vdots & & \vdots & \vdots & \vdots & & \vdots \\ 0 & 0 & 0 & \cdots & 1 & b_{n1} & b_{n2} & \cdots & b_{nn} \end{bmatrix}$$

The B matrix in this matrix is the desired inverse A^{-1}. If A does not have an inverse, then it is impossible to change the original matrix to this final form.

EXAMPLE 1

Find A^{-1} if $A = \begin{bmatrix} 2 & 4 \\ 3 & 5 \end{bmatrix}$.

Solution

First, form the matrix

$$\left[\begin{array}{cc:cc} 2 & 4 & 1 & 0 \\ 3 & 5 & 0 & 1 \end{array}\right]$$

Now multiply row 1 by $\frac{1}{2}$.

$$\left[\begin{array}{cc:cc} 1 & 2 & \frac{1}{2} & 0 \\ 3 & 5 & 0 & 1 \end{array}\right]$$

Next, add -3 times row 1 to row 2 to form a new row 2.

$$\left[\begin{array}{cc:cc} 1 & 2 & \frac{1}{2} & 0 \\ 0 & -1 & -\frac{3}{2} & 1 \end{array}\right]$$

Then multiply row 2 by -1.

$$\left[\begin{array}{cc:cc} 1 & 2 & \frac{1}{2} & 0 \\ 0 & 1 & \frac{3}{2} & -1 \end{array}\right]$$

Finally, add -2 times row 2 to row 1 to form a new row 1.

$$\left[\begin{array}{cc:cc} 1 & 0 & \boxed{\begin{array}{cc} -\frac{5}{2} & 2 \\ \frac{3}{2} & -1 \end{array}} \end{array}\right]$$

The matrix inside the box is A^{-1}; that is,

$$A^{-1} = \begin{bmatrix} -\frac{5}{2} & 2 \\ \frac{3}{2} & -1 \end{bmatrix}$$

This can be checked, as always, by showing that $AA^{-1} = A^{-1}A = I_2$.   ■

EXAMPLE 2   Find A^{-1} if $A = \begin{bmatrix} 1 & 1 & 2 \\ 2 & 3 & -1 \\ -3 & 1 & -2 \end{bmatrix}$.

Solution

Form the matrix $\begin{bmatrix} 1 & 1 & 2 & 1 & 0 & 0 \\ 2 & 3 & -1 & 0 & 1 & 0 \\ -3 & 1 & -2 & 0 & 0 & 1 \end{bmatrix}$.

Add -2 times row 1 to row 2, and add 3 times row 1 to row 3.

$$\begin{bmatrix} 1 & 1 & 2 & 1 & 0 & 0 \\ 0 & 1 & -5 & -2 & 1 & 0 \\ 0 & 4 & 4 & 3 & 0 & 1 \end{bmatrix}$$

Add -1 times row 2 to row 1, and add -4 times row 2 to row 3.

$$\begin{bmatrix} 1 & 0 & 7 & 3 & -1 & 0 \\ 0 & 1 & -5 & -2 & 1 & 0 \\ 0 & 0 & 24 & 11 & -4 & 1 \end{bmatrix}$$

Multiply row 3 by $\dfrac{1}{24}$.

$$\begin{bmatrix} 1 & 0 & 7 & 3 & -1 & 0 \\ 0 & 1 & -5 & -2 & 1 & 0 \\ 0 & 0 & 1 & \dfrac{11}{24} & -\dfrac{1}{6} & \dfrac{1}{24} \end{bmatrix}$$

Add -7 times row 3 to row 1, and add 5 times row 3 to row 2.

$$\begin{bmatrix} 1 & 0 & 0 & -\dfrac{5}{24} & \dfrac{1}{6} & -\dfrac{7}{24} \\ 0 & 1 & 0 & \dfrac{7}{24} & \dfrac{1}{6} & \dfrac{5}{24} \\ 0 & 0 & 1 & \dfrac{11}{24} & -\dfrac{1}{6} & \dfrac{1}{24} \end{bmatrix}$$

Therefore,

$$A^{-1} = \begin{bmatrix} -\dfrac{5}{24} & \dfrac{1}{6} & -\dfrac{7}{24} \\ \dfrac{7}{24} & \dfrac{1}{6} & \dfrac{5}{24} \\ \dfrac{11}{24} & -\dfrac{1}{6} & \dfrac{1}{24} \end{bmatrix}$$ Be sure to check this!

Systems of Equations

In Section 7.2 we used the concept of the multiplicative inverse to solve systems of two linear equations in two variables. This same technique can be applied to general systems of n linear equations in n variables. Let's consider one such example involving three equations in three variables.

EXAMPLE 3

Solve the system

$$\left(\begin{array}{c} x + y + 2z = -8 \\ 2x + 3y - z = 3 \\ -3x + y - 2z = 4 \end{array} \right)$$

Solution

If we let

$$A = \begin{bmatrix} 1 & 1 & 2 \\ 2 & 3 & -1 \\ -3 & 1 & -2 \end{bmatrix} \qquad X = \begin{bmatrix} x \\ y \\ z \end{bmatrix} \qquad \text{and} \qquad B = \begin{bmatrix} -8 \\ 3 \\ 4 \end{bmatrix}$$

then the given system can be represented by the matrix equation $AX = B$. Therefore, we know that $X = A^{-1}B$, so we need to find A^{-1} and the product $A^{-1}B$. The matrix A^{-1} was found in Example 2, so let's use that result and find $A^{-1}B$.

$$X = A^{-1}B = \begin{bmatrix} -\dfrac{5}{24} & \dfrac{1}{6} & -\dfrac{7}{24} \\ \dfrac{7}{24} & \dfrac{1}{6} & \dfrac{5}{24} \\ \dfrac{11}{24} & -\dfrac{1}{6} & \dfrac{1}{24} \end{bmatrix} \begin{bmatrix} -8 \\ 3 \\ 4 \end{bmatrix} = \begin{bmatrix} 1 \\ -1 \\ -4 \end{bmatrix}$$

The solution set of the given system is $\{(1, -1, -4)\}$. ∎

PROBLEM SET 7.3

For Problems 1–8, find $A + B$, $A - B$, $2A + 3B$, and $4A - 2B$.

1. $A = \begin{bmatrix} 2 & -1 & 4 \\ -2 & 0 & 5 \end{bmatrix}$, $B = \begin{bmatrix} -1 & 4 & -7 \\ 5 & -6 & 2 \end{bmatrix}$

2. $A = \begin{bmatrix} 3 & -6 \\ 2 & -1 \\ -4 & 5 \end{bmatrix}$, $B = \begin{bmatrix} 1 & 0 \\ 5 & -7 \\ -6 & 9 \end{bmatrix}$

3. $A = \begin{bmatrix} 2 & -1 & 4 & 12 \end{bmatrix}$, $B = \begin{bmatrix} -3 & -6 & 9 & -5 \end{bmatrix}$

4. $A = \begin{bmatrix} 3 \\ -9 \\ 7 \end{bmatrix}$, $B = \begin{bmatrix} -6 \\ 12 \\ 9 \end{bmatrix}$

5. $A = \begin{bmatrix} 3 & -2 & 1 \\ -1 & 4 & -7 \\ 0 & 5 & 9 \end{bmatrix}$, $B = \begin{bmatrix} 5 & -1 & -3 \\ 10 & -2 & 4 \\ 7 & 0 & 12 \end{bmatrix}$

6. $A = \begin{bmatrix} 7 & -4 \\ -5 & 9 \\ -1 & 2 \end{bmatrix}$, $B = \begin{bmatrix} 12 & 3 \\ -2 & -4 \\ -6 & 7 \end{bmatrix}$

7. $A = \begin{bmatrix} -1 & 0 \\ 2 & 3 \\ -5 & -4 \\ -7 & 11 \end{bmatrix}$, $B = \begin{bmatrix} 1 & 2 \\ -3 & 7 \\ 6 & -5 \\ 9 & -2 \end{bmatrix}$

8. $A = \begin{bmatrix} 0 & -1 & -2 \\ 3 & -4 & 6 \\ 5 & 4 & -9 \end{bmatrix}$, $B = \begin{bmatrix} 2 & 1 & -7 \\ -6 & 4 & 5 \\ 3 & -2 & -1 \end{bmatrix}$

For Problems 9–20, find AB and BA, whenever they exist.

9. $A = \begin{bmatrix} 2 & -1 \\ 0 & -4 \\ -5 & 3 \end{bmatrix}$, $B = \begin{bmatrix} 5 & -2 & 6 \\ -1 & 4 & -2 \end{bmatrix}$

10. $A = \begin{bmatrix} -2 & 3 & -1 \\ 7 & -4 & 5 \end{bmatrix}$, $B = \begin{bmatrix} 1 & -1 \\ -2 & 3 \\ -5 & -6 \end{bmatrix}$

11. $A = \begin{bmatrix} 2 & -1 & -3 \\ 0 & -4 & 7 \end{bmatrix}$, $B = \begin{bmatrix} 2 & 1 & -1 & 4 \\ 0 & -2 & 3 & 5 \\ -6 & 4 & -2 & 0 \end{bmatrix}$

12. $A = \begin{bmatrix} 3 & -1 & -4 \\ -5 & 2 & 2 \end{bmatrix}$, $B = \begin{bmatrix} 3 & -2 \\ -4 & -1 \end{bmatrix}$

13. $A = \begin{bmatrix} 1 & -1 & 2 \\ 0 & 1 & -2 \\ 3 & 1 & 4 \end{bmatrix}$, $B = \begin{bmatrix} 2 & 3 & -1 \\ 4 & 0 & 2 \\ -5 & 1 & -1 \end{bmatrix}$

14. $A = \begin{bmatrix} 1 & 0 & 1 \\ 0 & 1 & 1 \\ -1 & 2 & 3 \end{bmatrix}$, $B = \begin{bmatrix} -1 & -1 & 1 \\ 0 & 1 & 0 \\ 2 & -3 & 1 \end{bmatrix}$

15. $A = [2 \quad -1 \quad 3 \quad 4]$, $B = \begin{bmatrix} -1 \\ -3 \\ 2 \\ -4 \end{bmatrix}$

16. $A = \begin{bmatrix} -2 \\ 3 \\ -5 \end{bmatrix}$, $B = [3 \quad -4 \quad -5]$

17. $A = \begin{bmatrix} 2 \\ -7 \end{bmatrix}$, $B = \begin{bmatrix} 3 & -2 \\ 1 & 0 \\ -1 & 4 \end{bmatrix}$

18. $A = \begin{bmatrix} 3 & -2 & 2 & -4 \\ 1 & 0 & -1 & 2 \end{bmatrix}$, $B = \begin{bmatrix} 3 & -2 & 1 \\ -3 & 1 & 4 \\ 5 & 2 & 0 \\ -4 & -1 & -2 \end{bmatrix}$

19. $A = \begin{bmatrix} 3 \\ -4 \\ 2 \end{bmatrix}$, $B = [3 \quad -4]$

20. $A = [3 \quad -7]$, $B = \begin{bmatrix} 8 \\ -9 \end{bmatrix}$

For Problems 21–36, use the technique discussed in this section to find the multiplicative inverse (if it exists) of each matrix.

21. $\begin{bmatrix} 1 & 3 \\ 4 & 2 \end{bmatrix}$

22. $\begin{bmatrix} 1 & 2 \\ 2 & -3 \end{bmatrix}$

23. $\begin{bmatrix} 2 & 1 \\ 7 & 4 \end{bmatrix}$

24. $\begin{bmatrix} 3 & 7 \\ 2 & 5 \end{bmatrix}$

25. $\begin{bmatrix} -2 & 1 \\ 3 & -4 \end{bmatrix}$

26. $\begin{bmatrix} -3 & 1 \\ 3 & -2 \end{bmatrix}$

27. $\begin{bmatrix} 1 & 2 & 3 \\ 1 & 3 & 4 \\ 1 & 4 & 3 \end{bmatrix}$

28. $\begin{bmatrix} 1 & 3 & -2 \\ 1 & 4 & -1 \\ -2 & -7 & 5 \end{bmatrix}$

29. $\begin{bmatrix} 1 & -2 & 1 \\ -2 & 5 & 3 \\ 3 & -5 & 7 \end{bmatrix}$

30. $\begin{bmatrix} 1 & 4 & -2 \\ -3 & -11 & 1 \\ 2 & 7 & 3 \end{bmatrix}$

31. $\begin{bmatrix} 2 & 3 & -4 \\ 3 & -1 & -2 \\ 1 & -4 & 2 \end{bmatrix}$

32. $\begin{bmatrix} -2 & 2 & 3 \\ 1 & -1 & 0 \\ 0 & 1 & 4 \end{bmatrix}$

33. $\begin{bmatrix} 1 & 2 & 3 \\ -3 & -4 & 3 \\ 2 & 4 & -1 \end{bmatrix}$

34. $\begin{bmatrix} 1 & -2 & 3 \\ -1 & 3 & -2 \\ -2 & 6 & 1 \end{bmatrix}$

35. $\begin{bmatrix} 2 & 0 & 0 \\ 0 & 4 & 0 \\ 0 & 0 & 10 \end{bmatrix}$

36. $\begin{bmatrix} 1 & -3 & 5 \\ 0 & 1 & 2 \\ 0 & 0 & 1 \end{bmatrix}$

For Problems 37–46, use the method of matrix inverses to solve each system. The required multiplicative inverses were found in Problems 21–36.

37. $\begin{pmatrix} 2x + y = -4 \\ 7x + 4y = -13 \end{pmatrix}$

38. $\begin{pmatrix} 3x + 7y = -38 \\ 2x + 5y = -27 \end{pmatrix}$

39. $\begin{pmatrix} -2x + y = 1 \\ 3x - 4y = -14 \end{pmatrix}$

40. $\begin{pmatrix} -3x + y = -18 \\ 3x - 2y = 15 \end{pmatrix}$

41. $\begin{pmatrix} x + 2y + 3z = -2 \\ x + 3y + 4z = -3 \\ x + 4y + 3z = -6 \end{pmatrix}$

42. $\begin{pmatrix} x + 3y - 2z = 5 \\ x + 4y - z = 3 \\ -2x - 7y + 5z = -12 \end{pmatrix}$

43. $\begin{pmatrix} x - 2y + z = -3 \\ -2x + 5y + 3z = 34 \\ 3x - 5y + 7z = 14 \end{pmatrix}$

44. $\begin{pmatrix} x + 4y - 2z = 2 \\ -3x - 11y + z = -2 \\ 2x + 7y + 3z = -2 \end{pmatrix}$

45. $\begin{pmatrix} x + 2y + 3z = 2 \\ -3x - 4y + 3z = 0 \\ 2x + 4y - z = 4 \end{pmatrix}$

46. $\begin{pmatrix} x - 2y + 3z = -39 \\ -x + 3y - 2z = 40 \\ -2x + 6y + z = 45 \end{pmatrix}$

47. We can generate five systems of linear equations from the system

$$\begin{pmatrix} x + y + 2z = a \\ 2x + 3y - z = b \\ -3x + y - 2z = c \end{pmatrix}$$

by letting a, b, and c assume five different sets of values. Solve the system for each set of values. The inverse of the coefficient matrix of these systems is given in Example 2 of this section.

a. $a = 7, b = 1,$ and $c = -1$

b. $a = -7, b = 5,$ and $c = 1$

c. $a = -9, b = -8,$ and $c = 19$

d. $a = -1, b = -13,$ and $c = -17$

e. $a = -2, b = 0,$ and $c = -2$

THOUGHTS INTO WORDS

48. How would you describe row-by-column multiplication of matrices?

49. Give a step-by-step explanation of how to find the multiplicative inverse of the matrix $\begin{bmatrix} 1 & 3 \\ -2 & 4 \end{bmatrix}$ by using the technique of Section 7.3.

50. Explain how to find the multiplicative inverse of the matrix in Problem 49 by using the technique discussed in Section 7.2.

Further Investigations

51. Matrices can be used to code and decode messages. For example, suppose that we set up a one-to-one correspondence between the letters of the alphabet and the first 26 counting numbers, as follows.

A B C   Z
↕ ↕ ↕ ⋯ ↕
1 2 3   26

Now suppose that we want to code the message PLAY IT BY EAR. We can partition the letters of the message into groups of two. Because the last group will contain only one letter, let's arbitrarily stick in a Z to form a group of two. Let's also assign a number to each letter on the basis of the letter/number association we exhibited.

P L A Y I T B Y E A R Z
↕ ↕ ↕ ↑ ↑ ↕ ↕ ↕ ↕ ↕ ↑ ↕
16 12 1 25 9 20 2 25 5 1 18 26

Each pair of numbers can be recorded as columns in a 2×6 matrix B.

$$B = \begin{bmatrix} 16 & 1 & 9 & 2 & 5 & 18 \\ 12 & 25 & 20 & 25 & 1 & 26 \end{bmatrix}$$

Now let's choose a 2×2 matrix such that the matrix contains only integers and its inverse also contains only integers. For example, we can use $A = \begin{bmatrix} 3 & 1 \\ 5 & 2 \end{bmatrix}$; then $A^{-1} = \begin{bmatrix} 2 & -1 \\ -5 & 3 \end{bmatrix}$.

Next, let's find the product AB.

$$AB = \begin{bmatrix} 3 & 1 \\ 5 & 2 \end{bmatrix} \begin{bmatrix} 16 & 1 & 9 & 2 & 5 & 18 \\ 12 & 25 & 20 & 25 & 1 & 26 \end{bmatrix}$$

$$= \begin{bmatrix} 60 & 28 & 47 & 31 & 16 & 80 \\ 104 & 55 & 85 & 60 & 27 & 142 \end{bmatrix}$$

Now we have our coded message:

60   104   28   55   47   85   31   60   16   27   80   142

A person decoding the message would put the numbers back into a 2×6 matrix, multiply it on the left by A^{-1}, and convert the numbers back to letters.

Each of the following coded messages was formed by using the matrix $A = \begin{bmatrix} 2 & 3 \\ 1 & 2 \end{bmatrix}$. Decode each of the messages.

a. 68   40   77   51   78   49   23   15   29   19   85
52   41   27

b. 62   40   78   47   64   36   19   11   93   57   93
56   88   57

c. 64   36   58   37   63   36   21   13   75   47   63
36   38   23   118   72

d. 61   38   115   69   93   57   36   20   78   49   68
40   77   51   60   37   47   26   84   51   21   11

52. Suppose that the ordered pair (x, y) of a rectangular coordinate system is recorded as a 2×1 matrix and then multiplied on the left by the matrix $\begin{bmatrix} 1 & 0 \\ 0 & -1 \end{bmatrix}$. We would obtain

$$\begin{bmatrix} 1 & 0 \\ 0 & -1 \end{bmatrix} \begin{bmatrix} x \\ y \end{bmatrix} = \begin{bmatrix} x \\ -y \end{bmatrix}$$

The point $(x, -y)$ is an x-axis reflection of the point (x, y). Therefore, the matrix $\begin{bmatrix} 1 & 0 \\ 0 & -1 \end{bmatrix}$ performs an x-axis reflection. What type of geometric transformation is performed by each of the following matrices?

a. $\begin{bmatrix} -1 & 0 \\ 0 & 1 \end{bmatrix}$ **b.** $\begin{bmatrix} -1 & 0 \\ 0 & -1 \end{bmatrix}$

c. $\begin{bmatrix} 0 & -1 \\ 1 & 0 \end{bmatrix}$ [*Hint:* Check the slopes of lines through the origin.]

d. $\begin{bmatrix} 0 & 1 \\ -1 & 0 \end{bmatrix}$

 GRAPHING CALCULATOR ACTIVITIES

53. Use your calculator to check your answers for Problems 14, 18, 28, 30, 32, 34, 36, 42, 44, 46, and 47.

54. Use your calculator and the method of matrix inverses to solve each of the following systems. Be sure to check your solutions.

a. $\begin{cases} 2x - 3y + 4z = 54 \\ 3x + y - z = 32 \\ 5x - 4y + 3z = 58 \end{cases}$

b. $\begin{cases} 17x + 15y - 19z = 10 \\ 18x - 14y + 16z = 94 \\ 13x + 19y - 14z = -23 \end{cases}$

c. $\begin{cases} 1.98x + 2.49y + 3.15z = 45.72 \\ 2.29x + 1.95y + 2.75z = 42.05 \\ 3.15x + 3.20y + 1.85z = 42 \end{cases}$

d. $\begin{cases} x_1 + 2x_2 - 4x_3 + 7x_4 = -23 \\ 2x_1 - 3x_2 + 5x_3 - x_4 = -22 \\ 5x_1 + 4x_2 - 2x_3 - 8x_4 = 59 \\ 3x_1 - 7x_2 + 8x_3 + 9x_4 = -103 \end{cases}$

e. $\begin{cases} 2x_1 - x_2 + 3x_3 - 4x_4 + 12x_5 = 98 \\ x_1 + 2x_2 - x_3 - 7x_4 + 5x_5 = 41 \\ 3x_1 + 4x_2 - 7x_3 + 6x_4 - 9x_5 = -41 \\ 4x_1 - 3x_2 + x_3 - x_4 + x_5 = 4 \\ 7x_1 + 8x_2 - 4x_3 - 6x_4 - 6x_5 = 12 \end{cases}$

7.4   SYSTEMS INVOLVING LINEAR INEQUALITIES; LINEAR PROGRAMMING

Finding solution sets for **systems of linear inequalities** relies heavily on the graphing approach. (Recall that we discussed graphing of linear inequalities in Section 2.3.) The solution set of the system

$$\begin{pmatrix} x + y > 2 \\ x - y < 2 \end{pmatrix}$$

is the intersection of the solution sets of the individual inequalities. In Figure 7.1(a) we indicate the solution set for $x + y > 2$, and in Figure 7.1(b) we indicate the solution set for $x - y < 2$. The shaded region in Figure 7.1(c) represents the intersection of the two solution sets; therefore, it is the graph of the system. Remember that dashed lines are used to indicate that the points on the lines are not included in the solution set. In the following examples, we indicate only the final solution set for the system.

(a)

(b)

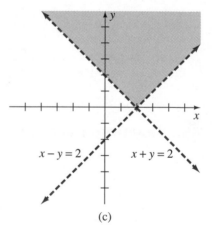

(c)

FIGURE 7.1

EXAMPLE 1

Solve the following system by graphing.

$$\begin{pmatrix} 2x - y \geq 4 \\ x + 2y < 2 \end{pmatrix}$$

Solution

The graph of $2x - y \geq 4$ consists of all points *on or below* the line $2x - y = 4$. The graph of $x + 2y < 2$ consists of all points *below* the line $x + 2y = 2$. The graph of the system is indicated by the shaded region in Figure 7.2. Note that all points in the shaded region are on or below the line $2x - y = 4$ and below the line $x + 2y = 2$.

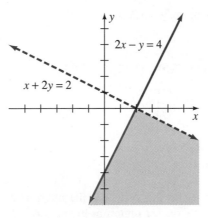

FIGURE 7.2

EXAMPLE 2

Solve the following system by graphing.

$$\begin{pmatrix} x \le 2 \\ y \ge -1 \end{pmatrix}$$

Solution

Remember that even though each inequality contains only one variable, we are working in a rectangular coordinate system involving ordered pairs. That is, the system could also be written

$$\begin{pmatrix} x + 0(y) \le 2 \\ 0(x) + y \ge -1 \end{pmatrix}$$

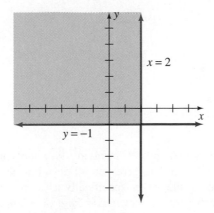

FIGURE 7.3

The graph of this system is the shaded region in Figure 7.3. Note that all points in the shaded region are *on or to the left* of the line $x = 2$ and *on or above* the line $y = -1$.

A system may contain more than two inequalities, as the next example illustrates.

EXAMPLE 3

Solve the following system by graphing.

$$\begin{pmatrix} x \geq 0 \\ y \geq 0 \\ 2x + 3y \leq 12 \\ 3x + y \leq 6 \end{pmatrix}$$

Solution

The solution set for the system is the intersection of the solution sets of the four inequalities. The shaded region in Figure 7.4 indicates the solution set for the system. Note that all points in the shaded region are *on or to the right* of the y axis, *on or above* the x axis, *on or below* the line $2x + 3y = 12$, and *on or below* the line $3x + y = 6$.

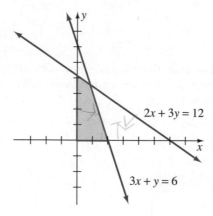

FIGURE 7.4 ■

Linear Programming: Another Look at Problem Solving

Throughout this text, problem solving is a unifying theme. Therefore, it seems appropriate at this time to give you a brief glimpse of an area of mathematics that was developed in the 1940s specifically as a problem-solving tool. Many applied problems involve the idea of *maximizing* or *minimizing* a certain function that is subject to various constraints; these can be expressed as linear inequalities. **Linear programming** was developed as one method for solving such problems.

    REMARK   The term *programming* refers to the distribution of limited resources in order to maximize or minimize a certain function, such as cost, profit, distance, and so on. Thus it does not mean the same thing it means in computer programming. The constraints that govern the distribution of resources determine the linear inequalities and equations; thus the term *linear programming* is used.

Before we introduce a linear programming type of problem, we need to extend one mathematical concept a bit. A **linear function in two variables** x and y is a function of the form $f(x, y) = ax + by + c$, where a, b, and c are real numbers. In other words, with each ordered pair (x, y) we associate a third number by the rule $ax + by + c$. For example, suppose the function f is described by $f(x, y) = 4x + 3y + 5$. Then $f(2, 1) = 4(2) + 3(1) + 5 = 16$.

First, let's take a look at some mathematical ideas that form the basis for solving a linear programming problem. Consider the shaded region in Figure 7.5 and the following linear functions in two variables.

$$f(x, y) = 4x + 3y + 5$$
$$f(x, y) = 2x + 7y - 1$$
$$f(x, y) = x - 2y$$

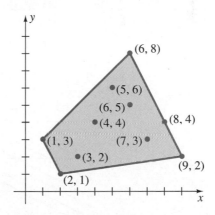

FIGURE 7.5

Suppose that we need to find the maximum value and the minimum value achieved by each of the functions in the indicated region. The following chart summarizes the values for the ordered pairs indicated in Figure 7.5. Note that for each function, the maximum and minimum values are obtained at vertices of the region.

We claim that for linear functions, maximum and minimum functional values are *always* obtained at vertices of the region. To substantiate this, let's consider the family of lines $x - 2y = k$, where k is an arbitrary constant. (We are now working only with the function $f(x, y) = x - 2y$). In slope–intercept form, $x - 2y = k$ becomes $y = \frac{1}{2}x - \frac{1}{2}k$; so we have a family of parallel lines each having a slope of $\frac{1}{2}$. In Figure 7.6 we sketched some of these lines, so that each line has at least one point in common with the given region. Note that $x - 2y$ reaches a minimum value of -10 at the vertex $(6, 8)$ and a maximum value of 5 at the vertex $(9, 2)$.

	ORDERED PAIRS	VALUE OF $f(x, y) = 4x + 3y + 5$	VALUE OF $f(x, y) = 2x + 7y - 1$	VALUE OF $f(x, y) = x - 2y$
Vertex	(2, 1)	16 *(minimum)*	10 *(minimum)*	0
	(3, 2)	23	19	−1
Vertex	(9, 2)	47	31	5 *(maximum)*
Vertex	(1, 3)	18	22	−5
	(7, 3)	42	34	1
	(4, 4)	33	35	−4
	(8, 4)	49	43	0
	(6, 5)	44	46	−4
	(5, 6)	43	51	−7
Vertex	(6, 8)	53 *(maximum)*	67 *(maximum)*	−10 *(minimum)*

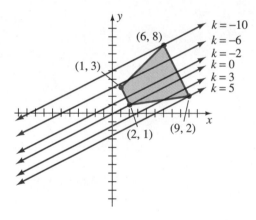

FIGURE 7.6

In general, suppose that f is a **linear function in two variables** x and y and that S is a region of the xy plane. If f attains a maximum (minimum) value in S, then that maximum (minimum) value is obtained at a vertex of S.

> **REMARK**   A subset of the xy plane is said to be **bounded** if there is a circle that contains all of its points; otherwise the subset is said to be **unbounded**. A bounded set will contain maximum and minimum values for a function, but an unbounded set may not contain such values.

Now we will consider two examples that illustrate a general graphing approach to solving a linear programming problem in two variables. The first example gives us the general makeup of such a problem; the second example will illustrate the type of setting from which the function and inequalities evolve.

EXAMPLE 4

Find the maximum value and the minimum value of the function $f(x, y) = 9x + 13y$ in the region determined by the following system of inequalities.

$$\begin{pmatrix} x \geq 0 \\ y \geq 0 \\ 2x + 3y \leq 18 \\ 2x + y \leq 10 \end{pmatrix}$$

Solution

First, let's graph the inequalities to determine the region, as indicated in Figure 7.7. (Such a region is called the **set of feasible solutions**, and the inequalities are referred to as **constraints**.) The point $(3, 4)$ is determined by solving the system

$$\begin{pmatrix} 2x + 3y = 18 \\ 2x + y = 10 \end{pmatrix}$$

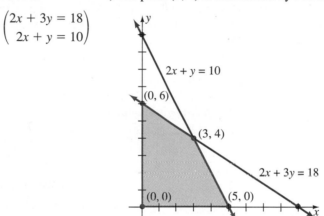

FIGURE 7.7

Next we can determine the values of the given function at the vertices of the region. (Such a function to be maximized or minimized is called the **objective function**.)

VERTICES	VALUE OF $f(x, y) = 9x + 13y$
$(0, 0)$	0 *(minimum)*
$(5, 0)$	45
$(3, 0)$	79 *(maximum)*
$(0, 6)$	18

A minimum value of 0 is obtained at $(0, 0)$, and a maximum value of 79 is obtained at $(3, 4)$. ■

PROBLEM 1

A company that manufactures gidgets and gadgets has the following production information available.

1. To produce a gidget requires 3 hours of working time on machine A and 1 hour on machine B.

2. To produce a gadget requires 2 hours on machine A and 1 hour on machine B.

3. Machine A is available for no more than 120 hours per week, and machine B is available for no more than 50 hours per week.

4. Gidgets can be sold at a profit of $3.75 each, and a profit of $3 can be realized on a gadget.

How many gidgets and how many gadgets should the company produce each week to maximize its profit? What would the maximum profit be?

Solution

Let x be the number of gidgets and y be the number of gadgets. Thus the profit function is $P(x, y) = 3.75x + 3y$. The constraints for the problem can be represented by the following inequalities.

$3x + 2y \leq 120$	Machine A is available for no more than 120 hours.
$x + y \leq 50$	Machine B is available for no more than 50 hours.
$x \geq 0$	The number of gidgets and gadgets must be
$y \geq 0$	represented by a nonnegative number.

When we graph these inequalities, we obtain the set of feasible solutions indicated by the shaded region in Figure 7.8. Next we find the value of the profit function at the vertices; this produces the chart that follows.

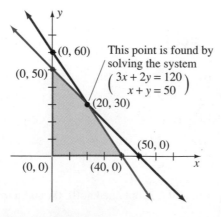

FIGURE 7.8

VERTICES	VALUE OF $P(x, y) = 3.75x + 3y$
$(0, 0)$	0
$(40, 0)$	150
$(20, 30)$	165 *(maximum)*
$(0, 50)$	150

Thus a maximum profit of $165 is realized by producing 20 gidgets and 30 gadgets. ▪

PROBLEM SET 7.4

For Problems 1–24, indicate the solution set for each system of inequalities by graphing the system and shading the appropriate region.

1. $\begin{pmatrix} x + y > 3 \\ x - y > 1 \end{pmatrix}$

2. $\begin{pmatrix} x - y < 2 \\ x + y < 1 \end{pmatrix}$

3. $\begin{pmatrix} x - 2y \le 4 \\ x + 2y > 4 \end{pmatrix}$

4. $\begin{pmatrix} 3x - y > 6 \\ 2x + y \le 4 \end{pmatrix}$

5. $\begin{pmatrix} 2x + 3y \le 6 \\ 3x - 2y \le 6 \end{pmatrix}$

6. $\begin{pmatrix} 4x + 3y \ge 12 \\ 3x - 4y \ge 12 \end{pmatrix}$

7. $\begin{pmatrix} 2x - y \ge 4 \\ x + 3y < 3 \end{pmatrix}$

8. $\begin{pmatrix} 3x - y < 3 \\ x + y \ge 1 \end{pmatrix}$

9. $\begin{pmatrix} x + 2y > -2 \\ x - y < -3 \end{pmatrix}$

10. $\begin{pmatrix} x - 3y < -3 \\ 2x - 3y > -6 \end{pmatrix}$

11. $\begin{pmatrix} y > x - 4 \\ y < x \end{pmatrix}$

12. $\begin{pmatrix} y \le x + 2 \\ y \ge x \end{pmatrix}$

13. $\begin{pmatrix} x - y > 2 \\ x - y > -1 \end{pmatrix}$

14. $\begin{pmatrix} x + y > 1 \\ x + y > 3 \end{pmatrix}$

15. $\begin{pmatrix} y \ge x \\ x > -1 \end{pmatrix}$

16. $\begin{pmatrix} y < x \\ y \le 2 \end{pmatrix}$

17. $\begin{pmatrix} y < x \\ y > x + 3 \end{pmatrix}$

18. $\begin{pmatrix} x \le 3 \\ y \le -1 \end{pmatrix}$

19. $\begin{pmatrix} y > -2 \\ x > 1 \end{pmatrix}$

20. $\begin{pmatrix} x + 2y > 4 \\ x + 2y < 2 \end{pmatrix}$

21. $\begin{pmatrix} x \ge 0 \\ y \ge 0 \\ x + y \le 4 \\ 2x + y \le 6 \end{pmatrix}$

22. $\begin{pmatrix} x \ge 0 \\ y \ge 0 \\ x - y \le 5 \\ 4x + 7y \le 28 \end{pmatrix}$

23. $\begin{pmatrix} x \ge 0 \\ y \ge 0 \\ 2x + y \le 4 \\ 2x - 3y \le 6 \end{pmatrix}$

24. $\begin{pmatrix} x \ge 0 \\ y \ge 0 \\ 3x + 5y \ge 15 \\ 5x + 3y \ge 15 \end{pmatrix}$

For Problems 25–28 (Figures 7.9 through 7.12), find the maximum value and the minimum value of the given function in the indicated region.

25. $f(x, y) = 3x + 5y$

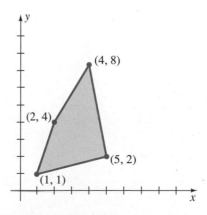

FIGURE 7.9

26. $f(x, y) = 8x + 3y$

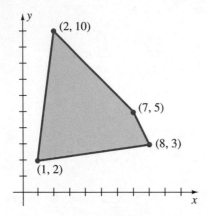

FIGURE 7.10

27. $f(x, y) = x + 4y$

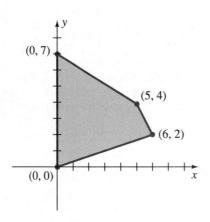

FIGURE 7.11

28. $f(x, y) = 2.5x + 3.5y$

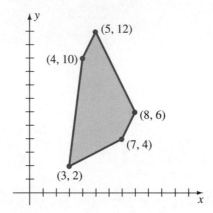

FIGURE 7.12

29. Maximze the function $f(x, y) = 3x + 7y$ in the region determined by the following constraints.

$$3x + 2y \leq 18$$
$$3x + 4y \geq 12$$
$$x \geq 0$$
$$y \geq 0$$

30. Maximize the function $f(x, y) = 1.5x + 2y$ in the region determined by the following constraints.

$$3x + 2y \leq 36$$
$$3x + 10y \leq 60$$
$$x \geq 0$$
$$y \geq 0$$

31. Maximize the function $f(x, y) = 40x + 55y$ in the region determined by the following constraints.

$$2x + y \leq 10$$
$$x + y \leq 7$$
$$2x + 3y \leq 18$$
$$x \geq 0$$
$$y \geq 0$$

32. Maximize the function $f(x, y) = 0.08x + 0.09y$ in the region determined by the following constraints.

$$x + y \le 8000$$
$$y \le \frac{1}{3}x$$
$$y \ge 500$$
$$x \le 7000$$
$$x \ge 0$$

33. Minimize the function $f(x, y) = 0.2x + 0.5y$ in the region determined by the following constraints.

$$2x + y \ge 12$$
$$2x + 5y \ge 20$$
$$x \ge 0$$
$$y \ge 0$$

34. Minimize the function $f(x, y) = 3x + 7y$ in the region determined by the following constraints.

$$x + y \ge 9$$
$$6x + 11y \ge 84$$
$$x \ge 0$$
$$y \ge 0$$

35. Maximize the function $f(x, y) = 9x + 2y$ in the region determined by the following constraints.

$$5y - 4x \le 20$$
$$4x + 5y \le 60$$
$$x \ge 0$$
$$x \le 10$$
$$y \ge 0$$

36. Maximize the function $f(x, y) = 3x + 4y$ in the region determined by the following constraints.

$$2y - x \le 6$$
$$x + y \le 12$$
$$x \ge 2$$
$$x \le 8$$
$$y \ge 0$$

For Problems 37–42, solve each linear programming problem by using the graphing method illustrated in Problem 1 on page 522.

37. Suppose that an investor wants to invest up to $10,000. She plans to buy one speculative type of stock and one conservative type. The speculative stock is paying a 12% return, and the conservative stock is paying a 9% return. She has decided to invest at least $2000 in the conservative stock and no more than $6000 in the speculative stock. Furthermore, she does not want the speculative investment to exceed the conservative one. How much should she invest at each rate to maximize her return?

38. A manufacturer of golf clubs makes a profit of $50 per set on a model A set and $45 per set on a model B set. Daily production of the model A clubs is between 30 and 50 sets, inclusive, and that of the model B clubs is between 10 and 20 sets, inclusive. The total daily production is not to exceed 50 sets. How many sets of each model should be manufactured per day to maximize the profit?

39. A company makes two types of calculators. Type A sells for $12, and type B sells for $10. It costs the company $9 to produce one type A calculator and $8 to produce one type B calculator. In one month, the company is equipped to produce between 200 and 300, inclusive, of the type A calculator and between 100 and 250, inclusive, of the type B calculator, but not more than 300 altogether. How many calculators of each type should be produced per month to maximize the difference between the total selling price and the total cost of production?

40. A manufacturer of small copiers makes a profit of $200 on a deluxe model and $250 on a standard model. The company wants to produce at least 50 deluxe models per week and at least 75 standard models per week. However, the weekly production is not to exceed 150 copiers. How many copiers of each kind should be produced in order to maximize the profit?

41. Products A and B are produced by a company according to the following production information.

a. To produce one unit of product A requires 1 hour of working time on machine I, 2 hours on machine II, and 1 hour on machine III.

b. To produce one unit of product B requires 1 hour of working time on machine I, one hour on machine II, and 3 hours on machine III.

c. Machine I is available for no more than 40 hours per week, machine II is available for no more than 40 hours per week, and machine III for no more than 60 hours per week.

d. Product A can be sold at a profit of $2.75 per unit and product B at a profit of $3.50 per unit.

How many units each of product A and product B should be produced per week to maximize profit?

42. Suppose that the company we refer to in Problem 1 also manufactures widgets and wadgets and has the following production information available.

a. To produce a widget requires 4 hours of working time on machine A and 2 hours on machine B.

b. To produce a wadget requires 5 hours of working time on machine A and 5 hours on machine B.

c. Machine A is available for no more than 200 hours per month, and machine B is available for no more than 150 hours per month.

d. Widgets can be sold at a profit of $7 each and wadgets at a profit of $8 each.

How many widgets and how many wadgets should be produced per month in order to maximize profit?

THOUGHTS INTO WORDS

43. Describe in your own words the process of solving a system of inequalities.

44. What is linear programming? Write a paragraph or two answering this question in a way that elementary algebra students could understand.

CHAPTER 7 SUMMARY

Be sure that you understand the following ideas pertaining to the algebra of matrices.

1. Matrices of the same dimension are added by adding elements in corresponding positions.

2. Matrix addition is a commutative and an associative operation.

3. Matrices of any specific dimension have an additive identity element, which is the matrix of that same dimension containing all zeros.

4. Every matrix A has an additive inverse, $-A$, which can be found by changing the sign of each element of A.

5. Matrices of the same dimension can be subtracted by the definition $A - B = A + (-B)$.

6. The scalar product of a real number k and a matrix A can be found by multiplying each element of A by k.

7. The following properties hold for scalar multiplication and matrix addition.

$$k(A + B) = kA + kB$$
$$(k + l)A = kA + lA$$
$$(kl)A = k(lA)$$

8. If A is an $m \times n$ matrix and B is an $n \times p$ matrix, then the product AB is an $m \times p$ matrix. The general term, c_{ij}, of the product matrix $C = AB$ is determined by the equation

$$c_{ij} = a_{i1}b_{1j} + a_{i2}b_{2j} + \cdots + a_{in}b_{nj}$$

9. Matrix multiplication is not a commutative operation, but it is an associative operation.

10. Matrix multiplication has two distributive properties:

$$A(B + C) = AB + AC \quad \text{and} \quad (A + B)C = AC + BC$$

11. The general multiplicative identity element, I_n, for square $n \times n$ matrices contains only 1s in the main diagonal and 0s elsewhere. For example,

$$I_2 = \begin{bmatrix} 1 & 0 \\ 0 & 1 \end{bmatrix} \quad \text{and} \quad I_3 = \begin{bmatrix} 1 & 0 & 0 \\ 0 & 1 & 0 \\ 0 & 0 & 1 \end{bmatrix}$$

12. If a square matrix A has a multiplicative inverse A^{-1}, then $AA^{-1} = A^{-1}A = I_n$.

13. The multiplicative inverse of the 2×2 matrix

$$A = \begin{bmatrix} a_{11} & a_{12} \\ a_{21} & a_{22} \end{bmatrix}$$

is

$$A^{-1} - \frac{1}{|A|} \begin{bmatrix} a_{22} & -a_{12} \\ -a_{21} & a_{11} \end{bmatrix}$$

for $|A| \neq 0$. If $|A| = 0$, then the matrix A has no inverse.

14. A general technique for finding the inverse of a square matrix, when it exists, is described on page 510.

15. The solution set of a system of n linear equations in n variables can be found by multiplying the inverse of the coefficient matrix times the column matrix consisting of the constant terms. For example, the solution set of the system

$$\begin{pmatrix} 2x + 3y - z = 4 \\ 3x - y + 2z = 5 \\ 5x - 7y - 4z = -1 \end{pmatrix}$$

can be found by the product

$$\begin{bmatrix} 2 & 3 & -1 \\ 3 & -1 & 2 \\ 5 & -7 & -4 \end{bmatrix}^{-1} \begin{bmatrix} 4 \\ 5 \\ -1 \end{bmatrix}$$

The solution set of a system of linear inequalities is the intersection of the solution sets of the individual inequalities. Such solution sets are easily determined by the graphing approach.

Linear programming problems deal with the idea of maximizing or minimizing a certain linear function that is subject to various constraints. The constraints are expressed as linear inequalities. Example 4 and Problem 1 (page 522) of Section 7.4 are a good summary of the general approach to linear programming problems in this chapter.

CHAPTER 7 REVIEW PROBLEM SET

For Problems 1–10, compute the indicated matrix, if it exists, using the following matrices.

$$A = \begin{bmatrix} 2 & -4 \\ -3 & 8 \end{bmatrix} \qquad B = \begin{bmatrix} 5 & -1 \\ 0 & 2 \end{bmatrix}$$

$$C = \begin{bmatrix} 3 & -1 \\ -2 & 4 \\ 5 & -6 \end{bmatrix} \qquad D = \begin{bmatrix} -2 & -1 & 4 \\ 5 & 0 & -3 \end{bmatrix},$$

$$E = \begin{bmatrix} 1 \\ -3 \\ -7 \end{bmatrix} \qquad F = \begin{bmatrix} 1 & -2 \\ 4 & -4 \\ 7 & -8 \end{bmatrix}$$

1. $A + B$ **2.** $B - A$ **3.** $C - F$

4. $2A + 3B$ **5.** $3C - 2F$ **6.** CD

7. DC **8.** $DC + AB$ **9.** DE

10. EF

11. Use A and B from the preceding problems and show that $AB \neq BA$.

12. Use C, D, and F from the preceding problems and show that $D(C + F) = DC + DF$.

13. Use C, D, and F from the preceding problems and show that $(C + F)D = CD + FD$.

For each matrix in Problems 14–23, find the multiplicative inverse, if it exists.

14. $\begin{bmatrix} 9 & 5 \\ 7 & 4 \end{bmatrix}$ **15.** $\begin{bmatrix} 9 & 4 \\ 7 & 3 \end{bmatrix}$

16. $\begin{bmatrix} -2 & 1 \\ 2 & 3 \end{bmatrix}$ **17.** $\begin{bmatrix} 4 & -6 \\ 2 & -3 \end{bmatrix}$

18. $\begin{bmatrix} -1 & -3 \\ -4 & -5 \end{bmatrix}$ **19.** $\begin{bmatrix} 0 & -3 \\ 7 & 6 \end{bmatrix}$

20. $\begin{bmatrix} 1 & -2 & 1 \\ 2 & -5 & 2 \\ -3 & 7 & 5 \end{bmatrix}$ **21.** $\begin{bmatrix} 1 & 3 & -2 \\ 4 & 13 & -7 \\ 5 & 16 & -8 \end{bmatrix}$

22. $\begin{bmatrix} -2 & 4 & 7 \\ 1 & -3 & 5 \\ 1 & -5 & 22 \end{bmatrix}$ **23.** $\begin{bmatrix} -1 & 2 & 3 \\ 2 & -5 & -7 \\ -3 & 5 & 11 \end{bmatrix}$

For Problems 24–28, use the multiplicative inverse matrix approach to solve each system. The required inverses were found in Problems 14–23.

24. $\begin{pmatrix} 9x + 5y = 12 \\ 7x + 4y = 10 \end{pmatrix}$ **25.** $\begin{pmatrix} -2x + y = -9 \\ 2x + 3y = 5 \end{pmatrix}$

26. $\begin{pmatrix} x - 2y + z = 7 \\ 2x - 5y + 2z = 17 \\ -3x + 7y + 5z = -32 \end{pmatrix}$

27. $\begin{pmatrix} x + 3y - 2z = -7 \\ 4x + 13y - 7z = -21 \\ 5x + 16y - 8z = -23 \end{pmatrix}$

28. $\begin{pmatrix} -x + 2y + 3z = 22 \\ 2x - 5y - 7z = -51 \\ -3x + 5y + 11z = 71 \end{pmatrix}$

For Problems 29–32, indicate the solution set for each system of linear inequalities by graphing the system and shading the appropriate region.

29. $\begin{pmatrix} 3x - 4y \geq 0 \\ 2x + 3y \leq 0 \end{pmatrix}$ **30.** $\begin{pmatrix} 3x - 2y < 6 \\ 2x - 3y < 6 \end{pmatrix}$

31. $\begin{pmatrix} x - 4y < 4 \\ 2x + y \geq 2 \end{pmatrix}$ **32.** $\begin{pmatrix} x \geq 0 \\ y \geq 0 \\ x + 2y \leq 4 \\ 2x - y \leq 4 \end{pmatrix}$

33. Maximize the function $f(x, y) = 8x + 5y$ in the region determined by the following constraints.

$$y \leq 4x$$
$$x + y \leq 5$$
$$x \geq 0$$
$$y \geq 0$$
$$x \leq 4$$

34. Maximize the function $f(x, y) = 2x + 7y$ in the region determined by the following constraints.

$$x \geq 0$$
$$y \geq 0$$
$$x + 2y \leq 16$$
$$x + y \leq 9$$
$$3x + 2y \leq 24$$

35. Maximize the function $f(x, y) = 7x + 5y$ in the region determined by the constraints of Problem 34.

36. Maximize the function $f(x, y) = 150x + 200y$ in the region determined by the constraints of Problem 34.

37. A manufacturer of electric ice cream freezers makes a profit of \$4.50 on a one-gallon freezer and a profit of \$5.25 on a two-gallon freezer. The company wants to produce at least 75 one-gallon and at least 100 two-gallon freezers per week. However, the weekly production is not to exceed a total of 250 free͏ͨ   ͏   many freezers of each type should be produce͏ ͏ ͏er week in order to maximize the profit?

CHAPTER 7 TEST

For Problems 1–10, compute the indicated matrix, if it exists, using the following matrices.

$$A = \begin{bmatrix} -1 & 3 \\ 4 & -2 \end{bmatrix} \qquad B = \begin{bmatrix} 3 & -2 \\ 4 & -1 \end{bmatrix} \qquad C = \begin{bmatrix} -3 \\ 5 \\ -6 \end{bmatrix}$$

$$D = \begin{bmatrix} 2 & -1 \\ 3 & -2 \\ 6 & 5 \end{bmatrix} \qquad E = \begin{bmatrix} 2 & -1 & 4 \\ 5 & 1 & -3 \end{bmatrix} \qquad F = \begin{bmatrix} -1 & 6 \\ 2 & -5 \\ 3 & 4 \end{bmatrix}$$

1. AB **2.** BA **3.** DE **4.** BC **5.** EC

6. $2A - B$ **7.** $3D + 2F$ **8.** $-3A - 2B$ **9.** EF **10.** $AB - EF$

For Problems 11–16, find the multiplicative inverse, if it exists.

11. $\begin{bmatrix} 3 & -2 \\ 5 & -3 \end{bmatrix}$ **12.** $\begin{bmatrix} -2 & 5 \\ 3 & -7 \end{bmatrix}$ **13.** $\begin{bmatrix} 1 & -3 \\ -2 & 8 \end{bmatrix}$

14. $\begin{bmatrix} 3 & 5 \\ 1 & 4 \end{bmatrix}$ **15.** $\begin{bmatrix} -2 & 2 & 3 \\ 1 & -1 & 0 \\ 0 & 1 & 4 \end{bmatrix}$ **16.** $\begin{bmatrix} 1 & -2 & 4 \\ 0 & 1 & 3 \\ 0 & 0 & 1 \end{bmatrix}$

For Problems 17–19, use the multiplicative inverse matrix approach to solve each system.

17. $\begin{pmatrix} 3x - 2y = 48 \\ 5x - 3y = 76 \end{pmatrix}$ **18.** $\begin{pmatrix} x - 3y = 36 \\ -2x + 8y = -100 \end{pmatrix}$ **19.** $\begin{pmatrix} 3x + 5y = 92 \\ x + 4y = 61 \end{pmatrix}$

20. Solve the system

$$\begin{pmatrix} -x + 3y + z = 1 \\ 2x + 5y = 3 \\ 3x + y - 2z = -2 \end{pmatrix}$$

where the inverse of the coefficient matrix is

$$\begin{bmatrix} -\dfrac{10}{9} & \dfrac{7}{9} & -\dfrac{5}{9} \\[2mm] \dfrac{4}{9} & -\dfrac{1}{9} & \dfrac{2}{9} \\[2mm] -\dfrac{13}{9} & \dfrac{10}{9} & -\dfrac{11}{9} \end{bmatrix}$$

21. Solve the system

$$\begin{pmatrix} x + y + 2z = 3 \\ 2x + 3y - z = 3 \\ -3x + y - 2z = 3 \end{pmatrix}$$

where the inverse of the coefficient matrix is

$$\begin{bmatrix} -\dfrac{5}{24} & \dfrac{1}{6} & -\dfrac{7}{24} \\[2mm] \dfrac{7}{24} & \dfrac{1}{6} & \dfrac{5}{24} \\[2mm] \dfrac{11}{24} & -\dfrac{1}{6} & \dfrac{1}{24} \end{bmatrix}$$

For Problems 22–24, indicate the solution set for each system of inequalities by graphing the system and shading the appropriate region.

22. $\begin{pmatrix} 2x - y > 4 \\ x + 3y < 3 \end{pmatrix}$ **23.** $\begin{pmatrix} 2x - 3y \leq 6 \\ x + 4y > 4 \end{pmatrix}$ **24.** $\begin{pmatrix} y \leq 2x - 2 \\ y \geq x + 1 \end{pmatrix}$

25. Maximize the function $f(x, y) = 500x + 350y$ in the region determined by the following constraints.

$$3x + 2y \leq 24$$
$$x + 2y \leq 16$$
$$x + y \leq 9$$
$$x \geq 0$$
$$y \geq 0$$

8

CONIC SECTIONS

Parabolic surfaces are used in the construction of satellite dishes.

Parabolas, circles, ellipses, and hyperbolas can be formed by causing a right circular conical surface and a plane to intersect, as shown in Figure 8.1; these figures are often referred to as **conic sections**. *The conic sections are not new to you. You did some graphing of circles, parabolas, ellipses, and hyperbolas in Chapters 2 and 3. At that time, however, except for the circle, we did not present any formal definitions or standard forms of equations. In Chapter 2 we developed the standard form for the equation of a circle, $(x - h)^2 + (y - k)^2 = r^2$. We used this equation to solve a variety of problems that pertain to circles. It is now time to study the other conic sections in the same manner. We will define each conic section and derive the standard form of an equation. Then we will use the standard forms to study specific conic sections.*

Circle Ellipse Parabola Hyperbola

FIGURE 8.1

8.1 PARABOLAS

We discussed parabolas as the graphs of quadratic functions in Sections 3.2 and 3.3. All parabolas in those sections had vertical lines as axes of symmetry. Furthermore, we did not state the definition for a parabola at that time. We shall now define a parabola and derive standard forms of equations for those that have either vertical or horizontal axes of symmetry.

> ### DEFINITION 8.1
>
> A **parabola** is the set of all points in a plane such that the distance of each point from a fixed point F (the **focus**) is equal to its distance from a fixed line d (the **directrix**) in the plane.

FIGURE 8.2

Using Definition 8.1, we can sketch a parabola by starting with a fixed line d and a fixed point F, not on d. Then a point P is on the parabola if and only if $PF = PP'$, where $\overline{PP'}$ is perpendicular to the directrix d (Figure 8.2). The dashed curved line in Figure 8.2 indicates the possible positions of P; it is the parabola. The line l, through F and perpendicular to the directrix d, is called the **axis of symmetry**. The point V, on the axis of symmetry halfway from F to the directrix d, is the vertex of the parabola.

We can derive a standard form for the equation of a parabola by superimposing coordinates on the plane such that the origin is at the vertex of the parabola and the y axis is the axis of symmetry (Figure 8.3). If the focus is at $(0, p)$, where $p \neq 0$,

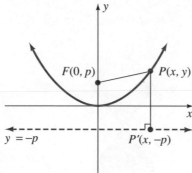

FIGURE 8.3

then the equation of the directrix is $y = -p$. Therefore, for any point P on the parabola, $PF = PP'$, and using the distance formula yields

$$\sqrt{(x - 0)^2 + (y - p)^2} = \sqrt{(x - x)^2 + (y + p)^2}$$

Squaring both sides and simplifying, we obtain

$$(x - 0)^2 + (y - p)^2 = (x - x)^2 + (y + p)^2$$
$$x^2 + y^2 - 2py + p^2 = y^2 + 2py + p^2$$
$$x^2 = 4py$$

Thus the **standard form for the equation of a parabola**, with its vertex at the origin and the y axis as its axis of symmetry, is

$$x^2 = 4py$$

If $p > 0$, the parabola opens upward; if $p < 0$, the parabola opens downward.

In Figure 8.4 the line segment $\overline{QP}$ is called the **latus rectum**. It contains the focus and is parallel to the directrix. Because $FP = PP' = |2p|$, the entire length of the latus rectum is $|4p|$ units. You will see in a moment how we can use this fact when graphing parabolas.

FIGURE 8.4

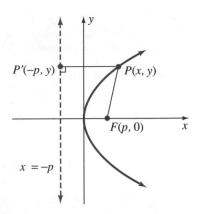

FIGURE 8.5

In a similar fashion, we can develop the standard form for the equation of a parabola with its vertex at the origin and the x axis as its axis of symmetry. By choosing a focus at $F(p, 0)$ and a directrix with an equation of $x = -p$ (see Figure 8.5), and by applying the definition of a parabola, we obtain the standard form for the equation:

$$y^2 = 4px$$

If $p > 0$, the parabola opens to the right, as in Figure 8.5; if $p < 0$, it opens to the left.

The concept of symmetry can be used to decide which of the two equations, $x^2 = 4py$ or $y^2 = 4px$, is to be used. The graph of $x^2 = 4py$ is symmetric with respect to the y axis because replacing x with $-x$ does not change the equation. Likewise, the graph of $y^2 = 4px$ is symmetric with respect to the x axis because replacing y with $-y$ leaves the equation unchanged. The following property summarizes our previous discussion.

PROPERTY 8.1

The graph of each of the following equations is a parabola that has its vertex at the origin and has the indicated focus, directrix, and symmetry.

1. $x^2 = 4py$ focus $(0, p)$, directrix $y = -p$, y-axis symmetry
2. $y^2 = 4px$ focus $(p, 0)$, directrix $x = -p$, x-axis symmetry

Now let's illustrate some uses of the equations $x^2 = 4py$ and $y^2 = 4px$.

EXAMPLE 1

Find the focus and directrix of the parabola $x^2 = -8y$ and sketch its graph.

Solution

Compare $x^2 = -8y$ to the standard form $x^2 = 4py$ and we have $4p = -8$. Therefore, $p = -2$ and the parabola opens downward. The focus is at $(0, -2)$ and the equation of the directrix is $y = -(-2) = 2$. The latus rectum is $|4p| = |-8| = 8$ units long. Therefore, the endpoints of the latus rectum are at $(4, -2)$ and $(-4, -2)$. The graph is sketched in Figure 8.6.

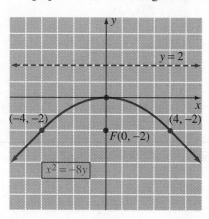

FIGURE 8.6

EXAMPLE 2

Write the equation of the parabola that is symmetric with respect to the y axis, has its vertex at the origin, and contains the point $P(6, 3)$.

Solution

The standard form of the parabola is $x^2 = 4py$. Because P is on the parabola, the ordered pair $(6, 3)$ must satisfy the equation. Therefore,

$$6^2 = 4p(3)$$
$$36 = 12p$$
$$3 = p$$

If $p = 3$, the equation becomes

$$x^2 = 4(3)y$$
$$x^2 = 12y$$

EXAMPLE 3

Find the focus and directrix of the parabola $y^2 = 6x$ and sketch its graph.

Solution

Compare $y^2 = 6x$ to the standard form $y^2 = 4px$; we see that $4p = 6$ and therefore $p = \dfrac{3}{2}$. Thus the focus is at $\left(\dfrac{3}{2}, 0\right)$ and the equation of the directrix is $x = -\dfrac{3}{2}$.

The parabola opens to the right. The latus rectum is $|4p| = |6| = 6$ units long. Therefore the endpoints of the latus rectum are at $\left(\frac{3}{2}, 3\right)$ and $\left(\frac{3}{2}, -3\right)$. The graph is sketched in Figure 8.7.

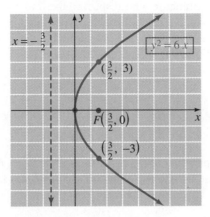

FIGURE 8.7

Other Parabolas

In much the same way, we can develop the standard form for an equation of a parabola that is symmetric with respect to a line parallel to a coordinate axis. In Figure 8.8 we have taken the vertex V at (h, k) and the focus F at $(h, k + p)$; the equation of the directrix is $y = k - p$. By the definition of a parabola, we know that $FP = PP'$. Therefore, applying the distance formula, we obtain

$$\sqrt{(x - h)^2 + (y - (k + p))^2} = \sqrt{(x - x)^2 + [y - (k - p)]^2}$$

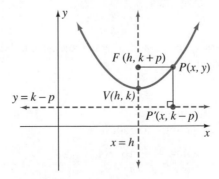

FIGURE 8.8

We leave it to the reader to show that this equation simplifies to

$$(x - h)^2 = 4p(y - k)$$

which is called the standard form for the equation of a parabola that has its vertex at (h, k) and is symmetric with respect to the line $x = h$. If $p > 0$, the parabola opens upward; if $p < 0$, the parabola opens downward.

In a similar fashion, we can show that the standard form for the equation of a parabola that has its vertex at (h, k) and is symmetric with respect to the line $y = k$ is

$$(y - k)^2 = 4p(x - h)$$

If $p > 0$, the parabola opens to the right; if $p < 0$, it opens to the left.

Let's summarize our discussion of parabolas that have lines of symmetry parallel to the x axis or y axis by stating the following property.

PROPERTY 8.2

The graph of each of the following equations is a parabola that has its vertex at (h, k) and has the indicated focus, directrix, and symmetry.

1. $(x - h)^2 = 4p(y - k)$ focus $(h, k + p)$, directrix $y = k - p$, line of symmetry $x = h$

2. $(y - k)^2 = 4p(x - h)$ focus $(h + p, k)$, directrix $x = h - p$, line of symmetry $y = k$

EXAMPLE 4

Find the vertex, focus, and directrix of the parabola $y^2 + 4y - 4x + 16 = 0$ and sketch its graph.

Solution

Write the given equation as $y^2 + 4y = 4x - 16$ and we can complete the square on the left side by adding 4 to both sides.

$$y^2 + 4y + 4 = 4x - 16 + 4$$
$$(y + 2)^2 = 4x - 12$$
$$(y + 2)^2 = 4(x - 3)$$

Now let's compare this final equation to the form $(y - k)^2 = 4p(x - h)$.

$$(y - (-2))^2 = 4(x - 3).$$

$$k \overset{\uparrow}{=} -2 \quad 4p = 4 \quad h = 3$$
$$p = 1$$

The vertex is at $(3, -2)$ and because $p > 0$, the parabola opens to the right and the focus is at $(4, -2)$. The equation of the directrix is $x = 2$. The latus rectum is $|4p| = |4| = 4$ units long and its endpoints are at $(4, 0)$ and $(4, -4)$. The graph is sketched in Figure 8.9.

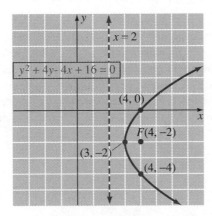

FIGURE 8.9 ■

REMARK If we were using a graphing calculator to graph the parabola in Example 4, then after the step $(y + 2)^2 = 4x - 12$, we would solve for y to obtain $y = -2 \pm \sqrt{4x - 12}$. Then we could enter the two functions $Y_1 = -2 + \sqrt{4x - 12}$ and $Y_2 = -2 - \sqrt{4x - 12}$ and obtain a figure that closely resembles Figure 8.9. (You are asked to do this in the Graphing Calculator Activities.) Some computer programs can graph the equation in Example 4 without changing its form.

EXAMPLE 5

Write the equation of the parabola if its focus is at $(-4, 1)$ and the equation of its directrix is $y = 5$.

Solution

Because the directrix is a horizontal line, we know that the equation of the parabola is of the form $(x - h)^2 = 4p(y - k)$. The vertex is halfway between the focus and the directrix, so the vertex is at $(-4, 3)$. This means that $h = -4$ and $k = 3$. The parabola opens downward because the focus is below the directrix, and the distance between the focus and the vertex is 2 units; thus $p = -2$. Substitute -4 for h, 3 for k, and -2 for p in the equation $(x - h)^2 = 4p(y - k)$ to obtain

$$(x - (-4))^2 = 4(-2)(y - 3)$$

which simplifies to

$$(x + 4)^2 = -8(y - 3)$$
$$x^2 + 8x + 16 = -8y + 24$$
$$x^2 + 8x + 8y - 8 = 0$$ ∎

REMARK For a problem such as Example 5, you may find it helpful to put the given information on a set of axes and draw a rough sketch of the parabola to help you with the analysis of the problem.

Parabolas possess various properties that make them very useful. For example, if a parabola is rotated about its axis, a parabolic surface is formed. The rays from a source of light placed at the focus of this surface reflect from the surface parallel to the axis. It is for this reason that parabolic reflectors are used on searchlights as in Figure 8.10. Likewise, rays of light coming into a parabolic surface parallel to the axis are reflected through the focus. This property of parabolas is useful in the design of mirrors for telescopes (see Figure 8.11) and in the construction of radar antennas.

FIGURE 8.10

FIGURE 8.11

A bullet fired into the air follows the curvature of a parabola if air resistance and other outside factors are ignored—in other words, if only the force of gravity is considered. (See Figure 8.12.)

FIGURE 8.12

PROBLEM SET 8.1

For Problems 1–22, find the vertex, focus, and directrix of the given parabola and sketch its graph.

1. $y^2 = 8x$ **2.** $y^2 = -4x$

3. $x^2 = -12y$ **4.** $x^2 = 8y$

5. $y^2 = -2x$ **6.** $y^2 = 6x$

7. $x^2 = 6y$ **8.** $x^2 = -7y$

9. $x^2 - 4y + 8 = 0$ **10.** $x^2 - 8y - 24 = 0$

11. $x^2 + 8y + 16 = 0$ **12.** $x^2 + 4y - 4 = 0$

13. $y^2 - 12x + 24 = 0$ **14.** $y^2 + 8x - 24 = 0$

15. $x^2 - 2x - 4y + 9 = 0$ **16.** $x^2 + 4x - 8y - 4 = 0$

17. $x^2 + 6x + 8y + 1 = 0$

18. $x^2 - 4x + 4y - 4 = 0$

19. $y^2 - 2y + 12x - 35 = 0$

20. $y^2 + 4y + 8x - 4 = 0$

21. $y^2 + 6y - 4x + 1 = 0$

22. $y^2 - 6y - 12x + 21 = 0$

For Problems 23–42, find an equation of the parabola that satisfies the given conditions.

23. Focus $(0, 3)$, directrix $y = -3$

24. Focus $\left(0, -\dfrac{1}{2}\right)$, directrix $y = \dfrac{1}{2}$

25. Focus $(-1, 0)$, directrix $x = 1$

26. Focus $(5, 0)$, directrix $x = 1$

27. Focus $(0, 1)$, directrix $y = 7$

28. Focus $(0, -2)$, directrix $y = -10$

29. Focus $(3, 4)$, directrix $y = -2$

30. Focus $(-3, -1)$, directrix $y = 7$

31. Focus $(-4, 5)$, directrix $x = 0$

32. Focus $(5, -2)$, directrix $x = -1$

33. Vertex $(0, 0)$, symmetric with respect to the x axis, and contains the point $(-3, 5)$

34. Vertex $(0, 0)$, symmetric with respect to the y axis, and contains the point $(-2, -4)$

35. Vertex $(0, 0)$, focus $\left(\dfrac{5}{2}, 0\right)$

36. Vertex $(0, 0)$, focus $\left(0, -\dfrac{7}{2}\right)$

37. Vertex $(7, 3)$, focus $(7, 5)$, and symmetric with respect to the line $x = 7$

38. Vertex $(-4, -6)$, focus $(-7, -6)$, and symmetric with respect to the line $y = -6$

39. Vertex $(8, -3)$, focus $(11, -3)$, and symmetric with respect to the line $y = -3$

40. Vertex $(-2, 9)$, focus $(-2, 5)$, and symmetric with respect to the line $x = -2$

41. Vertex $(-9, 1)$, symmetric with respect to the line $x = -9$, and contains the point $(-8, 0)$

42. Vertex $(6, -4)$, symmetric with respect to the line $y = -4$, and contains the point $(8, -3)$

For Problems 43–47, solve each problem.

43. One section of a suspension bridge hangs between two towers that are 40 feet above the surface and 300 feet

FIGURE 8.13

apart as in Figure 8.13. A cable strung between the tops of the two towers is in the shape of a parabola with its vertex 10 feet above the surface. With axes drawn as indicated in the figure, find the equation of the parabola.

44. Suppose that five equally spaced vertical cables are used to support the bridge in Figure 8.13. Find the total length of these supports.

45. Suppose that an arch is shaped like a parabola. It is 20 feet wide at the base and 100 feet high. How wide is the arch 50 feet above the ground?

46. A parabolic arch 27 feet high spans a parkway. How wide is the arch if the center section of the parkway, a section that is 50 feet wide, has a minimum clearance of 15 feet?

47. A parabolic arch spans a stream 200 feet wide. How high must the arch be above the stream to give a minimum clearance of 40 feet over a channel in the center that is 120 feet wide?

48. Give a step-by-step description of how you would go about graphing the parabola $x^2 - 2x - 4y - 7 = 0$.

49. Suppose that someone graphed the equation $y^2 - 6y - 2x + 11 = 0$ and obtained the graph in Figure 8.14. How do you know by looking at the equation that this graph is incorrect?

FIGURE 8.14

 GRAPHING CALCULATOR ACTIVITIES

50. The parabola determined by the equation $x^2 + 4x - 8y - 4 = 0$ (Problem 16) is easy to graph using a graphing calculator because it can be expressed as a function of x without much computation. Let's solve the equation for y.

$$8y = x^2 + 4x - 4$$
$$y = \frac{x^2 + 4x - 4}{8}$$

Use your graphing calculator to graph this function.

As noted in the Remark following Example 4, solving the equation $y^2 + 4y - 4x + 16 = 0$ for y produces two functions—namely, $Y_1 = -2 + \sqrt{4x - 12}$ and $Y_2 = -2 - \sqrt{4x - 12}$. Graph these two functions on the same set of axes. Your result should resemble Figure 8.9.

Use your graphing calculator to check your graphs for Problems 1–22.

8.2 ELLIPSES

Let's begin by defining the concept of an ellipse.

DEFINITION 8.2

An ellipse is the set of all points in a plane such that the sum of the distances of each point from two fixed points F and F' (the foci) in the plane is constant.

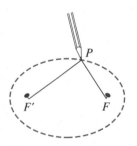

FIGURE 8.15

Using two thumbtacks, a piece of string, and a pencil, it is easy to draw an ellipse by satisfying the conditions of Definition 8.2. First, insert two thumbtacks in a piece of cardboard at points F and F' and fasten the ends of the piece of string to the thumbtacks, as in Figure 8.15. Then loop the string around the point of a pencil and hold the pencil so that the string is taut. Finally, move the pencil around the tacks, always keeping the string taut: You will draw an ellipse. The two points F and F' are the foci referred to in Definition 8.2, and the sum of the distances FP and $F'P$ is constant, because it represents the length of the piece of string. With the same piece of string, you can vary the shape of the ellipse by changing the positions of the foci. Moving F and F' farther apart will make the ellipse flatter. Likewise, moving F and F' closer together will cause the ellipse to resemble a circle. In fact, if $F = F'$, you will obtain a circle.

We can derive a standard form for the equation of an ellipse by superimposing coordinates on the plane such that the foci are on the x axis, equidistant from the origin (Figure 8.16). If F has coordinates $(c, 0)$, where $c > 0$, then F' has coordinates

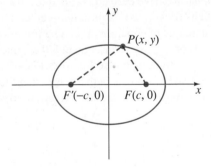

FIGURE 8.16

$(-c, 0)$, and the distance between F and F' is $2c$ units. We will let $2a$ represent the constant sum of $FP + F'P$. Note that $2a > 2c$ and therefore $a > c$. For any point P on the ellipse,

$$FP + F'P = 2a$$

Use the distance formula to write this as

$$\sqrt{(x-c)^2 + (y-0)^2} + \sqrt{(x+c)^2 + (y-0)^2} = 2a$$

Let's change the form of this equation to

$$\sqrt{(x-c)^2 + y^2} = 2a - \sqrt{(x+c)^2 + y^2}$$

and square both sides:

$$(x-c)^2 + y^2 = 4a^2 - 4a\sqrt{(x+c)^2 + y^2} + (x+c)^2 + y^2$$

This can be simplified to

$$a^2 + cx = a\sqrt{(x+c)^2 + y^2}$$

Again square both sides to produce

$$a^4 + 2a^2cx + c^2x^2 = a^2[(x+c)^2 + y^2]$$

which can be written in the form

$$x^2(a^2 - c^2) + a^2y^2 = a^2(a^2 - c^2)$$

Divide both sides by $a^2(a^2 - c^2)$, which leads to the form

$$\frac{x^2}{a^2} + \frac{y^2}{a^2 - c^2} = 1$$

Letting $b^2 = a^2 - c^2$, where $b > 0$, produces the equation

$$\frac{x^2}{a^2} + \frac{y^2}{b^2} = 1 \tag{1}$$

Because $c > 0$, $a > c$, and $b^2 = a^2 - c^2$, it follows that $a^2 > b^2$ and hence $a > b$. This equation that we have derived is called the **standard form for the equation of an ellipse** with its foci on the x axis and its center at the origin.

The x intercepts of equation (1) can be found by letting $y = 0$. Doing this produces $x^2/a^2 = 1$, or $x^2 = a^2$; consequently, the x intercepts are a and $-a$. The corresponding points on the graph (see Figure 8.17) are $A(a, 0)$ and $A'(-a, 0)$, and the line segment $\overline{A'A}$, which is of length $2a$, is called the **major axis** of the ellipse. The endpoints of the major axis are also referred to as the **vertices** of the ellipse.

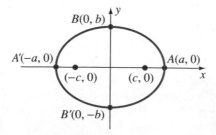

FIGURE 8.17

Similarly, letting $x = 0$ produces $y^2/b^2 = 1$ or $y^2 = b^2$; consequently, the y intercepts are b and $-b$. The corresponding points on the graph are $B(0, b)$ and $B'(0, -b)$, and the line segment $\overline{BB'}$, which is of length $2b$, is called the **minor axis**. Because $a > b$, **the major axis is always longer than the minor axis.** The point of intersection of the major and minor axes is called the **center** of the ellipse.

Let's summarize this discussion by stating the following property.

PROPERTY 8.3

The graph of the equation

$$\frac{x^2}{a^2} + \frac{y^2}{b^2} = 1$$

for $a^2 > b^2$, is an ellipse with the endpoints of its major axis (the vertices) at $(a, 0)$ and $(-a, 0)$ and the endpoints of its minor axis at $(0, b)$ and $(0, -b)$. The foci are at $(c, 0)$ and $(-c, 0)$, where $c^2 = a^2 - b^2$.

Note that replacing y with $-y$, or x with $-x$, or both x and y with $-x$ and $-y$, leaves the equation unchanged. Thus the graph of

$$\frac{x^2}{a^2} + \frac{y^2}{b^2} = 1$$

is symmetric with respect to the x axis, the y axis, and the origin.

EXAMPLE 1

Find the vertices, the endpoints of the minor axis, and the foci of the ellipse $4x^2 + 9y^2 = 36$, and sketch the ellipse.

Solution

The given equation can be changed to standard form by dividing both sides by 36.

$$\frac{4x^2}{36} + \frac{9y^2}{36} = \frac{36}{36}$$

$$\frac{x^2}{9} + \frac{y^2}{4} = 1$$

Therefore, $a^2 = 9$ and $b^2 = 4$; hence the vertices are at $(3, 0)$ and $(-3, 0)$, and the endpoints of the minor axis are at $(0, 2)$ and $(0, -2)$. Because $c^2 = a^2 - b^2$, we have

$$c^2 = 9 - 4 = 5$$

Thus the foci are at $\left(\sqrt{5}, 0\right)$ and $\left(\sqrt{-5}, 0\right)$. The ellipse is sketched in Figure 8.18.

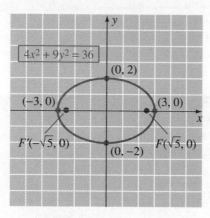

FIGURE 8.18

REMARK For a problem such as Example 1, it is not necessary to change to standard form to find the values for a and b. After all, $\pm a$ are the x intercepts and $\pm b$ are the y intercepts. These values can be found quite easily from the given form of the equation.

EXAMPLE 2

Find the equation of the ellipse with vertices at $(\pm 6, 0)$ and foci at $(\pm 4, 0)$.

Solution

From the given information, we know that $a = 6$ and $c = 4$. Therefore,

$$b^2 = a^2 - c^2 = 36 - 16 = 20$$

Substitute 36 for a^2 and 20 for b^2 in the standard form to produce

$$\frac{x^2}{36} + \frac{y^2}{20} = 1$$

Multiply both sides by 180 to get

$$5x^2 + 9y^2 = 180$$

Ellipses with Foci on the y Axis

We can develop a standard form for the equation of an ellipse with foci on the y axis in a similar fashion. The following property summarizes the results of such a development with the foci at $(0, c)$ and $(0, -c)$, where $c > 0$.

PROPERTY 8.4

The graph of the equation

$$\frac{x^2}{b^2} + \frac{y^2}{a^2} = 1$$

where $a^2 > b^2$, is an ellipse with the endpoints of its major axis (vertices) at $(0, a)$ and $(0, -a)$ and the endpoints of its minor axis at $(b, 0)$ and $(-b, 0)$. The foci are at $(0, c)$ and $(0, -c)$, where $c^2 = a^2 - b^2$.

From Properties 8.3 and 8.4 it is evident that an equation of an ellipse with its center at the origin and foci on a coordinate axis can be written in the form

$$\frac{x^2}{p} + \frac{y^2}{q} = 1 \qquad \text{or} \qquad qx^2 + py^2 = pq$$

where p and q are positive. If $p > q$, the major axis lies on the x axis, and if $q > p$, the major axis is on the y axis. It is not necessary to memorize these facts, because for any specific problem the endpoints of the major and minor axes are determined by the x and y intercepts. However, it is necessary to remember the relationship $c^2 = a^2 - b^2$.

EXAMPLE 3

Find the vertices, the endpoints of the minor axis, and the foci of the ellipse $18x^2 + 4y^2 = 36$, and sketch the ellipse.

Solution

To find the x intercepts, we let $y = 0$ and we obtain

$$18x^2 = 36$$
$$x^2 = 2$$
$$x = \pm\sqrt{2}$$

Similarly, to find the y intercepts, we let $x = 0$ and we obtain

$$4y^2 = 36$$
$$y^2 = 9$$
$$y = \pm 3$$

Because $3 > \sqrt{2}$, we know that $a = 3$ and $b = \sqrt{2}$. Therefore, the vertices are at $(0, 3)$ and $(0, -3)$, and the endpoints of the minor axes are at $(\sqrt{2}, 0)$ and $(-\sqrt{2}, 0)$. From the relationship $c^2 = a^2 - b^2$, we get

$$c^2 = 9 - 2 = 7$$

Thus the foci are at $\left(0, \sqrt{7}\right)$ and $\left(0, -\sqrt{7}\right)$. The ellipse is sketched in Figure 8.19.

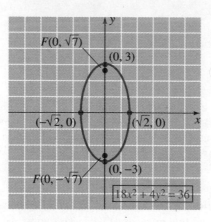

FIGURE 8.19

Other Ellipses

In the same way, we can develop the standard form for an equation of an ellipse that is symmetric with respect to a line parallel to a coordinate axis. We will not show such developments in this text, but Figures 8.20 and 8.21 indicate the basic facts we need in order to develop and use the resulting equations. Note that in each case, the center of the ellipse is at a point (h, k). Furthermore, the physical significance of a, b, and c is the same as before. However, these values are used relative to the center (h, k) to find the endpoints of the major and minor axes and to find the foci. Let's see how this works in a specific example.

FIGURE 8.20

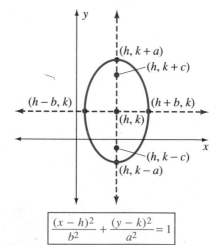

FIGURE 8.21

EXAMPLE 4

Find the vertices, the endpoints of the minor axis, and the foci of the ellipse $9x^2 + 54x + 4y^2 - 8y + 49 = 0$, and sketch the ellipse.

Solution

First, we need to change to standard form by completing the square on both x and y.

$$9(x^2 + 6x + \underline{}) + 4(y^2 - 2y + \underline{}) = -49$$
$$9(x^2 + 6x + 9) + 4(y^2 - 2y + 1) = -49 + 81 + 4$$
$$9(x + 3)^2 + 4(y - 1)^2 = 36$$
$$\frac{(x + 3)^2}{4} + \frac{(y - 1)^2}{9} = 1$$

Because $a > b$, this last equation is of the form

$$\frac{(x - h)^2}{b^2} + \frac{(y - k)^2}{a^2} = 1$$

where $h = -3$, $k = 1$, $a = 3$, and $b = 2$. Thus the endpoints of the major axis (vertices) are up three units and down three units from the center, $(-3, 1)$, so they are at $(-3, 4)$ and $(-3, -2)$. Likewise, the endpoints of the minor axis are two units to the right and two units to the left of the center. Thus they are at $(-1, 1)$ and $(-5, 1)$. From the relationship $c^2 = a^2 - b^2$, we get

$$c^2 = 9 - 4 = 5$$

Thus the foci are at $\left(-3, 1 + \sqrt{5}\right)$ and $\left(-3, 1 - \sqrt{5}\right)$. The ellipse is sketched in Figure 8.22.

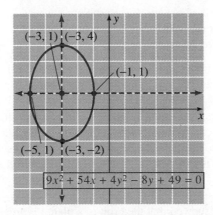

FIGURE 8.22

EXAMPLE 5

Write the equation of the ellipse that has vertices at $(-3, -5)$ and $(7, -5)$ and foci at $(-1, -5)$ and $(5, -5)$.

Solution

Because the vertices and foci are on the same horizontal line ($y = -5$), this ellipse has an equation of the form

$$\frac{(x - h)^2}{a^2} + \frac{(y - k)^2}{b^2} = 1$$

The center of the ellipse is at the midpoint of the major axis. Therefore,

$$h = \frac{-3 + 7}{2} = 2 \qquad \text{and} \qquad k = \frac{-5 + (-5)}{2} = -5$$

The distance between the center $(2, -5)$ and a vertex $(7, -5)$ is 5 units; thus $a = 5$. The distance between the center $(2, -5)$ and a focus $(5, -5)$ is 3 units; thus $c = 3$. Using the relationship $c^2 = a^2 - b^2$, we obtain

$$b^2 = a^2 - c^2 = 25 - 9 = 16$$

Now let's substitute 2 for h, -5 for k, 25 for a^2, and 16 for b^2 in the general form, and then we can simplify.

$$\frac{(x - 2)^2}{25} + \frac{(y + 5)^2}{16} = 1$$
$$16(x - 2)^2 + 25(y + 5)^2 = 400$$
$$16(x^2 - 4x + 4) + 25(y^2 + 10y + 25) = 400$$
$$16x^2 - 64x + 64 + 25y^2 + 250y + 625 = 400$$
$$16x^2 - 64x + 25y^2 + 250y + 289 = 0$$

∎

REMARK   Again, for a problem such as Example 5, it might be helpful to start by recording the given information on a set of axes and drawing a rough sketch of the figure.

Like parabolas, ellipses possess properties that make them very useful. For example, the elliptical surface formed by rotating an ellipse about its major axis has the following property: Light or sound waves emitted at one focus reflect off the surface and converge at the other focus. This is the principle behind "whispering galleries," such as the Rotunda of the Capitol Building in Washington, D.C. In such buildings, two people standing at two specific spots that are the foci of the elliptical ceiling can whisper and yet hear each other clearly, even though they may be quite far apart.

Ellipses also play an important role in astronomy. Johannes Kepler (1571–1630) showed that the orbit of a planet is an ellipse with the sun at one focus. For example, the orbit of the earth is elliptical but nearly circular; at the same time, the moon moves about the earth in an elliptical path (see Figure 8.23).

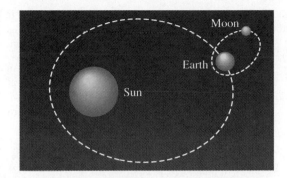

FIGURE 8.23

The arches for concrete bridges are sometimes elliptical. (One example is shown in Figure 8.25 in the next set of problems.) Also, elliptical gears are used in certain kinds of machinery that require a slow but powerful force at impact, such as a heavy-duty punch (see Figure 8.24).

FIGURE 8.24

PROBLEM SET 8.2

For Problems 1–22, find the vertices, the endpoints of the minor axis, and the foci of the given ellipse, and sketch its graph.

1. $\dfrac{x^2}{4} + \dfrac{y^2}{1} = 1$ **2.** $\dfrac{x^2}{16} + \dfrac{y^2}{1} = 1$

3. $\dfrac{x^2}{4} + \dfrac{y^2}{9} = 1$ **4.** $\dfrac{x^2}{4} + \dfrac{y^2}{16} = 1$

5. $9x^2 + 3y^2 = 27$ **6.** $4x^2 + 3y^2 = 36$

7. $2x^2 + 5y^2 = 50$ **8.** $5x^2 + 36y^2 = 180$

9. $12x^2 + y^2 = 36$ **10.** $8x^2 + y^2 = 16$

11. $7x^2 + 11y^2 = 77$ **12.** $4x^2 + y^2 = 12$

13. $4x^2 - 8x + 9y^2 - 36y + 4 = 0$

14. $x^2 + 6x + 9y^2 - 36y + 36 = 0$

15. $4x^2 + 16x + y^2 + 2y + 1 = 0$

16. $9x^2 - 36x + 4y^2 + 16y + 16 = 0$

17. $x^2 - 6x + 4y^2 + 5 = 0$

18. $16x^2 + 9y^2 + 36y - 108 = 0$

19. $9x^2 - 72x + 2y^2 + 4y + 128 = 0$

20. $5x^2 + 10x + 16y^2 + 160y + 325 = 0$

21. $2x^2 + 12x + 11y^2 - 88y + 172 = 0$

22. $9x^2 + 72x + y^2 + 6y + 135 = 0$

For Problems 23–36, find an equation of the ellipse that satisfies the given conditions.

23. Vertices $(\pm 5, 0)$, foci $(\pm 3, 0)$

24. Vertices $(\pm 4, 0)$, foci $(\pm 2, 0)$

25. Vertices $(0, \pm 6)$, foci $(0, \pm 5)$

26. Vertices $(0, \pm 3$, foci $(0, \pm 2)$

27. Vertices $(\pm 3, 0)$, length of minor axis is 2

28. Vertices $(0, \pm 5)$, length of minor axis is 4

29. Foci $(0, \pm 2)$, length of minor axis is 3

30. Foci $(\pm 1, 0)$, length of minor axis is 2

31. Vertices $(0, \pm 5)$, contains the point $(3, 2)$

32. Vertices $(\pm 6, 0)$, contains the point $(5, 1)$

33. Vertices $(5, 1)$ and $(-3, 1)$, foci $(3, 1)$ and $(-1, 1)$

34. Vertices $(2, 4)$ and $(2, -6)$, foci $(2, 3)$ and $(2, -5)$

35. Center $(0, 1)$ one focus at $(-4, 1)$, length of minor axis is 6

36. Center $(3, 0)$, one focus at $(3, 2)$, length of minor axis is 4

For Problems 37–40, solve each problem.

37. Find an equation of the set of points in a plane such that the sum of the distances between each point of the set and the points $(2, 0)$ and $(-2, 0)$ is 8 units.

38. Find an equation of the set of points in a plane such that the sum of the distances between each point of the set and the points $(0, 3)$ and $(0, -3)$ is 10 units.

39. An arch of the bridge shown in Figure 8.25 is semielliptical and the major axis is horizontal. The arch is 30 feet wide and 10 feet high. Find the height of the arch 10 feet from the center of the base.

F I G U R E   8 . 2 5

40. In Figure 8.25, how much clearance is there 10 feet from the bank?

THOUGHTS INTO WORDS

41. What type of figure is the graph of the equation $x^2 + 6x + 2y^2 - 20y + 59 = 0$? Explain your answer.

42. Suppose that someone graphed the equation $4x^2 - 16x + 9y^2 + 18y - 11 = 0$ and obtained the graph shown in Figure 8.26. How do you know by looking at the equation that this is an incorrect graph?

F I G U R E   8 . 2 6

GRAPHING CALCULATOR ACTIVITIES

43. Use your graphing calculator to check your graphs for Problems 13–22.

44. Use your graphing calculator to graph each of the following ellipses.

    a. $2x^2 - 40x + y^2 + 2y + 185 = 0$

    b. $x^2 - 4x + 2y^2 - 48y + 272 = 0$

    c. $4x^2 - 8x + y^2 - 4y - 136 = 0$

    d. $x^2 + 6x + 2y^2 + 56y + 301 = 0$

8.3   HYPERBOLAS

A hyperbola and an ellipse are similar by definition; however, an ellipse involves the *sum* of distances and a hyperbola involves the *difference* of distances.

> ### DEFINITION 8.3
>
> A **hyperbola** is the set of all points in a plane such that the difference of the distances of each point from two fixed points F and F' (the foci) in the plane is a positive constant.

Using Definition 8.3, we can sketch a hyperbola by starting with two fixed points F and F' as shown in Figure 8.27. Then we locate all points P such that $PF' - PF$ is a positive constant. Likewise, as shown in Figure 8.27, all points Q are located such that $QF - QF'$ is the same positive constant. The two dashed curved lines in Figure 8.27 make up the hyperbola. The two curves are sometimes referred to as the branches of the hyperbola.

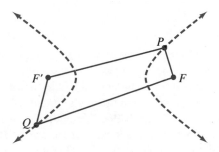

FIGURE 8.27

To develop a standard form for the equation of a hyperbola, let's superimpose coordinates on the plane such that the foci are located at $F(c, 0)$ and $F'(-c, 0)$, as indicated in Figure 8.28. Using the distance formula and setting $2a$ equal to the difference of the distances from any point P on the hyperbola to the foci, we have the following equation.

$$\left| \sqrt{(x - c)^2 + (y - 0)^2} - \sqrt{(x + c)^2 + (y - 0)^2} \right| = 2a$$

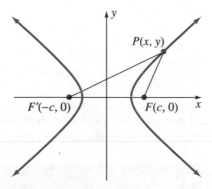

F I G U R E 8 . 2 8

(The absolute value sign is used to allow the point P to be on either branch of the hyperbola.) Using the same type of simplification procedure that we used for deriving the standard form for the equation of an ellipse, we find that this equation simplifies to

$$\frac{x^2}{a^2} - \frac{y^2}{c^2 - a^2} = 1$$

Letting $b^2 = c^2 - a^2$, where $b > 0$, we obtain the standard form

$$\frac{x^2}{a^2} - \frac{y^2}{b^2} = 1 \tag{1}$$

Equation (1) indicates that this hyperbola is symmetric with respect to both axes and the origin. Furthermore, by letting $y = 0$, we obtain $x^2/a^2 = 1$, or $x^2 = a^2$, so the x intercepts are a and $-a$. The corresponding points $A(a, 0)$ and $A'(-a, 0)$ are the **vertices** of the hyperbola, and the line segment AA' is called the **transverse axis**; it is of length $2a$ (see Figure 8.29). The midpoint of the transverse axis is called the **center** of the hyperbola; it is located at the origin. By letting $x = 0$ in equation (1), we obtain $-y^2/b^2 = 1$, or $y^2 = -b^2$. This implies that there are no y intercepts, as indicated in Figure 8.29.

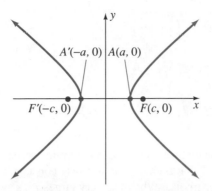

FIGURE 8.29

The following property summarizes the previous discussion.

PROPERTY 8.5

The graph of the equation

$$\frac{x^2}{a^2} - \frac{y^2}{b^2} = 1$$

is a hyperbola with vertices at $(a, 0)$ and $(-a, 0)$. The foci are at $(c, 0)$ and $(-c, 0)$, where $c^2 = a^2 + b^2$.

In conjunction with every hyperbola there are two intersecting lines that pass through the center of the hyperbola. These lines, referred to as *asymptotes*, are very helpful when we are sketching a hyperbola. Their equations are easily determined by using the following type of reasoning. Solving the equation

$$\frac{x^2}{a^2} - \frac{y^2}{b^2} = 1$$

for y produces $y = \pm\frac{b}{a}\sqrt{x^2 - a^2}$. From this form, it is evident that there are no points on the graph for $x^2 - a^2 < 0$—that is, if $-a < x < a$. However, there are points on the graph if $x \geq a$ or $x \leq -a$. If $x \geq a$, then $y = \pm\frac{b}{a}\sqrt{x^2 - a^2}$ can be written

$$y = \pm\frac{b}{a}\sqrt{x^2\left(1 - \frac{a^2}{x^2}\right)}$$

$$= \pm\frac{b}{a}\sqrt{x^2}\sqrt{1 - \frac{a^2}{x^2}}$$

$$= \pm\frac{b}{a}x\sqrt{1 - \frac{a^2}{x^2}}$$

Now suppose that we are going to determine some y values for very large values of x. (Remember that a and b are arbitrary constants; they have specific values for a particular hyperbola.) When x is very large, a^2/x^2 will be close to zero, so the radicand will be close to 1. Therefore, the y value will be close to either $(b/a)x$ or $-(b/a)x$. In other words, as x becomes larger and larger, the point $P(x, y)$ gets closer and closer to either the line $y = (b/a)x$ or the line $y = -(b/a)x$. A corresponding situation occurs when $x \leq a$. The lines with equations

$$y = \pm\frac{b}{a}x$$

are called the **asymptotes** of the hyperbola.

As we mentioned earlier, the asymptotes are very helpful for sketching hyperbolas. An easy way to sketch the asymptotes is first to plot the vertices $A(a, 0)$ and $\underline{A'}(-a, 0)$, and the points $B(0, b)$ and $B'(0, -b)$, as in Figure 8.30. The line segment $\overline{BB'}$ is of length $2b$ and is called the **conjugate axis** of the hyperbola. The horizontal line segments drawn through B and B', together with the vertical line segments drawn through A and A', form a rectangle. The diagonals of this rectangle have slopes b/a and $-(b/a)$. Therefore, by extending the diagonals, we obtain the asymptotes $y = (b/a)x$ and $y = -(b/a)x$. The two branches of the hyperbola can be sketched by using the asymptotes as guidelines, as shown in Figure 8.30.

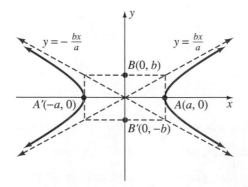

FIGURE 8.30

EXAMPLE 1

Find the vertices, the foci, and the equations of the asymptotes of the hyperbola $9x^2 - 4y^2 = 36$, and sketch the hyperbola.

Solution

Dividing both sides of the given equation by 36 and simplifying, we change the equation to the standard form.

$$\frac{x^2}{4} - \frac{y^2}{9} = 1$$

where $a^2 = 4$ and $b^2 = 9$. Hence $a = 2$ and $b = 3$. The vertices are $(\pm 2, 0)$ and the endpoints of the conjugate axis are $(0, \pm 3)$; these points determine the rectangle whose diagonals extend to become the asymptotes. Using $a = 2$ and $b = 3$, the equations of the asymptotes are $y = \frac{3}{2}x$ and $y = -\frac{3}{2}x$. Then, using the relationship $c^2 = a^2 + b^2$, we obtain $c^2 = 4 + 9 = 13$. Thus the foci are at $\left(\sqrt{13}, 0\right)$ and $\left(-\sqrt{13}, 0\right)$. Using the vertices and the asymptotes, we have sketched the hyperbola in Figure 8.31.

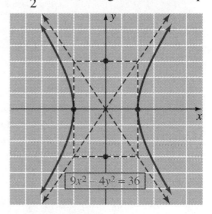

$9x^2 - 4y^2 = 36$

FIGURE 8.31 ■

EXAMPLE 2

Find the equation of the hyperbola with vertices at $(\pm 4, 0)$ and foci at $\left(\pm 2\sqrt{5}, 0\right)$.

Solution

From the given information, we know that $a = 4$ and $c = 2\sqrt{5}$. Then using the relationship $b^2 = c^2 - a^2$, we obtain

$$b^2 = \left(2\sqrt{5}\right)^2 - 4^2 = 20 - 16 = 4$$

Substituting 16 for a^2 and 4 for b^2 in the standard form produces

$$\frac{x^2}{16} - \frac{y^2}{4} = 1$$

Multiplying both sides of this equation by 16 yields

$$x^2 - 4y^2 = 16$$

■

Hyperbolas with Foci on the y Axis

In a similar fashion, we can develop a standard form for the equation of a hyperbola with foci on the y axis. The following property summarizes the results of such a development, where the foci are at $(0, c)$ and $(0, -c)$.

PROPERTY 8.6

The graph of the equation

$$\frac{y^2}{a^2} - \frac{x^2}{b^2} = 1$$

is a hyperbola with vertices at $(0, a)$ and $(0, -a)$. The foci are at $(0, c)$ and $(0, -c)$, where $c^2 = a^2 + b^2$.

For this type of hyperbola, the endpoints of the conjugate axis are at $(b, 0)$ and $(-b, 0)$. In this case we can find the asymptotes by extending the diagonals of the rectangle determined by the horizontal lines through the vertices and the vertical lines through the endpoints of the conjugate axis. The slopes of these diagonals are a/b and $-a/b$; thus the equations of these asymptotes are

$$y = \frac{a}{b}x \qquad \text{and} \qquad y = -\frac{a}{b}x$$

EXAMPLE 3

Find the vertices, the foci, and the equations of the asymptotes of the hyperbola $4y^2 - x^2 = 12$, and sketch the hyperbola.

Solution

Divide both sides of the given equation by 12 to change the equation to the standard form.

$$\frac{y^2}{3} - \frac{x^2}{12} = 1$$

where $a^2 = 3$ and $b^2 = 12$. Hence $a = \sqrt{3}$ and $b = 2\sqrt{3}$. The vertices, $\left(0, \pm\sqrt{3}\right)$, and the endpoints of the conjugate axis, $\left(\pm 2\sqrt{3}, 0\right)$, determine the rectangle whose diagonals extend to become the asymptotes. Using $a = 3$ and $b = 2\sqrt{3}$, the equations of the asymptotes are $y = \left(\sqrt{3}/2\sqrt{3}\right)x = \frac{1}{2}x$ and $y = -\frac{1}{2}x$. Then, using the relationship $c^2 = a^2 + b^2$, we obtain $c^2 = 3 + 12 = 15$. So the foci are at $\left(0, \sqrt{15}\right)$ and $\left(0, -\sqrt{15}\right)$. The hyperbola is sketched in Figure 8.32.

$$4y^2 - x^2 = 12$$

FIGURE 8.32 ■

Other Hyperbolas

In the same way, we can develop the standard form for an equation of a hyperbola that is symmetric with respect to a line parallel to a coordinate axis. We will not show such developments in this text but will simply state and use the results.

$$\frac{(x-h)^2}{a^2} - \frac{(y-k)^2}{b^2} = 1$$ A hyperbola with center at (h, k) and transverse axis on the horizontal line $y = k$

$$\frac{(y-k)^2}{a^2} - \frac{(x-h)^2}{b^2} = 1$$ A hyperbola with center at (h, k) and transverse axis on the vertical line $x = h$

The relationship $c^2 = a^2 + b^2$ still holds, and the physical significance of a, b, and c remains the same. However, these values are used relative to the center (h, k) to find the endpoints of the transverse and conjugate axes and to find the foci. Furthermore, the slopes of the asymptotes are as before, but these lines now contain the new center, (h, k). Let's see how all of this works in a specific example.

EXAMPLE 4

Find the vertices, the foci, and the equations of the asymptotes of the hyperbola $9x^2 - 36x - 16y^2 + 96y - 252 = 0$, and sketch the hyperbola.

Solution

First, we need to change to a standard form by completing the square on both x and y.

$$9(x^2 - 4x + \underline{\quad}) - 16(y^2 - 6y + \underline{\quad}) = 252$$
$$9(x^2 - 4x + 4) - 16(y^2 - 6y + 9) = 252 + 36 - 144$$
$$9(x - 2)^2 - 16(y - 3)^2 = 144$$
$$\frac{(x - 2)^2}{16} - \frac{(y - 3)^2}{9} = 1$$

The center is at (2, 3) and the transverse axis is on the line $y = 3$. Because $a^2 = 16$, we know that $a = 4$. Therefore, the vertices are four units to the right and four units to the left of the center, (2, 3), so they are at (6, 3) and (−2, 3). Likewise, because $b^2 = 9$, or $b = 3$, the endpoints of the conjugate axis are three units up and three units down from the center, so they are at (2, 6) and (2, 0). Using $a = 4$ and $b = 3$, the slopes of the asymptotes are $\frac{3}{4}$ and $-\frac{3}{4}$. Then, using the slopes, the center (2, 3), and the point–slope form for writing the equation of a line, we can determine the equations of the asymptotes to be $3x - 4y = -6$ and $3x + 4y = 18$. From the relationship $c^2 = a^2 + b^2$ we obtain $c^2 = 16 + 9 = 25$. Thus the foci are at (7, 3) and (−3, 3). The hyperbola is sketched in Figure 8.33.

$$9x^2 - 36x - 16y^2 + 96y - 252 = 0$$

F I G U R E   8 . 3 3　　　　　　　　　　　　　　　　　■

EXAMPLE 5

Find the equation of the hyperbola with vertices at (−4, 2) and (−4, −4) and with foci at (−4, 3) and (−4, −5).

Solution

Because the vertices and foci are on the same vertical line ($x = -4$), this hyperbola has an equation of the form

$$\frac{(y - k)^2}{a^2} - \frac{(x - h)^2}{b^2} = 1$$

The center of the hyperbola is at the midpoint of the transverse axis. Therefore,

$$h = \frac{-4 + (-4)}{2} = -4 \qquad \text{and} \qquad k = \frac{2 + (-4)}{2} = -1$$

The distance between the center, (−4, −1) and a vertex, (−4, 2) is three units, so $a = 3$. The distance between the center, (−4, −1), and a focus, (−4, 3), is four units, so $c = 4$. Then, using the relationship $c^2 = a^2 + b^2$, we obtain

$$b^2 = c^2 - a^2 = 16 - 9 = 7$$

Now we can substitute -4 for h, -1 for k, 9 for a^2, and 7 for b^2 in the general form and simplify.

$$\frac{(y+1)^2}{9} - \frac{(x+4)^2}{7} = 1$$

$$7(y+1)^2 - 9(x+4)^2 = 63$$

$$7(y^2 + 2y + 1) - 9(x^2 + 8x + 16) = 63$$

$$7y^2 + 14y + 7 - 9x^2 - 72x - 144 = 63$$

$$7y^2 + 14y - 9x^2 - 72x - 200 = 0 \qquad \blacksquare$$

The hyperbola also has numerous applications, including many you may not be aware of. For example, one method of artillery range-finding is based on the concept of a hyperbola. If each of two listening posts, P_1 and P_2 in Figure 8.34 records the time that an artillery blast is heard, then the difference between the times multiplied by the speed of sound gives the difference of the distances of the gun from the two fixed points. Thus the gun is located somewhere on the hyperbola whose foci are the two listening posts. By bringing in a third listening post, P_3, we can form another hyperbola with foci at P_2 and P_3. Then the location of the gun must be at one of the intersections of the two hyperbolas.

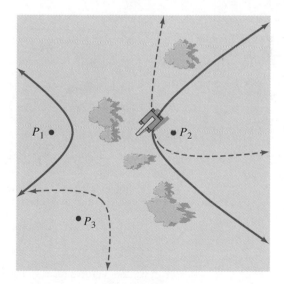

FIGURE 8.34

This same principle of intersecting hyperbolas is used in a long-range navigation system known as LORAN. Radar stations serve as the foci of the hyperbolas, and of course computers are used for the many calculations that are necessary to fix the location of a plane or ship. At the present time, LORAN is probably used mostly for coastal navigation in connection with small pleasure boats.

Some rather unique architectural creations have used the concept of a hyperbolic paraboloid, pictured in Figure 8.35. For example, the TWA building at Kennedy Airport is so designed. Some comets, upon entering the sun's gravitational field, follow a hyperbolic path, with the sun as one of the foci (see Figure 8.36).

FIGURE 8.35

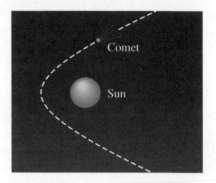

FIGURE 8.36

PROBLEM SET 8.3

For Problems 1–22, find the vertices, the foci, and the equations of the asymptotes, and sketch each hyperbola.

1. $\dfrac{x^2}{9} - \dfrac{y^2}{4} = 1$

2. $\dfrac{x^2}{4} - \dfrac{y^2}{16} = 1$

3. $\dfrac{y^2}{4} - \dfrac{x^2}{9} = 1$

4. $\dfrac{y^2}{16} - \dfrac{x^2}{4} = 1$

5. $9y^2 - 16x^2 = 144$

6. $4y^2 - x^2 = 4$

7. $x^2 - y^2 = 9$

8. $x^2 - y^2 = 1$

9. $5y^2 - x^2 = 25$

10. $y^2 - 2x^2 = 8$

11. $y^2 - 9x^2 = -9$

12. $16y^2 - x^2 = -16$

13. $4x^2 - 24x - 9y^2 - 18y - 9 = 0$

14. $9x^2 + 72x - 4y^2 - 16y + 92 = 0$

15. $y^2 - 4y - 4x^2 - 24x - 36 = 0$

16. $9y^2 + 54y - x^2 + 6x + 63 = 0$

17. $2x^2 - 8x - y^2 + 4 = 0$

18. $x^2 + 6x - 3y^2 = 0$

19. $y^2 + 10y - 9x^2 + 16 = 0$

20. $4y^2 - 16y - x^2 + 12 = 0$

21. $x^2 + 4x - y^2 - 4y - 1 = 0$

22. $y^2 + 8y - x^2 + 2x + 14 = 0$

For Problems 23–38, find an equation of the hyperbola that satisfies the given conditions.

23. Vertices $(\pm 2, 0)$, foci $(\pm 3, 0)$

24. Vertices $(\pm 1, 0)$, foci $(\pm 4, 0)$

25. Vertices $(0, \pm 3)$, foci $(0, \pm 5)$

26. Vertices $(0, \pm 2)$, foci $(0, \pm 6)$

27. Vertices $(\pm 1, 0)$, contains the point $(2, 3)$

28. Vertices $(0, \pm 1)$, contains the point $(-3, 5)$

29. Vertices $\left(0, \pm\sqrt{3}\right)$, length of conjugate axis is 4

30. Vertices $\left(\pm\sqrt{5}, 0\right)$, length of conjugate axis is 6

31. Foci $\left(\pm\sqrt{23}, 0\right)$, length of transverse axis is 8

32. Foci $\left(0, \pm 3\sqrt{2}\right)$, length of conjugate axis is 4

33. Vertices $(6, -3)$ and $(2, -3)$, foci $(7, -3)$ and $(1, -3)$

34. Vertices $(-7, -4)$ and $(-5, -4)$, foci $(-8, -4)$ and $(-4, -4)$

35. Vertices $(-3, 7)$ and $(-3, 3)$, foci $(-3, 9)$ and $(-3, 1)$

36. Vertices $(7, 5)$ and $(7, -1)$, foci $(7, 7)$ and $(7, -3)$

37. Vertices $(0, 0)$ and $(4, 0)$, foci $(5, 0)$ and $(-1, 0)$

38. Vertices $(0, 0)$ and $(0, -6)$, foci $(0, 2)$ and $(0, -8)$

For Problems 39–48, identify the graph of each of the equations as a straight line, a circle, a parabola, an ellipse, or a hyperbola. Do not sketch the graphs.

39. $x^2 - 7x + y^2 + 8y - 2 = 0$

40. $x^2 - 7x - y^2 + 8y - 2 = 0$

41. $5x - 7y = 9$

42. $4x^2 - x + y^2 + 2y - 3 = 0$

43. $10x^2 + y^2 = 8$

44. $-3x - 2y = 9$

45. $5x^2 + 3x - 2y^2 - 3y - 1 = 0$

46. $x^2 + y^2 - 3y - 6 = 0$

47. $x^2 - 3x + y - 4 = 0$

48. $5x + y^2 - 2y - 1 = 0$

49. What is the difference between the graphs of the equations $x^2 + y^2 = 0$ and $x^2 - y^2 = 0$?

50. What is the difference between the graphs of the equations $4x^2 + 9y^2 = 0$ and $9x^2 + 4y^2 = 0$?

51. A flashlight produces a "cone of light" that can be cut by the plane of a wall to illustrate the conic sections. Try shining a flashlight against a wall (stand within a couple of feet of the wall) at different angles to produce a circle, an ellipse, a parabola, and one branch of a hyperbola. (You may find it difficult to distinguish between a parabola and a branch of a hyperbola.) Write a paragraph explaining this experiment to someone else.

 GRAPHING CALCULATOR ACTIVITIES

52. Use a graphing calculator to check your graphs for Problems 13–22. Be sure to graph the asymptotes for each hyperbola.

53. Use a graphing calculator to check your answers for Problems 39–48.

8.4 **SYSTEMS INVOLVING NONLINEAR EQUATIONS**

In Chapters 6 and 7, we used several techniques to solve systems of linear equations. We will use two of those techniques in this section to solve some systems that contain at least one nonlinear equation. Furthermore, we will use our knowledge of graphing lines, circles, parabolas, ellipses, and hyperbolas to get a pictorial view of the systems. That will give us a basis for predicting approximate real number solutions if there are any. In other words, we have once again arrived at a topic that vividly illustrates the merging of mathematical ideas. Let's begin by considering a system that contains one linear and one nonlinear equation.

EXAMPLE 1

Solve the system $\begin{pmatrix} x^2 + y^2 = 13 \\ 3x + 2y = 0 \end{pmatrix}$.

Solution

From our previous graphing experiences, we should recognize that $x^2 + y^2 = 13$ is a circle and $3x + 2y = 0$ is a straight line. Thus the system can be pictured as in Figure 8.37. The graph indicates that the solution set of this system should consist of two ordered pairs of real numbers, which represent the points of intersection in the second and fourth quadrants.

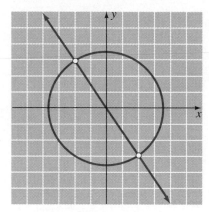

FIGURE 8.37

Now let's solve the system analytically by using the *substitution method*. Change the form of $3x + 2y = 0$ to $y = -3x/2$ and then substitute $-3x/2$ for y in the other equation to produce

$$x^2 + \left(-\frac{3x}{2}\right)^2 = 13$$

This equation can now be solved for x.

$$x^2 + \frac{9x^2}{4} = 13$$
$$4x^2 + 9x^2 = 52$$
$$13x^2 = 52$$
$$x^2 = 4$$
$$x = \pm 2$$

Substitute 2 for x and then -2 for x in the second equation of the system to produce two values for y.

$$3x + 2y = 0 \qquad\qquad 3x + 2y = 0$$
$$3(2) + 2y = 0 \qquad\quad 3(-2) + 2y = 0$$
$$2y = -6 \qquad\qquad\quad 2y = 6$$
$$y = -3 \qquad\qquad\quad y = 3$$

Therefore, the solution set of the system is $\{(2, -3), (-2, 3)\}$. ■

REMARK Don't forget that, as always, you can check the solutions by substituting them back into the original equations. Graphing the system permits you to approximate any possible real number solutions before solving the system. Then, after solving the system, you can use the graph again to check that the answers are reasonable.

EXAMPLE 2

Solve the system $\begin{pmatrix} x^2 + y^2 = 16 \\ y^2 - x^2 = 4 \end{pmatrix}$.

Solution

Graphing the system produces Figure 8.38. This figure indicates that there should be

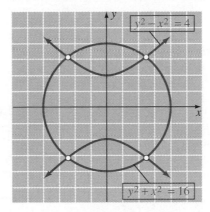

FIGURE 8.38

four ordered pairs of real numbers in the solution set of the system. Solving the system by using the *elimination method* works nicely. We can simply add the two equations, which eliminates the x's.

$$x^2 + y^2 = 16$$
$$\underline{-x^2 + y^2 = 4}$$
$$2y^2 = 20$$
$$y^2 = 10$$
$$y = \pm\sqrt{10}$$

Substituting $\sqrt{10}$ for y in the first equation yields

$$x^2 + y^2 = 16$$
$$x^2 + \left(\sqrt{10}\right)^2 = 16$$
$$x^2 + 10 = 16$$
$$x^2 = 6$$
$$x = \pm\sqrt{6}$$

Thus $\left(\sqrt{6}, \sqrt{10}\right)$ and $\left(-\sqrt{6}, \sqrt{10}\right)$ are solutions. Substituting $-\sqrt{10}$ for y in the first equation yields

$$x^2 + y^2 = 16$$
$$x^2 + \left(-\sqrt{10}\right)^2 = 16$$
$$x^2 + 10 = 16$$
$$x^2 = 6$$
$$x = \pm\sqrt{6}$$

Thus $\left(\sqrt{6}, -\sqrt{10}\right)$ and $\left(-\sqrt{6}, -\sqrt{10}\right)$ are also solutions. The solution set is $\left\{\left(-\sqrt{6}, \sqrt{10}\right), \left(-\sqrt{6}, -\sqrt{10}\right), \left(\sqrt{6}, \sqrt{10}\right), \left(\sqrt{6}, -\sqrt{10}\right)\right\}$. ∎

Sometimes a sketch of the graph of a system may not clearly indicate whether the system contains any real number solutions. The next example illustrates such a situation.

EXAMPLE 3

Solve the system $\begin{pmatrix} y = x^2 + 2 \\ 6x - 4y = -5 \end{pmatrix}$.

Solution

From previous graphing experiences, we recognize that $y = x^2 + 2$ is the basic parabola shifted upward two units and $6x - 4y = -5$ is a straight line (see Figure 8.39). Because of the close proximity of the curves, it is difficult to tell whether they intersect. In other words, the graph does not definitely indicate any real number solutions for the system.

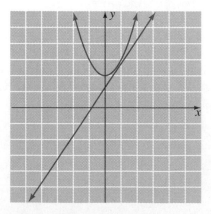

FIGURE 8.39

Let's solve the system by using the substitution method. We can substitute $x^2 + 2$ for y in the second equation, which produces two values for x.

$$6x - 4(x^2 + 2) = -5$$
$$6x - 4x^2 - 8 = -5$$
$$-4x^2 + 6x - 3 = 0$$
$$4x^2 - 6x + 3 = 0$$
$$x = \frac{6 \pm \sqrt{36 - 48}}{8}$$
$$x = \frac{6 \pm \sqrt{-12}}{8}$$
$$x = \frac{6 \pm 2i\sqrt{3}}{8}$$
$$x = \frac{3 \pm i\sqrt{3}}{4}$$

It is now obvious that the system has no real number solutions. That is, the line and the parabola do not intersect in the real number plane. However, there will be two pairs of complex numbers in the solution set. We can substitute $(3 + i\sqrt{3})/4$ for x in the first equation.

$$y = \left(\frac{3 + i\sqrt{3}}{4}\right)^2 + 2$$
$$= \frac{6 + 6i\sqrt{3}}{16} + 2$$
$$= \frac{6 + 6i\sqrt{3} + 32}{16}$$
$$= \frac{38 + 6i\sqrt{3}}{16} = \frac{19 + 3i\sqrt{3}}{8}$$

Likewise, we can substitute $(3 - i\sqrt{3})/4$ for x in the first equation.

$$y = \left(\frac{3 - i\sqrt{3}}{4}\right)^2 + 2$$
$$= \frac{6 - 6i\sqrt{3}}{16} + 2$$
$$= \frac{6 - 6i\sqrt{3} + 32}{16}$$
$$= \frac{38 - 6i\sqrt{3}}{16} = \frac{19 - 3i\sqrt{3}}{8}$$

The solution set is $\left\{\left(\dfrac{3 + i\sqrt{3}}{4}, \dfrac{19 + 3i\sqrt{3}}{8}\right), \left(\dfrac{3 - i\sqrt{3}}{4}, \dfrac{19 - 3i\sqrt{3}}{8}\right)\right\}$. ■

In Example 3 the use of a graphing utility may not, at first, indicate whether the system has any real number solutions. Suppose that we graph the system using a viewing rectangle such that $-15 \leq x \leq 15$ and $-10 \leq y \leq 10$. As shown in the display in Figure 8.40, we cannot tell whether the line and the parabola intersect.

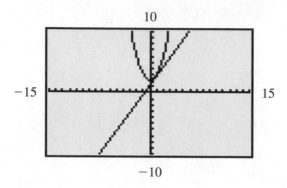

F I G U R E   8 . 4 0

However, if we change the viewing rectangle so that $0 \leq x \leq 2$ and $0 \leq y \leq 4$, as shown in Figure 8.41, then it becomes apparent that the two graphs do not intersect.

F I G U R E   8 . 4 1

EXAMPLE 4

Find the real number solutions for the system

$$\left(\begin{array}{l} y = \log_2(x - 3) - 2 \\ y = -\log_2 x \end{array} \right)$$

Solution

First, let's use a graphing calculator to obtain a graph of the system as shown in Figure 8.42. The two curves appear to intersect at approximately $x = 4$ and $y = -2$. To solve the system algebraically, we can equate the two expressions for y and solve the resulting equation for x.

F I G U R E 8 . 4 2

$$\log_2(x - 3) - 2 = -\log_2 x$$
$$\log_2 x + \log_2(x - 3) = 2$$
$$\log_2 x(x - 3) = 2$$

At this step we can either change to exponential form or rewrite 2 as $\log_2 4$.

$$\log_2 x(x - 3) = \log_2 4$$
$$x(x - 3) = 4$$
$$x^2 - 3x - 4 = 0$$
$$(x - 4)(x + 1) = 0$$

$$x - 4 = 0 \quad \text{or} \quad x + 1 = 0$$
$$x = 4 \quad \text{or} \quad x = -1$$

Because logarithms are not defined for negative numbers, -1 is discarded. Therefore, if $x = 4$, then

$$y = -\log_2 x$$

becomes

$$y = -\log_2 4$$
$$= -2$$

Therefore, the solution set is $\{(4, -2)\}$. ■

PROBLEM SET 8.4

For Problems 1–30, (a) graph the system so that approximate real number solutions (if there are any) can be predicted, and (b) solve the system by the substitution or elimination method.

1. $\begin{pmatrix} x^2 + y^2 = 5 \\ x + 2y = 5 \end{pmatrix}$

2. $\begin{pmatrix} x^2 + y^2 = 13 \\ 2x + 3y = 13 \end{pmatrix}$

3. $\begin{pmatrix} x^2 + y^2 = 26 \\ x + y = -4 \end{pmatrix}$

4. $\begin{pmatrix} x^2 + y^2 = 10 \\ x + y = -2 \end{pmatrix}$

5. $\begin{pmatrix} x^2 + y^2 = 2 \\ x - y = 4 \end{pmatrix}$

6. $\begin{pmatrix} x^2 + y^2 = 3 \\ x - y = -5 \end{pmatrix}$

7. $\begin{pmatrix} y = x^2 + 6x + 7 \\ 2x + y = -5 \end{pmatrix}$

8. $\begin{pmatrix} y = x^2 - 4x + 5 \\ y - x = 1 \end{pmatrix}$

9. $\begin{pmatrix} 2x + y = -2 \\ y = x^2 + 4x + 7 \end{pmatrix}$

10. $\begin{pmatrix} 2x + y = 0 \\ y = -x^2 + 2x - 4 \end{pmatrix}$

11. $\begin{pmatrix} y = x^2 - 3 \\ x + y = -4 \end{pmatrix}$

12. $\begin{pmatrix} y = -x^2 + 1 \\ x + y = 2 \end{pmatrix}$

13. $\begin{pmatrix} x^2 + 2y^2 = 9 \\ x - 4y = -9 \end{pmatrix}$

14. $\begin{pmatrix} 2x - y = 7 \\ 3x^2 + y^2 = 21 \end{pmatrix}$

15. $\begin{pmatrix} x + y = -3 \\ x^2 + 2y^2 - 12y - 18 = 0 \end{pmatrix}$

16. $\begin{pmatrix} 4x^2 + 9y^2 = 25 \\ 2x + 3y = 7 \end{pmatrix}$

17. $\begin{pmatrix} x - y = 2 \\ x^2 - y^2 = 16 \end{pmatrix}$

18. $\begin{pmatrix} x^2 - 4y^2 = 16 \\ 2y - x = 2 \end{pmatrix}$

19. $\begin{pmatrix} y = -x^2 + 3 \\ y = x^2 + 1 \end{pmatrix}$

20. $\begin{pmatrix} y = x^2 \\ y = x^2 - 4x + 4 \end{pmatrix}$

21. $\begin{pmatrix} y = x^2 + 2x - 1 \\ y = x^2 + 4x + 5 \end{pmatrix}$

22. $\begin{pmatrix} y = -x^2 + 1 \\ y = x^2 - 2 \end{pmatrix}$

23. $\begin{pmatrix} x^2 - y^2 = 4 \\ x^2 + y^2 = 4 \end{pmatrix}$

24. $\begin{pmatrix} 2x^2 + y^2 = 8 \\ x^2 + y^2 = 4 \end{pmatrix}$

25. $\begin{pmatrix} 8y^2 - 9x^2 = 6 \\ 8x^2 - 3y^2 = 7 \end{pmatrix}$

26. $\begin{pmatrix} 2x^2 + y^2 = 11 \\ x^2 - y^2 = 4 \end{pmatrix}$

27. $\begin{pmatrix} 2x^2 - 3y^2 = -1 \\ 2x^2 + 3y^2 = 5 \end{pmatrix}$

28. $\begin{pmatrix} 4x^2 + 3y^2 = 9 \\ y^2 - 4x^2 = 7 \end{pmatrix}$

29. $\begin{pmatrix} xy = 3 \\ 2x + 2y = 7 \end{pmatrix}$

30. $\begin{pmatrix} x^2 + 4y^2 = 25 \\ xy = 6 \end{pmatrix}$

For Problems 31–36, solve each system for all real number solutions.

31. $\begin{pmatrix} y = \log_3(x - 6) - 3 \\ y = -\log_3 x \end{pmatrix}$

32. $\begin{pmatrix} y = \log_{10}(x - 9) - 1 \\ y = -\log_{10} x \end{pmatrix}$

33. $\begin{pmatrix} y = e^x - 1 \\ y = 2e^{-x} \end{pmatrix}$

34. $\begin{pmatrix} y = 28 - 11e^x \\ y = -e^{2x} \end{pmatrix}$

35. $\begin{pmatrix} y = x^3 \\ y = x^3 + 2x^2 + 5x - 3 \end{pmatrix}$

36. $\begin{pmatrix} y = 3(4^x) - 8 \\ y = 4^{2x} - 2(4^x) - 4 \end{pmatrix}$

THOUGHTS INTO WORDS

37. What happens if you try to graph the system
$\begin{pmatrix} 7x^2 + 8y^2 = 36 \\ 11x^2 + 5y^2 = -4 \end{pmatrix}$?

38. For what value(s) of k will the line $x + y = k$ touch the ellipse $x^2 + 2y^2 = 6$ in one and only one point? Defend your answer.

39. The system
$$\begin{pmatrix} x^2 - 6x + y^2 - 4y + 4 = 0 \\ x^2 - 4x + y^2 + 8y - 5 = 0 \end{pmatrix}$$

represents two circles that intersect in two points. An equivalent system can be formed by replacing the second equation with the result of adding -1 times the first equation to the second equation. Thus we obtain the system

$$\begin{pmatrix} x^2 - 6x + y^2 - 4y + 4 = 0 \\ 2x + 12y - 9 = 0 \end{pmatrix}$$

Explain why the linear equation in this system is the equation of the common chord of the original two intersecting circles.

GRAPHING CALCULATOR ACTIVITIES

40. Graph the system of equations $\begin{pmatrix} y = x^2 + 2 \\ 6x - 4y = -5 \end{pmatrix}$ and use the trace and zoom features of your calculator to show that this system has no real number solutions.

41. Use a graphing calculator to graph the systems in Problems 31–36 and check the reasonableness of your answers to those problems.

For Problems 42–47, use a graphing calculator to approximate, to the nearest tenth, the real number solutions for each system of equations.

42. $\begin{pmatrix} y = e^x + 1 \\ y = x^3 + x^2 - 2x - 1 \end{pmatrix}$

43. $\begin{pmatrix} y = x^3 + 2x^2 - 3x + 2 \\ y = -x^3 - x^2 + 1 \end{pmatrix}$

44. $\begin{pmatrix} y = 2^x + 1 \\ y = 2^{-x} + 2 \end{pmatrix}$

45. $\begin{pmatrix} y = \ln(x - 1) \\ y = x^2 - 16x + 64 \end{pmatrix}$

46. $\begin{pmatrix} x = y^2 - 2y + 3 \\ x^2 + y^2 = 25 \end{pmatrix}$

47. $\begin{pmatrix} y^2 - x^2 = 16 \\ 2y^2 - x^2 = 8 \end{pmatrix}$

CHAPTER 8 SUMMARY

The following standard forms for the equations of conic sections were developed in this chapter.

Parabolas

$x^2 = 4py$ focus $(0, p)$, directrix $y = -p$, y-axis symmetry

$y^2 = 4px$ focus $(p, 0)$, directrix $x = -p$, x-axis symmetry

$(x - h)^2 = 4p(y - k)$ focus $(h, k + p)$, directrix $y = k - p$, symmetric with respect to the line $x = h$

$(y - k)^2 = 4p(x - h)$ focus $(h + p, k)$, directrix $x = h - p$, symmetric with respect to the line $y = k$

Ellipses

$\dfrac{x^2}{a^2} + \dfrac{y^2}{b^2} = 1$ center $(0, 0)$, vertices $(\pm a, 0)$, endpoints of minor axis $(0, \pm b)$, foci $(\pm c, 0)$, $c^2 = a^2 - b^2$, $a^2 > b^2$

$\dfrac{x^2}{b^2} + \dfrac{y^2}{a^2} = 1$ center $(0, 0)$, vertices $(0, \pm a)$, endpoints of minor axis $(\pm b, 0)$, foci $(0, \pm c)$, $c^2 = a^2 - b^2$, $a^2 > b^2$

$\dfrac{(x - h)^2}{a^2} + \dfrac{(y - k)^2}{b^2} = 1$ center (h, k), vertices $(h \pm a, k)$, endpoints of minor axis $(h, k \pm b)$, foci $(h \pm c, k)$, $c^2 = a^2 - b^2$, $a^2 > b^2$

$$\frac{(x - h)^2}{b^2} + \frac{(y - k)^2}{a^2} = 1$$

center (h, k), vertices $(h, k \pm a)$, endpoints of minor axis $(h \pm b, k)$, foci $(h, k \pm c)$, $c^2 = a^2 - b^2, a^2 > b^2$

Hyperbolas

$$\frac{x^2}{a^2} - \frac{y^2}{b^2} = 1$$

center $(0, 0)$, vertices $(\pm a, 0)$, endpoints of conjugate axis $(0, \pm b)$, foci $(\pm c, 0)$, $c^2 = a^2 + b^2$, asymptotes $y = \pm \dfrac{b}{a}x$

$$\frac{y^2}{a^2} - \frac{x^2}{b^2} = 1$$

center $(0, 0)$, vertices $(0, \pm a)$, endpoints of conjugate axis $(\pm b, 0)$, foci $(0, \pm c)$, $c^2 = a^2 + b^2$, asymptotes $y = \pm \dfrac{a}{b}x$

$$\frac{(x - h)^2}{a^2} - \frac{(y - k)^2}{b^2} = 1$$

center (h, k), vertices $(h \pm a, k)$, endpoints of conjugate axis $(h, k \pm b)$, foci $(h \pm c, k)$, $c^2 = a^2 + b^2$, asymptotes $y - k = \pm \dfrac{b}{a}(x - h)$

$$\frac{(y - k)^2}{a^2} - \frac{(x - h)^2}{b^2} = 1$$

center (h, k), vertices $(h, k \pm a)$, endpoints of conjugate axis $(h \pm b, k)$, foci $(h, k \pm c)$, $c^2 = a^2 + b^2$, asymptotes $y - k = \pm \dfrac{a}{b}(x - h)$

Systems that contain at least one nonlinear equation can often be solved by substitution or by the elimination method. Graphing the system will often provide a basis for predicting approximate real number solutions if there are any.

CHAPTER 8 REVIEW PROBLEM SET

For Problems 1–12, (a) identify the conic section as a parabola, an ellipse, or a hyperbola, (b) if it is a parabola, find its vertex, focus, and directrix; if it is an ellipse, find its vertices, the endpoints of its minor axis, and its foci; if it is a hyperbola, find its vertices, the endpoints of its conjugate axis, its foci, and its asymptotes, and (c) sketch each of the curves.

1. $x^2 + 2y^2 = 32$

2. $y^2 = -12x$

3. $3y^2 - x^2 = 9$

4. $2x^2 - 3y^2 = 18$

5. $5x^2 + 2y^2 = 20$

6. $x^2 = 2y$

7. $x^2 - 8x - 2y^2 + 4y + 10 = 0$

8. $9x^2 - 54x + 2y^2 + 8y + 71 = 0$

9. $y^2 - 2y + 4x + 9 = 0$

10. $x^2 + 2x + 8y + 25 = 0$

11. $x^2 + 10x + 4y^2 - 16y + 25 = 0$

12. $3y^2 + 12y - 2x^2 - 8x - 8 = 0$

For Problems 13–24, find the equation of the indicated conic section that satisfies the given conditions.

13. Parabola with vertex $(0, 0)$, focus $(-5, 0)$, directrix $x = 5$

14. Ellipse with vertices $(0, \pm 4)$, foci $\left(0, \pm \sqrt{15}\right)$

15. Hyperbola with vertices $\left(\pm \sqrt{2}, 0\right)$, length of conjugate axis 10

16. Ellipse with vertices $(\pm 2, 0)$, contains the point $(1, -2)$

17. Parabola with vertex $(0, 0)$, symmetric with respect to the y axis, contains the point $(2, 6)$

18. Hyperbola with vertices $(0, \pm 1)$, foci $\left(0, \pm \sqrt{10}\right)$

19. Ellipse with vertices $(6, 1)$, and $(6, 7)$, length of minor axis 2 units

20. Parabola with vertex $(4, -2)$, focus $(6, -2)$

21. Hyperbola with vertices $(-5, -3)$ and $(-5, -5)$, foci $(-5, -2)$ and $(-5, -6)$

22. Parabola with vertex $(-6, -3)$, symmetric with respect to the line $x = -6$, contains the point $(-5, -2)$

23. Ellipse with endpoints of minor axis $(-5, 2)$ and $(-5, -2)$, length of major axis 10 units

24. Hyperbola with vertices $(2, 0)$ and $(6, 0)$, length of conjugate axis 8 units

For Problems 25–30, (a) graph the system, and (b) solve the system by using the substitution or elimination method.

25. $\begin{pmatrix} x^2 + y^2 = 17 \\ x - 4y = -17 \end{pmatrix}$ **26.** $\begin{pmatrix} x^2 - y^2 = 8 \\ 3x - y = 8 \end{pmatrix}$

27. $\begin{pmatrix} x - y = 1 \\ y = x^2 + 4x + 1 \end{pmatrix}$ **28.** $\begin{pmatrix} 4x^2 - y^2 = 16 \\ 9x^2 + 9y^2 = 16 \end{pmatrix}$

29. $\begin{pmatrix} x^2 + 2y^2 = 8 \\ 2x^2 + 3y^2 = 12 \end{pmatrix}$ **30.** $\begin{pmatrix} y^2 - x^2 = 1 \\ 4x^2 + y^2 = 4 \end{pmatrix}$

CHAPTER 8 TEST

1. Find the focus of the parabola $x^2 = -20y$.

2. Find the vertex of the parabola $y^2 - 4y - 8x - 20 = 0$.

3. Find the equation of the directrix for the parabola $2y^2 = 24x$.

4. Find the focus of the parabola $y^2 = 24x$.

5. Find the vertex of the parabola $x^2 + 4x - 12y - 8 = 0$.

6. Find the equation of the directrix for the parabola $x^2 = -16y$.

7. Find the equation of the parabola that has its vertex at the origin, is symmetric with respect to the x axis, and contains the point $(-2, 4)$.

8. Find the equation of the parabola that has its vertex at $(3, 4)$ and its focus at $(3, 1)$.

9. Find the endpoints of the major axis of the ellipse $4x^2 + y^2 = 36$.

10. Find the length of the major axis of the ellipse $x^2 - 4x + 9y^2 - 18y + 4 = 0$.

11. Find the endpoints of the minor axis of the ellipse $9x^2 + 90x + 4y^2 - 8y + 193 = 0$.

12. Find the foci of the ellipse $x^2 + 4y^2 = 16$.

13. Find the center of the ellipse $3x^2 + 30x + y^2 - 16y + 79 = 0$.

14. Find the equation of the ellipse that has the endpoints of its major axis at $(0, \pm 10)$ and its foci at $(0, \pm 8)$.

15. Find the equation of the ellipse that has the endpoints of its major axis at $(2, -2)$ and $(10, -2)$ and the endpoints of its minor axis at $(6, 0)$ and $(6, -4)$.

16. Find the equations of the asymptotes of the hyperbola $4y^2 - 9x^2 = 32$.

17. Find the vertices of the hyperbola $y^2 - 6y - 3x^2 - 6x - 3 = 0$.

18. Find the foci of the hyperbola $5x^2 - 4y^2 = 20$.

19. Find the equation of the hyperbola that has its vertices at $(\pm 6, 0)$ and its foci at $(\pm 4\sqrt{3}, 0)$.

20. Find the equation of the hyperbola that has its vertices at $(0, 4)$ and $(-2, 4)$ and its foci at $(2, 4)$ and $(-4, 4)$.

21. How many real number solutions are there for the system $\left(\begin{array}{l} x^2 + y^2 = 16 \\ x^2 - 4y = 8 \end{array} \right)$?

22. Solve the system $\left(\begin{array}{l} x^2 + 4y^2 = 25 \\ \quad xy = 6 \end{array} \right)$.

For Problems 23–25, graph each conic section.

23. $y^2 + 4y + 8x - 4 = 0$

24. $9x^2 - 36x + 4y^2 + 16y + 16 = 0$

25. $x^2 + 6x - 3y^2 = 0$

9

SEQUENCES AND MATHEMATICAL INDUCTION

If you could get a job that pays only a penny the first day of your employment, but then doubles each succeeding day, by the 31st working day your salary would be $10,737,418.24.

*Suppose that an auditorium has 35 seats in the first row, 40 seats in the second row, 45 seats in the third row, and so on for ten rows. The numbers 35, 40, 45, 50, . . . , 80 represent the number of seats per row from row 1 through row 10. This list of numbers has a constant difference of 5 between any two successive numbers in the list; such a list is called an **arithmetic sequence**. (Used in this sense, the word* arithmetic *is pronounced with the accent on the syllable* met.*)*

*Suppose that a fungus culture growing under controlled conditions doubles in size each day. If today the size of the culture is 6 units, then the numbers 12, 24, 48, 96, 192 represent the size of the culture for the next 5 days. In this list of numbers, each number after the first is twice the previous number; such a list is called a **geometric sequence**. Arithmetic sequences and geometric sequences will be the center of our attention in this chapter.*

9.1 ARITHMETIC SEQUENCES

An **infinite sequence** is a function whose domain is the set of positive integers. For example, consider the function defined by the equation

$$f(n) = 5n + 1$$

where the domain is the set of positive integers. If we substitute the numbers of the domain in order, starting with 1, we can list the resulting ordered pairs:

$$(1, 6) \qquad (2, 11) \qquad (3, 16) \qquad (4, 21) \qquad (5, 26)$$

and so on. However, because we know we are using the domain of positive integers in order, starting with 1, there is no need to use ordered pairs. We can simply express the infinite sequence as

$$6, 11, 16, 21, 26, \ldots$$

Often the letter a is used to represent sequential functions, and the functional value of a at n is written a_n (this is read "a sub n") instead of $a(n)$. The sequence is then expressed

$$a_1, a_2, a_3, a_4, \ldots$$

where a_1 is the **first term**, a_2 is the **second term**, a_3 is the **third term**, and so on. The expression a_n, which defines the sequence, is called the **general term** of the sequence. Knowing the general term of a sequence enables us to find as many terms of the sequence as needed and also to find any specific terms. Consider the following example.

EXAMPLE 1

Find the first five terms of the sequence where $a_n = 2n^2 - 3$; find the 20th term.

Solution

The first five terms are generated by replacing n with 1, 2, 3, 4, and 5.

$$a_1 = 2(1)^2 - 3 = -1 \qquad a_2 = 2(2)^2 - 3 = 5$$
$$a_3 = 2(3)^2 - 3 = 15 \qquad a_4 = 2(4)^2 - 3 = 29$$
$$a_5 = 2(5)^2 - 3 = 47$$

The first five terms are thus $-1, 5, 15, 29$, and 47. The 20th term is

$$a_{20} = 2(20)^2 - 3 = 797$$

Arithmetic Sequences

An **arithmetic sequence** (also called an arithmetic progression) is a sequence that has a common difference between successive terms. The following are examples of arithmetic sequences.

$$1, 8, 15, 22, 29, \ldots$$
$$4, 7, 10, 13, 16, \ldots$$
$$4, 1, -2, -5, -8, \ldots$$
$$-1, -6, -11, -16, -21, \ldots$$

The common difference in the first sequence is 7. That is, $8 - 1 = 7$, $15 - 8 = 7$, $22 - 15 = 7$, $29 - 22 = 7$, and so on. The common differences for the next three sequences are 3, -3, and -5, respectively.

In a more general setting, we say that the sequence

$$a_1, a_2, a_3, a_4, \ldots, a_n, \ldots$$

is an arithmetic sequence if and only if there is a real number d such that

$$a_{k+1} - a_k = d$$

for every positive integer k. The number d is called the **common difference**.

From the definition we see that $a_{k+1} = a_k + d$. In other words, we can generate an arithmetic sequence that has a common difference of d by starting with a first term a_1 and then simply adding d to each successive term.

First term: a_1

Second term: $a_1 + d$

Third term: $a_1 + 2d$ $\qquad$ $(a_1 + d) + d = a_1 + 2d$

Fourth term: $a_1 + 3d$

$$\vdots$$

nth term: $a_1 + (n - 1)d$

Thus the **general term** of an arithmetic sequence is given by

$$a_n = a_1 + (n - 1)d$$

where a_1 is the first term and d is the common difference. This formula for the general term can be used to solve a variety of problems involving arithmetic sequences.

EXAMPLE 2

Find the general-term expression for the arithmetic sequence $6, 2, -2, -6, \ldots$.

Solution

The common difference, d, is $2 - 6 = -4$, and the first term, a_1, is 6. Substitute these values into $a_n = a_1 + (n - 1)d$ and simplify to obtain

$$\begin{aligned} a_n &= a_1 + (n - 1)d \\ &= 6 + (n - 1)(-4) \\ &= 6 - 4n + 4 \\ &= -4n + 10 \end{aligned}$$

■

EXAMPLE 3

Find the 40th term of the arithmetic sequence $1, 5, 9, 13, \ldots$.

Solution

Using $a_n = a_1 + (n - 1)d$, we obtain

$$\begin{aligned} a_{40} &= 1 + (40 - 1)4 \\ &= 1 + (39)(4) \\ &= 157 \end{aligned}$$

■

EXAMPLE 4

Find the first term of the arithmetic sequence where the fourth term is 26 and the ninth term is 61.

Solution

Using $a_n = a_1 + (n - 1)d$ with $a_4 = 26$ (the fourth term is 26) and $a_9 = 61$ (the ninth term is 61), we have

$$26 = a_1 + (4 - 1)d = a_1 + 3d$$
$$61 = a_1 + (9 - 1)d = a_1 + 8d$$

Solving the system of equations

$$\begin{pmatrix} a_1 + 3d = 26 \\ a_1 + 8d = 61 \end{pmatrix}$$

yields $a_1 = 5$ and $d - 7$. Thus the first term is 5.

■

Sums of Arithmetic Sequences

We often use sequences to solve problems, so we need to be able to find the sum of a certain number of terms of the sequence. Before we develop a general-sum formula for arithmetic sequences, let's consider an approach to a specific problem that we can then use in a general setting.

EXAMPLE 5

Find the sum of the first 100 positive integers.

Solution

We are being asked to find the sum of $1 + 2 + 3 + 4 + \cdots + 100$. Rather than adding in the usual way, lets find the sum in the following manner.

$$
\begin{array}{c}
1 + \quad 2 + \quad 3 + \quad 4 + \cdots + 100 \\
100 + \quad 99 + \quad 98 + \quad 97 + \cdots + \quad 1 \\
\hline
101 + 101 + 101 + 101 + \cdots + 101
\end{array}
$$

$$\frac{\overset{50}{\cancel{100}}(101)}{\cancel{2}} = 5050$$

Note that we simply wrote the indicated sum forward and backward, and then we added the results. In so doing, we produced 100 sums of 101, but half of them are repeats. For example, $100 + 1$ and $1 + 100$ are both counted in this process. Thus we divide the product $(100)(101)$ by 2, which yields the final result of 5050. ■

The *forward–backward* approach we used in Example 5 can be used to develop a formula for finding the sum of the first n terms of any arithmetic sequence. Consider an arithmetic sequence $a_1, a_2, a_3, a_4, \ldots, a_n$ with a common difference of d. Use S_n to represent the sum of the first n terms and proceed as follows.

$$S_n = a_1 + (a_1 + d) + (a_1 + 2d) + \cdots + (a_n - 2d) + (a_n - d) + a_n$$

Now write this sum in reverse.

$$S_n = a_n + (a_n - d) + (a_n - 2d) + \cdots + (a_1 + 2d) + (a_1 + d) + a_1$$

Add the two equations to produce

$$2S_n = (a_1 + a_n) + (a_1 + a_n) + (a_1 + a_n) + \cdots + (a_1 + a_n) + (a_1 + a_n) + (a_1 + a_n)$$

That is, we have n sums $a_1 + a_n$, so

$$2S_n = n(a_1 + a_n)$$

from which we obtain a **sum formula:**

$$S_n = \frac{n(a_1 + a_n)}{2}$$

Using the *n*th-term formula and/or the sum formula, we can solve a variety of problems involving arithmetic sequences.

EXAMPLE 6

Find the sum of the first 30 terms of the arithmetic sequence 3, 7, 11, 15, ...

Solution

Using $a_n = a_1 + (n - 1)d$, we can find the 30th term.

$$a_{30} = 3 + (30 - 1)4 = 3 + 29(4) = 119$$

Now we can use the sum formula.

$$S_{30} = \frac{30(3 + 119)}{2} = 1830$$

∎

EXAMPLE 7

Find the sum $7 + 10 + 13 + \cdots + 157$.

Solution

To use the sum formula, we need to know the number of terms. The *n*th-term formula will do that for us.

$$a_n = a_1 + (n - 1)d$$
$$157 = 7 + (n - 1)3$$
$$157 = 7 + 3n - 3$$
$$157 = 3n + 4$$
$$153 = 3n$$
$$51 = n$$

Now we can use the sum formula.

$$S_{51} = \frac{51(7 + 157)}{2} = 4182$$

∎

Keep in mind that we developed the sum formula for an arithmetic sequence by using the forward–backward technique, which we had previously used on a specific problem. Now that we have the sum formula, we have two choices when solving problems. We can either memorize the formula and use it or simply use the forward–backward technique. If you choose to use the formula and some day you forget it, don't panic. Just use the forward–backward technique. In other words, understanding the development of a formula often enables you to do problems even when you forget the formula itself.

Summation Notation

Sometimes a special notation is used to indicate the sum of a certain number of terms of a sequence. The capital Greek letter *sigma*, Σ, is used as a **summation symbol**. For example,

$$\sum_{i=1}^{5} a_i$$

represents the sum $a_1 + a_2 + a_3 + a_4 + a_5$. The letter i is frequently used as the **index of summation**; the letter i takes on all integer values from the lower limit to the upper limit, inclusive. Thus

$$\sum_{i=1}^{4} b_i = b_1 + b_2 + b_3 + b_4$$

$$\sum_{i=3}^{7} a_i = a_3 + a_4 + a_5 + a_6 + a_7$$

$$\sum_{i=1}^{15} i^2 = 1^2 + 2^2 + 3^2 + \cdots + 15^2$$

$$\sum_{i=1}^{n} a_i = a_1 + a_2 + a_3 + \cdots + a_n$$

If $a_1, a_2, a_3, \ldots$ represents an arithmetic sequence, we can now write the sum formula

$$\sum_{i=1}^{n} a_i = \frac{n}{2}(a_1 + a_n)$$

EXAMPLE 8

Find the sum $\sum_{i=1}^{50} (3i + 4)$.

Solution

This indicated sum means

$$\sum_{i=1}^{50} (3i + 4) = [3(1) + 4] + [3(2) + 4] + [3(3) + 4] + \cdots + [3(50) + 4]$$

$$= 7 + 10 + 13 + \cdots + 154$$

Because this is an indicated sum of an arithmetic sequence, we can use our sum formula.

$$S_{50} = \frac{50}{2}(7 + 154) = 4025$$

EXAMPLE 9

Find the sum $\sum\limits_{i=2}^{7} 2i^2$.

Solution

This indicated sum means

$$\sum_{i=2}^{7} 2i^2 = 2(2)^2 + 2(3)^2 + 2(4)^2 + 2(5)^2 + 2(6)^2 + 2(7)^2$$

$$= 8 + 18 + 32 + 50 + 72 + 98$$

This is not the indicated sum of an *arithmetic* sequence; therefore, let's simply add the numbers in the usual way. The sum is 278. ■

Example 9 suggests a word of caution. Be sure to analyze the sequence of numbers that is represented by the summation symbol. You may or may not be able to use a formula for adding the numbers.

PROBLEM SET 9.1

For Problems 1–10, write the first five terms of the sequence that has the indicated general term.

1. $a_n = 3n - 7$

2. $a_n = 5n - 2$

3. $a_n = -2n + 4$

4. $a_n = -4n + 7$

5. $a_n = 3n^2 - 1$

6. $a_n = 2n^2 - 6$

7. $a_n = n(n - 1)$

8. $a_n = (n + 1)(n + 2)$

9. $a_n = 2^{n+1}$

10. $a_n = 3^{n-1}$

11. Find the 15th and 30th terms of the sequence where $a_n = -5n - 4$.

12. Find the 20th and 50th terms of the sequence where $a_n = -n - 3$.

13. Find the 25th and 50th terms of the sequence where $a_n = (-1)^{n+1}$.

14. Find the 10th and 15th terms of the sequence where $a_n = -n^2 - 10$.

For Problems 15–24, find the general term (the *n*th term) for each arithmetic sequence.

15. $11, 13, 15, 17, 19, \ldots$

16. $7, 10, 13, 16, 19, \ldots$

17. $2, -1, -4, -7, -10, \ldots$

18. $4, 2, 0, -2, -4, \ldots$

19. $\dfrac{3}{2}, 2, \dfrac{5}{2}, 3, \dfrac{7}{2}, \ldots$

20. $0, \dfrac{1}{2}, 1, \dfrac{3}{2}, 2, \ldots$

21. $2, 6, 10, 14, 18, \ldots$

22. $2, 7, 12, 17, 22, \ldots$

23. $-3, -6, -9, -12, -15, \ldots$

24. $-4, -8, -12, -16, -20, \ldots$

For Problems 25–30, find the required term for each arithmetic sequence.

25. The 15th term of $3, 8, 13, 18, \ldots$

26. The 20th term of $4, 11, 18, 25, \ldots$

27. The 30th term of $15, 26, 37, 48, \ldots$

28. The 35th term of $9, 17, 25, 33, \ldots$

29. The 52nd term of $1, \dfrac{5}{3}, \dfrac{7}{3}, 3, \ldots$

30. The 47th term of $\dfrac{1}{2}, \dfrac{5}{4}, 2, \dfrac{11}{4}, \ldots$

For Problems 31–42, solve each problem.

31. If the 6th term of an arithmetic sequence is 12 and the 10th term is 16, find the first term.

32. If the 5th term of an arithmetic sequence is 14 and the 12th term is 42, find the first term.

33. If the 3rd term of an arithmetic sequence is 20 and the 7th term is 32, find the 25th term.

34. If the 5th term of an arithmetic sequence is −5 and the 15th term is −25, find the 50th term.

35. Find the sum of the first 50 terms of the arithmetic sequence 5, 7, 9, 11, 13, ...

36. Find the sum of the first 30 terms of the arithmetic sequence 0, 2, 4, 6, 8, ...

37. Find the sum of the first 40 terms of the arithmetic sequence 2, 6, 10, 14, 18, ...

38. Find the sum of the first 60 terms of the arithmetic sequence −2, 3, 8, 13, 18, ...

39. Find the sum of the first 75 terms of the arithmetic sequence 5, 2, −1, −4, −7, ...

40. Find the sum of the first 80 terms of the arithmetic sequence 7, 3, −1, −5, −9, ...

41. Find the sum of the first 50 terms of the arithmetic sequence $\frac{1}{2}, 1, \frac{3}{2}, 2, \frac{5}{2}, \ldots$.

42. Find the sum of the first 100 terms of the arithmetic sequence $-\frac{1}{3}, \frac{1}{3}, 1, \frac{5}{3}, \frac{7}{3}, \ldots$.

For Problems 43–50, find the indicated sum.

43. $1 + 5 + 9 + 13 + \cdots + 197$

44. $3 + 8 + 13 + 18 + \cdots + 398$

45. $2 + 8 + 14 + 20 + \cdots + 146$

46. $6 + 9 + 12 + 15 + \cdots + 93$

47. $(-7) + (-10) + (-13) + (-16) + \cdots + (-109)$

48. $(-5) + (-9) + (-13) + (-17) + \cdots + (-169)$

49. $(-5) + (-3) + (-1) + 1 + \cdots + 119$

50. $(-7) + (-4) + (-1) + 2 + \cdots + 131$

For Problems 51–58, solve each problem.

51. Find the sum of the first 200 odd whole numbers.

52. Find the sum of the first 175 positive even whole numbers.

53. Find the sum of all even numbers between 18 and 482, inclusive.

54. Find the sum of all odd numbers between 17 and 379, inclusive.

55. Find the sum of the first 30 terms of the arithmetic sequence with the general term $a_n = 5n - 4$.

56. Find the sum of the first 40 terms of the arithmetic sequence with the general term $a_n = 4n - 7$.

57. Find the sum of the first 25 terms of the arithmetic sequence with the general term $a_n = -4n - 1$.

58. Find the sum of the first 35 terms of the arithmetic sequence with the general term $a_n = -5n - 3$.

For Problems 59–70, find each sum.

59. $\displaystyle\sum_{i=1}^{45} (5i + 2)$

60. $\displaystyle\sum_{i=1}^{38} (3i + 6)$

61. $\displaystyle\sum_{i=1}^{30} (-2i + 4)$

62. $\displaystyle\sum_{i=1}^{40} (-3i + 3)$

63. $\displaystyle\sum_{i=4}^{32} (3i - 10)$

64. $\displaystyle\sum_{i=6}^{47} (4i - 9)$

65. $\displaystyle\sum_{i=10}^{20} 4i$

66. $\displaystyle\sum_{i=15}^{30} (-5i)$

67. $\displaystyle\sum_{i=1}^{5} i^2$

68. $\displaystyle\sum_{i=1}^{6} (i^2 + 1)$

69. $\displaystyle\sum_{i=3}^{8} (2i^2 + i)$

70. $\displaystyle\sum_{i=4}^{7} (3i^2 - 2)$

71. Before developing the formula $a_n = a_1 + (n - 1)d$, we stated the equation $a_{k+1} - a_k = d$. In your own words, explain what this equation says.

72. Explain how to find the sum $1 + 2 + 3 + 4 + \cdots + 175$ without using the sum formula.

73. Explain in words how to find the sum of the first n terms of an arithmetic sequence.

74. Explain how one can tell that a particular sequence is an arithmetic sequence.

Further Investigations

The general term of a sequence can consist of one expression for certain values of n and another expression (or expressions) for other values of n. That is, a **multiple description** of the sequence can be given. For example,

$$a_n = \begin{cases} 2n + 3 & \text{for } n \text{ odd} \\ 3n - 2 & \text{for } n \text{ even} \end{cases}$$

means that we use $a_n = 2n + 3$ for $n = 1, 3, 5, 7, \ldots$ and we use $a_n = 3n - 2$ for $n = 2, 4, 6, 8, \ldots$. The first six terms of this sequence are 5, 4, 9, 10, 13, and 16.

    For Problems 75–78, write the first six terms of each sequence.

75. $a_n = \begin{cases} 2n + 1 & \text{for } n \text{ odd} \\ 2n - 1 & \text{for } n \text{ even} \end{cases}$

76. $a_n = \begin{cases} \dfrac{1}{n} & \text{for } n \text{ odd} \\ n^2 & \text{for } n \text{ even} \end{cases}$

77. $a_n = \begin{cases} 3n + 1 & \text{for } n \leq 3 \\ 4n - 3 & \text{for } n > 3 \end{cases}$

78. $a_n = \begin{cases} 5n - 1 & \text{for } n \text{ a multiple of 3} \\ 2n & \text{otherwise} \end{cases}$

The multiple-description approach can also be used to give a **recursive description** for a sequence. A sequence is said to be **described recursively** if the first n terms are stated and then each succeeding term is defined as a function of one or more of the preceding terms. For example,

$$\begin{cases} a_1 = 2 \\ a_n = 2a_{n-1} & \text{for } n \geq 2 \end{cases}$$

means that the first term, a_1 is 2 and each succeeding term is 2 times the previous term. Thus the first six terms are 2, 4, 8, 16, 32, and 64.

For Problems 79–84, write the first six terms of each sequence.

79. $\begin{cases} a_1 = 4 \\ a_n = 3a_{n-1} & \text{for } n \geq 2 \end{cases}$

80. $\begin{cases} a_1 = 3 \\ a_n = a_{n-1} + 2 & \text{for } n \geq 2 \end{cases}$

81. $\begin{cases} a_1 = 1 \\ a_2 = 1 \\ a_n = a_{n-2} + a_{n-1} & \text{for } n \geq 3 \end{cases}$

82. $\begin{cases} a_1 = 2 \\ a_2 = 3 \\ a_n = 2a_{n-2} + 3a_{n-1} & \text{for } n \geq 3 \end{cases}$

83. $\begin{cases} a_1 = 3 \\ a_2 = 1 \\ a_n = (a_{n-1} - a_{n-2})^2 & \text{for } n \geq 3 \end{cases}$

84. $\begin{cases} a_1 = 1 \\ a_2 = 2 \\ a_3 = 3 \\ a_n = a_{n-1} + a_{n-2} + a_{n-3} & \text{for } n \geq 4 \end{cases}$

9.2 GEOMETRIC SEQUENCES

A **geometric sequence** or **geometric progression** is a sequence in which we obtain each term after the first by multiplying the preceding term by a common multiplier, called the **common ratio** of the sequence. We can find the common ratio of a geometric sequence by dividing any term (other than the first) by the preceding term. The following geometric sequences have common ratios of 3, 2, $\frac{1}{2}$, and -4, respectively.

$$1, 3, 9, 27, 81, \ldots$$
$$3, 6, 12, 24, 48, \ldots$$
$$16, 8, 4, 2, 1, \ldots$$
$$-1, 4, -16, 64, -256, \ldots$$

In a more general setting, we say that the sequence $a_1, a_2, a_3, \ldots, a_n, \ldots$ is a geometric sequence if and only if there is a nonzero real number r such that

$$a_{k+1} = ra_k$$

for every positive integer k. The nonzero real number r is called the common ratio of the sequence.

The previous equation can be used to generate a general geometric sequence that has a_1 as a first term and r as a common ratio. We can proceed as follows.

First term: a_1
Second term: $a_1 r$
Third term: $a_1 r^2$ $(a_1 r)(r) = a_1 r^2$
Fourth term: $a_1 r^3$

$$\vdots$$

nth term: $a_1 r^{n-1}$

Thus the **general term** of a geometric sequence is given by

$$a_n = a_1 r^{n-1}$$

where a_1 is the first term and r is the common ratio.

EXAMPLE 1

Find the general term for the geometric sequence $8, 16, 32, 64, \ldots$.

Solution

Using $a_n = a_1 r^{n-1}$, we obtain

$$a_n = 8(2)^{n-1} = (2^3)(2)^{n-1} = 2^{n+2}$$

■

EXAMPLE 2

Find the ninth term of the geometric sequence 27, 9, 3, 1, ...

Solution

Using $a_n = a_1 r^{n-1}$, we can find the ninth term as follows.

$$a_9 = 27\left(\frac{1}{3}\right)^{9-1} = 27\left(\frac{1}{3}\right)^8 = \frac{3^3}{3^8} = \frac{1}{3^5} = \frac{1}{243}$$

Sums of Geometric Sequences

As with arithmetic sequences, we often need to find the sum of a certain number of terms of a geometric sequence. Before we develop a general-sum formula for geometric sequences, let's consider an approach to a specific problem that we can then use in a general setting.

EXAMPLE 3

Find the sum $1 + 3 + 9 + 27 + \cdots + 6561$.

Solution

Let S represent the sum and proceed as follows.

$$S = 1 + 3 + 9 + 27 + \cdots + 6561 \qquad\qquad \textbf{(1)}$$
$$3S = 3 + 9 + 27 + \cdots + 6561 + 19683 \qquad \textbf{(2)}$$

Equation (2) is the result of multiplying equation (1) by the common ratio, 3. Subtracting equation (1) from equation (2) produces

$$2S = 19683 - 1 = 19682$$
$$S = 9841$$

Now let's consider a general geometric sequence $a_1, a_1 r, a_1 r^2, \ldots, a_1 r^{n-1}$. By applying a procedure similar to the one we used in Example 3, we can develop a formula for finding the sum of the first n terms of any geometric sequence. We let S_n represent the sum of the first n terms.

$$S_n = a_1 + a_1 r + a_1 r^2 + \cdots + a_1 r^{n-1} \qquad\qquad \textbf{(3)}$$

Next we multiply both sides of equation (3) by the common ratio r.

$$rS_n = a_1 r + a_1 r^2 + a_1 r^3 + \cdots + a_1 r^n \qquad\qquad \textbf{(4)}$$

We then subtract equation (3) from equation (4).

$$rS_n - S_n = a_1 r^n - a_1$$

When we apply the distributive property to the left side and then solve for S_n, we obtain

$$S_n(r - 1) = a_1 r^n - a_1$$
$$S_n = \frac{a_1 r^n - a_1}{r - 1}, \qquad r \neq 1$$

Therefore, the sum of the first n terms of a geometric sequence with a first term a_1 and a common ratio r is given by

$$S_n = \frac{a_1 r^n - a_1}{r - 1}, \qquad r \neq 1$$

EXAMPLE 4

Find the sum of the first eight terms of the geometric sequence $1, 2, 4, 8, \ldots$.

Solution

Use the sum formula to obtain

$$S_8 = \frac{1(2)^8 - 1}{2 - 1} = \frac{2^8 - 1}{1} = 255$$

∎

If the common ratio of a geometric sequence is less than 1, it may be more convenient to change the form of the sum formula. That is, the fraction

$$\frac{a_1 r^n - a_1}{r - 1}$$

can be changed to

$$\frac{a_1 - a_1 r^n}{1 - r}$$

by multiplying both the numerator and the denominator by -1. Thus, by using

$$S_n = \frac{a_1 - a_1 r^n}{1 - r}$$

we can sometimes avoid unnecessary work with negative numbers when $r < 1$, as the next example illustrates.

EXAMPLE 5

Find the sum $1 + \dfrac{1}{2} + \dfrac{1}{4} + \cdots + \dfrac{1}{256}$.

Solution A

To use the sum formula, we need to know the number of terms, which can be found by counting them or by applying the nth-term formula, as follows.

$$a_n = a_1 r^{n-1}$$
$$\frac{1}{256} = 1\left(\frac{1}{2}\right)^{n-1}$$
$$\left(\frac{1}{2}\right)^8 = \left(\frac{1}{2}\right)^{n-1}$$
$$8 = n - 1 \qquad \text{If } b^n = b^m, \text{ then } n = m.$$
$$9 = n$$

Now we use $n = 9$, $a_1 = 1$, and $r = \dfrac{1}{2}$ in the sum formula of the form

$$S_n = \frac{a_1 - a_1 r^n}{1 - r}$$

$$S_9 = \frac{1 - 1\left(\dfrac{1}{2}\right)^9}{1 - \dfrac{1}{2}} = \frac{1 - \dfrac{1}{512}}{\dfrac{1}{2}} = \frac{\dfrac{511}{512}}{\dfrac{1}{2}} = 1\frac{255}{256}$$ ∎

We can also do a problem like Example 5 without finding the number of terms; we use the general approach illustrated in Example 3. Solution B demonstrates this idea.

Solution B

Let S represent the desired sum.

$$S = 1 + \frac{1}{2} + \frac{1}{4} + \cdots + \frac{1}{256}$$

Multiply both sides by the common ratio, $\dfrac{1}{2}$.

$$\frac{1}{2} S = \frac{1}{2} + \frac{1}{4} + \frac{1}{8} + \cdots + \frac{1}{256} + \frac{1}{512}$$

Subtract the second equation from the first and solve for S.

$$\frac{1}{2}S = 1 - \frac{1}{512} = \frac{511}{512}$$

$$S = \frac{511}{256} = 1\frac{255}{256}$$ ∎

Summation notation can also be used to indicate the sum of a certain number of terms of a geometric sequence.

EXAMPLE 6

Find the sum $\displaystyle\sum_{i=1}^{10} 2^i$.

Solution

This indicated sum means

$$\sum_{i=1}^{10} 2^i = 2^1 + 2^2 + 2^3 + \cdots + 2^{10}$$

$$= 2 + 4 + 8 + \cdots + 1024$$

This is the indicated sum of a geometric sequence, so we can use the sum formula, with $a_1 = 2$, $r = 2$, and $n = 10$.

$$S_{10} = \frac{2(2)^{10} - 2}{2 - 1} = \frac{2(2^{10} - 1)}{1} = 2046 \qquad \blacksquare$$

The Sum of an Infinite Geometric Sequence

Let's take the formula

$$S_n = \frac{a_1 - a_1 r^n}{1 - r}$$

and rewrite the right side by applying the property

$$\frac{a - b}{c} = \frac{a}{c} - \frac{b}{c}$$

Thus we obtain

$$S_n = \frac{a_1}{1 - r} - \frac{a_1 r^n}{1 - r} \qquad\qquad (1)$$

Now let's examine the behavior of r^n for $|r| < 1$—that is, for $-1 < r < 1$. For example, suppose that $r = \frac{1}{2}$, then

$$r^2 = \left(\frac{1}{2}\right)^2 = \frac{1}{4} \qquad r^3 = \left(\frac{1}{2}\right)^3 = \frac{1}{8}$$

$$r^4 = \left(\frac{1}{2}\right)^4 = \frac{1}{16} \qquad r^4 = \left(\frac{1}{2}\right)^5 = \frac{1}{32}$$

and so on. We can make $\left(\frac{1}{2}\right)^n$ as close to zero as we please by choosing sufficiently large values for n. In general, for values of r such that $|r| < 1$, the expression r^n approaches zero as n gets larger and larger. Therefore, the fraction $a_1 r^n / (1 - r)$ in equation (1) approaches zero as n increases. We say that **the sum of the infinite geometric sequence is given by**

$$S_\infty = \frac{a_1}{1 - r}, \qquad |r| < 1$$

EXAMPLE 7

Find the sum of the infinite geometric sequence

$$1, \frac{1}{2}, \frac{1}{4}, \frac{1}{8}, \cdots$$

Solution

Because $a_1 = 1$ and $r = \dfrac{1}{2}$, we obtain

$$S_\infty = \frac{1}{1-\dfrac{1}{2}} = \frac{1}{\dfrac{1}{2}} = 2$$

When we state that $S_\infty = 2$ in Example 7, we mean that as we add more and more terms, the sum approaches 2. Observe what happens when we calculate the sum up to five terms.

First term: 1

Sum of first two terms: $1 + \dfrac{1}{2} = 1\dfrac{1}{2}$

Sum of first three terms: $1 + \dfrac{1}{2} + \dfrac{1}{4} = 1\dfrac{3}{4}$

Sum of first four terms: $1 + \dfrac{1}{2} + \dfrac{1}{4} + \dfrac{1}{8} = 1\dfrac{7}{8}$

Sum of first five terms: $1 + \dfrac{1}{2} + \dfrac{1}{4} + \dfrac{1}{8} + \dfrac{1}{16} = 1\dfrac{15}{16}$

If $|r| > 1$, the absolute value of r^n increases without bound as n increases. Consider the following two examples and note the unbounded growth of the absolute value of r^n.

Let $r = 3$. **Let $r = -2$.**

$r^2 = 3^2 = 9 \qquad r^2 = (-2)^2 = 4$

$r^3 = 3^3 = 27 \qquad r^3 = (-2)^3 = -8 \qquad |-8| = 8$

$r^4 = 3^4 = 81 \qquad r^4 = (-2)^4 = 16$

$r^5 = 3^5 = 243 \qquad r^5 = (-2)^5 = -32 \qquad |-32| = 32$

If $r = 1$, then $S_n = na_1$, and as n increases without bound, $|S_n|$ also increases without bound. If $r = -1$, then S_n will either be a_1 or 0. Therefore, we say that the sum of any infinite geometric sequence where $|r| \geq 1$ *does not exist*.

Repeating Decimals as Sums of Infinite Geometric Sequences

In Section 1.1, we defined rational numbers to be numbers that have either a terminating or a repeating decimal representation. For example,

$$2.23 \qquad 0.147 \qquad 0.\overline{3} \qquad 0.\overline{14} \qquad \text{and} \qquad 0.\overline{56}$$

are rational numbers. (Remember that $0.\overline{3}$ means 0.3333. . . .) Place value provides the basis for changing terminating decimals such as 2.23 and 0.147 to a/b form, where a and b are integers and $b \neq 0$.

$$2.23 = \frac{223}{100} \quad \text{and} \quad 0.147 = \frac{147}{1000}$$

However, changing repeating decimals to a/b form requires a different technique, and our work with sums of infinite geometric sequences provides the basis for one such approach. Consider the following examples.

EXAMPLE 8

Change $0.\overline{14}$ to a/b form, where a and b are integers and $b \neq 0$.

Solution

The repeating decimal $0.\overline{14}$ can be written as the indicated sum of an infinite geometric sequence with first term 0.14 and common ratio 0.01.

$$0.14 + 0.0014 + 0.000014 + \cdots$$

Using $S_\infty = a_1/(1 - r)$, we obtain

$$S_\infty = \frac{0.14}{1 - 0.01} = \frac{0.14}{0.99} = \frac{14}{99}$$

Thus $0.\overline{14} = \frac{14}{99}$. ∎

If the repeating block of digits does not begin immediately after the decimal point, as in $0.5\overline{6}$ we can make an adjustment in the technique we used in Example 8.

EXAMPLE 9

Change $0.5\overline{6}$ to a/b form, where a and b are integers and $b \neq 0$.

Solution

The repeating decimal $0.5\overline{6}$ can be written

$$(0.5) + (0.06 + 0.006 + 0.0006 + \cdots)$$

where

$$0.06 + 0.006 + 0.0006 + \cdots$$

is the indicated sum of the infinite geometric sequence with $a_1 = 0.06$ and $r = 0.1$. Therefore,

$$S_\infty = \frac{0.06}{1 - 0.1} = \frac{0.06}{0.9} = \frac{6}{90} = \frac{1}{15}$$

Now we can add 0.5 and $\frac{1}{15}$.

$$0.5\overline{6} = 0.5 + \frac{1}{15} = \frac{1}{2} + \frac{1}{15} = \frac{15}{30} + \frac{2}{30} = \frac{17}{30}$$ ∎

PROBLEM SET 9.2

For Problems 1–12, find the general term (the nth term) for each geometric sequence.

1. 3, 6, 12, 24, . . .

2. 2, 6, 18, 54, . . .

3. 3, 9, 27, 81, . . .

4. 2, 4, 8, 16, . . .

5. $\frac{1}{4}, \frac{1}{8}, \frac{1}{16}, \frac{1}{32}, \ldots$

6. 8, 4, 2, 1, . . .

7. 4, 16, 64, 256, . . .

8. $6, 2, \frac{2}{3}, \frac{2}{9}, \ldots$

9. 1, 0.3, 0.09, 0.027, . . .

10. 0.2, 0.04, 0.008, 0.0016, . . .

11. 1, -2, 4, -8, . . .

12. -3, 9, -27, 81, . . .

For Problems 13–20, find the required term for each geometric sequence.

13. The 8th term of $\frac{1}{2}$, 1, 2, 4, . . .

14. The 7th term of 2, 6, 18, 54, . . .

15. The 9th term of 729, 243, 81, 27, . . .

16. The 11th term of 768, 384, 192, 96, . . .

17. The 10th term of 1, -2, 4, -8, . . .

18. The 8th term of $-1, -\frac{3}{2}, -\frac{9}{4}, -\frac{27}{8}, \ldots$

19. The 8th term of $\frac{1}{2}, \frac{1}{6}, \frac{1}{18}, \frac{1}{54}, \ldots$

20. The 9th term of $\frac{16}{81}, \frac{8}{27}, \frac{4}{9}, \frac{2}{3}, \ldots$

For Problems 21–32, solve each problem.

21. Find the first term of the geometric sequence with 5th term $\frac{32}{3}$ and common ratio 2.

22. Find the first term of the geometric sequence with 4th term $\frac{27}{128}$ and common ratio $\frac{3}{4}$.

23. Find the common ratio of the geometric sequence with 3rd term 12 and 6th term 96.

24. Find the common ratio of the geometric sequence with 2nd term $\frac{8}{3}$ and 5th term $\frac{64}{81}$.

25. Find the sum of the first ten terms of the geometric sequence 1, 2, 4, 8, ...

26. Find the sum of the first seven terms of the geometric sequence 3, 9, 27, 81, ...

27. Find the sum of the first nine terms of the geometric sequence 2, 6, 18, 54, ...

28. Find the sum of the first ten terms of the geometric sequence 5, 10, 20, 40, ...

29. Find the sum of the first eight terms of the geometric sequence 8, 12, 18, 27, ...

30. Find the sum of the first eight terms of the geometric sequence 9, 12, 16, $\frac{64}{3}$, ...

31. Find the sum of the first ten terms of the geometric sequence -4, 8, -16, 32, ...

32. Find the sum of the first nine terms of the geometric sequence -2, 6, -18, 54, ...

For Problems 33–38, find each indicated sum.

33. $9 + 27 + 81 + \cdots + 729$

34. $2 + 8 + 32 + \cdots + 8192$

35. $4 + 2 + 1 + \cdots + \frac{1}{512}$

36. $1 + (-2) + 4 + \cdots + 256$

37. $(-1) + 3 + (-9) + \cdots + (-729)$

38. $16 + 8 + 4 + \cdots + \frac{1}{32}$

For Problems 39–44, find each indicated sum.

39. $\sum_{i=1}^{9} 2^{i-3}$

40. $\sum_{i=1}^{6} 3^i$

41. $\sum_{i=2}^{5} (-3)^{i+1}$

42. $\sum_{i=3}^{8} (-2)^{i-1}$ **43.** $\sum_{i=1}^{6} 3\left(\frac{1}{2}\right)^i$ **44.** $\sum_{i=1}^{5} 2\left(\frac{1}{3}\right)^i$

For Problems 45–56, find the sum of each infinite geometric sequence. If the sequence has no sum, so state.

45. $2, 1, \frac{1}{2}, \frac{1}{4}, \ldots$ **46.** $9, 3, 1, \frac{1}{3}, \ldots$

47. $1, \frac{2}{3}, \frac{4}{9}, \frac{8}{27}, \ldots$ **48.** $5, 3, \frac{9}{5}, \frac{27}{25}, \ldots$

49. $4, 8, 16, 32, \ldots$ **50.** $32, 16, 8, 4, \ldots$

51. $9, -3, 1, -\frac{1}{3}, \ldots$ **52.** $2, -6, 18, -54, \ldots$

53. $\frac{1}{2}, \frac{3}{8}, \frac{9}{32}, \frac{27}{128}, \ldots$ **54.** $4, -\frac{4}{3}, \frac{4}{9}, -\frac{4}{27}, \ldots$

55. $8, -4, 2, -1, \ldots$ **56.** $7, \frac{14}{5}, \frac{28}{25}, \frac{56}{125}, \ldots$

For Problems 57–68, change each repeating decimal to a/b form, where a and b are integers and $b \neq 0$. Express a/b in reduced form.

57. $0.\overline{3}$ **58.** $0.\overline{4}$ **59.** $0.\overline{26}$

60. $0.\overline{18}$ **61.** $0.\overline{123}$ **62.** $0.\overline{273}$

63. $0.2\overline{6}$ **64.** $0.4\overline{3}$ **65.** $0.21\overline{4}$

66. $0.37\overline{1}$ **67.** $2.\overline{3}$ **68.** $3.\overline{7}$

THOUGHTS INTO WORDS

69. Explain the difference between an arithmetic sequence and a geometric sequence.

70. What does it mean to say that the sum of the infinite geometric sequence $1, \frac{1}{2}, \frac{1}{4}, \frac{1}{8}, \cdots$ is 2?

71. What do we mean when we say that the infinite geometric sequence $1, 2, 4, 8, \cdots$ has no sum?

72. Why don't we discuss the sum of an infinite arithmetic sequence?

9.3   ANOTHER LOOK AT PROBLEM SOLVING

In the previous two sections, many of the exercises fell into one of the following four categories.

1. Find the nth term of an arithmetic sequence

$$a_n = a_1 + (n-1)d$$

2. Find the sum of the first n terms of an arithmetic sequence

$$S_n = \frac{n(a_1 + a_n)}{2}$$

3. Find the nth term of a geometric sequence

$$a_n = a_1 r^{n-1}$$

4. Find the sum of the first n terms of a geometric sequence

$$S_n = \frac{a_1 r^n - a_1}{r - 1}$$

In this section we want to use this knowledge of arithmetic sequences and geometric sequences to expand our problem-solving capabilities. Let's begin by restating some old problem-solving suggestions that continue to apply here; we will also consider some other suggestions that are directly related to problems that involve sequences of numbers. (We will indicate the new suggestions with an asterisk.)

Suggestions for Solving Word Problems

1. Read the problem carefully and make certain that you understand the meanings of all the words. Be especially alert for any technical terms used in the statement of the problem.

2. Read the problem a second time (perhaps even a third time) to get an overview of the situation being described and to determine the known facts, as well as what you are to find.

3. Sketch a figure, diagram, or chart that might be helpful in analyzing the problem.

*4. Write down the first few terms of the sequence to describe what is taking place in the problem. Be sure that you understand, term by term, what the sequence represents in the problem.

*5. Determine whether the sequence is arithmetic or geometric.

*6. Determine whether the problem is asking for a specific term of the sequence or for the sum of a certain number of terms.

*7. Carry out the necessary calculations and check your answer for reasonableness.

As we solve some problems, these suggestions will become more meaningful.

PROBLEM 1

Domenica started to work in 1975 at an annual salary of $14,500. She received a $1050 raise each year. What was her annual salary in 1984?

Solution

The following sequence represents her annual salary beginning in 1975.

$$14{,}500,\ 15{,}550,\ 16{,}600,\ 17{,}650,\ \ldots$$

This is an arithmetic sequence, with $a_1 = 14500$ and $d = 1050$. Because each term of the sequence represents her annual salary, we are looking for the tenth term.

$$a_{10} = 14{,}500 + (10 - 1)1050 = 14{,}500 + 9(1050) = 23{,}950$$

Her annual salary in 1984 was $23,950.   ∎

PROBLEM 2

An auditorium has 20 seats in the front row, 24 seats in the second row, 28 seats in the third row, and so on, for 15 rows. How many seats are there in the auditorium?

Solution

The following sequence represents the number of seats per row starting with the first row.

$$20, 24, 28, 32, \ldots$$

This is an arithmetic sequence, with $a_1 = 20$ and $d = 4$. Therefore, the 15th term, which represents the number of seats in the 15th row, is given by

$$a_{15} = 20 + (15 - 1)4 = 20 + 14(4) = 76$$

The total number of seats in the auditorium is represented by

$$20 + 24 + 28 + \cdots + 76$$

Use the sum formula for an arithmetic sequence to obtain

$$S_{15} = \frac{15}{2}(20 + 76) = 720$$

There are 720 seats in the auditorium. ∎

PROBLEM 3

Suppose that you save 25 cents the first day of a week, 50 cents the second day, and one dollar the third day and that you continue to double your savings each day. How much will you save on the seventh day? What will be your total savings for the week?

Solution

The following sequence represents your savings per day, expressed in cents.

$$25, 50, 100, \ldots$$

This is a geometric sequence, with $a_1 = 25$ and $r = 2$. Your savings on the seventh day is the seventh term of this sequence. Therefore, using $a_n = a_1 r^{n-1}$, we obtain

$$a_7 = 25(2)^6 = 1600$$

So you will save \$16 on the seventh day. Your total savings for the 7 days is given by

$$25 + 50 + 100 + \cdots + 1600$$

Use the sum formula for a geometric sequence to obtain

$$S_7 = \frac{25(2)^7 - 25}{2 - 1} = \frac{25(2^7 - 1)}{1} = 3175$$

Thus your savings for the entire week is \$31.75. ∎

PROBLEM 4

A pump is attached to a container for the purpose of creating a vacuum. For each stroke of the pump, $\frac{1}{4}$ of the air that remains in the container is removed. To the nearest tenth of a percent, how much of the air remains in the container after six strokes?

Solution

Let's draw a diagram to help with the analysis of this problem.

First stroke:	$\frac{1}{4}$ of the air is removed	$1 - \frac{1}{4} = \frac{3}{4}$ of the air remains
Second stroke:	$\frac{1}{4}\left(\frac{3}{4}\right) = \frac{3}{16}$ of the air is removed	$\frac{3}{4} - \frac{3}{16} = \frac{9}{16}$ of the air remains
Third stroke:	$\frac{1}{4}\left(\frac{9}{16}\right) = \frac{9}{64}$ of the air is removed	$\frac{9}{16} - \frac{9}{64} = \frac{27}{64}$ of the air remains

The diagram suggests two approaches to the problem.

Approach A   The sequence $\frac{1}{4}, \frac{3}{16}, \frac{9}{64}, \ldots$ represents, term by term, the fractional amount of air that is removed with each successive stroke. Therefore, we can find the total amount removed and subtract it from 100%. The sequence is geometric with $a_1 = \frac{1}{4}$ and $r = \frac{3}{4}$.

$$S_6 = \frac{\frac{1}{4} - \frac{1}{4}\left(\frac{3}{4}\right)^6}{1 - \frac{3}{4}} = \frac{\frac{1}{4}\left[1 - \left(\frac{3}{4}\right)^6\right]}{\frac{1}{4}}$$

$$= 1 - \frac{729}{4096} = \frac{3367}{4096} = 82.2\%$$

Therefore, $100\% - 82.2\% = 17.8\%$ of the air remains after six strokes.

Approach B   The sequence

$$\frac{3}{4}, \frac{9}{16}, \frac{27}{64}, \ldots$$

represents, term by term, the amount of air that remains in the container after each stroke. Therefore, when we find the sixth term of this geometric sequence, we will have the answer to the problem. Because $a_1 = \dfrac{3}{4}$ and $r = \dfrac{3}{4}$, we obtain

$$a_6 = \frac{3}{4}\left(\frac{3}{4}\right)^5 = \left(\frac{3}{4}\right)^6 = \frac{729}{4096} = 17.8\%$$

Therefore, 17.8% of the air remains after six strokes. ■

It will be helpful for you to take another look at the two approaches we used to solve Problem 4. Note in Approach B that finding the sixth term of the sequence produced the answer to the problem without any further calculations. In Approach A, we had to find the sum of six terms of the sequence and then subtract that amount from 100%. As we solve problems that involve sequences, we must understand what each particular sequence represents on a term-by-term basis.

PROBLEM SET 9.3

Use your knowledge of arithmetic sequences and geometric sequences to help solve Problems 1–28.

1. A man started to work in 1960 at an annual salary of $9500. He received a $700 raise each year. How much was his annual salary in 1981?

2. A woman started to work in 1970 at an annual salary of $13,400. She received a $900 raise per year. How much was her annual salary in 1985?

3. State University had an enrollment of 9600 students in 1960. Each year the enrollment increased by 150 students. What was the enrollment in 1973?

4. Math University had an enrollment of 12,800 students in 1977. Each year the enrollment decreased by 75 students. What was the enrollment in 1984?

5. The enrollment at University X is predicted to increase at the rate of 10% per year. If the enrollment for 1982 was 5000 students, find the predicted enrollment for 1986. Express your answer to the nearest whole number.

6. If you pay $12,000 for a car and it depreciates 20% per year, how much will it be worth in 5 years? Express your answer to the nearest dollar.

7. A tank contains 16,000 liters of water. Each day one-half of the water in the tank is removed and not replaced. How much water remains in the tank at the end of 7 days?

8. If the price of a pound of coffee is $3.20 and the projected rate of inflation is 5% per year, how much per pound should we expect coffee to cost in 5 years? Express your answer to the nearest cent.

9. A tank contains 5832 gallons of water. Each day one-third of the water in the tank is removed and not replaced. How much water remains in the tank at the end of 6 days?

10. A fungus culture growing under controlled conditions doubles in size each day. How many units will the culture contain after 7 days if it originally contains 4 units?

11. Sue is saving quarters. She saves 1 quarter the first day, 2 quarters the second day, 3 quarters the third day, and so on for 30 days. How much money will she have saved in 30 days?

12. Suppose you save a penny the first day of a month, 2 cents the second day, 3 cents the third day, and so on for 31 days. What will be your total savings for the 31 days?

13. Suppose you save a penny the first day of a month, 2 cents the second day, 4 cents the third day, and continue to double your savings each day. How much will you save on the 15th day of the month? How much will your total savings be for the 15 days?

14. Eric saved a nickel the first day of a month, a dime the second day, and 20 cents the third day and then continued to double this daily savings each day for 14 days. What was his daily savings on the 14th day? What was his total savings for the 14 days?

15. Ms. Bryan invested $1500 at 12% simple interest at the beginning of each year for a period of 10 years. Find the

total accumulated value of all the investments at the end of the 10-year period.

16. Mr. Woodley invested $1200 at 11% simple interest at the beginning of each year for a period of 8 years. Find the total accumulated value of all the investments at the end of the 8-year period.

17. An object falling from rest in a vacuum falls approximately 16 feet the first second, 48 feet the second second, 80 feet the third second, 112 feet the fourth second, and so on. How far will it fall in 11 seconds?

18. A raffle is organized so that the amount paid for each ticket is determined by the number on the ticket. The tickets are numbered with the consecutive odd whole numbers 1, 3, 5, 7, ... Each contestant pays as many cents as the number on the ticket drawn. How much money will the raffle take in if 1000 tickets are sold?

19. Suppose an element has a half-life of 4 hours. This means that if n grams of it exist at a specific time, then only $\frac{1}{2}n$ grams remain 4 hours later. If at a particular moment we have 60 grams of the element, how many grams of it will remain 24 hours later?

20. Suppose an element has a half-life of 3 hours. (See Problem 19 for a definition of half-life.) If at a particular moment we have 768 grams of the element, how many grams of it will remain 24 hours later?

21. A rubber ball is dropped from a height of 1458 feet, and at each bounce it rebounds one-third of the height from which it last fell. How far has the ball traveled by the time it strikes the ground for the sixth time?

22. A rubber ball is dropped from a height of 100 feet, and at each bounce it rebounds one-half of the height from which it last fell. What distance has the ball traveled up to the instant it hits the ground for the eighth time?

23. A pile of logs has 25 logs in the bottom layer, 24 logs in the next layer, 23 logs in the next layer, and so on, until the top layer has 1 log. How many logs are in the pile?

24. A well driller charges $9.00 per foot for the first 10 feet, $9.10 per foot for the next 10 feet, $9.20 per foot for the next 10 feet, and so on, at a price increase of $.10 per foot for succeeding intervals of 10 feet. How much does it cost to drill a well to a depth of 150 feet?

25. A pump is attached to a container for the purpose of creating a vacuum. For each stroke of the pump, $\frac{1}{3}$ of the air remaining in the container is removed. To the nearest tenth of a percent, how much of the air remains in the container after seven strokes?

26. Suppose that in Problem 25, each stroke of the pump removes $\frac{1}{2}$ of the air remaining in the container. What fractional part of the air has been removed after six strokes?

27. A tank contains 20 gallons of water. One-half of the water is removed and replaced with antifreeze. Then one-half of this mixture is removed and replaced with antifreeze. This process is continued eight times. How much water remains in the tank after the eighth replacement process?

28. The radiator of a truck contains 10 gallons of water. Suppose we remove 1 gallon of water and replace it with antifreeze. Then we remove 1 gallon of this mixture and replace it with antifreeze. This process is continued seven times. To the nearest tenth of a gallon, how much antifreeze is in the final mixture?

THOUGHTS INTO WORDS

29. Your friend solves Problem 6 as follows: If the car depreciates 20% per year, then at the end of 5 years it will have depreciated 100% and be worth zero dollars. How would you convince him that his reasoning is incorrect?

30. A contractor wants you to clear some land for a housing project. He anticipates that it will take 20 working days to do the job. He offers to pay you one of two ways: (1) a fixed amount of $3000 or (2) a penny the first day, 2 cents the second day, 4 cents the third day, and so on, doubling your daily wages for the 20 days. Which offer should you take and why?

9.4   MATHEMATICAL INDUCTION

Is $2^n > n$ for all positive integer values of n? In an attempt to answer this question we might proceed as follows.

If $n = 1$, then $2^n > n$ becomes $2^1 > 1$, a true statement.

If $n = 2$, then $2^n > n$ becomes $2^2 > 2$, a true statement.

If $n = 3$, then $2^n > n$ becomes $2^3 > 3$, a true statement.

We can continue in this way as long as we want, but obviously we can never show in this manner that $2^n > n$ for *every* positive integer n. However, we do have a form of proof, called **proof by mathematical induction**, that can be used to verify the truth of many mathematical statements involving positive integers. This form of proof is based on the following principle.

> ### Principle of Mathematical Induction
>
> Let P_n be a statement in terms of n, where n is a positive integer. If
>
> **1.** P_1 is true, and
>
> **2.** the truth of P_k implies the truth of P_{k+1} for every positive integer k
>
> then P_n is true for every positive integer n.

The principle of mathematical induction, a proof that some statement is true for all positive integers, consists of two parts. First, we must show that the statement is true for the positive integer 1. Then we must show that if the statement is true for some positive integer, then it follows that it is also true for the next positive integer. Let's illustrate what this means.

EXAMPLE I

Prove that $2^n > n$ for all positive integer values of n.

Proof

PART I   If $n = 1$, then $2^n > n$ becomes $2^1 > 1$, which is a true statement.

PART 2   We must prove that if $2^k > k$, then $2^{k+1} > k + 1$ for all positive integer values of k. In other words, we should be able to start with $2^k > k$ and from that deduce $2^{k+1} > k + 1$. This can be done as follows.

$$2^k > k$$
$$2(2^k) > 2(k) \qquad \text{Multiply both sides by 2.}$$
$$2^{k+1} > 2k$$

We know that $k \geq 1$ because we are working with positive integers. Therefore,

$$k + k \geq k + 1 \qquad \text{Add } k \text{ to both sides}$$
$$2k \geq k + 1$$

Because $2^{k+1} > 2k$ and $2k \geq k + 1$, by the transitive property we conclude that

$$2^{k+1} > k + 1$$

Therefore, using part 1 and part 2, we have proved that $2^n > n$ for *all* positive integers. ∎

It will be helpful for you to look back over the proof in Example 1. Note that in part 1 we established that $2^n > n$ is true for $n = 1$. Then in part 2 we established that if $2^n > n$ is true for any positive integer, then it must be true for the next consecutive positive integer. Therefore, because $2^n > n$ is true for $n = 1$, it must be true for $n = 2$. Likewise, if $2^n > n$ is true for $n = 2$, then it must be true for $n = 3$, and so on, for *all* positive integers.

We can depict proof by mathematical induction with dominoes. Suppose that in Figure 9.1 we have infinitely many dominoes lined up. If we can push the first

FIGURE 9.1

domino over (part 1 of a mathematical induction proof) and if the dominoes are spaced so that each time one falls over, it causes the next one to fall over (part 2 of a mathematical induction proof), then by pushing the first one over we will cause a chain reaction that will topple all of the dominoes (Figure 9.2).

FIGURE 9.2

Recall that in the first three sections of this chapter, we used a_n to represent the nth term of a sequence and S_n to represent the sum of the first n terms of a sequence. For example, if $a_n = 2n$, then the first three terms of the sequence are $a_1 = 2(1) = 2$, $a_2 = 2(2) = 4$, and $a_3 = 2(3) = 6$. Furthermore, the kth term is $a_k = 2(k) = 2k$ and the $(k + 1)$st term is $a_{k+1} = 2(k + 1) = 2k + 2$. Relative to this same sequence, we can state that $S_1 = 2$, $S_2 = 2 + 4 = 6$ and $S_3 = 2 + 4 + 6 = 12$.

There are numerous sum formulas for sequences that can be verified by mathematical induction. For such proofs, the following property of sequences is used.

$$S_{k+1} = S_k + a_{k+1}$$

This property states that **the sum of the first $k + 1$ terms is equal to the sum of the first k terms plus the $(k + 1)$st term.** Let's see how this can be used in a specific example.

EXAMPLE 2

Prove that $S_n = n(n + 1)$ for the sequence $a_n = 2n$, where n is any positive integer.

Proof

PART I   If $n = 1$, then $1(1 + 1) = 2$, and 2 is the first term of the sequence $a_n = 2n$, so $S_1 = a_1 = 2$.

PART 2   Now we need to prove that if $S_k = k(k + 1)$, then $S_{k+1} = (k + 1)(k + 2)$. Using the property $S_{k+1} = S_k + a_{k+1}$, we can proceed as follows.

$$\begin{aligned} S_{k+1} &= S_k + a_{k+1} \\ &= k(k + 1) + 2(k + 1) \\ &= (k + 1)(k + 2) \end{aligned}$$

Therefore, using part 1 and part 2, we have proved that $S_n = n(n + 1)$ will yield the correct sum for any number of terms of the sequence $a_n = 2n$. ∎

EXAMPLE 3

Prove that $S_n = 5n(n + 1)/2$ for the sequence $a_n = 5n$, where n is any positive integer.

Proof

PART I   Because $5(1)(1 + 1)/2 = 5$, and 5 is the first term of the sequence $a_n = 5n$, we have $S_1 = a_1 = 5$.

PART 2   We need to prove that if $S_k = 5k(k + 1)/2$, then $S_{k+1} = \dfrac{5(k + 1)(k + 2)}{2}$.

$$\begin{aligned} S_{k+1} &= S_k + a_{k+1} \\ &= \frac{5k(k + 1)}{2} + 5(k + 1) \\ &= \frac{5k(k + 1)}{2} + 5k + 5 \end{aligned}$$

$$= \frac{5k(k+1) + 2(5k+5)}{2}$$

$$= \frac{5k^2 + 5k + 10k + 10}{2}$$

$$= \frac{5k^2 + 15k + 10}{2}$$

$$= \frac{5(k^2 + 3k + 2)}{2}$$

$$= \frac{5(k+1)(k+2)}{2}$$

Therefore, using part 1 and part 2, we have proved that $S_n = 5n(n+1)/2$ yields the correct sum for any number of terms of the sequence $a_n = 5n$. ∎

EXAMPLE 4

Prove that $S_n = (4^n - 1)/3$ for the sequence $a_n = 4^{n-1}$, where n is any positive integer.

Proof

PART I Because $(4^1 - 1)/3 = 1$, and 1 is the first term of the sequence $a_n = 4^{n-1}$, we have $S_1 = a_1 = 1$.

PART 2 We need to prove that if $S_k = (4^k - 1)/3$, then $S_{k+1} = (4^{k+1} - 1)/3$.

$$S_{k+1} = S_k + a_{k+1}$$

$$= \frac{4^k - 1}{3} + 4^k$$

$$= \frac{4_k - 1 + 3(4^k)}{3}$$

$$= \frac{4^k + 3(4^k) - 1}{3}$$

$$= \frac{4^k(1 + 3) - 1}{3}$$

$$= \frac{4^k(4) - 1}{3}$$

$$= \frac{4^{k+1} - 1}{3}$$

Therefore, using part 1 and part 2, we have proved that $S_n = (4^n - 1)/3$ yields the correct sum for any number of terms of the sequence $a_n = 4^{n-1}$. ∎

As our final example of this section, let's consider a proof by mathematical induction involving the concept of divisibility.

EXAMPLE 5

Prove that for all positive integers n, the number $3^{2n} - 1$ is divisible by 8.

Proof

PART I   If $n = 1$, then $3^{2n} - 1$ becomes $3^{2(1)} - 1 = 3^2 - 1 = 8$, and of course 8 is divisible by 8.

PART 2   We need to prove that if $3^{2k} - 1$ is divisible by 8, then $3^{2k+2} - 1$ is divisible by 8 for all integer values of k. This can be verified as follows. If $3^{2k} - 1$ is divisible by 8, then for some integer x, we have $3^{2k} - 1 = 8x$. Therefore,

$$3^{2k} - 1 = 8x$$
$$3^{2k} = 1 + 8x$$
$$3^2(3^{2k}) = 3^2(1 + 8x) \qquad \text{Multiply both sides by } 3^2.$$
$$3^{2k+2} = 9(1 + 8x)$$
$$3^{2k+2} = 9 + 9(8x)$$
$$3^{2k+2} = 1 + 8 + 9(8x) \qquad 9 = 1 + 8$$
$$3^{2k+2} = 1 + 8(1 + 9x) \qquad \text{Apply distributive}$$
$$3^{2k+2} - 1 = 8(1 + 9x) \qquad \text{property to } 8 + 9(8x)$$

Therefore, $3^{2k+2} - 1$ is divisible by 8.

Thus, using part 1 and part 2, we have proved that $3^{2n} - 1$ is divisible by 8 for all positive integers n. ∎

We conclude this section with a few final comments about proof by mathematical induction. Every mathematical induction proof is a two-part proof, and both parts are absolutely necessary. There can be mathematical statements that hold for one or the other of the two parts but not for both. For example, $(a + b)^n = a^n + b^n$ is true for $n = 1$, but it is false for every positive integer greater than 1. Therefore, if we were to attempt a mathematical induction proof for $(a + b)^n = a^n + b^n$, we could establish part 1 but not part 2. Another example of this type is the statement that $n^2 - n + 41$ produces a prime number for all positive integer values of n. This statement is true for $n = 1, 2, 3, 4, \ldots, 40$, but it is false when $n = 41$ (because $41^2 - 41 + 41 = 41^2$, which is not a prime number).

It is also possible that part 2 of a mathematical induction proof can be established but not part 1. For example, consider the sequence $a_n = n$ and the sum formula $S_n = (n + 3)(n - 2)/2$. If $n = 1$, then $a_1 = 1$ but $S_1 = (4)(-1)/2 = -2$, so part 1 does not hold. However, it is possible to show that $S_k = (k + 3)(k - 2)/2$ implies $S_{k+1} = (k + 4)(k - 1)/2$. We will leave the details of this for you to do.

Finally, it is important to realize that some mathematical statements are true for all positive integers greater than some fixed positive integer other than 1. (Back in Figure 9.1, perhaps we cannot knock down the first four dominoes, whereas we can knock down the fifth domino and every one thereafter.) For example, we can prove by mathematical induction that $2^n > n^2$ for all positive integers $n > 4$. It requires a slight variation in the statement of the principle of mathematical induction. We will not concern ourselves with such problems in this text, but we want you to be aware of their existence.

PROBLEM SET 9.4

For Problems 1–10, use mathematical induction to prove each of the sum formulas for the indicated sequences. They are to hold for all positive integers n.

1. $S_n = \dfrac{n(n+1)}{2}$ for $a_n = n$

2. $S_n = n^2$ for $a_n = 2n - 1$

3. $S_n = \dfrac{n(3n+1)}{2}$ for $a_n = 3n - 1$

4. $S_n = \dfrac{n(5n+9)}{2}$ for $a_n = 5n + 2$

5. $S_n = 2(2^n - 1)$ for $a_n = 2^n$

6. $S_n = \dfrac{3(3^n - 1)}{2}$ for $a_n = 3^n$

7. $S_n = \dfrac{n(n+1)(2n+1)}{6}$ for $a_n = n^2$

8. $S_n = \dfrac{n^2(n+1)^2}{4}$ for $a_n = n^3$

9. $S_n = \dfrac{n}{n+1}$ for $a_n = \dfrac{1}{n(n+1)}$

10. $S_n = \dfrac{n(n+1)(n+2)}{3}$ for $a_n = n(n+1)$

In Problems 11–20, use mathematical induction to prove that each statement is true for all positive integers n.

11. $3^n \geq 2n + 1$

12. $4^n \geq 4n$

13. $n^2 \geq n$

14. $2^n \geq n + 1$

15. $4^n - 1$ is divisible by 3

16. $5^n - 1$ is divisible by 4

17. $6^n - 1$ is divisible by 5

18. $9^n - 1$ is divisible by 4

19. $n^2 + n$ is divisible by 2

20. $n^2 - n$ is divisible by 2

THOUGHTS INTO WORDS

21. How would you describe proof by mathematical induction?

22. Compare inductive reasoning to proof by mathematical induction.

CHAPTER 9 SUMMARY

There are four main topics in this chapter: arithmetic sequences, geometric sequences, problem solving, and mathematical induction.

Arithmetic Sequences

The sequence $a_1, a_2, a_3, a_4, \ldots$ is called **arithmetic** if and only if

$$a_{k+1} - a_k = d$$

for every positive integer k. In other words, there is a **common difference**, d, between successive terms.

The **general term** of an arithmetic sequence is given by the formula

$$a_n = a_1 + (n - 1)d$$

where a_1 is the first term, n is the number of terms, and d is the common difference.

The **sum** of the first n terms of an arithmetic sequence is given by the formula

$$S_n = \frac{n(a_1 + a_n)}{2}$$

Summation notation can be used to indicate the sum of a certain number of terms of a sequence. For example,

$$\sum_{i=1}^{5} 4^i = 4^1 + 4^2 + 4^3 + 4^4 + 4^5$$

Geometric Sequences

The sequence $a_1, a_2, a_3, a_4, \ldots$ is called **geometric** if and only if

$$a_{k+1} = ra_k$$

for every positive integer k. There is a **common ratio**, r, between successive terms.

The **general term** of a geometric sequence is given by the formula

$$a_n = a_1 r^{n-1}$$

where a_1 is the first term, n is the number of terms, and r is the common ratio.

The **sum** of the first n terms of a geometric sequence is given by the formula

$$S_n = \frac{a_1 r^n - a_1}{r - 1} \qquad r \neq 1$$

The **sum of an infinite geometric sequence** is given by the formula

$$S_\infty = \frac{a_1}{1 - r} \qquad \text{for } |r| < 1$$

If $|r| \geq 1$, the sequence has no sum.

Repeating decimals (such as $0.\overline{4}$) can be changed to a/b form, where a and b are integers and $b \neq 0$, by treating them as the sum of an infinite geometric sequence. For example, the repeating decimal $0.\overline{4}$ can be written $0.4 + 0.04 + 0.004 + 0.0004 + \cdots$.

Problem Solving

Many of the problem-solving suggestions offered earlier in this text are still appropriate when we are solving problems that deal with sequences. However, there are also some special suggestions pertaining to sequence problems.

1. Write down the first few terms of the sequence to describe what is taking place in the problem. Drawing a picture or diagram may help with this step.
2. Be sure that you understand, term by term, what the sequence represents in the problem.
3. Determine whether the sequence is arithmetic or geometric. (Those are the only kinds of sequences we are working with in this text.)
4. Determine whether the problem is asking for a specific term or for the sum of a certain number of terms.

Mathematical Induction

Proof by mathematical induction relies on the following **principle of induction**.

Let P_n be a statement in terms of n, where n is a positive integer. If

1. P_1 is true, and
2. The truth of P_k implies the truth of P_{k+1}, for every positive integer k

then P_n is true for every positive integer n.

CHAPTER 9 REVIEW PROBLEM SET

For Problems 1–10, find the general term (the *n*th term) for each sequence. These problems include both arithmetic sequences and geometric sequences.

1. $3, 9, 15, 21, \ldots$

2. $\dfrac{1}{3}, 1, 3, 9, \ldots$

3. $10, 20, 40, 80, \ldots$

4. $5, 2, -1, -4, \ldots$

5. $-5, -3, -1, 1, \ldots$

6. $9, 3, 1, \dfrac{1}{3}, \ldots$

7. $-1, 2, -4, 8, \ldots$

8. $12, 15, 18, 21, \ldots$

9. $\dfrac{2}{3}, 1, \dfrac{4}{3}, \dfrac{5}{3}, \ldots$

10. $1, 4, 16, 64, \ldots$

For Problems 11–16, find the required term of each of the sequences.

11. The 19th term of $1, 5, 9, 13, \ldots$

12. The 28th term of $-2, 2, 6, 10, \ldots$

13. The 9th term of $8, 4, 2, 1, \ldots$

14. The 8th term of $\dfrac{243}{32}, \dfrac{81}{16}, \dfrac{27}{8}, \dfrac{9}{4}, \ldots$

15. The 34th term of $7, 4, 1, -2, \ldots$

16. The 10th term of $-32, 16, -8, 4, \ldots$

For Problems 17–29, solve each problem.

17. If the 5th term of an arithmetic sequence is -19 and the 8th term is -34, find the common difference of the sequence.

18. If the 8th term of an arithmetic sequence is 37 and the 13th term is 57, find the 20th term.

19. Find the first term of a geometric sequence if the third term is 5 and the sixth term is 135.

20. Find the common ratio of a geometric sequence if the second term is $\dfrac{1}{2}$ and the sixth term is 8.

21. Find the sum of the first nine terms of the sequence 81, 27, 9, 3, ...

22. Find the sum of the first 70 terms of the sequence $-3, 0, 3, 6, \ldots$.

23. Find the sum of the first 75 terms of the sequence 5, 1, $-3, -7, \ldots$.

24. Find the sum of the first ten terms of the sequence where $a_n = 2^{5-n}$.

25. Find the sum of the first 95 terms of the sequence where $a_n = 7n + 1$.

26. Find the sum $5 + 7 + 9 + \cdots + 137$.

27. Find the sum $64 + 16 + 4 + \cdots + \dfrac{1}{64}$.

28. Find the sum of all even numbers between 8 and 384, inclusive.

29. Find the sum of all multiples of 3 between 27 and 276, inclusive.

For Problems 30–33, find each indicated sum.

30. $\displaystyle\sum_{i=1}^{45} (-2i + 5)$

31. $\displaystyle\sum_{i=1}^{5} i^3$

32. $\displaystyle\sum_{i=1}^{8} 2^{8-i}$

33. $\displaystyle\sum_{i=4}^{75} (3i - 4)$

For Problems 34–36, solve each problem.

34. Find the sum of the infinite geometric sequence 64, 16, 4, 1, ...

35. Change $0.\overline{36}$ to reduced a/b form, where a and b are integers and $b \neq 0$.

36. Change $0.4\overline{5}$ to reduced a/b form, where a and b are integers and $b \neq 0$.

Solve each of Problems 37–40 by using your knowledge of arithmetic sequences and geometric sequences.

37. Suppose that your savings account contains $3750 at the beginning of a year. If you withdrew $250 per month from the account, how much will it contain at the end of the year?

38. Sonya decides to start saving dimes. She plans to save 1 dime the first day of April, 2 dimes the second day, 3 dimes the third day, 4 dimes the fourth day, and so on for the 30 days of April. How much money will she save in April?

39. Nancy decides to start saving dimes. She plans to save 1 dime the first day of April, 2 dimes the second day, 4 dimes the third day, 8 dimes the fourth day, and so on for the first 15 days of April. How much will she save in 15 days?

40. A tank contains 61,440 gallons of water. Each day one-fourth of the water is drained out. How much water remains in the tank at the end of 6 days?

For Problems 41–43, show a mathematical induction proof.

41. Prove that $5^n > 5n - 1$ for all positive integer values of n.

42. Prove that $n^3 - n + 3$ is divisible by 3 for all positive integer values of n.

43. Prove that

$$S_n = \frac{n(n + 3)}{4(n + 1)(n + 2)}$$

is the sum formula for the sequence

$$a_n = \frac{1}{n(n + 1)(n + 2)}$$

where n is any positive integer.

CHAPTER 9 TEST

1. Find the 15th term of the sequence for which $a_n = -n^2 - 1$.

2. Find the fifth term of the sequence for which $a_n = 3(2)^{n-1}$.

3. Find the general term of the sequence $-3, -8, -13, -18, \ldots$.

4. Find the general term of the sequence $5, \dfrac{5}{2}, \dfrac{5}{4}, \dfrac{5}{8}, \ldots$.

5. Find the general term of the sequence $10, 16, 22, 28, \ldots$.

6. Find the 7th term of the sequence $8, 12, 18, 27, \ldots$.

7. Find the 75th term of the sequence $1, 4, 7, 10, \ldots$.

8. Find the number of terms in the sequence $7, 11, 15, \ldots, 243$.

9. Find the sum of the first 40 terms of the sequence $1, 4, 7, 10, \ldots$.

10. Find the sum of the first eight terms of the sequence $3, 6, 12, 24, \ldots$.

11. Find the sum of the first 45 terms of the sequence for which $a_n = 7n - 2$.

12. Find the sum of the first ten terms of the sequence for which $a_n = 3(2)^n$.

13. Find the sum of the first 150 positive even whole numbers.

14. Find the sum of the odd whole numbers between 11 and 193, inclusive.

15. Find the indicated sum $\displaystyle\sum_{i=1}^{50} (3i + 5)$.

16. Find the indicated sum $\displaystyle\sum_{i=1}^{10} (-2)^{i-1}$.

17. Find the sum of the infinite geometric sequence $3, \dfrac{3}{2}, \dfrac{3}{4}, \dfrac{3}{8}, \ldots$.

18. Find the sum of the infinite geometric sequence for which $a_n = 2\left(\dfrac{1}{3}\right)^{n+1}$.

19. Change $0.\overline{18}$ to reduced $\dfrac{a}{b}$ form, where a and b are integers and $b \neq 0$.

20. Change $0.2\overline{6}$ to reduced $\dfrac{a}{b}$ form, where a and b are integers and $b \neq 0$.

For Problems 21–23, solve each problem.

21. A tank contains 49,152 liters of gasoline. Each day three-fourths of the gasoline remaining in the tank is pumped out and not replaced. How much gasoline remains in the tank at the end of 7 days?

22. Suppose that you save a dime the first day of a month, $.20 the second day, and $.40 the third day and that you continue to double your savings per day for 15 days. Find the total amount that you will save at the end of 15 days.

23. A woman invests $350 at 12% simple interest at the beginning of each year for a period of 10 years. Find the total accumulated value of all the investments at the end of the 10-year period.

For Problems 24 and 25, show a mathematical induction proof.

24. $S_n = \dfrac{n(3n - 1)}{2}$ for $a_n = 3n - 2$.

25. $9^n - 1$ is divisible by 8 for all positive integer values for n.

10

COUNTING TECHNIQUES, PROBABILITY, AND THE BINOMIAL THEOREM

In a group of thirty people, there is approximately a 70% chance that at least two of them will have the same birthday (same month and same day of the month). In a group of 60 people, there is approximately a 99% chance that at least two of them will have the same birthday.

With an ordinary deck of 52 playing cards, there is 1 chance out of 54,145 that you will be dealt four aces in a five-card hand. The radio is predicting a 40% chance of locally severe thunderstorms by late afternoon. The odds in favor of the Cubs winning the pennant are 2 to 3. Suppose that in a box containing 50 light bulbs, 45 are good ones and 5 are burned out. If two bulbs are chosen at random, the probability of getting at least one good bulb is $\frac{243}{245}$. Historically, many basic probability concepts have been developed as a result of studying various games of chance. However, in recent years, probability applications have been surfacing at a phenomenal rate in a large variety of fields, such as physics, biology, psychology, economics, insurance, military science, manufacturing, and politics. It is our purpose in this chapter first to introduce some counting techniques and then to use those techniques to introduce some basic concepts of probability. The last section of the chapter will be devoted to the binomial theorem.

10.1 FUNDAMENTAL PRINCIPLE OF COUNTING

One very useful counting principle is referred to as the **fundamental principle of counting**. We will offer some examples, state the property, and then use it to solve a variety of counting problems. Let's consider two examples to lead up to the statement of the property.

PROBLEM 1

A woman has four skirts and five blouses. Assuming that each blouse can be worn with each skirt, how many different skirt–blouse outfits does she have?

Solution

For *each* of the four skirts she has a choice of five blouses. Therefore, she has $4(5) = 20$ different skirt–blouse outfits from which to choose. ∎

PROBLEM 2

Eric is shopping for a new bicycle and has two different models (5-speed or 10-speed) and four different colors (red, white, blue, or silver) from which to choose. How many different choices does he have?

Solution

His different choices can be counted with the help of a **tree diagram**.

For each of the two model choices, there are four choices of color. Altogether, then, Eric has 2(4) = 8 choices. ■

These two problems exemplify the following general principle.

Fundamental Principle of Counting

If one task can be accomplished in x different ways and, following this task, a second task can be accomplished in y different ways, then the first task followed by the second task can be accomplished in $x \cdot y$ different ways. (This counting principle can be extended to any finite number of tasks.)

As you apply the fundamental principle of counting, it is often helpful to analyze a problem systematically in terms of the tasks to be accomplished. Let's consider some examples.

PROBLEM 3

How many numbers of three different digits each can be formed by choosing from the digits 1, 2, 3, 4, 5 and 6?

Solution

Let's analyze this problem in terms of three tasks.

TASK I Choose the hundreds digit, for which there are six choices.

TASK 2 Now choose the tens digit, for which there are only five choices, because one digit was used in the hundreds place.

TASK 3 Now choose the units digit, for which there are only four choices, because two digits have been used for the other places.

Therefore, task 1 followed by task 2 followed by task 3 can be accomplished in (6)(5)(4) = 120 ways. In other words, there are 120 numbers of three different digits that can be formed by choosing from the six given digits. ■

Now look back over the solution for Problem 3 and think about each of the following questions.

1. Can we solve the problem by choosing the units digit first, then the tens digit, and finally the hundreds digit?

2. How many three-digit numbers can be formed from 1, 2, 3, 4, 5, and 6 if we do not require each number to have three *different* digits? (Your answer should be 216.)

3. Suppose that the digits from which to choose are 0, 1, 2, 3, 4, and 5. Now how many numbers of three different digits each can be formed, assuming that we do not want zero in the hundreds place? (Your answer should be 100.)

4. Suppose that we want to know the number of *even* numbers with three different digits each that can be formed by choosing from 1, 2, 3, 4, 5, and 6. How many are there? (Your answer should be 60.)

PROBLEM 4

Employee ID numbers at a certain factory consist of one capital letter followed by a three-digit number that contains no repeated digits. For example, A-014 is an ID number. How many such ID numbers can be formed? How many can be formed if repeated digits *are* allowed?

Solution

Again, let's analyze in terms of tasks to be completed.

TASK 1   Choose the letter part of the ID number: there are 26 choices.

TASK 2   Choose the first digit of the three-digit number: there are ten choices.

TASK 3   Choose the second digit: there are nine choices.

TASK 4   Choose the third digit: there are eight choices.

Therefore, applying the fundamental principle, we obtain $(26)(10)(9)(8) = 18,720$ possible ID numbers.

If repeat digits were allowed, then there would be $(26)(10)(10)(10) = 26,000$ possible ID numbers.   ■

PROBLEM 5

In how many ways can Al, Barb, Chad Dan, and Edna be seated in a row of five seats so that Al and Barb are seated side by side?

Solution

This problem can be analyzed in terms of three tasks.

TASK 1   Choose the two adjacent seats to be occupied by Al and Barb. An illustration such as Figure 10.1 helps us to see that there are four choices for the two adjacent seats.

FIGURE 10.1

TASK 2   Determine the number of ways in which Al and Barb can be seated. Because Al can be seated on the left and Barb on the right, or vice versa, there are two ways to seat Al and Barb for each pair of adjacent seats.

TASK 3   The remaining three people must be seated in the remaining three seats. This can be done in $(3)(2)(1) = 6$ different ways.

Therefore, by the fundamental principle, task 1 followed by task 2 followed by task 3 can be done in $(4)(2)(6) = 48$ ways.   ■

Suppose that in Problem 5, we wanted instead the number of ways in which the five people can sit so that Al and Barb are *not* side by side. We can determine this number by using either of two basically different techniques: (1) analyze and count the number of nonadjacent positions for Al and Barb, or (2) subtract the number of seating arrangements determined in Problem 5 from the total number of ways in which five people can be seated in five seats. Try doing this problem both ways and see whether you agree with the answer of 72 ways.

As you apply the fundamental principle of counting, you may find that for certain problems, simply thinking about an appropriate tree diagram is helpful, even though the size of the problem may make it inappropriate to write out the diagram in detail. Consider the following problem.

PROBLEM 6

Suppose that the undergraduate students in three departments—geography, history, and psychology—are to be classified according to sex and year in school. How many categories are needed?

Solution

Let's represent the various classifications symbolically as follows.

M:	Male	1.	Freshman	G:	Geography
F:	Female	2.	Sophomore	H:	History
		3.	Junior	P:	Psychology
		4.	Senior		

We can mentally picture a tree diagram such that each of the two sex classifications branches into four school-year classifications, which in turn branch into three department classifications. Thus we have $(2)(4)(3) = 24$ different categories.   ■

Another technique that works on certain problems involves what some people call the *back door* approach. For example, suppose we know that the classroom contains 50 seats. On some days, it may be easier to determine the number of students present by counting the number of empty seats and subtracting from 50 than by counting the number of students in attendance. (We suggested this back door approach as one way to count the nonadjacent seating arrangements in the discussion following Problem 5.) The next example further illustrates this approach.

PROBLEM 7

When rolling a pair of dice, in how many ways can we obtain a sum greater than 4?

Solution

For clarification purposes, let's use a red die and a white die. (It is not necessary to use different-colored dice, but it does help us analyze the different possible outcomes.) With a moment of thought, you will see that there are more ways to get a sum greater than 4 than there are ways to get a sum of 4 or less. Therefore, let's determine the number of possibilities for getting a sum of 4 or less; then we'll subtract that number from the total number of possible outcomes when rolling a pair of dice.

First, we can simply list and count the ways of getting a sum of 4 or less.

Red Die	White Die
1	1
1	2
1	3
2	1
2	2
3	1

There are six ways of getting a sum of 4 or less.

Second, because there are six possible outcomes on the red die and six possible outcomes on the white die, there is a total of $(6)(6) = 36$ possible outcomes when rolling a pair of dice.

Therefore, subtracting the number of ways of getting 4 or less from the total number of possible outcomes, we obtain $36 - 6 = 30$ ways of getting a sum greater than 4. ■

PROBLEM SET 10.1

Solve Problems 1–37.

1. If a woman has two skirts and ten blouses, how many different skirt–blouse combinations does she have?

2. If a man has eight shirts, five pairs of slacks, and three pairs of shoes, how many different shirt–slack–shoe combinations does he have?

3. In how many ways can four people be seated in a row of four seats?

4. How many numbers of two different digits can be formed by choosing from the digits 1, 2, 3, 4, 5, 6, and 7?

5. How many *even* numbers of three different digits can be formed by choosing from the digits 2, 3, 4, 5, 6, 7, 8, and 9?

6. How many *odd* numbers of four different digits can be formed by choosing from the digits 1, 2, 3, 4, 5, 6, 7, and 8?

7. Suppose that the students at a certain university are to be classified according to their college (College of Applied Science, College of Arts and Sciences, College of Business, College of Education, College of Fine Arts, College of Health and Physical Education), sex (female, male), and year in school (1, 2, 3, 4). How many categories are possible?

8. A medical researcher classifies subjects according to sex (female, male), smoking habits (smoker, nonsmoker), and weight (below average, average, above average). How many different combined classifications are used?

9. A pollster classifies voters according to sex (female, male), party affiliation (Democrat, Republican, Independent), and family income (below $10,000, $10,000–$19,999, $20,000–$29,999, $30,000–$39,999, $40,000–$49,999, $50,000 and above). How many combined classifications does the pollster use?

10. A couple is planning to have four children. How many ways can this happen in terms of boy–girl classification? (For example, *BBBG* indicates that the first three children are boys and the last is a girl.)

11. In how many ways can three officers—president, secretary, and treasurer—be selected from a club that has 20 members?

12. In how many ways can three officers—president, secretary, and treasurer—be selected from a club with 15 female and 10 male members, so that the president is female and the secretary and treasurer are male?

13. A disc jockey wants to play six songs once each in a half-hour program. How many different ways can he order these songs?

14. A state has agreed to have its automobile license plates consist of two letters followed by four digits. State officials do not want to repeat any letters or digits in any license numbers. How many different license plates will be available?

15. In how many ways can six people be seated in a row of six seats?

16. In how many ways can Al, Bob, Carl, Don, Ed, and Fern be seated in a row of six seats if Al and Bob want to sit side by side?

17. In how many ways can Amy, Bob, Cindy, Dan, and Elmer be seated in a row of five seats so that neither Amy nor Bob occupies an end seat?

18. In how many ways can Al, Bob, Carl, Don, Ed, and Fern be seated in a row of six seats if Al and Bob are not to be seated side by side? [*Hint:* Either Al and Bob will be seated side by side or they will not be seated side by side.]

19. In how many ways can Al, Bob, Carol, Dawn, and Ed be seated in a row of five chairs if Al is to be seated in the middle chair?

20. In how many ways can three letters be dropped in five mailboxes?

21. In how many ways can five letters be dropped in three mailboxes?

22. In how many ways can four letters be dropped in six mailboxes so that no two letters go in the same box?

23. In how many ways can six letters be dropped in four mailboxes so that no two letters go in the same box?

24. If five coins are tossed, in how many ways can they fall?

25. If three dice are tossed, in how many ways can they fall?

26. In how many ways can a sum less than ten be obtained when tossing a pair of dice?

27. In how many ways can a sum greater than five be obtained when tossing a pair of dice?

28. In how many ways can a sum greater than four be obtained when tossing three dice?

29. If no number contains repeated digits, how many numbers greater than 400 can be formed by choosing from the digits 2, 3, 4, and 5? [*Hint:* Consider both three-digit and four-digit numbers.]

30. If no number contains repeated digits, how many numbers greater than 5000 can be formed by choosing from the digits 1, 2, 3, 4, 5, and 6?

31. In how many ways can four boys and three girls be seated in a row of seven seats so that boys and girls occupy alternate seats?

32. In how many ways can three different mathematics books and four different history books be exhibited on a shelf so that all of the books in a subject area are side by side?

33. In how many ways can a true–false test of ten questions be answered?

34. If no number contains repeated digits, how many even numbers greater than 3000 can be formed by choosing from the digits 1, 2, 3, and 4?

35. If no number contains repeated digits, how many odd numbers greater than 40,000 can be formed by choosing from the digits 1, 2, 3, 4, and 5?

36. In how many ways can Al, Bob, Carol, Don, Ed, Faye, and George be seated in a row of seven seats so that Al, Bob, and Carol occupy consecutive seats in some order?

37. The license plates for a certain state consist of two letters followed by a four-digit number such that the first digit of the number is not zero. An example would be PK-2446.

  a. How many different license plates can be produced?

  b. How many different plates do not have a repeated letter?

  c. How many plates do not have any repeated digits in the number part of the plate?

  d. How many plates do not have a repeated letter and also do not have any repeated digits?

THOUGHTS INTO WORDS

38. How would you explain the fundamental principle of counting to a friend who missed class the day it was discussed?

39. Give two or three simple illustrations of the fundamental principle of counting.

40. Explain how you solved Problem 29.

10.2 PERMUTATIONS AND COMBINATIONS

As we develop the material in this section, factorial notation becomes very useful. The notation n! (which is read *n factorial*) is used with positive integers as follows.

$$1! = 1$$
$$2! = 2 \cdot 1 = 2$$
$$3! = 3 \cdot 2 \cdot 1 = 6$$
$$4! = 4 \cdot 3 \cdot 2 \cdot 1 = 24$$

Note that the factorial notation refers to an *indicated product*. In general, we write

$$n! = n(n - 1)(n - 2) \cdots 3 \cdot 2 \cdot 1$$

We also define $0! = 1$ so that certain formulas will be true for all nonnegative integers.

Now, as an introduction to the first concept of this section, let's consider a counting problem that closely resembles problems from the previous section.

PROBLEM 1

In how many ways can the three letters A, B, and C be arranged in a row?

Solution A

Certainly one approach to the problem is simply to list and count the arrangements.

ABC   ACB   BAC   BCA   CAB   CBA

There are six arrangements of the three letters.

Solution B

Another approach, one that can be generalized for more difficult problems, uses the fundamental principle of counting. Because there are three choices for the first letter of an arrangement, two choices for the second letter, and one choice for the third letter, there are $(3)(2)(1) = 6$ arrangements. ∎

Ordered arrangements are called **permutations**. In general, a permutation of a set of n elements is an ordered arrangement of the n elements; we will use the symbol $P(n, n)$ to denote the number of such permutations. For example, from Problem 1 we know that $P(3, 3) = 6$. Furthermore, by using the same basic approach as in Solution B of Problem 1, we can obtain

$$P(1, 1) = 1 = 1!$$
$$P(2, 2) = 2 \cdot 1 = 2!$$
$$P(4, 4) = 4 \cdot 3 \cdot 2 \cdot 1 = 4!$$
$$P(5, 5) = 5 \cdot 4 \cdot 3 \cdot 2 \cdot 1 = 5!$$

In general, the following formula becomes evident.

$$P(n, n) = n!$$

Now suppose that we are interested in the number of two-letter permutations that can be formed by choosing from the four letters A, B, C, and D. (Some examples of such permutations are AB, BA, AC, BC, and CB.) In other words, we want to find the number of two-element permutations that can be formed from a set of four elements. We denote this number by $P(4, 2)$. To find $P(4, 2)$, we can reason as follows. First, we can choose any one of the four letters to occupy the first position in the permutation, and then we can choose any one of the three remaining letters for the second position. Therefore, by the fundamental principle of counting, we have $(4)(3) = 12$ different two-letter permutations; that is, $P(4, 2) = 12$. By using a similar line of reasoning, we can determine the following numbers. (Make sure that you agree with each of these.)

$$P(4, 3) = 4 \cdot 3 \cdot 2 = 24$$
$$P(5, 2) = 5 \cdot 4 = 20$$
$$P(6, 4) = 6 \cdot 5 \cdot 4 \cdot 3 = 360$$
$$P(7, 3) = 7 \cdot 6 \cdot 5 = 210$$

In general, we say that **the number of r-element permutations that can be formed from a set of n elements is given by**

$$P(n, r) = \underbrace{n(n-1)(n-2) \cdots}_{r \text{ factors}}$$

Note that the indicated product for $P(n, r)$ begins with n. Thereafter each factor is 1 less than the previous one and there is a total of r factors. For example,

$$P(6, 2) = 6 \cdot 5 = 30$$
$$P(8, 3) = 8 \cdot 7 \cdot 6 = 336$$
$$P(9, 4) = 9 \cdot 8 \cdot 7 \cdot 6 = 3024$$

Let's consider two problems that illustrate the use of $P(n, n)$ and $P(n, r)$.

PROBLEM 2

In how many ways can five students be seated in a row of five seats?

Solution

The problem is asking for the number of five-element permutations that can be formed from a set of five elements. Thus we can apply $P(n, n) = n!$.

$$P(5, 5) = 5! = 5 \cdot 4 \cdot 3 \cdot 2 \cdot 1 = 120$$ ∎

PROBLEM 3

Suppose that seven people enter a swimming race. In how many ways can first, second, and third prizes be awarded?

Solution

This problem is asking for the number of three-element permutations that can be formed from a set of seven elements. Therefore, using the formula for $P(n, r)$, we obtain

$$P(7, 3) = 7 \cdot 6 \cdot 5 = 210$$ ∎

It should be evident that both Problem 2 and Problem 3 could have been solved by applying the fundamental principle of counting. In fact, the formulas for $P(n, n)$ and $P(n, r)$ do not really give us much additional problem-solving power. However, as we will see in a moment, they do provide the basis for developing a formula that is very useful as a problem-solving tool.

Permutations Involving Nondistinguishable Objects

Suppose we have two identical H's and one T in an arrangement such as HTH. If we switch the two identical H's, the newly formed arrangement, HTH, will not be distinguishable from the original. In other words, there are fewer distinguishable permutations of n elements when some of those elements are identical than when the n elements are distinctly different.

To see the effect of identical elements on the number of distinguishable permutations, let's look at some specific examples.

    2 identical H's   1 permutation (HH)

    2 different letters   2! permutations (HT, TH)

Therefore, having two different letters affects the number of permutations by a *factor of* 2!.

    3 identical H's   1 permutation (HHH)

    3 different letters   3! permutations

Therefore, having three different letters affects the number of permutations by *a factor* of 3!.

    4 identical H's   1 permutation (HHHH)

    4 different letters   4! permutations

Therefore, having four different letters affects the number of permutations by a *factor of* 4!.

Now let's solve a specific problem.

PROBLEM 4

How many distinguishable permutations can be formed from three identical H's and two identical T's?

Solution

If we had five distinctly different letters, we could form 5! permutations. But the three identical H's affect the number of distinguishable permutations by a factor of 3!, and the two identical T's affect the number of permutations by a factor of 2!. Therefore, we must divide 5! by 3! and 2!. Thus we obtain

$$\frac{5!}{(3!)(2!)} = \frac{5 \cdot \overset{2}{\cancel{4}} \cdot \cancel{3} \cdot \cancel{2} \cdot 1}{\cancel{3} \cdot \cancel{2} \cdot 1 \cdot \cancel{2} \cdot 1} = 10$$

distinguishable permutations of three H's and two T's. ■

The type of reasoning used in Problem 4 leads us to the following general counting technique. If there are n elements to be arranged, where there are r_1 of one kind, r_2 of another kind, r_3 of another kind, . . . , r_k of a kth kind, then the total number of distinguishable permutations is given by the expression

$$\frac{n!}{(r_1!)(r_2!)(r_3!) \cdots (r_k!)}$$

PROBLEM 5

How many different 11-letter permutations can be formed from the 11 letters of the word MISSISSIPPI?

Solution

Because there ar 4 I's, 4 S's, and 2 P's, we can form

$$\frac{11!}{(4!)(4!)(2!)} = \frac{11 \cdot 10 \cdot 9 \cdot 8 \cdot 7 \cdot 6 \cdot 5 \cdot 4 \cdot 3 \cdot 2 \cdot 1}{4 \cdot 3 \cdot 2 \cdot 1 \cdot 4 \cdot 3 \cdot 2 \cdot 1 \cdot 2 \cdot 1} = 34,650$$

distinguishable permutations. ■

Combinations (Subsets)

Permutations are *ordered* arrangements; however, *order* is often not a consideration. For example, suppose that we want to determine the number of three-person committees that can be formed from the five people: Al, Barb, Carol, Dawn, and Eric. Certainly the committee consisting of Al, Barb, and Eric is the same as the committee consisting of Barb, Eric, and Al. In other words, the order in which we choose or list the members is not important. Therefore, we are really dealing with subsets; that is, we are looking for the number of three-element subsets that can be formed from a set of five elements. Traditionally in this context, subsets have been called **combinations**. Stated another way, then, we are looking for the number of combinations of five things taken three at a time. In general, *r*-element subsets taken from a set of *n* elements are called **combinations of *n* things taken *r* at a time**. The symbol $C(n, r)$ denotes the number of these combinations.

Now let's restate that committee problem and show a detailed solution that can be generalized to handle a variety of problems dealing with combinations.

PROBLEM 6

How many three-person committees can be formed from the five people: Al, Barb, Carol, Dawn, and Eric?

Solution

Let's use the set {A, B, C, D, E} to represent the five people. Consider one possible three-person committee (subset), such as {A, B, C}; there are 3! permutations of these three letters. Now take another committee, such as {A, B, D}; there are also 3! permutations of these three letters. If we were to continue this process with all of the three-letter subsets that can be formed from the five letters, we would be counting all possible three-letter permutations of the five letters. That is, we would obtain $P(5, 3)$. Therefore, if we let $C(5, 3)$ represent the number of three-element subsets, then

$$(3!) \cdot C(5, 3) = P(5, 3)$$

Solving this equation for $C(5, 3)$ yields

$$C(5, 3) = \frac{P(5, 3)}{3!} = \frac{5 \cdot 4 \cdot 3}{3 \cdot 2 \cdot 1} = 10$$

So there are *ten* three-person committees that can be formed from the five people. ■

In general, $C(n, r)$ times $r!$ yields $P(n, r)$. Thus

$$(r!) \cdot C(n, r) = P(n, r)$$

and solving this equation for C(n, r) produces

$$C(n, r) = \frac{P(n, r)}{r!}$$

In other words, we can find the number of *combinations* of n things taken r at a time by dividing by $r!$ the number of permutations of n things taken r at a time. The following examples illustrate this idea.

$$C(2, 3) = \frac{P(7, 3)}{3!} = \frac{7 \cdot 6 \cdot 5}{3 \cdot 2 \cdot 1} = 35$$

$$C(9, 2) = \frac{P(9, 2)}{2!} = \frac{9 \cdot 8}{2 \cdot 1} = 36$$

$$C(10, 4) = \frac{P(10, 4)}{4!} = \frac{10 \cdot 9 \cdot 8 \cdot 7}{4 \cdot 3 \cdot 2 \cdot 1} = 210$$

PROBLEM 7

How many different five-card hands can be dealt from a deck of 52 playing cards?

Solution

Because the order in which the cards are dealt is not an issue, we are working with a combination (subset) problem. Thus, using the formula for $C(n, r)$, we obtain

$$C(52, 5) = \frac{P(52, 5)}{5!} = \frac{52 \cdot 51 \cdot 50 \cdot 49 \cdot 48}{5 \cdot 4 \cdot 3 \cdot 2 \cdot 1} = 2,598,960$$

There are 2,598,960 different five-card hands that can be dealt from a deck of 52 playing cards. ∎

Some counting problems, such as Problem 8, can be solved by using the fundamental principle of counting along with the combination formula.

PROBLEM 8

How many committees that consist of three women and two men can be formed from a group of five women and four men?

Solution

Let's think of this problem in terms of two tasks.

> **TASK I**   Choose a subset of three women from the five women. This can be done in

$$C(5, 3) = \frac{P(5, 3)}{3!} = \frac{5 \cdot 4 \cdot 3}{3 \cdot 2 \cdot 1} = 10 \text{ ways}$$

TASK 2   Choose a subset of two men from the four men. This can be done in

$$C(4, 2) = \frac{P(4, 2)}{2!} = \frac{4 \cdot 3}{2 \cdot 1} = 6 \text{ ways}$$

Task 1 followed by task 2 can be done in $(10)(6) = 60$ ways. Therefore, there are 60 committees consisting of three women and two men that can be formed. ■

Sometimes it takes a little thought to decide whether permutations or combinations should be used. Remember that **if order is to be considered, permutations should be used, but if order does not matter, then use combinations.** It is helpful to think of combinations as subsets.

PROBLEM 9

A small accounting firm has 12 computer programmers. Three of these people are to be promoted to systems analysts. In how many ways can the firm select the three people to be promoted?

Solution

Let's call the people A, B, C, D, E, F, G, H, I, J, K, and L. Suppose A, B, and C are chosen for promotion. Is this any different from choosing B, C, and A? Obviously not, so order does not matter and we are being asked a question about combinations. More specifically, we need to find the number of combinations of 12 people taken three at a time. Thus there are

$$C(12, 3) = \frac{P(12, 3)}{3!} = \frac{12 \cdot 11 \cdot 10}{3 \cdot 2 \cdot 1} = 220$$

different ways to choose the three people to be promoted. ■

PROBLEM 10

A club is to elect three officers—president, secretary, and treasurer—from a group of six people, all of whom are willing to serve in any office. How many different ways can the officers be chosen?

Solution

Let's call the candidates A, B, C, D, E, and F. Is electing A as president, B as secretary, and C as treasurer different from electing B as president, C as secretary, and A as treasurer? Obviously it is, and therefore we are working with permutations. Thus there are

$$P(6, 3) = 6 \cdot 5 \cdot 4 = 120$$

different ways of filling the offices. ■

PROBLEM SET 10.2

In Problems 1–12, evaluate each.

1. $P(5, 3)$
2. $P(8, 2)$

3. $P(6, 4)$
4. $P(9, 3)$

5. $C(7, 2)$
6. $C(8, 5)$

7. $C(10, 5)$
8. $C(12, 4)$

9. $C(15, 2)$
10. $P(5, 5)$

11. $C(5, 5)$
12. $C(11, 1)$

For Problems 13–44, solve each problem.

13. How many permutations of the four letters A, B, C, and D can be formed by using all the letters in each permutation?

14. In how many ways can six students be seated in a row of six seats?

15. How many three-person committees can be formed from a group of nine people?

16. How many two-card hands can be dealt from a deck of 52 playing cards?

17. How many three-letter permutations can be formed from the first eight letters of the alphabet (a) if repetitions are not allowed? (b) if repetitions are allowed?

18. In a seven-team baseball league, in how many ways can the top three positions in the final standings be filled?

19. In how many ways can the manager of a baseball team arrange his batting order of nine starters if he wants his best hitters in the top four positions?

20. In a baseball league of nine teams, how many games are needed to complete the schedule if each team plays 12 games with each other team?

21. How many committees consisting of four women and four men can be chosen from a group of seven women and eight men?

22. How many three-element subsets containing one vowel and two consonants can be formed from the set {a, b, c, d, e, f, g, h, i}?

23. Five associate professors are being considered for promotion to the rank of full professor, but only three will be promoted. How many different combinations of three could be promoted?

24. How many numbers of four different digits can be formed from the digits 1, 2, 3, 4, 5, 6, 7, 8, and 9 if each number must consist of two odd and two even digits?

25. How many three-element subsets containing the letter A can be formed from the set {A, B, C, D, E, F}?

26. How many four-person committees can be chosen from five women and three men if each committee must contain at least one man?

27. How many different seven-letter permutations can be formed from four identical H's and three identical T's?

28. How many different eight-letter permutations can be formed from six identical H's and two identical T's?

29. How many different nine-letter permutations can be formed from three identical A's, four identical B's, and two identical C's?

30. How many different ten-letter permutations can be formed from five identical A's, four identical B's, and one C?

31. How many different seven-letter permutations can be formed from the seven letters of the word ALGEBRA?

32. How many different 11-letter permutations can be formed from the 11 letters of the word MATHEMATICS?

33. In how many ways can x^4y^2 be written without using exponents? [*Hint:* One way is *xxxxyy.*]

34. In how many ways can $x^3y^4z^3$ be written without using exponents?

35. Ten basketball players are going to be divided into two teams of five players each for a game. In how many ways can this be done?

36. Ten basketball players are going to be divided into two teams of five in such a way that the two best players are on opposite teams. In how many ways can this be done?

37. A box contains nine good light bulbs and four defective bulbs. How many samples of three bulbs contain one defective bulb? How many samples of three bulbs contain *at least* one defective bulb?

38. How many five-person committees consisting of two juniors and three seniors can be formed from a group of six juniors and eight seniors?

39. In how many ways can six people be divided into two groups so that there are four in one group and two in the other? In how many ways can six people be divided into two groups of three each?

40. How many five-element subsets containing A and B can be formed from the set {A, B, C, D, E, F, G, H}?

41. How many four-element subsets containing A or B but not both A and B can be formed from the set {A, B, C, D, E, F, G}?

42. How many different five-person committees can be selected from nine people if two of those people refuse to serve together on a committee?

43. How many different line segments are determined by five points? By six points? By seven points? By *n* points?

44. *a.* How many five-card hands consisting of two kings and three aces can be dealt from a deck of 52 playing cards?

　　b. How many five-card hands consisting of three kings and two aces can be dealt from a deck of 52 playing cards?

　　c. How many five-card hands consisting of three cards of one face value and two cards of another face value can be dealt from a deck of 52 playing cards?

THOUGHTS INTO WORDS

45. Explain the difference between a permutation and a combination. Give an example of each one to illustrate your explanation.

46. Your friend is having difficulty distinguishing between permutations and combinations in problem-solving situations. What might you do to help her?

Further Investigations

47. In how many ways can six people be seated at a circular table? [*Hint:* Moving each person one place to the right (or left) does not create a new seating.]

48. The quantity $P(8, 3)$ can be expressed completely in factorial notation as follows.

$$P(8, 3) = \frac{P(8, 3) \cdot 5!}{5!} = \frac{(8 \cdot 7 \cdot 6)(5 \cdot 4 \cdot 3 \cdot 2 \cdot 1)}{5!} = \frac{8!}{5!}$$

Express each of the following in terms of factorial notation.

　a. $P(7, 3)$

　b. $P(9, 2)$

　c. $P(10, 7)$

　d. $P(n, r)$,　　$r \leq n$ and 0! is defined to be 1

49. Sometimes the formula

$$C(n, r) = \frac{n!}{r!(n - r)!}$$

is used to find the number of combinations of *n* things taken *r* at a time. Use the result from part (d) of Problem 48 and develop this formula.

50. Compute $C(7, 3)$ and $C(7, 4)$. Compute $C(8, 2)$ and $C(8, 6)$. Compute $C(9, 8)$ and $C(9, 1)$. Now argue that $C(n, r) = C(n, n - r)$ for $r \leq n$.

GRAPHING CALCULATOR ACTIVITIES

Before doing Problems 51–56 be sure that you can use your calculator to compute the number of permutations and combinations. Your calculator may possess a special sequence of keys for such computations. You may need to refer to your user's manual for this information.

51. Use your calculator to check your answers for Problems 1–12.

52. How many different five-card hands can be dealt from a deck of 52 playing cards?

53. How many different seven-card hands can be dealt from a deck of 52 playing cards?

54. How many different five-person committees can be formed from a group of 50 people?

55. How many different juries consisting of 11 people can be chosen from a group of 30 people?

56. How many seven-person committees consisting of three juniors and four seniors can be formed from 45 juniors and 53 seniors?

10.3   PROBABILITY

In order to introduce some terminology and notation, let's consider a simple experiment of tossing a regular six-sided die. There are six possible outcomes to this experiment: the 1, the 2, the 3, the 4, the 5, or the 6 will land up. This set of possible outcomes is called a *sample space*, and the individual elements of the sample space are called *sample points*. We will use S (sometimes with subscripts for identification purposes) to refer to a particular sample space of an experiment; then we will denote the number of sample points by $n(S)$. Thus, for the experiment of tossing a die, $S = \{1, 2, 3, 4, 5, 6\}$ and $n(S) = 6$

In general, the set of all possible outcomes of a given experiment is called the **sample space**, and the individual elements of the sample space are called **sample points**. (In this text we will be working only with sample spaces that are finite.)

Now suppose we are interested in some of the various possible outcomes in the die-tossing experiment. For example, we might be interested in the event *An even number comes up*. In this case we are satisfied if a 2, 4, or 6 appears on the top face of the die, and therefore the event *An even number comes up* is the subset $E = \{2, 4, 6\}$, where $n(E) = 3$. Perhaps, instead, we might be interested in the event *A multiple of 3 comes up*. This event determines the subset $F = \{3, 6\}$, where $n(F) = 2$.

In general, any subset of a sample space is called an **event** or an **event space.** If the event consists of exactly one element of the sample space, then it is called a **simple event**. Any nonempty event that is not simple is called a **compound event**. A compound event can be represented as the union of simple events.

It is now possible to give a very simple definition for *probability* as we want to use the term in this text.

> ## DEFINITION 10.1
>
> In an experiment where all possible outcomes in the sample space S are equally likely to occur, the **probability** of an event E is defined by
>
> $$P(E) = \frac{n(E)}{n(S)}$$
>
> where $n(E)$ denotes the number of elements in the event E and $n(S)$ denotes the number of elements in the sample space S.

Many probability problems can be solved by applying Definition 10.1. Such an approach requires that we be able to determine the number of elements in the sample space and the number of elements in the event space. For example, returning to the die-tossing experiment, the probability of getting an even number with one toss of the die is given by

$$P(E) = \frac{n(E)}{n(S)} = \frac{3}{6} = \frac{1}{2}$$

Let's consider two examples where the number of elements in both the sample space and the event space are quite easy to determine.

PROBLEM 1

A coin is tossed. Find the probability that a head turns up.

Solution

Let the sample space be $S = \{H, T\}$; then $n(S) = 2$. The event of a head turning up is the subset $E = \{H\}$, so $n(E) = 1$. Therefore, the probability of getting a head with one flip of a coin is given by

$$P(E) = \frac{n(E)}{n(S)} = \frac{1}{2}$$

■

PROBLEM 2

Two coins are tossed. What is the probability that *at least* one head will turn up?

Solution

For clarification purposes, let the coins be a penny and a nickel. The possible outcomes of this experiment are (1) a head on both coins (2) a head on the penny and a tail on the nickel (3) a tail on the penny and a head on the nickel, and (4) a tail on both coins. Using ordered-pair notation, where the first entry of a pair represents the penny and the second entry the nickel, we can write the sample space

$$S = \{(H, H), (H, T), (T, H), (T, T)\}$$

and $n(S) = 4$.

Let E be the event of getting at least one head. Thus $E = \{(H, H), (H, T), (T, H)\}$ and $n(E) = 3$. Therefore, the probability of getting at least one head with one toss of two coins is

$$P(E) = \frac{n(E)}{n(S)} = \frac{3}{4}$$

As you might expect, the counting techniques discussed in the first two sections of this chapter can frequently be used to solve probability problems.

PROBLEM 3

Four coins are tossed. Find the probability of getting three heads and one tail.

Solution

The sample space consists of the possible outcomes for tossing four coins. Because there are two things that can happen on each coin, by the fundamental principle of counting there are $2 \cdot 2 \cdot 2 \cdot 2 = 16$ possible outcomes for tossing four coins. Thus we know that $n(S) = 16$ without taking the time to list all of the elements. The event of getting three heads and one tail is the subset $E = \{(H, H, H, T), (H, H, T, H), (H, T, H, H), (T, H, H, H)\}$, where $n(E) = 4$. Therefore, the requested probability is

$$P(E) = \frac{n(E)}{n(S)} = \frac{4}{16} = \frac{1}{4}$$

PROBLEM 4

Al, Bob, Chad, Dawn, Eve, and Francis are randomly seated in a row of six chairs. What is the probability that Al and Bob are seated in the end seats?

Solution

The sample space consists of all possible ways of seating six people in six chairs, or, in other words, the permutations of six things taken six at a time. Thus $n(S) = P(6, 6) = 6! = 6 \cdot 5 \cdot 4 \cdot 3 \cdot 2 \cdot 1 = 720$.

The event space consists of all possible ways of seating the six people so that Al and Bob both occupy end seats. The number of these possibilities can be determined as follows.

> **TASK 1**   Put Al and Bob in the end seats. This can be done in two ways because Al can be on the left end and Bob on the right end, or vice versa.

> **TASK 2**   Put the other four people in the remaining four seats. This can be done in $4! = 4 \cdot 3 \cdot 2 \cdot 1 = 24$ different ways.

Therefore, task 1 followed by task 2 can be done in $(2)(24) = 48$ different ways, so $n(E) = 48$. Thus the requested probability is

$$P(E) = \frac{n(E)}{n(S)} = \frac{48}{720} = \frac{1}{15}$$

Note that in Problem 3, by using the fundamental principle of counting to determine the number of elements in the sample space, we did not actually have to list all of the elements. For the event space, we listed the elements and counted them in the usual way. In Problem 4 we used the permutation formula $P(n, n) = n!$ to determine the number of elements in the sample space, and then we used the fundamental principle to determine the number of elements in the event space. There are no definite rules about when to list the elements and when to apply some sort of counting technique. In general, we suggest that if you do not immediately see a counting pattern for a particular problem, you should begin the listing process. If a counting pattern then emerges as you are listing the elements, use the pattern at that time.

The combination (subset) formula we developed in Section 10.2, $C(n, r) = P(n, r)/r!$, is also a very useful tool for solving certain kinds of probability problems. The next three examples illustrate some problems of this type.

PROBLEM 5

A committee of three people is randomly selected from Alice, Barb, Chad, Dee, and Eric. What is the probability that Alice is on the committee?

Solution

The sample space, S, consists of all possible three-person committees that can be formed from the five people. Therefore,

$$n(S) = C(5, 3) = \frac{P(5, 3)}{3!} = \frac{5 \cdot 4 \cdot 3}{3 \cdot 2 \cdot 1} = 10$$

The event space, E, consists of all the three-person committees that have Alice as a member. Each of those committees contains Alice and two other people chosen from the four remaining people. Thus the number of such committees is $C(4, 2)$, so we obtain

$$n(E) = C(4, 2) = \frac{P(4, 2)}{2!} = \frac{4 \cdot 3}{2 \cdot 1} = 6$$

The requested probability is

$$P(E) = \frac{n(E)}{n(S)} = \frac{6}{10} = \frac{3}{5}$$ ∎

PROBLEM 6

A committee of four is chosen at random from a group of five seniors and four juniors. Find the probability that the committee will contain two seniors and two juniors.

Solution

The sample space, S, consists of all possible four-person committees that can be formed from the nine people. Thus

$$n(S) = C(9, 4) = \frac{P(9, 4)}{4!} = \frac{9 \cdot 8 \cdot 7 \cdot 6}{4 \cdot 3 \cdot 2 \cdot 1} = 126$$

The event space, E, consists of all four-person committees that contain two seniors and two juniors. They can be counted as follows.

> **TASK 1** Choose two seniors from the five available seniors in $C(5, 2) = 10$ ways.
>
> **TASK 2** Choose two juniors from the four available juniors in $C(4, 2) = 6$ ways.

Therefore, there are $10 \cdot 6 = 60$ committees consisting of two seniors and two juniors. The requested probability is

$$P(E) = \frac{n(E)}{n(S)} = \frac{60}{126} = \frac{10}{21} \qquad \blacksquare$$

PROBLEM 7

Eight coins are tossed. Find the probability of getting two heads and six tails.

Solution

Because either of two things can happen on each coin, the total number of possible outcomes, $n(S)$, is $2^8 = 256$.

We can select two coins, which are to fall heads, in $C(8,2) = 28$ ways. For each of these ways, there is only one way to select the other six coins that are to fall tails. Therefore, there are $28 \cdot 1 = 28$ ways of getting two heads and six tails, so $n(E) = 28$. The requested probability is

$$P(E) = \frac{n(E)}{n(S)} = \frac{28}{256} = \frac{7}{64} \qquad \blacksquare$$

PROBLEM SET 10.3

For Problems 1–4, *two* coins are tossed. Find the probability of tossing each of the following events.

1. One head and one tail

2. Two tails

3. At least one tail

4. No tails

For Problems 5–8, *three* coins are tossed. Find the probability of tossing each of the following events.

5. Three heads

6. Two heads and a tail

7. At least one head

8. Exactly one tail

For Problems 9–12, *four* coins are tossed. Find the probability of tossing each of the following events.

9. Four heads

10. Three heads and a tail

11. Two heads and two tails

12. At least one head

For Problems 13–16, *one* die is tossed. Find the probability of rolling each of the following events.

13. A multiple of 3

14. A prime number

15. An even number

16. A multiple of 7

For Problems 17–22, *two* dice are tossed. Find the probability of rolling each of the following events.

17. A sum of 6

18. A sum of 11

19. A sum less than 5

20. A 5 on exactly one die

21. A 4 on at least one die

22. A sum greater than 4

For Problems 23–26, *one* card is drawn from a standard deck of 52 playing cards. Find the probability of each of the following events.

23. A heart is drawn.

24. A king is drawn.

25. A spade or a diamond is drawn.

26. A red jack is drawn.

For Problems 27–30, suppose that 25 slips of paper numbered 1 to 25, inclusive, are put in a hat and then one is drawn out at random. Find the probability of each of the following events.

27. The slip with the 5 on it is drawn.

28. A slip with an even number on it is drawn.

29. A slip with a prime number on it is drawn.

30. A slip with a multiple of 6 on it is drawn.

For Problems 31–34, suppose that a committee of two boys is to be chosen at random from the five boys, Al, Bill, Carl, Dan, and Elmer. Find the probability of each of the following events.

31. Dan is on the committee.

32. Dan and Elmer are both on the committee.

33. Bill and Carl are not both on the committee.

34. Dan or Elmer, but not both of them, is on the committee.

For Problems 35–38, suppose that a five-person committee is selected at random from the eight people Al, Barb, Chad, Dominique, Eric, Fern, George, and Harriet. Find the probability of each of the following events.

35. Al and Barb are both on the committee.

36. George is not on the committee.

37. Either Chad or Dominique, but not both, is on the committee.

38. Neither Al nor Barb is on the committee.

For Problems 39–41, suppose that a box of ten items from a manufacturing company is known to contain two defective and eight nondefective items. A sample of three items is selected at random. Find the probability of each of the following events.

39. The sample contains all nondefective items.

40. The sample contains one defective and two nondefective items.

41. The sample contains two defective and one nondefective item.

For Problem 42–60, solve each problem.

42. A building has five doors. Find the probability that two people, entering the building at random, will choose the same door.

43. Bill, Carol, and Alice are to be seated at random in a row of three seats. Find the probability that Bill and Carol will be seated side by side.

44. April, Bill, Carl, and Denise are to be seated at random in a row of four chairs. What is the probability that April and Bill will occupy the end seats?

45. A committee of four girls is to be chosen at random from the five girls Alice, Becky, Candy, Dee, and Elaine. Find the probability that Elaine is not on the committee.

46. Three boys and two girls are to be seated at random in a row of five seats. What is the probability that the boys and girls will be in alternate seats?

47. Four different mathematics books and five different history books are randomly placed on a shelf. What is the probability that all of the books on a subject are side by side?

48. Each of three letters is to be mailed in any one of five different mailboxes. What is the probability that all will be mailed in the same mailbox?

49. Randomly form a four-digit number by using the digits 2, 3, 4, and 6 once each. What is the probability that the number formed is greater than 4000?

50. Randomly select one of the 120 permutations of the letters a, b, c, d, and e. Find the probability that in the chosen permutation, the letter a precedes the b (the a is to the left of the b).

51. A committee of four is chosen at random from a group of six women and five men. Find the probability that the committee contains two women and two men.

52. A committee of three is chosen at random from a group of four women and five men. Find the probability that the committee contains at least one man.

53. Ahmed, Bob, Carl, Dan, Ed, Frank, Gino, Harry, Julio, and Mike are randomly divided into two five-man teams for a basketball game. What is the probability that Ahmed, Bob, and Carl are on the same team?

54. Seven coins are tossed. Find the probability of getting four heads and three tails.

55. Nine coins are tossed. Find the probability of getting three heads and six tails.

56. Six coins are tossed. Find the probability of getting at least four heads.

57. Five coins are tossed. Find the probability of getting no more than three heads.

58. Each arrangement of the 11 letters of the word MISSIS-SIPPI is put on a slip of paper and placed in a hat. One slip is drawn at random from the hat. Find the probability that the slip contains an arrangement of the letters with the four S's at the beginning.

59. Each arrangement of the seven letters of the word OSMOSIS is put on a slip of paper and placed in a hat. One slip is drawn at random from the hat. Find the probability that the slip contains an arrangement of the letters with an O at the beginning and an O at the end.

60. Consider all possible arrangements of three identical H's and three identical T's. Suppose that one of these arrangements is selected at random. What is the probability that the selected arrangement has the three H's in consecutive positions?

THOUGHTS INTO WORDS

61. Explain the concepts of sample space and event space.

62. Why must probability answers fall between 0 and 1, inclusive? Give an example of a situation for which the probability is zero. Also give an example for which the probability is one.

Further Investigations

In Problem 7 of Section 10.2, we found that there are 2,598,960 different five-card hands that can be dealt from a hand of 52 playing cards. Therefore, probabilities for certain kinds of five-card poker hands can be calculated by using 2,598,960 as the number of elements in the sample space. For Problems 63–71, determine the number of different five-card poker hands of the indicated type that can be obtained.

63. A straight flush (five cards in sequence and of the same suit; aces are both low and high, so A2345 and 10JQKA are both acceptable)

64. Four of a kind (four of the same face value, such as four kings)

65. A full house (three cards of one face value and two cards of another face value)

66. A flush (five cards of the same suit but not in sequence)

67. A straight (five cards in sequence but not all of the same suit)

68. Three of a kind (three cards of one face value and two cards of two different face values)

69. Two pairs

70. Exactly one pair

71. No pairs

10.4   SOME PROPERTIES OF PROBABILITY; EXPECTED VALUES

There are several basic properties that are useful in the study of probability from both a theoretical and a computational viewpoint. We will discuss two of these properties at this time and some additional ones in the next section. The first property may seem to state the obvious, but still needs to be mentioned.

PROPERTY 10.1

For all events E,

$$0 \le P(E) \le 1$$

Property 10.1 simply states that probabilities must fall in the range from 0 to 1, inclusive. This seems reasonable because $P(E) = n(E)/n(S)$, and E is a subset of S. The next two examples illustrate circumstances where $P(E) = 0$ and $P(E) = 1$.

PROBLEM 1

Toss a regular six-sided die. What is the probability of getting a 7?

Solution

The sample space is $S = \{1, 2, 3, 4, 5, 6\}$, when $n(S) = 6$. The event space is $E = \varnothing$, so $n(E) = 0$. Therefore, the probability of getting a 7 is

$$P(E) = \frac{n(E)}{n(S)} = \frac{0}{6} = 0 \qquad \blacksquare$$

PROBLEM 2

What is the probability of getting a head or a tail with one flip of a coin?

Solution

The sample space is $S = \{H, T\}$ and the event space is $E = \{H, T\}$. Therefore, $n(S) = n(E) = 2$ and

$$P(E) = \frac{n(E)}{n(S)} = \frac{2}{2} = 1 \qquad \blacksquare$$

An event that has a probability of 1 is sometimes called **certain success**, and an event with a probability of zero is called a **certain failure**.

It should also be mentioned that Property 10.1 serves as a check for reasonableness of answers. In other words, when computing probabilities, we know that our answer must fall between 0 and 1, inclusive. Any other probability answer is simply not reasonable.

Complementary Events

Complementary events are complementary sets such that S, the sample space, serves as the universal set. The following examples illustrate this idea.

Sample Space	Event Space	Complement of Event Space
$S = \{1, 2, 3, 4, 5, 6\}$	$E = \{1, 2\}$	$E' = \{3, 4, 5, 6\}$
$S = \{H, T\}$	$E = \{T\}$	$E' = \{H\}$
$S = \{2, 3, 4, \ldots, 12\}$	$E = \{2, 3, 4\}$	$E' = \{5, 6, 7, \ldots, 12\}$
$S = \{1, 2, 3, \ldots, 25\}$	$E = \{3, 4, 5, \ldots, 25\}$	$E' = \{1, 2\}$

In each case, note that E' (the complement of E) consists of all elements of S that are *not* in E. Thus E and E' are called *complementary events*. Also note that for each example, $P(E) + P(E') = 1$. We can state the following general property.

PROPERTY 10.2

If E is any event of a sample space S, and E' is the complementary event, then

$$P(E) + P(E') = 1$$

From a computational viewpoint, Property 10.2 provides us with a double-barreled attack on some probability problems. That is, once we compute either $P(E)$ or $P(E')$, then we can determine the other one simply by subtracting from 1. For example, suppose that for a particular problem we can determine that $P(E) = \dfrac{3}{13}$. Then we immediately know that $P(E') = 1 - P(E) = 1 - \dfrac{3}{13} = \dfrac{10}{13}$. The following examples further illustrate the usefulness of Property 10.2.

PROBLEM 3

Two dice are tossed. Find the probability of getting a sum greater than 3.

Solution

 Let S be the familiar sample space of ordered pairs for this problem, where $n(S) = 36$. Let E be the event of obtaining a sum greater than 3. Then E' is the event of obtaining a sum less than or equal to 3; that is, $E' = \{(1, 1), (1, 2), (2, 1)\}$. Thus

$$P(E') = \frac{n(E')}{n(S)} = \frac{3}{36} = \frac{1}{12}$$

From this, we conclude that

$$P(E) = 1 - P(E') = 1 - \frac{1}{12} = \frac{11}{12}$$ ■

PROBLEM 4

Toss three coins and find the probability of getting at least one head.

Solution

The sample space, S, consists of all possible outcomes for tossing three coins. Using the fundamental principle of counting, we know that there are $(2)(2)(2) = 8$ outcomes, so $n(S) = 8$. Let E be the event of getting at least one head. Then E' is the complementary event of not getting any heads. The set E' is easy to list: $E' = \{(T, T, T)\}$. Thus $n(E') = 1$ and $P(E') = \dfrac{1}{8}$. From this, $P(E)$ can be determined to be

$$P(E) = 1 - P(E') = 1 - \frac{1}{8} = \frac{7}{8}$$

■

PROBLEM 5

A three-person committee is chosen at random from a group of five women and four men. Find the probability that the committee contains at least one woman.

Solution

Let the sample space, S, be the set of all possible three-person committees that can be formed from nine people. There are $C(9, 3) = 84$ such committees; therefore, $n(S) = 84$.

Let E be the event *The committee contains at least one woman*. Then E' is the complementary event, *The committee contains all men*. Thus E' consists of all three-man committees that can be formed from four men. There are $C(4, 3) = 4$ such committees; thus $n(E') = 4$. Therefore, we have

$$P(E') = \frac{n(E')}{n(S)} = \frac{4}{84} = \frac{1}{21}$$

which determines $P(E)$ to be

$$P(E) = 1 - P(E') = 1 - \frac{1}{21} = \frac{20}{21}$$

■

The concepts of **set intersection** and **set union** play an important role in the study of probability. If E and F are two events in a sample space S, then $E \cap F$ is the event consisting of all sample points of S that are in both E and F as indicated in Figure 10.2. Likewise, $E \cup F$ is the event consisting of all sample points of S that are in E or F, or both, as shown in Figure 10.3.

In Figure 10.4 there are 47 sample points in E, 38 sample points in F, and 15 sample points in $E \cap F$. How many sample points are there in $E \cup F$? Simply

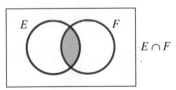

$E \cap F$

FIGURE 10.2

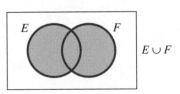

$E \cup F$

FIGURE 10.3

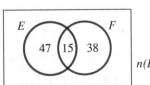

$n(E) = 47$
$n(F) = 38$
$n(E \cap F) = 15$

FIGURE 10.4

adding the number of points in E and F would result in counting the 15 points in $E \cap F$ twice. Therefore, 15 must be subtracted from the total number of points in E

and F, yielding $47 + 38 - 15 = 70$ points in $E \cup F$. We can state the following general counting property:

$$n(E \cup F) = n(E) + n(F) - n(E \cap F)$$

If we divide both sides of this equation by $n(S)$, we obtain the following probability property.

PROPERTY 10.3

For events E and F of a sample space S,

$$P(E \cup F) = P(E) + P(F) - P(E \cap F)$$

PROBLEM 6

What is the probability of getting an odd number or a prime number with one toss of a die?

Solution

Let $S = \{1, 2, 3, 4, 5, 6\}$ be the sample space, $E = \{1, 3, 5\}$ the event of getting an odd number, and $F = \{2, 3, 5\}$ the event of getting a prime number. Then $E \cap F = \{3, 5\}$, and using Property 11.3, we obtain

$$P(E \cup F) = \frac{3}{6} + \frac{3}{6} - \frac{2}{6} = \frac{4}{6} = \frac{2}{3}$$

∎

PROBLEM 7

Toss three coins. What is the probability of getting at least two heads or exactly one tail?

Solution

Using the fundamental principle of counting, we know that there are $2 \cdot 2 \cdot 2 = 8$ possible outcomes of tossing three coins; thus $n(S) = 8$. Let

$$E = \{(H, H, H), (H, H, T), (H, T, H), (T, H, H)\}$$

be the event of getting at least two heads, and let

$$F = \{(H, H, T), (H, T, H), (T, H, H)\}$$

be the event of getting exactly one tail. Then

$$E \cap F = \{(H, H, T), (H, T, H), (T, H, H)\}$$

and we can compute $P(E \cup F)$ as follows.

$$P(E \cup F) = P(E) + P(F) - P(E \cap F)$$
$$= \frac{4}{8} + \frac{3}{8} - \frac{3}{8}$$
$$= \frac{4}{8} = \frac{1}{2}$$

     In Property 10.3, if $E \cap F = \varnothing$, then the events E and F are said to be **mutually exclusive**. In other words, mutually exclusive events are events that cannot occur at the same time. For example, when we roll a die, the event of getting a 4 and the event of getting a 5 are mutually exclusive; they cannot both happen on the same roll. If $E \cap F = \varnothing$, then $P(E \cap F) = 0$ and Property 11.3 becomes $P(E \cup F) = P(E) + P(F)$ **for mutually exclusive events.**

PROBLEM 8

Suppose we have a jar that contains five white, seven green, and nine red marbles. If one marble is drawn at random from the jar, find the probability that it is white or green.

Solution

The events of drawing a white marble and drawing a green marble are mutually exclusive. Therefore, the probability of drawing a white or a green marble is

$$\frac{5}{12} + \frac{7}{21} = \frac{12}{21} = \frac{4}{7}$$

    Note that in the solution for Problem 8 we did not explicitly name and list the elements of the sample space or event spaces. It was obvious that the sample space contained 21 elements (21 marbles in the jar) and the event spaces contained five elements (five white marbles) and seven elements (seven green marbles). Thus it was unnecessary to name and list the sample space and event spaces.

PROBLEM 9

Suppose that the data in the following table represent the results of a survey of 1000 drivers after a holiday weekend.

	RAIN (R)	NO RAIN (R')	TOTAL
ACCIDENT (A)	35	10	45
NO ACCIDENT (A')	450	505	955
TOTAL	485	515	1000

If a person is selected at random, what is the probability that the person was in an accident or that it rained?

Solution

First, let's form a **probability table** by dividing each entry by 1000, the total number surveyed.

	RAIN (R)	**NO RAIN (R')**	**TOTAL**
ACCIDENT (A)	0.035	0.010	0.045
NO ACCIDENT (A')	0.450	0.505	0.955
TOTAL	0.485	0.515	1.000

Now we can use Property 10.3 and compute $P(A \cup R)$.

$$P(A \cup R) = P(A) + P(R) - P(A \cap R)$$
$$= 0.045 + 0.485 - 0.035$$
$$= 0.495$$

■

Expected Value

Suppose we toss a coin 500 times. We would expect to get approximately 250 heads. In other words, because the probability of getting a head with one toss of a coin is $\frac{1}{2}$, then in 500 tosses we should get approximately $500\left(\frac{1}{2}\right) = 250$ heads. The word *approximately* conveys a key idea. As we know from experience, it is possible to toss a coin several times and get all heads. However, with a large number of tosses, things should average out so that we get about an equal number of heads and tails.

As another example, consider the fact that the probability of getting a sum of 6 with one toss of a pair of dice is $\frac{5}{36}$. Therefore, if a pair of dice is tossed 360 times, we should expect to get a sum of 6 approximately $360\left(\frac{5}{36}\right) = 50$ times.

Let us now define the concept of *expected value*.

DEFINITION 10.2

If the k possible outcomes of an experiment are assigned the values x_1, $x_2, x_3, \ldots, x_k$, and if they occur with probabilities of $p_1, p_2, p_3, \ldots,$ p_k, respectively, then the **expected value** of the experiment (E_v) is given by

$$E_v = x_1 p_1 + x_2 p_2 + x_3 p_3 + \cdots + x_k p_k$$

The concept of expected value (also called **mathematical expectation**) is used in a variety of probability situations that deal with such things as fairness of games and decision making in business ventures. Let's consider some examples.

PROBLEM 10

Suppose that you buy one ticket in a lottery where 1000 tickets are sold. Furthermore, suppose that three prizes are awarded: one of $500, one of $300, and one of $100. What is your mathematical expectation?

Solution

Because you bought one ticket, the probability of your winning $500 is $\frac{1}{1000}$, that of your winning $300 is $\frac{1}{1000}$, and that of your winning $100 is $\frac{1}{1000}$. Multiplying each of these probabilities times the corresponding prize money and then adding the results yields your mathematical expectation.

$$E_v = \$500\left(\frac{1}{1000}\right) + \$300\left(\frac{1}{1000}\right) + \$100\left(\frac{1}{1000}\right)$$
$$= \$.50 + \$.30 + \$.10$$
$$= \$.90$$

In Problem 10, if you pay more than $.90 for a ticket, then it is not a **fair game** from your standpoint. If the price of the game is included in the calculation of the expected value, then a fair game is defined to be one where the expected value is zero.

PROBLEM 11

A player pays $5 to play a game where the probability of winning is $\frac{1}{5}$ and the probability of losing is $\frac{4}{5}$. If the player wins the game, he receives $25. Is this a fair game for the player?

Solution

Using Definition 10.2, let $x_1 = \$20$, which represents the $25 won minus the $5 paid to play, and let $x_2 = -\$5$, the amount paid to play the game. We are also given that $p_1 = \frac{1}{5}$ and $p_2 = \frac{4}{5}$. Thus the expected value is

$$E_v = \$20\left(\frac{1}{5}\right) + (-\$5)\left(\frac{4}{5}\right)$$
$$= \$4 - \$4$$
$$= 0$$

Because the expected value is zero, it is a fair game.

| PROBLEM 12 | Suppose you are interested in insuring a diamond ring for $2000 against theft. An insurance company charges a premium of $25 per year, claiming that there is a probability of 0.01 that the ring will be stolen during the year. What is your expected gain or loss if you take out the insurance? |

Solution

Using Definition 10.2, let $x_1 = \$1975$, which represents the $2000 minus the cost of the premium, $25, and let $x_2 = -\$25$. We also are given that $p_1 = 0.1$, so $p_2 = 1 - 0.01 = 0.99$. Thus the expected value is

$$
\begin{aligned}
E_v &= \$1975(0.01) + (-\$25)(0.99) \\
&= \$19.75 - \$24.75 \\
&= -\$5.00
\end{aligned}
$$

This means that if you insure with this company over many years and the circumstances remain the same, you will have an average net loss of $5 per year. ∎

PROBLEM SET 10.4

For Problems 1–4, *two* dice are tossed. Find the probability of rolling each of the following events.

1. A sum of 6

2. A sum greater than 2

3. A sum less than 8

4. A sum greater than 1

For Problems 5–8, *three* dice are tossed. Find the probability of rolling each of the following events.

5. A sum of 3

6. A sum greater than 4

7. A sum less than 17

8. A sum greater than 18

For Problems 9–12, *four* coins are tossed. Find the probability of getting each of the following events.

9. Four heads

10. Three heads and a tail

11. At least one tail

12. At least one head

For Problems 13–16, *five* coins are tossed. Find the probability of getting each of the following events.

13. Five tails

14. Four heads and a tail

15. At least one tail

16. At least two heads

For Problems 17–23, solve each problem.

17. Toss a pair of dice. What is the probability of not getting a double?

18. The probability that a certain horse will win the Kentucky Derby is $\dfrac{1}{20}$. What is the probability that it will lose the race?

19. One card is randomly drawn from a deck of 52 playing cards. What is the probability that it is not an ace?

20. Six coins are tossed. Find the probability of getting at least two heads.

21. A subset of two letters is chosen at random from the set {a, b, c, d, e, f, g, h, i}. Find the probability that the subset contains at least one vowel.

22. A two-person committee is chosen at random from a group of four men and three women. Find the probability that the committee contains at least one man.

23. A three-person committee is chosen at random from a group of seven women and five men. Find the probability that the committee contains at least one man.

For Problems 24–27, one die is tossed. Find the probability of rolling each of the following events.

24. A 3 or an odd number

25. A 2 or an odd number

26. An even number or a prime number

27. An odd number or a multiple of 3

For Problems 28–31, two dice are tossed. Find the probability of rolling each of the following events.

28. A double or a sum of 6

29. A sum of 10 or a sum greater than 8

30. A sum of 5 or a sum greater than 10

31. A double or a sum of 7

For Problems 32–56, solve each problem.

32. Two coins are tossed. Find the probability of getting exactly one head or at least one tail.

33. Three coins are tossed. Find the probability of getting at least two heads or exactly one tail.

34. A jar contains seven white, six blue, and ten red marbles. If one marble is drawn at random from the jar, find the probability that (a) the marble is white or blue; (b) the marble is white or red; (c) the marble is blue or red.

35. A coin and a die are tossed. Find the probability of getting a head on the coin or a 2 on the die.

36. A card is randomly drawn from a deck of 52 playing cards. Find the probability that it is a red card or a face card. (Jacks, queens, and kings are the face cards.)

37. The data in the following table represents the results of a survey of 1000 drivers after a holiday weekend.

	RAIN (R)	**NO RAIN (R′)**	**TOTAL**
ACCIDENT (A)	45	15	60
NO ACCIDENT (A′)	350	590	940
TOTAL	395	605	1000

If a person is selected at random from those surveyed, find the probability of each of the following events. (Express the probabilities in decimal form.)

a. The person was in an accident or it rained.

b. The person was not in an accident or it rained.

c. The person was not in an accident or it did not rain.

38. One hundred people were surveyed, and one question pertained to their educational background. The results of this question are given in the following table.

	FEMALE (F)	**MALE (F′)**	**TOTAL**
COLLEGE DEGREE (D)	30	20	50
NO COLLEGE DEGREE (D′)	15	35	50
TOTAL	45	55	100

If a person is selected at random from those surveyed, find the probability of each of the following events. Express the probabilities in decimal form.

a. The person is female or has a college degree.

b. The person is male or does not have a college degree.

c. The person is female or does not have a college degree.

39. In a recent election there were 1000 eligible voters. They were asked to vote on two issues, *A* and *B*. The results were as follows: 300 people voted for *A*, 400 people voted for *B*, and 175 voted for both *A* and *B*. If one person is chosen at random from the 1000 eligible voters, find the probability that the person voted for *A* or *B*.

40. A company has 500 employees among whom 200 are females, 15 are high-level executives, and 7 of the high-level executives are females. If one of the 500 employees is chosen at random, find the probability that the person chosen is female or a high-level executive.

41. A die is tossed 360 times. How many times would you expect to get a 6?

42. Two dice are tossed 360 times. How many times would you expect to get a sum of 5?

43. Two dice are tossed 720 times. How many times would you expect to get a sum greater than 9?

44. Four coins are tossed 80 times. How many times would you expect to get one head and three tails?

45. Four coins are tossed 144 times. How many times would you expect to get four tails?

46. Two dice are tossed 300 times. How many times would you expect to get a double?

47. Three coins are tossed 448 times. How many times would you expect to get three heads?

48. Suppose 5000 tickets are sold in a lottery. There are three prizes: The first is $1000, the second is $500, and the third is $100. What is the mathematical expectation of winning?

49. Your friend challenges you with the following game: You are to roll a pair of dice, and he will give you $5 if you roll a sum of 2 or 12, $2 if you roll a sum of 3 or 11, $1 if you roll a sum of 4 or 10. Otherwise you are to pay him $1. Should you play the game?

50. A contractor bids on a building project. There is a probability of 0.8 that he can show a profit of $30,000 and a probability of 0.2 that he will have to absorb a loss of $10,000. What is his mathematical expectation?

51. Suppose a person tosses two coins and receives $5 if 2 heads come up, receives $2 if 1 head and 1 tail come up,

and has to pay $2 if 2 tails come up. Is it a fair game for him?

52. A "wheel of fortune" is divided into four colors: red, white, blue, and yellow. The probability of the spinner landing on each of the colors and the money received is given by the following chart. The price to spin the wheel is $1.50. Is it a fair game?

COLOR	PROBABILITY OF LANDING ON THE COLOR	MONEY RECEIVED FOR LANDING ON THE COLOR
red	$\frac{4}{10}$	$.50
white	$\frac{3}{10}$	1.00
blue	$\frac{2}{10}$	2.00
yellow	$\frac{1}{10}$	5.00

53. A contractor estimates a probability of 0.7 of making $20,000 on a building project and a probability of 0.3 of losing $10,000 on the project. What is his mathematical expectation?

54. A farmer estimates his corn crop at 30,000 bushels. On the basis of past experience, he also estimates a probability of $\frac{3}{5}$ that he will make a profit of $.50 per bushel and a probability of $\frac{1}{5}$ of losing $.30 per bushel. What is his expected income from the corn crop?

55. Bill finds that the annual premium for insuring a stereo system for $2500 against theft is $75. If the probability that the set will be stolen during the year is 0.02, what is Bill's expected gain or loss by taking out the insurance?

56. Sandra finds that the annual premium for a $2000 insurance policy against the theft of a painting is $100. If the probability that the painting will be stolen during the year is 0.01, what is Sandra's expected gain or loss by taking out the insurance?

THOUGHTS INTO WORDS

57. If the probability of some event happening is 0.4, what is the probability of the event not happening? Explain your answer.

58. Explain each of the following concepts to a friend who missed class the day this section was discussed: using

complementary events to determine probabilities, using union and intersection of sets to determine probabilities, and using expected value to determine the fairness of a game.

Further Investigations

The term **odds** is sometimes used to express a probability statement. For example, we might say, "The odds in favor of the Cubs winning the pennant are 5 to 1," or "The odds against the Mets winning the pennant are 50 to 1." *Odds in favor* and *odds against* for equally likely outcomes can be defined as follows.

$$\text{Odds in favor} = \frac{\text{number of favorable outcomes}}{\text{number of unfavorable outcomes}}$$

$$\text{Odds against} = \frac{\text{number of unfavorable outcomes}}{\text{number of favorable outcomes}}$$

We have used the fractional form to define odds; however, in practice, the *to* vocabulary is commonly used. Thus the odds in favor of rolling a 4 with one roll of a die are usually stated as *1 to 5* instead of $\frac{1}{5}$. The odds against rolling a 4 are stated as *5 to 1*.

The *odds in favor of* statement about the Cubs means that there are 5 favorable outcomes compared to 1 unfavorable, or a total of 6 possible outcomes. So the *5 to 1 in favor of* statement also means that the probability of the Cubs winning the pennant is $\frac{5}{6}$. Likewise, the *50 to 1 against* statement about the Mets means that the probability that the Mets will not win the pennant is $\frac{50}{51}$.

Odds are usually stated in reduced form. For example, odds of 6 to 4 are usually stated as 3 to 2. Likewise, a fraction representing probability is reduced before being changed to a statement about odds.

59. What are the odds in favor of getting three heads with a toss of three coins?

60. What are the odds against getting four tails with a toss of four coins?

61. What are the odds against getting three heads and two tails with a toss of five coins?

62. What are the odds in favor of getting four heads and two tails with a toss of six coins?

63. What are the odds in favor of getting a sum of 5 with one toss of a pair of dice?

64. What are the odds against getting a sum greater than 5 with one toss of a pair of dice?

65. Suppose that one card is drawn at random from a deck of 52 playing cards. Find the odds against drawing a red card.

66. Suppose that one card is drawn at random from a deck of 52 playing cards. Find the odds in favor of drawing an ace or a king.

67. If $P(E) = \frac{4}{7}$ for some event E, find the odds in favor of E happening.

68. If $P(E) = \frac{5}{9}$ for some event E, find the odds against E happening.

69. Suppose that there is a predicted 40% chance of freezing rain. State the prediction in terms of the odds against getting freezing rain.

70. Suppose that there is a predicted 20% chance of thunderstorms. State the prediction in terms of the odds in favor of getting thunderstorms.

71. If the odds against an event happening are 5 to 2, find the probability that the event will occur.

72. The odds against Belly Dancer winning the fifth race are 20 to 9. What is the probability of Belly Dancer winning the fifth race?

73. The odds in favor of the Mets winning the pennant are stated as 7 to 5. What is the probability of the Mets winning the pennant?

74. The following chart contains some poker-hand probabilities. Complete the center column *Odds Against Being Dealt This Hand*. Note that fractions are reduced before being changed to odds.

5-CARD HAND	PROBABILITY OF BEING DEALT THIS HAND	ODDS AGAINST BEING DEALT THIS HAND
straight flush	$\dfrac{40}{2{,}598{,}960} = \dfrac{1}{64{,}974}$	64,973 to 1
four of a kind	$\dfrac{624}{2{,}598{,}960} =$	
full house	$\dfrac{3744}{2{,}598{,}960} =$	
flush	$\dfrac{5108}{2{,}598{,}960} =$	
straight	$\dfrac{10{,}200}{2{,}598{,}960} =$	
three of a kind	$\dfrac{54{,}912}{2{,}598{,}960} =$	
two pairs	$\dfrac{123{,}552}{2{,}598{,}960} =$	
one pair	$\dfrac{1{,}098{,}240}{2{,}598{,}960} =$	
no pairs	$\dfrac{1{,}302{,}540}{2{,}598{,}960} =$	

10.5 CONDITIONAL PROBABILITY; DEPENDENT AND INDEPENDENT EVENTS

Two events are often related in such a way that the probability of one of them may vary depending upon whether the other event has occurred. For example, the probability of rain may change drastically if additional information is obtained indicating a front moving through the area. Mathematically, the additional information about the front changes the sample space for the probability of rain.

In general, the probability of the occurence of an event E, given the occurrence of another event F, is called a **conditional probability** and is denoted $P(E|F)$. Let's look at a simple example and use it to motivate a definition for conditional probability.

What is the probability of rolling a prime number in one roll of a die? Let $S = \{1, 2, 3, 4, 5, 6\}$, so $n(S) = 6$; and let $E = \{2, 3, 5\}$, so $n(E) = 3$. Therefore,

$$P(E) = \frac{n(E)}{n(S)} = \frac{3}{6} = \frac{1}{2}$$

Next, what is the probability of rolling a prime number in one roll of a die, *given that an odd number has turned up?* Let $F = \{1, 3, 5\}$ be the new sample space of odd numbers. Then $n(F) = 3$. We are now interested in only that part of E (rolling a prime number) that is also in F, in other words, $E \cap F$. Therefore, because $E \cap F = \{3, 5\}$, the probability of E given F is

$$P(E|F) = \frac{n(E \cap F)}{n(F)} = \frac{2}{3}$$

When we divide both the numerator and the denominator of $n(E \cap F)/n(F)$ by $n(S)$, we obtain

$$\frac{\dfrac{n(E \cap F)}{n(S)}}{\dfrac{n(F)}{n(S)}} = \frac{P(E \cap F)}{P(F)}$$

Therefore, we can state the following general definition of the conditional probability of E given F for arbitrary events E and F.

DEFINITION 10.3   *Conditional Probability*

$$P(E|F) = \frac{P(E \cap F)}{P(F)}, \qquad P(F) \neq 0$$

In a problem in the previous section, the following probability table was formed relative to car accidents and weather conditions on a holiday weekend.

	RAIN (R)	NO RAIN (R′)	TOTAL
ACCIDENT (A)	0.035	0.010	0.045
NO ACCIDENT (A′)	0.450	0.505	0.955
TOTAL	0.485	0.515	1.000

Some conditional probabilities that can be calculated from the table follow.

$$P(A|R) = \frac{P(A \cap R)}{P(R)} = \frac{0.035}{0.485} = \frac{35}{485} = \frac{7}{97}$$

$$P(A'|R) = \frac{P(A' \cap R)}{P(R)} = \frac{0.450}{0.485} = \frac{450}{485} = \frac{90}{97}$$

$$P(A|R') = \frac{P(A \cap R')}{P(R')} = \frac{0.010}{0.515} = \frac{10}{515} = \frac{2}{103}$$

Note that the probability of an accident given that it was raining, $P(A\,|\,R)$, is greater than the probability of an accident given that it was not raining, $P(A\,|\,R')$. This seems reasonable.

PROBLEM 1

A die is tossed. Find the probability that a 4 came up if it is known that an even number turned up.

Solution

Let E be the event of rolling a 4, and let F be the event of rolling an even number. Therefore, $E = \{4\}$ and $F = \{2, 4, 6\}$, from which we obtain $E \cap F = \{4\}$. Using Definition 10.3, we obtain

$$P(E\,|\,F) = \frac{P(E \cap F)}{P(F)} = \frac{\dfrac{1}{6}}{\dfrac{3}{6}} = \frac{1}{3}$$ ■

PROBLEM 2

Suppose the probability that a student will enroll in a mathematics course is 0.45, the probability that he or she will enroll in a science course is 0.38, and the probability that he or she will enroll in both courses is 0.26. Find the probability that a student will enroll in a mathematics course, given that he or she is also enrolled in a science course. Also, find the probability that a student will enroll in a science course given that he or she is enrolled in mathematics.

Solution

Let M be the event *Will enroll in mathematics*, and let S be the event *Will enroll in science*. Therefore, using Definition 10.3, we obtain

$$P(M\,|\,S) = \frac{P(M \cap S)}{P(S)} = \frac{0.26}{0.38} = \frac{26}{38} = \frac{13}{19}$$

and

$$P(S\,|\,M) = \frac{P(S \cap M)}{P(M)} = \frac{0.26}{0.45} = \frac{26}{45}$$ ■

Independent and Dependent Events

Suppose that, when computing a conditional probability, we find that

$$P(E\,|\,F) = P(E)$$

This means that the probability of E is not affected by the occurrence or nonoccurence of F. In such a situation, we say that event E is *independent* of event F. It can be shown that if event E is independent of event F, then F is also independent of E; thus E and F are referred to as **independent events**. Furthermore, from the equations

$$P(E \mid F) = \frac{P(E \cap F)}{P(F)} \quad \text{and} \quad P(E \mid F) = P(E)$$

we see that

$$\frac{P(E \cap F)}{P(F)} = P(E)$$

which can be written

$$P(E \cap F) = P(E)P(F)$$

Therefore, we state the following general definition.

DEFINITION 10.4

Two events E and F are said to be **independent** if and only if

$$P(E \cap F) = P(E)P(F)$$

Two events that are not independent are called **dependent events**.

In the probability table preceding Problem 1, we see that $P(A) = 0.045$, $P(R) = 0.485$, and $P(A \cap R) = 0.035$. Because

$$P(A)P(R) = (0.045)(0.485) = 0.021825$$

and this does not equal $P(A \cap R)$, the events A (have a car accident) and R (rainy conditions) are not independent. This is not too surprising; we would certainly expect rainy conditions and automobile accidents to be related.

PROBLEM 3

Suppose we roll a white die and a red die. If we let E be the event *We roll a 4 on the white die* and we let F be the event *We roll a 6 on the red die*, are E and F independent events?

Solution

The sample space for rolling a pair of dice has $(6)(6) = 36$ elements. Using ordered pair notation, where the first entry represents the white die and the second entry the red die, we can list events E and F as follows.

$$E = \{(4, 1), (4, 2), (4, 3), (4, 4), (4, 5), (4, 6)\}$$
$$F = \{(1, 6), (2, 6), (3, 6), (4, 6), (5, 6), (6, 6)\}$$

Therefore, $E \cap F = \{(4, 6)\}$. Because $P(F) = \frac{1}{6}$, $P(E) = \frac{1}{6}$, and $P(E \cap F) = \frac{1}{36}$, we see that $P(E \cap F) = P(E)P(F)$, and the events E and F are independent. ∎

PROBLEM 4	Two coins are tossed. Let E be the event *Toss not more than one head*, and let F be the event *Toss at least one of each face*. Are these events independent?

Solution

The sample space has $(2)(2) = 4$ elements. The events E and F can be listed as follows.

$$E = \{(H, T), (T, H), (T, T)\}$$
$$F = \{(H, T), (T, H)\}$$

Therefore, $E \cap F = \{(H, T), (T, H)\}$. Because $P(E) = \dfrac{3}{4}$, $P(F) = \dfrac{1}{2}$, and $P(E \cap F) = \dfrac{1}{2}$, we see that $P(E \cap F) \neq P(E)P(F)$, so the events E and F are dependent.   ■

Sometimes the independence issue can be decided by the physical nature of the events in the problem. For instance, in Problem 3 it should seem evident that rolling a 4 on the white die is not affected by rolling a 6 on the red die. However, as in Problem 4, the description of the events may not clearly indicate whether the events are dependent.

From a problem-solving viewpoint, the following two statements are very helpful.

1. If E and F are independent events, then

$$P(E \cap F) = P(E)P(F)$$

(This property generalizes to any finite number of independent events.)

2. If E and F are dependent events, then

$$P(E \cap F) = P(E)P(F|E)$$

Let's analyze some problems using these ideas.

PROBLEM 5	A die is rolled three times. (This is equivalent to rolling three dice once each.) What is the probability of getting a 6 all three times?

Solution

The events of a 6 on the first roll, a 6 on the second roll, and a 6 on the third roll are independent events. Therefore, the probability of getting three 6's is

$$\left(\frac{1}{6}\right)\left(\frac{1}{6}\right)\left(\frac{1}{6}\right) = \frac{1}{216}$$

■

PROBLEM 6

A jar contains five white, seven green, and nine red marbles. If two marbles are drawn in succession *without replacement*, find the probability that both marbles are white.

Solution

Let E be the event of drawing a white marble on the first draw, and let F be the event of drawing a white marble on the second draw. Because the marble drawn first is not to be replaced before the second marble is drawn, we have dependent events. Therefore,

$$P(E \cap F) = P(E)P(F \mid E)$$
$$= \left(\frac{5}{21}\right)\left(\frac{4}{20}\right) = \frac{20}{420} = \frac{1}{21}$$

$P(F \mid E)$ means the probability of drawing a white marble on the second draw, given that a white marble was obtained on the first draw. ∎

The concept of *mutually exclusive events* may also enter the picture when we are working with independent or dependent events. Our final problems of this section illustrate this idea.

PROBLEM 7

A coin is tossed three times. Find the probability of getting two heads and one tail.

Solution

Two heads and one tail can be obtained in three different ways: (1) HHT (head on first toss, head on second toss, and tail on third toss), (2) HTH, and (3) THH. Thus we have three *mutually exclusive* events, each of which can be broken into *independent* events: first toss, second toss, and third toss. Therefore, the probability can be computed as follows

$$\left(\frac{1}{2}\right)\left(\frac{1}{2}\right)\left(\frac{1}{2}\right) + \left(\frac{1}{2}\right)\left(\frac{1}{2}\right)\left(\frac{1}{2}\right) + \left(\frac{1}{2}\right)\left(\frac{1}{2}\right)\left(\frac{1}{2}\right) = \frac{3}{8}$$ ∎

PROBLEM 8

A jar contains five white, seven green, and nine red marbles. If two marbles are drawn in succession *without replacement*, find the probability that one of them is white and the other is green.

Solution

The drawing of a white marble and a green marble can occur in two different ways: (1) by drawing a white first and then a green, and (2) by drawing a green first and then a white. Thus we have two mutually exclusive events, each of which is broken into two *dependent* events: first draw and second draw. Therefore, the probability can be computed as follows.

$$\left(\frac{5}{21}\right)\left(\frac{7}{20}\right) \qquad + \left(\frac{7}{21}\right)\left(\frac{5}{20}\right) = \frac{70}{420} = \frac{1}{6}$$

| White on first draw | Green on second draw | Green on first draw | White on second draw | ■ |

PROBLEM 9

Two cards are drawn in succession *with replacement* from a deck of 52 playing cards. Find the probability of drawing a jack and a queen.

Solution

Drawing a jack and a queen can occur two different ways: (1) a jack on the first draw and a queen on the second, and (2) a queen on the first draw and a jack on the second. Thus (1) and (2) are mutually exclusive events, and each is broken into the *independent* events of first draw and second draw with replacement. Therefore, the probability can be computed as follows.

$$\left(\frac{4}{52}\right)\left(\frac{4}{52}\right) \qquad + \left(\frac{4}{52}\right)\left(\frac{4}{52}\right) = \frac{32}{2704} = \frac{2}{169}$$

| Jack on first draw | Queen on second draw | Queen on first draw | Jack on second draw | ■ |

PROBLEM SET 10.5

For Problems 1–22, solve each problem.

1. A die is tossed. Find the probability that a 5 came up if it is known that an odd number came up.

2. A die is tossed. Find the probability that a prime number was obtained, given that an even number came up. Also find the probability that an even number came up, given that a prime number was obtained.

3. Two dice are rolled and someone indicates that the two numbers that come up are different. Find the probability that the sum of the two numbers is 6.

4. Two dice are rolled and someone indicates that the two numbers that come up are identical. Find the probability that the sum of the two numbers is 8.

5. One card is randomly drawn from a deck of 52 playing cards. Find the probability that it is a jack, given that the card is a face card. (We are considering jacks, queens, and kings as face cards.)

6. One card is randomly drawn from a deck of 52 playing cards. Find the probability that it is a spade, given the fact that it is a black card.

7. A coin and a die are tossed. Find the probability of getting a 5 on the die, given that a head comes up on the coin.

8. A family has three children. Assume that each child is as likely to be a boy as it is a girl. Find the probability that the family has three girls if it is known that the family has at least one girl.

9. The probability that a student will enroll in a mathematics course is 0.7, the probability that he or she will enroll in a history course is 0.3, and the probability that he or she will enroll in both mathematics and history is 0.2. Find

the probability that a student will enroll in mathematics, given that he or she is also enrolled in history. Also find the probability that a student will enroll in history, given that he or she is also enrolled in mathematics.

10. The following probability table contains data relative to car accidents and weather conditions on a holiday weekend.

	RAIN (R)	NO RAIN (R')	TOTAL
ACCIDENT (A)	0.025	0.015	0.040
NO ACCIDENT (A')	0.400	0.560	0.960
TOTAL	0.425	0.575	1.000

Find the probability that a person chosen at random from the survey was in an accident, given that it was raining. Also find the probability that a person was not in an accident, given that it was not raining.

11. One hundred people were surveyed, and one question pertained to their educational background. The responses to this question are given in the following table.

	FEMALE (F)	MALE (F')	TOTAL
COLLEGE DEGREE (D)	30	20	50
NO COLLEGE DEGREE (D')	15	35	50
TOTAL	45	55	100

Find the probability that a person chosen at random from the survey has a college degree, given that the person is female. Also find the probability that a person chosen is male, given that the person has a college degree.

12. In a recent election there were 1000 eligible voters. They were asked to vote on two issues, A and B. The results were as follows: 200 people voted for A, 400 people voted for B, and 50 people voted for both A and B. If one person is chosen at random from the 100 eligible voters, find the probability that the person voted for A, given that he or she voted for B. Also find the probability that the person voted for B, given that he or she voted for A.

13. A small company has 100 employees among whom 75 are males, 7 are administrators, and 5 of the administrators are males. If a person is chosen at random from the employees, find the probability that the person is an administrator, given that he is a male. Also find the probability that the person chosen is female, given that she is an administrator.

14. A survey claims that 80% of the households in a certain town have a color TV, 10% have a microwave oven, and 2% have both a color TV and a microwave oven. Find the probability that a randomly selected household will have a microwave oven, given that it has a color TV.

15. Consider a family of three children. Let *E* be the event *The first child is a boy*, and let *F* be the event *The family has exactly one boy*. Are events *E* and *F* dependent or independent?

16. Roll a white die and a green die. Let *E* be the event *Roll a 2 on the white die*, and let *F* be the event *Roll a 4 on the green die*. Are *E* and *F* dependent or independent events?

17. Toss 3 coins. Let *E* be the event *Toss not more that one head*, and let *F* be the event *Toss at least one of each face*. Are *E* and *F* dependent or independent events?

18. A card is drawn at random from a standard deck of 52 playing cards. Let *E* be the event *The card is a 2*, and let *F* be the event *The card is a 2 or a 3*. Are the events *E* and *F* dependent or independent?

19. A coin is tossed four times. Find the probability of getting three heads and one tail.

20. A coin is tossed five times. Find the probability of getting four heads and one tail.

21. Toss a pair of dice three times. Find the probability that a double is obtained on all three tosses.

22. Toss a pair of dice three times. Find the probability that each toss will produce a sum of 4.

For Problems 23–26, suppose that two cards are drawn in succession *without replacement* from a deck of 52 playing cards. Find the probability of each of the following events.

23. Both cards are 4's.

24. One card is an ace and one card is a king.

25. One card is a spade and one card is a diamond.

26. Both cards are black.

For Problems 27–30, suppose that two cards are drawn in succession *with replacement* from a deck of 52 playing cards. Find the probability of each of the following events.

27. Both cards are spades.

28. One card is an ace and one card is a king.

29. One card is the ace of spades and one card is the king of spades.

30. Both cards are red.

For Problems 31 and 32, solve each problem.

31. A person holds three kings from a deck of 52 playing cards. If the person draws two cards without replacement from the 49 cards remaining in the deck, find the probability of drawing the fourth king.

32. A person removes two aces and a king from a deck of 52 playing cards and draws, without replacement, two more cards from the deck. Find the probability that the person will draw two aces, or two kings, or an ace and a king.

For Problems 33–36, a bag contains five red and four white marbles. Two marbles are drawn in succession *with replacement*. Find the probability of each of the following events.

33. Both marbles drawn are red.

34. Both marbles drawn are white.

35. The first marble is red and the second marble is white.

36. At least one marble is red.

For Problems 37–40, a bag contains five white, four red, and four blue marbles. Two marbles are drawn in succession *with replacement*. Find the probability of each of the following events.

37. Both marbles drawn are white.

38. Both marbles drawn are red.

39. One red and one blue marble are drawn.

40. One white and one blue marble are drawn.

For Problems 41–44, a bag contains one red and two white marbles. Two marbles are drawn in succession *without replacement*. Find the probability of each of the following events.

41. One marble drawn is red and one marble drawn is white.

42. The first marble drawn is red and the second is white.

43. Both marbles drawn are white.

44. Both marbles drawn are red.

For Problems 45–48, a bag contains five red and 12 white marbles. Two marbles are drawn in succession *without replacement*. Find the probability of each of the following events.

45. Both marbles drawn are red.

46. Both marbles drawn are white.

47. One red and one white marble are drawn.

48. At least one marble drawn is red.

For Problems 49–52, a bag contains two red, three white, and four blue marbles. Two marbles are drawn in succession *without replacement*. Find the probability of each of the following events.

49. Both marbles drawn are white.

50. One marble drawn is white and one is blue.

51. Both marbles drawn are blue.

52. At least one red marble is drawn.

For Problems 53–56, a bag contains five white, one blue, and three red marbles. Three marbles are drawn in succession *with replacement*. Find the probability of each of the following events.

53. All three marbles drawn are blue.

54. One marble of each color is drawn.

55. One white and two red marbles are drawn.

56. One blue and two white marbles are drawn.

For Problems 57–60, a bag contains four white, one red, and two blue marbles, Three marbles are drawn in succession *without replacement*. Find the probability of each of the following events.

57. All three marbles drawn are white.

58. One red and two blue marbles are drawn.

59. One marble of each color is drawn.

60. One white and two red marbles are drawn.

For Problems 61 and 62, solve each problem.

61. Two boxes with red and white marbles are shown here. A marble is drawn at random from Box 1, and then a second marble is drawn from Box 2. Find the probability that both marbles drawn are white. Find the probability that both marbles drawn are red. Find the probability that one red and one white marble are drawn.

| 3 red |
| 4 white |
| Box 1 |

| 2 red |
| 1 white |
| Box 2 |

62. Three boxes containing red and white marbles are shown here. Randomly draw a marble from Box 1 and put it in Box 2. Then draw a marble from Box 2 and put it in Box 3. Then draw a marble from Box 3. What is the probability that the last marble drawn, from Box 3, is red? What is the probability that it is white?

| 2 red |
| 2 white |
| Box 1 |

| 3 red |
| 1 white |
| Box 2 |

| 3 white |
| Box 3 |

63. How would you explain the concept of conditional probability to a classmate who missed the discussion of this section?

64. How would you give a nontechnical description of conditional probability to an elementary algebra student?

65. Explain in your own words the concept of independent events.

66. Suppose that a bag contains two red and three white marbles. Furthermore, suppose that two marbles are drawn from the bag in succession *with replacement*. Explain how the following tree diagram can be used to determine that the probability of drawing two white marbles is $\frac{9}{25}$.

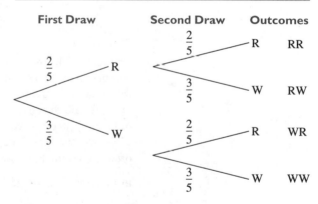

67. Explain how a tree diagram can be used to determine the probabilities for Problems 41–44.

10.6 BINOMIAL THEOREM

In Chapter 0 we developed a pattern for expanding binomials, using Pascal's triangle to determine the coefficients of each term. Now we will be more precise and develop a general formula, called the *binomial formula*. In other words, we want to develop a formula that will allow us to expand $(x + y)^n$, where n is any positive integer.

Let's begin, as we did in Chapter 0, by looking at some specific expansions, which can be verified by direct multiplication.

$$(x + y)^0 = 1$$
$$(x + y)^1 = x + y$$
$$(x + y)^2 = x^2 + 2xy + y^2$$
$$(x + y)^3 = x^3 + 3x^2y + 3xy^2 + y^3$$
$$(x + y)^4 = x^4 + 4x^3y + 6x^2y^2 + 4xy^3 + y^4$$
$$(x + y)^5 = x^5 + 5x^4y + 10x^3y^2 + 10x^2y^3 + 5xy^4 + y^5$$

First, note the pattern of the exponents for x and y on a term-by-term basis. The exponents of x begin with the exponent of the binomial and decrease by 1, term by term, until the last term has x^0, which is 1. The exponents of y begin with zero ($y^0 = 1$) and increase by 1, term by term, until the last term contains y to the power of the binomial. In other words, the variables in the expansion of $(x + y)^n$ have the following pattern.

$$x^n, \qquad x^{n-1}y, \qquad x^{n-2}y^2, \qquad x^{n-3}y^3, \qquad \ldots, \qquad xy^{n-1}, \qquad y^n$$

Note that for each term, the sum of the exponents of x and y is n.

Now let's look for a pattern for the coefficients by examining specifically the expansion of $(x + y)^5$.

$$(x + y)^5 = x^5 + 5x^4y^1 + 10x^3y^2 + 10x^2y^3 + 5x^1y^4 + 1y^5$$
$$\uparrow \qquad \uparrow \qquad \uparrow \qquad \uparrow \qquad \uparrow$$
$$C(5, 1) \quad C(5, 2) \quad C(5, 3) \quad C(5, 4) \quad C(5, 5)$$

As indicated by the arrows, the coefficients are numbers that arise as different-sized combinations of five things. To see why this happens, consider the coefficient for the term containing x^3y^2. The two y's (for y^2) come from two of the factors of $(x + y)$, and therefore the three x's (for x^3) must come from the other three factors of $(x + y)$. In other words, the coefficient is $C(5, 2)$.

We can now state a general expansion formula for $(x + y)^n$; this formula is often called the **binomial theorem**. But before stating it, let's make a small switch in notation. Instead of $C(n, r)$, we shall write $\binom{n}{r}$, which will prove to be a little more convenient at this time. The symbol $\binom{n}{r}$ still refers to the number of combinations of n things taken r at a time, but in this context it is often called a **binomial coefficient**.

Binomial Theorem

For any binomial $(x + y)$ and any natural number n,

$$(x + y)^n = x^n + \binom{n}{1}x^{n-1}y + \binom{n}{2}x^{n-2}y^2 + \cdots + \binom{n}{n}y^n$$

The binomial theorem can be proven by mathematical induction, but we will not do that in this text. Instead, we'll consider a few examples that put the binomial theorem to work.

EXAMPLE 1

Expand $(x + y)^7$.

Solution

$$(x + y)^7 = x^7 + \binom{7}{1}x^6y + \binom{7}{2}x^5y^2 + \binom{7}{3}x^4y^3 + \binom{7}{4}x^3y^4 + \binom{7}{5}x^2y^5 + \binom{7}{6}xy^6 + \binom{7}{7}y^7$$

$$= x^7 + 7x^6y + 21x^5y^2 + 35x^4y^3 + 35x^3y^4 + 21x^2y^5 + 7xy^6 + y^7 \qquad ■$$

EXAMPLE 2

Expand $(x - y)^5$.

Solution

We shall treat $(x - y)^5$ as $[x + (-y)]^5$

$$[x + (-y)]^5 = x^5 + \binom{5}{1}x^4(-y) + \binom{5}{2}x^3(-y)^2 + \binom{5}{3}x^2(-y)^3 + \binom{5}{4}x(-y)^4 + \binom{5}{5}(-y)^5$$

$$= x^5 - 5x^4y + 10x^3y^2 - 10x^2y^3 + 5xy^4 - y^5 \qquad ■$$

EXAMPLE 3

Expand $(2a + 3b)^4$.

Solution

Let $x = 2a$ and $y = 3b$ in the binomial theorem.

$$(2a + 3b)^4 = (2a)^4 + \binom{4}{1}(2a)^3(3b) + \binom{4}{2}(2a)^2(3b)^2 + \binom{4}{3}(2a)(3b)^3 + \binom{4}{4}(3b)^4$$

$$= 16a^4 + 96a^3b + 216a^2b^2 + 216ab^3 + 81b^4 \qquad ■$$

EXAMPLE 4

Expand $\left(a + \dfrac{1}{n}\right)^5$.

Solution

$$\left(a + \frac{1}{n}\right)^5 = a^5 + \binom{5}{1}a^4\left(\frac{1}{n}\right) + \binom{5}{2}a^3\left(\frac{1}{n}\right)^2 + \binom{5}{3}a^2\left(\frac{1}{n}\right)^3 + \binom{5}{4}a\left(\frac{1}{n}\right)^4 + \binom{5}{5}\left(\frac{1}{n}\right)^5$$

$$= a^5 + \frac{5a^4}{n} + \frac{10a^3}{n^2} + \frac{10a^2}{n^3} + \frac{5a}{n^4} + \frac{1}{n^5} \qquad ■$$

EXAMPLE 5

Expand $(x^2 - 2y^3)^6$.

Solution

$$
[x^2 + (-2y^3)]^6 = (x^2)^6 + \binom{6}{1}(x^2)^5(-2y^3) + \binom{6}{2}(x^2)^4(-2y^3)^2
$$
$$
+ \binom{6}{3}(x^2)^3(-2y^3)^3 + \binom{6}{4}(x^2)^2(-2y^3)^4
$$
$$
+ \binom{6}{5}(x^2)(-2y^3)^5 + \binom{6}{6}(-2y^3)^6
$$
$$
= x^{12} - 12x^{10}y^3 + 60x^8y^6 - 160x^6y^9 + 240x^4y^{12} - 192x^2y^{15} + 64y^{18} \quad \blacksquare
$$

Finding Specific Terms

Sometimes it is convenient to be able to write down the specific term of a binomial expansion without writing out the entire expansion. For example, suppose that we want the sixth term of the expansion $(x + y)^{12}$. We can proceed as follows: The sixth term will contain y^5. (Note in the binomial theorem that the **exponent of y is always one less than the number of the term.**) Because the sum of the exponents for x and y must be 12 (the exponent of the binomial), the sixth term will also contain x^7. The coefficient is $\binom{12}{5}$, where the 5 agrees with the exponent of y^5. Therefore, the sixth term of $(x + y)^{12}$ is

$$
\binom{12}{5}x^7y^5 = 792x^7y^5
$$

EXAMPLE 6

Find the fourth term of $(3a + 2b)^7$.

Solution

The fourth term will contain $(2b)^3$, and therefore it will also contain $(3a)^4$. The coefficient is $\binom{7}{3}$. Thus the fourth term is

$$
\binom{7}{3}(3a)^4(2b)^3 = (35)(81a^4)(8b^3) = 22{,}680a^4b^3 \quad \blacksquare
$$

EXAMPLE 7

Find the sixth term of $(4x - y)^9$.

Solution

The sixth term will contain $(-y)^5$, and therefore it will also contain $(4x)^4$. The coefficient is $\binom{9}{5}$. Thus the sixth term is

$$\binom{9}{5}(4x)^4(-y)^5 = (126)(256x^4)(-y^5)$$

$$= -32256x^4y^5 \qquad \blacksquare$$

PROBLEM SET 10.6

For Problems 1–26, expand and simplify each binomial.

1. $(x + y)^8$ **2.** $(x + y)^9$

3. $(x - y)^6$ **4.** $(x - y)^4$

5. $(a + 2b)^4$ **6.** $(3a + b)^4$

7. $(x - 3y)^5$ **8.** $(2x - y)^6$

9. $(2a - 3b)^4$ **10.** $(3a - 2b)^5$

11. $(x^2 + y)^5$ **12.** $(x + y^3)^6$

13. $(2x^2 - y^2)^4$ **14.** $(3x^2 - 2y^2)^5$

15. $(x + 3)^6$ **16.** $(x + 2)^7$

17. $(x - 1)^9$ **18.** $(x - 3)^4$

19. $\left(1 + \dfrac{1}{n}\right)^4$ **20.** $\left(2 + \dfrac{1}{n}\right)^5$

21. $\left(a - \dfrac{1}{n}\right)^6$ **22.** $\left(2a - \dfrac{1}{n}\right)^5$

23. $\left(1 + \sqrt{2}\right)^4$ **24.** $\left(2 + \sqrt{3}\right)^3$

25. $\left(3 - \sqrt{2}\right)^5$ **26.** $\left(1 - \sqrt{3}\right)^4$

For Problems 27–36, write the first four terms of each expansion.

27. $(x + y)^{12}$ **28.** $(x + y)^{15}$

29. $(x - y)^{20}$ **30.** $(a - 2b)^{13}$

31. $(x^2 - 2y^3)^{14}$ **32.** $(x^3 - 3y^2)^{11}$

33. $\left(a + \dfrac{1}{n}\right)^9$ **34.** $\left(2 - \dfrac{1}{n}\right)^6$

35. $(-x + 2y)^{10}$ **36.** $(-a - b)^{14}$

For Problems 37–46, find the specified term for each binomial expansion.

37. The fourth term of $(x + y)^8$

38. The seventh term of $(x + y)^{11}$

39. The fifth term of $(x - y)^9$

40. The fourth term of $(x - 2y)^6$

41. The sixth term of $(3a + b)^7$

42. The third term of $(2x - 5y)^5$

43. The eighth term of $(x^2 + y^3)^{10}$

44. The ninth term of $(a + b^3)^{12}$

45. The seventh term of $\left(1 - \dfrac{1}{n}\right)^{15}$

46. The eighth term of $\left(1 - \dfrac{1}{n}\right)^{13}$

THOUGHTS INTO WORDS

47. How would you explain binomial expansions to an elementary algebra student?

48. Explain how to find the fifth term of the expansion of $(2x + 3y)^9$ without writing out the entire expansion.

49. Is the tenth term of the expansion $(1 - 2)^{15}$ positive or negative? Explain how you determined the answer to this question.

Further Investigations

For Problems 50–53, expand and simplify each complex number.

52. $(2 - i)^6$

53. $(3 - 2i)^5$

50. $(1 + 2i)^5$

51. $(2 + i)^6$

CHAPTER 10 SUMMARY

We can summarize this chapter with three main topics: counting techniques, probability, and the binomial theorem.

Counting Techniques

The **fundamental principle of counting** states that if a first task can be accomplished in x ways and, following this task, a second task can be accomplished in y ways, then task 1 followed by task 2 can be accomplished in $x \cdot y$ ways. The principle extends to any finite number of tasks. As you solve problems involving the fundamental principle of counting, it is often helpful to analyze the problem in terms of the tasks to be completed.

Ordered arrangements are called **permutations**. The number of permutations of n things taken n at a time is given by

$$P(n, n) = n!$$

The number of r-element permutations that can be formed from a set of n elements is given by

$$P(n, r) = \underbrace{n(n - 1)(n - 2) \cdots}_{r \text{ factors}}$$

If there are n elements to be arranged, where there are r_1 of one kind, r_2 of another kind, r_3 of another kind, ... r_k of a kth kind, then the number of distinguishable permutations is given by

$$\frac{n!}{(r_1!)(r_2!)(r_3!) \cdots (r_k!)}$$

Combinations are subsets; the order in which the elements appear does not make a difference. The number of r-element combinations (subsets) that can be formed from a set of n elements is given by

$$C(n, r) = \frac{P(n, r)}{r!}$$

Does the order in which the elements appear make any difference? This is a key question to consider when trying to decide whether a particular problem involves permutations or combinations. If the answer to the question is yes, then it is a permutation problem; if the answer is no, then it is a combination problem. Don't forget that combinations are subsets.

Probability

In an experiment where all possible outcomes in the sample space S are equally likely to occur, the **probability** of an event E is defined by

$$P(E) = \frac{n(E)}{n(S)}$$

where $n(E)$ denotes the number of elements in the event E and $n(S)$ denotes the number of elements in the sample space S. The numbers $n(E)$ and $n(S)$ can often be determined by using one or more of the previously listed counting techniques. For all events E, it is always true that $0 \leq P(E) \leq 1$. That is, all probabilities fall in the range from 0 to 1, inclusive.

If E and E' are **complementary events**, then $P(E) + P(E') = 1$. Therefore, if we can calculate either $P(E)$ or $P(E')$, then we can find the other one by subtracting from 1.

For two events E and F, the probability of E or F is given by

$$P(E \cup F) = P(E) + P(F) - P(E \cap F)$$

If $E \cap F = \varnothing$, then E and F are **mutually exclusive events**.

The probability that an event E occurs, given that another event F has already occurred, is called **conditional probability**, and it is given by the equation

$$P(E|F) = \frac{P(E \cap F)}{P(F)}$$

Two events E and F are said to be **independent** if and only if

$$P(E \cap F) = P(E)P(F)$$

Two events that are not independent are called **dependent** events, and the probability of two dependent events is given by

$$P(E \cap F) = P(E)P(F|E)$$

The Binomial Theorem

For any binomial $(x + y)$ and any natural number n,

$$(x + y)^n = x^n + \binom{n}{1}x^{n-1}y + \binom{n}{2}x^{n-2}y^2 + \cdots + \binom{n}{n}y^n$$

Note the following patterns in a binomial expansion.

1. In each term, the sum of the exponents of x and y is n.

2. The exponents of x begin with the exponent of the binomial and decrease by 1, term by term, until the last term has x^0, which is 1. The exponents of y begin with zero ($y^0 = 1$) and increase by 1, term by term, until the last term contains y to the power of the binomial.

3. The coefficient of any term is given by $\binom{n}{r}$, where the value of r agrees with the exponent of y for that term. For example, if the term contains y^3, then the coefficient of that term is $\binom{n}{3}$.

4. The expansion of $(x + y)^n$ contains $n + 1$ terms.

CHAPTER 10 REVIEW PROBLEM SET

Problems 1–14 are counting type problems.

1. How many different arrangements of the letters A, B, C, D, E, and F can be made?

2. How many different nine-letter arrangements can be formed from the nine letters of the word APPARATUS?

3. How many odd numbers of three different digits each can be formed by choosing from the digits 1, 2, 3, 5, 7, 8, and 9?

4. In how many ways can Arlene, Brent, Carlos, Dave, Ernie, Frank, and Gladys be seated in a row of seven seats so that Arlene and Carlos are side by side?

5. In how many ways can a committee of three people be chosen from six people?

6. How many committees consisting of three men and two women can be formed from seven men and six women?

7. How many different five-card hands consisting of all hearts can be formed from a deck of 52 playing cards?

8. If no number contains repeated digits, how many numbers greater than 500 can be formed by choosing from the digits 2, 3, 4, 5, and 6?

9. How many three-person committees can be formed from four men and five women so that each committee contains at least one man?

10. How many different four-person committees can be formed from eight people if two particular people refuse to serve together on a committee?

11. How many four-element subsets containing A or B but not both A and B can be formed from the set {A, B, C, D, E, F, G, H}?

12. How many different six-letter permutations can be formed from four identical H's and two identical T's?

13. How many four-person committees consisting of two seniors, one sophomore, and one junior can be formed from three seniors, four juniors, and five sophomores?

14. In a baseball league of six teams, how many games are needed to complete a schedule if each team plays eight games with each other team?

Problems 15–35 pose some probability questions.

15. If three coins are tossed, find the probability of getting two heads and one tail.

16. If five coins are tossed, find the probability of getting three heads and two tails.

17. What is the probability of getting a sum of 8 with one roll of a pair of dice?

18. What is the probability of getting a sum more than 5 with one roll of a pair of dice?

19. Aimée, Brenda, Chuck, Dave and Eli are randomly seated in a row of five seats. Find the probability that Aimée and Chuck are not seated side by side.

20. Four girls and three boys are to be randomly seated in a row of seven seats. Find the probability that the girls and boys will be seated in alternate seats.

21. Six coins are tossed. Find the probability of getting at least two heads.

22. Two cards are randomly chosen from a deck of 52 playing cards. What is the probability that two jacks are drawn?

23. Each arrangement of the six letters of the word CYCLIC is put on a slip of paper and placed in a hat. One slip is drawn at random. Find the probability that the slip contains an arrangement with the Y at the beginning.

24. A committee of three is randomly chosen from one man and six women. What is the probability that the man is not on the committee?

25. A four-person committee is selected at random from the eight people Alice, Bob, Carl, Dee, Edna, Fred, Gina, and Hilda. Find the probability that Alice or Bob, but not both, is on the committee.

26. A committee of three is chosen at random from a group of five men and four women. Find the probability that the committee contains two men and one woman.

27. A committee of four is chosen at random from a group of six men and seven women. Find the probability that the committee contains at least one woman.

28. A bag contains five red and eight white marbles. Two marbles are drawn in succession *with replacement*. What is the probability that at least one red marble is drawn?

29. A bag contains four red, five white, and three blue marbles. Two marbles are drawn in succession *with replacement*. Find the probability that one red and one blue marble are drawn.

30. A bag contains four red and seven blue marbles. Two marbles are drawn in succession *without replacement*. Find the probability of drawing one red and one blue marble.

31. A bag contains three red, two white, and two blue marbles. Two marbles are drawn in succession *without replacement*. Find the probability of drawing at least one red marble.

32. Each of three letters is to be mailed in any one of four different mailboxes. What is the probability that all three letters will be mailed in the same mailbox?

33. The probability that a customer in a department store will buy a blouse is 0.15, the probability that she will buy a pair of shoes is 0.10, and the probability that she will buy both a blouse and a pair of shoes is 0.05. Find the probability that the customer will buy a blouse, given that she has already purchased a pair of shoes. Also find the probability that she will buy a pair of shoes, given that she has already purchased a blouse.

34. A survey of 500 employees of a company produced the following information.

EMPLOYMENT LEVEL	COLLEGE DEGREE	NO COLLEGE DEGREE
Managerial	45	5
Nonmanagerial	50	400

Find the probability that an employee chosen at random (a) is working in a managerial position, given that he or she has a college degree; and (b) has a college degree, given that he or she is working in a managerial position.

35. From a survey of 1000 college students, it was found that 450 of them owned cars, 700 of them owned stereos, and 200 of them owned both a car and a stereo. If a student is chosen at random from the 1000 students, find the probability that the student (a) owns a car, given the fact that he or she owns a stereo, and (b) owns a stereo, given the fact that he or she owns a car.

For Problems 36–41, expand each binomial and simplify.

36. $(x + 2y)^5$

37. $(x - y)^8$

38. $(a^2 - 3b^3)^4$

39. $\left(x + \dfrac{1}{n}\right)^6$

40. $\left(1 - \sqrt{2}\right)^5$

41. $(-a + b)^3$

42. Find the fourth term of the expansion of $(x - 2y)^{12}$.

43. Find the tenth term of the expansion of $(3a + b^2)^{13}$.

CHAPTER 10 TEST

For Problems 1–21, solve each problem.

1. In how many ways can Abdul, Barb, Corazon and Doug be seated in a row of 4 seats so that Abdul occupies an end seat?

2. How many even numbers of four different digits each can be formed by choosing from the digits 1, 2, 3, 5, 7, 8, and 9?

3. In how many ways can three letters be mailed in six mailboxes?

4. In a baseball league of ten teams, how many games are needed to complete the schedule if each team plays six games against each other team?

5. In how many ways can a sum greater than 5 be obtained when tossing a pair of dice?

6. In how many ways can six different mathematics books and three different biology books be placed on a shelf so that all of the books in a subject area are side by side?

7. How many four-element subsets containing A or B, but not both A and B, can be formed from the set {A, B, C, D, E, F, G}?

8. How many five-card hands consisting of two aces, two kings, and one queen can be dealt from a deck of 52 playing cards?

9. How many different nine-letter arrangements can be formed from the nine letters of the word SASSAFRAS?

10. How many committees consisting of four men and three women can be formed from a group of seven men and five women?

11. What is the probability of rolling a sum less than 9 with a pair of dice?

12. Six coins are tossed. Find the probability of getting three heads and three tails.

13. All possible numbers of three different digits each are formed from the digits 1, 2, 3, 4, 5, and 6. If one number is then chosen at random, find the probability that it is greater than 200.

14. A four-person committee is selected at random from Anwar, Barb, Chad, Dick, Edna, Fern, and Giraldo. What is the probability that neither Anwar nor Barb is on the committee?

15. From a group of three men and five women, a three-person committee is selected at random. Find the probability that the committee contains at least one man.

16. A box of 12 items is known to contain one defective and 11 nondefective items. If a sample of three items is selected at random, what is the probability that all three items are nondefective?

17. Five coins are tossed 80 times. How many times should you expect to get three heads and two tails?

18. Suppose 3000 tickets are sold in a lottery. There are three prizes: The first prize is $500, the second is $300, and the third is $100. What is the mathematical expectation of winning?

19. A bag contains seven white and 12 green marbles. Two marbles are drawn in succession, *with replacement*. Find the probability that one marble of each color is drawn.

20. A bag contains three white, five green, and seven blue marbles. Two marbles are drawn, *without replacement*. Find the probability that two green marbles are drawn.

21. In an election there were 2000 eligible voters. They were asked to vote on two issues, A and B. The results were as follows: 500 people voted for A, 800 people voted for B, and 250 people voted for both A and B. If one person is chosen at random from the 2000 eligible voters, find the probability that this person voted for A, given that he or she voted for B.

22. Expand and simplify $\left(2 - \dfrac{1}{n}\right)^6$

23. Expand and simplify $(3x + 2y)^5$.

24. Find the ninth term of the expansion of $\left(x - \dfrac{1}{2}\right)^{12}$.

25. Find the fifth term of the expansion of $(x + 3y)^7$.

COMMON LOGARITHMS

Table of Common Logarithms

N	0	1	2	3	4	5	6	7	8	9
1.0	.0000	.0043	.0086	.0128	.0170	.0212	.0253	.0294	.0334	.0374
1.1	.0414	.0453	.0492	.0531	.0569	.0607	.0645	.0682	.0719	.0755
1.2	.0792	.0828	.0864	.0899	.0934	.0969	.1004	.1038	.1072	.1106
1.3	.1139	.1173	.1206	.1239	.1271	.1303	.1335	.1367	.1399	.1430
1.4	.1461	.1492	.1523	.1553	.1584	.1614	.1644	.1673	.1703	.1732
1.5	.1761	.1790	.1818	.1847	.1875	.1903	.1931	.1959	.1987	.2014
1.6	.2041	.2068	.2095	.2122	.2148	.2175	.2201	.2227	.2253	.2279
1.7	.2304	.2330	.2355	.2380	.2405	.2430	.2455	.2480	.2504	.2529
1.8	.2553	.2577	.2601	.2625	.2648	.2672	.2695	.2718	.2742	.2765
1.9	.2788	.2810	.2833	.2856	.2878	.2900	.2923	.2945	.2967	.2989
2.0	.3010	.3032	.3054	.3075	.3096	.3118	.3139	.3160	.3181	.3201
2.1	.3222	.3243	.3263	.3284	.3304	.3324	.3345	.3365	.3385	.3404
2.2	.3424	.3444	.3464	.3483	.3502	.3522	.3541	.3560	.3579	.3598
2.3	.3617	.3636	.3655	.3674	.3692	.3711	.3729	.3747	.3766	.3784
2.4	.3802	.3820	.3838	.3856	.3874	.3892	.3909	.3927	.3945	.3962
2.5	.3979	.3997	.4014	.4031	.4048	.4065	.4082	.4099	.4116	.4133
2.6	.4150	.4166	.4183	.4200	.4216	.4232	.4249	.4265	.4281	.4298
2.7	.4314	.4330	.4346	.4362	.4378	.4393	.4409	.4425	.4440	.4456
2.8	.4472	.4487	.4502	.4518	.4533	.4548	.4564	.4579	.4594	.4609
2.9	.4624	.4639	.4654	.4669	.4683	.4698	.4713	.4728	.4742	.4757
3.0	.4771	.4786	.4800	.4814	.4829	.4843	.4857	.4871	.4886	.4900
3.1	.4914	.4928	.4942	.4955	.4969	.4983	.4997	.5011	.5024	.5038
3.2	.5051	.5065	.5079	.5092	.5105	.5119	.5132	.5145	.5159	.5172
3.3	.5185	.5198	.5211	.5224	.5237	.5250	.5263	.5276	.5289	.5302
3.4	.5315	.5328	.5340	.5353	.5366	.5378	.5391	.5403	.5416	.5428

Table of Common Logarithms (continued)

N	0	I	2	3	4	5	6	7	8	9
3.5	.5441	.5453	.5465	.5478	.5490	.5502	.5514	.5527	.5539	.5551
3.6	.5563	.5575	.5587	.5599	.5611	.5623	.5635	.5647	.5658	.5670
3.7	.5682	.5694	.5705	.5717	.5729	.5740	.5752	.5763	.5775	.5786
3.8	.5798	.5809	.5821	.5832	.5843	.5855	.5866	.5877	.5888	.5899
3.9	.5911	.5922	.5933	.5944	.5955	.5966	.5977	.5988	.5999	.6010
4.0	.6021	.6031	.6042	.6053	.6064	.6075	.6085	.6096	.6107	.6117
4.1	.6128	.6138	.6149	.6160	.6170	.6180	.6191	.6201	.6212	.6222
4.2	.6232	.6243	.6253	.6263	.6274	.6284	.6294	.6304	.6314	.6325
4.3	.6335	.6345	.6355	.6365	.6375	.6385	.6395	.6405	.6415	.6425
4.4	.6435	.6444	.6454	.6464	.6474	.6484	.6493	.6503	.6513	.6522
4.5	.6532	.6542	.6551	.6561	.6571	.6580	.6590	.6599	.6609	.6618
4.6	.6628	.6637	.6646	.6656	.6665	.6675	.6684	.6693	.6702	.6712
4.7	.6721	.6730	.6739	.6749	.6758	.6767	.6776	.6785	.6794	.6803
4.8	.6812	.6821	.6830	.6839	.6848	.6857	.6866	.6875	.6884	.6893
4.9	.6902	.6911	.6920	.6928	.6937	.6946	.6955	.6964	.6972	.6981
5.0	.6990	.6998	.7007	.7016	.7024	.7033	.7042	.7050	.7059	.7067
5.1	.7076	.7084	.7093	.7101	.7110	.7118	.7126	.7135	.7143	.7152
5.2	.7160	.7168	.7177	.7185	.7193	.7202	.7210	.7218	.7226	.7235
5.3	.7243	.7251	.7259	.7267	.7275	.7284	.7292	.7300	.7308	.7316
5.4	.7324	.7332	.7340	.7348	.7356	.7364	.7372	.7380	.7388	.7396
5.5	.7404	.7412	.7419	.7427	.7435	.7443	.7451	.7459	.7466	.7474
5.6	.7482	.7490	.7497	.7505	.7513	.7520	.7528	.7536	.7543	.7551
5.7	.7559	.7566	.7574	.7582	.7589	.7597	.7604	.7612	.7619	.7627
5.8	.7634	.7642	.7649	.7657	.7664	.7672	.7679	.7686	.7694	.7701
5.9	.7709	.7716	.7723	.7731	.7738	.7745	.7752	.7760	.7767	.7774
6.0	.7782	.7789	.7796	.7803	.7810	.7818	.7825	.7832	.7839	.7846
6.1	.7853	.7860	.7868	.7875	.7882	.7889	.7896	.7903	.7910	.7917
6.2	.7924	.7931	.7938	.7945	.7952	.7959	.7966	.7973	.7980	.7987
6.3	.7993	.8000	.8007	.8014	.8021	.8028	.8035	.8041	.8048	.8055
6.4	.8062	.8069	.8075	.8082	.8089	.8096	.8102	.8109	.8116	.8122
6.5	.8129	.8136	.8142	.8149	.8156	.8162	.8169	.8176	.8182	.8189
6.6	.8195	.8202	.8209	.8215	.8222	.8228	.8235	.8241	.8248	.8254
6.7	.8261	.8267	.8274	.8280	.8287	.8293	.8299	.8306	.8312	.8319
6.8	.8325	.8331	.8338	.8344	.8351	.8357	.8363	.8370	.8376	.8382
6.9	.8388	.8395	.8401	.8407	.8414	.8420	.8426	.8432	.8439	.8445
7.0	.8451	.8457	.8463	.8470	.8476	.8482	.8488	.8494	.8500	.8506
7.1	.8513	.8519	.8525	.8531	.8537	.8543	.8549	.8555	.8561	.8567
7.2	.8573	.8579	.8585	.8591	.8597	.8603	.8609	.8615	.8621	.8627
7.3	.8633	.8639	.8645	.8651	.8657	.8663	.8669	.8675	.8681	.8686
7.4	.8692	.8698	.8704	.8710	.8716	.8722	.8727	.8733	.8739	.8745

Table of Common Logarithms (continued)

N	0	1	2	3	4	5	6	7	8	9
7.5	.8751	.8756	.8762	.8768	.8774	.8779	.8785	.8791	.8797	.8802
7.6	.8808	.8814	.8820	.8825	.8831	.8837	.8842	.8848	.8854	.8859
7.7	.8865	.8871	.8876	.8882	.8887	.8893	.8899	.8904	.8910	.8915
7.8	.8921	.8927	.8932	.8938	.8943	.8949	.8954	.8960	.8965	.8971
7.9	.8976	.8982	.8987	.8993	.8998	.9004	.9009	.9015	.9020	.9025
8.0	.9031	.9036	.9042	.9047	.9053	.9058	.9063	.9069	.9074	.9079
8.1	.9085	.9090	.9096	.9101	.9106	.9112	.9117	.9122	.9128	.9133
8.2	.9138	.9143	.9149	.9154	.9159	.9165	.9170	.9175	.9180	.9186
8.3	.9191	.9196	.9201	.9206	.9212	.9217	.9222	.9227	.9232	.9238
8.4	.9243	.9248	.9253	.9258	.9263	.9269	.9274	.9279	.9284	.9289
8.5	.9294	.9299	.9304	.9309	.9315	.9320	.9325	.9330	.9335	.9340
8.6	.9345	.9350	.9355	.9360	.9365	.9370	.9375	.9380	.9385	.9390
8.7	.9395	.9400	.9405	.9410	.9415	.9420	.9425	.9430	.9435	.9440
8.8	.9445	.9450	.9455	.9460	.9465	.9469	.9474	.9479	.9484	.9489
8.9	.9494	.9499	.9504	.9509	.9513	.9518	.9523	.9528	.9533	.9538
9.0	.9542	.9547	.9552	.9557	.9562	.9566	.9571	.9576	.9581	.9586
9.1	.9590	.9595	.9600	.9605	.9609	.9614	.9619	.9624	.9628	.9633
9.2	.9638	.9643	.9647	.9652	.9657	.9661	.9666	.9671	.9675	.9680
9.3	.9685	.9689	.9694	.9699	.9703	.9708	.9713	.9717	.9722	.9727
9.4	.9731	.9736	.9741	.9745	.9750	.9754	.9759	.9763	.9768	.9773
9.5	.9777	.9782	.9786	.9791	.9795	.9800	.9805	.9809	.9814	.9818
9.6	.9823	.9827	.9832	.9836	.9841	.9845	.9850	.9854	.9859	.9863
9.7	.9868	.9872	.9877	.9881	.9886	.9890	.9894	.9899	.9903	.9908
9.8	.9912	.9917	.9921	.9926	.9930	.9934	.9939	.9943	.9948	.9952
9.9	.9956	.9961	.9965	.9969	.9974	.9978	.9983	.9987	.9991	.9996

Using a table to find a common logarithm is relatively easy, but it does require a little more effort than pushing a key on a calculator. Each number in the column headed n represents the first two significant digits of a number between 1 and 10, and each of the column headings 0 through 9 represents the third significant digit. To find the logarithm of a number such as 1.75, we look at the intersection of the row that contains 1.7 and the column headed 5. Thus we obtain

$$\log 1.75 = 0.2430$$

Similarly, we can find that

$$\log 2.09 = 0.3201 \quad \text{and} \quad \log 2.40 = 0.3802$$

Keep in mind that these values are also rounded to four decimal places.

Now suppose that we want to use the table to find the logarithm of a positive number greater than 10 or less than 1. To accomplish this, we represent the number in scientific notation and then apply the property $\log rs = \log r + \log s$. For example, to find $\log 134$, we can proceed as follows.

$$\log 134 = \log(1.34 \cdot 10^2)$$
$$= \log 1.34 + \log 10^2$$
$$= 0.1271 + 2 = 2.1271$$

By inspection, we know that the common logarithm of 10^2 is 2 (the exponent), and the common logarithm of 1.34 can be found in the table.

The decimal part (0.1271) of the logarithm 2.1271 is called the mantissa, and the integral part (2) is called the characteristic. Thus we can find the characteristic of a common logarithm by inspection (because it is the exponent of 10 when the number is written in scientific notation), and we can get the mantissa from a table. Let's consider two more examples.

$$\log 23.8 = \log (2.38 \cdot 10^1)$$
$$= \log 2.38 + \log 10^1$$
$$= 0.3766 + 1$$

From the table ↑   ↘ Exponent of 10

$$= 1.3766$$

$$\log 0.192 = \log(1.92 \cdot 10^{-1})$$
$$= \log 1.92 + \log 10^{-1}$$
$$= 0.2833 + (-1)$$

From the table ↑   ↖ Exponent of 10

$$= 0.2833 + (-1)$$

Note that in the last example, we expressed the logarithm of 0.192 as $0.2833 + (-1)$; we did not add 0.2833 and -1. This is normal procedure when using a table of common logarithms, because the mantissas given in the table are positive numbers. However, you should recognize that adding 0.2833 and -1 produces -0.7167, which agrees with the result obtained earlier with a calculator.

We can also use the table to find a number when given the common logarithm of the number. That is, given $\log x$, we can determine x from the table. Traditionally, x is referred to as the antilogarithm (abbreviated antilog) of $\log x$. Let's consider some examples.

EXAMPLE 1

Determine antilog 1.3365.

Solution

To find an antilogarithm, we simply reverse the process used before for finding a logarithm. Thus antilog 1.3365 means that 1 is the characteristic and 0.3365 the mantissa. We look for 0.3365 in the body of the common logarithm table, and we find that it is located at the intersection of the 2.1 row and the 7 column. Therefore, the antilogarithm is

$$2.17 \cdot 10^1 = 21.7$$

■

EXAMPLE 2

Determine antilog $[0.1523 + (-2)]$.

Solution

The mantissa, 0.1523, is located at the intersection of the 1.4 row and the 2 column. The characteristic is -2, so the antilogarithm is

$$1.42 \cdot 10^{-2} = 0.0142$$

■

EXAMPLE 3

Determine antilog -2.6038.

Solution

The mantissas given in a table are *positive* numbers. Thus we need to express -2.6038 in terms of a positive mantissa, and this can be done by adding and subtracting 3 as follows.

$$(-2.6038 + 3) - 3 = 0.3962 + (-3)$$

Now we can look for 0.3962 and find it at the intersection of the 2.4 row and the 9 column. Therefore, the antilogarithm is

$$2.49 \cdot 10^{-3} = 0.00249$$

■

Linear Interpolation

Now suppose that we want to determine log 2.774 from the table. Because the table contains only logarithms of numbers with, at most, three significant digits, we have a problem. However, by a process called linear interpolation, we can extend the capabilities of the table to include numbers with four significant digits.

First, let's consider a geometric basis of linear interpolation. Then we will use a systematic procedure for carrying out the necessary calculations. A portion of the graph of $y = \log x$, with the curvature exaggerated to help illustrate the principle involved, is shown in Figure A.1. The line segment that joins points P and Q is used to approximate the curve from P to Q. The actual value of log 2.744 is the ordinate of point C—that is, the length of $\overline{AC}$. This cannot be determined from the table. Instead we will use the ordinate of point B (the length of $\overline{AB}$) as an approximation for log 2.744.

Consider Figure A.2, where line segments $\overline{DB}$ and $\overline{EQ}$ are drawn perpendicular to $\overline{PE}$. The right triangles formed, $\triangle PDB$ and $\triangle PEQ$, are similar, and therefore the lengths of their corresponding sides are proportional. Thus we can write

$$\frac{PD}{PE} = \frac{DB}{EQ} \tag{1}$$

FIGURE A.1

FIGURE A.2

From Figure A.2 we see that

$$PD = 2.744 - 2.74 = 0.004$$

$$PE = 2.75 - 2.74 = 0.01$$

$$EQ = 0.4393 - 0.4378 = 0.0015$$

Therefore, proportion (1) becomes

$$\frac{0.004}{0.01} = \frac{DB}{0.0015}$$

Solving this proportion for *DB* yields

$$DB = 0.0006$$

Because $AB = AD + DB$, we have

$$AB = 0.4378 + 0.0006 = 0.4384$$

Thus we obtain log 2.744 = 0.4384.

Now let's suggest an abbreviated format for carrying out the calculations necessary to find log 2.744.

$$
\begin{array}{cc}
\mathbf{x} & \mathbf{Log\ x} \\[4pt]
4\left\{\begin{array}{c}2.740\\2.744\end{array}\right. & k\left\{\begin{array}{c}0.4378\\?\end{array}\right. \\[10pt]
\left.\begin{array}{c}\\2.750\end{array}\right\}10 & \left.\begin{array}{c}\\0.4393\end{array}\right\}0.0015
\end{array}
$$

Note that we have used 4 and 10 for the differences for values of *x* instead of 0.004 and 0.01, because the ratio $\dfrac{0.004}{0.01}$ equals $\dfrac{4}{10}$. Setting up a proportion and solving for *k* yields

$$\frac{4}{10} = \frac{k}{0.0015}$$

$$10k = 4(0.0015) = 0.0060$$

$$k = 0.0006$$

Thus log 2.744 = 0.4378 + 0.0006 = 0.4384.

Let's do another example to make sure the process is clear.

EXAMPLE 4

Find log 617.6

Solution

$$\log 617.6 = \log(6.176 \cdot 10^2)$$

$$= \log 6.176 + \log 10^2$$

Thus the characteristic is 2, and we can approximate the mantissa by using interpolation from the table as follows.

x	Log x

$$6\begin{cases} 6.170 \\ \\ 6.176 \\ 6.180 \end{cases}10 \qquad k\begin{cases} 0.7903 \\ \\ ? \\ 0.7910 \end{cases}0.007$$

$$\frac{6}{10} = \frac{k}{0.0007}$$

$$10k = 6(0.0007) = 0.0042$$

$$k = 0.00042 \approx 0.0004$$

Therefore, log 6.176 = 0.7903 + 0.0004 = 0.7907, and we can complete the solution for log 617.6 as follows.

$$\log 617.6 = \log(6.176 \cdot 10^2)$$

$$= \log 6.176 + \log 10^2$$

$$= 0.7907 + 2$$

$$= 2.7907$$

The process of linear interpolation can also be used to approximate an antilogarithm when the mantissa is between two values in the table. The following example illustrates this procedure.

EXAMPLE 5

Find antilog 1.6157.

Solution

From the table we see that the mantissa, 0.6157, is between 0.6149 and 0.6160. We can carry out the interpolation as follows.

$$h\begin{Bmatrix} 4.120 \\ ? \\ 4.130 \end{Bmatrix} 0.0101 \qquad 8\begin{Bmatrix} 0.6149 \\ 0.6157 \\ 0.6160 \end{Bmatrix} 11 \qquad \frac{0.0008}{0.0011} = \frac{8}{11}$$

$$\frac{h}{0.010} = \frac{8}{11}$$

$$h = 8(0.010) = 0.080$$

$$h = \frac{1}{11}(0.080) = 0.007 \quad \text{to the nearest thousandth}$$

Thus antilog $0.6157 = 4.120 + 0.007 = 4.127$. Therefore,

$$\text{antilog } 1.6157 = \text{antilog}(0.6157 + 1)$$
$$= 4.127 \cdot 10^1$$
$$= 41.27 \qquad\qquad\qquad\qquad\qquad \blacksquare$$

Computation with Common Logarithms

Let's first restate the basic properties of logarithms in terms of common logarithms (Remember that we are writing $\log x$ instead of $\log_{10} x$.)

If x and y are positive real numbers, then

    1. $\log xy = \log x + \log y$

    2. $\log \dfrac{x}{y} = \log x - \log y$

    3. $\log x^p = p \log x \qquad p$ is any real number.

The following two properties of equality that pertain to logarithms will also be used.

    4. If $x = y$ (x and y are positive), then $\log x = \log y$.

    5. If $\log x = \log y$, then $x = y$.

EXAMPLE 6

Find the product $(49.1)(876)$.

Solution

Let $N = (49.1)(876)$. By Property 4,

    $\log N = \log(49.1)(876)$

By Property 1,

    $\log N = \log 49.1 + \log 876$

From the table, we find that $\log 49.1 = 1.6911$ and that $\log 876 = 2.9425$. Thus

    $\log N = 1.6911 + 2.9425$
              $= 4.6336$

Therefore,

$$N = \text{antilog } 4.6336$$

By using linear interpolation, we can determine antilog 0.6336 to four significant digits. We obtain

$$N = \text{antilog } (0.6336 + 4)$$
$$= 4.301 \cdot 10^4$$
$$= 43,010$$

 Check   Using a calculator, we get

$$N = (49.1)(876) = 43011.6$$

EXAMPLE 7   Find the quotient $\dfrac{942}{64.8}$.

Solution

Let $N = \dfrac{94.2}{64.8}$. Therefore,

$$\log N = \log \frac{942}{64.8}$$
$$= \log 942 - \log 64.8 \qquad \log \frac{x}{y} - \log x - \log y$$
$$= 2.9741 - 1.8116 \qquad \text{from the table}$$
$$= 1.1625$$

Therefore,

$$N = \text{antilog } 1.1625$$
$$= \text{antilog}(0.01625 + 1)$$
$$= 1.454 \cdot 10^1$$
$$= 14.54$$

 Check   Using a calculator, we get

$$N = \frac{942}{64.8} = 14.537037$$

EXAMPLE 8   Evaluate $\dfrac{(571.4)(8.236)}{71.68}$.

Solution

Let $N = \dfrac{(571.4)(8.236)}{71.68}$. Therefore,

$$\log N = \log \frac{(571.4)(8.236)}{71.68}$$

$$= \log 571.4 + \log 8.236 - \log 71.68$$

$$= 2.7569 + 0.9157 - 1.8554 = 1.8172$$

Therefore,

$$N = \text{antilog } 1.8172$$

$$= \text{antilog}(0.8172 + 1)$$

$$= 6.564 \cdot 10^1 = 65.64$$

 Check   Using a calculator, we get

$$N = \frac{(571.4)(8.236)}{71.68} = 65.653605$$

EXAMPLE 9

Evaluate $\sqrt[3]{3770}$.

Solution

Let $N = \sqrt[3]{3770} = (3770)^{1/3}$. Therefore,

$$\log N = \log(3770)^{1/3}$$

$$= \frac{1}{3}\log 3770 \qquad \log x^p = p \log x$$

$$= \frac{1}{3}(3.5763)$$

$$= 1.1921$$

Therefore,

$$N = \text{antilog } 1.1921$$

$$= \text{antilog }(0.1921 + 1)$$

$$= 1.556 \cdot 10^1 = 15.56$$

 Check   Using a calculator, we get

$$N = \sqrt[3]{3370} = 15.563733$$

When using tables of logarithms, we sometimes must change the form of writing a logarithm so that the decimal part (mantissa) is positive. The next example illustrates this idea.

EXAMPLE 10

Find the quotient $\dfrac{1.73}{5.08}$.

Solution

Let $N = \dfrac{1.73}{5.08}$. Therefore,

$$\log N = \log \frac{1.73}{5.08}$$

$$= \log 1.73 - \log 5.08$$

$$= 0.2380 - 0.7059 = -0.4679$$

Now by adding 1 and subtracting 1, which changes the form but not the value, we obtain

$$\log N = -0.4679 + 1 - 1$$

$$= 0.5321 - 1$$

$$= 0.5321 + (-1)$$

Therefore,

$$N = \text{antilog}(0.5321 + (-1))$$

$$= 3.405 \cdot 10^{1} = 0.3405$$

 Check   Using a calculator, we get

$$N = \frac{1.73}{5.08} = 0.34055118$$

Sometimes it is also necessary to change the form of a logarithm so that a subsequent calculation will produce an **integer for the characteristic part of the logarithm.** Let's consider an example to illustrate this idea.

EXAMPLE 11

Evaluate $\sqrt[4]{0.0767}$.

Solution

Let $N = \sqrt[4]{0.0767} = (0.0767)^{1/4}$. Therefore,

$$\log N = \log(0.0767)^{1/4} = \frac{1}{4}\log 0.0767$$

$$= \frac{1}{4}(0.8848 + (-2))$$

$$= \frac{1}{4}(-2 + 0.8848)$$

At this stage we recognize that applying the distributive property will produce a nonintegral characteristic, $-\frac{1}{2}$. Therefore, let's add 4 and subtract 4 inside the parentheses, which will change the form as follows.

$$\log N = \frac{1}{4}(-2 + 0.8848 + 4 - 4)$$

$$= \frac{1}{4}(4 - 2 + 0.8848 - 4)$$

$$= \frac{1}{4}(2.8848 - 4)$$

Now, applying the distributive property, we obtain

$$\log N = \frac{1}{4}(2.8848) - \frac{1}{4}(4)$$

$$= 0.7212 - 1 = 0.7212 + (-1)$$

Therefore,

$$N = \text{antilog}[(0.7212 + (-1)]$$

$$= 5.262 \cdot 10^{-1} = 0.5262$$

Check Using a calculator, we get

$$N = \sqrt[4]{0.0767} = 0.5262816$$

PRACTICE EXERCISES

For Problems 1–8, use the table of common logarithms and linear interpolation to find each common logarithm.

1. log 4.327

2. log 27.43

3. log 128.9

4. log 3526

5. log 0.8761

6. log 0.07692

7. log 0.005186

8. log 0.0002558

For Problems 9–14, use the table of common logarithms and linear interpolation to find each antilogarithm to four significant digits.

9. antilog 0.4690

10. antilog 1.7971

11. antilog 2.1925

12. antilog 3.7225

13. antilog(0.5026 + (−1))

14. antilog(0.9397 + (−2))

For Problems 15–28, use common logarithms and linear interpolation to help evaluate each. Express your answers to four significant digits. Check your answers by using a calculator.

15. (294)(71.2)

16. (192.6)(4.017)

17. $\dfrac{23.4}{4.07}$

18. $\dfrac{718.5}{8.248}$

19. $(17.3)^5$

20. $(48.02)^3$

21. $\dfrac{(108)(76.2)}{13.4}$

22. $\dfrac{(126.3)(24.32)}{8.019}$

23. $\sqrt[5]{0.821}$

24. $\sqrt[4]{645.3}$

25. $(79.3)^{3/5}$

26. $(176.8)^{3/4}$

27. $\sqrt{\dfrac{(7.05)(18.7)}{0.521}}$

28. $\sqrt[3]{\dfrac{(41.3)(0.271)}{8.05}}$

NATURAL LOGARITHMS

The following table contains the natural logarithms for numbers between 0.1 and 10, inclusive, at intervals of 0.1. Be sure that you agree with the following values taken directly from the table.

$$\ln 1.6 = 0.4700$$
$$\ln 0.5 = -0.6931$$
$$\ln 4.8 = 1.5686$$
$$\ln 9.2 = 2.2192$$

Table of Natural Logarithms

n	$\ln n$	n	$\ln n$	n	$\ln n$	n	$\ln n$
0.1	−2.3026	2.6	0.9555	5.1	1.6292	7.6	2.0281
0.2	−1.6094	2.7	0.9933	5.2	1.6487	7.7	2.0412
0.3	−1.2040	2.8	1.0296	5.3	1.6677	7.8	2.0541
0.4	−0.9163	2.9	1.0647	5.4	1.6864	7.9	2.0669
0.5	−0.6931	3.0	1.0986	5.5	1.7047	8.0	2.0794
0.6	−0.5108	3.1	1.1314	5.6	1.7228	8.1	2.0919
0.7	−0.3567	3.2	1.1632	5.7	1.7405	8.2	2.1041
0.8	−0.2231	3.3	1.1939	5.8	1.7579	8.3	2.1163
0.9	−0.1054	3.4	1.2238	5.9	1.7750	8.4	2.1282
1.0	0.0000	3.5	1.2528	6.0	1.7918	8.5	2.1401
1.1	0.0953	3.6	1.2809	6.1	1.8083	8.6	2.1518
1.2	0.1823	3.7	1.3083	6.2	1.8245	8.7	2.1633
1.3	0.2624	3.8	1.3350	6.3	1.8405	8.8	2.1748
1.4	0.3365	3.9	1.3610	6.4	1.8563	8.9	2.1861
1.5	0.4055	4.0	1.3863	6.5	1.8718	9.0	2.1972

Table of Natural Logarithms (continued)

n	ln n	n	ln n	n	ln n	n	ln n
1.6	0.4700	4.1	1.4110	6.6	1.8871	9.1	2.2083
1.7	0.5306	4.2	1.4351	6.7	1.9021	9.2	2.2192
1.8	0.5878	4.3	1.4586	6.8	1.9169	9.3	2.2300
1.9	0.6419	4.4	1.4816	6.9	1.9315	9.4	2.2407
2.0	0.6931	4.5	1.5041	7.0	1.9459	9.5	2.2513
2.1	0.7419	4.6	1.5261	7.1	1.9601	9.6	2.2618
2.2	0.7885	4.7	1.5476	7.2	1.9741	9.7	2.2721
2.3	0.8329	4.8	1.5686	7.3	1.9879	9.8	2.2824
2.4	0.8755	4.9	1.5892	7.4	2.0015	9.9	2.2925
2.5	0.9163	5.0	1.6094	7.5	2.0149	10	2.3026

When using a table, we can approximate the natural logarithm of a positive number less than 0.1 or greater than 10 by using the property $\ln rs = \ln r + \ln s$ as follows.

$$\ln 190 = \ln(1.9 \cdot 10^2)$$
$$= \ln 1.9 + \ln 10^2$$
$$= \ln 1.9 + 2 \ln 10$$
$$= 0.6419 + 2(2.3026)$$

From the   From the
table   table

$$= 5.2471$$

$$\ln 0.0084 = \ln(8.4 \cdot 10^{-3})$$
$$= \ln 8.4 + \ln 10^{-3}$$
$$= \ln 8.4 + (-3)(\ln 10)$$
$$= 2.1282 - 3(2.3026)$$

From the   From the
table   table

$$= 2.1282 - 6.9078 = -4.7796$$

ANSWERS TO ODD-NUMBERED PROBLEMS AND ALL REVIEW PROBLEMS

CHAPTER 0

Problem Set 0.1 (page 12)

1. True   **3.** False   **5.** False   **7.** True
9. False   **11.** $\{46\}$ **13.** $\{0, -14, 46\}$
15. $\left\{\sqrt{5}, -\sqrt{2}, -\pi\right\}$ **17.** $\{0, -14\}$ **19.** $\subseteq$
21. $\subseteq$ **23.** $\nsubseteq$ **25.** $\subseteq$ **27.** $\nsubseteq$ **29.** $\subseteq$
31. $\nsubseteq$ **33.** $\{1\}$ **35.** $\{0, 1, 2, 3\}$
37. $\{\ldots -2, -1, 0, 1\}$ **39.** $\varnothing$ **41.** $\{0, 1, 2\}$
43. (a) 18   **(b)** 26   **(c)** 39   **(d)** 25
(e) 35   **(f)** 37
45. Commutative property of multiplication
47. Identity property of multiplication
49. Multiplication property of negative one
51. Distributive property
53. Commutative property of multiplication
55. Distributive property
57. Associative property of multiplication   **59.** -22
61. 100   **63.** -21 **65.** 8   **67.** 19
69. 66   **71.** -75 **73.** 34   **75.** 1
77. 11   **79.** 4

Problem Set 0.2 (page 23)

1. $\dfrac{1}{8}$ **3.** $-\dfrac{1}{1000}$ **5.** 27   **7.** 4

9. $-\dfrac{27}{8}$ **11.** 1   **13.** $\dfrac{16}{25}$ **15.** 4

17. $\dfrac{1}{100}$ or 0.01   **19.** $\dfrac{1}{100,000}$ or 0.00001

21. 81   **23.** $\dfrac{1}{16}$ **25.** $\dfrac{3}{4}$ **27.** $\dfrac{256}{25}$

29. $\dfrac{16}{25}$ **31.** $\dfrac{64}{81}$ **33.** 64

35. $\dfrac{1}{100,000}$ or 0.00001   **37.** $\dfrac{17}{72}$ **39.** $\dfrac{1}{6}$

41. $\dfrac{48}{19}$ **43.** $\dfrac{1}{x^4}$ **45.** $\dfrac{1}{a^2}$ **47.** $\dfrac{1}{a^6}$ **49.** $\dfrac{y^4}{x^3}$

51. $\dfrac{c^3}{a^3 b^6}$ **53.** $\dfrac{y^4}{4x^4}$ **55.** $\dfrac{x^4}{y^6}$ **57.** $\dfrac{9a^2}{4b^4}$

59. $\dfrac{1}{x^3}$ **61.** $\dfrac{a^3}{b}$ **63.** $-20x^4 y^5$ **65.** $-27x^3 y^9$

67. $\dfrac{8x^6}{27y^9}$ **69.** $-8x^6$ **71.** $\dfrac{6}{x^3 y}$ **73.** $\dfrac{6}{a^2 y^3}$

75. $\dfrac{4x^3}{y^5}$ **77.** $-\dfrac{5}{a^2 b}$ **79.** $\dfrac{1}{4x^2 y^4}$ **81.** $\dfrac{x+1}{x^2}$

83. $\dfrac{y - x^2}{x^2 y}$ **85.** $\dfrac{3b^3 + 2a^2}{a^2 b^3}$ **87.** $\dfrac{y^2 - x^2}{xy}$

89. $12x^{3a+1}$ **91.** 1   **93.** x^{2a} **95.** $-4y^{6b+2}$
97. x^b **99.** $(6.2)(10)^7$ **101.** $(4.12)(10)^{-4}$
103. 180,000   **105.** 0.0000023   **107.** 0.04
109. 30,000   **111.** 0.03

Problem Set 0.3 (page 30)

1. $14x^2 + x - 6$ **3.** $-x^2 - 4x - 9$
5. $6x - 11$ **7.** $6x^2 - 5x - 7$ **9.** $-x - 34$
11. $12x^3 y^2 + 15x^2 y^3$ **13.** $30a^4 b^3 - 24a^5 b^3 + 18a^4 b^4$
15. $x^2 + 20x + 96$ **17.** $n^2 - 16n + 48$
19. $sx + sy - tx - ty$ **21.** $6x^2 + 7x - 3$
23. $12x^2 - 37x + 21$ **25.** $x^2 + 8x + 16$
27. $4n^2 + 12n + 9$ **29.** $x^3 + x^2 - 14x - 24$
31. $6x^3 - x^2 - 11x + 6$ **33.** $x^3 + 2x^2 - 7x + 4$
35. $t^3 - 1$ **37.** $6x^3 + x^2 - 5x - 2$

39. $x^4 + 8x^3 + 15x^2 + 2x - 4$ **41.** $25x^2 - 4$

43. $x^4 - 10x^3 + 21x^2 + 20x + 4$ **45.** $4x^2 - 9y^2$

47. $x^3 + 15x^2 + 75x + 125$

49. $8x^3 + 12x^2 + 6x + 1$

51. $64x^3 - 144x^2 + 108x - 27$

53. $125x^3 - 150x^2y + 60xy^2 - 8y^3$

55. $a^7 + 7a^6b + 21a^5b^2 + 35a^4b^3 + 35a^3b^4 + 21a^2b^5 + 7ab^6 + b^7$

57. $x^5 - 5x^4y + 10x^3y^2 - 10x^2y^3 + 5xy^4 - y^5$

59. $x^4 + 8x^3y + 24x^2y^2 + 32xy^3 + 16y^4$

61. $64a^6 - 192a^5b + 240a^4b^2 - 160a^3b^3 + 60a^2b^4 - 12ab^5 + b^6$

63. $x^{14} + 7x^{12}y + 21x^{10}y^2 + 35x^8y^3 + 35x^6y^4 + 21x^4y^5 + 7x^2y^6 + y^7$

65. $32a^5 - 240a^4b + 720a^3b^2 - 1080a^2b^3 + 810ab^4 - 243b^5$ **67.** $3x^2 - 5x$

69. $-5a^4 + 4a^2 - 9a$ **71.** $5ab + 11a^2b^4$

73. $x^{2a} - y^{2b}$ **75.** $x^{2b} - 3x^b - 28$

77. $6x^{2b} + x^b - 2$ **79.** $x^{4a} - 2x^{2a} + 1$

81. $x^{3a} - 6x^{2a} + 12x^a - 8$

Problem Set 0.4 (page 40)

1. $2xy(3 - 4y)$ **3.** $(z + 3)(x + y)$

5. $(x + y)(3 + a)$ **7.** $(x - y)(a - b)$

9. $(3x + 5)(3x - 5)$ **11.** $(1 + 9n)(1 - 9n)$

13. $(x + 4 + y)(x + 4 - y)$

15. $(3s + 2t - 1)(3s - 2t + 1)$ **17.** $(x - 7)(x + 2)$

19. $(5 + x)(3 - x)$ **21.** Not factorable

23. $(3x - 5)(x - 2)$ **25.** $(5x + 1)(2x - 7)$

27. $(x - 2)(x^2 + 2x + 4)$

29. $(4x + 3y)(16x^2 - 12xy + 9y^2)$ **31.** $4(x^2 + 4)$

33. $x(x + 3)(x - 3)$ **35.** $(3a - 7)^2$

37. $2n(n^2 + 3n + 5)$ **39.** $(5x - 3)(2x + 9)$

41. $(6a - 1)^2$ **43.** $(4x - y)(2x + y)$

45. Not factorable   **47.** $2n(n^2 + 7n - 10)$

49. $4(x + 2)(x^2 - 2x + 4)$

51. $(x + 3)(x - 3)(x^2 + 5)$

53. $2y(x + 4)(x - 4)(x^2 + 3)$

55. $(a + b + c + d)(a + b - c - d)$

57. $(x + 4 + y)(x + 4 - y)$

59. $(x + y + 5)(x - y - 5)$ **61.** $(10x + 3)(6x - 5)$

63. $3x(7x - 4)(4x + 5)$ **65.** $(x^a + 4)(x^a - 4)$

67. $(x^n - y^n)(x^{2n} + x^ny^n + y^{2n})$

69. $(x^a + 4)(x^a - 7)$ **71.** $(2x^n - 5)(x^n + 6)$

73. $(x^{2n} + y^{2n})(x^n + y^n)(x^n - y^n)$

75. **(a)** $(x + 32)(x + 3)$ **(c)** $(x - 21)(x - 24)$

    (e) $(x + 28)(x + 32)$

Problem Set 0.5 (page 50)

1. $\dfrac{2x}{3}$ **3.** $\dfrac{7y^3}{9x}$ **5.** $\dfrac{a + 4}{a - 9}$ **7.** $\dfrac{x(2x + 7)}{y(x + 9)}$

9. $\dfrac{x^2 + xy + y^2}{x + 2y}$ **11.** $-\dfrac{2}{x + 1}$ **13.** $\dfrac{x}{2y^3}$

15. $-\dfrac{8x^3y^3}{15}$ **17.** $\dfrac{14}{27a}$ **19.** $5y$

21. $\dfrac{5(a + 3)}{a(a - 2)}$ **23.** $\dfrac{(x + 6y)^2(2x + 3y)}{y^3(x + 4y)}$

25. $\dfrac{3xy}{4(x + 6)}$ **27.** $\dfrac{x - 9}{42x^2}$ **29.** $\dfrac{8x + 5}{12}$

31. $\dfrac{7x}{24}$ **33.** $\dfrac{35b + 12a^3}{80a^2b^2}$ **35.** $\dfrac{12 + 9n - 10n^2}{12n^2}$

37. $\dfrac{9y + 8x - 12xy}{12xy}$ **39.** $\dfrac{13x + 14}{(2x + 1)(3x + 4)}$

41. $\dfrac{7x + 21}{x(x + 7)}$ **43.** $\dfrac{1}{a - 2}$ **45.** $\dfrac{1}{x + 1}$

47. $\dfrac{9x + 73}{(x + 3)(x + 7)(x + 9)}$

49. $\dfrac{3x^2 + 30x - 78}{(x + 1)(x - 1)(x + 8)(x - 2)}$

51. $\dfrac{-x^2 - x + 1}{(x + 1)(x - 1)}$ **53.** $\dfrac{-8}{(n^2 + 4)(n + 2)(n - 2)}$

55. $\dfrac{5x^2 + 16x + 5}{(x + 1)(x - 4)(x + 7)}$ **57.** **(a)** $\dfrac{5}{x - 1}$

(c) $\dfrac{5}{a - 3}$ **(e)** $x + 3$ **59.** $\dfrac{5y^2 - 3xy^2}{x^2y + 2x^2}$

61. $\dfrac{x + 1}{x - 1}$ **63.** $\dfrac{n - 1}{n + 1}$ **65.** $\dfrac{-6x - 4}{3x + 9}$

67. $\dfrac{x^2 + x + 1}{x + 1}$ **69.** $\dfrac{a^2 + 4a + 1}{4a + 1}$

71. $-\dfrac{2x + h}{x^2(x + h)^2}$ **73.** $-\dfrac{1}{(x + 1)(x + h + 1)}$

75. $-\dfrac{4}{(2x - 1)(2x + 2h - 1)}$ **77.** $\dfrac{y + x}{x^2y - xy^2}$

79. $\dfrac{x^2y^2 + 1}{y^2 - x}$

Problem Set 0.6 (page 61)

1. 9 **3.** 5 **5.** $\dfrac{6}{7}$ **7.** $-\dfrac{3}{2}$ **9.** $2\sqrt{6}$

11. $4\sqrt{7}$ **13.** $-6\sqrt{11}$ **15.** $\dfrac{3\sqrt{5}}{2}$

17. $2x\sqrt{3}$ **19.** $8x^2y^3\sqrt{y}$ **21.** $\dfrac{9y^3\sqrt{5x}}{7}$

23. $4\sqrt[3]{2}$ **25.** $2x\sqrt[3]{2x}$ **27.** $2x\sqrt[4]{3x}$

29. $\dfrac{2\sqrt{3}}{5}$ **31.** $\dfrac{\sqrt{14}}{4}$ **33.** $\dfrac{4\sqrt{15}}{5}$ **35.** $\dfrac{3\sqrt{2}}{7}$

37. $\dfrac{\sqrt{15}}{6x^2}$ **39.** $\dfrac{2\sqrt{15a}}{5ab}$ **41.** $\dfrac{3\sqrt[3]{2}}{2}$

43. $\dfrac{\sqrt[3]{18x^2y}}{3x}$ **45.** $12\sqrt{3}$ **47.** $3\sqrt{7}$

49. $\dfrac{11\sqrt{3}}{6}$ **51.** $-\dfrac{89\sqrt{2}}{30}$ **53.** $48\sqrt{6}$

55. $10\sqrt{6} + 8\sqrt{30}$ **57.** $3x\sqrt{6y} - 6\sqrt{2xy}$

59. $13 + 7\sqrt{3}$ **61.** $30 + 11\sqrt{6}$ **63.** 16

65. $x + 2\sqrt{xy} + y$ **67.** $a - b$ **69.** $3\sqrt{5} - 6$

71. $\sqrt{7} + \sqrt{3}$ **73.** $\dfrac{-2\sqrt{10} + 3\sqrt{14}}{43}$

75. $\dfrac{x + \sqrt{x}}{x - 1}$ **77.** $\dfrac{x - \sqrt{xy}}{x - y}$

79. $\dfrac{6x + 7\sqrt{xy} + 2y}{9x - 4y}$ **81.** $\dfrac{2}{\sqrt{2x + 2h} + \sqrt{2x}}$

83. $\dfrac{1}{\sqrt{x + h - 3} + \sqrt{x - 3}}$

Problem Set 0.7 (page 66)

1. 7   **3.** 8   **5.** -4 **7.** 2   **9.** 64

11. 0.001   **13.** $\dfrac{1}{32}$ **15.** 2   **17.** $15x^{7/12}$

19. $y^{5/12}$ **21.** $64x^{3/4}y^{3/2}$ **23.** $4x^{4/15}$

25. $\dfrac{7}{a^{1/12}}$ **27.** $\dfrac{16x^{4/3}}{81y}$ **29.** $\dfrac{y^{3/2}}{x}$ **31.** $8a^{9/2}x^2$

33. $\sqrt[4]{8}$ **35.** $\sqrt[12]{x^7}$ **37.** $xy\sqrt[4]{xy^3}$

39. $a\sqrt[12]{a^5b^{11}}$ **41.** $4\sqrt[6]{2}$ **43.** $\sqrt[6]{2}$ **45.** $\sqrt{2}$

47. $x\sqrt[12]{x^7}$ **49.** $\dfrac{5\sqrt[3]{x^2}}{x}$ **51.** $\dfrac{\sqrt[12]{x^3y^4}}{y}$

53. $\dfrac{\sqrt[20]{x^{15}y^8}}{y}$ **55.** $\dfrac{5\sqrt[12]{x^9y^8}}{4x}$

57. (a) $\sqrt[6]{2}$ (c) $\sqrt{x}$

63. (a) 13.391   (c) 2.702   (e) 4.304

65. $\dfrac{2x - 2}{(2x - 1)^{3/2}}$ **67.** $\dfrac{x}{(x^2 + 2x)^{3/2}}$ **69.** $\dfrac{4x}{(2x)^{4/3}}$

Problem Set 0.8 (page 74)

1. $13 + 8i$ **3.** $3 + 4i$ **5.** $-11 + i$

7. $-1 - 2i$ **9.** $-\dfrac{3}{20} + \dfrac{5}{12}i$ **11.** $\dfrac{7}{10} - \dfrac{11}{12}i$

13. $4 + 0i$ **15.** $3i$ **17.** $i\sqrt{19}$ **19.** $\dfrac{2}{3}i$

21. $2i\sqrt{2}$ **23.** $3i\sqrt{3}$ **25.** $3i\sqrt{6}$ **27.** $18i$

29. $12i\sqrt{2}$ **31.** -8 **33.** $-\sqrt{6}$

35. $-2\sqrt{5}$ **37.** $-2\sqrt{15}$ **39.** $-2\sqrt{14}$

41. 3   **43.** $\sqrt{6}$ **45.** -21 **47.** $8 + 12i$

49. $0 + 26i$ **51.** $53 - 26i$ **53.** $10 - 24i$

55. $-14 - 8i$ **57.** $-7 + 24i$ **59.** $-3 + 4i$

61. $113 + 0i$ **63.** $13 + 0i$ **65.** $-\dfrac{8}{13} + \dfrac{12}{13}i$

67. $1 - \dfrac{2}{3}i$ **69.** $0 - \dfrac{3}{2}i$ **71.** $\dfrac{22}{41} - \dfrac{7}{41}i$

73. $-1 + 2i$ **75.** $-\dfrac{17}{10} + \dfrac{1}{10}i$ **77.** $\dfrac{5}{13} - \dfrac{1}{13}i$

83. (a) $2 + 11i$ (c) $-11 + 2i$ (e) $-7 - 24i$

Chapter 0 Review Problem Set (page 78)

1. $\dfrac{1}{125}$ **2.** $-\dfrac{1}{81}$ **3.** $\dfrac{16}{9}$ **4.** $\dfrac{1}{9}$ **5.** -8

6. $\dfrac{3}{2}$ **7.** $-\dfrac{1}{2}$ **8.** $\dfrac{1}{6}$ **9.** 4   **10.** -8

11. $12x^2y$ **12.** $-30x^{7/6}$ **13.** $\dfrac{48}{a^{1/6}}$

14. $\dfrac{27y^{3/5}}{x^2}$ **15.** $\dfrac{4y^5}{x^5}$ **16.** $\dfrac{8y}{x^{7/12}}$ **17.** $\dfrac{16x^6}{y^6}$

18. $-\dfrac{a^3b^1}{3}$ **19.** $4x - 1$ **20.** $-3x + 8$

21. $12a - 19$ **22.** $20x^2 - 11x - 42$

23. $-12x^2 + 17x - 6$ **24.** $-35x^2 + 22x - 3$

25. $x^3 + x^2 - 19x - 28$ **26.** $2x^3 - x^2 + 10x + 6$

27. $25x^2 - 30x + 9$ **28.** $9x^2 + 42x + 49$

29. $8x^3 - 12x^2 + 6x - 1$

30. $27x^3 + 135x^2 + 225x + 125$

31. $x^4 + 2x^3 - 6x^2 - 22x - 15$

32. $2x^4 + 11x^3 - 16x^2 - 8x + 8$

33. $-4x^2y^3 + 8xy^2$ **34.** $-7y + 9xy^2$

35. $(3x + 2)(3x - 2)$ **36.** $3x(x + 5)(x - 8)$

37. $(2x + 5)^2$ **38.** $(x - y + 3)(x - y - 3)$

39. $(x - 2)(x - y)$

40. $(4x - 3y)(16x^2 + 12xy + 9y^2)$

41. $(3x - 4)(5x + 2)$ **42.** $3(x^3 + 12)$

43. Not factorable   **44.** $3(x + 2)(x^2 - 2x + 4)$

45. $(x + 3)(x - 3)(x + 2)(x - 2)$

46. $(2x - 1 - y)(2x - 1 + y)$ **47.** $\dfrac{2}{3y}$

48. $\dfrac{-5a^2}{3}$ **49.** $\dfrac{3x + 5}{x}$ **50.** $\dfrac{2(3x - 1)}{x^2 + 4}$

51. $\dfrac{29x - 10}{12}$ **52.** $\dfrac{x - 38}{15}$ **53.** $\dfrac{-6n + 15}{5n^2}$

54. $\dfrac{-3x - 16}{x(x + 7)}$ **55.** $\dfrac{3x^2 - 8x - 40}{(x + 4)(x - 4)(x - 10)}$

56. $\dfrac{8x - 4}{x(x + 2)(x - 2)}$ **57.** $\dfrac{3xy - 2x^2}{5y + 7x^2}$

58. $\dfrac{3x - 2}{4x + 3}$ **59.** $-\dfrac{6x + 3h}{x^2(x + h)^2}$ **60.** $\dfrac{12}{(x^2 + 2)^{3/2}}$

61. $20\sqrt{3}$ **62.** $6x\sqrt{6x}$ **63.** $2xy\sqrt[3]{4xy^2}$

64. $\sqrt{3}$ **65.** $\dfrac{\sqrt{10x}}{2y}$ **66.** $\dfrac{15 - 3\sqrt{2}}{23}$

67. $\dfrac{24 - 4\sqrt{6}}{15}$ **68.** $\dfrac{3x + 6\sqrt{xy}}{x - 4y}$ **69.** $\sqrt[6]{5^5}$

70. $\sqrt[12]{x^{11}}$ **71.** $x^2\sqrt[6]{x^5}$ **72.** $x\sqrt[10]{xy^9}$ **73.** $\sqrt[6]{5}$

74. $\dfrac{\sqrt[12]{x^{11}}}{x}$ **75.** $-11 - 6i$ **76.** $-1 - 2i$

77. $1 - 2i$ **78.** $21 + 0i$ **79.** $26 - 7i$

80. $-25 + 15i$ **81.** $-14 - 12i$ **82.** $29 + 0i$

83. $0 - \dfrac{5}{3}i$ **84.** $-\dfrac{6}{25} + \dfrac{17}{25}i$ **85.** $0 + i$

86. $-\dfrac{12}{29} - \dfrac{30}{29}i$ **87.** $10i$ **88.** $2i\sqrt{10}$

89. $16i\sqrt{5}$ **90.** -12 **91.** $-4\sqrt{3}$

92. $2\sqrt{2}$ **93.** 600,000,000   **94.** 800,000

Chapter 0 Test (page 80)

1. (a) $-\dfrac{1}{49}$ (b) $\dfrac{8}{27}$ (c) $\dfrac{8}{27}$ (d) $\dfrac{3}{4}$

2. $-\dfrac{15}{x^4 y^2}$ **3.** $-12x - 8$ **4.** $-30x^2 + 32x - 8$

5. $3x^3 - 4x^2 - 11x - 14$

6. $64x^3 - 48x^2 + 12x - 1$ **7.** $9x^3y + 12x^4y^2$

8. $3x(2x + 1)(3x - 4)$ **9.** $(x + 2)(6x - 5)$

10. $8(x + 2)(x^2 - 2x + 4)$

11. $(x - 2)(x + y)$ **12.** $\dfrac{21x^5}{20}$ **13.** $\dfrac{x + 3}{x^2 + 2x + 4}$

14. $\dfrac{n - 8}{12}$ **15.** $\dfrac{23x + 6}{6x(x - 3)(x + 2)}$ **16.** $\dfrac{8 - 13n}{2n^2}$

17. $\dfrac{2y^2 - 5xy}{3y^2 + 4x}$ **18.** $12x^2\sqrt{7x}$ **19.** $\dfrac{5\sqrt{2}}{6}$

20. $\dfrac{4\sqrt{3} + 3\sqrt{2}}{5}$ **21.** $2xy\sqrt[3]{6xy^2}$ **22.** $-4 - 3i$

23. $34 - 18i$ **24.** $85 + 0i$ **25.** $\dfrac{1}{10} + \dfrac{7}{10}i$

CHAPTER 1

Problem Set 1.1 (page 90)

1. $\{-2\}$ **3.** $\left\{-\dfrac{1}{2}\right\}$ **5.** $\{7\}$ **7.** $\left\{-\dfrac{3}{2}\right\}$

9. $\left\{-\dfrac{10}{3}\right\}$ **11.** $\{-10\}$ **13.** $\{-17\}$

15. $\left\{-\dfrac{21}{16}\right\}$ **17.** $\left\{\dfrac{3}{5}\right\}$ **19.** $\{-14\}$

21. $\{9\}$ **23.** $\left\{\dfrac{10}{7}\right\}$ **25.** $\{-10\}$

27. $\{1\}$ **29.** $\{-12\}$ **31.** $\{27\}$

33. $\left\{\dfrac{159}{5}\right\}$ **35.** $\{3\}$ **37.** $\{0\}$

39. $\left\{-\dfrac{2}{3}\right\}$ **41.** $\left\{\dfrac{1}{2}\right\}$ **43.** 14 and 9

45. 14, 16, and 18   **47.** 18 and 9

49. 10, 11, 12, and 13

51. 30°, 40°, and 110°

53. $24,000 for Renee, $20,000 for Kelly, $16,000 for Nina

55. 48 pennies, 21 nickels, and 11 dimes

57. 17 females and 26 males   **59.** 13 years old

61. Brad is 29 and Pedro is 23.

Problem Set 1.2 (page 101)

1. $\{1\}$ **3.** $\{9\}$ **5.** $\left\{\dfrac{10}{3}\right\}$ **7.** $\{4\}$

9. $\{14\}$ **11.** $\{9\}$ **13.** $\left\{\dfrac{1}{2}\right\}$ **15.** $\left\{\dfrac{1}{4}\right\}$

17. $\left\{\dfrac{2}{3}\right\}$ **19.** $\{-8\}$ **21.** $\varnothing$ **23.** $\{12\}$

25. $\{300\}$ **27.** $\{275\}$ **29.** $\left\{-\dfrac{66}{37}\right\}$ **31.** $\{6\}$

33. $w = \dfrac{P - 21}{2}$ **35.** $h = \dfrac{A - 2\pi r^2}{2\pi r}$

37. $F = \dfrac{9C + 160}{5}$ or $F = \dfrac{9}{5}C + 32$

39. $T = \dfrac{NC - NV}{C}$ **41.** $T = \dfrac{I + klt}{kl}$

43. $R_n = \dfrac{R_1 R_2}{R_1 + R_2}$ **45.** 17 and 81   **47.** $1050

49. $900 and $1350   **51.** 37 teachers and 740 students

53. $65   **55.** $950 per month   **57.** $32.20

59. $75   **61.** $30   **63.** 14 nickels and 29 dimes

65. 15 dimes, 45 quarters, and 10 half-dollars

67. $2000 at 9% and $3500 at 10%   **69.** $3500

71. 6 centimeters by 10 centimeters

CHAPTER 2

Problem Set 1.3 (page 112)

1. $\{-4, 7\}$ **3.** $\left\{-3, \dfrac{4}{3}\right\}$ **5.** $\left\{0, \dfrac{3}{2}\right\}$

7. $\left\{\pm\dfrac{2\sqrt{3}}{3}\right\}$ **9.** $\left\{\dfrac{-1 \pm 2\sqrt{5}}{2}\right\}$ **11.** $\left\{-\dfrac{5}{3}, \dfrac{2}{5}\right\}$

13. $\{2 \pm 2i\}$ **15.** $\left\{-\dfrac{7}{2}, \dfrac{1}{5}\right\}$ **17.** $\{4, 6\}$

19. $\left\{-5 \pm 3\sqrt{3}\right\}$ **21.** $\left\{\dfrac{3 \pm \sqrt{5}}{2}\right\}$

23. $\left\{-2 \pm i\sqrt{2}\right\}$ **25.** $\left\{\dfrac{-6 \pm \sqrt{46}}{2}\right\}$

27. $\{-16, 18\}$ **29.** $\left\{\dfrac{-5 \pm \sqrt{37}}{6}\right\}$ **31.** $\{-6. 9\}$

33. $\left\{-5, -\dfrac{1}{3}\right\}$ **35.** $\left\{1 \pm \sqrt{5}\right\}$ **37.** $\left\{\dfrac{3 \pm \sqrt{7}}{2}\right\}$

39. $\left\{\dfrac{3 \pm i\sqrt{19}}{2}\right\}$ **41.** $\left\{4 \pm 2\sqrt{3}\right\}$ **43.** $\left\{\dfrac{1}{2}\right\}$

45. $\left\{-\dfrac{3}{2}, \dfrac{1}{4}\right\}$ **47.** $\{-14, 12\}$ **49.** $\left\{\dfrac{3 \pm i\sqrt{47}}{4}\right\}$

51. $\left\{-1, \dfrac{5}{3}\right\}$ **53.** $\left\{\dfrac{-1 \pm \sqrt{2}}{2}\right\}$ **55.** $\left\{8 \pm 5\sqrt{2}\right\}$

57. $\left\{-10 \pm 5\sqrt{5}\right\}$ **59.** $\left\{\dfrac{1 \pm \sqrt{6}}{5}\right\}$

61. **(a)** One real solution   **(c)** One real solution
    (e) Two complex but nonreal solutions
    (g) Two unequal real solutions
63. 11 and 12   **65.** 12 feet
67. 10 meters and 24 meters
69. 8 inches by 14 inches
71. 7 meters wide and 18 meters long   **73.** 1 meter
75. 7 inches by 3 inches   **77.** 8 units

83. **(a)** $n = \dfrac{\sqrt{A\pi}}{\pi}$ **(c)** $t = \dfrac{\sqrt{2gs}}{g}$

    (e) $y = \dfrac{b\sqrt{x^2 - a^2}}{a}$

85. $k = \pm 4$
87. **(a)** $\{-1.359, 7.359\}$
    (c) $\{-10.280, 4.280\}$
    (e) $\{-0.259, -7.742\}$
    (g) $\{0.191, 1.309\}$
    (i) $\{-0.422, 5.922\}$

Problem Set 1.4 (page 125)

1. $\{-12\}$ **3.** $\left\{\dfrac{37}{15}\right\}$ **5.** $\{-2\}$ **7.** $\{-8, 1\}$

9. $\left\{\dfrac{6}{29}\right\}$ **11.** $\left\{n \mid n \neq \dfrac{3}{2} \text{ and } n \neq 3\right\}$

13. $\left\{-4, \dfrac{4}{3}\right\}$ **15.** $\left\{-\dfrac{1}{4}\right\}$ **17.** $\varnothing$ **19.** $\{-1\}$

21. 9 rows and 14 trees per row   **23.** $4\dfrac{1}{2}$ hours

25. 50 miles

27. 50 mph for the freight and 70 mph for the express
29. 3 liters
31. 3.5 liters of the 50% solution and 7 liters of the 80% solution

33. 5 quarts   **35.** $2\dfrac{2}{5}$ hours   **37.** 60 minutes

39. 9 minutes for Pat and 18 minutes for Mike
41. 8 hours   **43.** 7 golf balls
45. 60 hours

Problem Set 1.5 (page 133)

1. $\{-2, -1, 2\}$ **3.** $\left\{\pm i, \dfrac{3}{2}\right\}$

5. $\left\{-\dfrac{5}{4}, 0, \pm\dfrac{\sqrt{2}}{2}\right\}$ **7.** $\{0, 16\}$ **9.** $\{1\}$

11. $\{6\}$ **13.** $\{3\}$ **15.** $\varnothing$ **17.** $\{-15\}$

19. $\{9\}$ **21.** $\left\{\dfrac{2}{3}, 1\right\}$ **23.** $\{5\}$ **25.** $\{7\}$

27. $\{-2, -1\}$ **29.** $\{0\}$ **31.** $\{6\}$ **33.** $\{0, 4\}$

35. $\{\pm 1, \pm 2\}$ **37.** $\left\{\pm\dfrac{\sqrt{2}}{2}, \pm 2\right\}$

39. $\left\{\pm i\sqrt{5}, \pm\sqrt{7}\right\}$

41. $\left\{\pm\sqrt{2 + \sqrt{3}}, \pm\sqrt{2 - \sqrt{3}}\right\}$ **43.** $\{-125, 8\}$

45. $\left\{-\dfrac{8}{27}, \dfrac{27}{8}\right\}$ **47.** $\left\{-\dfrac{1}{6}, \dfrac{1}{2}\right\}$ **49.** $\{25, 36\}$

51. $\{4\}$
53. 12 inches   **55.** 320 feet
61. **(a)** $\{\pm 1.62, \pm.62\}$
    (c) $\{\pm 1.78, \pm.56\}$
    (e) $\{\pm 8.00, \pm 6.00\}$

Problem Set 1.6 (page 144)

1. $(-\infty, -2]$

3. $(1, 4)$

5. $(0, 2)$

7. $[-2, -1]$

9. $(-\infty, 1) \cup (3, \infty)$

11. $(-2, \infty)$

13. $[-5, 4]$ **15.** $\left(-1, \dfrac{3}{2}\right)$ **17.** $(-11, 13)$

19. $(-1, 5)$

21. $(-\infty, -2)$ **23.** $\left[-\frac{5}{3}, \infty\right)$ **25.** $[7, \infty)$

27. $\left(-\infty, \frac{17}{5}\right]$ **29.** $\left(-\infty, \frac{7}{3}\right)$ **31.** $[-20, \infty)$

33. $(300, \infty)$ **35.** $\left(\frac{1}{5}, \frac{7}{5}\right)$ **37.** $[1, 5]$

39. $(-4, 1)$ **41.** $(-\infty, -3) \cup (5, \infty)$

43. $[-1, 2]$ **45.** $(-\infty, -4) \cup \left(\frac{1}{3}, \infty\right)$

47. $\left[\frac{2}{5}, \frac{4}{3}\right]$ **49.** $\left(-\infty, \frac{1}{2}\right) \cup \left(\frac{1}{2}, \infty\right)$

51. $(-2, -1)$ **53.** $\left(-\infty, \frac{22}{3}\right]$

55. $(-4, 1) \cup (2, \infty)$ **57.** $(-\infty, -2] \cup \left[\frac{1}{2}, 5\right]$

59. $[-4, 0] \cup [6, \infty)$ **61.** $(-3, 2) \cup (2, \infty)$

63. Greater than 12%   **65.** 98 or better

67. Greater than or equal to 13.8 inches

69. Between $-4°$ and $23°$, inclusive

71. More than 250 miles

Problem Set 1.7 (page 153)

1. $(-\infty, -1) \cup (5, \infty)$ **3.** $\left(-2, \frac{1}{2}\right)$ **5.** $\left(\frac{1}{3}, 3\right]$

7. $[-3, -2)$ **9.** $(-\infty, -5) \cup (-2, \infty)$

11. $(-\infty, -5)$ **13.** $(-3, 2)$ **15.** $\{-4, 8\}$

17. $\left\{-\frac{13}{20}, \frac{3}{20}\right\}$ **19.** $\{-3, 4\}$ **21.** $\left\{-3, \frac{1}{3}\right\}$

23. $\varnothing$ **25.** $\left\{-\frac{10}{3}, 2\right\}$ **27.** $\left\{\frac{1}{4}, \frac{7}{4}\right\}$

29. $\left\{-\frac{2}{5}, 4\right\}$ **31.** $\{-2, 0\}$ **33.** $\{-1\}$

35. $(-6, 6)$ **37.** $(-\infty, -8) \cup (8, \infty)$

39. $(-\infty, \infty)$ **41.** $(-\infty, -2) \cup (8, \infty)$

43. $[-3, 4]$ **45.** $\left(-\infty, -\frac{11}{3}\right) \cup \left(\frac{7}{3}, \infty\right)$ **47.** $\varnothing$

49. $\left(-\frac{1}{2}, \frac{7}{2}\right)$ **51.** $(-\infty, \infty)$ **53.** $[-7, 3]$

55. $(-6, 0)$ **57.** $(-\infty, -6) \cup (-2, \infty)$

59. $(-\infty, 0] \cup [4, \infty)$ **61.** $(-6, 4)$

63. $\left(-\infty, \frac{5}{4}\right) \cup \left(\frac{7}{2}, \infty\right)$ **65.** $(-\infty, -3) \cup (-3, -1)$

67. $\left[-\frac{2}{5}, 0\right) \cup \left(0, \frac{2}{3}\right]$ **69.** $\left(-\infty, \frac{2}{5}\right] \cup \left[\frac{2}{3}, \infty\right)$

77. (a) $(-16, 6)$ **(c)** $(-\infty, -3) \cup (4, \infty)$
    (e) $(-3, 7)$

Chapter 1 Review Problem Set (page 156)

1. $\{-14\}$ **2.** $\{-19\}$ **3.** $\left\{\frac{10}{7}\right\}$ **4.** $\{200\}$

5. $\left\{-1, \frac{5}{3}\right\}$ **6.** $\left\{\frac{5}{4}, 6\right\}$ **7.** $(3 \pm i\}$

8. $\{-22, 18\}$ **9.** $\left\{-\frac{2}{5}, 0, \frac{1}{3}\right\}$ **10.** $\{-5\}$

11. $\left\{\frac{1}{2}, 6\right\}$ **12.** $\{\pm 3i, \pm\sqrt{5}\}$

13. $\left\{\pm\frac{\sqrt{5}}{5}, \pm\sqrt{2}\right\}$ **14.** $\left\{-1, 2, \frac{-5 \pm \sqrt{33}}{2}\right\}$

15. $\{2\}$ **16.** $\left\{-1, \frac{1}{2}\right\}$ **17.** $\{0\}$

18. $\left\{-\frac{6}{5}, \frac{8}{5}\right\}$ **19.** $\left\{\frac{2}{5}, 12\right\}$ **20.** $\left\{\frac{1}{4}, \frac{7}{4}\right\}$

21. $\{-\sqrt{2}, -1, \sqrt{2}\}$ **22.** $\left\{-64, \frac{27}{8}\right\}$

23. $(-8, \infty)$ **24.** $\left[-\frac{65}{4}, \infty\right)$ **25.** $\left(-\infty, -\frac{9}{2}\right)$

26. $(-\infty, 400]$ **27.** $[-2, 1]$ **28.** $\left(-\frac{2}{3}, 2\right)$

29. $(-3, 6)$ **30.** $(-\infty, -2] \cup [7, \infty)$

31. $(-\infty, -2) \cup (1, 4)$ **32.** $\left[-4, \frac{3}{2}\right)$

33. $\left(-\infty, \frac{1}{5}\right) \cup (2, \infty)$ **34.** $[-7, -3)$

35. $(-\infty, 4)$ **36.** $\left(-\infty, -\frac{1}{2}\right) \cup (2, \infty)$

37. $\left[-\frac{19}{3}, 3\right]$ **38.** $\left(-\frac{9}{2}, \frac{3}{2}\right)$

39. $(-1, 0) \cup \left(0, \frac{1}{3}\right)$ **40.** $\left(-\frac{3}{2}, \infty\right)$

41. 21, 23, and 25   **42.** 9 and 65

43. 7 centimeters by 12 centimeters

44. 13 nickels, 39 dimes, and 36 quarters   **45.** $20

46. 20 gallons

47. Rosie is 14 years old and her mother is 33 years old.

48. $350 at 9% and 450 at 12%   **49.** 95 or better

50. $10\frac{10}{11}$ minutes   **51.** $26\frac{2}{3}$ minutes

52. 40 shares at $15/share

53. 54 mph for Mike and 52 mph for Larry

54. Cindy 4 hours and Bill 6 hours

55. 15 centimeters and 20 centimeters

56. 5 inches by 7 inches

Chapter I Test (page 158)

1. $\{2\}$ **2.** $\left\{-\dfrac{3}{2}, \dfrac{1}{5}\right\}$ **3.** $\left\{-\dfrac{7}{5}, \dfrac{3}{5}\right\}$ **4.** $\{-1\}$

5. $\left\{\dfrac{1 + i\sqrt{31}}{4}\right\}$ **6.** $\{-4, -1\}$ **7.** $\{600\}$

8. $\left\{-1, \dfrac{11}{3}\right\}$ **9.** $\left\{\dfrac{1 \pm \sqrt{7}}{3}\right\}$ **10.** $\{-9, 0, 2\}$

11. $\left\{-\dfrac{6}{7}, 3\right\}$ **12.** $\{8\}$ **13.** $\varnothing$

14. $\left\{-\dfrac{1}{4}, \dfrac{2}{3}\right\}$ **15.** $(-\infty, -35]$ **16.** $(3, \infty)$

17. $\left(-1, \dfrac{7}{3}\right)$ **18.** $\left(-\infty, -\dfrac{11}{4}\right] \cup \left[\dfrac{1}{4}, \infty\right)$

19. $\left[-\dfrac{1}{2}, 5\right]$ **20.** $(-\infty, -2) \cup \left(\dfrac{1}{3}, \infty\right)$

21. $[-10, -6)$ **22.** $\dfrac{2}{3}$ of a cup

23. 15 miles per hour **24.** 150 shares
25. 9 centimeters by 14 centimeters

CHAPTER 2

Problem Set 2.1 (page 171)

1. 10 **3.** -5 **5.** 6 **7.** 15 **9.** 7

11. $\dfrac{1}{3}$ **13.** -7 **15.** $10; (6, 4)$

17. $\sqrt{13}; \left(2, -\dfrac{5}{2}\right)$ **19.** $3\sqrt{2}; \left(\dfrac{15}{2}, -\dfrac{11}{2}\right)$

21. $\dfrac{\sqrt{74}}{6}; \left(\dfrac{1}{12}, \dfrac{11}{12}\right)$ **23.** $(3, 5)$ **25.** $(2, 5)$

27. $\left(\dfrac{17}{8}, -7\right)$ **29.** $\left(4, \dfrac{25}{4}\right)$ **35.** $15 + 9\sqrt{5}$

39. 3 or -7 **41.** $(3, 8)$

43. Both midpoints are at $\left(\dfrac{7}{2}, \dfrac{5}{2}\right)$.

Problem Set 2.2 (page 181)

1.

3.

5.

7.

9.

11.

13.

The graph is the y-axis.

15.

17.

19.

21.

23.

25.

27.

29.

33.

35.

37.

39.

41.

43.

23. $2x + 3y = -1$ **25.** $y = -2$

27. $5x - 7y = -11$ **29.** $5x + 6y = 37$

31. $x + 5y = 14$ **33.** $y = -3$ **35.** $y = \frac{1}{2}x + 3$

37. $y = -\frac{3}{7}x + 2$ **39.** $y = 4x + \frac{3}{2}$

41. $y = -\frac{5}{6}x + \frac{1}{4}$ **43.** $5x - 4y = 20$

45. $x = -4$ **47.** $5x + 2y = 14$

49. $4x + y = -2$ **51.** Parallel

53. Perpendicular

55. Intersecting lines that are not perpendicular

57. Perpendicular   **59.** $m = \frac{2}{3}, b = -\frac{4}{3}$

61. $m = \frac{1}{2}, b = -\frac{7}{2}$ **63.** $m = -3, b = 0$

65. $m = \frac{7}{5}, b = -\frac{12}{5}$

67. (a)

(c)

(e)

73. $x - y = -7, x + 5y = -19, 2x + y = 16$

75. $9x + 8y = -2, 6x - 7y = 11, 15x + y = 9$

77. 250 feet   **79.** 19 centimeters

85. (a) $3x - y = 9$ **(c)** $2x + 7y = 0$

89. (a) $2x - y = 4$ **(c)** $5x + 2y = 2$

Problem Set 2.3 (page 195)

1. $\frac{3}{4}$ **3.** -5 **5.** 0 **7.** $-\frac{b}{a}$ **9.** $x = 7$

11. $x = -2$ **13.–17.** Answers will vary.

19. $x - 3y = -10$ **21.** $2x - y = 0$

Problem Set 2.4 (page 207)

1. $(4, -3); (-4, 3); (-4, -3)$

3. $(-6, 1); (6, -1); (6, 1)$

5. $(0, -4); (0, 4); (0, -4)$ **7.** y-axis

9. x axis   **11.** x axis, y axis, and origin

13. None   **15.** Origin   **17.** None   **19.** y axis

21.

23.

41.

43.

25.

27.

45.

47.

29.

31.

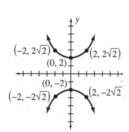

Problem Set 2.5 (page 218)

1. $x^2 + y^2 - 4x - 6y - 12 = 0$
3. $x^2 + y^2 + 2x + 10y + 17 = 0$
5. $x^2 + y^2 - 6x = 0$ 7. $x^2 + y^2 - 49 = 0$
9. $(3, 5), r = 2$ 11. $(-5, -7), r = 1$
13. $(5, 0), r = 5$ 15. $(0, 0), r = 2\sqrt{2}$
17. $\left(\dfrac{1}{2}, 1\right), r = 2$
19. $x^2 + y^2 - 6x - 6y - 67 = 0$
21. $x^2 + y^2 - 14x + 14y + 49 = 0$
23. $x^2 + y^2 + 6x - 10y + 9 = 0$ and $x^2 + y^2 + 6x + 10y + 9 = 0$

33.

35.

25.

27.

37.

39.

29.

31.

33.

35.

15.

16.

37.

39.

17.

18.

41.

43.

19.

20.

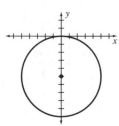

47. **(a)** $(1, 4), r = 3$ **(c)** $(-6, -4), r = 8$
    (e) $(0, 6), r = 9$

Chapter 2 Review Problem Set (page 222)

1. 5   **2.** -5 **3.** $\left(9, \dfrac{1}{3}\right)$ **4.** $(-2, 6)$

7. x axis   **8.** None   **9.** x axis, y axis, and origin
10. y axis   **11.** Origin   **12.** y axis

13.

14.

21.

22.

23. -5 **24.** $\dfrac{5}{7}$ **25.** $3x + 4y = 29$

26. $2x - y = -4$ **27.** $4x + 3y = -4$
28. $x + 2y = 3$
29. $x^2 + y^2 - 10x + 12y + 60 = 0$
30. $x^2 + y^2 - 4x - 6y - 4 = 0$
31. $x^2 + y^2 + 10x - 24y = 0$
32. $x^2 + y^2 + 8x + 8y + 16 = 0$

Chapter 2 Test (page 223)

1. 8   **2.** $(-4, 6)$ **3.** $(6, -4)$ **4.** $-\dfrac{4}{9}$

5. $\dfrac{2}{7}$ **6.** $3x + 4y = -12$

7. $11x - 3y = 23$ **8.** $x - 5y = -21$

9. $7x - 4y = 1$ **10.** $x + 0y = -2$

11. $x^2 + 6x + y^2 + 12y + 29 = 0$

12. $x^2 - 4x + y^2 - 8y - 20 = 0$

13. $x^2 - 8x + y^2 + 6y = 0$

14. Center at $(-8, 5)$ and a radius of length 3 units

15. 5, $\sqrt{37}$, and $2\sqrt{5}$ units   **16.** 1 and 5

17. $\pm\sqrt{3}$ **18.** 6 units   **19.** $y = \pm\dfrac{3}{4}x$

20. (a) x axis   **(b)** origin   **(c)** y axis
(d) x axis, y axis, and origin

21.

22.

23.

24.

25.

Chapters 0, 1, and 2 Cumulative Review Problem Set (page 225)

1. $\dfrac{1}{27}$ **2.** $-\dfrac{1}{16}$ **3.** $\dfrac{9}{4}$ **4.** $-\dfrac{2}{3}$ **5.** 9

6. $\dfrac{9}{16}$ **7.** $\dfrac{20}{x^2y^3}$ **8.** $-\dfrac{56a}{b}$ **9.** $4x^4y^2$

10. $\dfrac{5y^2}{x^4}$ **11.** $\dfrac{x^{1/3}}{17y^{7/4}}$ **12.** $\dfrac{4a^8}{b^{14}}$ **13.** $-30\sqrt{2}$

14. $6xy\sqrt{3x}$ **15.** $2xy^2\sqrt[3]{7xy}$ **16.** $\dfrac{3\sqrt{6}}{10}$

17. $\dfrac{\sqrt{21xy}}{7y}$ **18.** $-\dfrac{5\left(\sqrt{2} + 3\right)}{7}$

19. $\dfrac{6\sqrt{14} + 3\sqrt{42}}{2}$ **20.** $\dfrac{4x - 12\sqrt{xy}}{x - 9y}$

21. $\dfrac{3x^3y^2}{8}$ **22.** $-\dfrac{3b^3}{8a^3}$ **23.** $\dfrac{5x + 1}{x}$

24. $\dfrac{21x + 5}{24}$ **25.** $\dfrac{10 - 3n}{6n^2}$

26. $\dfrac{5x^2 + 18x + 27}{(x + 9)(x - 3)(x + 3)}$ **27.** $\left\{-\dfrac{23}{4}\right\}$ **28.** $\{3\}$

29. $\left\{\dfrac{3}{7}\right\}$ **30.** $\left\{\pm\dfrac{2}{3}\right\}$ **31.** $\{-4, 0, 2\}$

32. $\left\{\dfrac{3}{7}, 4\right\}$ **33.** $\{\pm 4i, \pm 1\}$ **34.** $\left\{-\dfrac{1}{5}, 1\right\}$

35. $\left\{\dfrac{3 \pm \sqrt{17}}{4}\right\}$ **36.** $\left\{\dfrac{-13 \pm \sqrt{205}}{2}\right\}$ **37.** $\{1\}$

38. $\left\{\dfrac{1 \pm 2i}{2}\right\}$ **39.** $\left(-\infty, \dfrac{1}{8}\right)$ **40.** $\left[-\dfrac{5}{9}, \infty\right)$

41. $(-\infty, 250]$ **42.** $(-\infty, -8) \cup (3, \infty)$

43. $\left(-\dfrac{3}{2}, \dfrac{1}{3}\right)$ **44.** $\left(-3, \dfrac{1}{2}\right) \cup (4, \infty)$

45. $\left(-1, \dfrac{2}{3}\right]$ **46.** $(1, 7]$

47. $\left(-\infty, -\dfrac{4}{3}\right) \cup (2, \infty)$ **48.** $\left(-\dfrac{9}{5}, 3\right)$

49.   **50.**

51.   **52.**

53.   **54.**

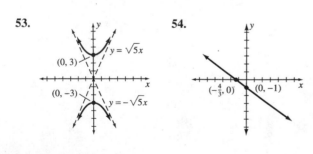

55. $(-7, 4); r = 3$ **56.** $3x - 4y = -26$
57. $(-2, 6)$ **58.** $4x + 3y = 25$
59. $28.60; 31.43
60. $3000 at 5% and $4500 at 6%
61. Length of 8 inches and width of 4.5 inches

62. The side is 20 centimeters long and the altitude is 8 centimeters long.   **63.** 16 milliliters
64. 30 miles   **65.** 3 hours

CHAPTER 3

Problem Set 3.1 (page 235)

1. $f(3) = -1; f(5) = -5; f(-2) = 9$
3. $g(3) = -20; g(-1) = -8; g(-4) = -41$
5. $h(3) = \dfrac{5}{4}; h(4) = \dfrac{23}{12}; h\left(-\dfrac{1}{2}\right) = -\dfrac{13}{12}$
7. $f(5) = 3; f\left(\dfrac{1}{2}\right) = 0; f(23) = 3\sqrt{5}$
9. $f(4) = 4; f(10) = 10; f(-3) = 9; f(-5) = 25$
11. $f(3) = 6; f(5) = 10; f(-3) = 6; f(-5) = 10$
13. $f(2) = 1; f(0) = 0; f\left(-\dfrac{1}{2}\right) = 0; f(-4) = -1$
15. -7 **17.** $-2a - h + 4$ **19.** $6a + 3h - 1$
21. $3a^2 + 3ah + h^2 - 2a - h + 2$
23. $-\dfrac{2}{(a-1)(a+h-1)}$ **25.** $-\dfrac{2a+h}{a^2(a+h)^2}$
27. Yes   **29.** No   **31.** Yes   **33.** Yes
35. $D = \left\{x \,|\, x \geq \dfrac{4}{3}\right\}$
     $R = \{f(x) \,|\, f(x) \geq 0\}$
37. $D = \{x \,|\, x \text{ is any real number}\}$
     $R = \{f(x) \,|\, f(x) \geq -2\}$
39. $D = \{x \,|\, x \text{ is any real number}\}$
     $R = \{f(x) \,|\, f(x) \text{ is any nonnegative real number}\}$
41. $D = \{x \,|\, x \text{ is any nonnegative real number}\}$
     $R = \{f(x) \,|\, f(x) \text{ is any nonpositive real number}\}$
43. $D = \{x \,|\, x \neq -2\}$
45. $D = \left\{x \,|\, x \neq \dfrac{1}{2} \text{ and } x \neq -4\right\}$
47. $D = \{x \,|\, x \neq 2 \text{ and } x \neq -2\}$
49. $D = \{x \,|\, x \neq -3 \text{ and } x \neq 4\}$
51. $D = \left\{x \,|\, x \neq -\dfrac{5}{2} \text{ and } x \neq \dfrac{1}{3}\right\}$
53. $(-\infty, -4] \cup [4, \infty)$ **55.** $(-\infty, \infty)$
57. $(-\infty, -5] \cup [8, \infty)$ **59.** $\left(-\infty, -\dfrac{5}{2}\right] \cup \left[\dfrac{7}{4}, \infty\right)$
61. 12.57; 28.27; 452.39; 907.92   **63.** 48; 64; 48; 0
65. $55; $60; $67.50; $75
67. 125.66; 301.59; 804.25   **69.** Odd
71. Neither   **73.** Neither   **75.** Even
77. Odd

Problem Set 3.2 (page 249)

1.

3.

21.

23.

5.

7.

25.

27.

9.

11.

29.

31.

13.

15.

33.

35.

17.

19.

37.

39.

41.

43.

13.

15.

45.

17.

19.

Problem Set 3.3 (page 261)

1.

3.

5.

7.

9.

11.

21. 3 and 5; $(4, -1)$ **23.** 6 and 8; $(7, -2)$

25. 4 and 6; $(5, 1)$

27. $7 + \sqrt{5}$ and $7 - \sqrt{5}$; $(7, 5)$

29. No x intercepts; $\left(\dfrac{9}{2}, -\dfrac{3}{4}\right)$

31. $\dfrac{1 + \sqrt{5}}{2}$ and $\dfrac{1 - \sqrt{5}}{2}$; $\left(\dfrac{1}{2}, 5\right)$

33. 70

35. 144

37. 25 and 25

39. 60 meters by 60 meters

41. 1100 subscribers at \$13.75 per month

47. 75 feet

Problem Set 3.4 (page 274)

1.

3.

5.

7.

25.

27.

9.

11.

29.

13.

15.

31. (a)

(c)

17.

19.

21.

23.

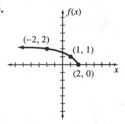

Problem Set 3.5 (page 281)

1. $8x - 2; -2x - 6; 15x^2 - 14x - 8; \dfrac{3x - 4}{5x + 2}$

3. $x^2 - 7x + 3; x^2 - 5x + 5; -x^3 + 5x^2 + 2x - 4;$
$\dfrac{x^2 - 6x + 4}{-x - 1}$

5. $2x^2 + 3x - 6; -5x + 4; x^4 + 3x^3 - 10x^2 + x + 5;$
$\dfrac{x^2 - x - 1}{x^2 + 4x - 5}$

7. $\sqrt{x - 1} + \sqrt{x}; \sqrt{x - 1} - \sqrt{x}; \sqrt{x^2 - x};$
$\dfrac{\sqrt{x(x - 1)}}{x}$

9. $(f \circ g)(x) = 6x - 2, D = \{\text{all reals}\}$
$(g \circ f)(x) = 6x - 1, D = \{\text{all reals}\}$

11. $(f \circ g)(x) = 10x + 2, D = \{\text{all reals}\}$
$(g \circ f)(x) = 10x - 5, D = \{\text{all reals}\}$

13. $(f \circ g)(x) = 3x^2 + 7, D = \{\text{all reals}\}$
$(g \circ f)(x) = 9x^2 + 24x + 17, D = \{\text{all reals}\}$

15. $(f \circ g)(x) = 3x^2 + 9x - 16, D = \{\text{all reals}\}$
$(g \circ f)(x) = 9x^2 - 15x, D = \{\text{all reals}\}$

17. $(f \circ g)(x) = \dfrac{1}{2x + 7}, D = \left\{x \mid x \neq -\dfrac{7}{2}\right\}$

$(g \circ f)(x) = \dfrac{7x + 2}{x}, D = \{x \mid x \neq 0\}$

19. $(f \circ g)(x) = \sqrt{3x - 3}, D = \{x \mid x \geq 1\}$
$(g \circ f)(x) = 3\sqrt{x - 2} - 1, D = \{x \mid x \geq 2\}$

21. $(f \circ g)(x) = \dfrac{x}{2 - x}, D = \{x \mid x \neq 0 \text{ and } x \neq 2\}$

$(g \circ f)(x) = 2x - 2, D = \{x \mid x \neq 1\}$

23. $(f \circ g)(x) = 2\sqrt{x - 1} + 1, D = \{x \mid x \geq 1\}$
$(g \circ f)(x) = \sqrt{2x}, D = \{x \mid x \geq 0\}$

25. $(f \circ g)(x) = x, D = \{x \mid x \neq 0\}$
$(g \circ f)(x) = x, D = \{x \mid x \neq 1\}$

27. $4; 50$ **29.** $9; 0$ **31.** $\sqrt{11}; 5$

8.

9.

10.

11.

Problem Set 3.6 (page 288)

1. $y = kx^2$ **3.** $A = klw$ **5.** $V = \dfrac{k}{p}$

7. $V - khr^2$ **9.** 24 **11.** $\dfrac{22}{7}$ **13.** $\dfrac{1}{2}$

15. 7 **17.** 6 **19.** 8 **21.** 96
23. 5 hours   **25.** 2 seconds   **27.** 24 days
29. 28 **31.** $\$2400$
37. 2.8 seconds
39. 1.4

12.

13.
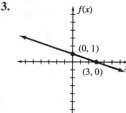

Chapter 3 Review Problem Set (page 293)

1. $7; 4; 32$ **2.** **(a)** -5 **(b)** $4a + 2h - 1$
(c) $-6a - 3h + 2$

3. The domain is the set of all real numbers, and the range is the set of all real numbers greater than or equal to 5.

4. The domain is the set of all real numbers except $\dfrac{1}{2}$

and -4.

5. $(-\infty, 2] \cup [5, \infty)$

6.

7.

14.

15.

16. $x^2 - 2x; -x^2 + 6x + 6; 2x^3 - 5x^2 - 18x - 9;$
$\dfrac{2x + 3}{x^2 - 4x - 3}$

17. $(f \circ g)(x) = -6x + 12; D = \{\text{all reals}\}$
$(g \circ f)(x) = -6x + 25; D = \{\text{all reals}\}$

18. $(f \circ g)(x) = 25x^2 - 40x + 11; D = \{\text{all reals}\}$
$(g \circ f)(x) = 5x^2 - 29; D = \{\text{all reals}\}$

19. $(f \circ g)(x) = \sqrt{x - 3}; D = \{x \mid x \geq 3\}$
$(g \circ f)(x) = \sqrt{x - 5} + 2; D = \{x \mid x \geq 5\}$

20. $(f \circ g)(x) = \dfrac{x + 2}{-3x - 5}; D = \left\{x \mid x \neq -2 \text{ and } x \neq -\dfrac{5}{3}\right\}$

$(g \circ f)(x) = \dfrac{x - 3}{2x - 5}; D = \left\{x \mid x \neq 3 \text{ and } x \neq \dfrac{5}{2}\right\}$

21. **(a)** Neither   **(b)** Odd
   (c) Even   **(d)** Neither

22. $f(5) = 23; f(0) = -2; f(-3) = 13$

23. $f(g(6)) = -2; g(f(-2)) = 0$

24. $(f \circ g)(1) = 1; (g \circ f)(-3) = 5$

25. 2 and 8

26. 112 students

27. 9

28. 441

29. 128 pounds

30. 15 hours

Chapter 3 Test (page 295)

1. $\dfrac{11}{6}$ **2.** 11   **3.** $6a + 3h + 2$

4. $\left\{x \mid x \neq -4 \text{ and } x \neq \dfrac{1}{2}\right\}$ **5.** $\left\{x \mid x \leq \dfrac{5}{3}\right\}$

6. $(f + g)(x) = 2x^2 + 2x - 6; (f - g)(x) = -2x^2 + 4x + 4; (f \cdot g)(x) = 6x^3 - 5x^2 - 14x + 5$

7. $(f \circ g)(x) = -21x - 2$ **8.** $(g \circ f)(x) = 8x^2 + 38x + 48$

9. $(f \circ g)(x) = \dfrac{3x}{2 - 2x}$

10. 12; 7   **11.** **a.** Even   **b.** Odd   **c.** Neither
   d. Even

12. $\{x \mid x \neq 0 \text{ and } x \neq 1\}$ **13.** 18; 10; 0

14. $(f \cdot g)(x) = x^3 + 4x^2 - 11x + 6; \left(\dfrac{f}{g}\right)(x) = x + 6$

15. 6 and 54   **16.** 15   **17.** -4 **18.** \$96

19. The graph of $g(x) = (x - 6)^3 - 4$ is the graph of $f(x) = x^3$ translated six units to the right and four units downward.

20. The graph of $g(x) = -|x| + 8$ is the graph of $f(x) = |x|$ reflected across the x axis and then translated eight units upward.

21. The graph of $g(x) = -\sqrt{x + 5} + 7$ is the graph of $f(x) = \sqrt{x}$ reflected across the x axis and then translated five units to the left and seven units upward.

22.

23.

24.

25.

CHAPTER 4

Problem Set 4.1 (page 305)

1. $\{3\}$ **3.** $\{3\}$ **5.** $\{4\}$ **7.** $\{2\}$

9. $\{-2\}$ **11.** $\left\{\dfrac{5}{3}\right\}$ **13.** $\left\{\dfrac{3}{2}\right\}$

15. $\left\{\dfrac{4}{9}\right\}$ **17.** $\left\{\dfrac{4}{3}\right\}$ **19.** $\left\{\dfrac{2}{3}\right\}$

21.

23.

41.

25.

27.

Problem Set 4.2 (page 315)

1. **(a)** \$.67   **(c)** \$2.31   **(e)** \$12,623
   (g) \$803   **3.** \$384.66   **5.** \$480.31
7. \$2479.35   **9.** \$1816.70   **11.** \$1356.59
13. \$22,553.65   **15.** \$567.63   **17.** \$1422.36
19. \$8963.38   **21.** \$17,547.35   **23.** \$23,558.88
25. 5.9%   **27.** 8.06%
29. 8.25% compounded quarterly
31. 50 grams; 37 grams   **33.** 2226; 3320; 7389
35. 2000
37. **(a)** 6.5 pounds per square inch
   (c) 13.6 pounds per square inch
39. **(a)** Approximately 100 times brighter
   (c) Approximately 10 billion times brighter

29.

31.

41.

43.

33.

35.

45.

37.

39.

49.

	8%	10%	12%	14%
5 years	\$1492	1649	1822	2014
10 years	2226	2718	3320	4055
15 years	3320	4482	6050	8166
20 years	4953	7389	11,023	16,445
25 years	7389	12,182	20,086	33,115

51.

	8%	10%	12%	14%
Compounded annually	$2159	2594	3106	3707
Compounded semiannually	2191	2653	3207	3870
Compounded quarterly	2208	2685	3262	3959
Compounded monthly	2220	2707	3300	4022
Compounded continuously	2226	2718	3320	4055

53.

55.

51. $f^{-1}(x) = \frac{1}{3}x$

53. $f^{-1}(x) = \frac{x-1}{2}$

55. $f^{-1}(x) = \frac{x+2}{x}$ for $x > 0$

57. $f^{-1}(x) = \sqrt{x+4}$ for $x \geq -4$

59. Increasing on $[0, \infty)$ and decreasing on $(-\infty, 0]$
61. Decreasing on $(-\infty, \infty)$
63. Increasing on $(-\infty, -2]$ and decreasing on $[-2, \infty)$
65. Increasing on $(-\infty, -4]$ and increasing on $[-4, \infty)$

71. (a) $f^{-1}(x) = \frac{x+9}{3}$

(c) $f^{-1}(x) = -x + 1$

(e) $f^{-1}(x) = -\frac{1}{5}x$

Problem Set 4.3 (page 327)
1. Yes   **3.** No   **5.** Yes   **7.** Yes
9. Yes   **11.** No   **13.** No
15. Domain of f: $\{1, 2, 5\}$
Range of f: $\{5, 9, 21\}$
$f^{-1} = \{(5, 1), (9, 2), (21, 5)\}$
Domain of f^{-1}: $\{5, 9, 21\}$
Range of f^{-1}: $\{1, 2, 5\}$
17. Domain of f: $\{0, 2, -1, -2\}$
Range of f: $\{0, 8, -1, -8\}$
f^{-1}: $\{(0, 0), (8, 2), (-1, -1), (-8, -2)\}$
Domain of f^{-1}: $\{0, 8, -1, -8\}$
Range of f^{-1}: $\{0, 2, -1, -2\}$
27. No   **29.** Yes   **31.** No   **33.** Yes
35. Yes   **37.** $f^{-1}(x) = x + 4$

39. $f^{-1}(x) = \frac{-x-4}{3}$ **41.** $f^{-1}(x) = \frac{12x+10}{9}$

43. $f^{-1}(x) = -\frac{3}{2}x$ **45.** $f^{-1}(x) = x^2$ for $x \geq 0$

47. $f^{-1}(x) = \sqrt{x-4}$ for $x \geq 4$

49. $f^{-1}(x) = \frac{1}{x-1}$ for $x > 1$

Problem Set 4.4 (page 338)
1. $\log_3 9 = 2$ **3.** $\log_5 125 = 3$

5. $\log_2\left(\frac{1}{16}\right) = -4$ **7.** $\log_{10} 0.01 = -2$

9. $2^6 = 64$ **11.** $10^{-1} = .1$ **13.** $2^{-4} = \frac{1}{16}$

15. 2   **17.** −1   **19.** 1   **21.** $\frac{1}{2}$ **23.** $\frac{1}{2}$

25. $-\frac{1}{8}$ **27.** 7   **29.** 0   **31.** {25}

33. {32}   **35.** {9}   **37.** {1}   **39.** 5.1293

41. 6.9657   **43.** 1.4037   **45.** 7.4512

47. 6.3219   **49.** −0.3791   **51.** 0.5766

53. 2.1531   **55.** 0.3949

57. $\log_b x + \log_b y + \log_b z$ **59.** $2 \log_b x + 3 \log_b y$

61. $\frac{1}{2} \log_b x + \frac{1}{2} \log_b y$ **63.** $\frac{1}{2} \log_b x - \frac{1}{2} \log_b y$

65. $\log_b\left(\dfrac{xy}{z}\right)$ **67.** $\log_b\left(\dfrac{x}{yz}\right)$ **69.** $\log_b\left(x\sqrt{y}\right)$

71. $\log_b\left(\dfrac{x^2\sqrt{x-1}}{(2x+5)^4}\right)$ **73.** $\left\{\dfrac{9}{4}\right\}$ **75.** {25}

77. {4}   **79.** $\left\{\dfrac{19}{8}\right\}$ **81.** {9}   **83.** {1}

49.

51.

53.

Problem Set 4.5 (page 346)

1. 0.8597   **3.** 1.7179   **5.** 3.5071

7. −0.1373   **9.** −3.4685   **11.** 411.43

13. 90095   **15.** 79.543   **17.** 0.048440

19. 0.0064150   **21.** 1.6094   **23.** 3.4843

25. 6.0638   **27.** −0.7765   **29.** −3.4609

31. 1.6034   **33.** 3.1346   **35.** 108.56

37. 0.48268   **39.** 0.035994

41.

43.

45.

47.

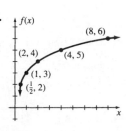

Problem Set 4.6 (page 356)

1. {3.17}   **3.** {2.99}   **5.** {1.81}   **7.** {1.41}

9. {1.41}   **11.** {3.10}   **13.** {1.82}

15. {7.84}   **17.** {10.32}   **19.** {2}   **21.** $\left\{\dfrac{29}{8}\right\}$

23. $\left\{\dfrac{-1 + \sqrt{65}}{2}\right\}$ **25.** $\left\{\sqrt{2}\right\}$ **27.** {6}

29. {1, 100}   **31.** 2.402   **33.** 0.461

35. 2.657   **37.** 1.211   **39.** 7.9 years

41. 12.2 years   **43.** 11.8%   **45.** 6.6 years

47. 1.5 hours   **49.** 34.7 years   **51.** 6.7

53. Approximately 8 times   **55.** 5.8   **57.** 10

65. $x = \log\left(y \pm \sqrt{y^2 - 1}\right)$

Chapter 4 Review Problem Set (page 362)

1. 32   **2.** −125   **3.** 81   **4.** 3   **5.** −2

6. $\frac{1}{3}$ **7.** $\frac{1}{4}$ **8.** −5   **9.** 1   **10.** 12

11. {5}   **12.** $\left\{\dfrac{1}{9}\right\}$ **13.** $\left\{\dfrac{7}{2}\right\}$ **14.** {3.40}

15. {8}   **16.** $\left\{\dfrac{1}{11}\right\}$ **17.** {1.95}   **18.** {1.41}

19. {1.56}   **20.** {20}   **21.** {10^{100}}   **22.** {2}

23. $\left\{\dfrac{11}{2}\right\}$ **24.** {0}   **25.** 0.3680   **26.** 1.3222

27. 1.4313   **28.** 0.5634

29. (a) $\log_b x - 2 \log_b y$ **(b)** $\frac{1}{4} \log_b x + \frac{1}{2} \log_b y$

    (c) $\frac{1}{2} \log_b x - 3 \log_b y$

30. (a) $\log_b x^3 y^2$ **(b)** $\log_b \left(\dfrac{\sqrt{y}}{x^4} \right)$

    (c) $\log_b \left(\dfrac{\sqrt{xy}}{z^2} \right)$ **31.** 1.58   **32.** 0.63

33. 3.79   **34.** -2.12

35.

36.

37.

38.

39.

40.

41.

42.

43. \$2219.91   **44.** \$4797.55   **45.** \$15,999.31
46. Yes   **47.** No   **48.** Yes   **49.** Yes

50. $f^{-1}(x) = \dfrac{x - 5}{4}$ **51.** $f^{-1}(x) = \dfrac{-x - 7}{3}$

52. $f^{-1}(x) = \dfrac{6x + 2}{5}$ **53.** $f^{-1}(x) = \sqrt{-2 - x}$

54. Increasing on $(-\infty, 4]$ and decreasing on $[4, \infty)$
55. Increasing on $[3, \infty)$
56. Approximately 5.3 years
57. Approximately 12.1 years
58. Approximately 8.7%   **59.** 61,070; 67,493; 74,591
60. Approximately 4.8 hours   **61.** 133 grams
62. 8.1

Chapter 4 Test (page 364)

1. $\dfrac{1}{2}$ **2.** 1   **3.** 1   **4.** -1 **5.** $\{-3\}$

6. $\left\{ -\dfrac{3}{2} \right\}$ **7.** $\left\{ \dfrac{8}{3} \right\}$ **8.** $\{243\}$ **9.** $\{2\}$

10. $\left\{ \dfrac{2}{5} \right\}$ **11.** 4.1919   **12.** 0.2031

13. 0.7325   **14.** $\{5.17\}$ **15.** $\{10.29\}$

16. 4.0069   **17.** $f^{-1}(x) = -\dfrac{1}{3}x - 2$

18. $f^{-1}(x) = \dfrac{3}{2}x + \dfrac{9}{10}$ **19.** Yes   **20.** \$6342.08

21. 13.5 years   **22.** 7.8 hours   **23.** 4813 grams
24.

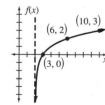

25.

CHAPTER 5

Problem Set 5.1 (page 372)

1. $Q: 4x + 5$ **3.** $Q: t^2 + 2t - 4$
   $R: 0$ $R: 0$
5. $Q: 2x + 5$ **7.** $Q: 3x - 4$
   $R: 1$ $R: 3x - 1$
9. $Q: 5y - 1$ **11.** $Q: 4a + 6$
   $R: -8y - 2$ $R: 7a - 19$
13. $Q: 3x + 4y$ **15.** $Q: 3x + 4$
   $R: 0$ $R: 0$
17. $Q: x + 6$ **19.** $Q: 4x - 3$
   $R: 14$ $R: 2$

21. $Q: x^2 - 1$
$R: 0$

23. $Q: 3x^3 - 4x^2 + 6x - 13$
$R: 12$

25. $Q: x^2 - 2x - 3$
$R: 0$

27. $Q: x^3 + 7x^2 + 21x + 56$
$R: 167$

29. $Q: x^2 + 3x + 2$
$R: 0$

31. $Q: x^4 + x^3 + x^2 + x + 1$
$R: 0$

33. $Q: x^4 + x^3 + x^2 + x + 1$
$R: 2$

35. $Q: 2x^2 + 2x - 3$
$R: \dfrac{9}{2}$

37. $Q: 4x^3 + 2x^2 - 4x - 2$
$R: 0$

Problem Set 5.2 (page 376)

1. $f(2) = -2$ **3.** $f(-4) = -105$
5. $f(-2) = 9$ **7.** $f(6) = 74$ **9.** $f(3) = 200$
11. $f(-1) = 5$ **13.** $f(7) = -5$

15. $f(-2) = -27$ **17.** $f\left(\dfrac{1}{2}\right) = -2$ **19.** Yes

21. Yes   **23.** No   **25.** Yes   **27.** Yes
29. $(x + 2)(x + 6)(x - 1)$
31. $(x - 3)(2x - 1)(3x + 2)$ **33.** $(x + 1)^2(x - 4)$
35. $k = 6$ **37.** $k = -30$
39. Let $f(x) = x^{12} - 4096$; then $f(-2) = 0$; therefore, $x + 2$ is a factor of $f(x)$.
41. Let $f(x) = x^n - 1$. Because $1^n = 1$ for all positive integral values of n, then $f(1) = 0$ and $x - 1$ is a factor.
43. **(a)** Let $f(x) = x^n - y^n$. Therefore, $f(y) = y^n - y^n = 0$ and $x - y$ is a factor of $f(x)$.
  (c) Let $f(x) = x^n + y^n$. Therefore, $f(-y) = (-y)^n + y^n = -y^n + y^n = 0$ when n is odd, and $x - (-y) = x + y$ is a factor of $f(x)$.
47. $f(1 + i) = 2 + 6i$
51. **(a)** $f(4) = 137; f(-5) = 11; f(7) = 575$
  (c) $f(4) = -79; f(5) = -162; f(-3) = 110$

Problem Set 5.3 (page 388)

1. $\{-2, -1, 2\}$ **3.** $\left\{-\dfrac{3}{2}, \dfrac{1}{3}, 1\right\}$

5. $\left\{-7, \dfrac{2}{3}, 2\right\}$ **7.** $\{-1, 4\}$ **9.** $\{-3, 1, 2, 4\}$

11. $\{-2, 1 \pm \sqrt{7}\}$ **13.** $\left\{-\dfrac{2}{3}, 1, \pm\sqrt{2}\right\}$

15. $\left\{-\dfrac{4}{3}, 0, \dfrac{1}{2}, 3\right\}$ **17.** $\{-1, 2, 1 \pm i\}$

19. $\left\{-1, \dfrac{3}{2}, 2, \pm i\right\}$

27. **(a)** $\{-4, -2, 1\}$ **(c)** $\left\{-4, -2, \dfrac{3}{2}\right\}$

29. Two positive *or* two nonreal complex solutions
31. One negative and two nonreal complex solutions
33. One positive and two negative *or* one positive and two nonreal complex solutions
35. One negative and two positive and two nonreal complex solutions *or* one negative and four nonreal complex solutions
37. One positive and one negative and four nonreal complex solutions
39. $x^4 + 2x^3 - 9x^2 - 2x + 8 = 0$
41. $12x^3 - 37x^2 - 3x + 18 = 0$
43. $x^4 + 12x^3 + 54x^2 + 108x + 81 = 0$
45. $x^3 + 13x + 34 = 0$
47. $x^4 + 4x^3 + 14x^2 + 4x + 13 = 0$

Problem Set 5.4 (page 400)

1.

3.

5.

7.

9.

11.

13.

15.

33.

17.

19.

35. **(a)** 60   **(c)** $f(x) > 0$ for $(-4, 3) \cup (5, \infty)$
$f(x) < 0$ for $(-\infty, -4) \cup (3, 5)$

37. **(a)** 432   **(c)** $f(x) > 0$ for $(-3, 4) \cup (4, \infty)$
$f(x) < 0$ for $(-\infty, -3)$

39. **(a)** 8
(c) $f(x) > 0$ for $(-\infty, -2) \cup (-2, 1) \cup (2, \infty)$
$f(x) < 0$ for $(1, 2)$

41. **(a)** 512   **(c)** $f(x) > 0$ for $(-2, 4) \cup (4, \infty)$
$f(x) < 0$ for $(-\infty, -2)$

45. **(a)** 1.6   **(c)** 4.4   **(e)** -1.4

51. **(a)** -2, 1, and 4; $f(x) > 0$ for $(-2, 1) \cup (4, \infty)$ and
$f(x) < 0$ for $(-\infty, -2) \cup (1, 4)$
(c) 2 and 3; $f(x) > 0$ for $(3, \infty)$ and $f(x) < 0$ for $(2, 3)$
$\cup (-\infty, 2)$
(e) -3, -1, and 2; $f(x) > 0$ for $(-\infty, -3) \cup (2, \infty)$ and
$f(x) < 0$ for $(-3, -1) \cup (-1, 2)$

53. **(a)** -3, 3; $(0.5, 3.1)$, $(-1.8, 10.1)$
(c) -2.2, 2.2; $(-1.4, -8.0)$, $(0.0, -4.0)$, $(1.4, 8.0)$

55. 32 units

21.

23.

25.

27.

Problem Set 5.5 (page 411)

1.

3.

5.

7.

29.

31.

9.

11.

Problem Set 5.6 (page 420)

1.

3.

13.

15.

5.

7.

17.

19.

9.

11.

21.

13.

15.

25. (a)

(c)

17.

19.

Problem Set 5.7 (page 426)

1. $\dfrac{4}{x-2} + \dfrac{7}{x+1}$ **3.** $\dfrac{3}{x+1} - \dfrac{5}{x-1}$

5. $\dfrac{1}{3x-1} + \dfrac{6}{2x+3}$ **7.** $\dfrac{2}{x-1} + \dfrac{3}{x+2} - \dfrac{4}{x-3}$

9. $\dfrac{-1}{x} + \dfrac{2}{2x-1} - \dfrac{3}{4x+1}$ **11.** $\dfrac{2}{x-2} + \dfrac{5}{(x-2)^2}$

13. $\dfrac{4}{x} + \dfrac{7}{x^2} - \dfrac{10}{x+3}$ **15.** $\dfrac{-3}{x^2+1} - \dfrac{2}{x-4}$

17. $\dfrac{3}{x+2} - \dfrac{2}{(x+2)^2} + \dfrac{1}{(x+2)^3}$

19. $\dfrac{2}{x} + \dfrac{3x+5}{x^2-x+3}$ **21.** $\dfrac{2x}{x^2+1} + \dfrac{3-x}{(x^2+1)^2}$

Chapter 5 Review Problem Set (page 429)

1. $Q: 3x^2 - 5x + 4$
  $R: 4$

2. $Q: 2a - 1$
  $R: 5$

3. $Q: 3x^2 - x + 5$
  $R: 3$

4. $Q: 5x^2 - 3x - 3$
  $R: 16$

5. $Q: -2x^3 + 9x^2 - 38x + 151$
  $R: -605$

6. $Q: -3x^3 + 9x^2 - 32x + 96$
  $R: -279$

7. $f(1) = 1$

8. $f(-3) = -197$ **9.** $f(-2) = 20$

10. $f(8) = 0$ **11.** Yes **12.** No **13.** Yes

14. Yes **15.** $\{-3, 1, 5\}$ **16.** $\left\{-\dfrac{7}{2}, -1, \dfrac{5}{4}\right\}$

17. $\{1, 2, 1 \pm 5i\}$ **18.** $\{-2, 3 \pm \sqrt{7}\}$

19. Two positive and two negative solutions *or* two positive and two nonreal complex solutions *or* two negative and two nonreal complex solutions *or* four nonreal complex solutions.

20. One negative and four nonreal complex solutions

21.

22.

23.

24.

25.

26.

27.

28.

29. $\dfrac{1}{x} - \dfrac{2}{x^2} + \dfrac{4}{x+2}$ **30.** $\dfrac{3x+1}{x^2+4} - \dfrac{5}{2x-1}$

Chapter 5 Test (page 431)

1. $Q: 2x^2 - 3x - 4$
  $R: 0$

2. $Q: 3x^3 - x^2 - 2x - 6$
  $R: 3$

3. $Q: 4x^3 + 8x^2 + 9x + 17$
  $R: 38$

4. -24 **5.** 5 **6.** 39 **7.** No **8.** No

9. Yes **10.** No

11. One positive, one negative, and two nonreal complex solutions

12. $-7, 0,$ and $\dfrac{2}{3}$ **13.** $x = -3$ **14.** $f(x) = 5$

15. y axis **16.** Origin **17.** $f(x) = 4x - 3$

18. $\dfrac{3}{2x-1} + \dfrac{4}{x-6}$ **19.** $-\dfrac{1}{x} + \dfrac{2}{x+2} - \dfrac{2}{(x+2)^2}$

20. $\dfrac{1}{x+1} + \dfrac{2x-2}{x^2-x+3}$

21.

22.

23.

24.

25.

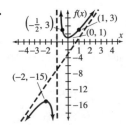

Chapters 0–5 Cumulative Review Problem Set (page 433)

1. $\dfrac{64}{27}$ **2.** $-\dfrac{2}{3}$ **3.** $-\dfrac{1}{25}$ **4.** 16   **5.** $\dfrac{1}{27}$

6. 3   **7.** -4 **8.** -5 **9.** 16   **10.** 3

11. $(-\infty, -6] \cup \left[\dfrac{1}{2}, \infty\right)$

12. $(f \circ g)(-2) = 26$ and $(g \circ f)(3) = 59$

13. $(f \circ g)(x) = -2x + 8$ and $D = \{x \mid x \neq 4\}$

$(g \circ f)(x) = -\dfrac{x}{4x+2}$ and

$D = \left\{x \mid x \neq 0 \text{ and } x \neq -\dfrac{1}{2}\right\}$

14. $f^{-1}(x) = \dfrac{-x+7}{2}$ **15.** $2a + h + 7$

16. $f(9) = 33$ **17.** $3x^4 + 9x^3 + 2x^2 - x - 2$

18. No   **19.** 5.64   **20.** $(-3, 2)$ and $r = 3$

21. $x + 3y = 2$ **22.** $4x + 3y = 5$ **23.** 16 units

24. $y = \pm\dfrac{1}{3}x$ **25.** 12   **26.** $\dfrac{2}{7}$ **27.** \$784

28. 8.7 years

29. 10 nickels, 15 dimes, and 32 quarters   **30.** \$125

31. $1\dfrac{1}{3}$ quarts   **32.** 45 miles   **33.** 4 hours

34. $\left\{\dfrac{3}{5}\right\}$ **35.** $\left\{\dfrac{-13 \pm \sqrt{193}}{2}\right\}$ **36.** $\{-7, 0, 2\}$

37. $\left\{-\dfrac{5}{2}, -1, \dfrac{2}{3}\right\}$ **38.** $\left\{-1, \dfrac{5}{2}\right\}$ **39.** $\{0\}$

40. $\{-1\}$ **41.** $\left\{\dfrac{2}{3}\right\}$ **42.** $\{3\}$

43. $\{\pm 3i, \pm\sqrt{6}\}$ **44.** $\left\{-\dfrac{13}{2}, 4\right\}$

45. $\left\{1, 2, \dfrac{-1 \pm i\sqrt{11}}{2}\right\}$

46. $(-\infty, -5)$

47. $\left[-\dfrac{11}{17}, \infty\right)$ **48.** $(-3, 6)$

49. $[-3, 1] \cup [2, \infty)$ **50.** $\left(-\infty, -\dfrac{5}{2}\right) \cup \left(\dfrac{7}{2}, \infty\right)$

51. $\left[-\dfrac{10}{3}, 2\right]$ **52.** $\left(-\infty, \dfrac{3}{4}\right] \cup (2, \infty)$

53. $(-\infty, 4) \cup \left(\dfrac{15}{2}, \infty\right)$

54.

55.

64.

56.

57.

CHAPTER 6

Problem Set 6.1 (page 444)

1. $\{(7, 9)\}$ **3.** $\{(-4, 7)\}$ **5.** $\{(6, 3)\}$

7. $a = -3$ and $b = -4$

9. $\left\{\left(k, \frac{2}{3}k - \frac{4}{3}\right)\right\}$, a dependent system

11. $u = 5$ and $t = 7$ **13.** $\{(2, -5)\}$

15. $\varnothing$, an inconsistent system   **17.** $\left\{\left(-\frac{3}{4}, -\frac{6}{5}\right)\right\}$

a19. $\{(3, -4)\}$ **21.** $\{(2, 8)\}$ **23.** $\{(-1, -5)\}$

25. $\varnothing$, an inconsistent system

27. $a = \frac{5}{27}$ and $b = -\frac{26}{27}$ **29.** $x = -6$ and $t = 12$

31. $\left\{\left(-\frac{1}{2}, \frac{1}{3}\right)\right\}$ **33.** $\left\{\left(\frac{13}{22}, \frac{2}{11}\right)\right\}$ **35.** $\{(-4, 2)\}$

37. $\{(5, 5)\}$ **39.** $\varnothing$, an inconsistent system

41. $\{(12, -24)\}$ **43.** $t = 8$ and $u = 3$

45. $\{(200, 800)\}$ **47.** $\{(400, 800)\}$ **49.** $\{(3.5, 7)\}$

51. 17 and 36   **53.** 15°, 75°   **55.** 72

57. 34   **59.** 8 single rooms and 15 double rooms

61. 2500 student tickets and 500 nonstudent tickets

63. $500 at 9% and $1500 at 11%

65. 3 miles per hour

67. $1.25 per tennis ball and $1.75 per golf ball

69. 30 five-dollar bills and 18 ten-dollar bills

75. $\{(4, 6)\}$ **77.** $\{(2, -3)\}$ **79.** $\left\{\left(\frac{1}{4}, -\frac{2}{3}\right)\right\}$

58.

59.

60.

61.

62.

63.

Problem Set 6.2 (page 454)

1. $\{(-4, -2, 3)\}$ **3.** $\{(-2, 5, 2)\}$

5. $\{(4, -1, -2)\}$ **7.** $\{(3, 1, 2)\}$

9. $[(-1, 3, 5)\}$ **11.** $\{(-2, -1, 3)\}$

13. $\{(0, 2, 4)\}$ **15.** $\{(4, -1, -2)\}$

17. $\{(-4, 0, -1)\}$ **19.** $\{(2, 2, -3)\}$

21. 4 pounds of pecans, 4 pounds of almonds, and 12 pounds of peanuts

23. 7 nickels, 13 dimes, and 22 quarters

25. $40°, 60°$, and $80°$

27. $500 at 12%, $1000 at 13%, and $1500 at 14%

29. 50 of type A, 75 of type B, and 150 of type C

31. $x^2 + y^2 - 4x - 20y - 2 = 0$

Problem Set 6.3 (page 464)

1. Yes   **3.** Yes   **5.** No   **7.** No

9. Yes   **11.** $\{(-1, -5)\}$ **13.** $\{(3, -6)\}$

15. $\varnothing$ **17.** $\{(-2, -9)\}$ **19.** $\{(-1, -2, 3)\}$

21. $\{(3, -1, 4)\}$ **23.** $\{(0, -2, 4)\}$

25. $\{(-7k + 8, -5k + 7, k)\}$ **27.** $\{(-4, -3, -2)\}$

29. $\{(4, -1, -2)\}$ **31.** $\{(1, -1, 2, -3)\}$

33. $\{(2, 1, 3, -2)\}$ **35.** $\{(-2, 4, -3, 0)\}$

37. $\varnothing$ **39.** $\{(-3k + 5, -1, -4k + 2, k)\}$

41. $\{(-3k + 9, k, 2, -3)\}$

45. $\{(17k - 6, 10k - 5, k)\}$

47. $\left\{\left(-\dfrac{1}{2}k + \dfrac{34}{11}, \dfrac{1}{2}k - \dfrac{5}{11}, k\right)\right\}$ **49.** $\varnothing$

Problem Set 6.4 (page 474)

1. 22   **3.** -29 **5.** 20   **7.** 5   **9.** -2

11. $-\dfrac{2}{3}$ **13.** -25 **15.** 58   **17.** 39

19. -12 **21.** -41 **23.** -8 **25.** 1088

27. -140 **29.** 81   **31.** 146

33. Property 6.3   **35.** Property 6.2

37. Property 6.4   **39.** Property 6.3

41. Property 6.5

Problem Set 6.5 (page 482)

1. $\{(1, 4)\}$ **3.** $\{(3, -5)\}$ **5.** $\{(2, -1)\}$

7. $\varnothing$ **9.** $\left\{\left(-\dfrac{1}{4}, \dfrac{2}{3}\right)\right\}$ **11.** $\left\{\left(\dfrac{2}{17}, \dfrac{52}{17}\right)\right\}$

13. $\{(9, -2)\}$ **15.** $\left\{\left(2, -\dfrac{5}{7}\right)\right\}$ **17.** $\{(0, 2, -3)\}$

19. $\{(2, 6, 7)\}$ **21.** $\{(4, -4, 5)\}$

23. $\{(-1, 3, -4)\}$ **25.** Infinitely many solutions

27. $\left\{\left(-2, \dfrac{1}{2}, -\dfrac{2}{3}\right)\right\}$ **29.** $\left\{\left(3, \dfrac{1}{2}, -\dfrac{1}{3}\right)\right\}$

31. $(-4, 6, 0)$ **37.** $(0, 0, 0)$

39. Infinitely many solutions

Chapter 6 Review Problem Set (page 486)

1. $\{(3, -7)\}$ **2.** $\{(-1, -3)\}$ **3.** $\{(0, -4)\}$

4. $\left\{\left(\dfrac{23}{3}, -\dfrac{14}{3}\right)\right\}$ **5.** $\{(4, -6)\}$

6. $\left\{\left(-\dfrac{6}{7}, -\dfrac{15}{7}\right)\right\}$ **7.** $\{(-1, 2, -5)\}$

8. $\{(2, -3, -1)\}$ **9.** $\{(5, -4)\}$ **10.** $\{(2, 7)\}$

11. $\{(-2, 2, -1)\}$ **12.** $\{(0, -1, 2)\}$

13. $\{(-3, -1)\}$ **14.** $\{(4, 6)\}$ **15.** $\{(2, -3, -4)\}$

16. $\{(-1, 2, -5)\}$ **17.** $\{(5, -5)\}$

18. $\{(-12, 12)\}$ **19.** $\left\{\left(\dfrac{5}{7}, \dfrac{4}{7}\right)\right\}$

20. $\{(-10, -7)\}$ **21.** $\{(1, 1, -4)\}$

22. $\{(-4, 0, 1)\}$ **23.** $\varnothing$ **24.** $\{(-2, -4, 6)\}$

25. -34 **26.** 13   **27.** -40 **28.** 16

29. 51   **30.** 125   **31.** 72

32. $900 at 10% and $1600 at 12%

33. 20 nickels, 32 dimes, and 54 quarters

34. $25°, 45°$, and $110°$

Chapter 6 Test (page 488)

1. III   **2.** I   **3.** III   **4.** II

5. 8   **6.** $-\dfrac{7}{12}$ **7.** -18 **8.** 112

9. Infinitely many   **10.** $\{(-2, 4)\}$

11. $\{(3, -1)\}$ **12.** $x = -12$ **13.** $y = -\dfrac{13}{11}$

14. No   **15.** $x = 14$ **16.** $y = 13$

17. Infinitely many   **18.** None

19. $\left\{\left(\dfrac{11}{5}, 6, -3\right)\right\}$ **20.** $\{(-2, -1, 0)\}$

21. $x = 1$ **22.** $y = 4$ **23.** 52

24. 2 liters   **25.** 22 quarters

CHAPTER 7

Problem Set 7.1 (page 497)

1. $\begin{bmatrix} 3 & -5 \\ 8 & 3 \end{bmatrix}$ **3.** $\begin{bmatrix} -2 & 21 \\ -7 & 2 \end{bmatrix}$ **5.** $\begin{bmatrix} -2 & 1 \\ -3 & 19 \end{bmatrix}$

7. $\begin{bmatrix} -1 & -5 \\ 2 & 3 \end{bmatrix}$ **9.** $\begin{bmatrix} -12 & -14 \\ -18 & -20 \end{bmatrix}$

11. $\begin{bmatrix} 2 & -11 \\ -7 & 0 \end{bmatrix}$

13. $AB = \begin{bmatrix} 4 & -6 \\ 8 & -12 \end{bmatrix}, BA = \begin{bmatrix} -5 & 5 \\ 3 & -3 \end{bmatrix}$

15. $AB = \begin{bmatrix} -5 & -18 \\ -4 & 42 \end{bmatrix}, BA = \begin{bmatrix} 19 & -39 \\ -16 & 18 \end{bmatrix}$

17. $AB = \begin{bmatrix} 14 & -28 \\ 7 & -14 \end{bmatrix}, BA = \begin{bmatrix} 0 & 0 \\ 0 & 0 \end{bmatrix}$

19. $AB = \begin{bmatrix} -14 & -7 \\ -12 & -1 \end{bmatrix}, BA = \begin{bmatrix} -2 & -3 \\ -32 & -13 \end{bmatrix}$

21. $AB = \begin{bmatrix} 1 & 0 \\ 0 & 1 \end{bmatrix}, BA = \begin{bmatrix} 1 & 0 \\ 0 & 1 \end{bmatrix}$

23. $AB = \begin{bmatrix} 0 & -\frac{5}{3} \\ \frac{17}{6} & -3 \end{bmatrix}, BA = \begin{bmatrix} 0 & -\frac{17}{6} \\ \frac{5}{3} & -3 \end{bmatrix}$

25. $AB = \begin{bmatrix} 1 & 0 \\ 0 & 1 \end{bmatrix}, BA = \begin{bmatrix} 1 & 0 \\ 0 & 1 \end{bmatrix}$

27. $AB = \begin{bmatrix} 3 & -2 \\ 4 & 5 \end{bmatrix}, BA = \begin{bmatrix} 5 & 4 \\ -2 & 3 \end{bmatrix}$

29. $AD = \begin{bmatrix} 1 & 1 \\ 9 & 9 \end{bmatrix}, DA = \begin{bmatrix} 3 & 7 \\ 3 & 7 \end{bmatrix}$

49. $A^2 = \begin{bmatrix} -1 & -4 \\ 8 & 7 \end{bmatrix}, A^3 = \begin{bmatrix} -9 & -11 \\ 22 & 13 \end{bmatrix}$

Problem Set 7.2 (page 504)

1. $\begin{bmatrix} 3 & -7 \\ -2 & 5 \end{bmatrix}$

3. $\begin{bmatrix} -5 & 8 \\ 2 & -3 \end{bmatrix}$

5. $\begin{bmatrix} -\frac{2}{5} & \frac{1}{5} \\ \frac{3}{10} & \frac{1}{10} \end{bmatrix}$

7. Does not exist

9. $\begin{bmatrix} -\frac{5}{7} & \frac{2}{7} \\ -\frac{4}{7} & \frac{3}{7} \end{bmatrix}$

11. $\begin{bmatrix} -\frac{3}{5} & \frac{1}{5} \\ 1 & 0 \end{bmatrix}$

13. $\begin{bmatrix} -\frac{4}{5} & \frac{3}{5} \\ \frac{1}{5} & -\frac{2}{5} \end{bmatrix}$

15. $\begin{bmatrix} 2 & -\frac{5}{3} \\ 1 & -\frac{2}{3} \end{bmatrix}$

17. $\begin{bmatrix} \frac{1}{2} & \frac{1}{2} \\ \frac{1}{2} & -\frac{1}{2} \end{bmatrix}$

19. $\begin{bmatrix} 30 \\ 36 \end{bmatrix}$

21. $\begin{bmatrix} 0 \\ 5 \end{bmatrix}$

23. $\begin{bmatrix} -4 \\ 13 \end{bmatrix}$ **25.** $\begin{bmatrix} -4 \\ -13 \end{bmatrix}$ **27.** $\{(2, 3)\}$

29. $\{(-2, 5)\}$ **31.** $\{(0, -1)\}$ **33.** $\{(-1, -1)\}$

35. $\{(4, 7)\}$ **37.** $\left\{\left(-\frac{1}{3}, \frac{1}{2}\right)\right\}$ **39.** $\{(-9, 20)\}$

Problem Set 7.3 (page 512)

1. $\begin{bmatrix} 1 & 3 & -3 \\ 3 & -6 & 7 \end{bmatrix}$; $\begin{bmatrix} 3 & -5 & 11 \\ -7 & 6 & 3 \end{bmatrix}$; $\begin{bmatrix} 1 & 10 & -13 \\ 11 & -18 & 16 \end{bmatrix}$; $\begin{bmatrix} 10 & -12 & 30 \\ -18 & 12 & 16 \end{bmatrix}$

3. $[-1 \quad -7 \quad 13 \quad 7]$; $[5 \quad 5 \quad -5 \quad 17]$; $[-5 \quad -20 \quad 35 \quad 9]$; $[14 \quad 8 \quad -2 \quad 58]$

5. $\begin{bmatrix} 8 & -3 & -2 \\ 9 & 2 & -3 \\ 7 & 5 & 21 \end{bmatrix}$; $\begin{bmatrix} -2 & -1 & 4 \\ -11 & 6 & -11 \\ -7 & 5 & -3 \end{bmatrix}$; $\begin{bmatrix} 21 & -7 & -7 \\ 28 & 2 & 2 \\ 21 & 10 & 54 \end{bmatrix}$; $\begin{bmatrix} 2 & -6 & 10 \\ -24 & 20 & -36 \\ -14 & 20 & 12 \end{bmatrix}$

7. $\begin{bmatrix} 0 & 2 \\ -1 & 10 \\ 1 & -9 \\ 2 & 9 \end{bmatrix}$; $\begin{bmatrix} -2 & -2 \\ 5 & -4 \\ -11 & 1 \\ -16 & 13 \end{bmatrix}$; $\begin{bmatrix} 1 & 6 \\ -5 & 27 \\ 8 & -23 \\ 13 & 16 \end{bmatrix}$; $\begin{bmatrix} -6 & -4 \\ 14 & -2 \\ -32 & -6 \\ -46 & 48 \end{bmatrix}$

9. $AB = \begin{bmatrix} 11 & -8 & 14 \\ 4 & -16 & 8 \\ -28 & 22 & -36 \end{bmatrix}$; $BA = \begin{bmatrix} -20 & 21 \\ 8 & -21 \end{bmatrix}$

11. $AB = \begin{bmatrix} 22 & -8 & 1 & 3 \\ -42 & 36 & -26 & -20 \end{bmatrix}$; BA does not exist

13. $AB = \begin{bmatrix} -12 & 5 & -5 \\ 14 & -2 & 4 \\ -10 & 13 & -5 \end{bmatrix}$; $BA = \begin{bmatrix} -1 & 0 & -6 \\ 10 & -2 & 16 \\ -8 & 5 & -16 \end{bmatrix}$

15. $AB = [-9]$; $BA = \begin{bmatrix} -2 & 1 & -3 & -4 \\ -6 & 3 & -9 & -12 \\ 4 & -2 & 6 & 8 \\ -8 & 4 & -12 & -16 \end{bmatrix}$

17. AB does not exist; $BA = \begin{bmatrix} 20 \\ 2 \\ -30 \end{bmatrix}$

19. $AB = \begin{bmatrix} 9 & -12 \\ -12 & 16 \\ 6 & -8 \end{bmatrix}$; BA does not exist.

21. $\begin{bmatrix} -\dfrac{1}{5} & \dfrac{3}{10} \\ \dfrac{2}{5} & -\dfrac{1}{10} \end{bmatrix}$ **23.** $\begin{bmatrix} 4 & -1 \\ -7 & 2 \end{bmatrix}$

25. $\begin{bmatrix} -\dfrac{4}{5} & -\dfrac{1}{5} \\ -\dfrac{3}{5} & -\dfrac{2}{5} \end{bmatrix}$ **27.** $\begin{bmatrix} \dfrac{7}{2} & -3 & \dfrac{1}{2} \\ -\dfrac{1}{2} & 0 & \dfrac{1}{2} \\ -\dfrac{1}{2} & 1 & -\dfrac{1}{2} \end{bmatrix}$

29. $\begin{bmatrix} -50 & -9 & 11 \\ -23 & -4 & 5 \\ 5 & 1 & -1 \end{bmatrix}$ **31.** Does not exist

33. $\begin{bmatrix} \dfrac{4}{7} & -1 & -\dfrac{9}{7} \\ -\dfrac{3}{14} & \dfrac{1}{2} & \dfrac{6}{7} \\ \dfrac{2}{7} & 0 & -\dfrac{1}{7} \end{bmatrix}$ **35.** $\begin{bmatrix} \dfrac{1}{2} & 0 & 0 \\ 0 & \dfrac{1}{4} & 0 \\ 0 & 0 & \dfrac{1}{10} \end{bmatrix}$

37. $\{(-3, 2)\}$ **39.** $\{(2, 5)\}$ **41.** $\{(-1, -2, 1)\}$
43. $\{(-2, 3, 5)\}$ **45.** $\{(-4, 3, 0)\}$
47. **(a)** $\{(-1, 2, 3)\}$ **(c)** $\{(-5, 0, -2)\}$
(e) $\{(1, -1, -1)\}$
49. **(a)** y-axis reflection
(c) 90° counterclockwise rotation

Problem Set 7.4 (page 523)

1.

3.

5.

7.

9.

11.

13.

15.

17. $\varnothing$

19.

21.

23.

25. Minimum of 8 and maximum of 52
27. Minimum of 0 and maximum of 28 **29.** 63
31. 340 **33.** 2 **35.** 98
37. $5000 at 9% and $5000 at 12%
39. 300 of type A and 200 of type B
41. 12 units of A and 16 units of B

Chapter 7 Review Problem Set (page 528)

1. $\begin{bmatrix} 7 & -5 \\ -3 & 10 \end{bmatrix}$ **2.** $\begin{bmatrix} 3 & 3 \\ 3 & -6 \end{bmatrix}$ **3.** $\begin{bmatrix} 2 & 1 \\ -6 & 8 \\ -2 & 2 \end{bmatrix}$

4. $\begin{bmatrix} 19 & -11 \\ -6 & 22 \end{bmatrix}$ **5.** $\begin{bmatrix} 7 & 1 \\ -14 & 20 \\ 1 & -2 \end{bmatrix}$

6. $\begin{bmatrix} -11 & -3 & 15 \\ 24 & 2 & -20 \\ -40 & -5 & 38 \end{bmatrix}$ **7.** $\begin{bmatrix} 16 & -26 \\ 0 & 13 \end{bmatrix}$

8. $\begin{bmatrix} 26 & -36 \\ -15 & 32 \end{bmatrix}$ **9.** $\begin{bmatrix} -27 \\ 26 \end{bmatrix}$

10. *EF* does not exist.

14. $\begin{bmatrix} 4 & -5 \\ -7 & 9 \end{bmatrix}$

15. $\begin{bmatrix} -3 & 4 \\ 7 & -9 \end{bmatrix}$ **16.** $\begin{bmatrix} -\dfrac{3}{8} & \dfrac{1}{8} \\ \dfrac{1}{4} & \dfrac{1}{4} \end{bmatrix}$

17. Inverse does not exist. **18.** $\begin{bmatrix} \dfrac{5}{7} & -\dfrac{3}{7} \\ -\dfrac{4}{7} & \dfrac{1}{7} \end{bmatrix}$

19. $\begin{bmatrix} \dfrac{2}{7} & \dfrac{1}{7} \\ -\dfrac{1}{3} & 0 \end{bmatrix}$ **20.** $\begin{bmatrix} \dfrac{39}{8} & -\dfrac{17}{8} & -\dfrac{1}{8} \\ 2 & -1 & 0 \\ \dfrac{1}{8} & \dfrac{1}{8} & \dfrac{1}{8} \end{bmatrix}$

21. $\begin{bmatrix} 8 & -8 & 5 \\ -3 & 2 & -1 \\ -1 & -1 & 1 \end{bmatrix}$ **22.** Inverse does not exist.

23. $\begin{bmatrix} -\dfrac{20}{3} & -\dfrac{7}{3} & \dfrac{1}{3} \\ -\dfrac{1}{3} & -\dfrac{2}{3} & -\dfrac{1}{3} \\ -\dfrac{5}{3} & -\dfrac{1}{3} & \dfrac{1}{3} \end{bmatrix}$ **24.** $\{(-2, 6)\}$

25. $\{(4, -1)\}$ **26.** $\{(2, -3, -1)\}$
27. $\{(-3, 2, 5)\}$ **28.** $\{(-4, 3, 4)\}$

29.   **30.**

31.   **32.**

33. 37   **34.** 56   **35.** 57   **36.** 1700
37. 75 one-gallon and 175 two-gallon freezers

Chapter 7 Test (page 530)

1. $\begin{bmatrix} 9 & -1 \\ 4 & -6 \end{bmatrix}$ **2.** $\begin{bmatrix} -11 & 13 \\ -8 & 14 \end{bmatrix}$ **3.** $\begin{bmatrix} -1 & -3 & 11 \\ -4 & -5 & 18 \\ 37 & -1 & 9 \end{bmatrix}$

4. Does not exist **5.** $\begin{bmatrix} -35 \\ 8 \end{bmatrix}$ **6.** $\begin{bmatrix} -5 & 8 \\ 4 & -3 \end{bmatrix}$

7. $\begin{bmatrix} 4 & 9 \\ 13 & -16 \\ 24 & 23 \end{bmatrix}$ **8.** $\begin{bmatrix} -3 & -5 \\ -20 & 8 \end{bmatrix}$ **9.** $\begin{bmatrix} 8 & 33 \\ -12 & 13 \end{bmatrix}$

10. $\begin{bmatrix} 1 & -34 \\ 16 & -19 \end{bmatrix}$ **11.** $\begin{bmatrix} -3 & 2 \\ -5 & 3 \end{bmatrix}$ **12.** $\begin{bmatrix} 7 & 5 \\ 3 & 2 \end{bmatrix}$

13. $\begin{bmatrix} 4 & \dfrac{3}{2} \\ 1 & \dfrac{1}{2} \end{bmatrix}$ **14.** $\begin{bmatrix} \dfrac{4}{7} & -\dfrac{5}{7} \\ -\dfrac{1}{7} & \dfrac{3}{7} \end{bmatrix}$

15. $\begin{bmatrix} -\dfrac{4}{3} & -\dfrac{5}{3} & 1 \\ -\dfrac{4}{3} & -\dfrac{8}{3} & 1 \\ \dfrac{1}{3} & \dfrac{2}{3} & 0 \end{bmatrix}$ **16.** $\begin{bmatrix} 1 & 2 & -10 \\ 0 & 1 & -3 \\ 0 & 0 & 1 \end{bmatrix}$

17. $\{(8, -12)\}$
18. $\{(-6, -14)\}$ **19.** $(\{9, 13\})$
20. $\left\{ \left(\dfrac{7}{3}, -\dfrac{1}{3}, \dfrac{13}{3} \right) \right\}$ **21.** $\{(-1, 2, 1)\}$

22.

23.

24.

25. 4050

CHAPTER 8

Problem Set 8.1 (page 541)

1. $V(0, 0), F(2, 0),$
$x = -2$

3. $V(0, 0), F(0, -3),$
$y = 3$

5. $V(0, 0), F\left(-\dfrac{1}{2}, 0\right),$
$x = \dfrac{1}{2}$

7. $V(0, 0), F\left(0, \dfrac{3}{2}\right),$
$y = -\dfrac{3}{2}$

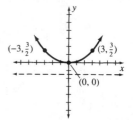

9. $V(0, 2), F(0, 3),$
$y = 1$

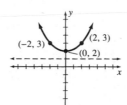

11. $V(0, -2),$
$F(0, -4),$
$y = 0$

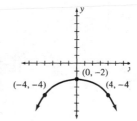

13. $V(2, 0), F(5, 0),$
$x = -1$

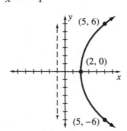

15. $V(1, 2), F(1, 3),$
$y = 1$

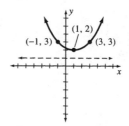

17. $V(-3, 1),$
$F(-3, -1),$
$y = 3$

19. $V(3, 1), F(0, 1),$
$x = 6$

21. $V(-2, -3)$,
$F(-1, -3)$,
$x = -3$

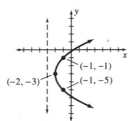

23. $x^2 = 12y$ **25.** $y^2 = -4x$
27. $x^2 + 12y - 48 = 0$
29. $x^2 - 6x - 12y + 21 = 0$
31. $y^2 - 10y + 8x + 41 = 0$ **33.** $3y^2 = -25x$
35. $y^2 = 10x$ **37.** $x^2 - 14x - 8y + 73 = 0$
39. $y^2 + 6y - 12x + 105 = 0$
41. $x^2 + 18x + y + 80 = 0$
43. $x^2 = 750(y - 10)$ **45.** $10\sqrt{2}$ feet
47. 62.5 feet

Problem Set 8.2 (page 551)

For Problems 1–22, the foci are indicated above the graph and the vertices and endpoints of the minor axes are indicated on the graph.

1. $F(\sqrt{3}, 0)$,
$F'(-\sqrt{3}, 0)$

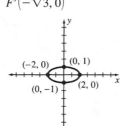

3. $F(0, \sqrt{5})$,
$F'(0, -\sqrt{5})$

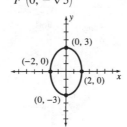

5. $F(0, \sqrt{6})$,
$F'(0, -\sqrt{6})$

7. $F(\sqrt{15}, 0)$,
$F'(-\sqrt{15}, 0)$

9. $F(0, \sqrt{33})$,
$F'(0, -\sqrt{33})$

11. $F(2, 0)$,
$F'(-2, 0)$

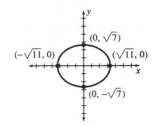

13. $F(1 + \sqrt{5}, 2)$,
$F'(1 - \sqrt{5}, 2)$

15. $F(-2, -1 + 2\sqrt{3})$,
$F'(-2, -1 - 2\sqrt{3})$

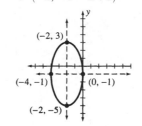

17. $F(3 + \sqrt{3}, 0)$,
$F'(3 - \sqrt{3}, 0)$

19. $F(4, -1 + \sqrt{7})$,
$F'(4, -1 - \sqrt{7})$

21. $F(0, 4), F'(-6, 4)$

23. $16x^2 + 25y^2 = 400$ **25.** $36x^2 + 11y^2 = 396$
27. $x^2 + 9y^2 = 9$ **29.** $100x^2 + 36y^2 = 225$
31. $7x^2 + 3y^2 = 75$
33. $3x^2 - 6x + 4y^2 - 8y - 41 = 0$

35. $9x^2 + 25y^2 - 50y - 200 = 0$

37. $3x^2 + 4y^2 = 48$ **39.** $\dfrac{10\sqrt{5}}{3}$ feet

Problem Set 8.3 (page 562)

For Problems 1–22, the foci and equations of the asymptotes are indicated above the graphs. The vertices are given on the graphs.

1. $F(\sqrt{13}, 0)$,
$F'(-\sqrt{13}, 0)$,
$y = \pm\dfrac{2}{3}x$

3. $F(0, \sqrt{13})$,
$F'(0, -\sqrt{13})$,
$y = \pm\dfrac{2}{3}x$

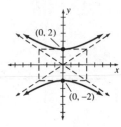

5. $F(0, 5)$,
$F'(0, -5)$,
$y = \pm\dfrac{4}{3}x$

7. $F(3\sqrt{2}, 0)$,
$F'(-3\sqrt{2}, 0)$,
$y = \pm x$

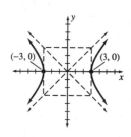

9. $F(0, \sqrt{30})$,
$F'(0, -\sqrt{30})$,
$y = \pm\dfrac{\sqrt{5}}{5}x$

11. $F(\sqrt{10}, 0)$,
$F'(-\sqrt{10}, 0)$,
$y = \pm 3x$

13. $F(3 + \sqrt{13}, -1)$,
$F'(3 - \sqrt{13}, -1)$
$2x - 3y = 9$ and
$2x + 3y = 3$

15. $F(-3, 2 + \sqrt{5})$,
$F'(-3, 2 - \sqrt{5})$
$2x - y = -8$ and
$2x + y = -4$

17. $F(2 + \sqrt{6}, 0)$,
$F'(2 - \sqrt{6}, 0)$
$\sqrt{2}x - y = 2\sqrt{2}$ and
$\sqrt{2}x + y = 2\sqrt{2}$

19. $F(0, -5 + \sqrt{10})$,
$F'(0, -5 - \sqrt{10})$
$3x - y = 5$ and
$3x + y = -5$

21. $F(-2 + \sqrt{2}, -2)$,
$F'(-2 - \sqrt{2}, -2)$
$x - y = 0$ and
$x + y = -4$

23. $5x^2 - 4y^2 = 20$ **25.** $16y^2 - 9x^2 = 144$
27. $3x^2 - y^2 = 3$ **29.** $4y^2 - 3x^2 = 12$
31. $7x^2 - 16y^2 = 112$
33. $5x^2 - 40x - 4y^2 - 24y + 24 = 0$
35. $3y^2 - 30y - x^2 - 6x + 54 = 0$
37. $5x^2 - 20x - 4y^2 = 0$ **39.** Circle
41. Straight line   **43.** Ellipse   **45.** Hyperbola
47. Parabola

Problem Set 8.4 (page 570)

1. $\{(1, 2)\}$ **3.** $\{(1, -5), (-5, 1)\}$

5. $\{(2 + i\sqrt{3}, -2 + i\sqrt{3}), (2 - i\sqrt{3}, -2 - i\sqrt{3})\}$

7. $\{(-6, 7), (-2, -1)\}$ **9.** $\{(-3, 4)\}$

11. $\left\{\left(\dfrac{-1 + i\sqrt{3}}{2}, \dfrac{-7 - i\sqrt{3}}{2}\right), \left(\dfrac{-1 - i\sqrt{3}}{2}, \dfrac{-7 + i\sqrt{3}}{2}\right)\right\}$

13. $\{(-1, 2)\}$ **15.** $\{(-6, 3), (-2, -1)\}$

17. $\{(5, 3)\}$ **19.** $\{(1, 2), (-1, 2)\}$ **21.** $\{(-3, 2)\}$

23. $\{(2, 0), (-2, 0)\}$

25. $\{(\sqrt{2}, \sqrt{3}), (\sqrt{2}, -\sqrt{3}), (-\sqrt{2}, \sqrt{3}), (-\sqrt{2}, -\sqrt{3})\}$

27. $\{(1, 1), (1, -1), (-1, 1), (-1, -1)\}$

29. $\left\{\left(2, \dfrac{3}{2}\right), \left(\dfrac{3}{2}, 2\right)\right\}$ **31.** $\{(9, -2)\}$

33. $\{(\ln 2, 1)\}$ **35.** $\left\{\left(\dfrac{1}{2}, \dfrac{1}{8}\right), (-3, -27)\right\}$

43. $\{(-2.3, 7.4)\}$ **45.** $\{(6.7, 1.7), (9.5, 2.1)\}$

47. None

Chapter 8 Review Problem Set (page 572)

1. $F(4, 0), F'(-4, 0)$

2. $F(-3, 0)$

3. $F(0, 2\sqrt{3}),$
$F'(0, -2\sqrt{3})$
$y = \pm\dfrac{\sqrt{3}}{3}x$

4. $F(\sqrt{15}, 0),$
$F'(-\sqrt{15}, 0)$
$y = \pm\dfrac{\sqrt{6}}{3}x$

5. $F(0, \sqrt{6}),$
$F'(0, -\sqrt{6})$

6. $F\left(0, \dfrac{1}{2}\right)$

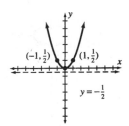

7. $F(4 + \sqrt{6}, 1), F'(4 - \sqrt{6}, 1)$
$\sqrt{2}x - 2y = 4\sqrt{2} - 2 \text{ and } \sqrt{2}x + 2y = 4\sqrt{2} + 2$

8. $F(3, -2 + \sqrt{7}), F'(3, -2 - \sqrt{7})$

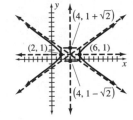

9. $F(-3, 1), x = -1$

10. $F(-1, -5), y = -1$

11. $F\left(-5 + 2\sqrt{3}, 2\right), F'\left(-5 - 2\sqrt{3}, 2\right)$

12. $F\left(-2, -2 + \sqrt{10}\right), F'\left(-2, -2 - \sqrt{10}\right)$
$\sqrt{6}x - 3y = 6 - 2\sqrt{6}$ and $\sqrt{6}x + 3y = -6 - 2\sqrt{6}$

13. $y^2 = -20x$ **14.** $y^2 + 16x^2 = 16$
15. $25x^2 - 2y^2 = 50$ **16.** $4x^2 + 3y^2 = 16$
17. $3x^2 = 2y$ **18.** $9y^2 - x^2 = 9$
19. $9x^2 - 108x + y^2 - 8y + 331 = 0$
20. $y^2 + 4y - 8x + 36 = 0$
21. $3y^2 + 24y - x^2 - 10x + 20 = 0$
22. $x^2 + 12x - y + 33 = 0$
23. $4x^2 + 40x + 25y^2 = 0$
24. $4x^2 - 32x - y^2 + 48 = 0$ **25.** $\{(-1, 4)\}$
26. $\{(3, 1)\}$ **27.** $\{(-1, -2), (-2, -3)\}$
28. $\left\{\left(\dfrac{4\sqrt{2}}{3}, \dfrac{4}{3}i\right), \left(\dfrac{4\sqrt{2}}{3}, -\dfrac{4}{3}i\right), \left(-\dfrac{4\sqrt{2}}{3}, \dfrac{4}{3}i\right),\right.$
$\left.\left(-\dfrac{4\sqrt{2}}{3}, -\dfrac{4}{3}i\right)\right\}$ **29.** $\{(0, 2), (0, -2)\}$

30. $\left\{\left(\dfrac{\sqrt{15}}{5}, \dfrac{2\sqrt{10}}{5}\right), \left(\dfrac{\sqrt{15}}{5}, -\dfrac{2\sqrt{10}}{5}\right),\right.$
$\left.\left(-\dfrac{\sqrt{15}}{5}, \dfrac{2\sqrt{10}}{5}\right), \left(-\dfrac{\sqrt{15}}{5}, -\dfrac{2\sqrt{10}}{5}\right)\right\}$

Chapter 8 Test (page 574)
1. $(0, -5)$ **2.** $(-3, 2)$ **3.** $x = -3$
4. $(6, 0)$ **5.** $(-2, -1)$ **6.** $y = 4$
7. $y^2 + 8x = 0$ **8.** $x^2 - 6x + 12y - 39 = 0$
9. $(0, 6)$ and $(0, -6)$ **10.** six units
11. $(-7, 1)$ and $(-3, 1)$ **12.** $\left(-2\sqrt{3}, 0\right)$ and $\left(2\sqrt{3}, 0\right)$

13. $(-5, 2)$ **14.** $25x^2 + 9y^2 = 900$
15. $x^2 - 12x + 4y^2 + 8y + 36 = 0$ **16.** $y = \pm\dfrac{3}{2}x$
17. $(-1, 6)$ and $(-1, 0)$ **18.** $(\pm 3, 0)$
19. $x^2 - 3y^2 = 36$
20. $8x^2 + 16x - y^2 + 8y - 16 = 0$
21. 3 **22.** $\left\{(3, 2), (-3, -2), \left(4, \dfrac{3}{2}\right), \left(-4, -\dfrac{3}{2}\right)\right\}$

23.

24.

25.
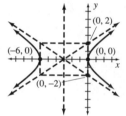

CHAPTER 9

Problem Set 9.1 (page 583)
1. $-4, -1, 2, 5, 8$ **3.** $2, 0, -2, -4, -6$
5. $2, 11, 26, 47, 74$ **7.** $0, 2, 6, 12, 20$
9. $4, 8, 16, 32, 64$ **11.** $a_{15} = -79; a_{30} = -154$
13. $a_{25} = 1; a_{50} = -1$ **15.** $a_n = 2n + 9$
17. $a_n = -3n + 5$ **19.** $a_n = \dfrac{n + 2}{2}$
21. $a_n = 4n - 2$ **23.** $a_n = -3n$ **25.** 73
27. 334 **29.** 35 **31.** 7 **33.** 86
35. 2700 **37.** 3200 **39.** -7950 **41.** 637.5
43. 4950 **45.** 1850 **47.** -2030 **49.** 3591
51. 40,000 **53.** 58,250 **55.** 2205
57. -1325 **59.** 5265 **61.** -810 **63.** 1276
65. 660 **67.** 55 **69.** 431
75. $3, 3, 7, 7, 11, 11$ **77.** $4, 7, 10, 13, 17, 21$
79. $4, 12, 36, 108, 324, 972$ **81.** $1, 1, 2, 3, 5, 8$
83. $3, 1, 4, 9, 25, 256$

Problem Set 9.2 (page 593)

1. $a_n = 3(2)^{n-1}$ **3.** $a_n = 3^n$ **5.** $a_n = \left(\dfrac{1}{2}\right)^{n+1}$

7. $a_n = 4^n$ **9.** $a_n = (0.3)^{n-1}$

11. $a_n = (-2)^{n-1}$ **13.** 64   **15.** $\dfrac{1}{9}$

17. -512 **19.** $\dfrac{1}{4374}$ **21.** $\dfrac{2}{3}$ **23.** 2

25. 1023   **27.** 19,682   **29.** $394\dfrac{1}{16}$

31. 1364   **33.** 1089   **35.** $7\dfrac{511}{512}$ **37.** -547

39. $127\dfrac{3}{4}$ **41.** 540   **43.** $2\dfrac{61}{64}$ **45.** 4

47. 3   **49.** No sum   **51.** $\dfrac{27}{4}$ **53.** 2

55. $\dfrac{16}{3}$ **57.** $\dfrac{1}{3}$ **59.** $\dfrac{26}{99}$ **61.** $\dfrac{41}{333}$

63. $\dfrac{4}{15}$ **65.** $\dfrac{106}{495}$ **67.** $\dfrac{7}{3}$

Problem Set 9.3 (page 598)

1. $24,200   **3.** 11,550   **5.** 7320
7. 125 liters   **9.** 512 gallons   **11.** $116.25
13. $163.84; $327.67   **15.** $24,900

17. 1936 feet   **19.** $\dfrac{15}{16}$ of a gram   **21.** 2910 feet

23. 325 logs   **25.** 5.9%   **27.** $\dfrac{5}{64}$ of a gallon

Problem Set 9.4 (page 605)

These problems are proofs by mathematical induction and require class discussion.

Chapter 9 Review Problem Set (page 608)

1. $a_n = 6n - 3$ **2.** $a_n = 3^{n-2}$ **3.** $a_n = 5 \cdot 2^n$
4. $a_n = -3n + 8$ **5.** $a_n = 2n - 7$
6. $a_n = 3^{3-n}$ **7.** $a_n = -(-2)^{n-1}$
8. $a_n = 3n + 9$ **9.** $a_n = \dfrac{n+1}{3}$

10. $a_n = 4^{n-1}$ **11.** 73   **12.** 106   **13.** $\dfrac{1}{32}$

14. $\dfrac{4}{9}$ **15.** -92 **16.** $\dfrac{1}{16}$ **17.** -5

18. 85   **19.** $\dfrac{5}{9}$ **20.** 2 or -2 **21.** $121\dfrac{40}{81}$

22. 7035   **23.** $-10,725$ **24.** $31\dfrac{31}{32}$

25. 32,015   **26.** 4757   **27.** $85\dfrac{21}{64}$

28. 37,044   **29.** 12,726   **30.** -1845

31. 225   **32.** 255   **33.** 8244   **34.** $85\dfrac{1}{3}$

35. $\dfrac{4}{11}$ **36.** $\dfrac{41}{90}$ **37.** $750   **38.** $46.50

39. $3276.70   **40.** 10,935 gallons

Chapter 9 Test (page 610)

1. -226 **2.** 48   **3.** $a_n = -5n + 2$
4. $a_n = 5(2)^{1-n}$ **5.** $a_n = 6n + 4$ **6.** $\dfrac{729}{8}$, or $91\dfrac{1}{8}$
7. 223   **8.** 60 terms   **9.** 2380   **10.** 765
11. 7155   **12.** 6138   **13.** 22,650   **14.** 9384
15. 4075   **16.** -341 **17.** 6   **18.** $\dfrac{1}{3}$

19. $\dfrac{2}{11}$ **20.** $\dfrac{4}{15}$ **21.** 3 liters   **22.** $3276.70

23. $5810   **24. and 25.** Instructor supplies proof.

CHAPTER 10

Problem Set 10.1 (page 617)

1. 20   **3.** 24   **5.** 168   **7.** 48   **9.** 36
11. 6840   **13.** 720   **15.** 720   **17.** 36
19. 24   **21.** 243   **23.** Impossible   **25.** 216
27. 26   **29.** 36   **31.** 144   **33.** 1024
35. 30   **37. (a)** 6,084,000   **(c)** 3,066,336

Problem Set 10.2 (page 626)

1. 60   **3.** 360   **5.** 21   **7.** 252   **9.** 105
11. 1   **13.** 24   **15.** 84   **17. (a)** 336
19. 2880   **21.** 2450   **23.** 10   **25.** 10
27. 35   **29.** 1260   **31.** 2520   **33.** 15
35. 126   **37.** 144; 202   **39.** 15; 10   **41.** 20

43. 10; 15; 21; $\dfrac{n(n-1)}{2}$ **47.** 120

53. 133,784,560   **55.** 54,627,300

Problem Set 10.3 (page 632)

1. $\dfrac{1}{2}$ **3.** $\dfrac{3}{4}$ **5.** $\dfrac{1}{8}$ **7.** $\dfrac{7}{8}$ **9.** $\dfrac{1}{16}$

11. $\dfrac{3}{8}$ **13.** $\dfrac{1}{3}$ **15.** $\dfrac{1}{2}$ **17.** $\dfrac{5}{36}$ **19.** $\dfrac{1}{6}$

21. $\dfrac{11}{36}$ **23.** $\dfrac{1}{4}$ **25.** $\dfrac{1}{2}$ **27.** $\dfrac{1}{25}$ **29.** $\dfrac{9}{25}$

31. $\dfrac{2}{5}$ **33.** $\dfrac{9}{10}$ **35.** $\dfrac{5}{14}$ **37.** $\dfrac{15}{28}$ **39.** $\dfrac{7}{15}$

41. $\dfrac{1}{15}$ **43.** $\dfrac{2}{3}$ **45.** $\dfrac{1}{5}$ **47.** $\dfrac{1}{63}$ **49.** $\dfrac{1}{2}$

51. $\dfrac{5}{11}$ **53.** $\dfrac{1}{6}$ **55.** $\dfrac{21}{128}$ **57.** $\dfrac{13}{16}$

59. $\dfrac{1}{21}$ **63.** 40   **65.** 3744   **67.** 10,200

69. 123,552   **71.** 1,302,540

Problem Set 10.4 (page 642)

1. $\dfrac{5}{36}$ **3.** $\dfrac{7}{12}$ **5.** $\dfrac{1}{216}$ **7.** $\dfrac{53}{54}$ **9.** $\dfrac{1}{16}$

11. $\dfrac{15}{16}$ **13.** $\dfrac{1}{32}$ **15.** $\dfrac{31}{32}$ **17.** $\dfrac{5}{6}$ **19.** $\dfrac{12}{13}$

21. $\dfrac{7}{12}$ **23.** $\dfrac{37}{44}$ **25.** $\dfrac{2}{3}$ **27.** $\dfrac{2}{3}$ **29.** $\dfrac{5}{18}$

31. $\dfrac{1}{3}$ **33.** $\dfrac{1}{2}$ **35.** $\dfrac{7}{12}$

37. (a) 0.410   (c) 0.955   **39.** 0.525
41. 60   **43.** 120   **45.** 9   **47.** 56
49. It is a fair game.   **51.** Yes   **53.** $11,000
55. −$25   **59.** 1 to 7   **61.** 11 to 5
63. 1 to 8   **65.** 1 to 1   **67.** 4 to 3

69. 3 to 2   **71.** $\dfrac{2}{7}$ **73.** $\dfrac{7}{12}$

Problem Set 10.5 (page 652)

1. $\dfrac{1}{3}$ **3.** $\dfrac{2}{15}$ **5.** $\dfrac{1}{3}$ **7.** $\dfrac{1}{6}$ **9.** $\dfrac{2}{3};\dfrac{2}{7}$

11. $\dfrac{2}{3};\dfrac{2}{5}$ **13.** $\dfrac{1}{5};\dfrac{2}{7}$ **15.** Dependent

17. Independent   **19.** $\dfrac{1}{4}$ **21.** $\dfrac{1}{216}$ **23.** $\dfrac{1}{221}$

25. $\dfrac{13}{102}$ **27.** $\dfrac{1}{16}$ **29.** $\dfrac{1}{1352}$ **31.** $\dfrac{2}{49}$

33. $\dfrac{25}{81}$ **35.** $\dfrac{20}{81}$ **37.** $\dfrac{25}{169}$ **39.** $\dfrac{32}{169}$

41. $\dfrac{2}{3}$ **43.** $\dfrac{1}{3}$ **45.** $\dfrac{5}{68}$ **47.** $\dfrac{15}{34}$ **49.** $\dfrac{1}{12}$

51. $\dfrac{1}{6}$ **53.** $\dfrac{1}{729}$ **55.** $\dfrac{5}{27}$ **57.** $\dfrac{4}{35}$

59. $\dfrac{8}{35}$ **61.** $\dfrac{4}{21};\dfrac{2}{7};\dfrac{11}{21}$

Problem Set 10.6 (page 659)

1. $x^8 + 8x^7y + 28x^6y^2 + 56x^5y^3 + 70x^4y^4 + 56x^3y^5 + 28x^2y^6 + 8xy^7 + y^8$
3. $x^6 - 6x^5y + 15x^4y^2 - 20x^3y^3 + 15x^2y^4 - 6xy^5 + y^6$
5. $a^4 + 8a^3b + 24a^2b^2 + 32ab^3 + 16b^4$
7. $x^5 - 15x^4y + 90x^3y^2 - 270x^2y^3 + 405xy^4 - 243y^5$
9. $16a^4 - 96a^3b + 216a^2b^2 - 216ab^3 + 81b^4$
11. $x^{10} + 5x^8y + 10x^6y^2 + 10x^4y^3 + 5x^2y^4 + y^5$
13. $16x^8 - 32x^6y^2 + 24x^4y^4 - 8x^2y^6 + y^8$
15. $x^6 + 18x^5 + 135x^4 + 540x^3 + 1215x^2 + 1458x + 729$
17. $x^9 - 9x^8 + 36x^7 - 84x^6 + 126x^5 - 126x^4 + 84x^3 - 36x^2 + 9x - 1$

19. $1 + \dfrac{4}{n} + \dfrac{6}{n^2} + \dfrac{4}{n^3} + \dfrac{1}{n^4}$

21. $a^6 - \dfrac{6a^5}{n} + \dfrac{15a^4}{n^2} - \dfrac{20a^3}{n^3} + \dfrac{15a^2}{n^4} - \dfrac{6a}{n^5} + \dfrac{1}{n^6}$

23. $17 + 12\sqrt{2}$ **25.** $843 - 589\sqrt{2}$
27. $x^{12} + 12x^{11}y + 66x^{10}y^2 + 220x^9y^3$
29. $x^{20} - 20x^{19}y + 190x^{18}y^2 - 1140x^{17}y^3$
31. $x^{28} - 28x^{26}y^3 + 364x^{24}y^6 - 2912x^{22}y^9$

33. $a^9 + \dfrac{9a^8}{n} + \dfrac{36a^7}{n^2} + \dfrac{84a^6}{n^3}$

35. $x^{10} - 20x^9y + 180x^8y^2 - 960x^7y^3$ **37.** $56x^5y^3$
39. $126x^5y^4$ **41.** $189a^2b^5$ **43.** $120x^6y^{21}$
45. $\dfrac{5005}{n^6}$ **51.** $-117 + 44i$ **53.** $-597 - 122i$

Chapter 10 Review Problem Set (page 662)

1. 720   **2.** 30,240   **3.** 150   **4.** 1440
5. 20   **6.** 525   **7.** 1287   **8.** 264
9. 74   **10.** 55   **11.** 40   **12.** 15

13. 60   **14.** 120   **15.** $\dfrac{3}{8}$ **16.** $\dfrac{5}{16}$

17. $\dfrac{5}{36}$ **18.** $\dfrac{13}{18}$ **19.** $\dfrac{3}{5}$ **20.** $\dfrac{1}{35}$ **21.** $\dfrac{57}{64}$

22. $\dfrac{1}{221}$ **23.** $\dfrac{1}{6}$ **24.** $\dfrac{4}{7}$ **25.** $\dfrac{4}{7}$ **26.** $\dfrac{10}{21}$

27. $\dfrac{140}{143}$ **28.** $\dfrac{105}{169}$ **29.** $\dfrac{1}{6}$ **30.** $\dfrac{28}{55}$

31. $\dfrac{5}{7}$ **32.** $\dfrac{1}{16}$ **33.** $\dfrac{1}{2};\dfrac{1}{3}$

34. (a) $\dfrac{9}{19}$ (b) $\dfrac{9}{10}$ **35.** (a) $\dfrac{2}{7}$ (b) $\dfrac{4}{9}$

36. $x^5 + 10x^4y + 40x^3y^2 + 80x^2y^3 + 80xy^4 + 32y^5$
37. $x^8 - 8x^7y + 28x^6y^2 - 56x^5y^3 + 70x^4y^4 - 56x^3y^5 + 28x^2y^6 - 8xy^7 + y^8$
38. $a^8 - 12a^6b^3 + 54a^4b^6 - 108a^2b^9 + 81b^{12}$

39. $x^6 + \dfrac{6x^5}{n} + \dfrac{15x^4}{n^2} + \dfrac{20x^3}{n^3} + \dfrac{15x^2}{n^4} + \dfrac{6x}{n^5} + \dfrac{1}{n^6}$

40. $41 - 29\sqrt{2}$ **41.** $-a^3 + 3a^2b - 3ab^2 + b^3$

42. $-1760x^9y^3$ **43.** $57915a^4b^{18}$

Chapter 10 Test (page 665)

1. 12   **2.** 240   **3.** 216   **4.** 270

5. 26   **6.** 8640   **7.** 20   **8.** 144

9. 2520   **10.** 350   **11.** $\dfrac{13}{18}$ **12.** $\dfrac{5}{16}$

13. $\dfrac{5}{6}$ **14.** $\dfrac{1}{7}$ **15.** $\dfrac{23}{28}$ **16.** $\dfrac{3}{4}$

17. 25 times   **18.** \$.30   **19.** $\dfrac{168}{361}$

20. $\dfrac{2}{21}$ **21.** $\dfrac{5}{16}$

22. $64 - \dfrac{192}{n} + \dfrac{240}{n^2} - \dfrac{160}{n^3} + \dfrac{60}{n^4} - \dfrac{12}{n^5} + \dfrac{1}{n^6}$

23. $243x^5 + 810x^4y + 1080x^3y^2 + 720x^2y^3 + 240xy^4 + 32y^5$

24. $\dfrac{495}{256}x^4$ **25.** $2835x^3y^4$

INDEX

PHOTO CREDITS

IN-BOOK SURVEY

At Brooks/Cole, we are excited about creating new types of learning materials that are interactive, three-dimensional, and fun to use. To guide us in our publishing/development process, we hope that you'll take just a few moments to fill out the survey below. Your answers can help us make decisions that will allow us to produce a wide variety of videos, CD-ROMs, and Internet-based learning systems to complement standard textbooks. If you're interested in working with us as a student Beta-tester, be sure to fill in your name, telephone number, and address. We look forward to hearing from you!

In addition to books, which of the following learning tools do you currently use in your counseling/human services/social work courses?

_____ **Video** _____ in class _____ school library _____ own VCR

_____ **CD-ROM** _____ in class _____ in lab _____ own computer

_____ **Macintosh disks** _____ in class _____ in lab _____ own computer

_____ **Windows disks** _____ in class _____ in lab _____ own computer

_____ **Internet** _____ in class _____ in lab _____ own computer

How often do you access the Internet? _____

My own home computer is:

_____ Macintosh _____ DOS _____ Windows _____ Windows 95

The computer I use in class for counseling/human services/social work courses is:

_____ Macintosh _____ DOS _____ Windows _____ Windows 95

If you are NOT currently using multimedia materials in your counseling/human services/social work courses, but can see ways that video, CD-ROM, Internet, or other technologies could enhance your learning, please comment below:

_____

_____

_____

Other comments (optional): _____

_____

Name _____ Telephone _____

Address _____

School _____

Professor/Course_____

You can fax this form to us at (408) 375-6414; e-mail to: info@brookscole.com; or detach, fold, secure, and mail.

TO THE OWNER OF THIS BOOK:

I hope that you have found *College Algebra,* Fourth Edition, useful. So that this book can be improved in a future edition, would you take the time to complete this sheet and return it? Thank you.

School and address: _____

Department: _____

Instructor's name: _____

1. What I like most about this book is: _____

_____

_____

2. What I like least about this book is: _____

_____

_____

3. My general reaction to this book is: _____

_____

4. The name of the course in which I used this book is: _____

_____

5. Were all of the chapters of the book assigned for you to read? _____

   If not, which ones weren't? _____

 6. In the space below, or on a separate sheet of paper, please write specific suggestions for improving this book and anything else you'd care to share about your experience in using the book.

_____

_____

_____

_____

_____

Optional:

Your name: _____ Date: _____

May Brooks/Cole quote you, either in promotion for *College Algebra,* Fourth Edition, or in future publishing ventures?

Yes: _____ No: _____

Sincerely,

Jerome E. Kaufmann

area A width w base b volume V

perimeter P surface area S circumference C area of base B

length l altitude (height) h radius r slant height s

Rectangle

$A = lw \qquad P = 2l + 2w$

Triangle

$A = \frac{1}{2}bh$

Square

$A = s^2 \qquad P = 4s$

Parallelogram

$A = bh$

Trapezoid

$A = \frac{1}{2}h(b_1 + b_2)$

Circle

$A = \pi r^2 \qquad C = 2\pi r$

30°–60° Right Triangle

Right Triangle

$a^2 + b^2 = c^2$

Isosceles Right Triangle

Right Circular Cylinder

$V = \pi r^2 h \qquad S = 2\pi r^2 + 2\pi rh$

Sphere

$S = 4\pi r^2 \qquad V = \frac{4}{3}\pi r^3$

Right Circular Cone

$V = \frac{1}{3}\pi r^2 h \qquad S = \pi r^2 + \pi rs$

Pyramid

$V = \frac{1}{3}Bh$

Prism

$V = Bh$